area A
perimeter P
length l

width w
surface area S
altitude (height) h

base ... me V
circ... of base B
radius ... slant height s

Rectangle

$$A = lw \qquad P = 2l + 2w$$

Triangle

$$A = \frac{1}{2}bh$$

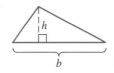

Square

$$A = s^2 \qquad P = 4s$$

Parallelogram

$$A = bh$$

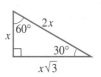

Trapezoid

$$A = \frac{1}{2}h(b_1 + b_2)$$

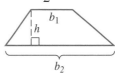

Circle

$$A = \pi r^2 \qquad C = 2\pi r$$

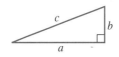

30°–60° Right Triangle

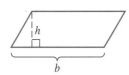

Right Triangle

$$a^2 + b^2 = c^2$$

Isosceles Right Triangle

Right Circular Cylinder

$$V = \pi r^2 h \qquad S = 2\pi r^2 + 2\pi rh$$

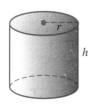

Sphere

$$S = 4\pi r^2 \qquad V = \frac{4}{3}\pi r^3$$

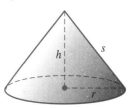

Right Circular Cone

$$V = \frac{1}{3}\pi r^2 h \qquad S = \pi r^2 + \pi rs$$

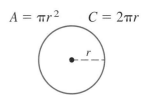

Pyramid

$$V = \frac{1}{3}Bh$$

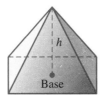

Prism

$$V = Bh$$

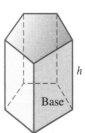

SEVENTH EDITION

Intermediate Algebra

JEROME E. KAUFMANN

KAREN L. SCHWITTERS
Seminole Community College

THOMSON
BROOKS/COLE

Australia • Canada • Mexico • Singapore • Spain
United Kingdom • United States

THOMSON

BROOKS/COLE

Mathematics Editor: *Jennifer Huber*
Assistant Editor: *Rebecca Subity*
Editorial Assistant: *Carrie Dodson*
Technology Project Manager: *Star Mackenzie*
Marketing Manager: *Leah Thomson*
Marketing Assistant: *Jessica Perry*
Advertising Project Manager: *Bryan Vann*
Project Manager, Editorial Production: *Kirk Bomont*
Print/Media Buyer: *Kris Waller*
Production Service: *Susan Graham*

Text Designer: *Carolyn Deacy*
Photo Researcher: *Sarah Evertson*
Copy Editor: *Connie Day*
Illustrators: *Network Graphics and
    Scientific Illustrators*
Cover Designer: *Roger Knox*
Cover Image: *Rowan Moore*
Cover Printer: *Lehigh Press*
Compositor: *G & S Typesetting, Inc.*
Printer: *Quebecor World–Taunton*

For more information about our products, contact us at:
Thomson Learning Academic Resource Center
1-800-423-0563
For permission to use material from this text, contact
us by:
Phone: 1-800-730-2214   **Fax:** 1-800-730-2215
Web: http://www.thomsonrights.com

Library of Congress Control Number: 2003106055

ISBN 0-534-40050-7
Instructor's Edition: ISBN 0-534-40437-5

Brooks/Cole — Thomson Learning
10 Davis Drive
Belmont, CA 94002–3098
USA

Asia
Thomson Learning
5 Shenton Way #01-01
UIC Building
Singapore 068808

Australia/New Zealand
Thomson Learning
102 Dodds Street
Southbank, Victoria 3006
Australia

Canada
Nelson
1120 Birchmount Road
Toronto, Ontario M1K 5G4
Canada

Europe/Middle East/Africa
Thomson Learning
High Holborn House
50/51 Bedford Row
London WC1R 4LR
United Kingdom

Latin America
Thomson Learning
Seneca, 53
Colonia Polanco
11560 Mexico D.F.
Mexico

Spain
Paraninfo
Calle/Magallanes, 25
28015 Madrid, Spain

Contents

CHAPTER 4

CHAPTER 5

CHAPTER 6

CHAPTER 11

Exponential and Logarithmic Functions   553

CHAPTER 12

Sequences and Series   599

Preface

When preparing *Intermediate Algebra, Seventh Edition*, we attempted to preserve the features that made the previous editions successful while incorporating improvements suggested by reviewers.

This text is written for college students who need an algebra course that bridges the gap between elementary algebra and the more advanced courses in pre-calculus mathematics. It covers topics that are usually classified as intermediate algebra topics.

The basic concepts of intermediate algebra are presented in a simple, straightforward way. Algebraic ideas are developed in a logical sequence and in an easy-to-read manner without excessive formalism. Concepts are developed through examples, reinforced continuously through additional examples, and then applied in a variety of problem-solving situations.

New in This Edition

- In Chapter 1 more examples and problems involving operations with rational numbers in common fraction form have been added. More steps have been added to the explanations.
- An Appendix A has been added to review prime number factorizations and operations of fractions. This review is intended to reacquaint students with the basic rules of adding, subtracting, multiplying, and dividing fractions.
- In Section 2.3 two approaches to solving equations involving decimals are shown, and in Section 2.4 there are more examples of application problems.
- In Chapter 3 care has been taken to provide ample problems of factoring trinomials with a leading coefficient other than 1, and these are at the appropriate level of difficulty for intermediate algebra.
- Synthetic division has been removed from the appendix and is now incorporated into Section 4.5. The material is at the end of the section and at the end of the problem set, so that it can easily be considered an optional topic.

- Throughout Chapter 4 more explanation and more steps have been added to the examples.
- In Section 5.3 the concept of simplifying radicals containing variables — without using the assumption of nonnegative variables — is placed in the Further Investigations section of the problem set.
- In Section 6.5 solving equations with rational exponents has been added to the Further Investigations section of the problem set.
- Chapter 3 in the previous edition is now Chapter 7. This allows for a later introduction of graphing when students have gained some algebraic maturity. Now the chapters are arranged so that a student can progress continuously through graphing lines (Chapter 7), graphing conic sections (Chapter 8), and graphing functions and their translations (Chapter 9). The graphing calculator is introduced in Chapter 7. Throughout the text the graphing calculator material is placed at the end of a section or at the end of a problem set so that using the graphing calculator is optional.
- In Section 7.5 the midpoint of a line segment and finding the equation of a perpendicular bisector of a line segment have been added to the Further Investigations section of the problem set.
- In Chapter 9 domain and range is presented in both interval notation and set notation. Answers in the back of the book are given in both formats.

Other Special Features

- A common thread runs throughout the book: namely, *learn a skill;* next, *use the skill to help solve equations and inequalities;* and then *use equations and inequalities to solve word problems.* This thread influenced some of the decisions we made in preparing this text.
  1. Approximately 450 word problems are scattered throughout the text. These problems deal with a large variety of applications that show the connection between mathematics and the real world.
  2. Many problem-solving suggestions are offered throughout the text, with special discussions in several sections. When appropriate, different methods for solving the same problem are shown. The problem-solving suggestions are demonstrated in more than 85 worked-out examples.
  3. Newly acquired skills are used as soon as possible to solve equations and inequalities, which are, in turn, used to solve word problems. Therefore, the concept of solving equations and inequalities is introduced early and is developed throughout the text. The concepts of factoring, solving equations, and solving word problems are tied together in Chapter 3.
- As recommended by the American Mathematical Association of Two-Year Colleges, many basic geometric concepts are integrated in a problem-solving setting. *Intermediate Algebra, Seventh Edition* contains approximately 20 worked-out examples and 100 problems that connect algebra, geometry, and real-world applications. Specific discussions of geometric concepts are contained in the following sections:

Section 2.2: Complementary and supplementary angles; the sum of the measures of the angles of a triangle equals 180°

Section 2.4: Area and volume formulas

Section 3.4: More on area and volume formulas, perimeter, and circumference formulas

Section 3.7: The Pythagorean theorem

Section 6.2: More on the Pythagorean theorem, including work with isosceles right triangles and 30°– 60° right triangles

- Specific graphing ideas (intercepts, symmetry, restrictions, asymptotes, and transformations) are introduced and used throughout Chapters 7 and 8. In Section 8.3, the work with parabolas from Chapter 7 is used to develop definitions for translations, reflections, stretchings, and shrinkings. These transformations are then applied to the graphs of

$$f(x) = x^3 \qquad f(x) = \frac{1}{x} \qquad f(x) = \sqrt{x} \qquad \text{and} \qquad f(x) = |x|$$

- All answers for Chapter Review Problem Sets, Chapter Tests, and Cumulative Review Problem Sets appear in the back of the text.

Additional Comments about Some of the Chapters

- Chapter 1 is written so that it can be covered quickly, and on an individual basis if so desired, by those who need only a brief review of some basic arithmetic and algebraic concepts. Appendix A is for students needing a more through review of operations with fractions.
- Chapter 2 presents an early introduction to the heart of the intermediate algebra course. Problem solving and the solving of equations and inequalities are introduced early so they can be used as unifying themes throughout the text.
- Chapter 6 is organized to give students the opportunity to learn, on a day-by-day basis, different techniques for solving quadratic equations. The process of completing the square is treated as a viable equation-solving tool for certain types of quadratic equations. The emphasis on completing the square in this setting pays dividends in Chapter 8 when we graph parabolas, circles, ellipses, and hyperbolas. Section 6.5 offers some guidance as to when to use a particular technique for solving quadratic equations. In addition, the often-overlooked relationships involving the sum and product of roots are discussed and used as an effective checking procedure.
- Chapter 8 is written on the premise that intermediate algebra students should become very familiar with straight lines, parabolas, and circles and should be exposed in only a limited fashion to ellipses and hyperbolas.
- Chapter 9 is devoted entirely to functions; our treatment of this topic does not jump back and forth between functions and relations that are not functions. This chapter includes some work with the composition of functions and the use of linear and quadratic functions in problem-solving situations.

- Chapter 10 contains the various techniques for solving systems of linear equations. It is organized so that instructors can use as much of the chapter as they need for their particular course. Section 10.2 presents the elimination-by-addition method, which emphasizes equivalent systems and sets the stage for future work with matrices.
- Chapter 11 presents a modern-day version of the concepts of exponents and logarithms. The emphasis is on making the concepts and their applications understood. The calculator is used as a computational tool.

Ancillaries

For the Instructor

Annotated Instructor's Edition. This special version of the complete student text contains a Resource Integration Guide with answers printed next to all respective exercises. Graphs, tables, and other answers appear in a special answer section at the back of the text.

Test Bank. The *Test Bank* includes eight tests per chapter as well as three final exams. The tests are made up of a combination of multiple-choice, free-response, true/false, and fill-in-the-blank questions.

Complete Solutions Manual. The *Complete Solutions Manual* provides worked-out solutions to all of the problems in the text.

BCA Instructor Version. With a balance of efficiency and high-performance simplicity and versatility, *Brooks/Cole Assessment* gives you the power to transform the learning and teaching experience. *BCA Instructor Version* is made up of two components, *BCA Testing* and *BCA Tutorial*. *BCA Testing* is a revolutionary, Internet-ready, text-specific testing suite that allows instructors to customize exams and track student progress. *BCA Testing* offers full algorithmic generation of problems and free response mathematics. *BCA Tutorial* is a text-specific, interactive tutorial software program that is delivered via the Web (at http://bca.brookscole.com) and is offered in both student and instructor versions. The tracking program built into the instructor version of the software enables instructors to carefully monitor student progress. Results flow automatically to your gradebook and you can easily communicate to individuals, sections, or entire courses.

Text-Specific Videotapes. These text-specific videotape sets, available at no charge to qualified adopters of the text, feature 10- to 20-minute problem-solving lessons that cover each section of every chapter.

MyCourse 2.0. Ask us about our new free online course builder! Whether you want only the easy-to-use tools to build it or the content to furnish it, Brooks/Cole offers you a simple solution for a custom course Web site that allows you to assign, track, and report on student progress; load your syllabus; and more. Contact your Brooks/Cole representative for details, or visit http://mycourse.thomsonlearning.com

For the Student

Student Solutions Manual. The *Student Solutions Manual* provides worked-out solutions to the odd-numbered problems, and all chapter review, chapter test, and cumulative review problems in the text.

Web site (http://mathematics.brookscole.com). When you adopt a Thomson-Brooks/Cole mathematics text, you and your students will have access to a variety of teaching and learning resources. This Web site features everything from book-specific resources to newsgroups.

InfoTrac College Edition is automatically packaged FREE with every new copy of this text! *InfoTrac College Edition* is a world-class, online university library that offers the full text of articles from almost 4000 scholarly and popular publications; it is updated daily and goes back as far as 22 years. Both adopters and their students receive unlimited access for four months.

BCA Tutorial Student Version. This text-specific, interactive tutorial software is delivered via the Web (at http://bca.brookscole.com). It is browser-based, making it an intuitive mathematical guide, even for students with little technological proficiency. So sophisticated, it's simple, *BCA Tutorial* allows students to work with real math notation in real time, providing instant analysis and feedback. *BCA Tutorial Student Version* is also available on CD Rom for those students who want to use the tutorial locally on their computers and not via the Internet.

Interactive Video Skillbuilder CD-ROM. Think of it as portable office hours! The *Interactive Video Skillbuilder CD-ROM* contains more than eight hours of video instruction. The problems worked during each video lesson are shown next to the viewing screen so that students can try working them out before watching the solution. Examples in the book that are taught on video are identified by this logo in the margin.

WebTutor Toolbox for WebCT (0-534-27488-9)
WebTutor Toolbox for Blackboard (0-534-27489-7)
Preloaded with content and available free via PIN code when packaged with this text, *WebTutor ToolBox for WebCT and Blackboard* pairs the content of this text's rich Book Companion Web Site with the sophisticated course management functionality of a WebCT and Blackboard product. You can assign materials (including online quizzes) and have the results flow automatically to your gradebook. ToolBox is ready to use as soon as you log on — or, you can customize its preloaded content by uploading images and other resources, adding Web links, or creating your own practice materials. Students only have access to student resources on the Web site. Instructors can enter a PIN code for access to password-protected Instructor Resources.

Vmentor. An ideal solution for homework help and tutoring. When you bundle *vMentor* with your favorite Brooks/Cole mathematics text, your students will have access, via the Web, to highly qualified tutors with thorough knowledge of our textbooks. When students get stuck on a particular problem or concept, they need only log on to *vMentor*, where they can talk to *vMentor* tutors who will skillfully guide them through the problem, using the whiteboard for illustration. To take a live test-drive of *BCA* or *vMentor*, visit us online at http://bca.brookscole.com.

Explorations in Beginning and Intermediate Algebra Using the TI-82/83/83-Plus/85/86 Graphing Calculator, Third Edition (0-534-40644-0)
Deborah J. Cochener and Bonnie M. Hodge, both of Austin Peay State University
This user-friendly workbook improves students' understanding and their retention of algebra concepts through a series of activities and guided explorations using the graphing calculator. An ideal supplement for any beginning or intermediate algebra course, *Explorations in Beginning and Intermediate Algebra, Third Edition* is an ideal tool for integrating technology without sacrificing course content. By clearly and succinctly teaching keystrokes, class time is devoted to investigations instead of how to use a graphing calculator.

The Math Student's Guide to the TI-83 Graphing Calculator (0-534-37802-1)
The Math Student's Guide to the TI-86 Graphing Calculator (0-534-37801-3)
The Math Student's Guide to the TI-83 Plus Graphing Calculator (0-534-42021-4)
The Math Student's Guide to the TI-89 Graphing Calculator (0-534-42022-2)
Trish Cabral of Butte College
These videos are designed for students who are new to the graphing calculator or for those who would like to brush up on their skills. Each instructional graphing calculator videotape covers basic calculations, the custom menu, graphing, advanced graphing, matrix operations, trigonometry, parametric equations, polar coordinates, calculus, Statistics I and one-variable data, and Statistics II with linear regression. These wonderful tools are each 105 minutes in length and cover all of the important functions of a graphing calculator.

Mastering Mathematics: How to Be a Great Math Student, Third Edition (0-534-34947-1)
Richard Manning Smith, Bryant College
Providing solid tips for every stage of study, *Mastering Mathematics* stresses the importance of a positive attitude and gives students the tools to succeed in their math course.

Beginning and Intermediate Algebra Activities Manual
Instructor Edition (0-534-35356-8); Student Edition (0-534-35355-X)
Debbie Garrison, Judy Jones, and Jolene Rhodes, all of Valencia Community College
Designed as a stand-alone supplement for any beginning or intermediate algebra text, *Activities in Beginning & Intermediate Algebra* is a collection of activities written to incorporate the recommendations from the NCTM and from AMATYC's Crossroads. Activities can be used during class or in a laboratory setting to introduce, teach, or reinforce a topic.

Conquering Math Anxiety: A Self-Help Workbook, Second Edition (0-534-38634-2)
Cynthia Arem, Pima Community College
A comprehensive workbook that provides a variety of exercises and worksheets along with detailed explanations of methods to help "math-anxious" students deal

with and overcome math fears. This edition now comes with a free relaxation CD-ROM and a detailed list of Internet resources.

Active Arithmetic and Algebra: Activities for Prealgebra and Beginning Algebra (0-534-36771-2)
Judy Jones, Valencia Community College
This activities manual includes a variety of approaches to learning mathematical concepts. Sixteen activities, including puzzles, games, data collection, graphing, and writing activities are included.

Math Facts: Survival Guide to Basic Mathematics, Second Edition (0-534-94734-4)
Algebra Facts: Survival Guide to Basic Algebra (0-534-19986-0)
Theodore John Szymanski, Tompkins-Cortland Community College
This booklet gives easy access to the most crucial concepts and formulas in algebra. Although it is bound, this booklet is structured to work like flash cards.

Acknowledgments

We would like to take this opportunity to thank the following people who served as reviewers for this text:

Tracy M. Boone
Pennsylvania State University-Altoona

Cynthia Fleck
Wright State University

Kay Haralson
Austin Peay State University

Susan Kutryb
Hudson Valley Community College

Sandra Mayo
Los Angeles Mission College

Kathryn T. McClellan
Tarrant County Junior College-Northeast

Jamie McGill
East Tennessee State University

Reed Parr
Salt Lake Community College

C. L. Pinchback
University of Central Arkansas

William Radulovich
Florida Community College at Jacksonville

Mary Voxman
University of Idaho

Dennis W. Watson
Clark College

Jerome E. Kaufmann
Karen L. Schwitters

Numbers from the set of integers are used to express temperatures that are below 0°F.

© Alden Pellett/The Image Works

Basic Concepts and Properties

The temperature at 6 p.m. was $-3°F$. By 11 p.m. the temperature had dropped another $5°F$. We can use the **numerical expression** $-3 - 5$ to determine the temperature at 11 p.m.

Justin has p pennies, n nickels, and d dimes in his pocket. The **algebraic expression** $p + 5n + 10d$ represents that amount of money in cents.

Algebra is often described as a **generalized arithmetic**. That description may not tell the whole story, but it does convey an important idea: A good understanding of arithmetic provides a sound basis for the study of algebra. In this chapter we use the concepts of **numerical expression** and **algebraic expression** to review some ideas from arithmetic and to begin the transition to algebra. Be sure that you thoroughly understand the basic concepts we review in this first chapter.

1

1.1   Sets, Real Numbers, and Numerical Expressions

In arithmetic, we use symbols such as 6, $\dfrac{2}{3}$, 0.27, and π to represent numbers. The symbols $+$, $-$, $\cdot$, and $\div$ commonly indicate the basic operations of addition, subtraction, multiplication, and division, respectively. Thus we can form specific **numerical expressions.** For example, we can write the indicated sum of six and eight as $6 + 8$.

In algebra, the concept of a variable provides the basis for generalizing arithmetic ideas. For example, by using x and y to represent any numbers, we can use the expression $x + y$ to represent the indicated sum of any two numbers. The x and y in such an expression are called **variables** and the phrase $x + y$ is called an **algebraic expression.**

We can extend to algebra many of the notational agreements we make in arithmetic, with a few modifications. The following chart summarizes the notational agreements that pertain to the four basic operations.

Operation	Arithmetic	Algebra	Vocabulary
Addition	$4 + 6$	$x + y$	The **sum** of x and y
Subtraction	$14 - 10$	$a - b$	The **difference** of a and b
Multiplication	$7 \cdot 5$ or 7×5	$a \cdot b$, $a(b)$, $(a)b$, $(a)(b)$, or ab	The **product** of a and b
Division	$8 \div 4$, $\dfrac{8}{4}$, or $4\overline{)8}$	$x \div y$, $\dfrac{x}{y}$, or $y\overline{)x}$	The **quotient** of x and y

Note the different ways to indicate a product, including the use of parentheses. The ab form is the simplest and probably the most widely used form. Expressions such as abc, $6xy$, and $14xyz$ all indicate multiplication. We also call your attention to the various forms that indicate division. In algebra, we usually use the fractional form $\dfrac{x}{y}$, although the other forms do serve a purpose at times.

Use of Sets

We can use some of the basic vocabulary and symbolism associated with the concept of sets in the study of algebra. A **set** is a collection of objects and the objects are called **elements** or **members** of the set. In arithmetic and algebra the elements of a set are usually numbers.

The use of set braces, { }, to enclose the elements (or a description of the elements) and the use of capital letters to name sets provide a convenient way to communicate about sets. For example, we can represent a set A, which consists of the vowels of the alphabet, in any of the following ways:

$A = \{$vowels of the alphabet$\}$ Word description

$A = \{a, e, i, o, u\}$ List or roster description

$A = \{x | x$ is a vowel$\}$ Set builder notation

We can modify the listing approach if the number of elements is quite large. For example, all of the letters of the alphabet can be listed as

$\{a, b, c, \ldots, z\}$

We simply begin by writing enough elements to establish a pattern; then the three dots indicate that the set continues in that pattern. The final entry indicates the last element of the pattern. If we write

$\{1, 2, 3, \ldots\}$

the set begins with the counting numbers 1, 2, and 3. The three dots indicate that it continues in a like manner forever; there is no last element. A set that consists of no elements is called the **null set** (written $\varnothing$).

Set builder notation combines the use of braces and the concept of a variable. For example, $\{x | x$ is a vowel$\}$ is read "the set of all x such that x is a vowel." Note that the vertical line is read "such that." We can use set builder notation to describe the set $\{1, 2, 3, \ldots\}$ as $\{x | x > 0$ and x is a whole number$\}$.

We use the symbol $\in$ to denote set membership. Thus if $A = \{a, e, i, o, u\}$, we can write $e \in A$, which we read as "e is an element of A." The slash symbol, $/$, is commonly used in mathematics as a negation symbol. For example, $m \notin A$ is read as "m is not an element of A."

Two sets are said to be *equal* if they contain exactly the same elements. For example,

$\{1, 2, 3\} = \{2, 1, 3\}$

because both sets contain the same elements; the order in which the elements are written doesn't matter. The slash mark through the equality symbol denotes "is not equal to." Thus if $A = \{1, 2, 3\}$ and $B = \{1, 2, 3, 4\}$, we can write $A \neq B$, which we read as "set A is not equal to set B."

Real Numbers

We refer to most of the algebra that we will study in this text as the **algebra of real numbers.** This simply means that the variables represent real numbers. Therefore, it is necessary for us to be familiar with the various terms that are used to classify different types of real numbers.

$\{1, 2, 3, 4, \ldots\}$	Natural numbers, counting numbers, positive integers
$\{0, 1, 2, 3, \ldots\}$	Whole numbers, nonnegative integers
$\{\ldots -3, -2, -1\}$	Negative integers
$\{\ldots -3, -2, -1, 0\}$	Nonpositive integers
$\{\ldots -3, -2, -1, 0, 1, 2, 3, \ldots\}$	Integers

We define a **rational number** as any number that can be expressed in the form $\frac{a}{b}$, where a and b are integers and b is not zero. The following are examples of rational numbers.

$$-\frac{3}{4}, \quad \frac{2}{3}, \quad -4, \quad 0, \quad 0.3, \quad 6\frac{1}{2}$$

-4 because $-4 = \frac{-4}{1} = \frac{4}{-1}$ 0 because $0 = \frac{0}{1} = \frac{0}{2} = \frac{0}{3} = \ldots$

0.3 because $0.3 = \frac{3}{10}$ $6\frac{1}{2}$ because $6\frac{1}{2} = \frac{13}{2}$

We can also define a rational number in terms of a decimal representation. Before doing so, let's review the different possibilities for decimal representations. We can classify decimals as **terminating, repeating,** or **nonrepeating.** Some examples follow.

$$\begin{bmatrix} 0.3 \\ 0.46 \\ 0.789 \\ 0.6234 \end{bmatrix} \text{ Terminating decimals} \qquad \begin{bmatrix} 0.6666\ldots \\ 0.141414\ldots \\ 0.694694694\ldots \\ 0.2317171717\ldots \\ 0.5417283283283\ldots \end{bmatrix} \text{ Repeating decimals}$$

$$\begin{bmatrix} 0.276314583\ldots \\ 0.21411811161111\ldots \\ 0.673183329333\ldots \end{bmatrix} \text{ Nonrepeating decimals}$$

A repeating decimal has a block of digits that repeats indefinitely. This repeating block of digits may be of any number of digits and may or may not begin immediately after the decimal point. A small horizontal bar (overbar) is commonly used to indicate the repeat block. Thus $0.6666\ldots$ is written as $0.\overline{6}$, and $0.2317171717\ldots$ is written as $0.23\overline{17}$.

In terms of decimals, we define a **rational number** as a number that has either a terminating or a repeating decimal representation. The following examples illustrate some rational numbers written in $\frac{a}{b}$ form and in decimal form.

$$\frac{3}{4} = 0.75 \qquad \frac{3}{11} = 0.\overline{27} \qquad \frac{1}{8} = 0.125 \qquad \frac{1}{7} = 0.\overline{142857} \qquad \frac{1}{3} = 0.\overline{3}$$

We define an **irrational number** as a number that *cannot* be expressed in $\dfrac{a}{b}$ form, where a and b are integers and b is not zero. Furthermore, an irrational number has a nonrepeating and nonterminating decimal representation. Some examples of irrational numbers and a partial decimal representation for each follow.

$$\sqrt{2} = 1.414213562373095\ldots \quad \sqrt{3} = 1.73205080756887\ldots$$

$$\pi = 3.14159265358979\ldots$$

The entire set of **real numbers** is composed of the rational numbers along with the irrationals. Every real number is either a rational number or an irrational number. The following tree diagram summarizes the various classifications of the real number system.

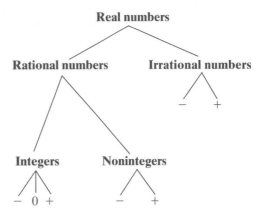

We can trace any real number down through the diagram as follows:

7 is real, rational, an integer, and positive.

$-\dfrac{2}{3}$ is real, rational, noninteger, and negative.

$\sqrt{7}$ is real, irrational, and positive.

0.38 is real, rational, noninteger, and positive.

REMARK: We usually refer to the set of nonnegative integers, $\{0, 1, 2, 3, \ldots\}$, as the set of **whole numbers** and to the set of positive integers, $\{1, 2, 3, \ldots\}$, as the set of **natural numbers.** The set of whole numbers differs from the set of natural numbers by the inclusion of the number zero.

The concept of subset is convenient to use at this time. A set A is a **subset** of a set B if and only if every element of A is also an element of B. This is written as $A \subseteq B$ and read as "A is a subset of B." For example, if $A = \{1, 2, 3\}$ and $B = \{1, 2, 3, 5, 9\}$, then $A \subseteq B$ because every element of A is also an element of B. The slash mark again denotes negation, so if $A = \{1, 2, 5\}$ and $B = \{2, 4, 7\}$, we can say that A is not a subset of B by writing $A \not\subseteq B$. Figure 1.1 represents the subset

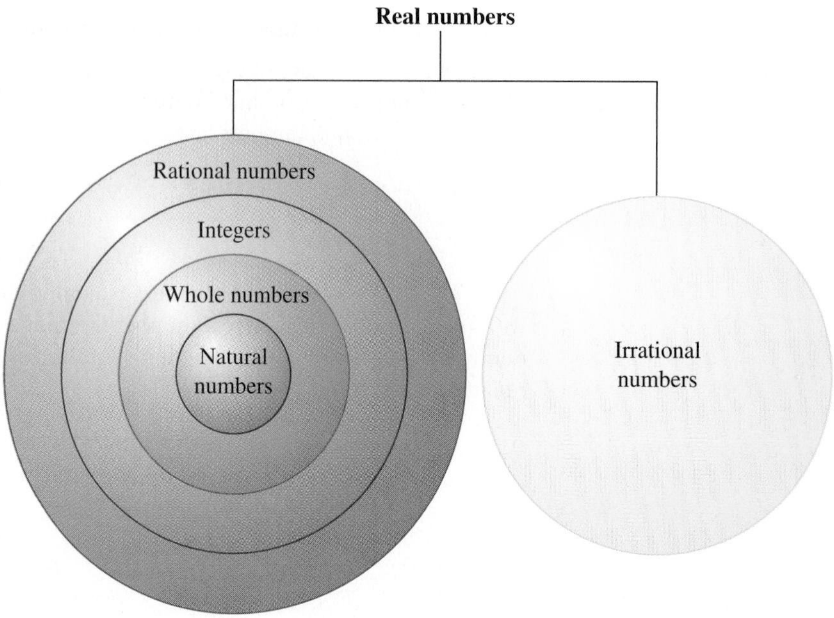

Real numbers

Figure 1.1

relationships for the set of real numbers. Refer to Figure 1.1 as you study the following statements that use subset vocabulary and subset symbolism.

   1. The set of whole numbers is a subset of the set of integers.

$$\{0, 1, 2, 3, \ldots\} \subseteq \{\ldots, -2, -1, 0, 1, 2, \ldots\}$$

   2. The set of integers is a subset of the set of rational numbers.

$$\{\ldots, -2, -1, 0, 1, 2, \ldots\} \subseteq \{x | x \text{ is a rational number}\}$$

   3. The set of rational numbers is a subset of the set of real numbers.

$$\{x | x \text{ is a rational number}\} \subseteq \{y | y \text{ is a real number}\}$$

Equality

The relation **equality** plays an important role in mathematics — especially when we are manipulating real numbers and algebraic expressions that represent real numbers. An equality is a statement in which two symbols, or groups of symbols, are names for the same number. The symbol $=$ is used to express an equality. Thus we can write

$$6 + 1 = 7 \qquad 18 - 2 = 16 \qquad 36 \div 4 = 9$$

(The symbol $\neq$ means *is not equal to.*) The following four basic properties of equality are self-evident, but we do need to keep them in mind. (We will expand this list in Chapter 2 when we work with solutions of equations.)

Properties of Equality

Reflexive Property

For any real number a,
$$a = a$$

EXAMPLES: $14 = 14$ $x = x$ $a + b = a + b$

Symmetric Property

For any real numbers a and b,
   if $a = b$, then $b = a$

EXAMPLES: If $13 + 1 = 14$, then $14 = 13 + 1$.
   If $3 = x + 2$, then $x + 2 = 3$.

Transitive Property

For any real numbers a, b, and c,
   if $a = b$ and $b = c$,   then $a = c$

EXAMPLES: If $3 + 4 = 7$ and $7 = 5 + 2$, then $3 + 4 = 5 + 2$.
   If $x + 1 = y$ and $y = 5$, then $x + 1 = 5$.

Substitution Property

For any real numbers a and b: If $a = b$, then a may be replaced by b, or b may be replaced by a, in any statement without changing the meaning of the statement.

EXAMPLES: If $x + y = 4$ and $x = 2$, then $2 + y = 4$.
   If $a - b = 9$ and $b = 4$, then $a - 4 = 9$.

Numerical Expressions

Let's conclude this section by *simplifying some numerical expressions* that involve whole numbers. When simplifying numerical expressions, we perform the operations in the following order. Be sure that you agree with the result in each example.

1. Perform the operations inside the symbols of inclusion (parentheses, brackets, and braces) and above and below each fraction bar. Start with the innermost inclusion symbol.
2. Perform all multiplications and divisions in the order in which they appear from left to right.
3. Perform all additions and subtractions in the order in which they appear from left to right.

EXAMPLE 1   Simplify $20 + 60 \div 10 \cdot 2$

Solution

First do the division.

$$20 + 60 \div 10 \cdot 2 = 20 + 6 \cdot 2$$

Next do the multiplication.

$$20 + 6 \cdot 2 = 20 + 12$$

Then do the addition.

$$20 + 12 = 32$$

Thus $20 + 60 \div 10 \cdot 2$ simplifies to 32.

EXAMPLE 2   Simplify $7 \cdot 4 \div 2 \cdot 3 \cdot 2 \div 4$.

Solution

The multiplications and divisions are to be done from left to right in the order in which they appear.

$$
\begin{aligned}
7 \cdot 4 \div 2 \cdot 3 \cdot 2 \div 4 &= 28 \div 2 \cdot 3 \cdot 2 \div 4 \\
&= 14 \cdot 3 \cdot 2 \div 4 \\
&= 42 \cdot 2 \div 4 \\
&= 84 \div 4 \\
&= 21
\end{aligned}
$$

Thus $7 \cdot 4 \div 2 \cdot 3 \cdot 2 \div 4$ simplifies to 21.

EXAMPLE 3   Simplify $5 \cdot 3 + 4 \div 2 - 2 \cdot 6 - 28 \div 7$.

Solution

First we do the multiplications and divisions in the order in which they appear. Then we do the additions and subtractions in the order in which they appear. Our work may take on the following format.

$$5 \cdot 3 + 4 \div 2 - 2 \cdot 6 - 28 \div 7 = 15 + 2 - 12 - 4 = 1$$

EXAMPLE 4   Simplify $(4 + 6)(7 + 8)$.

Solution

We use the parentheses to indicate the *product* of the quantities $4 + 6$ and $7 + 8$. We perform the additions inside the parentheses first and then multiply.

$$(4 + 6)(7 + 8) = (10)(15) = 150$$

EXAMPLE 5

Simplify $(3 \cdot 2 + 4 \cdot 5)(6 \cdot 8 - 5 \cdot 7)$.

Solution

First we do the multiplications inside the parentheses.

$$(3 \cdot 2 + 4 \cdot 5)(6 \cdot 8 - 5 \cdot 7) = (6 + 20)(48 - 35)$$

Then we do the addition and subtraction inside the parentheses.

$$(6 + 20)(48 - 35) = (26)(13)$$

Then we find the final product.

$$(26)(13) = 338$$ ∎

EXAMPLE 6

Simplify $6 + 7[3(4 + 6)]$.

Solution

We use brackets for the same purposes as parentheses. In such a problem we need to simplify *from the inside out;* that is, we perform the operations in the innermost parentheses first. We thus obtain

$$6 + 7[3(4 + 6)] = 6 + 7[3(10)]$$
$$= 6 + 7[30]$$
$$= 6 + 210$$
$$= 216$$ ∎

EXAMPLE 7

Simplify $\dfrac{6 \cdot 8 \div 4 - 2}{5 \cdot 4 - 9 \cdot 2}$.

Solution

First we perform the operations above and below the fraction bar. Then we find the final quotient.

$$\frac{6 \cdot 8 \div 4 - 2}{5 \cdot 4 - 9 \cdot 2} = \frac{48 \div 4 - 2}{20 - 18} = \frac{12 - 2}{2} = \frac{10}{2} = 5$$ ∎

REMARK: With parentheses we could write the problem in Example 7 as $(6 \cdot 8 \div 4 - 2) \div (5 \cdot 4 - 9 \cdot 2)$.

PROBLEM SET 1.1

For Problems 1–10, identify each statement as true or false.

1. Every irrational number is a real number.

2. Every rational number is a real number.

3. If a number is real, then it is irrational.

4. Every real number is a rational number.

5. All integers are rational numbers.

6. Some irrational numbers are also rational numbers.

7. Zero is a positive integer.

8. Zero is a rational number.

9. All whole numbers are integers.

10. Zero is a negative integer.

For Problems 11–18, from the list $0, 14, \frac{2}{3}, \pi, \sqrt{7}, -\frac{11}{14},$ $2.34, 3.\overline{21}, 6\frac{7}{8}, -\sqrt{17}, -19,$ and $-2.6,$ identify each of the following.

11. The whole numbers

12. The natural numbers

13. The rational numbers

14. The integers

15. The nonnegative integers

16. The irrational numbers

17. The real numbers

18. The nonpositive integers

For Problems 19–28, use the following set designations.

$N = \{x \mid x \text{ is a natural number}\}$
$Q = \{x \mid x \text{ is a rational number}\}$
$W = \{x \mid x \text{ is a whole number}\}$
$H = \{x \mid x \text{ is an irrational number}\}$
$I = \{x \mid x \text{ is an integer}\}$
$R = \{x \mid x \text{ is a real number}\}$

Place $\subseteq$ or $\nsubseteq$ in each blank to make a true statement.

19. R _____ N **20.** N _____ R

21. I _____ Q **22.** N _____ I

23. Q _____ H **24.** H _____ Q

25. N _____ W **26.** W _____ I

27. I _____ N **28.** I _____ W

For Problems 29–32, classify the real number by tracing through the diagram in the text (see page 5).

29. -8 **30.** 0.9

31. $-\sqrt{2}$ **32.** $\frac{5}{6}$

For Problems 33–42, list the elements of each set. For example, the elements of $\{x \mid x \text{ is a natural number less than 4}\}$ can be listed as $\{1, 2, 3\}$.

33. $\{x \mid x \text{ is a natural number less than 3}\}$

34. $\{x \mid x \text{ is a natural number greater than 3}\}$

35. $\{n \mid n \text{ is a whole number less than 6}\}$

36. $\{y \mid y \text{ is an integer greater than } -4\}$

37. $\{y \mid y \text{ is an integer less than 3}\}$

38. $\{n \mid n \text{ is a positive integer greater than } -7\}$

39. $\{x \mid x \text{ is a whole number less than 0}\}$

40. $\{x \mid x \text{ is a negative integer greater than } -3\}$

41. $\{n \mid n \text{ is a nonnegative integer less than 5}\}$

42. $\{n \mid n \text{ is a nonpositive integer greater than 3}\}$

For Problems 43–50, replace each question mark to make the given statement an application of the indicated property of equality. For example, $16 = ?$ becomes $16 = 16$ because of the reflexive property of equality.

43. If $y = x$ and $x = -6,$ then $y = ?$ (Transitive property of equality)

44. $5x + 7 = ?$ (Reflexive property of equality)

45. If $n = 2$ and $3n + 4 = 10,$ then $3(?) + 4 = 10$ (Substitution property of equality)

46. If $y = x$ and $x = z + 2,$ then $y = ?$ (Transitive property of equality)

47. If $4 = 3x + 1,$ then $? = 4$ (Symmetric property of equality)

48. If $t = 4$ and $s + t = 9,$ then $s + ? = 9$ (Substitution property of equality)

49. $5x = ?$ (Reflexive property of equality)

50. If $5 = n + 3,$ then $n + 3 = ?$ (Symmetric property of equality)

For Problems 51–74, simplify each of the numerical expressions.

51. $16 + 9 - 4 - 2 + 8 - 1$

52. $18 + 17 - 9 - 2 + 14 - 11$

53. $9 \div 3 \cdot 4 \div 2 \cdot 14$

54. $21 \div 7 \cdot 5 \cdot 2 \div 6$

55. $7 + 8 \cdot 2$

56. $21 - 4 \cdot 3 + 2$

57. $9 \cdot 7 - 4 \cdot 5 - 3 \cdot 2 + 4 \cdot 7$

58. $6 \cdot 3 + 5 \cdot 4 - 2 \cdot 8 + 3 \cdot 2$

59. $(17 - 12)(13 - 9)(7 - 4)$

60. $(14 - 12)(13 - 8)(9 - 6)$

61. $13 + (7 - 2)(5 - 1)$

62. $48 - (14 - 11)(10 - 6)$

63. $(5 \cdot 9 - 3 \cdot 4)(6 \cdot 9 - 2 \cdot 7)$

64. $(3 \cdot 4 + 2 \cdot 1)(5 \cdot 2 + 6 \cdot 7)$

65. $7[3(6 - 2)] - 64$

66. $12 + 5[3(7 - 4)]$

67. $[3 + 2(4 \cdot 1 - 2)][18 - (2 \cdot 4 - 7 \cdot 1)]$

68. $3[4(6 + 7)] + 2[3(4 - 2)]$

69. $14 + 4\left(\dfrac{8 - 2}{12 - 9}\right) - 2\left(\dfrac{9 - 1}{19 - 15}\right)$

70. $12 + 2\left(\dfrac{12 - 2}{7 - 2}\right) - 3\left(\dfrac{12 - 9}{17 - 14}\right)$

71. $[7 + 2 \cdot 3 \cdot 5 - 5] \div 8$

72. $[27 - (4 \cdot 2 + 5 \cdot 2)][(5 \cdot 6 - 4) - 20]$

73. $\dfrac{3 \cdot 8 - 4 \cdot 3}{5 \cdot 7 - 34} + 19$

74. $\dfrac{4 \cdot 9 - 3 \cdot 5 - 3}{18 - 12}$

75. You must of course be able to do calculations like those in Problems 51–74 both with and without a calculator. Furthermore, different types of calculators handle the priority-of-operations issue in different ways. Be sure you can do Problems 51–74 with *your* calculator.

■ ■ ■ **Thoughts into words**

76. Explain in your own words the difference between the reflexive property of equality and the symmetric property of equality.

77. Your friend keeps getting an answer of 30 when simplifying $7 + 8(2)$. What mistake is he making and how would you help him?

78. Do you think $3\sqrt{2}$ is a rational or an irrational number? Defend your answer.

79. Explain why every integer is a rational number but not every rational number is an integer.

80. Explain the difference between $1.\overline{3}$ and 1.3.

1.2 Operations with Real Numbers

Before we review the four basic operations with real numbers, let's briefly discuss some concepts and terminology we commonly use with this material. It is often helpful to have a geometric representation of the set of real numbers as indicated in Figure 1.2. Such a representation, called the **real number line,** indicates a one-

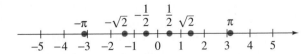

Figure 1.2

to-one correspondence between the set of real numbers and the points on a line. In other words, to each real number there corresponds one and only one point on the line, and to each point on the line there corresponds one and only one real number. The number associated with each point on the line is called the **coordinate** of the point.

Many operations, relations, properties, and concepts pertaining to real numbers can be given a geometric interpretation on the real number line. For example, the addition problem $(-1) + (-2)$ can be depicted on the number line as in Figure 1.3.

$$(-1) + (-2) = -3$$

Figure 1.3

Figure 1.4

The inequality relations also have a geometric interpretation. The statement $a > b$ (which is read "a is greater than b") means that a is to the right of b, and the statement $c < d$ (which is read "c is less than d") means that c is to the left of d as shown in Figure 1.4. The symbol $\leq$ means *is less than or equal to,* and the symbol $\geq$ means *is greater than or equal to.*

The property $-(-x) = x$ can be represented on the number line by following the sequence of steps shown in Figure 1.5.

1. Choose a point having a coordinate of x.
2. Locate its opposite, written as $-x$, on the other side of zero.
3. Locate the opposite of $-x$, written as $-(-x)$, on the other side of zero.

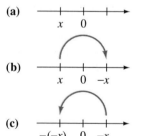

Figure 1.5

Therefore, we conclude that **the opposite of the opposite of any real number is the number itself,** and we symbolically express this by $-(-x) = x$.

REMARK: The symbol -1 can be read "negative one," "the negative of one," "the opposite of one," or "the additive inverse of one." The opposite-of and additive-inverse-of terminology is especially meaningful when working with variables. For example, the symbol $-x$, which is read "the opposite of x" or "the additive inverse of x," emphasizes an important issue. Because x can be any real number, $-x$ (the opposite of x) can be zero, positive, or negative. If x is positive, then $-x$ is negative. If x is negative, then $-x$ is positive. If x is zero, then $-x$ is zero.

Absolute Value

We can use the concept of **absolute value** to describe precisely how to operate with positive and negative numbers. Geometrically, the absolute value of any number is the distance between the number and zero on the number line. For example, the absolute value of 2 is 2. The absolute value of -3 is 3. The absolute value of 0 is 0 (see Figure 1.6).

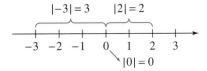

Figure 1.6

Symbolically, absolute value is denoted with vertical bars. Thus we write

$$|2| = 2 \qquad |-3| = 3 \qquad |0| = 0$$

More formally, we define the concept of absolute value as follows:

DEFINITION 1.1

For all real numbers a,

    1. If $a \geq 0$, then $|a| = a$.
    2. If $a < 0$, then $|a| = -a$.

According to Definition 1.1, we obtain

$$|6| = 6 \qquad\qquad \text{By applying part 1 of Definition 1.1}$$

$$|0| = 0 \qquad\qquad \text{By applying part 1 of Definition 1.1}$$

$$|-7| = -(-7) = 7 \qquad \text{By applying part 2 of Definition 1.1}$$

Note that the absolute value of a positive number is the number itself, but the absolute value of a negative number is its opposite. Thus the absolute value of any number except zero is positive, and the absolute value of zero is zero. Together, these facts indicate that the absolute value of any real number is equal to the absolute value of its opposite. We summarize these ideas in the following properties.

Properties of Absolute Value

The variables a and b represent any real number.

    1. $|a| \geq 0$
    2. $|a| = |-a|$
    3. $|a - b| = |b - a|$ $a - b$ and $b - a$ are opposites of each other.

Adding Real Numbers

We can use various physical models to describe the addition of real numbers. For example, profits and losses pertaining to investments: A loss of $25.75 (written as -25.75) on one investment, along with a profit of $22.20 (written as 22.20) on a second investment, produces an overall loss of $3.55. Thus $(-25.75) + 22.20 = -3.55$. Think in terms of profits and losses for each of the following examples.

$$50 + 75 = 125 \qquad\qquad 20 + (-30) = -10$$

$$-4.3 + (-6.2) = -10.5 \qquad -27 + 43 = 16$$

$$\frac{7}{8} + \left(-\frac{1}{4}\right) = \frac{5}{8} \qquad\qquad -3\frac{1}{2} + \left(-3\frac{1}{2}\right) = -7$$

Though all problems that involve addition of real numbers could be solved using the profit-loss interpretation, it is sometimes convenient to have a more precise description of the addition process. For this purpose we use the concept of absolute value.

Addition of Real Numbers

Two Positive Numbers   The sum of two positive real numbers is the sum of their absolute values.

Two Negative Numbers   The sum of two negative real numbers is the opposite of the sum of their absolute values.

One Positive and One Negative Number   The sum of a positive real number and a negative real number can be found by subtracting the smaller absolute value from the larger absolute value and giving the result the sign of the original number that has the larger absolute value. If the two numbers have the same absolute value, then their sum is 0.

Zero and Another Number   The sum of 0 and any real number is the real number itself.

Now consider the following examples in terms of the previous description of addition. These examples include operations with rational numbers in common fraction form. If you need a review on operations with fractions see appendix A.

$$(-6) + (-8) = -(|-6| + |-8|) = -(6 + 8) = -14$$

$$(-1.6) + (-7.7) = -(|-1.6| + |-7.7|) = -(1.6 + 7.7) = -9.3$$

$$6\frac{3}{4} + \left(-2\frac{1}{2}\right) = \left(\left|6\frac{3}{4}\right| - \left|-2\frac{1}{2}\right|\right) = \left(6\frac{3}{4} - 2\frac{1}{2}\right) = \left(6\frac{3}{4} - 2\frac{2}{4}\right) = 4\frac{1}{4}$$

$$14 + (-21) = -(|-21| - |14|) = -(21 - 14) = -7$$

$$-72.4 + 72.4 = 0 \qquad 0 + (-94) = -94$$

Subtracting Real Numbers

We can describe the subtraction of real numbers in terms of addition.

Subtraction of Real Numbers

If a and b are real numbers, then

$$a - b = a + (-b)$$

It may be helpful for you to read $a - b = a + (-b)$ as "a minus b is equal to a plus the opposite of b." In other words, every subtraction problem can be changed to an equivalent addition problem. Consider the following examples.

$$7 - 9 = 7 + (-9) = -2, \qquad -5 - (-13) = -5 + 13 = 8$$

$$6.1 - (-14.2) = 6.1 + 14.2 = 20.3, \qquad -16 - (-11) = -16 + 11 = -5$$

$$-\frac{7}{8} - \left(-\frac{1}{4}\right) = -\frac{7}{8} + \frac{1}{4} = -\frac{7}{8} + \frac{2}{8} = -\frac{5}{8}$$

It should be apparent that addition is a key operation. To simplify numerical expressions that involve addition and subtraction, we can first change all subtractions to additions and then perform the additions.

E X A M P L E   1

Simplify $7 - 9 - 14 + 12 - 6 + 4$.

Solution

$$7 - 9 - 14 + 12 - 6 + 4 = 7 + (-9) + (-14) + 12 + (-6) + 4$$

$$= -6$$ ∎

EXAMPLE 2 Simplify $-2\dfrac{1}{8} + \dfrac{3}{4} - \left(-\dfrac{3}{8}\right) - \dfrac{1}{2}$

Solution

$$-2\frac{1}{8} + \frac{3}{4} - \left(-\frac{3}{8}\right) - \frac{1}{2} = -2\frac{1}{8} + \frac{3}{4} + \frac{3}{8} + \left(-\frac{1}{2}\right)$$

$$= -\frac{17}{8} + \frac{6}{8} + \frac{3}{8} + \left(-\frac{4}{8}\right) \quad \text{Change to equivalent fractions with a common denominator.}$$

$$= -\frac{12}{8} = -\frac{3}{2}$$ ∎

It is often helpful to convert subtractions to additions *mentally*. In the next two examples, the work shown in the dashed boxes could be done in your head.

EXAMPLE 3 Simplify $4 - 9 - 18 + 13 - 10$.

Solution

$$4 - 9 - 18 + 13 - 10 = \boxed{4 + (-9) + (-18) + 13 + (-10)}$$

$$= -20$$ ∎

EXAMPLE 4 Simplify $\left(\dfrac{2}{3} - \dfrac{1}{5}\right) - \left(\dfrac{1}{2} - \dfrac{7}{10}\right)$

Solution

$$\left(\frac{2}{3} - \frac{1}{5}\right) - \left(\frac{1}{2} - \frac{7}{10}\right) = \boxed{\left[\frac{2}{3} + \left(-\frac{1}{5}\right)\right] - \left[\frac{1}{2} + \left(-\frac{7}{10}\right)\right]}$$

$$= \left[\frac{10}{15} + \left(-\frac{3}{15}\right)\right] - \left[\frac{5}{10} + \left(-\frac{7}{10}\right)\right]$$

Within the brackets, change to equivalent fractions with a common denominator.

$$= \left(\frac{7}{15}\right) - \left(-\frac{2}{10}\right)$$

$$= \boxed{\left(\frac{7}{15}\right) + \left(+\frac{2}{10}\right)}$$

$$= \frac{14}{30} + \left(+\frac{6}{30}\right) \quad \text{Change to equivalent fractions with a common denominator.}$$

$$= \frac{20}{30} = \frac{2}{3}$$ ∎

Multiplying Real Numbers

We can interpret the multiplication of whole numbers as repeated addition. For example, $3 \cdot 2$ means three 2s; thus $3 \cdot 2 = 2 + 2 + 2 = 6$. This same repeated-addition interpretation of multiplication can be used to find the product of a positive number and a negative number, as shown by the following examples.

$$2(-3) = -3 + (-3) = -6, \qquad 3(-2) = -2 + (-2) + (-2) = -6$$

$$4(-1.2) = -1.2 + (-1.2) + (-1.2) + (-1.2) = -4.8$$

$$3\left(-\frac{1}{8}\right) = -\frac{1}{8} + \left(-\frac{1}{8}\right) + \left(-\frac{1}{8}\right) = -\frac{3}{8}$$

When we are multiplying whole numbers, the order in which we multiply two factors does not change the product. For example, $2(3) = 6$ and $3(2) = 6$. Using this idea, we can handle a negative number times a positive number as follows:

$$(-2)(3) = (3)(-2) = (-2) + (-2) + (-2) = -6$$

$$(-3)(4) = (4)(-3) = (-3) + (-3) + (-3) + (-3) = -12$$

$$\left(-\frac{3}{7}\right)(2) = (2)\left(-\frac{3}{7}\right) = -\frac{3}{7} + \left(-\frac{3}{7}\right) = -\frac{6}{7}$$

Finally, let's consider the product of two negative integers. The following pattern using integers helps with the reasoning.

$$4(-2) = -8 \qquad 3(-2) = -6 \qquad 2(-2) = -4$$

$$1(-2) = -2 \qquad 0(-2) = 0 \qquad (-1)(-2) = ?$$

To continue this pattern, the product of -1 and -2 has to be 2. In general, this type of reasoning helps us realize that the product of any two negative real numbers is a positive real number. Using the concept of absolute value, we can describe the **multiplication of real numbers** as follows:

Multiplication of Real Numbers

1. The product of two positive or two negative real numbers is the product of their absolute values.
2. The product of a positive real number and a negative real number (either order) is the opposite of the product of their absolute values.
3. The product of zero and any real number is zero.

The following examples illustrate this description of multiplication. Again, the steps shown in the dashed boxes are usually performed mentally.

$$(-6)(-7) = |-6| \cdot |-7| = 6 \cdot 7 = 42$$

$$(8)(-9) = -(|8| \cdot |-9|) = -(8 \cdot 9) = -72$$

$$\left(-\frac{3}{4}\right)\left(\frac{1}{3}\right) = -\left(\left|-\frac{3}{4}\right| \cdot \left|\frac{1}{3}\right|\right) = -\left(\frac{3}{4} \cdot \frac{1}{3}\right) = -\frac{1}{4}$$

$$(-14.3)(0) = 0$$

The previous examples illustrated a step-by-step process for multiplying real numbers. In practice, however, the key is to remember that the product of two positive or two negative numbers is positive and that the product of a positive number and a negative number (either order) is negative.

Dividing Real Numbers

The relationship between multiplication and division provides the basis for dividing real numbers. For example, we know that $8 \div 2 = 4$ because $2 \cdot 4 = 8$. In other words, the quotient of two numbers can be found by looking at a related multiplication problem. In the following examples, we used this same type of reasoning to determine some quotients that involve integers.

$$\frac{6}{-2} = -3 \quad \text{because } (-2)(-3) = 6$$

$$\frac{-12}{3} = -4 \quad \text{because } (3)(-4) = -12$$

$$\frac{-18}{-2} = 9 \quad \text{because } (-2)(9) = -18$$

$$\frac{0}{-5} = 0 \quad \text{because } (-5)(0) = 0$$

$$\frac{-8}{0} \text{ is undefined} \qquad \text{Remember that division by zero is undefined!}$$

A precise description for **division of real numbers** follows.

Division of Real Numbers

1. The quotient of two positive or two negative real numbers is the quotient of their absolute values.
2. The quotient of a positive real number and a negative real number or of a negative real number and a positive real number is the opposite of the quotient of their absolute values.
3. The quotient of zero and any nonzero real number is zero.
4. The quotient of any nonzero real number and zero is undefined.

The following examples illustrate this description of division. Again, for practical purposes, the key is to remember whether the quotient is positive or negative.

$$\frac{-16}{-4} = \frac{|-16|}{|-4|} = \frac{16}{4} = 4 \qquad \frac{28}{-7} = -\left(\frac{|28|}{|-7|}\right) = -\left(\frac{28}{7}\right) = -4$$

$$\frac{-3.6}{4} = -\left(\frac{|-3.6|}{|4|}\right) = -\left(\frac{3.6}{4}\right) = -0.9 \qquad \frac{0}{\frac{7}{8}} = 0$$

Now let's simplify some numerical expressions that involve the four basic operations with real numbers. Remember that multiplications and divisions are done first, from left to right, before additions and subtractions are performed.

EXAMPLE 5   Simplify $-2\frac{1}{3} + 4\left(-\frac{2}{3}\right) - (-5)\left(-\frac{1}{3}\right)$

Solution

$$-2\frac{1}{3} + 4\left(-\frac{2}{3}\right) - (-5)\left(-\frac{1}{3}\right) = -2\frac{1}{3} + \left(-\frac{8}{3}\right) + \left(-\frac{5}{3}\right)$$

$$= -\frac{7}{3} + \left(-\frac{8}{3}\right) + \left(-\frac{5}{3}\right) \quad \text{Change to improper fraction.}$$

$$= -\frac{20}{3} \qquad \blacksquare$$

EXAMPLE 6   Simplify $-24 \div 4 + 8(-5) - (-5)(3)$.

Solution

$$-24 \div 4 + 8(-5) - (-5)(3) = -6 + (-40) - (-15)$$

$$= -6 + (-40) + 15$$

$$= -31 \qquad \blacksquare$$

EXAMPLE 7   Simplify $-7.3 - 2[-4.6(6 - 7)]$.

Solution

$$-7.3 - 2[-4.6(6 - 7)] = -7.3 - 2[-4.6(-1)] = -7.3 - 2[4.6]$$

$$= -7.3 - 9.2$$

$$= -7.3 + (-9.2)$$

$$= -16.5 \qquad \blacksquare$$

EXAMPLE 8   Simplify $[3(-7) - 2(9)][5(-7) + 3(9)]$.

Solution

$$[3(-7) - 2(9)][5(-7) + 3(9)] = [-21 - 18][-35 + 27]$$

$$= [-39][-8]$$

$$= 312 \qquad \blacksquare$$

PROBLEM SET 1.2

For Problems 1–50, perform the following operations with real numbers.

1. $8 + (-15)$

2. $9 + (-18)$

3. $(-12) + (-7)$

4. $(-7) + (-14)$

5. $-8 - 14$

6. $-17 - 9$

7. $9 - 16$

8. $8 - 22$

9. $(-9)(-12)$

10. $(-6)(-13)$

11. $(5)(-14)$

12. $(-17)(4)$

13. $(-56) \div (-4)$

14. $(-81) \div (-3)$

15. $\dfrac{-112}{16}$

16. $\dfrac{-75}{5}$

17. $-2\dfrac{3}{8} + 5\dfrac{7}{8}$

18. $-1\dfrac{1}{5} + 3\dfrac{4}{5}$

19. $4\dfrac{1}{3} - \left(-1\dfrac{1}{6}\right)$

20. $1\dfrac{1}{12} - \left(-5\dfrac{3}{4}\right)$

21. $\left(-\dfrac{1}{3}\right)\left(\dfrac{2}{5}\right)$

22. $(-8)\left(\dfrac{1}{3}\right)$

23. $\dfrac{1}{2} \div \left(-\dfrac{1}{8}\right)$

24. $\dfrac{2}{3} \div \left(-\dfrac{1}{6}\right)$

25. $0 \div (-14)$

26. $(-19) \div 0$

27. $(-21) \div 0$

28. $0 \div (-11)$

29. $-21 - 39$

30. $-23 - 38$

31. $-17.3 + 12.5$

32. $-16.3 + 19.6$

33. $21.42 - 7.29$

34. $2.73 - 8.14$

35. $-21.4 - (-14.9)$

36. $-32.6 - (-9.8)$

37. $(5.4)(-7.2)$

38. $(-8.5)(-3.3)$

39. $\dfrac{-1.2}{-6}$

40. $\dfrac{-6.3}{0.7}$

41. $\left(-\dfrac{1}{3}\right) + \left(-\dfrac{3}{4}\right)$

42. $-\dfrac{5}{6} + \dfrac{3}{8}$

43. $-\dfrac{3}{2} - \left(-\dfrac{3}{4}\right)$

44. $\dfrac{5}{8} - \dfrac{11}{12}$

45. $-\dfrac{2}{3} - \dfrac{7}{9}$

46. $\dfrac{5}{6} - \left(-\dfrac{2}{9}\right)$

47. $\left(-\dfrac{3}{4}\right)\left(\dfrac{4}{5}\right)$

48. $\left(\dfrac{1}{2}\right)\left(-\dfrac{4}{5}\right)$

49. $\dfrac{3}{4} \div \left(-\dfrac{1}{2}\right)$

50. $\left(-\dfrac{5}{6}\right) \div \left(-\dfrac{7}{8}\right)$

For Problems 51–90, simplify each numerical expression.

51. $9 - 12 - 8 + 5 - 6$

52. $6 - 9 + 11 - 8 - 7 + 14$

53. $-21 + (-17) - 11 + 15 - (-10)$

54. $-16 - (-14) + 16 + 17 - 19$

55. $7\dfrac{1}{8} - \left(2\dfrac{1}{4} - 3\dfrac{7}{8}\right)$

56. $-4\dfrac{3}{5} - \left(1\dfrac{1}{5} - 2\dfrac{3}{10}\right)$

57. $16 - 18 + 19 - [14 - 22 - (31 - 41)]$

58. $-19 - [15 - 13 - (-12 + 8)]$

59. $[14 - (16 - 18)] - [32 - (8 - 9)]$

60. $[-17 - (14 - 18)] - [21 - (-6 - 5)]$

61. $4\dfrac{1}{12} - \dfrac{1}{2}\left(\dfrac{1}{3}\right)$

62. $-\dfrac{4}{5} - \dfrac{1}{2}\left(-\dfrac{3}{5}\right)$

63. $-5 + (-2)(7) - (-3)(8)$

64. $-9 - 4(-2) + (-7)(6)$

65. $\dfrac{2}{5}\left(-\dfrac{3}{4}\right) - \left(-\dfrac{1}{2}\right)\left(\dfrac{3}{5}\right)$

66. $-\dfrac{2}{3}\left(\dfrac{1}{4}\right) + \left(-\dfrac{1}{3}\right)\left(\dfrac{5}{4}\right)$

67. $(-6)(-9) + (-7)(4)$

68. $(-7)(-7) - (-6)(4)$

69. $3(5 - 9) - 3(-6)$

70. $7(8 - 9) + (-6)(4)$

71. $(6 - 11)(4 - 9)$

72. $(7 - 12)(-3 - 2)$

73. $-6(-3 - 9 - 1)$

74. $-8(-3 - 4 - 6)$

75. $56 \div (-8) - (-6) \div (-2)$

76. $-65 \div 5 - (-13)(-2) + (-36) \div 12$

77. $-3[5 - (-2)] - 2(-4 - 9)$

78. $-2(-7 + 13) + 6(-3 - 2)$

79. $\dfrac{-6 + 24}{-3} + \dfrac{-7}{-6 - 1}$

80. $\dfrac{-12 + 20}{-4} + \dfrac{-7 - 11}{-9}$

81. $14.1 - (17.2 - 13.6)$

82. $-9.3 - (10.4 + 12.8)$

83. $3(2.1) - 4(3.2) - 2(-1.6)$

84. $5(-1.6) - 3(2.7) + 5(6.6)$

85. $7(6.2 - 7.1) - 6(-1.4 - 2.9)$

86. $-3(2.2 - 4.5) - 2(1.9 + 4.5)$

87. $\dfrac{2}{3} - \left(\dfrac{3}{4} - \dfrac{5}{6}\right)$

88. $-\dfrac{1}{2} - \left(\dfrac{3}{8} + \dfrac{1}{4}\right)$

89. $3\left(\dfrac{1}{2}\right) + 4\left(\dfrac{2}{3}\right) - 2\left(\dfrac{5}{6}\right)$

90. $2\left(\dfrac{3}{8}\right) - 5\left(\dfrac{1}{2}\right) + 6\left(\dfrac{3}{4}\right)$

91. Use a calculator to check your answers for Problems 51–86.

92. A scuba diver was 32 feet below sea level when he noticed that his partner had his extra knife. He ascended 13 feet to meet his partner and then continued to dive down for another 50 feet. How far below sea level is the diver?

93. Jeff played 18 holes of golf on Saturday. On each of 6 holes he was 1 under par, on each of 4 holes he was 2 over par, on 1 hole he was 3 over par, on each of 2 holes he shot par, and on each of 5 holes he was 1 over par. How did he finish relative to par?

94. After dieting for 30 days, Ignacio has lost 18 pounds. What number describes his average weight change per day?

95. Michael bet $5 on each of the 9 races at the racetrack. His only winnings were $28.50 on one race. How much did he win (or lose) for the day?

96. Max bought a piece of trim molding that measured $11\dfrac{3}{8}$ feet in length. Because of defects in the wood, he had to trim $1\dfrac{5}{8}$ feet off one end and he also had to remove $\dfrac{3}{4}$ of a foot off the other end. How long was the piece of molding after he trimmed the ends?

97. Natasha recorded the daily gains or losses for her company stock for a week. On Monday it gained 1.25 dollars; on Tuesday it gained 0.88 dollars; on Wednesday it lost 0.50 dollars; on Thursday it lost 1.13 dollars; on Friday it gained 0.38 dollars. What was the net gain (or loss) for the week?

98. On a summer day in Florida, the afternoon temperature was 96°F. After a thunderstorm, the temperature dropped 8°F. What would be the temperature if the sun came back out and the temperature rose 5°F?

99. In an attempt to lighten a dragster, the racing team exchanged two rear wheels for wheels that each weighed 15.6 pounds less. They also exchanged the crankshaft for one that weighed 4.8 pounds less. They changed the rear axle for one that weighed 23.7 pounds less but had to add an additional roll bar that weighed 10.6 pounds. If they wanted to lighten the dragster by 50 pounds, did they meet their goal?

100. A large corporation has five divisions. Two of the divisions had earnings of $2,300,000 each. The other three divisions had a loss of $1,450,000, a loss of $640,000, and a gain of $1,850,000, respectively. What was the net gain (or loss) of the corporation for the year?

■ ■ ■ **Thoughts into words**

101. Explain why $\dfrac{0}{8} = 0$, but $\dfrac{8}{0}$ is undefined.

102. The following simplification problem is incorrect. The answer should be -11. Find and correct the error.

$$8 \div (-4)(2) - 3(4) \div 2 + (-1) = (-2)(2) - 12 \div 1$$
$$= -4 - 12$$
$$= -16$$

1.3   Properties of Real Numbers and the Use of Exponents

At the beginning of this section we will list and briefly discuss some of the basic properties of real numbers. Be sure that you understand these properties, for they not only facilitate manipulations with real numbers but also serve as the basis for many algebraic computations.

Closure Property for Addition

If a and b are real numbers, then $a + b$ is a unique real number.

Closure Property for Multiplication

If a and b are real numbers, then ab is a unique real number.

We say that the set of real numbers is *closed* with respect to addition and also with respect to multiplication. That is, the sum of two real numbers is a unique real number, and the product of two real numbers is a unique real number. We use the word *unique* to indicate *exactly one*.

Commutative Property of Addition

If a and b are real numbers, then

$$a + b = b + a$$

Commutative Property of Multiplication

If a and b are real numbers, then

$$ab = ba$$

We say that addition and multiplication are commutative operations. This means that the order in which we add or multiply two numbers does not affect the result. For example, $6 + (-8) = (-8) + 6$ and $(-4)(-3) = (-3)(-4)$. It is also important to realize that subtraction and division are *not* commutative operations; order *does* make a difference. For example, $3 - 4 = -1$ but $4 - 3 = 1$. Likewise, $2 \div 1 = 2$ but $1 \div 2 = \dfrac{1}{2}$.

Associative Property of Addition

If a, b, and c are real numbers, then

$$(a + b) + c = a + (b + c)$$

Associative Property of Multiplication

If a, b, and c are real numbers, then

$$(ab)c = a(bc)$$

Addition and multiplication are **binary operations.** That is, we add (or multiply) two numbers at a time. The associative properties apply if more than two numbers are to be added or multiplied; they are grouping properties. For example, $(-8 + 9) + 6 = -8 + (9 + 6)$; changing the grouping of the numbers does not affect the final sum. This is also true for multiplication, which is illustrated by $[(-4)(-3)](2) = (-4)[(-3)(2)]$. Subtraction and division are *not* associative operations. For example, $(8 - 6) - 10 = -8$, but $8 - (6 - 10) = 12$. An example showing that division is not associative is $(8 \div 4) \div 2 = 1$, but $8 \div (4 \div 2) = 4$.

Identity Property of Addition

If a is any real number, then

$$a + 0 = 0 + a = a$$

Zero is called the identity element for addition. This merely means that the sum of any real number and zero is identically the same real number. For example, $-87 + 0 = 0 + (-87) = -87$.

Identity Property of Multiplication

If a is any real number, then

$$a(1) = 1(a) = a$$

We call 1 the identity element for multiplication. The product of any real number and 1 is identically the same real number. For example, $(-119)(1) = (1)(-119) = -119$.

Additive Inverse Property

For every real number a, there exists a unique real number $-a$ such that

$$a + (-a) = -a + a = 0$$

The real number $-a$ is called the **additive inverse of a** or the **opposite of a**. For example, 16 and -16 are additive inverses, and their sum is 0. The additive inverse of 0 is 0.

Multiplication Property of Zero

If a is any real number, then

$$(a)(0) = (0)(a) = 0$$

The product of any real number and zero is zero. For example, $(-17)(0) = 0(-17) = 0$.

Multiplication Property of Negative One

If a is any real number, then

$$(a)(-1) = (-1)(a) = -a$$

The product of any real number and -1 is the opposite of the real number. For example, $(-1)(52) = (52)(-1) = -52$.

Multiplicative Inverse Property

For every nonzero real number a, there exists a unique real number $\dfrac{1}{a}$ such that

$$a\left(\frac{1}{a}\right) = \frac{1}{a}(a) = 1$$

The number $\dfrac{1}{a}$ is called the **multiplicative inverse of a** or the **reciprocal of a.**
For example, the reciprocal of 2 is $\dfrac{1}{2}$ and $2\left(\dfrac{1}{2}\right) = \dfrac{1}{2}(2) = 1$. Likewise, the reciprocal of $\dfrac{1}{2}$ is $\dfrac{1}{\frac{1}{2}} = 2$. Therefore, 2 and $\dfrac{1}{2}$ are said to be reciprocals (or multiplicative inverses) of each other. Because division by zero is undefined, zero does not have a reciprocal.

Distributive Property

If a, b, and c are real numbers, then

$$a(b + c) = ab + ac$$

The distributive property ties together the operations of addition and multiplication. We say that **multiplication distributes over addition.** For example, $7(3 + 8) = 7(3) + 7(8)$. Because $b - c = b + (-c)$, it follows that **multiplication also distributes over subtraction.** This can be expressed symbolically as $a(b - c) = ab - ac$. For example, $6(8 - 10) = 6(8) - 6(10)$.

The following examples illustrate the use of the properties of real numbers to facilitate certain types of manipulations.

EXAMPLE 1

Simplify $[74 + (-36)] + 36$.

Solution

In such a problem, it is much more advantageous to group -36 and 36.

$$[74 + (-36)] + 36 = 74 + [(-36) + 36]$$ By using the associative
$$= 74 + 0 = 74$$ property for addition   ∎

EXAMPLE 2

Simplify $[(-19)(25)](-4)$.

Solution

It is much easier to group 25 and -4. Thus

$$[(-19)(25)](-4) = (-19)[(25)(-4)]$$ By using the associative
$$= (-19)(-100)$$ property for multiplication
$$= 1900$$ ∎

EXAMPLE 3

Simplify $17 + (-14) + (-18) + 13 + (-21) + 15 + (-33)$.

Solution

We could add in the order in which the numbers appear. However, because addition is commutative and associative, we could change the order and group in any convenient way. For example, we could add all of the positive integers and add all of the negative integers and then find the sum of these two results. It might be convenient to use the vertical format as follows:

$$
\begin{array}{rrr}
 & -14 & \\
17 & -18 & \\
13 & -21 & -86 \\
\underline{15} & \underline{-33} & \underline{45} \\
45 & -86 & -41
\end{array}
$$

∎

EXAMPLE 4

Simplify $-25(-2 + 100)$.

Solution

For this problem, it might be easiest to apply the distributive property first and then simplify.

$$-25(-2 + 100) = (-25)(-2) + (-25)(100)$$
$$= 50 + (-2500)$$
$$= -2450$$ ∎

EXAMPLE 5   Simplify $(-87)(-26 + 25)$.

Solution

For this problem, it would be better not to apply the distributive property but instead to add the numbers inside the parentheses first and then find the indicated product.

$$(-87)(-26 + 25) = (-87)(-1)$$

$$= 87$$

EXAMPLE 6   Simplify $3.7(104) + 3.7(-4)$.

Solution

Remember that the distributive property allows us to change from the form $a(b + c)$ to $ab + ac$ or from the form $ab + ac$ to $a(b + c)$. In this problem, we want to use the latter change. Thus

$$3.7(104) + 3.7(-4) = 3.7[104 + (-4)]$$

$$= 3.7(100)$$

$$= 370$$

Examples 4, 5, and 6 illustrate an important issue. Sometimes the form $a(b + c)$ is more convenient, but at other times the form $ab + ac$ is better. In these cases, as well as in the cases of other properties, you should *think first* and decide whether or not the properties can be used to make the manipulations easier.

Exponents

Exponents are used to indicate repeated multiplication. For example, we can write $4 \cdot 4 \cdot 4$ as 4^3, where the "raised 3" indicates that 4 is to be used as a factor 3 times. The following general definition is helpful.

DEFINITION 1.2

If n is a positive integer and b is any real number, then

$$b^n = \underbrace{bbb \cdots b}_{n \text{ factors of } b}$$

We refer to the b as the **base** and to n as the **exponent.** The expression b^n can be read "b to the nth power." We commonly associate the terms *squared* and *cubed* with exponents of 2 and 3, respectively. For example, b^2 is read "b squared" and b^3 as "b cubed." An exponent of 1 is usually not written, so b^1 is written as b. The following examples illustrate Definition 1.2.

$$2^3 = 2 \cdot 2 \cdot 2 = 8 \qquad \left(\frac{1}{2}\right)^5 = \frac{1}{2} \cdot \frac{1}{2} \cdot \frac{1}{2} \cdot \frac{1}{2} \cdot \frac{1}{2} = \frac{1}{32}$$

$$3^4 = 3 \cdot 3 \cdot 3 \cdot 3 = 81 \qquad (0.7)^2 = (0.7)(0.7) = 0.49$$

$$-5^2 = -(5 \cdot 5) = -25 \qquad (-5)^2 = (-5)(-5) = 25$$

Please take special note of the last two examples. Note that $(-5)^2$ means that -5 is the base and is to be used as a factor twice. However, -5^2 means that 5 is the base and that after it is squared, we take the opposite of that result.

Simplifying numerical expressions that contain exponents creates no trouble if we keep in mind that exponents are used to indicate repeated multiplication. Let's consider some examples.

E X A M P L E   7

Simplify $3(-4)^2 + 5(-3)^2$.

Solution

$$3(-4)^2 + 5(-3)^2 = 3(16) + 5(9) \qquad \text{Find the powers.}$$

$$= 48 + 45$$

$$= 93$$

E X A M P L E   8

Simplify $(2 + 3)^2$

Solution

$$(2 + 3)^2 = (5)^2 \qquad \qquad \text{Add inside the parentheses before}$$
$$\qquad \qquad \qquad \qquad \text{applying the exponent.}$$

$$= 25 \qquad \qquad \qquad \text{Square the 5.}$$

E X A M P L E   9

Simplify $[3(-1) - 2(1)]^3$.

Solution

$$[3(-1) - 2(1)]^3 = [-3 - 2]^3$$

$$= [-5]^3$$

$$= -125$$

E X A M P L E   1 0

Simplify $4\left(\dfrac{1}{2}\right)^3 - 3\left(\dfrac{1}{2}\right)^2 + 6\left(\dfrac{1}{2}\right) + 2$.

Solution

$$4\left(\frac{1}{2}\right)^3 - 3\left(\frac{1}{2}\right)^2 + 6\left(\frac{1}{2}\right) + 2 = 4\left(\frac{1}{8}\right) - 3\left(\frac{1}{4}\right) + 6\left(\frac{1}{2}\right) + 2$$

$$= \frac{1}{2} - \frac{3}{4} + 3 + 2$$

$$= \frac{19}{4}$$

PROBLEM SET 1.3

For Problems 1–14, state the property that justifies each of the statements. For example, $3 + (-4) = (-4) + 3$ because of the commutative property of addition.

1. $[6 + (-2)] + 4 = 6 + [(-2) + 4]$

2. $x(3) = 3(x)$

3. $42 + (-17) = -17 + 42$

4. $1(x) = x$

5. $-114 + 114 = 0$

6. $(-1)(48) = -48$

7. $-1(x + y) = -(x + y)$

8. $-3(2 + 4) = -3(2) + (-3)(4)$

9. $12yx = 12xy$

10. $[(-7)(4)](-25) = (-7)[4(-25)]$

11. $7(4) + 9(4) = (7 + 9)4$

12. $(x + 3) + (-3) = x + [3 + (-3)]$

13. $[(-14)(8)](25) = (-14)[8(25)]$

14. $\left(\dfrac{3}{4}\right)\left(\dfrac{4}{3}\right) = 1$

For Problems 15–26, simplify each numerical expression. Be sure to take advantage of the properties whenever they can be used to make the computations easier.

15. $36 + (-14) + (-12) + 21 + (-9) - 4$

16. $-37 + 42 + 18 + 37 + (-42) - 6$

17. $[83 + (-99)] + 18$

18. $[63 + (-87)] + (-64)$

19. $(25)(-13)(4)$

20. $(14)(25)(-13)(4)$

21. $17(97) + 17(3)$

22. $-86[49 + (-48)]$

23. $14 - 12 - 21 - 14 + 17 - 18 + 19 - 32$

24. $16 - 14 - 13 - 18 + 19 + 14 - 17 + 21$

25. $(-50)(15)(-2) - (-4)(17)(25)$

26. $(2)(17)(-5) - (4)(13)(-25)$

For Problems 27–54, simplify each of the numerical expressions.

27. $2^3 - 3^3$

28. $3^2 - 2^4$

29. $-5^2 - 4^2$

30. $-7^2 + 5^2$

31. $(-2)^3 - 3^2$

32. $(-3)^3 + 3^2$

33. $3(-1)^3 - 4(3)^2$

34. $4(-2)^3 - 3(-1)^4$

35. $7(2)^3 + 4(-2)^3$

36. $-4(-1)^2 - 3(2)^3$

37. $-3(-2)^3 + 4(-1)^5$

38. $5(-1)^3 - (-3)^3$

39. $(-3)^2 - 3(-2)(5) + 4^2$

40. $(-2)^2 - 3(-2)(6) - (-5)^2$

41. $2^3 + 3(-1)^3(-2)^2 - 5(-1)(2)^2$

42. $-2(3)^2 - 2(-2)^3 - 6(-1)^5$

43. $(3 + 4)^2$

44. $(4 - 9)^2$

45. $[3(-2)^2 - 2(-3)^2]^3$

46. $[-3(-1)^3 - 4(-2)^2]^2$

47. $2(-1)^3 - 3(-1)^2 + 4(-1) - 5$

48. $(-2)^3 + 2(-2)^2 - 3(-2) - 1$

49. $2^4 - 2(2)^3 - 3(2)^2 + 7(2) - 10$

50. $3(-3)^3 + 4(-3)^2 - 5(-3) + 7$

51. $3\left(\dfrac{1}{2}\right)^4 - 2\left(\dfrac{1}{2}\right)^3 + 5\left(\dfrac{1}{2}\right)^2 - 4\left(\dfrac{1}{2}\right) + 1$

52. $4(0.1)^2 - 6(0.1) + 0.7$

53. $-\left(\dfrac{2}{3}\right)^2 + 5\left(\dfrac{2}{3}\right) - 4$

54. $4\left(\dfrac{1}{3}\right)^3 + 3\left(\dfrac{1}{3}\right)^2 + 2\left(\dfrac{1}{3}\right) + 6$

C **55.** Use your calculator to check your answers for Problems 27–52.

C For Problems 56–64, use your calculator to evaluate each numerical expression.

56. 2^{10}

57. 3^7

58. $(-2)^8$

59. $(-2)^{11}$

60. -4^9

61. -5^6

62. $(3.14)^3$

63. $(1.41)^4$

64. $(1.73)^5$

The symbol, **C**, signals a problem that requires a calculator.

■ ■ ■ Thoughts into words

65. State, in your own words, the multiplication property of negative one.

66. Explain how the associative and commutative properties can help simplify $[(25)(97)](-4)$.

67. Your friend keeps getting an answer of 64 when simplifying -2^6. What mistake is he making, and how would you help him?

68. Write a sentence explaining in your own words how to evaluate the expression $(-8)^2$. Also write a sentence explaining how to evaluate -8^2.

69. For what natural numbers n does $(-1)^n = -1$? For what natural numbers n does $(-1)^n = 1$? Explain your answers.

70. Is the set $\{0, 1\}$ closed with respect to addition? Is the set $\{0, 1\}$ closed with respect to multiplication? Explain your answers.

1.4 Algebraic Expressions

Algebraic expressions such as

$$2x, \quad 8xy, \quad 3xy^2, \quad -4a^2b^3c, \quad \text{and} \quad z$$

are called **terms**. A term is an indicated product that may have any number of factors. The variables involved in a term are called **literal factors** and the numerical factor is called the **numerical coefficient.** Thus, in $8xy$, the x and y are literal factors and 8 is the numerical coefficient. The numerical coefficient of the term $-4a^2bc$ is -4. Because $1(z) = z$, the numerical coefficient of the term z is understood to be 1. Terms that have the same literal factors are called **similar terms** or **like terms.** Some examples of similar terms are

$$3x \quad \text{and} \quad 14x \qquad\qquad 5x^2 \quad \text{and} \quad 18x^2$$

$$7xy \quad \text{and} \quad -9xy \qquad\qquad 9x^2y \quad \text{and} \quad -14x^2y$$

$$2x^3y^2, \quad 3x^3y^2, \quad \text{and} \quad -7x^3y^2$$

By the symmetric property of equality, we can write the distributive property as

$$ab + ac = a(b + c)$$

Then the commutative property of multiplication can be applied to change the form to

$$ba + ca = (b + c)a$$

This latter form provides the basis for simplifying algebraic expressions by **combining similar terms.** Consider the following examples.

$$3x + 5x = (3 + 5)x \qquad\qquad -6xy + 4xy = (-6 + 4)xy$$
$$= 8x \qquad\qquad\qquad\qquad = -2xy$$
$$5x^2 + 7x^2 + 9x^2 = (5 + 7 + 9)x^2 \qquad 4x - x = 4x - 1x$$
$$= 21x^2 \qquad\qquad\qquad = (4 - 1)x = 3x$$

More complicated expressions might require that we first rearrange the terms by applying the commutative property for addition.

$$7x + 2y + 9x + 6y = 7x + 9x + 2y + 6y$$
$$= (7 + 9)x + (2 + 6)y \qquad \text{Distributive property}$$
$$= 16x + 8y$$

$$6a - 5 - 11a + 9 = 6a + (-5) + (-11a) + 9$$
$$= 6a + (-11a) + (-5) + 9 \qquad \text{Commutative property}$$
$$= (6 + (-11))a + 4 \qquad\qquad \text{Distributive property}$$
$$= -5a + 4$$

As soon as you thoroughly understand the various simplifying steps, you may want to do the steps mentally. Then you could go directly from the given expression to the simplified form, as follows:

$$14x + 13y - 9x + 2y = 5x + 15y$$
$$3x^2y - 2y + 5x^2y + 8y = 8x^2y + 6y$$
$$-4x^2 + 5y^2 - x^2 - 7y^2 = -5x^2 - 2y^2$$

Applying the distributive property to remove parentheses and then to combine similar terms sometimes simplifies an algebraic expression (as the next examples illustrate).

$$4(x + 2) + 3(x + 6) = 4(x) + 4(2) + 3(x) + 3(6)$$
$$= 4x + 8 + 3x + 18$$
$$= 4x + 3x + 8 + 18$$
$$= (4 + 3)x + 26$$
$$= 7x + 26$$

$$-5(y + 3) - 2(y - 8) = -5(y) - 5(3) - 2(y) - 2(-8)$$
$$= -5y - 15 - 2y + 16$$
$$= -5y - 2y - 15 + 16$$
$$= -7y + 1$$

$$5(x - y) - (x + y) = 5(x - y) - 1(x + y) \qquad \text{Remember, } -a = -1(a).$$
$$= 5(x) - 5(y) - 1(x) - 1(y)$$
$$= 5x - 5y - 1x - 1y$$
$$= 4x - 6y$$

When we are multiplying two terms such as 3 and $2x$, the associative property for multiplication provides the basis for simplifying the product.

$$3(2x) = (3 \cdot 2)x = 6x$$

This idea is put to use in the following example.

$$3(2x + 5y) + 4(3x + 2y) = 3(2x) + 3(5y) + 4(3x) + 4(2y)$$
$$= 6x + 15y + 12x + 8y$$
$$= 6x + 12x + 15y + 8y$$
$$= 18x + 23y$$

After you are sure of each step, a more simplified format may be used, as the following examples illustrate.

$$5(a + 4) - 7(a + 3) = 5a + 20 - 7a - 21 \qquad \text{Be careful with this sign.}$$
$$= -2a - 1$$
$$3(x^2 + 2) + 4(x^2 - 6) = 3x^2 + 6 + 4x^2 - 24$$
$$= 7x^2 - 18$$
$$2(3x - 4y) - 5(2x - 6y) = 6x - 8y - 10x + 30y$$
$$= -4x + 22y$$

Evaluating Algebraic Expressions

An algebraic expression takes on a numerical value whenever each variable in the expression is replaced by a real number. For example, if x is replaced by 5 and y by 9, the algebraic expression $x + y$ becomes the numerical expression $5 + 9$, which simplifies to 14. We say that $x + y$ has a value of 14 when x equals 5 and y equals 9. If $x = -3$ and $y = 7$, then $x + y$ has a value of $-3 + 7 = 4$. The following examples illustrate the process of finding a value of an algebraic expression. We commonly refer to the process as **evaluating algebraic expressions.**

EXAMPLE 1   Find the value of $3x - 4y$ when $x = 2$ and $y = -3$.

Solution

$$3x - 4y = 3(2) - 4(-3), \quad \text{when } x = 2 \text{ and } y = -3$$
$$= 6 + 12$$
$$= 18$$

E X A M P L E   2

Evaluate $x^2 - 2xy + y^2$ for $x = -2$ and $y = -5$.

Solution

$$x^2 - 2xy + y^2 = (-2)^2 - 2(-2)(-5) + (-5)^2, \quad \text{when } x = -2 \text{ and } y = -5$$

$$= 4 - 20 + 25$$

$$= 9$$

E X A M P L E   3

Evaluate $(a + b)^2$ for $a = 6$ and $b = -2$.

Solution

$$(a + b)^2 = [6 + (-2)]^2, \quad \text{when } a = 6 \text{ and } b = -2$$

$$= (4)^2$$

$$= 16$$

E X A M P L E   4

Evaluate $(3x + 2y)(2x - y)$ for $x = 4$ and $y = -1$.

Solution

$$(3x + 2y)(2x - y) = [3(4) + 2(-1)][2(4) - (-1)] \quad \text{when } x = 4$$
$$\text{and } y = -1$$

$$= (12 - 2)(8 + 1)$$

$$= (10)(9)$$

$$= 90$$

E X A M P L E   5

Evaluate $7x - 2y + 4x - 3y$ for $x = -\dfrac{1}{2}$ and $y = \dfrac{2}{3}$.

Solution

Let's first simplify the given expression.

$$7x - 2y + 4x - 3y = 11x - 5y$$

Now we can substitute $-\dfrac{1}{2}$ for x and $\dfrac{2}{3}$ for y.

$$11x - 5y = 11\left(-\frac{1}{2}\right) - 5\left(\frac{2}{3}\right)$$

$$= -\frac{11}{2} - \frac{10}{3}$$

$$= -\frac{33}{6} - \frac{20}{6} \qquad \text{Change to equivalent fractions}$$
$$\text{with a common denominator.}$$

$$= -\frac{53}{6}$$

EXAMPLE 6

Evaluate $2(3x + 1) - 3(4x - 3)$ for $x = -6.2$.

Solution

Let's first simplify the given expression.

$$2(3x + 1) - 3(4x - 3) = 6x + 2 - 12x + 9$$
$$= -6x + 11$$

Now we can substitute -6.2 for x.

$$-6x + 11 = -6(-6.2) + 11$$
$$= 37.2 + 11$$
$$= 48.2$$

EXAMPLE 7

Evaluate $2(a^2 + 1) - 3(a^2 + 5) + 4(a^2 - 1)$ for $a = 10$.

Solution

Let's first simplify the given expression.

$$2(a^2 + 1) - 3(a^2 + 5) + 4(a^2 - 1) = 2a^2 + 2 - 3a^2 - 15 + 4a^2 - 4$$
$$= 3a^2 - 17$$

Substituting $a = 10$, we obtain

$$3a^2 - 17 = 3(10)^2 - 17$$
$$= 3(100) - 17$$
$$= 300 - 17$$
$$= 283$$

Translating from English to Algebra

To use the tools of algebra to solve problems, we must be able to translate from English to algebra. This translation process requires that we recognize key phrases in the English language that translate into algebraic expressions (which involve the operations of addition, subtraction, multiplication, and division). Some of these key phrases and their algebraic counterparts are listed in the following table. The variable n represents the number being referred to in each phrase. When translating, remember that the commutative property holds only for the operations of addition and multiplication. Therefore, order will be crucial to algebraic expressions that involve subtraction and division.

English phrase	Algebraic expression
Addition	
The sum of a number and 4	$n + 4$
7 more than a number	$n + 7$
A number plus 10	$n + 10$
A number increased by 6	$n + 6$
8 added to a number	$n + 8$
Subtraction	
14 minus a number	$14 - n$
12 less than a number	$n - 12$
A number decreased by 10	$n - 10$
The difference between a number and 2	$n - 2$
5 subtracted from a number	$n - 5$
Multiplication	
14 times a number	$14n$
The product of 4 and a number	$4n$
$\dfrac{3}{4}$ of a number	$\dfrac{3}{4}n$
Twice a number	$2n$
Multiply a number by 12	$12n$
Division	
The quotient of 6 and a number	$\dfrac{6}{n}$
The quotient of a number and 6	$\dfrac{n}{6}$
A number divided by 9	$\dfrac{n}{9}$
The ratio of a number and 4	$\dfrac{n}{4}$
Mixture of operations	
4 more than three times a number	$3n + 4$
5 less than twice a number	$2n - 5$
3 times the sum of a number and 2	$3(n + 2)$
2 more than the quotient of a number and 12	$\dfrac{n}{12} + 2$
7 times the difference of 6 and a number	$7(6 - n)$

An English statement may not always contain a key word such as *sum, difference, product,* or *quotient.* Instead, the statement may describe a physical situation, and from this description we must deduce the operations involved. Some suggestions for handling such situations are given in the following examples.

EXAMPLE 8

Sonya can type 65 words per minute. How many words will she type in m minutes?

Solution

The total number of words typed equals the product of the rate per minute and the number of minutes. Therefore, Sonya should be able to type $65m$ words in m minutes. ■

EXAMPLE 9

Russ has n nickels and d dimes. Express this amount of money in cents.

Solution

Each nickel is worth 5 cents and each dime is worth 10 cents. We represent the amount in cents by $5n + 10d$. ■

EXAMPLE 10

The cost of a 50-pound sack of fertilizer is d dollars. What is the cost per pound for the fertilizer?

Solution

We calculate the cost per pound by dividing the total cost by the number of pounds. We represent the cost per pound by $\dfrac{d}{50}$. ■

The English statement we want to translate into algebra may contain some geometric ideas. Tables 1.1 and 1.2 contain some of the basic relationships that pertain to linear measurement in the English and metric systems, respectively.

Table 1.1 English system

12 inches = 1 foot
3 feet = 1 yard
1760 yards = 1 mile
5280 feet = 1 mile

Table 1.2 Metric system

1 kilometer = 1000 meters
1 hectometer = 100 meters
1 dekameter = 10 meters
1 decimeter = 0.1 meter
1 centimeter = 0.01 meter
1 millimeter = 0.001 meter

EXAMPLE 11

The distance between two cities is k kilometers. Express this distance in meters.

Solution

Because 1 kilometer equals 1000 meters, the distance in meters is represented by $1000k$. ■

E X A M P L E   1 2   The length of a rope is y yards and f feet. Express this length in inches.

Solution

Because 1 foot equals 12 inches and 1 yard equals 36 inches, the length of the rope in inches can be represented by $36y + 12f$. ∎

E X A M P L E   1 3   The length of a rectangle is l centimeters and the width is w centimeters. Express the perimeter of the rectangle in meters.

Solution

A sketch of the rectangle may be helpful (Figure 1.7).

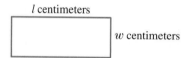

l centimeters

w centimeters

Figure 1.7

The perimeter of a rectangle is the sum of the lengths of the four sides. Thus the perimeter in centimeters is $l + w + l + w$, which simplifies to $2l + 2w$. Now, because 1 centimeter equals 0.01 meter, the perimeter, in meters, is $0.01(2l + 2w)$. This could also be written as $\dfrac{2l + 2w}{100} = \dfrac{2(l + w)}{100} = \dfrac{l + w}{50}$. ∎

PROBLEM SET 1.4

Simplify the algebraic expressions in Problems 1–14 by combining similar terms.

1. $-7x + 11x$

2. $5x - 8x + x$

3. $5a^2 - 6a^2$

4. $12b^3 - 17b^3$

5. $4n - 9n - n$

6. $6n + 13n - 15n$

7. $4x - 9x + 2y$

8. $7x - 9y - 10x - 13y$

9. $-3a^2 + 7b^2 + 9a^2 - 2b^2$ **10.** $-xy + z - 8xy - 7z$

11. $15x - 4 + 6x - 9$

12. $5x - 2 - 7x + 4 - x - 1$

13. $5a^2b - ab^2 - 7a^2b$

14. $8xy^2 - 5x^2y + 2xy^2 + 7x^2y$

Simplify the algebraic expressions in Problems 15–34 by removing parentheses and combining similar terms.

15. $3(x + 2) + 5(x + 3)$

16. $5(x - 1) + 7(x + 4)$

17. $-2(a - 4) - 3(a + 2)$

18. $-7(a + 1) - 9(a + 4)$

19. $3(n^2 + 1) - 8(n^2 - 1)$

20. $4(n^2 + 3) + (n^2 - 7)$

21. $-6(x^2 - 5) - (x^2 - 2)$

22. $3(x + y) - 2(x - y)$

23. $5(2x + 1) + 4(3x - 2)$

24. $5(3x - 1) + 6(2x + 3)$

25. $3(2x - 5) - 4(5x - 2)$

26. $3(2x - 3) - 7(3x - 1)$

27. $-2(n^2 - 4) - 4(2n^2 + 1)$

28. $-4(n^2 + 3) - (2n^2 - 7)$

29. $3(2x - 4y) - 2(x + 9y)$

30. $-7(2x - 3y) + 9(3x + y)$

31. $3(2x - 1) - 4(x + 2) - 5(3x + 4)$

32. $-2(x - 1) - 5(2x + 1) + 4(2x - 7)$

33. $-(3x - 1) - 2(5x - 1) + 4(-2x - 3)$

34. $4(-x - 1) + 3(-2x - 5) - 2(x + 1)$

Evaluate the algebraic expressions in Problems 35–57 for the given values of the variables.

35. $3x + 7y$, $x = -1$ and $y = -2$

36. $5x - 9y$, $x = -2$ and $y = 5$

37. $4x^2 - y^2$, $x = 2$ and $y = -2$

38. $3a^2 + 2b^2$, $a = 2$ and $b = 5$

39. $2a^2 - ab + b^2$, $a = -1$ and $b = -2$

40. $-x^2 + 2xy + 3y^2$, $x = -3$ and $y = 3$

41. $2x^2 - 4xy - 3y^2$, $x = 1$ and $y = -1$

42. $4x^2 + xy - y^2$, $x = 3$ and $y = -2$

43. $3xy - x^2y^2 + 2y^2$, $x = 5$ and $y = -1$

44. $x^2y^3 - 2xy + x^2y^2$, $x = -1$ and $y = -3$

45. $7a - 2b - 9a + 3b$, $a = 4$ and $b = -6$

46. $-4x + 9y - 3x - y$, $x = -4$ and $y = 7$

47. $(x - y)^2$, $x = 5$ and $y = -3$

48. $2(a + b)^2$, $a = 6$ and $b = -1$

49. $-2a - 3a + 7b - b$, $a = -10$ and $b = 9$

50. $3(x - 2) - 4(x + 3)$, $x = -2$

51. $-2(x + 4) - (2x - 1)$, $x = -3$

52. $-4(2x - 1) + 7(3x + 4)$, $x = 4$

53. $2(x - 1) - (x + 2) - 3(2x - 1)$, $x = -1$

54. $-3(x + 1) + 4(-x - 2) - 3(-x + 4)$, $x = -\dfrac{1}{2}$

55. $3(x^2 - 1) - 4(x^2 + 1) - (2x^2 - 1)$, $x = \dfrac{2}{3}$

56. $2(n^2 + 1) - 3(n^2 - 3) + 3(5n^2 - 2)$, $n = \dfrac{1}{4}$

57. $5(x - 2y) - 3(2x + y) - 2(x - y)$, $x = \dfrac{1}{3}$ and $y = -\dfrac{3}{4}$

C For Problems 58–63, use your calculator and evaluate each of the algebraic expressions for the indicated values. Express the final answers to the nearest tenth.

58. πr^2, $\pi = 3.14$ and $r = 2.1$

59. πr^2, $\pi = 3.14$ and $r = 8.4$

60. $\pi r^2 h$, $\pi = 3.14, r = 1.6$, and $h = 11.2$

61. $\pi r^2 h$, $\pi = 3.14, r = 4.8$, and $h = 15.1$

62. $2\pi r^2 + 2\pi rh$, $\pi = 3.14, r = 3.9$, and $h = 17.6$

63. $2\pi r^2 + 2\pi rh$, $\pi = 3.14, r = 7.8$, and $h = 21.2$

For Problems 64–78, translate each English phrase into an algebraic expression and use n to represent the unknown number.

64. The sum of a number and 4

65. A number increased by 12

66. A number decreased by 7

67. Five less than a number

68. A number subtracted from 75

69. The product of a number and 50

70. One-third of a number

71. Four less than one-half of a number

72. Seven more than three times a number

73. The quotient of a number and 8

74. The quotient of 50 and a number

75. Nine less than twice a number

76. Six more than one-third of a number

77. Ten times the difference of a number and 6

78. Twelve times the sum of a number and 7

For Problems 79–99, answer the question with an algebraic expression.

79. Brian is n years old. How old will he be in 20 years?

80. Crystal is n years old. How old was she 5 years ago?

81. Pam is t years old and her mother is 3 less than twice as old as Pam. What is the age of Pam's mother?

82. The sum of two numbers is 65 and one of the numbers is x. What is the other number?

83. The difference of two numbers is 47 and the smaller number is n. What is the other number?

84. The product of two numbers is 98 and one of the numbers is n. What is the other number?

85. The quotient of two numbers is 8 and the smaller number is y. What is the other number?

86. The perimeter of a square is c centimeters. How long is each side of the square?

87. The perimeter of a square is m meters. How long, in centimeters, is each side of the square?

88. Jesse has n nickels, d dimes, and q quarters in his bank. How much money, in cents, does he have in his bank?

89. Tina has c cents, which is all in quarters. How many quarters does she have?

90. If n represents a whole number, what represents the next larger whole number?

The symbol, ⬛, signals a problem that requires a calculator.

91. If n represents an odd integer, what represents the next larger odd integer?

92. If n represents an even integer, what represents the next larger even integer?

93. The cost of a 5-pound box of candy is c cents. What is the price per pound?

94. Larry's annual salary is d dollars. What is his monthly salary?

95. Mila's monthly salary is d dollars. What is her annual salary?

96. The perimeter of a square is i inches. What is the perimeter expressed in feet?

97. The perimeter of a rectangle is y yards and f feet. What is the perimeter expressed in feet?

98. The length of a line segment is d decimeters. How long is the line segment expressed in meters?

99. The distance between two cities is m miles. How far is this, expressed in feet?

⬛ **100.** Use your calculator to check your answers for Problems 35–54.

⬛ ⬛ ⬛ Thoughts into words

101. Explain the difference between simplifying a numerical expression and evaluating an algebraic expression.

102. How would you help someone who is having difficulty expressing n nickels and d dimes in terms of cents?

103. When asked to write an algebraic expression for "8 more than a number," you wrote $x + 8$ and another student wrote $8 + x$. Are both expressions correct? Explain your answer.

104. When asked to write an algebraic expression for "6 less than a number," you wrote $x - 6$ and another student wrote $6 - x$. Are both expressions correct? Explain your answer.

CHAPTER 1

SUMMARY

(1.1) A **set** is a collection of objects; the objects are called **elements** or **members** of the set. Set A is a subset of set B if and only if every member of A is also a member of B. The sets of **natural numbers, whole numbers, integers, rational numbers,** and **irrational numbers** are all subsets of the set of **real numbers.**

We can evaluate **numerical expressions** by performing the operations in the following order.

1. Perform the operations inside the parentheses and above and below fraction bars.

2. Find all powers or convert them to indicated multiplication.

3. Perform all multiplications and divisions in the order in which they appear from left to right.

4. Perform all additions and subtractions in the order in which they appear from left to right.

(1.2) The **absolute value** of a real number a is defined as follows:

1. If $a \geq 0$, then $|a| = a$.

2. If $a < 0$, then $|a| = -a$.

Operations with Real Numbers

Addition

1. The sum of two positive real numbers is the sum of their absolute values.

2. The sum of two negative real numbers is the opposite of the sum of their absolute values.

3. The sum of one positive and one negative number is found as follows:

   a. If the positive number has the larger absolute value, then the sum is the difference of their absolute values when the smaller absolute value is subtracted from the larger absolute value.

   b. If the negative number has the larger absolute value, then the sum is the opposite of the difference of their absolute values when the smaller absolute value is subtracted from the larger absolute value.

Subtraction

Applying the principle that $a - b = a + (-b)$ changes every subtraction problem to an equivalent addition problem. Then the rules for addition can be followed.

Multiplication

1. The product of two positive numbers or two negative real numbers is the product of their absolute values.

2. The product of one positive and one negative real number is the opposite of the product of their absolute values.

Division

1. The quotient of two positive numbers or two negative real numbers is the quotient of absolute values.

2. The quotient of one positive and one negative real number is the opposite of the quotient of their absolute values.

(1.3) The following basic properties of real numbers help with numerical manipulations and serve as a basis for algebraic computations.

Closure properties

$$a + b \text{ is a real number}$$
$$ab \text{ is a real number}$$

Commutative properties

$$a + b = b + a$$
$$ab = ba$$

Associative properties

$$(a + b) + c = a + (b + c)$$
$$(ab)c = a(bc)$$

Identity properties

$$a + 0 = 0 + a = a$$
$$a(1) = 1(a) = a$$

Additive inverse property

$$a + (-a) = (-a) + a = 0$$

Multiplication property of zero

$$a(0) = 0(a) = 0$$

Multiplication property of negative one

$$-1(a) = a(-1) = -a$$

Multiplicative inverse property

$$a\left(\frac{1}{a}\right) = \left(\frac{1}{a}\right)a = 1$$

Distributive properties

$$a(b + c) = ab + ac$$
$$a(b - c) = ab - ac$$

(1.4) Algebraic expressions such as

$$2x, \quad 8xy, \quad 3xy^2, \quad -4a^2b^3c, \quad \text{and} \quad z$$

are called **terms**. A term is an indicated product and may have any number of factors. We call the variables in a term the **literal factors,** and we call the numerical factor the **numerical coefficient.** Terms that have the same literal factors are called **similar** or **like terms.**

The distributive property in the form $ba + ca = (b + c)a$ serves as the basis for **combining similar terms.** For example,

$$3x^2y + 7x^2y = (3 + 7)x^2y = 10x^2y$$

To translate English phrases into algebraic expressions, we must be familiar with the key phrases that signal whether we are to find a sum, difference, product, or quotient.

CHAPTER 1   REVIEW PROBLEM SET

1. From the list $0, \sqrt{2}, \frac{3}{4}, -\frac{5}{6}, 8\frac{1}{3}, -\sqrt{3}, -8, 0.34, .2\bar{3},$

   67, and $\frac{9}{7}$, identify each of the following.

   a. The natural numbers
   b. The integers
   c. The nonnegative integers
   d. The rational numbers
   e. The irrational numbers

For Problems 2–10, state the property of equality or the property of real numbers that justifies each of the statements. For example, $6(-7) = -7(6)$ because of the commutative property for multiplication; and if $2 = x + 3$, then $x + 3 = 2$ is true because of the symmetric property of equality.

2. $7 + (3 + (-8)) = (7 + 3) + (-8)$

3. If $x = 2$ and $x + y = 9$, then $2 + y = 9$.

4. $-1(x + 2) = -(x + 2)$

5. $3(x + 4) = 3(x) + 3(4)$

6. $[(17)(4)](25) = (17)[(4)(25)]$

7. $x + 3 = 3 + x$

8. $3(98) + 3(2) = 3(98 + 2)$

9. $\left(\frac{3}{4}\right)\left(\frac{4}{3}\right) = 1$

10. If $4 = 3x - 1$, then $3x - 1 = 4$.

For Problems 11–22, simplify each of the numerical expressions.

11. $-8\frac{1}{4} + \left(-4\frac{5}{8}\right) - \left(-6\frac{3}{8}\right)$

12. $9\frac{1}{3} - 12\frac{1}{2} + \left(-4\frac{1}{6}\right) - \left(-1\frac{1}{6}\right)$

13. $-8(2) - 16 \div (-4) + (-2)(-2)$

14. $4(-3) - 12 \div (-4) + (-2)(-1) - 8$

15. $-3(2 - 4) - 4(7 - 9) + 6$

16. $[48 + (-73)] + 74$

17. $[5(-2) - 3(-1)][-2(-1) + 3(2)]$

18. $-4^2 - 2^3$

19. $(-2)^4 + (-1)^3 - 3^2$

20. $2(-1)^2 - 3(-1)(2) - 2^2$

21. $[4(-1) - 2(3)]^2$

22. $3 - [-2(3 - 4)] + 7$

For Problems 23–32, simplify each of the algebraic expressions by combining similar terms.

23. $3a^2 - 2b^2 - 7a^2 - 3b^2$

24. $4x - 6 - 2x - 8 + x + 12$

25. $\frac{1}{5}ab^2 - \frac{3}{10}ab^2 + \frac{2}{5}ab^2 + \frac{7}{10}ab^2$

26. $-\frac{2}{3}x^2y - \left(-\frac{3}{4}x^2y\right) - \frac{5}{12}x^2y - 2x^2y$

27. $3(2n^2 + 1) + 4(n^2 - 5)$

28. $-2(3a - 1) + 4(2a + 3) - 5(3a + 2)$

29. $-(n - 1) - (n + 2) + 3$

30. $3(2x - 3y) - 4(3x + 5y) - x$

31. $4(a - 6) - (3a - 1) - 2(4a - 7)$

32. $-5(x^2 - 4) - 2(3x^2 + 6) + (2x^2 - 1)$

For Problems 33–42, evaluate each of the algebraic expressions for the given values of the variables.

33. $-5x + 4y$ for $x = \frac{1}{2}$ and $y = -1$

34. $3x^2 - 2y^2$ for $x = \frac{1}{4}$ and $y = -\frac{1}{2}$

35. $-5(2x - 3y)$ for $x = 1$ and $y = -3$

36. $(3a - 2b)^2$ for $a = -2$ and $b = 3$

37. $a^2 + 3ab - 2b^2$ for $a = 2$ and $b = -2$

38. $3n^2 - 4 - 4n^2 + 9$ for $n = 7$

39. $3(2x - 1) + 2(3x + 4)$ for $x = 1.2$

40. $-4(3x - 1) - 5(2x - 1)$ for $x = -2.3$

41. $2(n^2 + 3) - 3(n^2 + 1) + 4(n^2 - 6)$ for $n = -\frac{2}{3}$

42. $5(3n - 1) - 7(-2n + 1) + 4(3n - 1)$ for $n = \frac{1}{2}$

For Problems 43–50, translate each English phrase into an algebraic expression and use n to represent the unknown number.

43. Four increased by twice a number

44. Fifty subtracted from three times a number

45. Six less than two-thirds of a number

46. Ten times the difference of a number and 14

47. Eight subtracted from five times a number

48. The quotient of a number and three less than the number

49. Three less than five times the sum of a number and 2

50. Three-fourths of the sum of a number and 12

For Problems 51–60, answer the question with an algebraic expression.

51. The sum of two numbers is 37 and one of the numbers is n. What is the other number?

52. Yuriko can type w words in an hour. What is her typing rate per minute?

53. Harry is y years old. His brother is 7 years less than twice as old as Harry. How old is Harry's brother?

54. If n represents a multiple of 3, what represents the next largest multiple of 3?

55. Celia has p pennies, n nickels, and q quarters. How much, in cents, does Celia have?

56. The perimeter of a square is i inches. How long, in feet, is each side of the square?

57. The length of a rectangle is y yards and the width is f feet. What is the perimeter of the rectangle expressed in inches?

58. The length of a piece of wire is d decimeters. What is the length expressed in centimeters?

59. Joan is f feet and i inches tall. How tall is she in inches?

60. The perimeter of a rectangle is 50 centimeters. If the rectangle is c centimeters long, how wide is it?

CHAPTER 1   *TEST*

1. State the property of equality that justifies writing $x + 4 = 6$ for $6 = x + 4$.

2. State the property of real numbers that justifies writing $5(10 + 2)$ as $5(10) + 5(2)$.

For Problems 3–11, simplify each numerical expression.

3. $-4 - (-3) + (-5) - 7 + 10$

4. $7 - 8 - 3 + 4 - 9 - 4 + 2 - 12$

5. $5\left(-\dfrac{1}{3}\right) - 3\left(-\dfrac{1}{2}\right) + 7\left(-\dfrac{2}{3}\right) + 1$

6. $(-6) \cdot 3 \div (-2) - 8 \div (-4)$

7. $-\dfrac{1}{2}(3 - 7) - \dfrac{2}{5}(2 - 17)$

8. $[48 + (-93)] + (-49)$

9. $3(-2)^3 + 4(-2)^2 - 9(-2) - 14$

10. $[2(-6) + 5(-4)][-3(-4) - 7(6)]$

11. $[-2(-3) - 4(2)]^5$

12. Simplify $6x^2 - 3x - 7x^2 - 5x - 2$ by combining similar terms.

13. Simplify $3(3n - 1) - 4(2n + 3) + 5(-4n - 1)$ by removing parentheses and combining similar terms.

For Problems 14–20, evaluate each algebraic expression for the given values of the variables.

14. $-7x - 3y$ for $x = -6$ and $y = 5$

15. $3a^2 - 4b^2$ for $a = -\dfrac{3}{4}$ and $b = \dfrac{1}{2}$

16. $6x - 9y - 8x + 4y$ for $x = \dfrac{1}{2}$ and $y = -\dfrac{1}{3}$

17. $-5n^2 - 6n + 7n^2 + 5n - 1$ for $n = -6$

18. $-7(x - 2) + 6(x - 1) - 4(x + 3)$ for $x = 3.7$

19. $-2xy - x + 4y$ for $x = -3$ and $y = 9$

20. $4(n^2 + 1) - (2n^2 + 3) - 2(n^2 + 3)$ for $n = -4$

For Problems 21 and 22, translate the English phrase into an algebraic expression using n to represent the unknown number.

21. Thirty subtracted from six times a number

22. Four more than three times the sum of a number and 8

For Problems 23–25, answer each question with an algebraic expression.

23. The product of two numbers is 72 and one of the numbers is n. What is the other number?

24. Tao has n nickels, d dimes, and q quarters. How much money, in cents, does she have?

25. The length of a rectangle is x yards and the width is y feet. What is the perimeter of the rectangle expressed in feet?

Most shoppers take advantage of the discounts offered by retailers. When making decisions about purchases, it is beneficial to be able to compute the sale prices.

© James Leynse /CORBIS-SABA

Equations, Inequalities, and Problem Solving

A retailer of sporting goods bought a putter for $18. He wants to price the putter to make a profit of 40% of the selling price. What price should he mark on the putter? The equation $s = 18 + 0.4s$ can be used to determine that the putter should be sold for $30.

Throughout this text, we develop algebraic skills, use these skills to help solve equations and inequalities, and then use equations and inequalities to solve applied problems. In this chapter, we review and expand concepts that are important to the development of problem-solving skills.

43

InfoTrac Project

Do a subject guide search on Euclid. Find an article that lists the three main elements that Euclid defines as the basic parts of geometry. Apply these three elements in solving the following word problem. Marcus traveled 4 hours and 15 minutes at the posted rate of speed. Because of bad road conditions, he had to reduce his speed by 20 mph for the next 3 hours and 24 minutes. If he traveled a total of 467.5 miles, what was the posted rate of speed? Don't forget to turn your minutes into parts of an hour. Make up a word problem for another student to solve.

2.1   Solving First-Degree Equations

In Section 1.1, we stated that an equality (equation) is a statement where two symbols, or groups of symbols, are names for the same number. It should be further stated that an equation may be true or false. For example, the equation $3 + (-8) = -5$ is true, but the equation $-7 + 4 = 2$ is false.

Algebraic equations contain one or more variables. The following are examples of algebraic equations.

$$3x + 5 = 8 \qquad 4y - 6 = -7y + 9 \qquad x^2 - 5x - 8 = 0$$

$$3x + 5y = 4 \qquad x^3 + 6x^2 - 7x - 2 = 0$$

An algebraic equation such as $3x + 5 = 8$ is neither true nor false as it stands, and we often refer to it as an "open sentence." Each time that a number is substituted for x, the algebraic equation $3x + 5 = 8$ becomes a numerical statement that is true or false. For example, if $x = 0$, then $3x + 5 = 8$ becomes $3(0) + 5 = 8$, which is a false statement. If $x = 1$, then $3x + 5 = 8$ becomes $3(1) + 5 = 8$, which is a true statement. **Solving an equation** refers to the process of finding the number (or numbers) that make(s) an algebraic equation a true numerical statement. We call such numbers the **solutions** or **roots** of the equation, and we say that they **satisfy** the equation. We call the set of all solutions of an equation its **solution set.** Thus {1} is the solution set of $3x + 5 = 8$.

In this chapter, we will consider techniques for solving **first-degree equations in one variable.** This means that the equations contain only one variable and that this variable has an exponent of 1. The following are examples of first-degree equations in one variable.

$$3x + 5 = 8 \qquad\qquad \frac{2}{3}y + 7 = 9$$

$$7a - 6 = 3a + 4 \qquad \frac{x - 2}{4} = \frac{x - 3}{5}$$

Equivalent equations are equations that have the same solution set. For example,

1. $3x + 5 = 8$
2. $3x = 3$
3. $x = 1$

are all equivalent equations because {1} is the solution set of each.

The general procedure for solving an equation is to continue replacing the given equation with equivalent but simpler equations until we obtain an equation of the form *variable = constant* or *constant = variable*. Thus, in the example above, $3x + 5 = 8$ was simplified to $3x = 3$, which was further simplified to $x = 1$, from which the solution set {1} is obvious.

To solve equations we need to use the various properties of equality. In addition to the reflexive, symmetric, transitive, and substitution properties we listed in Section 1.1, the following properties of equality play an important role.

Addition Property of Equality

For all real numbers a, b, and c,

$$a = b \quad \text{if and only if } a + c = b + c$$

Multiplication Property of Equality

For all real numbers a, b, and c, where $c \neq 0$,

$$a = b \quad \text{if and only if } ac = bc$$

The addition property of equality states that when the same number is added to both sides of an equation, an equivalent equation is produced. The multiplication property of equality states that we obtain an equivalent equation whenever we multiply both sides of an equation by the same *nonzero* real number. The following examples demonstrate the use of these properties to solve equations.

EXAMPLE 1

Solve $2x - 1 = 13$.

Solution

$$2x - 1 = 13$$

$$2x - 1 + 1 = 13 + 1 \qquad \text{Add 1 to both sides.}$$

$$2x = 14$$

$$\frac{1}{2}(2x) = \frac{1}{2}(14) \qquad \text{Multiply both sides by } \frac{1}{2}.$$

$$x = 7$$

The solution set is {7}. ■

To check an apparent solution, we can substitute it into the original equation and see if we obtain a true numerical statement.

 Check

$$2x - 1 = 13$$

$$2(7) - 1 \overset{?}{=} 13$$

$$14 - 1 \overset{?}{=} 13$$

$$13 = 13$$

Now we know that $\{7\}$ is the solution set of $2x - 1 = 13$. We will not show our checks for every example in this text, but do remember that checking is a way to detect arithmetic errors.

EXAMPLE 2 Solve $-7 = -5a + 9$.

Solution

$$-7 = -5a + 9$$

$$-7 + (-9) = 5a + 9 + (-9) \qquad \text{Add } -9 \text{ to both sides.}$$

$$-16 = -5a$$

$$-\frac{1}{5}(-16) = -\frac{1}{5}(-5a) \qquad \text{Multiply both sides by } -\frac{1}{5}.$$

$$\frac{16}{5} = a$$

The solution set is $\left\{\dfrac{16}{5}\right\}$. ∎

Note that in Example 2 the final equation is $\dfrac{16}{5} = a$ instead of $a = \dfrac{16}{5}$. Technically, the symmetric property of equality (if $a = b$, then $b = a$) would permit us to change from $\dfrac{16}{5} = a$ to $a = \dfrac{16}{5}$, but such a change is not necessary to determine that the solution is $\dfrac{16}{5}$. Note that we could use the symmetric property at the very beginning to change $-7 = -5a + 9$ to $-5a + 9 = -7$; some people prefer having the variable on the left side of the equation.

Let's clarify another point. We stated the properties of equality in terms of only two operations, addition and multiplication. We could also include the operations of subtraction and division in the statements of the properties. That is, we could think in terms of subtracting the same number from both sides of an equation and also in terms of dividing both sides of an equation by the same nonzero number. For example, in the solution of Example 2, we could subtract 9 from both sides rather than adding -9 to both sides. Likewise, we could divide both sides by -5 instead of multiplying both sides by $-\dfrac{1}{5}$.

EXAMPLE 3 Solve $7x - 3 = 5x + 9$.

Solution

$$7x - 3 = 5x + 9$$

$$7x - 3 + (-5x) = 5x + 9 + (-5x) \qquad \text{Add } -5x \text{ to both sides.}$$

$$2x - 3 = 9$$

$$2x - 3 + 3 = 9 + 3 \qquad \text{Add 3 to both sides.}$$

$$2x = 12$$

$$\frac{1}{2}(2x) = \frac{1}{2}(12) \qquad \text{Multiply both sides by } \frac{1}{2}.$$

$$x = 6$$

The solution set is $\{6\}$.

E X A M P L E   4

Solve $4(y - 1) + 5(y + 2) = 3(y - 8)$.

Solution

$$4(y - 1) + 5(y + 2) = 3(y - 8)$$

$$4y - 4 + 5y + 10 = 3y - 24 \qquad \text{Remove parentheses by applying the distributive property.}$$

$$9y + 6 = 3y - 24 \qquad \text{Simplify the left side by combining similar terms.}$$

$$9y + 6 + (-3y) = 3y - 24 + (-3y) \qquad \text{Add } -3y \text{ to both sides.}$$

$$6y + 6 = -24$$

$$6y + 6 + (-6) = -24 + (-6) \qquad \text{Add } -6 \text{ to both sides.}$$

$$6y = -30$$

$$\frac{1}{6}(6y) = \frac{1}{6}(-30) \qquad \text{Multiply both sides by } \frac{1}{6}.$$

$$y = -5$$

The solution set is $\{-5\}$.

We can summarize the process of solving first-degree equations in one variable as follows:

STEP 1 Simplify both sides of the equation as much as possible.

STEP 2 Use the addition property of equality to isolate a term that contains the variable on one side of the equation and a constant on the other side.

STEP 3 Use the multiplication property of equality to make the coefficient of the variable 1. That is, multiply both sides of the equation by the reciprocal of the numerical coefficient of the variable. The solution set should now be obvious.

STEP 4 Check each solution by substituting it in the original equation and verifying that the resulting numerical statement is true.

Use of Equations to Solve Problems

To use the tools of algebra to solve problems, we must be able to translate back and forth between the English language and the language of algebra. More specifically, we need to translate English sentences into algebraic equations. Such translations allow us to use our knowledge of equation solving to solve word problems. Let's consider an example.

PROBLEM 1   If we subtract 27 from three times a certain number, the result is 18. Find the number.

Solution

Let n represent the number to be found. The sentence "If we subtract 27 from three times a certain number, the result is 18" translates into the equation $3n - 27 = 18$. Solving this equation, we obtain

$$3n - 27 = 18$$

$$3n = 45 \qquad \text{Add 27 to both sides.}$$

$$n = 15 \qquad \text{Multiply both sides by } \frac{1}{3}.$$

The number to be found is 15.   ■

We often refer to the statement "Let n represent the number to be found" as **declaring the variable.** We need to choose a letter to use as a variable and indicate what it represents for a specific problem. This may seem like an insignificant idea, but as the problems become more complex, the process of declaring the variable becomes even more important. Furthermore, it is true that you could probably solve a problem such as Problem 1 without setting up an algebraic equation. However, as problems increase in difficulty, the translation from English to algebra becomes a key issue. Therefore, even with these relatively easy problems, we suggest that you concentrate on the translation process.

The next example involves the use of integers. Remember that the set of integers consists of $\{\ldots -2, -1, 0, 1, 2, \ldots\}$. Furthermore, the integers can be classified as even, $\{\ldots -4, -2, 0, 2, 4, \ldots\}$, or odd, $\{\ldots -3, -1, 1, 3, \ldots\}$.

PROBLEM 2   The sum of three consecutive integers is 13 greater than twice the smallest of the three integers. Find the integers.

Solution

Because consecutive integers differ by 1, we will represent them as follows: Let n represent the smallest of the three consecutive integers; then $n + 1$ represents the second largest, and $n + 2$ represents the largest.

The sum of the three
consecutive integers   13 greater than twice the smallest

$$n + (n + 1) + (n + 2) = 2n + 13$$
$$3n + 3 = 2n + 13$$
$$n = 10$$

The three consecutive integers are 10, 11, and 12.

To check our answers for Problem 2, we must determine whether or not they satisfy the conditions stated in the original problem. Because 10, 11, and 12 are consecutive integers whose sum is 33, and because twice the smallest plus 13 is also 33 (2(10) +13 = 33), we know that our answers are correct. (Remember, in checking a result for a word problem, it is *not* sufficient to check the result in the equation set up to solve the problem; the equation itself may be in error!)

In the two previous problems, the equation formed was almost a direct translation of a sentence in the statement of the problem. Now let's consider a situation where we need to think in terms of a guideline not explicitly stated in the problem.

P R O B L E M   3

Khoa received a car repair bill for $106. This included $23 for parts, $22 per hour for each hour of labor, and $6 for taxes. Find the number of hours of labor.

Solution

See Figure 2.1. Let h represent the number of hours of labor. Then $22h$ represents the total charge for labor.

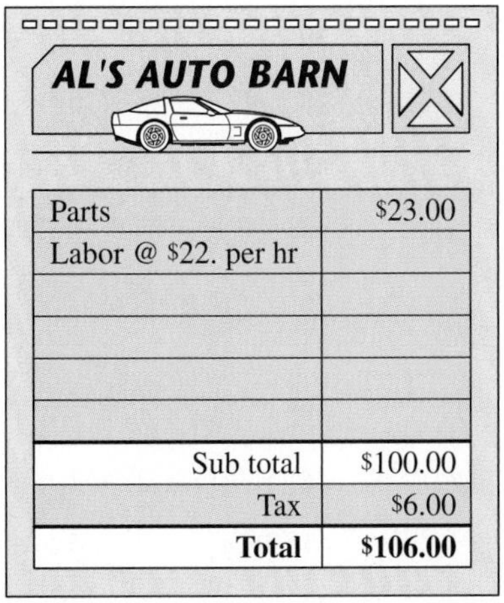

AL'S AUTO BARN

Parts	$23.00
Labor @ $22. per hr	
Sub total	$100.00
Tax	$6.00
Total	**$106.00**

Figure 2.1

We can use a guideline of *charge for parts plus charge for labor plus tax equals the total bill* to set up the following equation.

Parts   Labor   Tax   Total bill

$$23 \ + \ 22h \ + \ 6 \ = \ 106$$

Solving this equation, we obtain

$$22h + 29 = 106$$

$$22h = 77$$

$$h = 3\frac{1}{2}$$

Khoa was charged for $3\frac{1}{2}$ hours of labor.   ■

PROBLEM SET 2.1

For problems 1–50, solve each equation.

1. $3x + 4 = 16$

2. $4x + 2 = 22$

3. $5x + 1 = -14$

4. $7x + 4 = -31$

5. $-x - 6 = 8$

6. $8 - x = -2$

7. $4y - 3 = 21$

8. $6y - 7 = 41$

9. $3x - 4 = 15$

10. $5x + 1 = 12$

11. $-4 = 2x - 6$

12. $-14 = 3a - 2$

13. $-6y - 4 = 16$

14. $-8y - 2 = 18$

15. $4x - 1 = 2x + 7$

16. $9x - 3 = 6x + 18$

17. $5y + 2 = 2y - 11$

18. $9y + 3 = 4y - 10$

19. $3x + 4 = 5x - 2$

20. $2x - 1 = 6x + 15$

21. $-7a + 6 = -8a + 14$

22. $-6a - 4 = -7a + 11$

23. $5x + 3 - 2x = x - 15$

24. $4x - 2 - x = 5x + 10$

25. $6y + 18 + y = 2y + 3$

26. $5y + 14 + y = 3y - 7$

27. $4x - 3 + 2x = 8x - 3 - x$

28. $x - 4 - 4x = 6x + 9 - 8x$

29. $6n - 4 - 3n = 3n + 10 + 4n$

30. $2n - 1 - 3n = 5n - 7 - 3n$

31. $4(x - 3) = -20$

32. $3(x + 2) = -15$

33. $-3(x - 2) = 11$

34. $-5(x - 1) = 12$

35. $5(2x + 1) = 4(3x - 7)$

36. $3(2x - 1) = 2(4x + 7)$

37. $5x - 4(x - 6) = -11$

38. $3x - 5(2x + 1) = 13$

39. $-2(3x - 1) - 3 = -4$

40. $-6(x - 4) - 10 = -12$

41. $-2(3x + 5) = -3(4x + 3)$

42. $-(2x - 1) = -5(2x + 9)$

43. $3(x - 4) - 7(x + 2) = -2(x + 18)$

44. $4(x - 2) - 3(x - 1) = 2(x + 6)$

45. $-2(3n - 1) + 3(n + 5) = -4(n - 4)$

46. $-3(4n + 2) + 2(n - 6) = -2(n + 1)$

47. $3(2a - 1) - 2(5a + 1) = 4(3a + 4)$

48. $4(2a + 3) - 3(4a - 2) = 5(4a - 7)$

49. $-2(n - 4) - (3n - 1) = -2 + (2n - 1)$

50. $-(2n - 1) + 6(n + 3) = -4 - (7n - 11)$

For Problems 51–66, use an algebraic approach to solve each problem.

51. If 15 is subtracted from three times a certain number, the result is 27. Find the number.

52. If 1 is subtracted from seven times a certain number, the result is the same as if 31 is added to three times the number. Find the number.

53. Find three consecutive integers whose sum is 42.

54. Find four consecutive integers whose sum is −118.

55. Find three consecutive odd integers such that three times the second minus the third is 11 more than the first.

56. Find three consecutive even integers such that four times the first minus the third is 6 more than twice the second.

57. The difference of two numbers is 67. The larger number is 3 less than six times the smaller number. Find the numbers.

58. The sum of two numbers is 103. The larger number is 1 more than five times the smaller number. Find the numbers.

59. Angelo is paid double time for each hour he works over 40 hours in a week. Last week he worked 46 hours and earned $572. What is his normal hourly rate?

60. Suppose that a plumbing repair bill, not including tax, was $130. This included $25 for parts and an amount for 5 hours of labor. Find the hourly rate that was charged for labor.

61. Suppose that Maria has 150 coins consisting of pennies, nickels, and dimes. The number of nickels she has is 10 less than twice the number of pennies; the number of dimes she has is 20 less than three times the number of pennies. How many coins of each kind does she have?

62. Hector has a collection of nickels, dimes, and quarters totaling 122 coins. The number of dimes he has is 3 more than four times the number of nickels, and the number of quarters he has is 19 less than the number of dimes. How many coins of each kind does he have?

63. The selling price of a ring is $750. This represents $150 less than three times the cost of the ring. Find the cost of the ring.

64. In a class of 62 students, the number of females is one less than twice the number of males. How many females and how many males are there in the class?

65. An apartment complex contains 230 apartments each having one, two, or three bedrooms. The number of two-bedroom apartments is 10 more than three times the number of three-bedroom apartments. The number of one-bedroom apartments is twice the number of two-bedroom apartments. How many apartments of each kind are in the complex?

66. Barry sells bicycles on a salary-plus-commission basis. He receives a monthly salary of $300 and a commission of $15 for each bicycle that he sells. How many bicycles must he sell in a month to have a total monthly income of $750?

■ ■ ■ Thoughts into words

67. Explain the difference between a numerical statement and an algebraic equation.

68. Are the equations $7 = 9x - 4$ and $9x - 4 = 7$ equivalent equations? Defend your answer.

69. Suppose that your friend shows you the following solution to an equation.

$$17 = 4 - 2x$$

$$17 + 2x = 4 - 2x + 2x$$

$$17 + 2x = 4$$

$$17 + 2x - 17 = 4 - 17$$

$$2x = -13$$

$$x = \frac{-13}{2}$$

Is this a correct solution? What suggestions would you have in terms of the method used to solve the equation?

70. Explain in your own words what it means to declare a variable when solving a word problem.

71. Make up an equation whose solution set is the null set and explain why this is the solution set.

72. Make up an equation whose solution set is the set of all real numbers and explain why this is the solution set.

■ ■ ■ **Further investigations**

73. Solve each of the following equations.

(a) $5x + 7 = 5x - 4$

(b) $4(x - 1) = 4x - 4$

(c) $3(x - 4) = 2(x - 6)$

(d) $7x - 2 = -7x + 4$

(e) $2(x - 1) + 3(x + 2) = 5(x - 7)$

(f) $-4(x - 7) = -2(2x + 1)$

74. Verify that for any three consecutive integers, the sum of the smallest and largest is equal to twice the middle integer. [*Hint:* Use n, $n + 1$, and $n + 2$ to represent the three consecutive integers.]

75. Verify that no four consecutive integers can be found such that the product of the smallest and largest is equal to the product of the other two integers.

2.2   Equations Involving Fractional Forms

To solve equations that involve fractions, it is usually easiest to begin by **clearing the equation of all fractions**. This can be accomplished by multiplying both sides of the equation by the least common multiple of all the denominators in the equation. Remember that the least common multiple of a set of whole numbers is the smallest nonzero whole number that is divisible by each of the numbers. For example, the least common multiple of 2, 3, and 6 is 12. When working with fractions, we refer to the least common multiple of a set of denominators as the **least common denominator** (LCD). Let's consider some equations involving fractions.

E X A M P L E   1   Solve $\dfrac{1}{2}x + \dfrac{2}{3} = \dfrac{3}{4}$.

Solution

$$\frac{1}{2}x + \frac{2}{3} = \frac{3}{4}$$

$$12\left(\frac{1}{2}x + \frac{2}{3}\right) = 12\left(\frac{3}{4}\right) \qquad \text{Multiply both sides by 12, which is the LCD of 2, 3, and 4.}$$

$$12\left(\frac{1}{2}x\right) + 12\left(\frac{2}{3}\right) = 12\left(\frac{3}{4}\right) \qquad \text{Apply the distributive property to the left side.}$$

$$6x + 8 = 9$$

$$6x = 1$$

$$x = \frac{1}{6}$$

The solution set is $\left\{\dfrac{1}{6}\right\}$.

Check

$$\frac{1}{2}x + \frac{2}{3} = \frac{3}{4}$$

$$\frac{1}{2}\left(\frac{1}{6}\right) + \frac{2}{3} \overset{?}{=} \frac{3}{4}$$

$$\frac{1}{12} + \frac{2}{3} \overset{?}{=} \frac{3}{4}$$

$$\frac{1}{12} + \frac{8}{12} \overset{?}{=} \frac{3}{4}$$

$$\frac{9}{12} \overset{?}{=} \frac{3}{4}$$

$$\frac{3}{4} = \frac{3}{4}$$ ∎

EXAMPLE 2   Solve $\dfrac{x}{2} + \dfrac{x}{3} = 10$.

Solution

$$\frac{x}{2} + \frac{x}{3} = 10 \qquad \text{Recall that } \frac{x}{2} = \frac{1}{2}x.$$

$$6\left(\frac{x}{2} + \frac{x}{3}\right) = 6(10) \qquad \text{Multiply both sides by the LCD.}$$

$$6\left(\frac{x}{2}\right) + 6\left(\frac{x}{3}\right) = 6(10) \qquad \begin{array}{l}\text{Apply the distributive property to} \\ \text{the left side.}\end{array}$$

$$3x + 2x = 60$$

$$5x = 60$$

$$x = 12$$

The solution set is {12}. ∎

As you study the examples in this section, pay special attention to the steps shown in the solutions. There are no hard and fast rules as to which steps should be performed mentally; this is an individual decision. When you solve problems, show enough steps to allow the flow of the process to be understood and to minimize the chances of making careless computational errors.

EXAMPLE 3   Solve $\dfrac{x-2}{3} + \dfrac{x+1}{8} = \dfrac{5}{6}$.

 Solution

$$\frac{x-2}{3} + \frac{x+1}{8} = \frac{5}{6}$$

$$24\left(\frac{x-2}{3} + \frac{x+1}{8}\right) = 24\left(\frac{5}{6}\right) \qquad \text{Multiply both sides by the LCD.}$$

$$24\left(\frac{x-2}{3}\right) + 24\left(\frac{x+1}{8}\right) = 24\left(\frac{5}{6}\right) \qquad \text{Apply the distributive property to the left side.}$$

$$8(x-2) + 3(x+1) = 20$$

$$8x - 16 + 3x + 3 = 20$$

$$11x - 13 = 20$$

$$11x = 33$$

$$x = 3$$

The solution set is {3}. ■

EXAMPLE 4 Solve $\dfrac{3t-1}{5} - \dfrac{t-4}{3} = 1$.

Solution

$$\frac{3t-1}{5} - \frac{t-4}{3} = 1$$

$$15\left(\frac{3t-1}{5} - \frac{t-4}{3}\right) = 15(1) \qquad \text{Multiply both sides by the LCD.}$$

$$15\left(\frac{3t-1}{5}\right) - 15\left(\frac{t-4}{3}\right) = 15(1) \qquad \text{Apply the distributive property to the left side.}$$

$$3(3t-1) - 5(t-4) = 15$$

$$9t - 3 - 5t + 20 = 15 \qquad \text{Be careful with this sign!}$$

$$4t + 17 = 15$$

$$4t = -2$$

$$t = -\frac{2}{4} = -\frac{1}{2} \qquad \text{Reduce!}$$

The solution set is $\left\{-\dfrac{1}{2}\right\}$. ■

Problem Solving

As we expand our skills for solving equations, we also expand our capabilities for solving word problems. There is no one definite procedure that will ensure success at solving word problems, but the following suggestions can be helpful.

Suggestions for Solving Word Problems

1. Read the problem carefully and make certain that you understand the meanings of all of the words. Be especially alert for any technical terms used in the statement of the problem.
2. Read the problem a second time (perhaps even a third time) to get an overview of the situation being described. Determine the known facts as well as what is to be found.
3. Sketch any figure, diagram, or chart that might be helpful in analyzing the problem.
4. Choose a meaningful variable to represent an unknown quantity in the problem (perhaps t, if time is an unknown quantity) and represent any other unknowns in terms of that variable.
5. Look for a guideline that you can use to set up an equation. A guideline might be a formula, such as *distance equals rate times time*, or a statement of a relationship, such as "The sum of the two numbers is 28."
6. Form an equation containing the variable that translates the conditions of the guideline from English to algebra.
7. Solve the equation and use the solution to determine all facts requested in the problem.
8. Check all answers back into the **original statement of the problem.**

Keep these suggestions in mind as we continue to solve problems. We will elaborate on some of these suggestions at different times throughout the text. Now let's consider some problems.

PROBLEM 1

Find a number such that three-eighths of the number minus one-half of it is 14 less than three-fourths of the number.

Solution

Let n represent the number to be found.

$$\frac{3}{8}n - \frac{1}{2}n = \frac{3}{4}n - 14$$

$$8\left(\frac{3}{8}n - \frac{1}{2}n\right) = 8\left(\frac{3}{4}n - 14\right)$$

$$8\left(\frac{3}{8}n\right) - 8\left(\frac{1}{2}n\right) = 8\left(\frac{3}{4}n\right) - 8(14)$$

$$3n - 4n = 6n - 112$$

$$-n = 6n - 112$$

$$-7n = -112$$

$$n = 16$$

The number is 16. Check it!

PROBLEM 2

The width of a rectangular parking lot is 8 feet less than three-fifths of the length. The perimeter of the lot is 400 feet. Find the length and width of the lot.

Solution

Let l represent the length of the lot. Then $\frac{3}{5}l - 8$ represents the width (Figure 2.2).

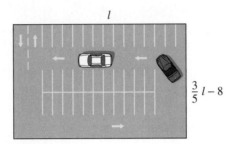

Figure 2.2

A guideline for this problem is the formula *the perimeter of a rectangle equals twice the length plus twice the width* $(P = 2l + 2w)$. Use this formula to form the following equation.

$$P = 2l + 2w$$

$$400 = 2l + 2\left(\frac{3}{5}l - 8\right)$$

Solving this equation, we obtain

$$400 = 2l + \frac{6l}{5} - 16$$

$$5(400) = 5\left(2l + \frac{6l}{5} - 16\right)$$

$$2000 = 10l + 6l - 80$$

$$2000 = 16l - 80$$

$$2080 = 16l$$

$$130 = l.$$

The length of the lot is 130 feet and the width is $\frac{3}{5}(130) - 8 = 70$ feet.

In Problems 1 and 2, note the use of different letters as variables. It is helpful to choose a variable that has significance for the problem you are working on. For example, in Problem 2 the choice of l to represent the length seems natural and meaningful. (Certainly this is another matter of personal preference, but you might consider it.)

In Problem 2 a geometric relationship ($P = 2l + 2w$) serves as a guideline for setting up the equation. The following geometric relationships pertaining to angle measure may also serve as guidelines.

1. Complementary angles are two angles the sum of whose measures is 90°.
2. Supplementary angles are two angles the sum of whose measures is 180°.
3. The sum of the measures of the three angles of a triangle is 180°.

PROBLEM 3

One of two complementary angles is 6° larger than one-half of the other angle. Find the measure of each of the angles.

Solution

Let a represent the measure of one of the angles. Then $\frac{1}{2}a + 6$ represents the measure of the other angle. Because they are complementary angles, the sum of their measures is 90°.

$$a + \left(\frac{1}{2}a + 6\right) = 90$$

$$2a + a + 12 = 180$$

$$3a + 12 = 180$$

$$3a = 168$$

$$a = 56$$

If a = 56, then $\frac{1}{2}a + 6$ becomes $\frac{1}{2}(56) + 6 = 34$. The angles have measures of 34° and 56°. ∎

PROBLEM 4

Dominic's present age is 10 years more than Michele's present age. In 5 years Michele's age will be three-fifths of Dominic's age. What are their present ages?

Solution

Let x represent Michele's present age. Then Dominic's age will be represented by $x + 10$. In 5 years, everyone's age is increased by 5 years, so we need to add 5 to Michele's present age and 5 to Dominic's present age to represent their ages in 5 years. Therefore, in 5 years Michele's age will be represented by $x + 5$, and Dominic's age will be represented by $x + 15$. Thus we can set up the equation reflecting the fact that in 5 years, Michele's age will be three-fifths of Dominic's age.

$$x + 5 = \frac{3}{5}(x + 15)$$

$$5(x + 5) = 5\left[\frac{3}{5}(x + 15)\right]$$

$$5x + 25 = 3(x + 15)$$

$$5x + 25 = 3x + 45$$

$$2x + 25 = 45$$

$$2x = 20$$

$$x = 10$$

Because x represents Michele's present age, we know her age is 10. Dominic's present age is represented by $x + 10$, so his age is 20.   ■

Keep in mind that the problem-solving suggestions offered in this section simply outline a general algebraic approach to solving problems. You will add to this list throughout this course and in any subsequent mathematics courses that you take. Furthermore, you will be able to pick up additional problem-solving ideas from your instructor and from fellow classmates as you discuss problems in class. Always be on the alert for any ideas that might help you become a better problem solver.

PROBLEM SET 2.2

For Problems 1–40, solve each equation.

1. $\dfrac{3}{4}x = 9$

2. $\dfrac{2}{3}x = -14$

3. $\dfrac{-2x}{3} = \dfrac{2}{5}$

4. $\dfrac{-5x}{4} = \dfrac{7}{2}$

5. $\dfrac{n}{2} - \dfrac{2}{3} = \dfrac{5}{6}$

6. $\dfrac{n}{4} - \dfrac{5}{6} = \dfrac{5}{12}$

7. $\dfrac{5n}{6} - \dfrac{n}{8} = \dfrac{-17}{12}$

8. $\dfrac{2n}{5} - \dfrac{n}{6} = \dfrac{-7}{10}$

9. $\dfrac{a}{4} - 1 = \dfrac{a}{3} + 2$

10. $\dfrac{3a}{7} - 1 = \dfrac{a}{3}$

11. $\dfrac{h}{4} + \dfrac{h}{5} = 1$

12. $\dfrac{h}{6} + \dfrac{3h}{8} = 1$

13. $\dfrac{h}{2} - \dfrac{h}{3} + \dfrac{h}{6} = 1$

14. $\dfrac{3h}{4} + \dfrac{2h}{5} = 1$

15. $\dfrac{x - 2}{3} + \dfrac{x + 3}{4} = \dfrac{11}{6}$

16. $\dfrac{x + 4}{5} + \dfrac{x - 1}{4} = \dfrac{37}{10}$

17. $\dfrac{x + 2}{2} - \dfrac{x - 1}{5} = \dfrac{3}{5}$

18. $\dfrac{2x + 1}{3} - \dfrac{x + 1}{7} = -\dfrac{1}{3}$

19. $\dfrac{n + 2}{4} - \dfrac{2n - 1}{3} = \dfrac{1}{6}$

20. $\dfrac{n - 1}{9} - \dfrac{n + 2}{6} = \dfrac{3}{4}$

21. $\dfrac{y}{3} + \dfrac{y - 5}{10} = \dfrac{4y + 3}{5}$

22. $\dfrac{y}{3} + \dfrac{y - 2}{8} = \dfrac{6y - 1}{12}$

23. $\dfrac{4x - 1}{10} - \dfrac{5x + 2}{4} = -3$

24. $\dfrac{2x - 1}{2} - \dfrac{3x + 1}{4} = \dfrac{3}{10}$

25. $\dfrac{2x - 1}{8} - 1 = \dfrac{x + 5}{7}$

26. $\dfrac{3x + 1}{9} + 2 = \dfrac{x - 1}{4}$

27. $\dfrac{2a - 3}{6} + \dfrac{3a - 2}{4} + \dfrac{5a + 6}{12} = 4$

28. $\dfrac{3a - 1}{4} + \dfrac{a - 2}{3} - \dfrac{a - 1}{5} = \dfrac{21}{20}$

29. $x + \dfrac{3x - 1}{9} - 4 = \dfrac{3x + 1}{3}$

30. $\dfrac{2x + 7}{8} + x - 2 = \dfrac{x - 1}{2}$

31. $\dfrac{x + 3}{2} + \dfrac{x + 4}{5} = \dfrac{3}{10}$

32. $\dfrac{x - 2}{5} - \dfrac{x - 3}{4} = -\dfrac{1}{20}$

33. $n + \dfrac{2n - 3}{9} - 2 = \dfrac{2n + 1}{3}$

34. $n - \dfrac{3n + 1}{6} - 1 = \dfrac{2n + 4}{12}$

35. $\dfrac{3}{4}(t - 2) - \dfrac{2}{5}(2t - 3) = \dfrac{1}{5}$

36. $\dfrac{2}{3}(2t + 1) - \dfrac{1}{2}(3t - 2) = 2$

37. $\dfrac{1}{2}(2x - 1) - \dfrac{1}{3}(5x + 2) = 3$

38. $\dfrac{2}{5}(4x - 1) + \dfrac{1}{4}(5x + 2) = -1$

39. $3x - 1 + \dfrac{2}{7}(7x - 2) = -\dfrac{11}{7}$

40. $2x + 5 + \dfrac{1}{2}(6x - 1) = -\dfrac{1}{2}$

For Problems 41–58, use an algebraic approach to solve each problem.

41. Find a number such that one-half of the number is 3 less than two-thirds of the number.

42. One-half of a number plus three-fourths of the number is 2 more than four-thirds of the number. Find the number.

43. Suppose that the width of a certain rectangle is 1 inch more than one-fourth of its length. The perimeter of the rectangle is 42 inches. Find the length and width of the rectangle.

44. Suppose that the width of a rectangle is 3 centimeters less than two-thirds of its length. The perimeter of the rectangle is 114 centimeters. Find the length and width of the rectangle.

45. Find three consecutive integers such that the sum of the first plus one-third of the second plus three-eighths of the third is 25.

46. Lou is paid $1\dfrac{1}{2}$ times his normal hourly rate for each hour he works over 40 hours in a week. Last week he worked 44 hours and earned $276. What is his normal hourly rate?

47. A board 20 feet long is cut into two pieces such that the length of one piece is two-thirds of the length of the other piece. Find the length of the shorter piece of board.

48. Jody has a collection of 116 coins consisting of dimes, quarters, and silver dollars. The number of quarters is 5 less than three-fourths of the number of dimes. The number of silver dollars is 7 more than five-eighths of the number of dimes. How many coins of each kind are in her collection?

49. The sum of the present ages of Angie and her mother is 64 years. In eight years Angie will be three-fifths as old as her mother at that time. Find the present ages of Angie and her mother.

50. Annilee's present age is two-thirds of Jessie's present age. In 12 years the sum of their ages will be 54 years. Find their present ages.

51. Sydney's present age is one-half of Marcus's present age. In 12 years, Sydney's age will be five-eighths of Marcus's age. Find their present ages.

52. The sum of the present ages of Ian and his brother is 45. In 5 years, Ian's age will be five-sixths of his brother's age. Find their present ages.

53. Aura took three biology exams and has an average score of 88. Her second exam score was 10 points better than her first, and her third exam score was 4 points better than her second exam. What were her three exam scores?

54. The average of the salaries of Tim, Maida, and Aaron is $24,000 per year. Maida earns $10,000 more than Tim, and Aaron's salary is $2000 more than twice Tim's salary. Find the salary of each person.

55. One of two supplementary angles is 4° more than one-third of the other angle. Find the measure of each of the angles.

56. If one-half of the complement of an angle plus three-fourths of the supplement of the angle equals 110°, find the measure of the angle.

57. If the complement of an angle is 5° less than one-sixth of its supplement, find the measure of the angle.

58. In $\triangle ABC$, angle B is 8° less than one-half of angle A and angle C is 28° larger than angle A. Find the measures of the three angles of the triangle.

■ ■ ■ **Thoughts into words**

59. Explain why the solution set of the equation $x + 3 = x + 4$ is the null set.

60. Explain why the solution set of the equation $\dfrac{x}{3} + \dfrac{x}{2} = \dfrac{5x}{6}$ is the entire set of real numbers.

61. Why must potential answers to word problems be checked back into the original statement of the problem?

62. Suppose your friend solved the problem "Find two consecutive odd integers whose sum is 28" like this:

$$x + x + 1 = 28$$

$$2x = 27$$

$$x = \frac{27}{2} = 13\frac{1}{2}$$

She claims that $13\frac{1}{2}$ will check in the equation. Where has she gone wrong and how would you help her?

2.3   Equations Involving Decimals and Problem Solving

In solving equations that involve fractions, usually the procedure is to clear the equation of all fractions. For solving equations that involve decimals, there are two commonly used procedures. One procedure is to keep the numbers in decimal form and solve the equation by applying the properties. Another procedure is to multiply both sides of the equation by an appropriate power of 10 to clear the equation of all decimals. Which technique to use depends on your personal preference and on the complexity of the equation. The following examples demonstrate both techniques.

EXAMPLE 1   Solve $0.2x + 0.24 = 0.08x + 0.72$

Solution

Let's clear the decimals by multiplying both sides of the equation by 100.

$$0.2x + 0.24 = 0.08x + 0.72$$

$$100(0.2x + 0.24) = 100(0.08x + 0.72)$$

$$100(0.2x) + 100(0.24) = 100(0.08x) + 100(0.72)$$

$$20x + 24 = 8x + 72$$

$$12x + 24 = 72$$

$$12x = 48$$

$$x = 4$$

 Check

$$0.2x + 0.24 = 0.08x + 0.72$$

$$0.2(4) + 0.24 \overset{?}{=} 0.08(4) + 0.72$$

$$0.8 + 0.24 \overset{?}{=} 0.32 + 0.72$$

$$1.04 = 1.04$$

The solution set is {4}.

Solve $0.07x + 0.11x = 3.6$

Solution

Let's keep this problem in decimal form.

$$0.07x + 0.11x = 3.6$$

$$0.18x = 3.6$$

$$x = \frac{3.6}{0.18}$$

$$x = 20$$

 Check

$$0.07x + 0.11x = 3.6$$

$$0.07(20) + 0.11(20) \overset{?}{=} 3.6$$

$$1.4 + 2.2 \overset{?}{=} 3.6$$

$$3.6 = 3.6$$

The solution set is {20}.

EXAMPLE   3

Solve $s = 1.95 + 0.35s$

Solution

Let's keep this problem in decimal form.

$$s = 1.95 + 0.35s$$

$$s + (-0.35s) = 1.95 + 0.35s + (-0.35s)$$

$$0.65s = 1.95 \qquad \text{Remember, } s = 1.00s.$$

$$s = \frac{1.95}{0.65}$$

$$s = 3$$

The solution set is $\{3\}$. Check it!

EXAMPLE 4

Solve $0.12x + 0.11(7000 - x) = 790$

Solution

Let's clear the decimals by multiplying both sides of the equation by 100.

$$0.12x + 0.11(7000 - x) = 790$$

$$100[0.12x + 0.11(7000 - x)] = 100(790) \qquad \text{Multiply both sides by 100.}$$

$$100(0.12x) + 100[0.11(7000 - x)] = 100(790)$$

$$12x + 11(7000 - x) = 79{,}000$$

$$12x + 77{,}000 - 11x = 79{,}000$$

$$x + 77{,}000 = 79{,}000$$

$$x = 2000$$

The solution set is $\{2000\}$.

Back to Problem Solving

We can solve many consumer problems with an algebraic approach. For example, let's consider some discount sale problems involving the relationship *original selling price minus discount equals discount sale price*.

Original selling price − Discount = Discount sale price

PROBLEM 1

Karyl bought a dress at a 35% discount sale for $32.50. What was the original price of the dress?

Solution

Let p represent the original price of the dress. Using the discount sale relationship as a guideline, we find that the problem translates into an equation as follows:

Original selling price	Minus	Discount	Equals	Discount sale price
p	−	$(35\%)(p)$	=	$32.50

Switching this equation to decimal form and solving, we obtain

$$p - (35\%)(p) = 32.50$$

$$(65\%)(p) = 32.50$$

$$0.65p = 32.50$$

$$p = 50$$

The original price of the dress was $50.

■

PROBLEM 2

A pair of jogging shoes that was originally priced at $25 is on sale for 20% off. Find the discount sale price of the shoes.

Solution

Let s represent the discount sale price.

Original price	Minus	Discount	Equals	Sale price
$25	−	(20%)($25)	=	s

Solving this equation we obtain

$$25 - (20\%)(25) = s$$

$$25 - (0.2)(25) = s$$

$$25 - 5 = s$$

$$20 = s$$

The shoes are on sale for $20.

■

REMARK: Keep in mind that if an item is on sale for 35% off, then the purchaser will pay 100% − 35% = 65% of the original price. Thus, in Problem 1, you could begin with the equation $0.65p = 32.50$. Likewise, in Problem 2, you could start with the equation $s = 0.8(25)$.

Another basic relationship that pertains to consumer problems is *selling price equals cost plus profit*. We can state profit (also called markup, markon, and margin of profit) in different ways. Profit may be stated as a percent of the selling price, as a percent of the cost, or simply in terms of dollars and cents. We shall consider some problems for which the profit is calculated either as a percent of the cost or as a percent of the selling price.

Selling price = Cost + Profit

PROBLEM 3

A retailer has some shirts that cost $20 each. She wants to sell them at a profit of 60% of the cost. What selling price should be marked on the shirts?

Solution

Let s represent the selling price. Use the relationship *selling price equals cost plus profit* as a guideline.

Selling price	Equals	Cost	Plus	Profit
↓		↓		↓
s	$=$	$\$20$	$+$	$(60\%)(\$20)$

Solving this equation yields

$$s = 20 + (60\%)(20)$$

$$s = 20 + (0.6)(20)$$

$$s = 20 + 12$$

$$s = 32$$

The selling price should be $32.   ■

REMARK: A profit of 60% of the cost means that the selling price is 100% of the cost plus 60% of the cost, or 160% of the cost. Thus, in Problem 3 we could solve the equation $s = 1.6(20)$.

PROBLEM 4

A retailer of sporting goods bought a putter for $18. He wants to price the putter such that he will make a profit of 40% of the selling price. What price should he mark on the putter?

Solution

Let s represent the selling price.

Selling price	Equals	Cost	Plus	Profit
↓		↓		↓
s	$=$	$\$18$	$+$	$(40\%)(s)$

Solving this equation yields

$$s = 18 + (40\%)(s)$$

$$s = 18 + 0.4s$$

$$0.6s = 18$$

$$s = 30$$

The selling price should be $30.   ■

PROBLEM 5

If a maple tree costs a landscaper $55.00 and he sells it for $80.00, what is his rate of profit based on the cost? Round the rate to the nearest tenth of a percent.

Solution

Let r represent the rate of profit and use the following guideline.

Selling price	Equals	Cost	Plus	Profit
80.00	=	55.00	+	$r(55.00)$
25.00	=	$r(55.00)$		
$\dfrac{25.00}{55.00}$	=	r		
0.455	$\approx$	r		

To change the answer to a percent, multiply 0.455 by 100. Thus his rate of profit is 45.5%. ■

We can solve certain types of investment and money problems by using an algebraic approach. Consider the following examples.

PROBLEM 6

Erick has 40 coins, consisting only of dimes and nickels, worth $3.35. How many dimes and how many nickels does he have?

Solution

Let x represent the number of dimes. Then the number of nickels can be represented by the total number of coins minus the number of dimes. Hence $40 - x$ represents the number of nickels. Because we know the amount of money Erick has, we need to multiply the number of each coin by its value. Use the following guideline.

Money from the dimes	Plus	Money from the nickels	Equals	Total money	
$0.10x$	+	$0.05(40 - x)$	=	3.35	
$10x$	+	$5(40 - x)$	=	335	Multiply both sides by 100.
$10x$	+	$200 - 5x$	=	335	
		$5x + 200$	=	335	
		$5x$	=	135	
		x	=	27	

The number of dimes is 27 and the number of nickels is $40 - x = 13$. So, Erick has 27 dimes and 13 nickels. ■

PROBLEM 7   A man invests $8000, part of it at 11% and the remainder at 12%. His total yearly interest from the two investments is $930. How much did he invest at each rate?

Solution

Let x represent the amount he invested at 11%. Then $8000 - x$ represents the amount he invested at 12%. Use the following guideline.

Interest earned from 11% investment	+	Interest earned from 12% investment	=	Total amount of interest earned
↓		↓		↓
$(11\%)(x)$	+	$(12\%)(8000 - x)$	=	$930

Solving this equation yields

$$(11\%)(x) + (12\%)(8000 - x) = 930$$

$$0.11x + 0.12(8000 - x) = 930$$

$$11x + 12(8000 - x) = 93{,}000 \qquad \text{Multiply both sides by 100.}$$

$$11x + 96{,}000 - 12x = 93{,}000$$

$$-x + 96{,}000 = 93{,}000$$

$$-x = -3000$$

$$x = 3000$$

Therefore, $3000 was invested at 11% and $8000 - $3000 = $5000 was invested at 12%. ■

Don't forget to check word problems; determine whether the answers satisfy the conditions stated in the *original* problem. A check for Problem 7 follows.

Check

We claim that $3000 is invested at 11% and $5000 at 12% and this satisfies the condition that $8000 is invested. The $3000 at 11% produces $330 of interest, and the $5000 at 12% produces $600. Therefore, the interest from the investments is $930. The conditions of the problem are satisfied, and our answers are correct.

PROBLEM SET 2.3

For Problems 1–28, solve each equation.

1. $0.14x = 2.8$

2. $1.6x = 8$

3. $0.09y = 4.5$

4. $0.07y = 0.42$

5. $n + 0.4n = 56$

6. $n - 0.5n = 12$

7. $s = 9 + 0.25s$

8. $s = 15 + 0.4s$

9. $s = 3.3 + 0.45s$

10. $s = 2.1 + 0.6s$

11. $0.11x + 0.12(900 - x) = 104$

12. $0.09x + 0.11(500 - x) = 51$

13. $0.08(x + 200) = 0.07x + 20$

14. $0.07x = 152 - 0.08(2000 - x)$

15. $0.12t - 2.1 = 0.07t - 0.2$

16. $0.13t - 3.4 = 0.08t - 0.4$

17. $0.92 + 0.9(x - 0.3) = 2x - 5.95$

18. $0.3(2n - 5) = 11 - 0.65n$

19. $0.1d + 0.11(d + 1500) = 795$

20. $0.8x + 0.9(850 - x) = 715$

21. $0.12x + 0.1(5000 - x) = 560$

22. $0.10t + 0.12(t + 1000) = 560$

23. $0.09(x + 200) = 0.08x + 22$

24. $0.09x = 1650 - 0.12(x + 5000)$

25. $0.3(2t + 0.1) = 8.43$

26. $0.5(3t + 0.7) = 20.6$

27. $0.1(x - 0.1) - 0.4(x + 2) = -5.31$

28. $0.2(x + 0.2) + 0.5(x - 0.4) = 5.44$

For Problems 29–50, use an algebraic approach to solve each problem.

29. Judy bought a coat at a 20% discount sale for $72. What was the original price of the coat?

30. Jim bought a pair of slacks at a 25% discount sale for $24. What was the original price of the slacks?

31. Find the discount sale price of a $64 item that is on sale for 15% off.

32. Find the discount sale price of a $72 item that is on sale for 35% off.

33. A retailer has some skirts that cost $30 each. She wants to sell them at a profit of 60% of the cost. What price should she charge for the skirts?

34. The owner of a pizza parlor wants to make a profit of 70% of the cost for each pizza sold. If it costs $2.50 to make a pizza, at what price should each pizza be sold?

35. If a ring costs a jeweler $200, at what price should it be sold to yield a profit of 50% on the selling price?

36. If a head of lettuce costs a retailer $0.32, at what price should it be sold to yield a profit of 60% on the selling price?

37. If a pair of shoes costs a retailer $24 and he sells them for $39.60, what is his rate of profit based on the cost?

38. A retailer has some skirts that cost her $45 each. If she sells them for $83.25 per skirt, find her rate of profit based on the cost.

39. If a computer costs an electronics dealer $300 and she sells them for $800, what is her rate of profit based on the selling price?

40. A textbook costs a bookstore $45 and the store sells it for $60. Find the rate of profit based on the selling price.

41. Mitsuko's salary for next year is $34,775. This represents a 7% increase over this year's salary. Find Mitsuko's present salary.

42. Don bought a car for $15,794, with 6% tax included. What was the price of the car without the tax?

43. Eva invested a certain amount of money at 10% interest and $1500 more than that amount at 11%. Her total yearly interest was $795. How much did she invest at each rate?

44. A total of $4000 was invested, part of it at 8% interest and the remainder at 9%. If the total yearly interest amounted to $350, how much was invested at each rate?

45. A sum of $95,000 is split between two investments, one paying 6% and the other 9%. If the total yearly interest amounted to $7290, how much was invested at 9%?

46. If $1500 is invested at 6% interest, how much money must be invested at 9% so that the total return for both investments is $301.50?

47. Suppose that Javier has a handful of coins, consisting of pennies, nickels, and dimes, worth $2.63. The number of nickels is 1 less than twice the number of pennies, and the number of dimes is 3 more than the number of nickels. How many coins of each kind does he have?

48. Sarah has a collection of nickels, dimes, and quarters worth $15.75. She has 10 more dimes than nickels and twice as many quarters as dimes. How many coins of each kind does she have?

49. A collection of 70 coins consisting of dimes, quarters, and half-dollars has a value of $17.75. There are three times as many quarters as dimes. Find the number of each kind of coin.

50. Abby has 37 coins, consisting only of dimes and quarters, worth $7.45. How many dimes and how many quarters does she have?

■ ■ ■ Thoughts into words

51. Go to Problem 39 and calculate the rate of profit based on cost. Compare the rate of profit based on cost to the rate of profit based on selling price. From a consumer's viewpoint, would you prefer that a retailer figure his profit on the basis of the cost of an item or on the basis of its selling price? Explain your answer.

52. Is a 10% discount followed by a 30% discount the same as a 30% discount followed by a 10% discount? Justify your answer.

53. What is wrong with the following solution and how should it be done?

$$1.2x + 2 = 3.8$$
$$10(1.2x) + 2 = 10(3.8)$$
$$12x + 2 = 38$$
$$12x = 36$$
$$x = 3$$

■ ■ ■ Further investigations

For Problems 54–63, solve each equation and express the solutions in decimal form. Be sure to check your solutions. Use your calculator whenever it seems helpful.

54. $1.2x + 3.4 = 5.2$

55. $0.12x - 0.24 = 0.66$

56. $0.12x + 0.14(550 - x) = 72.5$

57. $0.14t + 0.13(890 - t) = 67.95$

58. $0.7n + 1.4 = 3.92$

59. $0.14n - 0.26 = 0.958$

60. $0.3(d + 1.8) = 4.86$

61. $0.6(d - 4.8) = 7.38$

62. $0.8(2x - 1.4) = 19.52$

63. $0.5(3x + 0.7) = 20.6$

64. The following formula can be used to determine the selling price of an item when the profit is based on a percent of the selling price.

$$\text{Selling price} = \frac{\text{Cost}}{100\% - \text{Percent of profit}}$$

Show how this formula is developed.

65. A retailer buys an item for $90, resells it for $100, and claims that she is making only a 10% profit. Is this claim correct?

66. Is a 10% discount followed by a 20% discount equal to a 30% discount? Defend your answer.

2.4   Formulas

To find the distance traveled in 4 hours at a rate of 55 miles per hour, we multiply the rate times the time; thus the distance is $55(4) = 220$ miles. We can state the rule *distance equals rate times time* as a formula: $d = rt$. Formulas are rules we state in symbolic form, usually as equations.

Formulas are typically used in two different ways. At times a formula is solved for a specific variable when we are given the numerical values for the other variables. This is much like evaluating an algebraic expression. At other times we need to change the form of an equation by solving for one variable in terms of the other variables. Throughout our work on formulas, we will use the properties of equality and the techniques we have previously learned for solving equations. Let's consider some examples.

E X A M P L E   1

If we invest P dollars at r percent for t years, the amount of simple interest i is given by the formula $i = Prt$. Find the amount of interest earned by \$500 at 7% for 2 years.

Solution

By substituting \$500 for P, 7% for r, and 2 for t, we obtain

$$i = Prt$$

$$i = (500)(7\%)(2)$$

$$i = (500)(0.07)(2)$$

$$i = 70$$

Thus we earn \$70 in interest.

E X A M P L E   2

If we invest P dollars at a simple rate of r percent, then the amount A accumulated after t years is given by the formula $A = P + Prt$. If we invest \$500 at 8%, how many years will it take to accumulate \$600?

Solution

Substituting \$500 for P, 8% for r, and \$600 for A, we obtain

$$A = P + Prt$$

$$600 = 500 + 500(8\%)(t)$$

Solving this equation for t yields

$$600 = 500 + 500(0.08)(t)$$

$$600 = 500 + 40t$$

$$100 = 40t$$

$$2\frac{1}{2} = t$$

It will take $2\frac{1}{2}$ years to accumulate \$600.

When we are using a formula, it is sometimes convenient first to change its form. For example, suppose we are to use the *perimeter* formula for a rectangle ($P = 2l + 2w$) to complete the following chart.

Perimeter (*P*)	32	24	36	18	56	80	
Length (*l*)	10	7	14	5	15	22	All in centimeters
Width (*w*)	?	?	?	?	?	?	

Because w is the unknown quantity, it would simplify the computational work if we first solved the formula for w in terms of the other variables as follows:

$$P = 2l + 2w$$

$$P - 2l = 2w \qquad \text{Add } -2l \text{ to both sides.}$$

$$\frac{P - 2l}{2} = w \qquad \text{Multiply both sides by } \frac{1}{2}.$$

$$w = \frac{P - 2l}{2} \qquad \text{Apply the symmetric property of equality.}$$

Now for each value for P and l, we can easily determine the corresponding value for w. Be sure you agree with the following values for w: 6, 5, 4, 4, 13, and 18. Likewise we can also solve the formula $P = 2l + 2w$ for l in terms of P and w. The result would be $l = \frac{P - 2w}{2}$.

Let's consider some other often-used formulas and see how we can use the properties of equality to alter their forms. Here we will be solving a formula for a specified variable in terms of the other variables. The key is to isolate the term that contains the variable being solved for. Then, by appropriately applying the multiplication property of equality, we will solve the formula for the specified variable. Throughout this section, we will identify formulas when we first use them. (Some geometric formulas are also given on the endsheets.)

E X A M P L E   3   Solve $A = \dfrac{1}{2}bh$ for h (area of a triangle).

Solution

$$A = \frac{1}{2}bh$$

$$2A = bh \qquad \text{Multiply both sides by 2.}$$

$$\frac{2A}{b} = h \qquad \text{Multiply both sides by } \frac{1}{b}.$$

$$h = \frac{2A}{b} \qquad \text{Apply the symmetric property of equality.} \qquad \blacksquare$$

E X A M P L E   4   Solve $A = P + Prt$ for t.

Solution

$$A = P + Prt$$

$$A - P = Prt \qquad \text{Add } -P \text{ to both sides.}$$

$$\frac{A - P}{Pr} = t$$ 　　　　Multiply both sides by $\frac{1}{Pr}$.

$$t = \frac{A - P}{Pr}$$ 　　　　Apply the symmetric property of equality.　■

EXAMPLE 5

Solve $A = P + Prt$ for P.

Solution

$$A = P + Prt$$

$$A = P(1 + rt)$$ 　　　Apply the distributive property to the right side.

$$\frac{A}{1 + rt} = P$$ 　　　Multiply both sides by $\frac{1}{1 + rt}$.

$$P = \frac{A}{1 + rt}$$ 　　　Apply the symmetric property of equality.　■

EXAMPLE 6

Solve $A = \frac{1}{2}h(b_1 + b_2)$ for b_1 (area of a trapezoid).

Solution

$$A = \frac{1}{2}h(b_1 + b_2)$$

$$2A = h(b_1 + b_2)$$ 　　　Multiply both sides by 2.

$$2A = hb_1 + hb_2$$ 　　　Apply the distributive property to right side.

$$2A - hb_2 = hb_1$$ 　　　Add $-hb_2$ to both sides.

$$\frac{2A - hb_2}{h} = b_1$$ 　　　Multiply both sides by $\frac{1}{h}$.

$$b_1 = \frac{2A - hb_2}{h}$$ 　　　Apply the symmetric property of equality.　■

In order to isolate the term containing the variable being solved for, we will apply the distributive property in different ways. In Example 5 you *must* use the distributive property to change from the form $P + Prt$ to $P(1 + rt)$. However, in Example 6 we used the distributive property to change $h(b_1 + b_2)$ to $hb_1 + hb_2$. In both problems the key is to isolate the term that contains the variable being solved for, so that an appropriate application of the multiplication property of equality will produce the desired result. Also note the use of subscripts to identify the two bases of a trapezoid. Subscripts enable us to use the same letter b to identify the bases, but b_1 represents one base and b_2 the other.

Sometimes we are faced with equations such as $ax + b = c$, where x is the variable and $a, b,$ and c are referred to as *arbitrary constants.* Again we can use the properties of equality to solve the equation for x as follows:

$$ax + b = c$$

$$ax = c - b \qquad \text{Add } -b \text{ to both sides.}$$

$$x = \frac{c - b}{a} \qquad \text{Multiply both sides by } \frac{1}{a}.$$

In Chapter 7, we will be working with equations such as $2x - 5y = 7$, which are called equations of two variables in x and y. Often we need to change the form of such equations by solving for one variable in terms of the other variable. The properties of equality provide the basis for doing this.

EXAMPLE 7   Solve $2x - 5y = 7$ for y in terms of x.

Solution

$$2x - 5y = 7$$

$$-5y = 7 - 2x \qquad \text{Add } -2x \text{ to both sides.}$$

$$y = \frac{7 - 2x}{-5} \qquad \text{Multiply both sides by } -\frac{1}{5}.$$

$$y = \frac{2x - 7}{5} \qquad \begin{array}{l}\text{Multiply the numerator and denominator of the frac-}\\ \text{tion on the right by } -1. \text{ (This final step is not ab-}\\ \text{solutely necessary, but usually we prefer to have a}\\ \text{positive number as a denominator.)} \end{array} \quad\blacksquare$$

Equations of two variables may also contain arbitrary constants. For example, the equation $\frac{x}{a} + \frac{y}{b} = 1$ contains the variables x and y and the arbitrary constants a and b.

EXAMPLE 8   Solve the equation $\frac{x}{a} + \frac{y}{b} = 1$ for x.

Solution

$$\frac{x}{a} + \frac{y}{b} = 1$$

$$ab\left(\frac{x}{a} + \frac{y}{b}\right) = ab(1) \qquad \text{Multiply both sides by } ab.$$

$$bx + ay = ab$$

$$bx = ab - ay \qquad \text{Add } -ay \text{ to both sides.}$$

$$x = \frac{ab - ay}{b} \qquad \text{Multiply both sides by } \frac{1}{b}. \qquad \blacksquare$$

REMARK: Traditionally, equations that contain more than one variable, such as those in Examples 3–8, are called **literal equations.** As illustrated, it is sometimes necessary to solve a literal equation for one variable in terms of the other variable(s).

Formulas and Problem Solving

We often use formulas as guidelines for setting up an appropriate algebraic equation when solving a word problem. Let's consider an example to illustrate this point.

P R O B L E M   1

How long will it take $500 to double itself if we invest it at 8% simple interest?

Solution

For $500 to grow into $1000 (double itself), it must earn $500 in interest. Thus we let t represent the number of years it will take $500 to earn $500 in interest. Now we can use the formula $i = Prt$ as a guideline.

$$i = Prt$$
$$\downarrow \quad \downarrow \downarrow \downarrow$$
$$500 = 500(8\%)(t)$$

Solving this equation, we obtain

$$500 = 500(0.08)(t)$$
$$1 = 0.08t$$
$$100 = 8t$$
$$12\frac{1}{2} = t$$

It will take $12\frac{1}{2}$ years. $\qquad \blacksquare$

Sometimes we use formulas in the analysis of a problem but not as the main guideline for setting up the equation. For example, uniform motion problems involve the formula $d = rt$, but the main guideline for setting up an equation for such problems is usually a statement about times, rates, or distances. Let's consider an example to demonstrate.

P R O B L E M   2

Mercedes starts jogging at 5 miles per hour. One-half hour later, Karen starts jogging on the same route at 7 miles per hour. How long will it take Karen to catch Mercedes?

Solution

First, let's sketch a diagram and record some information (Figure 2.3).

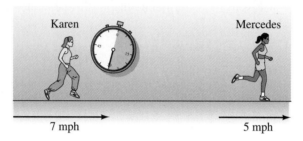

Figure 2.3

If we let t represent Karen's time, then $t + \dfrac{1}{2}$ represent Mercedes' time. We can use the statement *Karen's distance equals Mercedes' distance* as a guideline.

$$\underset{\substack{\downarrow \\ 7t}}{\text{Karen's distance}} \quad = \quad \underset{\substack{\downarrow \\ 5\left(t + \dfrac{1}{2}\right)}}{\text{Mercedes' distance}}$$

Solving this equation, we obtain

$$7t = 5t + \frac{5}{2}$$

$$2t = \frac{5}{2}$$

$$t = \frac{5}{4}$$

Karen should catch Mercedes in $1\dfrac{1}{4}$ hours.

REMARK: An important part of problem solving is the ability to sketch a meaningful figure that can be used to record the given information and help in the analysis of the problem. Our sketches were done by professional artists for aesthetic purposes. Your sketches can be very roughly drawn as long as they depict the situation in a way that helps you analyze the problem.

Note that in the solution of Problem 2 we used a figure and a simple arrow diagram to record and organize the information pertinent to the problem. Some people find it helpful to use a chart for that purpose. We shall use a chart in Problem 3. Keep in mind that we are not trying to dictate a particular approach; you decide what works best for you.

P R O B L E M   3   Two trains leave a city at the same time, one traveling east and the other traveling west. At the end of $9\frac{1}{2}$ hours, they are 1292 miles apart. If the rate of the train traveling east is 8 miles per hour faster than the rate of the other train, find their rates.

Solution

If we let r represent the rate of the westbound train, then $r + 8$ represents the rate of the eastbound train. Now we can record the times and rates in a chart and then use the distance formula ($d = rt$) to represent the distances.

	Rate	Time	Distance ($d = rt$)
Westbound train	r	$9\frac{1}{2}$	$\frac{19}{2}r$
Eastbound train	$r + 8$	$9\frac{1}{2}$	$\frac{19}{2}(r + 8)$

Because the distance that the westbound train travels plus the distance that the eastbound train travels equals 1292 miles, we can set up and solve the following equation.

$$\frac{\text{Eastbound}}{\text{distance}} + \frac{\text{Westbound}}{\text{distance}} = \frac{\text{Miles}}{\text{apart}}$$

$$\frac{19r}{2} + \frac{19(r + 8)}{2} = 1292$$

$$19r + 19(r + 8) = 2584$$

$$19r + 19r + 152 = 2584$$

$$38r = 2432$$

$$r = 64$$

The westbound train travels at a rate of 64 miles per hour, and the eastbound train travels at a rate of $64 + 8 = 72$ miles per hour.   ■

Now let's consider a problem that is often referred to as a mixture problem. There is no basic formula that applies to all of these problems, but we suggest that you think in terms of a pure substance, which is often helpful in setting up a guideline. Also keep in mind that the phrase "a 40% solution of some substance" means that the solution contains 40% of that particular substance and 60% of something else mixed with it. For example, a 40% salt solution contains 40% salt and the other 60% is something else, probably water. Now let's illustrate what we mean by suggesting that you think in terms of a pure substance.

P R O B L E M   4

Bryan's Pest Control stocks a 7% solution of insecticide for lawns and also a 15% solution. How many gallons of each should be mixed to produce 40 gallons that is 12% insecticide?

Solution

The key idea in solving such a problem is to recognize the following guideline.

$$\begin{pmatrix} \text{Amount of insecticide} \\ \text{in the 7\% solution} \end{pmatrix} + \begin{pmatrix} \text{Amount of insecticide} \\ \text{in the 15\% solution} \end{pmatrix} = \begin{pmatrix} \text{Amount of insecticide in} \\ \text{40 gallons of 15\% solution} \end{pmatrix}$$

Let x represent the gallons of 7% solution. Then $40 - x$ represents the gallons of 15% solution. The guideline translates into the following equation.

$$(7\%)(x) + (15\%)(40 - x) = (12\%)(40)$$

Solving this equation yields

$$0.07x + 0.15(40 - x) = 0.12(40)$$

$$0.07x + 6 - 0.15x = 4.8$$

$$-0.08x + 6 = 4.8$$

$$-0.08x = -1.2$$

$$x = 15$$

Thus 15 gallons of 7% solution and $40 - x = 25$ gallons of 15% solution need to be mixed to obtain 40 gallons of 12% solution. ■

P R O B L E M   5

How many liters of pure alcohol must we add to 20 liters of a 40% solution to obtain a 60% solution?

Solution

The key idea in solving such a problem is to recognize the following guideline.

$$\begin{pmatrix} \text{Amount of pure} \\ \text{alcohol in the} \\ \text{original solution} \end{pmatrix} + \begin{pmatrix} \text{Amount of} \\ \text{pure alcohol} \\ \text{to be added} \end{pmatrix} = \begin{pmatrix} \text{Amount of pure} \\ \text{alcohol in the} \\ \text{final solution} \end{pmatrix}$$

Let l represent the number of liters of pure alcohol to be added, and the guideline translates into the following equation.

$$(40\%)(20) + l = 60\%(20 + l)$$

Solving this equation yields

$$0.4(20) + l = 0.6(20 + l)$$

$$8 + l = 12 + 0.6l$$

$$0.4l = 4$$

$$l = 10$$

We need to add 10 liters of pure alcohol. (Remember to check this answer back into the original statement of the problem.) ■

PROBLEM SET 2.4

1. Solve $i = Prt$ for i, given that $P = \$300$, $r = 8\%$, and $t = 5$ years.

2. Solve $i = Prt$ for i, given that $P = \$500$, $r = 9\%$, and $t = 3\frac{1}{2}$ years.

3. Solve $i = Prt$ for t, given that $P = \$400$, $r = 11\%$, and $i = \$132$.

4. Solve $i = Prt$ for t, given that $P = \$250$, $r = 12\%$, and $i = \$120$.

5. Solve $i = Prt$ for r, given that $P = \$600$, $t = 2\frac{1}{2}$ years, and $i = \$90$. Express r as a percent.

6. Solve $i = Prt$ for r, given that $P = \$700$, $t = 2$ years, and $i = \$126$. Express r as a percent.

7. Solve $i = Prt$ for P, given that $r = 9\%$, $t = 3$ years, and $i = \$216$.

8. Solve $i = Prt$ for P, given that $r = 8\frac{1}{2}\%$, $t = 2$ years, and $i = \$204$.

9. Solve $A = P + Prt$ for A, given that $P = \$1000$, $r = 12\%$, and $t = 5$ years.

10. Solve $A = P + Prt$ for A, given that $P = \$850$, $r = 9\frac{1}{2}\%$, and $t = 10$ years.

11. Solve $A = P + Prt$ for r, given that $A = \$1372$, $P = \$700$, and $t = 12$ years. Express r as a percent.

12. Solve $A = P + Prt$ for r, given that $A = \$516$, $P = \$300$, and $t = 8$ years. Express r as a percent.

13. Solve $A = P + Prt$ for P, given that $A = \$326$, $r = 7\%$, and $t = 9$ years.

14. Solve $A = P + Prt$ for P, given that $A = \$720$, $r = 8\%$, and $t = 10$ years.

15. Use the formula $A = \frac{1}{2}h(b_1 + b_2)$ and complete the following chart.

A	98	104	49	162	$16\frac{1}{2}$	$38\frac{1}{2}$	Square feet
h	14	8	7	9	3	11	Feet
b_1	8	12	4	16	4	5	Feet
b_2	?	?	?	?	?	?	Feet

A = area, h = height, b_1 = one base, b_2 = other base

16. Use the formula $P = 2l + 2w$ and complete the following chart. (You may want to change the form of the formula.)

P	28	18	12	34	68	Centimeters
w	6	3	2	7	14	Centimeters
l	?	?	?	?	?	Centimeters

P = perimeter, w = width, l = length

Solve each of the following for the indicated variable.

17. $V = Bh$ for h (Volume of a prism)

18. $A = lw$ for l (Area of a rectangle)

19. $V = \pi r^2 h$ for h (Volume of a circular cylinder)

20. $V = \dfrac{1}{3}Bh$ for B (Volume of a pyramid)

21. $C = 2\pi r$ for r (Circumference of a circle)

22. $A = 2\pi r^2 + 2\pi rh$ for h (Surface area of a circular cylinder)

23. $I = \dfrac{100M}{C}$ for C (Intelligence quotient)

24. $A = \dfrac{1}{2}h(b_1 + b_2)$ for h (Area of a trapezoid)

25. $F = \dfrac{9}{5}C + 32$ for C (Celsius to Fahrenheit)

26. $C = \dfrac{5}{9}(F - 32)$ for F (Fahrenheit to Celsius)

For Problems 27–36, solve each equation for x.

27. $y = mx + b$

28. $\dfrac{x}{a} + \dfrac{y}{b} = 1$

29. $y - y_1 = m(x - x_1)$

30. $a(x + b) = c$

31. $a(x + b) = b(x - c)$

32. $x(a - b) = m(x - c)$

33. $\dfrac{x - a}{b} = c$

34. $\dfrac{x}{a} - 1 = b$

35. $\dfrac{1}{3}x + a = \dfrac{1}{2}b$

36. $\dfrac{2}{3}x - \dfrac{1}{4}a = b$

For Problems 37–46, solve each equation for the indicated variable.

37. $2x - 5y = 7$ for x

38. $5x - 6y = 12$ for x

39. $-7x - y = 4$ for y

40. $3x - 2y = -1$ for y

41. $3(x - 2y) = 4$ for x

42. $7(2x + 5y) = 6$ for y

43. $\dfrac{y - a}{b} = \dfrac{x + b}{c}$ for x

44. $\dfrac{x - a}{b} = \dfrac{y - a}{c}$ for y

45. $(y + 1)(a - 3) = x - 2$ for y

46. $(y - 2)(a + 1) = x$ for y

Solve each of Problems 47–62 by setting up and solving an appropriate algebraic equation.

47. Suppose that the length of a certain rectangle is 2 meters less than four times its width. The perimeter of the rectangle is 56 meters. Find the length and width of the rectangle.

48. The perimeter of a triangle is 42 inches. The second side is 1 inch more than twice the first side, and the third side is 1 inch less than three times the first side. Find the lengths of the three sides of the triangle.

49. How long will it take $500 to double itself at 9% simple interest?

50. How long will it take $700 to triple itself at 10% simple interest?

51. How long will it take P dollars to double itself at 9% simple interest?

52. How long will it take P dollars to triple itself at 10% simple interest?

53. Two airplanes leave Chicago at the same time and fly in opposite directions. If one travels at 450 miles per hour and the other at 550 miles per hour, how long will it take for them to be 4000 miles apart?

54. Look at Figure 2.4. Tyrone leaves city A on a moped traveling toward city B at 18 miles per hour. At the same time, Tina leaves city B on a bicycle traveling toward city A at 14 miles per hour. The distance between the two cities is 112 miles. How long will it take before Tyrone and Tina meet?

55. Juan starts walking at 4 miles per hour. An hour and a half later, Cathy starts jogging along the same route at

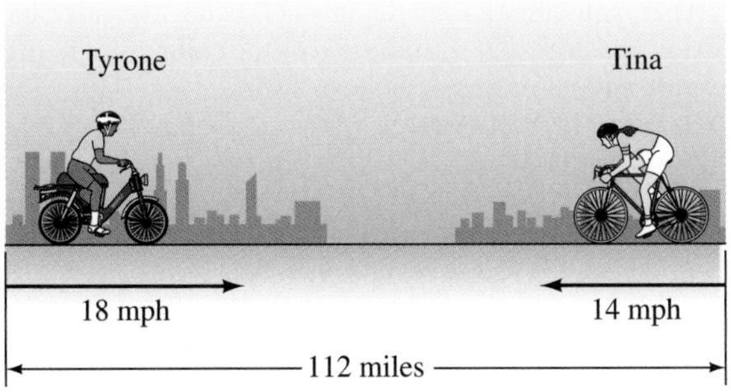

Figure 2.4

6 miles per hour. How long will it take Cathy to catch up with Juan?

56. A car leaves a town at 60 kilometers per hour. How long will it take a second car, traveling at 75 kilometers per hour, to catch the first car if it leaves 1 hour later?

57. Bret started on a 70-mile bicycle ride at 20 miles per hour. After a time he became a little tired and slowed down to 12 miles per hour for the rest of the trip. The entire trip of 70 miles took $4\frac{1}{2}$ hours. How far had Bret ridden when he reduced his speed to 12 miles per hour?

58. How many gallons of a 12%-salt solution must be mixed with 6 gallons of a 20%-salt solution to obtain a 15%-salt solution?

59. Suppose that you have a supply of a 30% solution of alcohol and a 70% solution of alcohol. How many quarts of each should be mixed to produce 20 quarts that is 40% alcohol?

60. How many cups of grapefruit juice must be added to 40 cups of punch that is 5% grapefruit juice to obtain a punch that is 10% grapefruit juice?

61. How many milliliters of pure acid must be added to 150 milliliters of a 30% solution of acid to obtain a 40% solution?

62. A 16-quart radiator contains a 50% solution of antifreeze. How much needs to be drained out and replaced with pure antifreeze to obtain a 60% antifreeze solution?

■ ■ ■ **Thoughts into words**

63. Some people subtract 32 and then divide by 2 to estimate the change from a Fahrenheit reading to a Celsius reading. Why does this give an estimate and how good is the estimate?

64. One of your classmates analyzes Problem 56 as follows: "The first car has traveled 60 kilometers before the second car starts. Because the second car travels 15 kilometers per hour faster, it will take $\frac{60}{15} = 4$ hours for the second car to overtake the first car." How would you react to this analysis of the problem?

65. Summarize the new ideas relative to problem solving that you have acquired thus far in this course.

■ ■ ■ **Further investigations**

For Problems 66–73, use your calculator to help solve each formula for the indicated variable.

66. Solve $i = Prt$ for i, given that $P = \$875, r = 12\frac{1}{2}\%$, and $t = 4$ years.

67. Solve $i = Prt$ for i, given that $P = \$1125, r = 13\frac{1}{4}\%$, and $t = 4$ years.

68. Solve $i = Prt$ for t, given that $i = \$453.25, P = \$925$, and $r = 14\%$.

69. Solve $i = Prt$ for t, given that $i = \$243.75, P = \$1250$, and $r = 13\%$.

70. Solve $i = Prt$ for r, given that $i = \$356.50, P = \$1550$, and $t = 2$ years. Express r as a percent.

71. Solve $i = Prt$ for r, given that $i = \$159.50, P = \$2200$, and $t = 0.5$ of a year. Express r as a percent.

72. Solve $A = P + Prt$ for P, given that $A = \$1423.50$, $r = 9\frac{1}{2}\%$, and $t = 1$ year.

73. Solve $A = P + Prt$ for P, given that $A = \$2173.75$, $r = 8\frac{3}{4}\%$, and $t = 2$ years.

74. If you have access to computer software that includes spreadsheets, go to Problems 15 and 16. You should be able to enter the given information in rows. Then, when you enter a formula in a cell below the information and drag that formula across the columns, the software should produce all the answers.

2.5   Inequalities

We listed the basic inequality symbols in Section 1.2. With these symbols we can make various **statements of inequality:**

$a < b$ means a is less than b.

$a \leq b$ means a is less than or equal to b.

$a > b$ means a is greater than b.

$a \geq b$ means a is greater than or equal to b.

Here are some examples of **numerical statements of inequality:**

$$7 + 8 > 10 \qquad -4 + (-6) \geq -10$$

$$-4 > -6 \qquad 7 - 9 \leq -2$$

$$7 - 1 < 20 \qquad 3 + 4 > 12$$

$$8(-3) < 5(-3) \qquad 7 - 1 < 0$$

Note that only $3 + 4 > 12$ and $7 - 1 < 0$ are *false;* the other six are *true* numerical statements.

Algebraic inequalities contain one or more variables. The following are examples of algebraic inequalities.

$$x + 4 > 8 \qquad 3x + 2y \leq 4$$
$$3x - 1 < 15 \qquad x^2 + y^2 + z^2 \geq 7$$
$$y^2 + 2y - 4 \geq 0$$

An algebraic inequality such as $x + 4 > 8$ is neither true nor false as it stands, and we call it an **open sentence.** For each numerical value we substitute for x, the algebraic inequality $x + 4 > 8$ becomes a numerical statement of inequality that is true or false. For example, if $x = -3$, then $x + 4 > 8$ becomes $-3 + 4 > 8$, which is false. If $x = 5$, then $x + 4 > 8$ becomes $5 + 4 > 8$, which is true. **Solving an inequality** is the process of finding the numbers that make an algebraic inequality a true numerical statement. We call such numbers the *solutions* of the inequality; the solutions *satisfy* the inequality.

The general process for solving inequalities closely parallels the process for solving equations. We continue to replace the given inequality with equivalent, but simpler, inequalities. For example,

$$3x + 4 > 10 \qquad\qquad\qquad \textbf{(1)}$$

$$3x > 6 \qquad\qquad\qquad \textbf{(2)}$$

$$x > 2 \qquad\qquad\qquad \textbf{(3)}$$

are all equivalent inequalities; that is, they all have the same solutions. By inspection we see that the solutions for (3) are all numbers greater than *2.* Thus (1) has the same solutions.

The exact procedure for simplifying inequalities so that we can determine the solutions is based primarily on two properties. The first of these is the addition property of inequality.

Addition Property of Inequality

For all real numbers *a, b,* and *c,*

$$a > b \quad \text{if and only if} \quad a + c > b + c$$

The addition property of inequality states that we can add any number to both sides of an inequality to produce an equivalent inequality. We have stated the property in terms of $>$, but analogous properties exist for $<$, $\geq$, and $\leq$.

Before we state the multiplication property of inequality let's look at some numerical examples.

$2 < 5$	Multiply both sides by 4	$4(2) < 4(5)$	$8 < 20$
$-3 > -7$	Multiply both sides by 2	$2(-3) > 2(-7)$	$-6 > -14$
$-4 < 6$	Multiply both sides by 10	$10(-4) < 10(6)$	$-40 < 60$
$4 < 8$	Multiply both sides by –3	$-3(4) > -3(8)$	$-12 > -24$
$3 > -2$	Multiply both sides by –4	$-4(3) < -4(-2)$	$-12 < 8$
$-4 < -1$	Multiply both sides by –2	$-2(-4) > -2(-1)$	$8 > 2$

Notice in the first three examples that when we multiply both sides of an inequality by a *positive number*, we get an inequality of the *same sense.* That means that if the original inequality is *less than*, then the new inequality is *less than*; and if the original inequality is *greater than*, then the new inequality is *greater than*. The last three examples illustrate that when we multiply both sides of an inequality by a *negative number* we get an inequality of the *opposite sense.*

We can state the multiplication property of inequality as follows.

Multiplication Property of Inequality

(a) For all real numbers *a, b,* and *c,* with $c > 0$,

$$a > b \quad \text{if and only if} \quad ac > bc$$

(b) For all real numbers *a, b,* and *c,* with $c < 0$,

$$a > b \quad \text{if and only if} \quad ac < bc$$

Similar properties hold if we reverse each inequality or if we replace $>$ with $\geq$ and $<$ with $\leq$. For example, if $a \leq b$ and $c < 0$, then $ac \geq bc$.

Now let's use the addition and multiplication properties of inequality to help solve some inequalities.

E X A M P L E   1   Solve $3x - 4 > 8$.

Solution

$$3x - 4 > 8$$

$$3x - 4 + 4 > 8 + 4 \qquad \text{Add 4 to both sides.}$$

$$3x > 12$$

$$\frac{1}{3}(3x) > \frac{1}{3}(12) \qquad \text{Multiply both sides by } \frac{1}{3}.$$

$$x > 4$$

The solution set is $\{x \mid x > 4\}$. (Remember that we read the set $\{x \mid x > 4\}$ as "the set of all x such that x is greater than 4.")   ∎

In Example 1, once we obtained the simple inequality $x > 4$, the solution set $\{x \mid x > 4\}$ became obvious. We can also express solution sets for inequalities on a number line graph. Figure 2.5 shows the graph of the solution set for Example 1. The left-hand parenthesis at 4 indicates that 4 is *not* a solution, and the thickened appearance of the part of the line to the right of 4 indicates that all numbers greater than 4 are solutions.

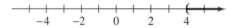

Figure 2.5

It is also convenient to express solution sets of inequalities using **interval notation**. For example, the notation $(4, \infty)$ also refers to the set of real numbers greater than 4. As in Figure 2.5, the left-hand parenthesis indicates that 4 is not to be included. The infinity symbol, ∞, along with the right-hand parenthesis, indicates that there is no right-hand endpoint. Following is a partial list of interval notations, along with the sets of graphs they represent (Figure 2.6). We will add to this list in the next section.

Set	Graph	Interval notation
$\{x \mid x > a\}$		(a, ∞)
$\{x \mid x \geq a\}$	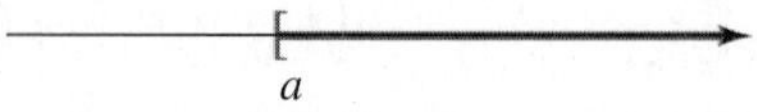	$[a, \infty)$
$\{x \mid x < b\}$	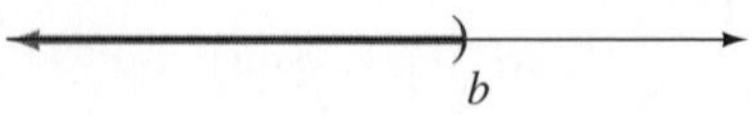	$(-\infty, b)$
$\{x \mid x \leq b\}$		$(-\infty, b]$

Figure 2.6

Note the use of square brackets to indicate the inclusion of endpoints. From now on, we will express the solution sets of inequalities using interval notation.

EXAMPLE 2

Solve $-2x + 1 > 5$ and graph the solutions.

Solution

$$-2x + 1 > 5$$

$$-2x + 1 + (-1) > 5 + (-1) \qquad \text{Add } -1 \text{ to both sides.}$$

$$-2x > 4$$

$$-\frac{1}{2}(-2x) < -\frac{1}{2}(4) \qquad \text{Multiply both sides by } -\frac{1}{2}.$$

$$x < -2 \qquad \text{Note that the sense of the inequality has been reversed.}$$

The solution set is $(-\infty, -2)$, which can be illustrated on a number line as in Figure 2.7.

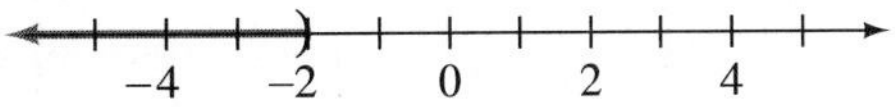

Figure 2.7

Checking solutions for an inequality presents a problem. Obviously, we cannot check all of the infinitely many solutions for a particular inequality. However, by checking at least one solution, especially when the multiplication property has been used, we might catch the common mistake of forgetting to change the sense of an inequality. In Example 2 we are claiming that all numbers less than -2 will satisfy the original inequality. Let's check one such number, say -4.

$$-2x + 1 > 5$$

$$-2(-4) + 1 \overset{?}{>} 5 \quad \text{when } x = -4$$

$$8 + 1 \overset{?}{>} 5$$

$$9 > 5$$

Thus -4 satisfies the original inequality. Had we forgotten to switch the sense of the inequality when both sides were multiplied by $-\frac{1}{2}$, our answer would have been $x > -2$, and we would have detected such an error by the check.

Many of the same techniques used to solve equations, such as removing parentheses and combining similar terms, may be used to solve inequalities. However, we must be extremely careful when using the multiplication property of inequality. Study each of the following examples very carefully. The format we used highlights the major steps of a solution.

EXAMPLE 3   Solve $-3x + 5x - 2 \geq 8x - 7 - 9x$.

Solution

$$-3x + 5x - 2 \geq 8x - 7 - 9x$$

$$2x - 2 \geq -x - 7 \qquad \text{Combine similar terms on both sides.}$$

$$3x - 2 \geq -7 \qquad \text{Add } x \text{ to both sides.}$$

$$3x \geq -5 \qquad \text{Add 2 to both sides.}$$

$$\frac{1}{3}(3x) \geq \frac{1}{3}(-5) \qquad \text{Multiply both sides by } \frac{1}{3}.$$

$$x \geq -\frac{5}{3}$$

The solution set is $\left[-\dfrac{5}{3}, \infty\right)$.

EXAMPLE 4   Solve $-5(x - 1) \leq 10$ and graph the solutions.

Solution

$$-5(x - 1) \leq 10$$

$$-5x + 5 \leq 10 \qquad \text{Apply the distributive property on the left.}$$

$$-5x \leq 5 \qquad \text{Add } -5 \text{ to both sides.}$$

$$-\frac{1}{5}(-5x) \geq -\frac{1}{5}(5) \qquad \text{Multiply both sides by } -\frac{1}{5}\text{, which reverses the inequality.}$$

$$x \geq -1$$

The solution set is $[-1, \infty)$ and it can be graphed as in Figure 2.8.

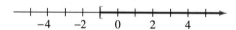

Figure 2.8

EXAMPLE 5   Solve $4(x - 3) > 9(x + 1)$.

Solution

$$4(x - 3) > 9(x + 1)$$

$$4x - 12 > 9x + 9 \qquad \text{Apply the distributive property.}$$

$$-5x - 12 > 9 \qquad \text{Add } -9x \text{ to both sides.}$$

$$-5x > 21 \qquad \text{Add 12 to both sides.}$$

$$-\frac{1}{5}(-5x) < -\frac{1}{5}(21)$$

Multiply both sides by $-\frac{1}{5}$, which reverses the inequality.

$$x < -\frac{21}{5}$$

The solution set is $\left(-\infty, -\frac{21}{5}\right)$.

The next example will solve the inequality without indicating the justification for each step. Be sure that you can supply the reasons for the steps.

E X A M P L E   6   Solve $3(2x + 1) - 2(2x + 5) < 5(3x - 2)$.

Solution

$$3(2x + 1) - 2(2x + 5) < 5(3x - 2)$$

$$6x + 3 - 4x - 10 < 15x - 10$$

$$2x - 7 < 15x - 10$$

$$-13x - 7 < -10$$

$$-13x < -3$$

$$-\frac{1}{13}(-13x) > -\frac{1}{13}(-3)$$

$$x > \frac{3}{13}$$

The solution set is $\left(\frac{3}{13}, \infty\right)$.

PROBLEM SET 2.5

For Problems 1–8, express the given inequality in interval notation and sketch a graph of the interval.

1. $x > 1$

2. $x > -2$

3. $x \geq -1$

4. $x \geq 3$

5. $x < -2$

6. $x < 1$

7. $x \leq 2$

8. $x \leq 0$

For Problems 9–16, express each interval as an inequality using the variable x. For example, we can express the interval $[5, \infty)$ as $x \geq 5$.

9. $(-\infty, 4)$

10. $(-\infty, -2)$

11. $(-\infty, -7]$

12. $(-\infty, 9]$

13. $(8, \infty)$

14. $(-5, \infty)$

15. $[-7, \infty)$

16. $[10, \infty)$

For Problems 17–40, solve each of the inequalities and graph the solution set on a number line.

17. $x - 3 > -2$

18. $x + 2 < 1$

19. $-2x \geq 8$

20. $-3x \leq -9$

21. $5x \leq -10$

22. $4x \geq -4$

23. $2x + 1 < 5$

24. $2x + 2 > 4$

25. $3x - 2 > -5$

26. $5x - 3 < -3$

27. $-7x - 3 \le 4$

28. $-3x - 1 \ge 8$

29. $2 + 6x > -10$

30. $1 + 6x > -17$

31. $5 - 3x < 11$

32. $4 - 2x < 12$

33. $15 < 1 - 7x$

34. $12 < 2 - 5x$

35. $-10 \le 2 + 4x$

36. $-9 \le 1 + 2x$

37. $3(x + 2) > 6$

38. $2(x - 1) < -4$

39. $5x + 2 \ge 4x + 6$

40. $6x - 4 \le 5x - 4$

For Problems 41–70, solve each inequality and express the solution set using interval notation.

41. $2x - 1 > 6$

42. $3x - 2 < 12$

43. $-5x - 2 < -14$

44. $5 - 4x > -2$

45. $-3(2x + 1) \ge 12$

46. $-2(3x + 2) \le 18$

47. $4(3x - 2) \ge -3$

48. $3(4x - 3) \le -11$

49. $6x - 2 > 4x - 14$

50. $9x + 5 < 6x - 10$

51. $2x - 7 < 6x + 13$

52. $2x - 3 > 7x + 22$

53. $4(x - 3) \le -2(x + 1)$

54. $3(x - 1) \ge -(x + 4)$

55. $5(x - 4) - 6(x + 2) < 4$

56. $3(x + 2) - 4(x - 1) < 6$

57. $-3(3x + 2) - 2(4x + 1) \ge 0$

58. $-4(2x - 1) - 3(x + 2) \ge 0$

59. $-(x - 3) + 2(x - 1) < 3(x + 4)$

60. $3(x - 1) - (x - 2) > -2(x + 4)$

61. $7(x + 1) - 8(x - 2) < 0$

62. $5(x - 6) - 6(x + 2) < 0$

63. $-5(x - 1) + 3 > 3x - 4 - 4x$

64. $3(x + 2) + 4 < -2x + 14 + x$

65. $3(x - 2) - 5(2x - 1) \ge 0$

66. $4(2x - 1) - 3(3x + 4) \ge 0$

67. $-5(3x + 4) < -2(7x - 1)$

68. $-3(2x + 1) > -2(x + 4)$

69. $-3(x + 2) > 2(x - 6)$

70. $-2(x - 4) < 5(x - 1)$

■ ■ ■ **Thoughts into words**

71. Do the *less than* and *greater than* relations possess a symmetric property similar to the symmetric property of equality? Defend your answer.

72. Give a step-by-step description of how you would solve the inequality $-3 > 5 - 2x$.

73. How would you explain to someone why it is necessary to reverse the inequality symbol when multiplying both sides of an inequality by a negative number?

■ ■ ■ **Further investigations**

74. Solve each of the following inequalities.

(a) $5x - 2 > 5x + 3$

(b) $3x - 4 < 3x + 7$

(c) $4(x + 1) < 2(2x + 5)$

(d) $-2(x - 1) > 2(x + 7)$

(e) $3(x - 2) < -3(x + 1)$

(f) $2(x + 1) + 3(x + 2) < 5(x - 3)$

2.6   More on Inequalities and Problem Solving

When we discussed solving equations that involve fractions, we found that **clearing the equation of all fractions** is frequently an effective technique. To accomplish this, we multiply both sides of the equation by the least common denominator of all the denominators in the equation. This same basic approach also works very well with inequalities that involve fractions, as the next examples demonstrate.

E X A M P L E   1   Solve $\dfrac{2}{3}x - \dfrac{1}{2}x > \dfrac{3}{4}$.

Solution

$$\frac{2}{3}x - \frac{1}{2}x > \frac{3}{4}$$

$$12\left(\frac{2}{3}x - \frac{1}{2}x\right) > 12\left(\frac{3}{4}\right) \qquad \text{Multiply both sides by 12, which is the LCD of 3, 2, and 4.}$$

$$12\left(\frac{2}{3}x\right) - 12\left(\frac{1}{2}x\right) > 12\left(\frac{3}{4}\right) \qquad \text{Apply the distributive property.}$$

$$8x - 6x > 9$$

$$2x > 9$$

$$x > \frac{9}{2}$$

The solution set is $\left(\dfrac{9}{2}, \infty\right)$.   ∎

E X A M P L E   2   Solve $\dfrac{x + 2}{4} + \dfrac{x - 3}{8} < 1$.

Solution

$$\frac{x + 2}{4} + \frac{x - 3}{8} < 1$$

$$8\left(\frac{x + 2}{4} + \frac{x - 3}{8}\right) < 8(1) \qquad \text{Multiply both sides by 8, which is the LCD of 4 and 8.}$$

$$8\left(\frac{x + 2}{4}\right) + 8\left(\frac{x - 3}{8}\right) < 8(1)$$

$$2(x + 2) + (x - 3) < 8$$

$$2x + 4 + x - 3 < 8$$

$$3x + 1 < 8$$

$$3x < 7$$

$$x < \frac{7}{3}$$

The solution set is $\left(-\infty, \dfrac{7}{3}\right)$.   ■

EXAMPLE 3   Solve $\dfrac{x}{2} - \dfrac{x - 1}{5} \geq \dfrac{x + 2}{10} - 4$.

Solution

$$\frac{x}{2} - \frac{x - 1}{5} \geq \frac{x + 2}{10} - 4$$

$$10\left(\frac{x}{2} - \frac{x - 1}{5}\right) \geq 10\left(\frac{x + 2}{10} - 4\right)$$

$$10\left(\frac{x}{2}\right) - 10\left(\frac{x - 1}{5}\right) \geq 10\left(\frac{x + 2}{10}\right) - 10(4)$$

$$5x - 2(x - 1) \geq x + 2 - 40$$

$$5x - 2x + 2 \geq x - 38$$

$$3x + 2 \geq x - 38$$

$$2x + 2 \geq -38$$

$$2x \geq -40$$

$$x \geq -20$$

The solution set is $[-20, \infty)$.   ■

The idea of **clearing all decimals** also works with inequalities in much the same way as it does with equations. We can multiply both sides of an inequality by an appropriate power of 10 and then proceed to solve in the usual way. The next two examples illustrate this procedure.

EXAMPLE 4   Solve $x \geq 1.6 + 0.2x$.

Solution

$$x \geq 1.6 + 0.2x$$

$$10(x) \geq 10(1.6 + 0.2x) \qquad \text{Multiply both sides by 10.}$$

$$10x \geq 16 + 2x$$

$$8x \geq 16$$

$$x \geq 2$$

The solution set is $[2, \infty)$.   ■

EXAMPLE 5   Solve $0.08x + 0.09(x + 100) \geq 43$.

Solution

$$0.08x + 0.09(x + 100) \geq 43$$

$$100(0.08x + 0.09(x + 100)) \geq 100(43) \qquad \text{Multiply both sides by 100.}$$

$$8x + 9(x + 100) \geq 4300$$

$$8x + 9x + 900 \geq 4300$$

$$17x + 900 \geq 4300$$

$$17x \geq 3400$$

$$x \geq 200$$

The solution set is $[200, \infty)$.   ■

Compound Statements

We use the words "and" and "or" in mathematics to form **compound statements.**
The following are examples of compound numerical statements that use "and." We
call such statements **conjunctions.** We agree to call a conjunction true only if all of
its component parts are true. Statements 1 and 2 below are true, but statements 3,
4, and 5 are false.

1. $3 + 4 = 7$ and $-4 < -3$.   True
2. $-3 < -2$ and $-6 > -10$.   True
3. $6 > 5$ and $-4 < -8$.   False
4. $4 < 2$ and $0 < 10$.   False
5. $-3 + 2 = 1$ and $5 + 4 = 8$.   False

We call compound statements that use "or" **disjunctions.** The following are
examples of disjunctions that involve numerical statements.

6. $0.14 > 0.13$ or $0.235 < 0.237$.   True

7. $\dfrac{3}{4} > \dfrac{1}{2}$ or $-4 + (-3) = 10$.   True

8. $-\dfrac{2}{3} > \dfrac{1}{3}$ or $(0.4)(0.3) = 0.12$.   True

9. $\dfrac{2}{5} < -\dfrac{2}{5}$ or $7 + (-9) = 16$.   False

A disjunction is true if at least one of its component parts is true. In other words,
disjunctions are false only if all of the component parts are false. Thus statements
6, 7, and 8 are true, but statement 9 is false.

Now let's consider finding solutions for some compound statements that involve algebraic inequalities. Keep in mind that our previous agreements for labeling conjunctions and disjunctions true or false form the basis for our reasoning.

EXAMPLE 6   Graph the solution set for the conjunction $x > -1$ and $x < 3$.

Solution

The key word is "and," so we need to satisfy both inequalities. Thus all numbers between -1 and 3 are solutions, and we can indicate this on a number line as in Figure 2.9.

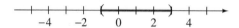

Figure 2.9

Using interval notation, we can represent the interval enclosed in parentheses in Figure 2.9 by $(-1, 3)$. Using set builder notation we can express the same interval as $\{x \mid -1 < x < 3\}$, where the statement $-1 < x < 3$ is read "Negative one is less than x, and x is less than three." In other words, x is between -1 and 3.   ■

Example 6 represents another concept that pertains to sets. The set of all elements common to two sets is called the **intersection** of the two sets. Thus in Example 6, we found the intersection of the two sets $\{x \mid x > -1\}$ and $\{x \mid x < 3\}$ to be the set $\{x \mid -1 < x < 3\}$. In general, we define the intersection of two sets as follows:

> **DEFINITION 2.1**
>
> The **intersection** of two sets A and B (written $A \cap B$) is the set of all elements that are in both A and in B. Using set builder notation, we can write
>
> $$A \cap B = \{x \mid x \in A \text{ and } x \in B\}$$

EXAMPLE 7   Solve the conjunction $3x + 1 > -5$ *and* $2x + 5 > 7$, and graph its solution set on a number line.

Solution

First, let's simplify both inequalities.

$$
\begin{array}{ccc}
3x + 1 > -5 & \text{and} & 2x + 5 > 7 \\
3x > -6 & \text{and} & 2x > 2 \\
x > -2 & \text{and} & x > 1
\end{array}
$$

Because this is a conjunction, we must satisfy both inequalities. Thus all numbers greater than 1 are solutions, and the solution set is $(1, \infty)$. We show the graph of the solution set in Figure 2.10.

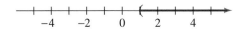

Figure 2.10

We can solve a conjunction such as $3x + 1 > -3$ and $3x + 1 < 7$, in which the same algebraic expression (in this case $3x + 1$) is contained in both inequalities, by using the **compact form** $-3 < 3x + 1 < 7$ as follows:

$$-3 \; < \; 3x + 1 \; < 7$$

$$-4 \; < \; 3x \; < \; 6 \qquad \text{Add } -1 \text{ to the left side, middle, and right side.}$$

$$-\frac{4}{3} \; < \; x \; < 2 \qquad \text{Multiply through by } \frac{1}{3}.$$

The solution set is $\left(-\dfrac{4}{3}, 2 \right)$.

The word *and* ties the concept of a conjunction to the set concept of intersection. In a like manner, the word *or* links the idea of a disjunction to the set concept of **union.** We define the union of two sets as follows:

DEFINITION 2.2

The **union** of two sets A and B (written $A \cup B$) is the set of all elements that are in A or in B, or in both. Using set builder notation, we can write

$$A \cup B = \{x | x \in A \text{ or } x \in B\}$$

EXAMPLE 8

Graph the solution set for the disjunction $x < -1$ or $x > 2$, and express it using interval notation.

Solution

The key word is "or," so all numbers that satisfy either inequality (or both) are solutions. Thus all numbers less than -1, along with all numbers greater than 2, are the solutions. The graph of the solution set is shown in Figure 2.11.

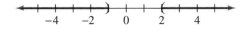

Figure 2.11

Using interval notation and the set concept of union, we can express the solution set as $(-\infty, -1) \cup (2, \infty)$.   ∎

Example 8 illustrates that in terms of set vocabulary, the solution set of a disjunction is the union of the solution sets of the component parts of the disjunction. Note that there is no compact form for writing $x < -1$ or $x > 2$ or for any disjunction.

E X A M P L E   9   Solve the disjunction $2x - 5 < -11$ or $5x + 1 \geq 6$, and graph its solution set on a number line.

Solution

First, let's simplify both inequalities.

$$2x - 5 < -11 \quad \text{or} \quad 5x + 1 \geq 6$$
$$2x < -6 \quad \text{or} \quad 5x \geq 5$$
$$x < -3 \quad \text{or} \quad x \geq 1$$

This is a disjunction, and all numbers less than -3, along with all numbers greater than or equal to 1, will satisfy it. Thus the solution set is $(-\infty, -3) \cup [1, \infty)$. Its graph is shown in Figure 2.12.

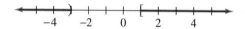

Figure 2.12

In summary, to solve a compound sentence involving an inequality, proceed as follows:

1. Solve separately each inequality in the compound sentence.
2. If it is a conjunction, the solution set is the intersection of the solution sets of each inequality.
3. If it is a disjunction, the solution set is the union of the solution sets of each inequality.

The following agreements on the use of interval notation (Figure 2.13) should be added to the list in Figure 2.6.

Set	Graph	Interval notation
$\{x \mid a < x < b\}$	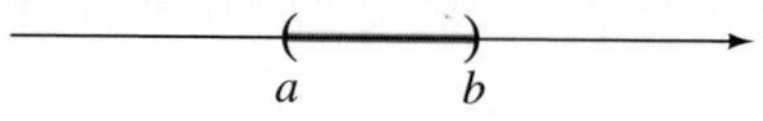	(a, b)
$\{x \mid a \leq x < b\}$	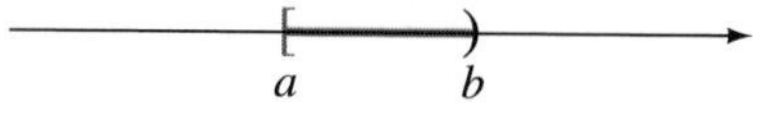	$[a, b)$
$\{x \mid a < x \leq b\}$	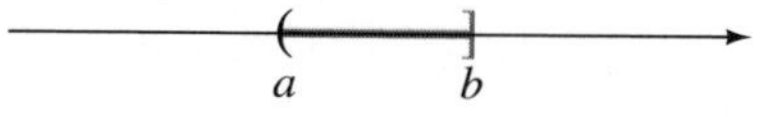	$(a, b]$
$\{x \mid a \leq x \leq b\}$	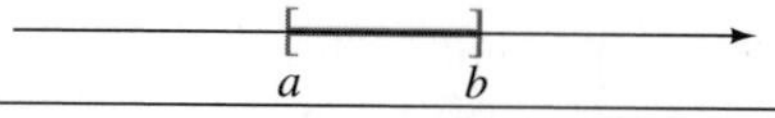	$[a, b]$

Figure 2.13

Problem Solving

We will conclude this section with some word problems that contain inequality statements.

PROBLEM 1

Sari had scores of 94, 84, 86, and 88 on her first four exams of the semester. What score must she obtain on the fifth exam to have an average of 90 or better for the five exams?

Solution

Let s represent the score Sari needs on the fifth exam. Because the average is computed by adding all scores and dividing by the number of scores, we have the following inequality to solve.

$$\frac{94 + 84 + 86 + 88 + s}{5} \geq 90$$

Solving this inequality, we obtain

$$\frac{352 + s}{5} \geq 90$$

$$5\left(\frac{352 + s}{5}\right) \geq 90 \qquad \text{Multiply both sides by 5.}$$

$$352 + s \geq 450$$

$$s \geq 98$$

Sari must receive a score of 98 or better. ∎

PROBLEM 2

An investor has $1000 to invest. Suppose she invests $500 at 8% interest. At what rate must she invest the other $500 so that the two investments together yield more than $100 of yearly interest?

Solution

Let r represent the unknown rate of interest. We can use the following guideline to set up an inequality.

$$\begin{array}{ccccc} \text{Interest from} & + & \text{Interest from } r & > & \$100 \\ \text{8\% investment} & & \text{percent investment} & & \\ \downarrow & & \downarrow & & \downarrow \\ (8\%)(\$500) & + & r(\$500) & > & \$100 \end{array}$$

Solving this inequality yields

$$40 + 500r > 100$$

$$500r > 60$$

$$r > \frac{60}{500}$$

$$r > 0.12 \qquad \text{Change to a decimal.}$$

She must invest the other \$500 at a rate greater than 12%. ■

PROBLEM 3   If the temperature for a 24-hour period ranged between 41°F and 59°F, inclusive (that is, $41 \leq F \leq 59$), what was the range in Celsius degrees?

Solution

Use the formula $F = \frac{9}{5}C + 32$, to solve the following compound inequality.

$$41 \leq \frac{9}{5}C + 32 \leq 59$$

Solving this yields

$$9 \leq \frac{9}{5}C \leq 27 \qquad\qquad \text{Add } -32.$$

$$\frac{5}{9}(9) \leq \frac{5}{9}\left(\frac{9}{5}C\right) \leq \frac{5}{9}(27) \qquad \text{Multiply by } \frac{5}{9}.$$

$$5 \leq C \leq 15$$

The range was between 5°C and 15°C, inclusive. ■

PROBLEM SET 2.6

For Problems 1–18, solve each of the inequalities and express the solution sets in interval notation.

1. $\frac{2}{5}x + \frac{1}{3}x > \frac{44}{15}$

2. $\frac{1}{4}x - \frac{4}{3}x < -13$

3. $x - \frac{5}{6} < \frac{x}{2} + 3$

4. $x + \frac{2}{7} > \frac{x}{2} - 5$

5. $\frac{x-2}{3} + \frac{x+1}{4} \geq \frac{5}{2}$

6. $\frac{x-1}{3} + \frac{x+2}{5} \leq \frac{3}{5}$

7. $\frac{3-x}{6} + \frac{x+2}{7} \leq 1$

8. $\frac{4-x}{5} + \frac{x+1}{6} \geq 2$

9. $\frac{x+3}{8} - \frac{x+5}{5} \geq \frac{3}{10}$

10. $\frac{x-4}{6} - \frac{x-2}{9} \leq \frac{5}{18}$

11. $\frac{4x-3}{6} - \frac{2x-1}{12} < -2$

12. $\frac{3x+2}{9} - \frac{2x+1}{3} > -1$

13. $0.06x + 0.08(250 - x) \geq 19$

14. $0.08x + 0.09(2x) \geq 130$

15. $0.09x + 0.1(x + 200) > 77$

16. $0.07x + 0.08(x + 100) > 38$

17. $x \geq 3.4 + 0.15x$

18. $x \geq 2.1 + 0.3x$

For Problems 19–34, graph the solution set for each compound inequality and express the solution sets in interval notation.

19. $x > -1$ and $x < 2$

20. $x > 1$ and $x < 4$

21. $x \leq 2$ and $x > -1$

22. $x \leq 4$ and $x \geq -2$

23. $x > 2$ or $x < -1$

24. $x > 1$ or $x < -4$

25. $x \leq 1$ or $x > 3$

26. $x < -2$ or $x \geq 1$

27. $x > 0$ and $x > -1$

28. $x > -2$ and $x > 2$

29. $x < 0$ and $x > 4$

30. $x > 1$ or $x < 2$

31. $x > -2$ or $x < 3$

32. $x > 3$ and $x < -1$

33. $x > -1$ or $x > 2$

34. $x < -2$ or $x < 1$

For Problems 35–44, solve each compound inequality and graph the solution sets. Express the solution sets in interval notation.

35. $x - 2 > -1$ and $x - 2 < 1$

36. $x + 3 > -2$ and $x + 3 < 2$

37. $x + 2 < -3$ or $x + 2 > 3$

38. $x - 4 < -2$ or $x - 4 > 2$

39. $2x - 1 \geq 5$ and $x > 0$

40. $3x + 2 > 17$ and $x \geq 0$

41. $5x - 2 < 0$ and $3x - 1 > 0$

42. $x + 1 > 0$ and $3x - 4 < 0$

43. $3x + 2 < -1$ or $3x + 2 > 1$

44. $5x - 2 < -2$ or $5x - 2 > 2$

For Problems 45–56, solve each compound inequality using the compact form. Express the solution sets in interval notation.

45. $-3 < 2x + 1 < 5$

46. $-7 < 3x - 1 < 8$

47. $-17 \leq 3x - 2 \leq 10$

48. $-25 \leq 4x + 3 \leq 19$

49. $1 < 4x + 3 < 9$

50. $0 < 2x + 5 < 12$

51. $-6 < 4x - 5 < 6$

52. $-2 < 3x + 4 < 2$

53. $-4 \leq \dfrac{x - 1}{3} \leq 4$

54. $-1 \leq \dfrac{x + 2}{4} \leq 1$

55. $-3 < 2 - x < 3$

56. $-4 < 3 - x < 4$

For Problems 57–67, solve each problem by setting up and solving an appropriate inequality.

57. Suppose that Lance has $500 to invest. If he invests $300 at 9% interest, at what rate must he invest the remaining $200 so that the two investments yield more than $47 in yearly interest?

58. Mona invests $100 at 8% yearly interest. How much does she have to invest at 9% so that the total yearly interest from the two investments exceeds $26?

59. The average height of the two forwards and the center of a basketball team is 6 feet and 8 inches. What must the average height of the two guards be so that the team average is at least 6 feet and 4 inches?

60. Thanh has scores of 52, 84, 65, and 74 on his first four math exams. What score must he make on the fifth exam to have an average of 70 or better for the five exams?

61. Marsha bowled 142 and 170 in her first two games. What must she bowl in the third game to have an average of at least 160 for the three games?

62. Candace had scores of 95, 82, 93, and 84 on her first four exams of the semester. What score must she obtain on the fifth exam to have an average of 90 or better for the five exams?

63. Suppose that Derwin shot rounds of 82, 84, 78, and 79 on the first four days of a golf tournament. What must he shoot on the fifth day of the tournament to average 80 or less for the five days?

64. The temperatures for a 24-hour period ranged between $-4°F$ and $23°F$, inclusive. What was the range in Celsius degrees? $\left(\text{Use F} = \dfrac{9}{5}C + 32. \right)$

65. Oven temperatures for baking various foods usually range between 325°F and 425°F, inclusive. Express this range in Celsius degrees. (Round answers to the nearest degree.)

66. A person's intelligence quotient (I) is found by dividing mental age (M), as indicated by standard tests, by chronological age (C) and then multiplying this ratio by 100. The formula $I = \dfrac{100M}{C}$ can be used. If the I range of a group of 11-year-olds is given by $80 \leq I \leq 140$, find the range of the mental age of this group.

67. Repeat Problem 66 for an I range of 70 to 125, inclusive, for a group of 9-year-olds.

■ ■ ■ **Thoughts into words**

68. Explain the difference between a conjunction and a disjunction. Give an example of each (outside the field of mathematics).

69. How do you know by inspection that the solution set of the inequality $x + 3 > x + 2$ is the entire set of real numbers?

70. Find the solution set for each of the following compound statements, and in each case explain your reasoning.

(a) $x < 3$ and $5 > 2$
(b) $x < 3$ or $5 > 2$
(c) $x < 3$ and $6 < 4$
(d) $x < 3$ or $6 < 4$

2.7   Equations and Inequalities Involving Absolute Value

In Section 1.2, we defined the absolute value of a real number by

$$|a| = \begin{cases} a, & \text{if } a \geq 0 \\ -a, & \text{if } a < 0 \end{cases}$$

We also interpreted the absolute value of any real number to be the distance between the number and zero on a number line. For example, $|6| = 6$ translates to 6 units between 6 and 0. Likewise, $|-8| = 8$ translates to 8 units between -8 and 0.

The interpretation of absolute value as distance on a number line provides a straightforward approach to solving a variety of equations and inequalities involving absolute value. First, let's consider some equations.

E X A M P L E   1   Solve $|x| = 2$.

Solution

Think in terms of distance between the number and zero, and you will see that x must be 2 or -2. That is, the equation $|x| = 2$ is equivalent to

$$x = -2 \quad \text{or} \quad x = 2$$

The solution set is $\{-2, 2\}$.   ■

E X A M P L E   2

Solve $|x + 2| = 5$.

Solution

The number, $x + 2$, must be -5 or 5. Thus $|x + 2| = 5$ is equivalent to

$$x + 2 = -5 \qquad \text{or} \qquad x + 2 = 5$$

Solving each equation of the disjunction yields

$$x + 2 = -5 \qquad \text{or} \qquad x + 2 = 5$$
$$x = -7 \qquad\quad \text{or} \qquad x = 3$$

The solution set is $\{-7, 3\}$.

 Check

$$|x + 2| = 5 \qquad\qquad |x + 2| = 5$$
$$|-7 + 2| \stackrel{?}{=} 5 \qquad\qquad |3 + 2| \stackrel{?}{=} 5$$
$$|-5| \stackrel{?}{=} 5 \qquad\qquad |5| \stackrel{?}{=} 5$$
$$5 = 5 \qquad\qquad\quad 5 = 5$$

The following general property should seem reasonable from the distance interpretation of absolute value.

PROPERTY 2.1

$|x| = k$ is equivalent to $x = -k$ or $x = k$, where k is a positive number.

Example 3 demonstrates our format for solving equations of the form $|x| = k$.

E X A M P L E   3

Solve $|5x + 3| = 7$.

Solution

$$|5x + 3| = 7$$
$$5x + 3 = -7 \qquad \text{or} \qquad 5x + 3 = 7$$
$$5x = -10 \qquad \text{or} \qquad 5x = 4$$
$$x = -2 \qquad \text{or} \qquad x = \frac{4}{5}$$

The solution set is $\left\{-2, \dfrac{4}{5}\right\}$. Check these solutions!

The distance interpretation for absolute value also provides a good basis for solving some inequalities that involve absolute value. Consider the following examples.

E X A M P L E   4   Solve $|x| < 2$ and graph the solution set.

Solution

The number, x, must be less than two units away from zero. Thus $|x| < 2$ is equivalent to

$$x > -2 \quad \text{and} \quad x < 2$$

The solution set is $(-2, 2)$ and its graph is shown in Figure 2.14.

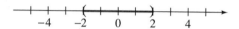

Figure 2.14

E X A M P L E   5   Solve $|x + 3| < 1$ and graph the solutions.

Solution

Let's continue to think in terms of distance on a number line. The number, $x + 3$, must be less than one unit away from zero. Thus $|x + 3| < 1$ is equivalent to

$$x + 3 > -1 \quad \text{and} \quad x + 3 < 1$$

Solving this conjunction yields

$$x + 3 > -1 \quad \text{and} \quad x + 3 < 1$$
$$x > -4 \quad \text{and} \quad x < -2$$

The solution set is $(-4, -2)$ and its graph is shown in Figure 2.15.

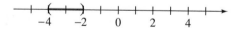

Figure 2.15

Take another look at Examples 4 and 5. The following general property should seem reasonable.

PROPERTY 2.2

$|x| < k$ is equivalent to $x > -k$ and $x < k$, where k is a positive number.

Remember that we can write a conjunction such as $x > -k$ and $x < k$ in the compact form $-k < x < k$. The compact form provides a very convenient format for solving inequalities such as $|3x - 1| < 8$, as Example 6 illustrates.

EXAMPLE 6   Solve $|3x - 1| < 8$ and graph the solutions.

Solution

$$|3x - 1| < 8$$

$$-8 < 3x - 1 < 8$$

$$-7 < 3x < 9 \qquad \text{Add 1 to left side, middle, and right side.}$$

$$\frac{1}{3}(-7) < \frac{1}{3}(3x) < \frac{1}{3}(9) \qquad \text{Multiply through by } \frac{1}{3}.$$

$$-\frac{7}{3} < x < 3$$

The solution set is $\left(-\frac{7}{3}, 3 \right)$ and its graph is shown in Figure 2.16.

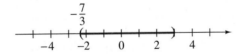

Figure 2.16

The distance interpretation also clarifies a property that pertains to *greater than* situations involving absolute value. Consider the following examples.

EXAMPLE 7   Solve $|x| > 1$ and graph the solutions.

Solution
The number, x, must be more than one unit away from zero. Thus $|x| > 1$ is equivalent to

$$x < -1 \qquad \text{or} \qquad x > 1$$

The solution set is $(-\infty, -1) \cup (1, \infty)$ and its graph is shown in Figure 2.17.

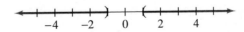

Figure 2.17

EXAMPLE 8

Solve $|x - 1| > 3$ and graph the solutions.

Solution

The number, $x - 1$, must be more than three units away from zero. Thus $|x - 1| > 3$ is equivalent to

$$x - 1 < -3 \quad \text{or} \quad x - 1 > 3$$

Solving this disjunction yields

$$x - 1 < -3 \quad \text{or} \quad x - 1 > 3$$
$$x < -2 \quad \text{or} \quad x > 4$$

The solution set is $(-\infty, -2) \cup (4, \infty)$ and its graph is shown in Figure 2.18.

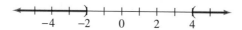

Figure 2.18

Examples 7 and 8 illustrate the following general property.

PROPERTY 2.3

$|x| > k$ is equivalent to $x < -k$ or $x > k$, where k is a positive number.

Therefore, solving inequalities of the form $|x| > k$ can take the format shown in Example 9.

EXAMPLE 9

Solve $|3x - 1| > 2$ and graph the solutions.

Solution

$$|3x - 1| > 2$$
$$3x - 1 < -2 \quad \text{or} \quad 3x - 1 > 2$$
$$3x < -1 \quad \text{or} \quad 3x > 3$$
$$x < -\frac{1}{3} \quad \text{or} \quad x > 1$$

The solution set is $\left(-\infty, -\frac{1}{3}\right) \cup (1, \infty)$ and its graph is shown in Figure 2.19.

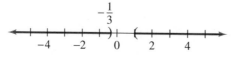

Figure 2.19

Properties 2.1, 2.2, and 2.3 provide the basis for solving a variety of equations and inequalities that involve absolute value. However, if at any time you become doubtful about what property applies, don't forget the distance interpretation. Furthermore, note that in each of the properties, k is a positive number. If k is a nonpositive number, we can determine the solution sets by inspection, as indicated by the following examples.

$|x + 3| = 0$ has a solution of $x = -3$ because the number $x + 3$ has to be 0. The solution set of $|x + 3| = 0$ is $\{-3\}$.

$|2x - 5| = -3$ has no solutions because the absolute value (distance) cannot be negative. The solution set is $\varnothing$, the null set.

$|x - 7| < -4$ has no solutions because we cannot obtain an absolute value less than -4. The solution set is $\varnothing$.

$|2x - 1| > -1$ is satisfied by all real numbers because the absolute value of $(2x - 1)$, regardless of what number is substituted for x, will always be greater than -1. The solution set is the set of all real numbers, which we can express in interval notation as $(-\infty, \infty)$.

PROBLEM SET 2.7

For Problems 1–14, solve each inequality and graph the solutions.

1. $|x| < 5$

2. $|x| < 1$

3. $|x| \leq 2$

4. $|x| \leq 4$

5. $|x| > 2$

6. $|x| > 3$

7. $|x - 1| < 2$

8. $|x - 2| < 4$

9. $|x + 2| \leq 4$

10. $|x + 1| \leq 1$

11. $|x + 2| > 1$

12. $|x + 1| > 3$

13. $|x - 3| \geq 2$

14. $|x - 2| \geq 1$

For Problems 15–50, solve each equation and inequality.

15. $|x - 1| = 8$

16. $|x + 2| = 9$

17. $|x - 2| > 6$

18. $|x - 3| > 9$

19. $|x + 3| < 5$

20. $|x + 1| < 8$

21. $|2x - 4| = 6$

22. $|3x - 4| = 14$

23. $|2x - 1| \leq 9$

24. $|3x + 1| \leq 13$

25. $|4x + 2| \geq 12$

26. $|5x - 2| \geq 10$

27. $|3x + 4| = 11$

28. $|5x - 7| = 14$

29. $|4 - 2x| = 6$

30. $|3 - 4x| = 8$

31. $|2 - x| > 4$

32. $|4 - x| > 3$

33. $|1 - 2x| < 2$

34. $|2 - 3x| < 5$

35. $|5x + 9| \leq 16$

36. $|7x - 6| \geq 22$

37. $\left| x - \dfrac{3}{4} \right| = \dfrac{2}{3}$

38. $\left| x + \dfrac{1}{2} \right| = \dfrac{3}{5}$

39. $|-2x + 7| \leq 13$

40. $|-3x - 4| \leq 15$

41. $\left| \dfrac{x - 3}{4} \right| < 2$

42. $\left| \dfrac{x + 2}{3} \right| < 1$

43. $\left| \dfrac{2x + 1}{2} \right| > 1$

44. $\left| \dfrac{3x - 1}{4} \right| > 3$

45. $|2x - 3| + 2 = 5$

46. $|3x - 1| - 1 = 9$

47. $|x + 7| - 3 \geq 4$

48. $|x - 2| + 4 \geq 10$

49. $|2x - 1| + 1 \leq 6$

50. $|4x + 3| - 2 \leq 5$

For Problems 51–60, solve each equation and inequality *by inspection.*

51. $|2x + 1| = -4$

52. $|5x - 1| = -2$

53. $|3x - 1| > -2$

54. $|4x + 3| < -4$

55. $|5x - 2| = 0$

56. $|3x - 1| = 0$

57. $|4x - 6| < -1$

58. $|x + 9| > -6$

59. $|x + 4| < 0$

60. $|x + 6| > 0$

■ ■ ■ **Thoughts into words**

61. Explain how you would solve the inequality $|2x + 5| > -3$.

62. Why is 2 the only solution for $|x - 2| \le 0$?

63. Explain how you would solve the equation $|2x - 3| = 0$.

■ ■ ■ **Further investigations**

Consider the equation $|x| = |y|$. This equation will be a true statement if x is equal to y or if x is equal to the opposite of y. Use the following format, $x = y$ or $x = -y$, to solve the equations in Problems 64–69.

For Problems 64–69, solve each equation.

64. $|3x + 1| = |2x + 3|$

65. $|-2x - 3| = |x + 1|$

66. $|2x - 1| = |x - 3|$

67. $|x - 2| = |x + 6|$

68. $|x + 1| = |x - 4|$

69. $|x + 1| = |x - 1|$

70. Use the definition of absolute value to help prove Property 2.1.

71. Use the definition of absolute value to help prove Property 2.2.

72. Use the definition of absolute value to help prove Property 2.3.

SUMMARY

(2.1) Solving an algebraic equation refers to the process of finding the number (or numbers) that make(s) the algebraic equation a true numerical statement. We call such numbers the **solutions** or **roots** of the equation that **satisfy** the equation. We call the set of all solutions of an equation the **solution set.** The general procedure for solving an equation is to continue replacing the given equation with **equivalent but simpler** equations until we arrive at one that can be solved by inspection. Two properties of equality play an important role in the process of solving equations.

Addition Property of Equality $a = b$ if and only if $a + c = b + c$.

Multiplication Property of Equality For $c \neq 0$, $a = b$ if and only if $ac = bc$.

(2.2) To solve an equation involving fractions, first **clear the equation of all fractions.** It is usually easiest to begin by multiplying both sides of the equation by the least common multiple of all of the denominators in the equation (by the least common denominator, or LCD).

Keep the following suggestions in mind as you solve word problems.

1. Read the problem carefully.
2. Sketch any figure, diagram, or chart that might be helpful.
3. Choose a meaningful variable.
4. Look for a guideline.
5. Form an equation or inequality.
6. Solve the equation or inequality.
7. Check your answers.

(2.3) To solve equations that contain decimals, you can clear the equation of all decimals by multiplying both sides by an appropriate power of 10, or you can keep the problem in decimal form and perform the calculations with decimals.

(2.4) We use equations to put rules in symbolic form; we call these rules **formulas.**

We can solve a formula such as $P = 2l + 2w$ for $l \left(l = \dfrac{P - 2w}{2} \right)$ or for $w \left(w = \dfrac{P - 2l}{2} \right)$ by applying the addition and multiplication properties of equality.

We often use formulas as **guidelines** for solving word problems.

(2.5) Solving an algebraic inequality refers to the process of finding the numbers that make the algebraic inequality a true numerical statement. We call such numbers the **solutions,** and we call the set of all solutions the **solution set.**

The general procedure for solving an inequality is to continue replacing the given inequality with **equivalent, but simpler,** inequalities until we arrive at one that we can solve by inspection. The following properties form the basis for solving algebraic inequalities.

1. $a > b$ if and only if $a + c > b + c$. (Addition property)
2. **a.** For $c > 0$, $a > b$ if and only if $ac > bc$.
   b. For $c < 0$, $a > b$ if and only if $ac < bc$. (Multiplication properties)

(2.6) To solve compound sentences that involve inequalities, we proceed as follows:

1. Solve separately each inequality in the compound sentence.
2. If it is a **conjunction,** the solution set is the **intersection** of the solution sets of each inequality.
3. If it is a **disjunction,** the solution set is the **union** of the solution sets of each inequality.

We define the intersection and union of two sets as follows.

Intersection $A \cap B = \{x | x \in A \quad and \quad x \in B\}$

Union $A \cup B = \{x | x \in A \quad or \quad x \in B\}$

The following are some examples of solution sets that we examined in Sections 2.5 and 2.6 (Figure 2.20).

Solution set	Graph	Interval notation
$\{x \mid x > 1\}$	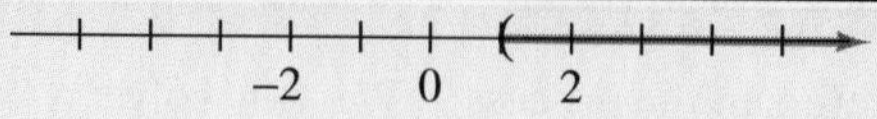	$(1, \infty)$
$\{x \mid x \geq 2\}$	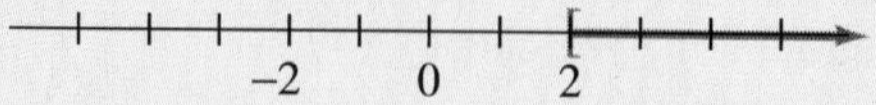	$[2, \infty)$
$\{x \mid x < 0\}$	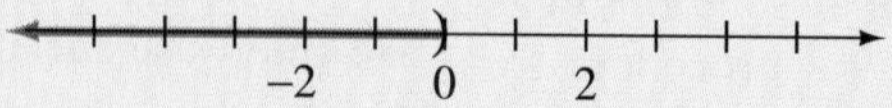	$(-\infty, 0)$
$\{x \mid x \leq -1\}$	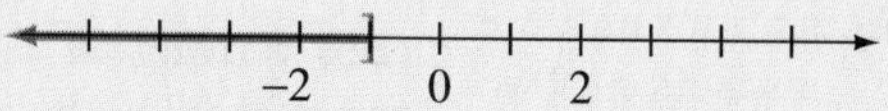	$(-\infty, -1]$
$\{x \mid -2 < x \leq 2\}$	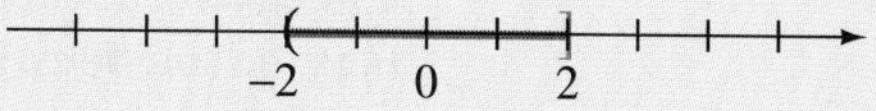	$(-2, 2]$
$\{x \mid x \leq -1 \text{ or } x > 1\}$	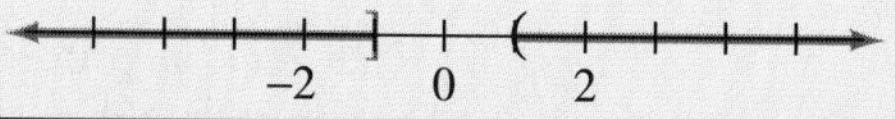	$(-\infty, -1] \cup (1, \infty)$

Figure 2.20

(2.7) We can interpret the **absolute value** of a number on the number line as the distance between that number and zero. The following properties form the basis for solving equations and inequalities involving absolute value.

1. $|x| = k$ is equivalent to $x = -k$ or $x = k$
2. $|x| < k$ is equivalent to $x > -k$ and $x < k$ $\left. \right\} \quad k > 0$
3. $|x| > k$ is equivalent to $x < -k$ or $x > k$

CHAPTER 2   REVIEW PROBLEM SET

For Problems 1–15, solve each of the equations.

1. $5(x - 6) = 3(x + 2)$

2. $2(2x + 1) - (x - 4) = 4(x + 5)$

3. $-(2n - 1) + 3(n + 2) = 7$

4. $2(3n - 4) + 3(2n - 3) = -2(n + 5)$

5. $\dfrac{3t - 2}{4} = \dfrac{2t + 1}{3}$

6. $\dfrac{x + 6}{5} + \dfrac{x - 1}{4} = 2$

7. $1 - \dfrac{2x - 1}{6} = \dfrac{3x}{8}$

8. $\dfrac{2x + 1}{3} + \dfrac{3x - 1}{5} = \dfrac{1}{10}$

9. $\dfrac{3n - 1}{2} - \dfrac{2n + 3}{7} = 1$

10. $|3x - 1| = 11$

11. $0.06x + 0.08\,(x + 100) = 15$

12. $0.4(t - 6) = 0.3(2t + 5)$

13. $0.1(n + 300) = 0.09n + 32$

14. $0.2(x - 0.5) - 0.3(x + 1) = 0.4$

15. $|2n + 3| = 4$

For Problems 16–20, solve each equation for x.

16. $ax - b = b + 2$

17. $ax = bx + c$

18. $m(x + a) = p(x + b)$

19. $5x - 7y = 11$

20. $\dfrac{x - a}{b} = \dfrac{y + 1}{c}$

For Problems 21–24, solve each of the formulas for the indicated variable.

21. $A = \pi r^2 + \pi rs$ for s

22. $A = \dfrac{1}{2}h(b_1 + b_2)$ for b_2

23. $S_n = \dfrac{n(a_1 + a_2)}{2}$ for n.

24. $\dfrac{1}{R} = \dfrac{1}{R_1} + \dfrac{1}{R_2}$ for R

For Problems 25–36, solve each of the inequalities.

25. $5x - 2 \geq 4x - 7$

26. $3 - 2x < -5$

27. $2(3x - 1) - 3(x - 3) > 0$

28. $3(x + 4) \leq 5(x - 1)$

29. $\dfrac{5}{6}n - \dfrac{1}{3}n < \dfrac{1}{6}$

30. $\dfrac{n - 4}{5} + \dfrac{n - 3}{6} > \dfrac{7}{15}$

31. $s \geq 4.5 + 0.25s$

32. $0.07x + 0.09(500 - x) \geq 43$

33. $|2x - 1| < 11$

34. $|3x + 1| > 10$

35. $-3(2t - 1) - (t + 2) > -6(t - 3)$

36. $\dfrac{2}{3}(x - 1) + \dfrac{1}{4}(2x + 1) < \dfrac{5}{6}(x - 2)$

For Problems 37–44, graph the solutions of each compound inequality.

37. $x > -1$ and $x < 1$

38. $x > 2$ or $x \leq -3$

39. $x > 2$ and $x > 3$

40. $x < 2$ or $x > -1$

41. $2x + 1 > 3$ or $2x + 1 < -3$

42. $2 \leq x + 4 \leq 5$

43. $-1 < 4x - 3 \leq 9$

44. $x + 1 > 3$ and $x - 3 < -5$

Solve each of Problems 45–56 by setting up and solving an appropriate equation or inequality.

45. The width of a rectangle is 2 meters more than one-third of the length. The perimeter of the rectangle is 44 meters. Find the length and width of the rectangle.

46. A total of $500 was invested, part of it at 7% interest and the remainder at 8%. If the total yearly interest from both investments amounted to $38, how much was invested at each rate?

47. Susan's average score for her first three psychology exams is 84. What must she get on the fourth exam so that her average for the four exams is 85 or better?

48. Find three consecutive integers such that the sum of one-half of the smallest and one-third of the largest is one less than the other integer.

49. Pat is paid time-and-a-half for each hour he works over 36 hours in a week. Last week he worked 42 hours for a total of $472.50. What is his normal hourly rate?

50. Marcela has a collection of nickels, dimes, and quarters worth $24.75. The number of dimes is 10 more than twice the number of nickels, and the number of quarters is 25 more than the number of dimes. How many coins of each kind does she have?

51. If the complement of an angle is one-tenth of the supplement of the angle, find the measure of the angle.

52. A retailer has some sweaters that cost her $38 each. She wants to sell them at a profit of 20% of her cost. What price should she charge for the sweaters?

53. How many pints of a 1% hydrogen peroxide solution should be mixed with a 4% hydrogen peroxide solution to obtain 10 pints of a 2% hydrogen peroxide solution?

54. Gladys leaves a town driving at a rate of 40 miles per hour. Two hours later, Reena leaves from the same place traveling the same route. She catches Gladys in 5 hours and 20 minutes. How fast was Reena traveling?

55. In $1\frac{1}{4}$ hours more time, Rita, riding her bicycle at 12 miles per hour, rode 2 miles farther than Sonya, who was riding her bicycle at 16 miles per hour. How long did each girl ride?

56. How many cups of orange juice must be added to 50 cups of a punch that is 10% orange juice to obtain a punch that is 20% orange juice?

CHAPTER 2

TEST

For Problems 1–10, solve each equation.

1. $5x - 2 = 2x - 11$

2. $6(n - 2) - 4(n + 3) = -14$

3. $-3(x + 4) = 3(x - 5)$

4. $3(2x - 1) - 2(x + 5) = -(x - 3)$

5. $\dfrac{3t - 2}{4} = \dfrac{5t + 1}{5}$

6. $\dfrac{5x + 2}{3} - \dfrac{2x + 4}{6} = -\dfrac{4}{3}$

7. $|4x - 3| = 9$

8. $\dfrac{1 - 3x}{4} + \dfrac{2x + 3}{3} = 1$

9. $2 - \dfrac{3x - 1}{5} = -4$

10. $0.05x + 0.06(1500 - x) = 83.5$

11. Solve $\dfrac{2}{3}x - \dfrac{3}{4}y = 2$ for y.

12. Solve $S = 2\pi r(r + h)$ for h.

For Problems 13–20, solve each inequality and express the solution set using interval notation.

13. $7x - 4 > 5x - 8$

14. $-3x - 4 \le x + 12$

15. $2(x - 1) - 3(3x + 1) \ge -6(x - 5)$

16. $\dfrac{3}{5}x - \dfrac{1}{2}x < 1$

17. $\dfrac{x - 2}{6} - \dfrac{x + 3}{9} > -\dfrac{1}{2}$

18. $0.05x + 0.07(800 - x) \ge 52$

19. $|6x - 4| < 10$

20. $|4x + 5| \ge 6$

For Problems 21–25, solve each problem by setting up and solving an appropriate equation or inequality.

21. Dela bought a dress at a 20% discount sale for $57.60. Find the original price of the dress.

22. The length of a rectangle is 1 centimeter more than three times its width. If the perimeter of the rectangle is 50 centimeters, find the length of the rectangle.

23. How many cups of grapefruit juice must be added to 30 cups of a punch that is 8% grapefruit juice to obtain a punch that is 10% grapefruit juice?

24. Rex has scores of 85, 92, 87, 88, and 91 on the first five exams. What score must he make on the sixth exam to have an average of 90 or better for all six exams?

25. If the complement of an angle is $\dfrac{2}{11}$ of the supplement of the angle, find the measure of the angle.

Polynomials

A quadratic equation can be solved to determine the width of a uniform strip trimmed off both sides and ends of a sheet of paper to obtain a specified area for the sheet of paper.

A strip of uniform width cut off of both sides and both ends of an 8-inch by 11-inch sheet of paper must reduce the size of the paper to an area of 40 square inches. Find the width of the strip. With the equation $(11 - 2x)(8 - 2x) = 40$, you can determine that the strip should be 1.5 inches wide.

The main object of this text is to help you develop algebraic skills, use these skills to solve equations and inequalities, and use equations and inequalities to solve word problems. The work in this chapter will focus on a class of algebraic expressions called **polynomials.**

© Tony Freeman /PhotoEdit

InfoTrac Project Do a subject guide search on the metric system. Find a periodical article on why we should learn the metric system. Write a brief summary of the article. If the length of a game room is 5 meters less than twice the width, what is the length if the area is 117 square meters?

3.1 Polynomials: Sums and Differences

Recall that algebraic expressions such as $5x$, $-6y^2$, $7xy$, $14a^2b$, and $-17ab^2c^3$ are called terms. A **term** is an indicated product and may contain any number of factors. The variables in a term are called **literal factors** and the numerical factor is called the **numerical coefficient.** Thus in $7xy$, the x and y are literal factors, 7 is the numerical coefficient, and the term is in two variables (x and y).

Terms that contain variables with only whole numbers as exponents are called **monomials.** The previously listed terms, $5x$, $-6y^2$, $7xy$, $14a^2b$, and $-17ab^2c^3$, are all monomials. (We shall work later with some algebraic expressions, such as $7x^{-1}y^{-1}$ and $6a^{-2}b^{-3}$, that are not monomials.)

The **degree** of a monomial is the sum of the exponents of the literal factors.

$7xy$ is of degree 2.

$14a^2b$ is of degree 3.

$-17ab^2c^3$ is of degree 6.

$5x$ is of degree 1.

$-6y^2$ is of degree 2.

If the monomial contains only one variable, then the exponent of the variable is the degree of the monomial. The last two examples illustrate this point. We say that any nonzero constant term is of degree zero.

A **polynomial** is a monomial or a finite sum (or difference) of monomials. Thus

$$4x^2, \qquad 3x^2 - 2x - 4, \qquad 7x^4 - 6x^3 + 4x^2 + x - 1,$$

$$3x^2y - 2xy^2, \qquad \frac{1}{5}a^2 - \frac{2}{3}b^2, \qquad \text{and} \qquad 14$$

are examples of polynomials. In addition to calling a polynomial with one term a **monomial,** we also classify polynomials with two terms as **binomials** and those with three terms as **trinomials.**

The **degree of a polynomial** is the degree of the term with the highest degree in the polynomial. The following examples illustrate some of this terminology.

The polynomial $4x^3y^4$ is a monomial in two variables of degree 7.

The polynomial $4x^2y - 2xy$ is a binomial in two variables of degree 3.

The polynomial $9x^2 - 7x + 1$ is a trinomial in one variable of degree 2.

Combining Similar Terms

Remember that *similar terms*, or *like terms*, are terms that have the same literal factors. In the preceding chapters, we have frequently simplified algebraic expressions by combining similar terms, as the next examples illustrate.

$$2x + 3y + 7x + 8y = \boxed{2x + 7x + 3y + 8y}$$
$$= \boxed{(2 + 7)x + (3 + 8)y}$$
$$= 9x + 11y$$

Steps in dashed boxes are usually done mentally.

$$4a - 7 - 9a + 10 = \boxed{4a + (-7) + (-9a) + 10}$$
$$= \boxed{4a + (-9a) + (-7) + 10}$$
$$= \boxed{(4 + (-9))a + (-7) + 10}$$
$$= -5a + 3$$

Both addition and subtraction of polynomials rely on basically the same ideas. The commutative, associative, and distributive properties provide the basis for rearranging, regrouping, and combining similar terms. Let's consider some examples.

E X A M P L E   1   Add $4x^2 + 5x + 1$ and $7x^2 - 9x + 4$.

Solution

We generally use the horizontal format for such work. Thus

$$(4x^2 + 5x + 1) + (7x^2 - 9x + 4) = (4x^2 + 7x^2) + (5x - 9x) + (1 + 4)$$
$$= 11x^2 - 4x + 5 \qquad ∎$$

E X A M P L E   2   Add $5x - 3$, $3x + 2$, and $8x + 6$.

Solution

$$(5x - 3) + (3x + 2) + (8x + 6) = (5x + 3x + 8x) + (-3 + 2 + 6)$$
$$= 16x + 5 \qquad ∎$$

E X A M P L E   3   Find the indicated sum: $(-4x^2y + xy^2) + (7x^2y - 9xy^2) + (5x^2y - 4xy^2)$.

Solution

$$(-4x^2y + xy^2) + (7x^2y - 9xy^2) + (5x^2y - 4xy^2)$$
$$= (-4x^2y + 7x^2y + 5x^2y) + (xy^2 - 9xy^2 - 4xy^2)$$
$$= 8x^2y - 12xy^2 \qquad ∎$$

The idea of subtraction as adding the opposite extends to polynomials in general. Hence the expression $a - b$ is equivalent to $a + (-b)$. We can form the opposite of a polynomial by taking the opposite of each term. For example, the opposite of $3x^2 - 7x + 1$ is $-3x^2 + 7x - 1$. We express this in symbols as

$$-(3x^2 - 7x + 1) = -3x^2 + 7x - 1$$

Now consider the following subtraction problems.

E X A M P L E   4

Subtract $3x^2 + 7x - 1$ from $7x^2 - 2x - 4$.

Solution

Use the horizontal format to obtain

$$(7x^2 - 2x - 4) - (3x^2 + 7x - 1) = (7x^2 - 2x - 4) + (-3x^2 - 7x + 1)$$
$$= (7x^2 - 3x^2) + (-2x - 7x) + (-4 + 1)$$
$$= 4x^2 - 9x - 3 \qquad ■$$

E X A M P L E   5

Subtract $-3y^2 + y - 2$ from $4y^2 + 7$.

Solution

Because subtraction is not a commutative operation, be sure to perform the subtraction in the correct order.

$$(4y^2 + 7) - (-3y^2 + y - 2) = (4y^2 + 7) + (3y^2 - y + 2)$$
$$= (4y^2 + 3y^2) + (-y) + (7 + 2)$$
$$= 7y^2 - y + 9 \qquad ■$$

The next example demonstrates the use of the vertical format for this work.

E X A M P L E   6

Subtract $4x^2 - 7xy + 5y^2$ from $3x^2 - 2xy + y^2$.

Solution

$$\begin{array}{l} 3x^2 - 2xy + y^2 \\ \underline{4x^2 - 7xy + 5y^2} \end{array}$$
Note which polynomial goes on the bottom and how the similar terms are aligned.

Now we can mentally form the opposite of the bottom polynomial and add.

$$\begin{array}{l} 3x^2 - 2xy + y^2 \\ \underline{4x^2 - 7xy + 5y^2} \\ -x^2 + 5xy - 4y^2 \end{array}$$
The opposite of $4x^2 - 7xy + 5y^2$ is $-4x^2 + 7xy - 5y^2$. ■

We can also use the distributive property and the properties $a = 1(a)$ and $-a = -1(a)$ when adding and subtracting polynomials. The next examples illustrate this approach.

EXAMPLE 7 Perform the indicated operations: $(5x - 2) + (2x - 1) - (3x + 4)$.

Solution

$$
\begin{aligned}
(5x - 2) + (2x - 1) - (3x + 4) &= 1(5x - 2) + 1(2x - 1) - 1(3x + 4) \\
&= 1(5x) - 1(2) + 1(2x) - 1(1) - 1(3x) - 1(4) \\
&= 5x - 2 + 2x - 1 - 3x - 4 \\
&= 5x + 2x - 3x - 2 - 1 - 4 \\
&= 4x - 7 \quad ■
\end{aligned}
$$

We can do some of the steps mentally and simplify our format, as shown in the next two examples.

EXAMPLE 8 Perform the indicated operations: $(5a^2 - 2b) - (2a^2 + 4) + (-7b - 3)$.

Solution

$$
\begin{aligned}
(5a^2 - 2b) - (2a^2 + 4) + (-7b - 3) &= 5a^2 - 2b - 2a^2 - 4 - 7b - 3 \\
&= 3a^2 - 9b - 7 \quad ■
\end{aligned}
$$

EXAMPLE 9 Simplify $(4t^2 - 7t - 1) - (t^2 + 2t - 6)$.

Solution

$$
\begin{aligned}
(4t^2 - 7t - 1) - (t^2 + 2t - 6) &= 4t^2 - 7t - 1 - t^2 - 2t + 6 \\
&= 3t^2 - 9t + 5 \quad ■
\end{aligned}
$$

Remember that a polynomial in parentheses preceded by a negative sign can be written without the parentheses by replacing each term with its opposite. Thus in Example 9, $-(t^2 + 2t - 6) = -t^2 - 2t + 6$. Finally, let's consider a simplification problem that contains grouping symbols within grouping symbols.

EXAMPLE 10 Simplify $7x + [3x - (2x + 7)]$.

Solution

$$
\begin{aligned}
7x + [3x - (2x + 7)] &= 7x + [3x - 2x - 7] \qquad \text{Remove the innermost} \\
&\qquad\qquad\qquad\qquad\qquad\quad \text{parentheses first.} \\
&= 7x + [x - 7] \\
&= 7x + x - 7 \\
&= 8x - 7 \quad ■
\end{aligned}
$$

Sometimes we encounter polynomials in a geometric setting. For example, we can find a polynomial that represents the total surface area of the rectangular solid in Figure 3.1 as follows:

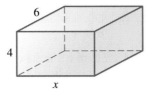

Figure 3.1

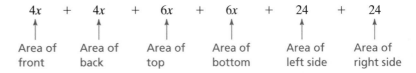

Simplifying $4x + 4x + 6x + 6x + 24 + 24$, we obtain the polynomial $20x + 48$, which represents the total surface area of the rectangular solid. Furthermore, by evaluating the polynomial $20x + 48$ for different positive values of x, we can determine the total surface area of any rectangular solid for which two dimensions are 4 and 6. The following chart contains some specific rectangular solids.

x	4 by 6 by x rectangular solid	Total surface area $(20x + 48)$
2	4 by 6 by 2	$20(2) + 48 = 88$
4	4 by 6 by 4	$20(4) + 48 = 128$
5	4 by 6 by 5	$20(5) + 48 = 148$
7	4 by 6 by 7	$20(7) + 48 = 188$
12	4 by 6 by 12	$20(12) + 48 = 288$

PROBLEM SET 3.1

For Problems 1–10, determine the degree of the given polynomials.

1. $7xy + 6y$

2. $-5x^2y^2 - 6xy^2 + x$

3. $-x^2y + 2xy^2 - xy$

4. $5x^3y^2 - 6x^3y^3$

5. $5x^2 - 7x - 2$

6. $7x^3 - 2x + 4$

7. $8x^6 + 9$

8. $5y^6 + y^4 - 2y^2 - 8$

9. -12

10. $7x - 2y$

For Problems 11–20, add the given polynomials.

11. $3x - 7$ and $7x + 4$

12. $9x + 6$ and $5x - 3$

13. $-5t - 4$ and $-6t + 9$

14. $-7t + 14$ and $-3t - 6$

15. $3x^2 - 5x - 1$ and $-4x^2 + 7x - 1$

16. $6x^2 + 8x + 4$ and $-7x^2 - 7x - 10$

17. $12a^2b^2 - 9ab$ and $5a^2b^2 + 4ab$

18. $15a^2b^2 - ab$ and $-20a^2b^2 - 6ab$

19. $2x - 4, -7x + 2,$ and $-4x + 9$

20. $-x^2 - x - 4, 2x^2 - 7x + 9,$ and $-3x^2 + 6x - 10$

For Problems 21–30, subtract the polynomials using the horizontal format.

21. $5x - 2$ from $3x + 4$

22. $7x + 5$ from $2x - 1$

23. $-4a - 5$ from $6a + 2$

24. $5a + 7$ from $-a - 4$

25. $3x^2 - x + 2$ from $7x^2 + 9x + 8$

26. $5x^2 + 4x - 7$ from $3x^2 + 2x - 9$

27. $2a^2 - 6a - 4$ from $-4a^2 + 6a + 10$

28. $-3a^2 - 6a + 3$ from $3a^2 + 6a - 11$

29. $2x^3 + x^2 - 7x - 2$ from $5x^3 + 2x^2 + 6x - 13$

30. $6x^3 + x^2 + 4$ from $9x^3 - x - 2$

For Problems 31–40, subtract the polynomials using the vertical format.

31. $5x - 2$ from $12x + 6$

32. $3x - 7$ from $2x + 1$

33. $-4x + 7$ from $-7x - 9$

34. $-6x - 2$ from $5x + 6$

35. $2x^2 + x + 6$ from $4x^2 - x - 2$

36. $4x^2 - 3x - 7$ from $-x^2 - 6x + 9$

37. $x^3 + x^2 - x - 1$ from $-2x^3 + 6x^2 - 3x + 8$

38. $2x^3 - x + 6$ from $x^3 + 4x^2 + 1$

39. $-5x^2 + 6x - 12$ from $2x - 1$

40. $2x^2 - 7x - 10$ from $-x^3 - 12$

For Problems 41–46, perform the operations as described.

41. Subtract $2x^2 - 7x - 1$ from the sum of $x^2 + 9x - 4$ and $-5x^2 - 7x + 10$.

42. Subtract $4x^2 + 6x + 9$ from the sum of $-3x^2 - 9x + 6$ and $-2x^2 + 6x - 4$.

43. Subtract $-x^2 - 7x - 1$ from the sum of $4x^2 + 3$ and $-7x^2 + 2x$.

44. Subtract $-4x^2 + 6x - 3$ from the sum of $-3x + 4$ and $9x^2 - 6$.

45. Subtract the sum of $5n^2 - 3n - 2$ and $-7n^2 + n + 2$ from $-12n^2 - n + 9$.

46. Subtract the sum of $-6n^2 + 2n - 4$ and $4n^2 - 2n + 4$ from $-n^2 - n + 1$.

For Problems 47–56, perform the indicated operations.

47. $(5x + 2) + (7x - 1) + (-4x - 3)$

48. $(-3x + 1) + (6x - 2) + (9x - 4)$

49. $(12x - 9) - (-3x + 4) - (7x + 1)$

50. $(6x + 4) - (4x - 2) - (-x - 1)$

51. $(2x^2 - 7x - 1) + (-4x^2 - x + 6) + (-7x^2 - 4x - 1)$

52. $(5x^2 + x + 4) + (-x^2 + 2x + 4) + (-14x^2 - x + 6)$

53. $(7x^2 - x - 4) - (9x^2 - 10x + 8) + (12x^2 + 4x - 6)$

54. $(-6x^2 + 2x + 5) - (4x^2 + 4x - 1) + (7x^2 + 4)$

55. $(n^2 - 7n - 9) - (-3n + 4) - (2n^2 - 9)$

56. $(6n^2 - 4) - (5n^2 + 9) - (6n + 4)$

For Problems 57–70, simplify by removing the inner parentheses first and working outward.

57. $3x - [5x - (x + 6)]$

58. $7x - [2x - (-x - 4)]$

59. $2x^2 - [-3x^2 - (x^2 - 4)]$

60. $4x^2 - [-x^2 - (5x^2 - 6)]$

61. $-2n^2 - [n^2 - (-4n^2 + n + 6)]$

62. $-7n^2 - [3n^2 - (-n^2 - n + 4)]$

63. $[4t^2 - (2t + 1) + 3] - [3t^2 + (2t - 1) - 5]$

64. $-(3n^2 - 2n + 4) - [2n^2 - (n^2 + n + 3)]$

65. $[2n^2 - (2n^2 - n + 5)] + [3n^2 + (n^2 - 2n - 7)]$

66. $3x^2 - [4x^2 - 2x - (x^2 - 2x + 6)]$

67. $[7xy - (2x - 3xy + y)] - [3x - (x - 10xy - y)]$

68. $[9xy - (4x + xy - y)] - [4y - (2x - xy + 6y)]$

69. $[4x^3 - (2x^2 - x - 1)] - [5x^3 - (x^2 + 2x - 1)]$

70. $[x^3 - (x^2 - x + 1)] - [-x^3 + (7x^2 - x + 10)]$

71. Find a polynomial that represents the perimeter of each of the following figures (Figures 3.2, 3.3, and 3.4).

(a)

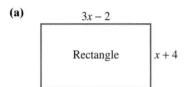

$3x - 2$

Rectangle $x + 4$

Figure 3.2

(b)

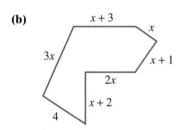

$x + 3$

$3x$

x

$x + 1$

$2x$

$x + 2$

4

Figure 3.3

(c)

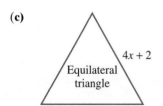

$4x + 2$

Equilateral triangle

Figure 3.4

72. Find a polynomial that represents the total surface area of the rectangular solid in Figure 3.5.

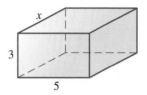

x

3

5

Figure 3.5

Now use that polynomial to determine the total surface area of each of the following rectangular solids.

(a) 3 by 5 by $\underline{4}$ **(b)** 3 by 5 by $\underline{7}$

(c) 3 by 5 by $\underline{11}$ **(d)** 3 by 5 by $\underline{13}$

73. Find a polynomial that represents the total surface area of the right circular cylinder in Figure 3.6. Now use that polynomial to determine the total surface area of each of the following right circular cylinders that have a base with a radius of 4. Use 3.14 for π, and express the answers to the nearest tenth.

(a) $h = 5$ **(b)** $h = 7$

(c) $h = 14$ **(d)** $h = 18$

4

h

Figure 3.6

■ ■ ▨ **Thoughts into words**

74. Explain how to subtract the polynomial $-3x^2 + 2x - 4$ from $4x^2 + 6$.

75. Is the sum of two binomials always another binomial? Defend your answer.

76. Explain how to simplify the expression

$$7x - [3x - (2x - 4) + 2] - x$$

3.2 Products and Quotients of Monomials

Suppose that we want to find the product of two monomials such as $3x^2y$ and $4x^3y^2$. To proceed, use the properties of real numbers and keep in mind that exponents indicate repeated multiplication.

$$(3x^2y)(4x^3y^2) = (3 \cdot x \cdot x \cdot y)(4 \cdot x \cdot x \cdot x \cdot y \cdot y)$$
$$= 3 \cdot 4 \cdot x \cdot x \cdot x \cdot x \cdot x \cdot y \cdot y \cdot y$$
$$= 12x^5y^3$$

You can use such an approach to find the product of any two monomials. However, there are some basic properties of exponents that make the process of multiplying monomials a much easier task. Let's consider each of these properties and illustrate its use when multiplying monomials. The following examples demonstrate the first property.

$$x^2 \cdot x^3 = (x \cdot x)(x \cdot x \cdot x) = x^5$$
$$a^4 \cdot a^2 = (a \cdot a \cdot a \cdot a)(a \cdot a) = a^6$$
$$b^3 \cdot b^4 = (b \cdot b \cdot b)(b \cdot b \cdot b \cdot b) = b^7$$

In general,

$$b^n \cdot b^m = \underbrace{(b \cdot b \cdot b \cdot \ldots b)}_{\substack{n \text{ factors} \\ \text{of } b}}\underbrace{(b \cdot b \cdot b \cdot \ldots b)}_{\substack{m \text{ factors} \\ \text{of } b}}$$

$$= \underbrace{b \cdot b \cdot b \cdot \ldots b}_{(n + m) \text{ factors of } b}$$

$$= b^{n+m}$$

We can state the first property as follows:

PROPERTY 3.1

If b is any real number, and n and m are positive integers, then

$$b^n \cdot b^m = b^{n+m}$$

Property 3.1 says that to find the product of two positive integral powers of the same base, we add the exponents and use this sum as the exponent of the common base.

$$x^7 \cdot x^8 = x^{7+8} = x^{15} \qquad\qquad y^6 \cdot y^4 = y^{6+4} = y^{10}$$

$$2^3 \cdot 2^8 = 2^{3+8} = 2^{11} \qquad\qquad (-3)^4 \cdot (-3)^5 = (-3)^{4+5} = (-3)^9$$

$$\left(\frac{2}{3}\right)^7 \cdot \left(\frac{2}{3}\right)^5 = \left(\frac{2}{3}\right)^{5+7} = \left(\frac{2}{3}\right)^{12}$$

The following examples illustrate the use of Property 3.1, along with the commutative and associative properties of multiplication, to form the basis for multiplying monomials. The steps enclosed in the dashed boxes could be performed mentally.

EXAMPLE 1

$$(3x^2y)(4x^3y^2) = 3 \cdot 4 \cdot x^2 \cdot x^3 \cdot y \cdot y^2$$
$$= 12x^{2+3}y^{1+2}$$
$$= 12x^5y^3$$

EXAMPLE 2

$$(-5a^3b^4)(7a^2b^5) = -5 \cdot 7 \cdot a^3 \cdot a^2 \cdot b^4 \cdot b^5$$
$$= -35a^{3+2}b^{4+5}$$
$$= -35a^5b^9$$

EXAMPLE 3

$$\left(\frac{3}{4}xy\right)\left(\frac{1}{2}x^5y^6\right) = \frac{3}{4} \cdot \frac{1}{2} \cdot x \cdot x^5 \cdot y \cdot y^6$$
$$= \frac{3}{8}x^{1+5}y^{1+6}$$
$$= \frac{3}{8}x^6y^7$$

EXAMPLE 4

$$(-ab^2)(-5a^2b) = (-1)(-5)(a)(a^2)(b^2)(b)$$
$$= 5a^{1+2}b^{2+1}$$
$$= 5a^3b^3$$

EXAMPLE 5

$$(2x^2y^2)(3x^2y)(4y^3) = 2 \cdot 3 \cdot 4 \cdot x^2 \cdot x^2 \cdot y^2 \cdot y \cdot y^3$$
$$= 24x^{2+2}y^{2+1+3}$$
$$= 24x^4y^6$$

The following examples demonstrate another useful property of exponents.

$$(x^2)^3 = x^2 \cdot x^2 \cdot x^2 = x^{2+2+2} = x^6$$
$$(a^3)^2 = a^3 \cdot a^3 = a^{3+3} = a^6$$
$$(b^4)^3 = b^4 \cdot b^4 \cdot b^4 = b^{4+4+4} = b^{12}$$

In general,

$$(b^n)^m = \underbrace{b^n \cdot b^n \cdot b^n \cdot \ldots b^n}_{m \text{ factors of } b^n}$$

adding m of these

$$= b^{\overbrace{n+n+n+\cdots+n}}$$

$$= b^{mn}$$

We can state this property as follows:

PROPERTY 3.2

If b is any real number, and m and n are positive integers, then

$$(b^n)^m = b^{mn}$$

The following examples show how Property 3.2 is used to find "the power of a power."

$$(x^4)^5 = x^{5(4)} = x^{20} \qquad (y^6)^3 = y^{3(6)} = y^{18}$$

$$(2^3)^7 = 2^{7(3)} = 2^{21}$$

A third property of exponents pertains to raising a monomial to a power. Consider the following examples, which we use to introduce the property.

$$(3x)^2 = (3x)(3x) = 3 \cdot 3 \cdot x \cdot x = 3^2 \cdot x^2$$

$$(4y^2)^3 = (4y^2)(4y^2)(4y^2) = 4 \cdot 4 \cdot 4 \cdot y^2 \cdot y^2 \cdot y^2 = (4)^3(y^2)^3$$

$$(-2a^3b^4)^2 = (-2a^3b^4)(-2a^3b^4) = (-2)(-2)(a^3)(a^3)(b^4)(b^4)$$

$$= (-2)^2(a^3)^2(b^4)^2$$

In general,

$$(ab)^n = \underbrace{(ab)(ab)(ab) \cdot \ldots (ab)}_{n \text{ factors of } ab}$$

$$= (\underbrace{a \cdot a \cdot a \cdot a \cdot \ldots a}_{\substack{n \text{ factors} \\ \text{of } a}})(\underbrace{b \cdot b \cdot b \cdot \ldots b}_{\substack{n \text{ factors} \\ \text{of } b}})$$

$$= a^n b^n$$

We can formally state Property 3.3 as follows:

PROPERTY 3.3

If a and b are real numbers and n is a positive integer, then

$$(ab)^n = a^n b^n$$

Property 3.3 and Property 3.2 form the basis for raising a monomial to a power, as in the next examples.

E X A M P L E   6

$(x^2y^3)^4 = (x^2)^4(y^3)^4$ Use $(ab)^n = a^n b^n$.

$= x^8 y^{12}$ Use $(b^n)^m = b^{mn}$.

E X A M P L E   7

$(3a^5)^3 = (3)^3(a^5)^3$

$= 27a^{15}$

E X A M P L E   8

$(-2xy^4)^5 = (-2)^5(x)^5(y^4)^5$

$= -32x^5 y^{20}$

Dividing Monomials

To develop an effective process for dividing by a monomial, we need yet another property of exponents. This property is a direct consequence of the definition of an exponent. Study the following examples.

$$\frac{x^4}{x^3} = \frac{x \cdot x \cdot x \cdot x}{x \cdot x \cdot x} = x \qquad \frac{x^3}{x^3} = \frac{x \cdot x \cdot x}{x \cdot x \cdot x} = 1$$

$$\frac{a^5}{a^2} = \frac{a \cdot a \cdot a \cdot a \cdot a}{a \cdot a} = a^3 \qquad \frac{y^5}{y^5} = \frac{y \cdot y \cdot y \cdot y \cdot y}{y \cdot y \cdot y \cdot y \cdot y} = 1$$

$$\frac{y^8}{y^4} = \frac{y \cdot y \cdot y \cdot y \cdot y \cdot y \cdot y \cdot y}{y \cdot y \cdot y \cdot y} = y^4$$

We can state the general property as follows:

PROPERTY 3.4

If b is any nonzero real number, and m and n are positive integers, then

1. $\dfrac{b^n}{b^m} = b^{n-m}$,   when $n > m$

2. $\dfrac{b^n}{b^m} = 1$,   when $n = m$

Applying Property 3.4 to the previous examples yields

$$\frac{x^4}{x^3} = x^{4-3} = x^1 = x \qquad \frac{x^3}{x^3} = 1$$

$$\frac{a^5}{a^2} = a^{5-2} = a^3 \qquad \frac{y^5}{y^5} = 1$$

$$\frac{y^8}{y^4} = y^{8-4} = y^4$$

(We will discuss the situation when $n < m$ in a later chapter.)

Property 3.4, along with our knowledge of dividing integers, provides the basis for dividing monomials. The following examples demonstrate the process.

$$\frac{24x^5}{3x^2} = 8x^{5-2} = 8x^3 \qquad\qquad \frac{-36a^{13}}{-12a^5} = 3a^{13-5} = 3a^8$$

$$\frac{-56x^9}{7x^4} = -8x^{9-4} = -8x^5 \qquad \frac{72b^5}{8b^5} = 9 \quad \left(\frac{b^5}{b^5} = 1\right)$$

$$\frac{48y^7}{-12y} = -4y^{7-1} = -4y^6 \qquad \frac{12x^4y^7}{2x^2y^4} = 6x^{4-2}y^{7-4} = 6x^2y^3$$

PROBLEM SET 3.2

For Problems 1–36, find each product.

1. $(4x^3)(9x)$

2. $(6x^3)(7x^2)$

3. $(-2x^2)(6x^3)$

4. $(2xy)(-4x^2y)$

5. $(-a^2b)(-4ab^3)$

6. $(-8a^2b^2)(-3ab^3)$

7. $(x^2yz^2)(-3xyz^4)$

8. $(-2xy^2z^2)(-x^2y^3z)$

9. $(5xy)(-6y^3)$

10. $(-7xy)(4x^4)$

11. $(3a^2b)(9a^2b^4)$

12. $(-8a^2b^2)(-12ab^5)$

13. $(m^2n)(-mn^2)$

14. $(-x^3y^2)(xy^3)$

15. $\left(\frac{2}{5}xy^2\right)\left(\frac{3}{4}x^2y^4\right)$

16. $\left(\frac{1}{2}x^2y^6\right)\left(\frac{2}{3}xy\right)$

17. $\left(-\frac{3}{4}ab\right)\left(\frac{1}{5}a^2b^3\right)$

18. $\left(-\frac{2}{7}a^2\right)\left(\frac{3}{5}ab^3\right)$

19. $\left(-\frac{1}{2}xy\right)\left(\frac{1}{3}x^2y^3\right)$

20. $\left(\frac{3}{4}x^4y^5\right)(-x^2y)$

21. $(3x)(-2x^2)(-5x^3)$

22. $(-2x)(-6x^3)(x^2)$

23. $(-6x^2)(3x^3)(x^4)$

24. $(-7x^2)(3x)(4x^3)$

25. $(x^2y)(-3xy^2)(x^3y^3)$

26. $(xy^2)(-5xy)(x^2y^4)$

27. $(-3y^2)(-2y^2)(-4y^5)$

28. $(-y^3)(-6y)(-8y^4)$

29. $(4ab)(-2a^2b)(7a)$

30. $(3b)(-2ab^2)(7a)$

31. $(-ab)(-3ab)(-6ab)$

32. $(-3a^2b)(-ab^2)(-7a)$

33. $\left(\frac{2}{3}xy\right)(-3x^2y)(5x^4y^5)$

34. $\left(\frac{3}{4}x\right)(-4x^2y^2)(9y^3)$

35. $(12y)(-5x)\left(-\frac{5}{6}x^4y\right)$

36. $(-12x)(3y)\left(-\frac{3}{4}xy^6\right)$

For Problems 37–58, raise each monomial to the indicated power.

37. $(3xy^2)^3$

38. $(4x^2y^3)^3$

39. $(-2x^2y)^5$

40. $(-3xy^4)^3$

41. $(-x^4y^5)^4$

42. $(-x^5y^2)^4$

43. $(ab^2c^3)^6$

44. $(a^2b^3c^5)^5$

45. $(2a^2b^3)^6$

46. $(2a^3b^2)^6$

47. $(9xy^4)^2$

48. $(8x^2y^5)^2$

49. $(-3ab^3)^4$

50. $(-2a^2b^4)^4$

51. $-(2ab)^4$

52. $-(3ab)^4$

53. $-(xy^2z^3)^6$

54. $-(xy^2z^3)^8$

55. $(-5a^2b^2c)^3$

56. $(-4abc^4)^3$

57. $(-xy^4z^2)^7$

58. $(-x^2y^4z^5)^5$

For Problems 59–74, find each quotient.

59. $\dfrac{9x^4y^5}{3xy^2}$

60. $\dfrac{12x^2y^7}{6x^2y^3}$

61. $\dfrac{25x^5y^6}{-5x^2y^4}$

62. $\dfrac{56x^6y^4}{-7x^2y^3}$

63. $\dfrac{-54ab^2c^3}{-6abc}$

64. $\dfrac{-48a^3bc^5}{-6a^2c^4}$

65. $\dfrac{-18x^2y^2z^6}{xyz^2}$

66. $\dfrac{-32x^4y^5z^8}{x^2yz^3}$

67. $\dfrac{a^3b^4c^7}{-abc^5}$

68. $\dfrac{-a^4b^5c}{a^2b^4c}$

69. $\dfrac{-72x^2y^4}{-8x^2y^4}$

70. $\dfrac{-96x^4y^5}{12x^4y^4}$

71. $\dfrac{14ab^3}{-14ab}$

72. $\dfrac{-12abc^2}{12bc}$

73. $\dfrac{-36x^3y^5}{2y^5}$

74. $\dfrac{-48xyz^2}{2xz}$

For Problems 75–90, find each product. Assume that the variables in the exponents represent positive integers. For example,

$$(x^{2n})(x^{3n}) = x^{2n+3n} = x^{5n}$$

75. $(2x^n)(3x^{2n})$

76. $(3x^{2n})(x^{3n-1})$

77. $(a^{2n-1})(a^{3n+4})$

78. $(a^{5n-1})(a^{5n+1})$

79. $(x^{3n-2})(x^{n+2})$

80. $(x^{n-1})(x^{4n+3})$

81. $(a^{5n-2})(a^3)$

82. $(x^{3n-4})(x^4)$

83. $(2x^n)(-5x^n)$

84. $(4x^{2n-1})(-3x^{n+1})$

85. $(-3a^2)(-4a^{n+2})$

86. $(-5x^{n-1})(-6x^{2n+4})$

87. $(x^n)(2x^{2n})(3x^2)$

88. $(2x^n)(3x^{3n-1})(-4x^{2n+5})$

89. $(3x^{n-1})(x^{n+1})(4x^{2-n})$

90. $(-5x^{n+2})(x^{n-2})(4x^{3-2n})$

91. Find a polynomial that represents the total surface area of the rectangular solid in Figure 3.7. Also find a polynomial that represents the volume.

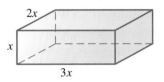

Figure 3.7

92. Find a polynomial that represents the total surface area of the rectangular solid in Figure 3.8. Also find a polynomial that represents the volume.

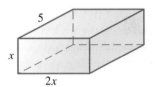

Figure 3.8

93. Find a polynomial that represents the area of the shaded region in Figure 3.9. The length of a radius of the larger circle is r units, and the length of a radius of the smaller circle is 6 units.

Figure 3.9

■ ■ ■ **Thoughts into words**

94. How would you convince someone that $x^6 \div x^2$ is x^4 and not x^3?

95. Your friend simplifies $2^3 \cdot 2^2$ as follows:

$$2^3 \cdot 2^2 = 4^{3+2} = 4^5 = 1024$$

What has she done incorrectly and how would you help her?

3.3 Multiplying Polynomials

We usually state the distributive property as $a(b + c) = ab + ac$; however, we can extend it as follows:

$$a(b + c + d) = ab + ac + ad$$

$$a(b + c + d + e) = ab + ac + ad + ae \qquad \text{etc.}$$

We apply the commutative and associative properties, the properties of exponents, and the distributive property together to find the product of a monomial and a polynomial. The following examples illustrate this idea.

EXAMPLE 1

$$3x^2(2x^2 + 5x + 3) = 3x^2(2x^2) + 3x^2(5x) + 3x^2(3)$$
$$= 6x^4 + 15x^3 + 9x^2$$ ■

EXAMPLE 2

$$-2xy(3x^3 - 4x^2y - 5xy^2 + y^3) = -2xy(3x^3) - (-2xy)(4x^2y)$$
$$-(-2xy)(5xy^2) + (-2xy)(y^3)$$
$$= -6x^4y + 8x^3y^2 + 10x^2y^3 - 2xy^4$$ ■

Now let's consider the product of two polynomials neither of which is a monomial. Consider the following examples.

EXAMPLE 3

$$(x + 2)(y + 5) = x(y + 5) + 2(y + 5)$$
$$= x(y) + x(5) + 2(y) + 2(5)$$
$$= xy + 5x + 2y + 10$$ ■

Note that each term of the first polynomial is multiplied by each term of the second polynomial.

EXAMPLE 4

$$(x - 3)(y + z + 3) = x(y + z + 3) - 3(y + z + 3)$$
$$= xy + xz + 3x - 3y - 3z - 9$$ ■

Multiplying polynomials often produces similar terms that can be combined to simplify the resulting polynomial.

EXAMPLE 5

$$(x + 5)(x + 7) = x(x + 7) + 5(x + 7)$$
$$= x^2 + 7x + 5x + 35$$
$$= x^2 + 12x + 35$$ ■

E X A M P L E   6

$$(x - 2)(x^2 - 3x + 4) = x(x^2 - 3x + 4) - 2(x^2 - 3x + 4)$$
$$= x^3 - 3x^2 + 4x - 2x^2 + 6x - 8$$
$$= x^3 - 5x^2 + 10x - 8$$

■

In Example 6, we are claiming that

$$(x - 2)(x^2 - 3x + 4) = x^3 - 5x^2 + 10x - 8$$

for all real numbers. In addition to going back over our work, how can we verify such a claim? Obviously, we cannot try all real numbers, but trying at least one number gives us a partial check. Let's try the number 4.

$$(x - 2)(x^2 - 3x + 4) = (4 - 2)(4^2 - 3(4) + 4)$$
$$= 2(16 - 12 + 4)$$
$$= 2(8)$$
$$= 16$$

$$x^3 - 5x^2 + 10x - 8 = 4^3 - 5(4)^2 + 10(4) - 8$$
$$= 64 - 80 + 40 - 8$$
$$= 16$$

E X A M P L E   7

$$(3x - 2y)(x^2 + xy - y^2) = 3x(x^2 + xy - y^2) - 2y(x^2 + xy - y^2)$$
$$= 3x^3 + 3x^2y - 3xy^2 - 2x^2y - 2xy^2 + 2y^3$$
$$= 3x^3 + x^2y - 5xy^2 + 2y^3$$

■

It helps to be able to find the product of two binomials without showing all of the intermediate steps. This is quite easy to do with the *three-step shortcut pattern* demonstrated by Figures 3.10 and 3.11 in the following examples.

E X A M P L E   8

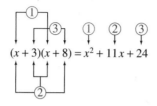

Figure 3.10

STEP ①. Multiply $x \cdot x$.

STEP ②. Multiply $3 \cdot x$ and $8 \cdot x$ and combine.

STEP ③. Multiply $3 \cdot 8$.

■

E X A M P L E   9

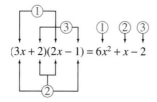

Figure 3.11

Now see if you can use the pattern to find the following products.

$$(x + 2)(x + 6) = ?$$

$$(x - 3)(x + 5) = ?$$

$$(2x + 5)(3x + 7) = ?$$

$$(3x - 1)(4x - 3) = ?$$

Your answers should be $x^2 + 8x + 12$, $x^2 + 2x - 15$, $6x^2 + 29x + 35$, and $12x^2 - 13x + 3$. Keep in mind that this shortcut pattern applies only to finding the product of two binomials.

We can use exponents to indicate repeated multiplication of polynomials. For example, $(x + 3)^2$ means $(x + 3)(x + 3)$, and $(x + 4)^3$ means $(x + 4)(x + 4) \cdot (x + 4)$. To square a binomial, we can simply write it as the product of two equal binomials and apply the shortcut pattern. Thus

$$(x + 3)^2 = (x + 3)(x + 3) = x^2 + 6x + 9$$

$$(x - 6)^2 = (x - 6)(x - 6) = x^2 - 12x + 36 \qquad \text{and}$$

$$(3x - 4)^2 = (3x - 4)(3x - 4) = 9x^2 - 24x + 16$$

When squaring binomials, be careful not to forget the middle term. That is to say, $(x + 3)^2 \neq x^2 + 3^2$; instead, $(x + 3)^2 = x^2 + 6x + 9$.

When multiplying binomials, there are some special patterns that you should recognize. We can use these patterns to find products, and later we will use some of them when factoring polynomials.

P A T T E R N

$$(a + b)^2 = (a + b)(a + b) = a^2 \quad + \quad 2ab \quad + \quad b^2$$

Square of first term of binomial

+ Twice the product of the two terms of binomial

+ Square of second term of binomial

Examples

$$(x + 4)^2 = x^2 + 8x + 16$$

$$(2x + 3y)^2 = 4x^2 + 12xy + 9y^2$$

$$(5a + 7b)^2 = 25a^2 + 70ab + 49b^2$$

PATTERN

$$(a - b)^2 = (a - b)(a - b) = a^2 \quad - \quad 2ab \quad + \quad b^2$$

Square of   Twice the   Square of
first term   − product of   + second term
of binomial   the two terms   of binomial
                   of binomial

Examples

$$(x - 8)^2 = x^2 - 16x + 64$$

$$(3x - 4y)^2 = 9x^2 - 24xy + 16y^2$$

$$(4a - 9b)^2 = 16a^2 - 72ab + 81b^2$$

PATTERN

$$(a + b)(a - b) = a^2 \quad - \quad b^2$$

Square of   Square of
first term   − second term
of binomials   of binomials

Examples

$$(x + 7)(x - 7) = x^2 - 49$$

$$(2x + y)(2x - y) = 4x^2 - y^2$$

$$(3a - 2b)(3a + 2b) = 9a^2 - 4b^2$$

Now suppose that we want to cube a binomial. One approach is as follows:

$$
\begin{aligned}
(x + 4)^3 &= (x + 4)(x + 4)(x + 4) \\
&= (x + 4)(x^2 + 8x + 16) \\
&= x(x^2 + 8x + 16) + 4(x^2 + 8x + 16) \\
&= x^3 + 8x^2 + 16x + 4x^2 + 32x + 64 \\
&= x^3 + 12x^2 + 48x + 64
\end{aligned}
$$

Another approach is to cube a general binomial and then use the resulting pattern.

PATTERN

$$
\begin{aligned}
(a + b)^3 &= (a + b)(a + b)(a + b) \\
&= (a + b)(a^2 + 2ab + b^2) \\
&= a(a^2 + 2ab + b^2) + b(a^2 + 2ab + b^2) \\
&= a^3 + 2a^2b + ab^2 + a^2b + 2ab^2 + b^3 \\
&= a^3 + 3a^2b + 3ab^2 + b^3
\end{aligned}
$$

Let's use the pattern $(a + b)^3 = a^3 + 3a^2b + 3ab^2 + b^3$ to cube the binomial $x + 4$.

$$(x + 4)^3 = x^3 + 3x^2(4) + 3x(4)^2 + 4^3$$
$$= x^3 + 12x^2 + 48x + 64$$

Because $a - b = a + (-b)$, we can easily develop a pattern for cubing $a - b$.

PATTERN

$$(a - b)^3 = [a + (-b)]^3$$
$$= a^3 + 3a^2(-b) + 3a(-b)^2 + (-b)^3$$
$$= a^3 - 3a^2b + 3ab^2 - b^3$$

Now let's use the pattern $(a - b)^3 = a^3 - 3a^2b + 3ab^2 - b^3$ to cube the binomial $3x - 2y$.

$$(3x - 2y)^3 = (3x)^3 - 3(3x)^2(2y) + 3(3x)(2y)^2 - (2y)^3$$
$$= 27x^3 - 54x^2y + 36xy^2 - 8y^3$$

Finally, we need to realize that if the patterns are forgotten or do not apply, then we can revert to applying the distributive property.

$$(2x - 1)(x^2 - 4x + 6) = 2x(x^2 - 4x + 6) - 1(x^2 - 4x + 6)$$
$$= 2x^3 - 8x^2 + 12x - x^2 + 4x - 6$$
$$= 2x^3 - 9x^2 + 16x - 6$$

Back to the Geometry Connection

As you might expect, there are geometric interpretations for many of the algebraic concepts we present in this section. We will give you the opportunity to make some of these connections between algebra and geometry in the next problem set. Let's conclude this section with a problem that allows us to use some algebra and geometry.

EXAMPLE 10

A rectangular piece of tin is 16 inches long and 12 inches wide as shown in Figure 3.12. From each corner a square piece x inches on a side is cut out. The flaps are then turned up to form an open box. Find polynomials that represent the volume and outside surface area of the box.

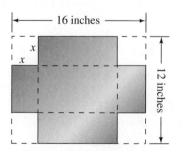

Figure 3.12

Solution

The length of the box will be $16 - 2x$, the width $12 - 2x$, and the height x. With the volume formula $V = lwh$, the polynomial $(16 - 2x)(12 - 2x)(x)$, which simplifies to $4x^3 - 56x^2 + 192x$, represents the volume.

The outside surface area of the box is the area of the original piece of tin minus the four corners that were cut off. Therefore, the polynomial $16(12) - 4x^2$, or $192 - 4x^2$, represents the outside surface area of the box. ■

REMARK: Recall that in Section 3.1 we found the total surface area of a rectangular solid by adding the areas of the sides, top, and bottom. Use this approach for the open box in Example 10 to check our answer of $192 - 4x^2$. Keep in mind that the box has no top.

PROBLEM SET 3.3

For Problems 1–74, find each indicated product. Remember the shortcut for multiplying binomials and the other special patterns we discussed in this section.

1. $2xy(5xy^2 + 3x^2y^3)$

2. $3x^2y(6y^2 - 5x^2y^4)$

3. $-3a^2b(4ab^2 - 5a^3)$

4. $-7ab^2(2b^3 - 3a^2)$

5. $8a^3b^4(3ab - 2ab^2 + 4a^2b^2)$

6. $9a^3b(2a - 3b + 7ab)$

7. $-x^2y(6xy^2 + 3x^2y^3 - x^3y)$

8. $-ab^2(5a + 3b - 6a^2b^3)$

9. $(a + 2b)(x + y)$

10. $(t - s)(x + y)$

11. $(a - 3b)(c + 4d)$

12. $(a - 4b)(c - d)$

13. $(x + 6)(x + 10)$

14. $(x + 2)(x + 10)$

15. $(y - 5)(y + 11)$

16. $(y - 3)(y + 9)$

17. $(n + 2)(n - 7)$

18. $(n + 3)(n - 12)$

19. $(x + 6)(x - 6)$

20. $(t + 8)(t - 8)$

21. $(x - 6)^2$

22. $(x - 2)^2$

23. $(x - 6)(x - 8)$

24. $(x - 3)(x - 13)$

25. $(x + 1)(x - 2)(x - 3)$

26. $(x - 1)(x + 4)(x - 6)$

27. $(x - 3)(x + 3)(x - 1)$

28. $(x - 5)(x + 5)(x - 8)$

29. $(t + 9)^2$

30. $(t + 13)^2$

31. $(y - 7)^2$

32. $(y - 4)^2$

33. $(4x + 5)(x + 7)$

34. $(6x + 5)(x + 3)$

35. $(3y - 1)(3y + 1)$

36. $(5y - 2)(5y + 2)$

37. $(7x - 2)(2x + 1)$

38. $(6x - 1)(3x + 2)$

39. $(1 + t)(5 - 2t)$

40. $(3 - t)(2 + 4t)$

41. $(3t + 7)^2$

42. $(4t + 6)^2$

43. $(2 - 5x)(2 + 5x)$

44. $(6 - 3x)(6 + 3x)$

45. $(7x - 4)^2$

46. $(5x - 7)^2$

47. $(6x + 7)(3x - 10)$

48. $(4x - 7)(7x + 4)$

49. $(2x - 5y)(x + 3y)$

50. $(x - 4y)(3x + 7y)$

51. $(5x - 2a)(5x + 2a)$

52. $(9x - 2y)(9x + 2y)$

53. $(t + 3)(t^2 - 3t - 5)$

54. $(t - 2)(t^2 + 7t + 2)$

55. $(x - 4)(x^2 + 5x - 4)$

56. $(x + 6)(2x^2 - x - 7)$

57. $(2x - 3)(x^2 + 6x + 10)$

58. $(3x + 4)(2x^2 - 2x - 6)$

59. $(4x - 1)(3x^2 - x + 6)$

60. $(5x - 2)(6x^2 + 2x - 1)$

61. $(x^2 + 2x + 1)(x^2 + 3x + 4)$

62. $(x^2 - x + 6)(x^2 - 5x - 8)$

63. $(2x^2 + 3x - 4)(x^2 - 2x - 1)$

64. $(3x^2 - 2x + 1)(2x^2 + x - 2)$

65. $(x + 2)^3$

66. $(x + 1)^3$

67. $(x - 4)^3$

68. $(x - 5)^3$

69. $(2x + 3)^3$

70. $(3x + 1)^3$

71. $(4x - 1)^3$

72. $(3x - 2)^3$

73. $(5x + 2)^3$

74. $(4x - 5)^3$

For Problems 75–84, find the indicated products. Assume all variables that appear as exponents represent positive integers.

75. $(x^n - 4)(x^n + 4)$

76. $(x^{3a} - 1)(x^{3a} + 1)$

77. $(x^a + 6)(x^a - 2)$

78. $(x^a + 4)(x^a - 9)$

79. $(2x^n + 5)(3x^n - 7)$

80. $(3x^n + 5)(4x^n - 9)$

81. $(x^{2a} - 7)(x^{2a} - 3)$

82. $(x^{2a} + 6)(x^{2a} - 4)$

83. $(2x^n + 5)^2$

84. $(3x^n - 7)^2$

85. Explain how Figure 3.13 can be used to demonstrate geometrically that $(x + 2)(x + 6) = x^2 + 8x + 12$.

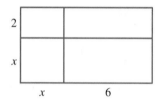

Figure 3.13

86. Find a polynomial that represents the sum of the areas of the two rectangles shown in Figure 3.14.

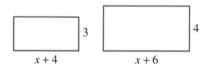

Figure 3.14

87. Find a polynomial that represents the area of the shaded region in Figure 3.15.

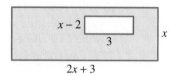

Figure 3.15

88. Explain how Figure 3.16 can be used to demonstrate geometrically that $(x + 7)(x - 3) = x^2 + 4x - 21$.

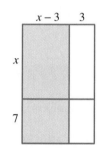

Figure 3.16

89. A square piece of cardboard is 16 inches on a side. A square piece x inches on a side is cut out from each corner. The flaps are then turned up to form an open box. Find polynomials that represent the volume and outside surface area of the box.

■ ■ ■ **Thoughts into words**

90. How would you simplify $(2^3 + 2^2)^2$? Explain your reasoning.

91. Describe the process of multiplying two polynomials.

92. Determine the number of terms in the product of $(x + y)$ and $(a + b + c + d)$ without doing the multiplication. Explain how you arrived at your answer.

■ ■ ■ Further investigations

93. We have used the following two multiplication patterns.

$(a + b)^2 = a^2 + 2ab + b^2$

$(a + b)^3 = a^3 + 3a^2b + 3ab^2 + b^3$

By multiplying, we can extend these patterns as follows:

$(a + b)^4 = a^4 + 4a^3b + 6a^2b^2 + 4ab^3 + b^4$

$(a + b)^5 = a^5 + 5a^4b + 10a^3b^2 + 10a^2b^3 + 5a^4 + b^5$

On the basis of these results, see if you can determine a pattern that will enable you to complete each of the following without using the long-multiplication process.

(a) $(a + b)^6$ **(b)** $(a + b)^7$

(c) $(a + b)^8$ **(d)** $(a + b)^9$

94. Find each of the following indicated products. These patterns will be used again in Section 3.5.

(a) $(x - 1)(x^2 + x + 1)$ **(b)** $(x + 1)(x^2 - x + 1)$

(c) $(x + 3)(x^2 - 3x + 9)$ **(d)** $(x - 4)(x^2 + 4x + 16)$

(e) $(2x - 3)(4x^2 + 6x + 9)$

(f) $(3x + 5)(9x^2 - 15x + 25)$

95. Some of the product patterns can be used to do arithmetic computations mentally. For example, let's use the pattern $(a + b)^2 = a^2 + 2ab + b^2$ to compute 31^2 mentally. Your thought process should be "$31^2 = (30 + 1)^2 = 30^2 + 2(30)(1) + 1^2 = 961$." Compute each

of the following numbers mentally, and then check your answers.

(a) 21^2 **(b)** 41^2 **(c)** 71^2

(d) 32^2 **(e)** 52^2 **(f)** 82^2

96. Use the pattern $(a - b)^2 = a^2 - 2ab + b^2$ to compute each of the following numbers mentally, and then check your answers.

(a) 19^2 **(b)** 29^2 **(c)** 49^2

(d) 79^2 **(e)** 38^2 **(f)** 58^2

97. Every whole number with a units digit of 5 can be represented by the expression $10x + 5$, where x is a whole number. For example, $35 = 10(3) + 5$ and $145 = 10(14) + 5$. Now let's observe the following pattern when squaring such a number.

$(10x + 5)^2 = 100x^2 + 100x + 25$

$$= \overline{100x(x + 1) + 25}$$

The pattern inside the dashed box can be stated as "add 25 to the product of x, $x + 1$, and 100." Thus, to compute 35^2 mentally, we can think "$35^2 = 3(4)(100) + 25 = 1225$." Compute each of the following numbers mentally, and then check your answers.

(a) 15^2 **(b)** 25^2 **(c)** 45^2

(d) 55^2 **(e)** 65^2 **(f)** 75^2

(g) 85^2 **(h)** 95^2 **(i)** 105^2

3.4 Factoring: Use of the Distributive Property

Recall that 2 and 3 are said to be *factors* of 6 because the product of 2 and 3 is 6. Likewise, in an indicated product such as $7ab$, the 7, a, and b are called factors of the product. If a positive integer greater than 1 has no factors that are positive integers other than itself and 1, then it is called a **prime number.** Thus the prime numbers less than 20 are 2, 3, 5, 7, 11, 13, 17, and 19. A positive integer greater than 1 that is not a prime number is called a **composite number.** The composite numbers less than 20 are 4, 6, 8, 9, 10, 12, 14, 15, 16, and 18. Every composite number is the product of prime numbers. Consider the following examples.

$$4 = 2 \cdot 2 \qquad\qquad 63 = 3 \cdot 3 \cdot 7$$

$$12 = 2 \cdot 2 \cdot 3 \qquad 121 = 11 \cdot 11$$

$$35 = 5 \cdot 7$$

The indicated product form that contains only prime factors is called the **prime factorization form** of a number. Thus the prime factorization form of 63 is $3 \cdot 3 \cdot 7$. We also say that the number has been **completely factored** when it is in the prime factorization form.

In general, factoring is the reverse of multiplication. Previously, we have used the distributive property to find the product of a monomial and a polynomial, as in the next examples.

$$3(x + 2) = 3(x) + 3(2) = 3x + 6$$

$$5(2x - 1) = 5(2x) - 5(1) = 10x - 5$$

$$x(x^2 + 6x - 4) = x(x^2) + x(6x) - x(4) = x^3 + 6x^2 - 4x$$

We shall also use the distributive property [in the form $ab + ac = a(b + c)$] to reverse the process — that is, to factor a given polynomial. Consider the following examples. (The steps in the dashed boxes can be done mentally.)

$$3x + 6 = \overline{3(x) + 3(2)} = 3(x + 2),$$

$$10x - 5 = \overline{5(2x) - 5(1)} = 5(2x - 1),$$

$$x^3 + 6x^2 - 4x = \overline{x(x^2) + x(6x) - x(4)} = x(x^2 + 6x - 4)$$

Note that in each example a given polynomial has been factored into the product of a monomial and a polynomial. Obviously, polynomials could be factored in a variety of ways. Consider some factorizations of $3x^2 + 12x$.

$$3x^2 + 12x = 3x(x + 4) \qquad \text{or} \qquad 3x^2 + 12x = 3(x^2 + 4x) \qquad \text{or}$$

$$3x^2 + 12x = x(3x + 12) \qquad \text{or} \qquad 3x^2 + 12x = \frac{1}{2}(6x^2 + 24x)$$

We are, however, primarily interested in the first of the previous factorization forms, which we refer to as the **completely factored form.** A polynomial with integral coefficients is in completely factored form if

1. It is expressed as a product of polynomials with *integral coefficients,* and
2. No polynomial, other than a monomial, within the factored form can be further factored into polynomials with integral coefficients.

Do you see why only the first of the above factored forms of $3x^2 + 12x$ is said to be in completely factored form? In each of the other three forms, the polynomial inside the parentheses can be factored further. Moreover, in the last form, $\frac{1}{2}(6x^2 + 24x)$, the condition of using only integral coefficients is violated.

The factoring process that we discuss in this section, $ab + ac = a(b + c)$, is often referred to as **factoring out the highest common monomial factor.** The key idea in this process is to recognize the monomial factor that is common to all terms. For example, we observe that each term of the polynomial $2x^3 + 4x^2 + 6x$ has a factor of $2x$. Thus we write

$$2x^3 + 4x^2 + 6x = 2x(\qquad)$$

and insert within the parentheses the appropriate polynomial factor. We determine the terms of this polynomial factor by dividing each term of the original polynomial by the factor of $2x$. The final, completely factored form is

$$2x^3 + 4x^2 + 6x = 2x(x^2 + 2x + 3)$$

The following examples further demonstrate this process of factoring out the highest common monomial factor.

$$12x^3 + 16x^2 = 4x^2(3x + 4) \qquad 6x^2y^3 + 27xy^4 = 3xy^3(2x + 9y)$$

$$8ab - 18b = 2b(4a - 9) \qquad 8y^3 + 4y^2 = 4y^2(2y + 1)$$

$$30x^3 + 42x^4 - 24x^5 = 6x^3(5 + 7x - 4x^2)$$

Note that in each example, the common monomial factor itself is not in a completely factored form. For example, $4x^2(3x + 4)$ is not written as $2 \cdot 2 \cdot x \cdot x \cdot (3x + 4)$.

Sometimes there may be a common binomial factor rather than a common monomial factor. For example, each of the two terms of the expression $x(y + 2) + z(y + 2)$ has a binomial factor of $(y + 2)$. Thus we can factor $(y + 2)$ from each term, and our result is

$$x(y + 2) + z(y + 2) = (y + 2)(x + z)$$

Consider a few more examples that involve a common binomial factor.

$$a^2(b + 1) + 2(b + 1) = (b + 1)(a^2 + 2)$$

$$x(2y - 1) - y(2y - 1) = (2y - 1)(x - y)$$

$$x(x + 2) + 3(x + 2) = (x + 2)(x + 3)$$

It may be that the original polynomial exhibits no apparent common monomial or binomial factor, which is the case with $ab + 3a + bc + 3c$. However, by factoring a from the first two terms and c from the last two terms, we get

$$ab + 3a + bc + 3c = a(b + 3) + c(b + 3)$$

Now a common binomial factor of $(b + 3)$ is obvious, and we can proceed as before.

$$a(b + 3) + c(b + 3) = (b + 3)(a + c)$$

We refer to this factoring process as **factoring by grouping.** Let's consider a few more examples of this type.

$$ab^2 - 4b^2 + 3a - 12 = b^2(a - 4) + 3(a - 4)$$ Factor b^2 from the first two terms and 3 from the last two terms.

$$= (a - 4)(b^2 + 3)$$ Factor common binomial from both terms.

$$x^2 - x + 5x - 5 = x(x - 1) + 5(x - 1)$$ Factor x from the first two terms and 5 from the last two terms.

$$= (x - 1)(x + 5)$$ Factor common binomial from both terms.

$$x^2 + 2x - 3x - 6 = x(x + 2) - 3(x + 2)$$ Factor x from the first two terms and -3 from the last two terms.

$$= (x + 2)(x - 3)$$ Factor common binomial factor from both terms.

It may be necessary to rearrange some terms before applying the distributive property. Terms that contain common factors need to be grouped together, and this may be done in more than one way. The next example illustrates this idea.

$$4a^2 - bc^2 - a^2b + 4c^2 = 4a^2 - a^2b + 4c^2 - bc^2$$
$$= a^2(4 - b) + c^2(4 - b)$$
$$= (4 - b)(a^2 + c^2) \quad \text{or}$$
$$4a^2 - bc^2 - a^2b + 4c^2 = 4a^2 + 4c^2 - bc^2 - a^2b$$
$$= 4(a^2 + c^2) - b(c^2 + a^2)$$
$$= 4(a^2 + c^2) - b(a^2 + c^2)$$
$$= (a^2 + c^2)(4 - b)$$

Equations and Problem Solving

One reason why factoring is an important algebraic skill is that it extends our techniques for solving equations. Each time we examine a factoring technique, we will then use it to help solve certain types of equations.

We need another property of equality before we consider some equations where the highest-common-factor technique is useful. Suppose that the product of two numbers is zero. Can we conclude that at least one of these numbers must itself be zero? Yes. Let's state a property that formalizes this idea. Property 3.5, along with the highest-common-factor pattern, provides us with another technique for solving equations.

> **PROPERTY 3.5**
>
> Let a and b be real numbers. Then
>
> $$ab = 0 \quad \text{if and only if } a = 0 \text{ or } b = 0$$

EXAMPLE 1

Solve $x^2 + 6x = 0$.

Solution

$$x^2 + 6x = 0$$

$$x(x + 6) = 0 \qquad \text{Factor the left side.}$$

$$x = 0 \quad \text{or} \quad x + 6 = 0 \qquad ab = 0 \text{ if and only if } a = 0 \text{ or } b = 0$$

$$x = 0 \quad \text{or} \quad x = -6$$

Thus both 0 and -6 will satisfy the original equation, and the solution set is $\{-6, 0\}$. ∎

EXAMPLE 2

Solve $a^2 = 11a$.

Solution

$$a^2 = 11a$$

$$a^2 - 11a = 0 \qquad \text{Add } -11a \text{ to both sides.}$$

$$a(a - 11) = 0 \qquad \text{Factor the left side.}$$

$$a = 0 \quad \text{or} \quad a - 11 = 0 \qquad ab = 0 \text{ if and only if } a = 0 \text{ or } b = 0$$

$$a = 0 \quad \text{or} \quad a = 11$$

The solution set is $\{0, 11\}$. ∎

REMARK: Note that in Example 2 we did *not* divide both sides of the equation by a. This would cause us to lose the solution of 0.

EXAMPLE 3

Solve $3n^2 - 5n = 0$.

Solution

$$3n^2 - 5n = 0$$

$$n(3n - 5) = 0$$

$$n = 0 \quad \text{or} \quad 3n - 5 = 0$$

$$n = 0 \quad \text{or} \quad 3n = 5$$

$$n = 0 \qquad \text{or} \qquad n = \frac{5}{3}$$

The solution set is $\left\{0, \frac{5}{3}\right\}$. ■

E X A M P L E   4

Solve $3ax^2 + bx = 0$ for x.

Solution

$$3ax^2 + bx = 0$$

$$x(3ax + b) = 0$$

$$x = 0 \qquad \text{or} \qquad 3ax + b = 0$$

$$x = 0 \qquad \text{or} \qquad 3ax = -b$$

$$x = 0 \qquad \text{or} \qquad x = -\frac{b}{3a}$$

The solution set is $\left\{0, -\frac{b}{3a}\right\}$. ■

Many of the problems that we solve in the next few sections have a geometric setting. Some basic geometric figures, along with appropriate formulas, are listed in the inside front cover of this text. You may need to refer to them to refresh your memory.

P R O B L E M   1

The area of a square is three times its perimeter. Find the length of a side of the square.

Solution

Let s represent the length of a side of the square (Figure 3.17). The area is represented by s^2 and the perimeter by $4s$. Thus

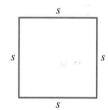

Figure 3.17

$$s^2 = 3(4s) \qquad \text{The area is to be three times the perimeter.}$$

$$s^2 = 12s$$

$$s^2 - 12s = 0$$

$$s(s - 12) = 0$$

$$s = 0 \qquad \text{or} \qquad s = 12$$

Because 0 is not a reasonable solution, it must be a 12-by-12 square. (Be sure to check this answer in the original statement of the problem!) ■

P R O B L E M   2

Suppose that the volume of a right circular cylinder is numerically equal to the total surface area of the cylinder. If the height of the cylinder is equal to the length of a radius of the base, find the height.

Solution

Because $r = h$, the formula for volume $V = \pi r^2 h$ becomes $V = \pi r^3$ and the formula for the total surface area $S = 2\pi r^2 + 2\pi r h$ becomes $S = 2\pi r^2 + 2\pi r^2$, or $S = 4\pi r^2$. Therefore, we can set up and solve the following equation.

$$\pi r^3 = 4\pi r^2$$

$$\pi r^3 - 4\pi r^2 = 0$$

$$\pi r^2(r - 4) = 0$$

$$\pi r^2 = 0 \quad \text{or} \quad r - 4 = 0$$

$$r = 0 \quad \text{or} \quad r = 4$$

Zero is not a reasonable answer, therefore the height must be 4 units.   ■

PROBLEM SET 3.4

For Problems 1–10, classify each number as prime or composite.

1. 63

2. 81

3. 59

4. 83

5. 51

6. 69

7. 91

8. 119

9. 71

10. 101

For Problems 11–20, factor each of the composite numbers into the product of prime numbers. For example, $30 = 2 \cdot 3 \cdot 5$.

11. 28

12. 39

13. 44

14. 49

15. 56

16. 64

17. 72

18. 84

19. 87

20. 91

For Problems 21–46, factor completely.

21. $6x + 3y$

22. $12x + 8y$

23. $6x^2 + 14x$

24. $15x^2 + 6x$

25. $28y^2 - 4y$

26. $42y^2 - 6y$

27. $20xy - 15x$

28. $27xy - 36y$

29. $7x^3 + 10x^2$

30. $12x^3 - 10x^2$

31. $18a^2b + 27ab^2$

32. $24a^3b^2 + 36a^2b$

33. $12x^3y^4 - 39x^4y^3$

34. $15x^4y^2 - 45x^5y^4$

35. $8x^4 + 12x^3 - 24x^2$

36. $6x^5 - 18x^3 + 24x$

37. $5x + 7x^2 + 9x^4$

38. $9x^2 - 17x^4 + 21x^5$

39. $15x^2y^3 + 20xy^2 + 35x^3y^4$

40. $8x^5y^3 - 6x^4y^5 + 12x^2y^3$

41. $x(y + 2) + 3(y + 2)$

42. $x(y - 1) + 5(y - 1)$

43. $3x(2a + b) - 2y(2a + b)$

44. $5x(a - b) + y(a - b)$

45. $x(x + 2) + 5(x + 2)$

46. $x(x - 1) - 3(x - 1)$

For Problems 47–64, factor by grouping.

47. $ax + 4x + ay + 4y$

48. $ax - 2x + ay - 2y$

49. $ax - 2bx + ay - 2by$

50. $2ax - bx + 2ay - by$

51. $3ax - 3bx - ay + by$

52. $5ax - 5bx - 2ay + 2by$

53. $2ax + 2x + ay + y$

54. $3bx + 3x + by + y$

55. $ax^2 - x^2 + 2a - 2$

56. $ax^2 - 2x^2 + 3a - 6$

57. $2ac + 3bd + 2bc + 3ad$

58. $2bx + cy + cx + 2by$

59. $ax - by + bx - ay$

60. $2a^2 - 3bc - 2ab + 3ac$

61. $x^2 + 9x + 6x + 54$

62. $x^2 - 2x + 5x - 10$

63. $2x^2 + 8x + x + 4$

64. $3x^2 + 18x - 2x - 12$

For Problems 65–80, solve each of the equations.

65. $x^2 + 7x = 0$

66. $x^2 + 9x = 0$

67. $x^2 - x = 0$

68. $x^2 - 14x = 0$

69. $a^2 = 5a$

70. $b^2 = -7b$

71. $-2y = 4y^2$

72. $-6x = 2x^2$

73. $3x^2 + 7x = 0$

74. $-4x^2 + 9x = 0$

75. $4x^2 = 5x$

76. $3x = 11x^2$

77. $x - 4x^2 = 0$

78. $x - 6x^2 = 0$

79. $12a = -a^2$

80. $-5a = -a^2$

For Problems 81–86, solve each equation for the indicated variable.

81. $5bx^2 - 3ax = 0$ for x

82. $ax^2 + bx = 0$ for x

83. $2by^2 = -3ay$ for y

84. $3ay^2 = by$ for y

85. $y^2 - ay + 2by - 2ab = 0$ for y

86. $x^2 + ax + bx + ab = 0$ for x

For Problems 87–96, set up an equation and solve each of the following problems.

87. The square of a number equals seven times the number. Find the number.

88. Suppose that the area of a square is six times its perimeter. Find the length of a side of the square.

89. The area of a circular region is numerically equal to three times the circumference of the circle. Find the length of a radius of the circle.

90. Find the length of a radius of a circle such that the circumference of the circle is numerically equal to the area of the circle.

91. Suppose that the area of a circle is numerically equal to the perimeter of a square and that the length of a radius of the circle is equal to the length of a side of the square. Find the length of a side of the square. Express your answer in terms of π.

92. Find the length of a radius of a sphere such that the surface area of the sphere is numerically equal to the volume of the sphere.

93. Suppose that the area of a square lot is twice the area of an adjoining rectangular plot of ground. If the rectangular plot is 50 feet wide and its length is the same as the length of a side of the square lot, find the dimensions of both the square and the rectangle.

94. The area of a square is one-fourth as large as the area of a triangle. One side of the triangle is 16 inches long, and the altitude to that side is the same length as a side of the square. Find the length of a side of the square.

95. Suppose that the volume of a sphere is numerically equal to twice the surface area of the sphere. Find the length of a radius of the sphere.

96. Suppose that a radius of a sphere is equal in length to a radius of a circle. If the volume of the sphere is numerically equal to four times the area of the circle, find the length of a radius for both the sphere and the circle.

■ ■ ■ **Thoughts into words**

97. Is $2 \cdot 3 \cdot 5 \cdot 7 \cdot 11 + 7$ a prime or a composite number? Defend your answer.

98. Suppose that your friend factors $36x^2y + 48xy^2$ as follows:

$$36x^2y + 48xy^2 = (4xy)(9x + 12y)$$

$$= (4xy)(3)(3x + 4y)$$

$$= 12xy(3x + 4y)$$

Is this a correct approach? Would you have any suggestion to offer your friend?

99. Your classmate solves the equation $3ax + bx = 0$ for x as follows:

$$3ax + bx = 0$$

$$3ax = -bx$$

$$x = \frac{-bx}{3a}$$

How should he know that the solution is incorrect? How would you help him obtain the correct solution?

■ ■ ■ **Further investigations**

100. The total surface area of a right circular cylinder is given by the formula $A = 2\pi r^2 + 2\pi rh$, where r represents the radius of a base and h represents the height of the cylinder. For computational purposes, it may be more convenient to change the form of the right side of the formula by factoring it.

$$A = 2\pi r^2 + 2\pi rh$$
$$= 2\pi r(r + h)$$

Use $A = 2\pi r(r + h)$ to find the total surface area of each of the following cylinders. Also, use $\dfrac{22}{7}$ as an approximation for π.

(a) $r = 7$ centimeters and $h = 12$ centimeters

(b) $r = 14$ meters and $h = 20$ meters

(c) $r = 3$ feet and $h = 4$ feet

(d) $r = 5$ yards and $h = 9$ yards

For Problems 101–106, factor each expression. Assume that all variables that appear as exponents represent positive integers.

101. $2x^{2a} - 3x^a$

102. $6x^{2a} + 8x^a$

103. $y^{3m} + 5y^{2m}$

104. $3y^{5m} - y^{4m} - y^{3m}$

105. $2x^{6a} - 3x^{5a} + 7x^{4a}$

106. $6x^{3a} - 10x^{2a}$

3.5   Factoring: Difference of Two Squares and Sum or Difference of Two Cubes

In Section 3.3, we examined some special multiplication patterns. One of these patterns was

$$(a + b)(a - b) = a^2 - b^2$$

This same pattern, viewed as a factoring pattern, is referred to as the difference of two squares.

Difference of Two Squares

$$a^2 - b^2 = (a + b)(a - b)$$

Applying the pattern is fairly simple, as these next examples demonstrate. Again the steps in dashed boxes are usually performed mentally.

$$x^2 - 16 = (x)^2 - (4)^2 = (x + 4)(x - 4)$$
$$4x^2 - 25 = (2x)^2 - (5)^2 = (2x + 5)(2x - 5)$$
$$16x^2 - 9y^2 = (4x)^2 - (3y)^2 = (4x + 3y)(4x - 3y)$$
$$1 - a^2 = (1)^2 - (a)^2 = (1 + a)(1 - a)$$

Multiplication is commutative, so the order of writing the factors is not important. For example, $(x + 4)(x - 4)$ can also be written as $(x - 4)(x + 4)$.

You must be careful not to assume an analogous factoring pattern for the *sum* of two squares; *it does not exist*. For example, $x^2 + 4 \neq (x + 2)(x + 2)$ because $(x + 2)(x + 2) = x^2 + 4x + 4$. We say that a polynomial such as $x^2 + 4$ is a **prime polynomial** or that it is not factorable using integers.

Sometimes the difference-of-two-squares pattern can be applied more than once, as the next examples illustrate.

$$x^4 - y^4 = (x^2 + y^2)(x^2 - y^2) = (x^2 + y^2)(x + y)(x - y)$$
$$16x^4 - 81y^4 = (4x^2 + 9y^2)(4x^2 - 9y^2) = (4x^2 + 9y^2)(2x + 3y)(2x - 3y)$$

It may also be that the squares are other than simple monomial squares, as in the next three examples.

$$(x + 3)^2 - y^2 = ((x + 3) + y)((x + 3) - y) = (x + 3 + y)(x + 3 - y)$$
$$4x^2 - (2y + 1)^2 = (2x + (2y + 1))(2x - (2y + 1))$$
$$= (2x + 2y + 1)(2x - 2y - 1)$$
$$(x - 1)^2 - (x + 4)^2 = ((x - 1) + (x + 4))((x - 1) - (x + 4))$$
$$= (x - 1 + x + 4)(x - 1 - x - 4)$$
$$= (2x + 3)(-5)$$

It is possible to apply both the technique of factoring out a common monomial factor and the pattern of the difference of two squares to the same problem. In general, it is best to look first for a common monomial factor. Consider the following examples.

$$2x^2 - 50 = 2(x^2 - 25) \qquad\qquad 9x^2 - 36 = 9(x^2 - 4)$$
$$= 2(x + 5)(x - 5) \qquad\qquad = 9(x + 2)(x - 2)$$
$$48y^3 - 27y = 3y(16y^2 - 9)$$
$$= 3y(4y + 3)(4y - 3)$$

Word of Caution   The polynomial $9x^2 - 36$ can be factored as follows:

$$9x^2 - 36 = (3x + 6)(3x - 6)$$
$$= 3(x + 2)(3)(x - 2)$$
$$= 9(x + 2)(x - 2)$$

However, when one takes this approach, there seems to be a tendency to stop at the step $(3x + 6)(3x - 6)$. Therefore, remember the suggestion to *look first for a common monomial factor*.

The following examples should help you summarize all of the factoring techniques we have considered thus far.

$$7x^2 + 28 = 7(x^2 + 4)$$

$$4x^2y - 14xy^2 = 2xy(2x - 7y)$$

$$x^2 - 4 = (x + 2)(x - 2)$$

$$18 - 2x^2 = 2(9 - x^2) = 2(3 + x)(3 - x)$$

$y^2 + 9$ is not factorable using integers.

$5x + 13y$ is not factorable using integers.

$$x^4 - 16 = (x^2 + 4)(x^2 - 4) = (x^2 + 4)(x + 2)(x - 2)$$

Sum and Difference of Two Cubes

As we pointed out before, there exists no sum-of-squares pattern analogous to the difference-of-squares factoring pattern. That is, a polynomial such as $x^2 + 9$ is not factorable using integers. However, patterns do exist for both the sum and the difference of two cubes. These patterns are as follows:

Sum and Difference of Two Cubes
$a^3 + b^3 = (a + b)(a^2 - ab + b^2)$
$a^3 - b^3 = (a - b)(a^2 + ab + b^2)$

Note how we apply these patterns in the next four examples.

$$x^3 + 27 = (x)^3 + (3)^3 = (x + 3)(x^2 - 3x + 9)$$

$$8a^3 + 125b^3 = (2a)^3 + (5b)^3 = (2a + 5b)(4a^2 - 10ab + 25b^2)$$

$$x^3 - 1 = (x)^3 - (1)^3 = (x - 1)(x^2 + x + 1)$$

$$27y^3 - 64x^3 = (3y)^3 - (4x)^3 = (3y - 4x)(9y^2 + 12xy + 16x^2)$$

Equations and Problem Solving

Remember that each time we pick up a new factoring technique we also develop more power for solving equations. Let's consider how we can use the difference-of-two-squares factoring pattern to help solve certain types of equations.

EXAMPLE 1 Solve $x^2 = 16$.

Solution

$$x^2 = 16$$

$$x^2 - 16 = 0$$

$$(x + 4)(x - 4) = 0$$

$$x + 4 = 0 \quad \text{or} \quad x - 4 = 0$$

$$x = -4 \quad \text{or} \quad x = 4$$

The solution set is $\{-4, 4\}$. (Be sure to check these solutions in the original equation!) ■

EXAMPLE 2 Solve $9x^2 = 64$.

Solution

$$9x^2 = 64$$

$$9x^2 - 64 = 0$$

$$(3x + 8)(3x - 8) = 0$$

$$3x + 8 = 0 \quad \text{or} \quad 3x - 8 = 0$$

$$3x = -8 \quad \text{or} \quad 3x = 8$$

$$x = -\frac{8}{3} \quad \text{or} \quad x = \frac{8}{3}$$

The solution set is $\left\{-\dfrac{8}{3}, \dfrac{8}{3}\right\}$. ■

EXAMPLE 3 Solve $7x^2 - 7 = 0$.

Solution

$$7x^2 - 7 = 0$$

$$7(x^2 - 1) = 0$$

$$x^2 - 1 = 0 \qquad \text{Multiply both sides by } \frac{1}{7}.$$

$$(x + 1)(x - 1) = 0$$

$$x + 1 = 0 \quad \text{or} \quad x - 1 = 0$$

$$x = -1 \quad \text{or} \quad x = 1$$

The solution set is $\{-1, 1\}$. ■

In the previous examples we have been using the property $ab = 0$ if and only if $a = 0$ or $b = 0$. This property can be extended to any number of factors whose product is zero. Thus for three factors, the property could be stated $abc = 0$ if and only if $a = 0$ or $b = 0$ or $c = 0$. The next two examples illustrate this idea.

E X A M P L E   4

Solve $x^4 - 16 = 0$.

Solution

$$x^4 - 16 = 0$$
$$(x^2 + 4)(x^2 - 4) = 0$$
$$(x^2 + 4)(x + 2)(x - 2) = 0$$
$$x^2 + 4 = 0 \quad \text{or} \quad x + 2 = 0 \quad \text{or} \quad x - 2 = 0$$
$$x^2 = -4 \quad \text{or} \quad x = -2 \quad \text{or} \quad x = 2$$

The solution set is $\{-2, 2\}$. (Because no real numbers, when squared, will produce -4, the equation $x^2 = -4$ yields no additional real number solutions.) ■

E X A M P L E   5

Solve $x^3 - 49x = 0$.

Solution

$$x^3 - 49x = 0$$
$$x(x^2 - 49) = 0$$
$$x(x + 7)(x - 7) = 0$$
$$x = 0 \quad \text{or} \quad x + 7 = 0 \quad \text{or} \quad x - 7 = 0$$
$$x = 0 \quad \text{or} \quad x = -7 \quad \text{or} \quad x = 7$$

The solution set is $\{-7, 0, 7\}$. ■

The more we know about solving equations, the better we are at solving word problems.

P R O B L E M   1

The combined area of two squares is 40 square centimeters. Each side of one square is three times as long as a side of the other square. Find the dimensions of each of the squares.

Solution

Let s represent the length of a side of the smaller square. Then $3s$ represents the length of a side of the larger square (Figure 3.18).

$$s^2 + (3s)^2 = 40$$

$$s^2 + 9s^2 = 40$$

$$10s^2 = 40$$

$$s^2 = 4$$

$$s^2 - 4 = 0$$

$$(s + 2)(s - 2) = 0$$

$$s + 2 = 0 \quad \text{or} \quad s - 2 = 0$$

$$s = -2 \quad \text{or} \quad s = 2$$

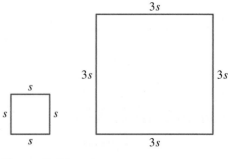

Figure 3.18

Because s represents the length of a side of a square, the solution -2 has to be disregarded. Thus the length of a side of the small square is 2 centimeters, and the large square has sides of length $3(2) = 6$ centimeters. ∎

PROBLEM SET 3.5

For Problems 1–20, use the difference-of-squares pattern to factor each of the following.

1. $x^2 - 1$

2. $x^2 - 9$

3. $16x^2 - 25$

4. $4x^2 - 49$

5. $9x^2 - 25y^2$

6. $x^2 - 64y^2$

7. $25x^2y^2 - 36$

8. $x^2y^2 - a^2b^2$

9. $4x^2 - y^4$

10. $x^6 - 9y^2$

11. $1 - 144n^2$

12. $25 - 49n^2$

13. $(x + 2)^2 - y^2$

14. $(3x + 5)^2 - y^2$

15. $4x^2 - (y + 1)^2$

16. $x^2 - (y - 5)^2$

17. $9a^2 - (2b + 3)^2$

18. $16s^2 - (3t + 1)^2$

19. $(x + 2)^2 - (x + 7)^2$

20. $(x - 1)^2 - (x - 8)^2$

For Problems 21–44, factor each of the following polynomials completely. Indicate any that are not factorable using integers. Don't forget to look first for a common monomial factor.

21. $9x^2 - 36$

22. $8x^2 - 72$

23. $5x^2 + 5$

24. $7x^2 + 28$

25. $8y^2 - 32$

26. $5y^2 - 80$

27. $a^3b - 9ab$

28. $x^3y^2 - xy^2$

29. $16x^2 + 25$

30. $x^4 - 16$

31. $n^4 - 81$

32. $4x^2 + 9$

33. $3x^3 + 27x$

34. $20x^3 + 45x$

35. $4x^3y - 64xy^3$

36. $12x^3 - 27xy^2$

37. $6x - 6x^3$

38. $1 - 16x^4$

39. $1 - x^4y^4$

40. $20x - 5x^3$

41. $4x^2 - 64y^2$

42. $9x^2 - 81y^2$

43. $3x^4 - 48$

44. $2x^5 - 162x$

For Problems 45–56, use the sum-of-two-cubes or the difference-of-two-cubes pattern to factor each of the following.

45. $a^3 - 64$

46. $a^3 - 27$

47. $x^3 + 1$

48. $x^3 + 8$

49. $27x^3 + 64y^3$

50. $8x^3 + 27y^3$

51. $1 - 27a^3$

52. $1 - 8x^3$

53. $x^3y^3 - 1$

54. $125x^3 + 27y^3$

55. $x^6 - y^6$

56. $x^6 + y^6$

For Problems 57–70, find all real number solutions for each equation.

57. $x^2 - 25 = 0$

58. $x^2 - 1 = 0$

59. $9x^2 - 49 = 0$

60. $4y^2 = 25$

61. $8x^2 - 32 = 0$

62. $3x^2 - 108 = 0$

63. $3x^3 = 3x$

64. $4x^3 = 64x$

65. $20 - 5x^2 = 0$

66. $54 - 6x^2 = 0$

67. $x^4 - 81 = 0$

68. $x^5 - x = 0$

69. $6x^3 + 24x = 0$

70. $4x^3 + 12x = 0$

For Problems 71–80, set up an equation and solve each of the following problems.

71. The cube of a number equals nine times the same number. Find the number.

72. The cube of a number equals the square of the same number. Find the number.

73. The combined area of two circles is 80π square centimeters. The length of a radius of one circle is twice the length of a radius of the other circle. Find the length of the radius of each circle.

74. The combined area of two squares is 26 square meters. The sides of the larger square are five times as long as

the sides of the smaller square. Find the dimensions of each of the squares.

75. A rectangle is twice as long as it is wide, and its area is 50 square meters. Find the length and the width of the rectangle.

76. Suppose that the length of a rectangle is one and one-third times as long as its width. The area of the rectangle is 48 square centimeters. Find the length and width of the rectangle.

77. The total surface area of a right circular cylinder is 54π square inches. If the altitude of the cylinder is twice the length of a radius, find the altitude of the cylinder.

78. The total surface area of a right circular cone is 108π square feet. If the slant height of the cone is twice the length of a radius of the base, find the length of a radius.

79. The sum of the areas of a circle and a square is $(16\pi + 64)$ square yards. If a side of the square is twice the length of a radius of the circle, find the length of a side of the square.

80. The length of an altitude of a triangle is one-third the length of the side to which it is drawn. If the area of the triangle is 6 square centimeters, find the length of that altitude.

■ ■ ■ **Thoughts into words**

81. Explain how you would solve the equation $4x^3 = 64x$.

82. What is wrong with the following factoring process?

$25x^2 - 100 = (5x + 10)(5x - 10)$

How would you correct the error?

83. Consider the following solution:

$$6x^2 - 24 = 0$$

$$6(x^2 - 4) = 0$$

$$6(x + 2)(x - 2) = 0$$

$6 = 0$ or $x + 2 = 0$ or $x - 2 = 0$

$6 = 0$ or $x = -2$ or $x = 2$

The solution set is $\{-2, 2\}$.

Is this a correct solution? Would you have any suggestion to offer the person who used this approach?

3.6 Factoring Trinomials

One of the most common types of factoring used in algebra is the expression of a trinomial as the product of two binomials. To develop a factoring technique, we first look at some multiplication ideas. Let's consider the product $(x + a)(x + b)$ and use the distributive property to show how each term of the resulting trinomial is formed.

$$(x + a)(x + b) = x(x + b) + a(x + b)$$
$$= x(x) + x(b) + a(x) + a(b)$$
$$= x^2 + (a + b)x + ab$$

Note that the coefficient of the middle term is the sum of a and b and that the last term is the product of a and b. These two relationships can be used to factor trinomials. Let's consider some examples.

EXAMPLE 1

Factor $x^2 + 8x + 12$.

Solution

We need to complete the following with two integers whose sum is 8 and whose product is 12.

$$x^2 + 8x + 12 = (x + \underline{\quad})(x + \underline{\quad})$$

The possible pairs of factors of 12 are 1(12), 2(6), and 3(4). Because $6 + 2 = 8$, we can complete the factoring as follows:

$$x^2 + 8x + 12 = (x + 6)(x + 2)$$

To check our answer, we find the product of $(x + 6)$ and $(x + 2)$. ■

EXAMPLE 2

Factor $x^2 - 10x + 24$.

Solution

We need two integers whose product is 24 and whose sum is -10. Let's use a small table to organize our thinking.

Factors	Product of the factors	Sum of the factors
$(-1)(-24)$	24	-25
$(-2)(-12)$	24	-14
$(-3)(-8)$	24	-11
$(-4)(-6)$	24	-10

The bottom line contains the numbers that we need. Thus

$$x^2 - 10x + 24 = (x - 4)(x - 6)$$

■

EXAMPLE 3

Factor $x^2 + 7x - 30$.

Solution

We need two integers whose product is -30 and whose sum is 7.

Factors	Product of the factors	Sum of the factors
$(-1)(30)$	-30	29
$(1)(-30)$	-30	-29
$(2)(-15)$	-30	-13
$(-2)(15)$	-30	13
$(-3)(10)$	-30	7

No need to search any further

The numbers that we need are -3 and 10, and we can complete the factoring.

$$x^2 + 7x - 30 = (x + 10)(x - 3)$$

■

EXAMPLE 4

Factor $x^2 + 7x + 16$.

Solution

We need two integers whose product is 16 and whose sum is 7.

Factors	Product of the factors	Sum of the factors
$(1)(16)$	16	17
$(2)(8)$	16	10
$(4)(4)$	16	8

We have exhausted all possible pairs of factors of 16 and no two factors have a sum of 7, so we conclude that $x^2 + 7x + 16$ *is not factorable using integers.*   ■

The tables in Examples 2, 3, and 4 were used to illustrate one way of organizing your thoughts for such problems. Normally you would probably factor such problems mentally without taking the time to formulate a table. Note, however, that in Example 4 the table helped us to be absolutely sure that we tried all the possibilities. Whether or not you use the table, keep in mind that the key ideas are the product and sum relationships.

E X A M P L E   5   Factor $n^2 - n - 72$.

Solution

Note that the coefficient of the middle term is -1. Hence we are looking for two integers whose product is -72, and because their sum is -1, the absolute value of the negative number must be 1 larger than the positive number. The numbers are -9 and 8, and we can complete the factoring.

$$n^2 - n - 72 = (n - 9)(n + 8)$$ ■

E X A M P L E   6   Factor $t^2 + 2t - 168$.

Solution

We need two integers whose product is -168 and whose sum is 2. Because the absolute value of the constant term is rather large, it might help to look at it in prime factored form.

$$168 = 2 \cdot 2 \cdot 2 \cdot 3 \cdot 7$$

Now we can mentally form two numbers by using all of these factors in different combinations. Using two 2s and a 3 in one number and the other 2 and the 7 in the second number produces $2 \cdot 2 \cdot 3 = 12$ and $2 \cdot 7 = 14$. The coefficient of the middle term of the trinomial is 2, so we know that we must use 14 and -12. Thus we obtain

$$t^2 + 2t - 168 = (t + 14)(t - 12)$$ ■

Trinomials of the Form $ax^2 + bx + c$

We have been factoring trinomials of the form $x^2 + bx + c$—that is, trinomials where the coefficient of the squared term is 1. Now let's consider factoring trinomials where the coefficient of the squared term is not 1. First, let's illustrate an informal trial-and-error technique that works quite well for certain types of trinomials. This technique is based on our knowledge of multiplication of binomials.

E X A M P L E   7   Factor $2x^2 + 11x + 5$.

Solution

By looking at the first term, $2x^2$, and the positive signs of the other two terms, we know that the binomials are of the form

$$(x + \underline{\hspace{1em}})(2x + \underline{\hspace{1em}})$$

Because the factors of the last term, 5, are 1 and 5, we have only the following two possibilities to try.

$$(x + 1)(2x + 5) \quad \text{or} \quad (x + 5)(2x + 1)$$

By checking the middle term formed in each of these products, we find that the second possibility yields the correct middle term of $11x$. Therefore,

$$2x^2 + 11x + 5 = (x + 5)(2x + 1)$$ ■

EXAMPLE 8   Factor $10x^2 - 17x + 3$.

Solution

First, observe that $10x^2$ can be written as $x \cdot 10x$ or $2x \cdot 5x$. Second, because the middle term of the trinomial is negative and the last term is positive, we know that the binomials are of the form

$$(x - \underline{\quad})(10x - \underline{\quad}) \quad \text{or} \quad (2x - \underline{\quad})(5x - \underline{\quad})$$

The factors of the last term, 3, are 1 and 3, so the following possibilities exist.

$$(x - 1)(10x - 3) \quad (2x - 1)(5x - 3)$$
$$(x - 3)(10x - 1) \quad (2x - 3)(5x - 1)$$

By checking the middle term formed in each of these products, we find that the product $(2x - 3)(5x - 1)$ yields the desired middle term of $-17x$. Therefore,

$$10x^2 - 17x + 3 = (2x - 3)(5x - 1)$$ ■

EXAMPLE 9   Factor $4x^2 + 6x + 9$.

Solution

The first term, $4x^2$, and the positive signs of the middle and last terms indicate that the binomials are of the form

$$(x + \underline{\quad})(4x + \underline{\quad}) \quad \text{or} \quad (2x + \underline{\quad})(2x + \underline{\quad}).$$

Because the factors of 9 are 1 and 9 or 3 and 3, we have the following five possibilities to try.

$$(x + 1)(4x + 9) \quad (2x + 1)(2x + 9)$$
$$(x + 9)(4x + 1) \quad (2x + 3)(2x + 3)$$
$$(x + 3)(4x + 3)$$

When we try all of these possibilities we find that none of them yields a middle term of $6x$. Therefore, $4x^2 + 6x + 9$ is not factorable using integers. ■

By now it is obvious that factoring trinomials of the form $ax^2 + bx + c$ can be tedious. The key idea is to organize your work so that you consider all possibilities.

We suggested one possible format in the previous three examples. As you practice such problems, you may come across a format of your own. Whatever works best for you is the right approach.

There is another, more systematic technique that you may wish to use with some trinomials. It is an extension of the technique we used at the beginning of this section. To see the basis of this technique, let's look at the following product.

$$(px + r)(qx + s) = px(qx) + px(s) + r(qx) + r(s)$$
$$= (pq)x^2 + (ps + rq)x + rs$$

Note that the product of the coefficient of the x^2 term and the constant term is $pqrs$. Likewise, the product of the two coefficients of x, ps and rq, is also $pqrs$. Therefore, when we are factoring the trinomial $(pq)x^2 + (ps + rq)x + rs$, the two coefficients of x must have a sum of $(ps) + (rq)$ and a product of $pqrs$. Let's see how this works in some examples.

E X A M P L E   1 0

Factor $6x^2 - 11x - 10$

Solution

First, multiply the coefficient of the x^2 term, 6, and the constant term, -10.

$$(6)(-10) = -60$$

Now find two integers whose sum is -11 and whose product is -60. The integers 4 and -15 satisfy these conditions.

Then rewrite the original problem, expressing the middle term as a sum of terms with these factors of -60 as their coefficients.

$$6x^2 - 11x - 10 = 6x^2 + 4x - 15x - 10$$

After rewriting the problem, we can factor by grouping — that is, factoring $2x$ from the first two terms and -5 from the last two terms.

$$6x^2 + 4x - 15x - 10 = 2x(3x + 2) - 5(3x + 2)$$

Now a common binomial factor of $(3x + 2)$ is obvious, and we can proceed as follows:

$$2x(3x + 2) - 5(3x + 2) = (3x + 2)(2x - 5)$$

Thus $6x^2 - 11x - 10 = (3x + 2)(2x - 5)$. ■

E X A M P L E   1 1

Factor $4x^2 - 29x + 30$

Solution

First, multiply the coefficient of the x^2 term, 4, and the constant term, 30.

$$(4)(30) = 120$$

Now find two integers whose sum is -29 and whose product is 120. The integers -24 and -5 satisfy these conditions.

Then rewrite the original problem, expressing the middle term as a sum of terms with these factors of 120 as their coefficients.

$$4x^2 - 29x + 30 = 4x^2 - 24x - 5x + 30$$

After rewriting the problem, we can factor by grouping — that is, factoring $4x$ from the first two terms and -5 from the last two terms.

$$4x^2 - 29x - 5x + 30 = 4x(x - 6) - 5(x - 6)$$

Now a common binomial factor of $(x - 6)$ is obvious, and we can proceed as follows:

$$4x(x - 6) - 5(x - 6) = (x - 6)(4x - 5)$$

Thus $4x^2 - 29x + 30 = (x - 6)(4x - 5)$. ■

The technique presented in Examples 10 and 11 has concrete steps to follow. Examples 7 through 9 were factored by trial-and-error technique. Both of the techniques we used have their strengths and weaknesses. Which technique to use depends on the complexity of the problem and on your personal preference. The more that you work with both techniques, the more comfortable you will feel using them.

Summary of Factoring Techniques

Before we summarize our work with factoring techniques, let's look at two more special factoring patterns. In Section 3.3 we used the following two patterns to square binomials.

$$(a + b)^2 = a^2 + 2ab + b^2 \qquad \text{and} \qquad (a - b)^2 = a^2 - 2ab + b^2$$

These patterns can also be used for factoring purposes.

$$a^2 + 2ab + b^2 = (a + b)^2 \qquad \text{and} \qquad a^2 - 2ab + b^2 = (a - b)^2$$

The trinomials on the left sides are called **perfect-square trinomials;** they are the result of squaring a binomial. We can always factor perfect-square trinomials using the usual techniques for factoring trinomials. However, they are easily recognized by the nature of their terms. For example, $4x^2 + 12x + 9$ is a perfect-square trinomial because

1. The first term is a perfect square. $(2x)^2$
2. The last term is a perfect square. $(3)^2$
3. The middle term is twice the product of the quantities $2(2x)(3)$
   being squared in the first and last terms.

Likewise, $9x^2 - 30x + 25$ is a perfect-square trinomial because

1. The first term is a perfect square. $(3x)^2$
2. The last term is a perfect square. $(5)^2$
3. The middle term is the negative of twice the product of $-2(3x)(5)$
   the quantities being squared in the first and last terms.

Once we know that we have a perfect-square trinomial, the factors follow immediately from the two basic patterns. Thus

$$4x^2 + 12x + 9 = (2x + 3)^2 \qquad 9x^2 - 30x + 25 = (3x - 5)^2$$

Here are some additional examples of perfect-square trinomials and their factored forms.

$$x^2 + 14x + 49 = (x)^2 + 2(x)(7) + (7) = (x + 7)^2$$
$$n^2 - 16n + 64 = (n)^2 - 2(n)(8) + (8)^2 = (n - 8)^2$$
$$36a^2 + 60ab + 25b^2 = (6a)^2 + 2(6a)(5b) + (5b)^2 = (6a + 5b)^2$$
$$16x^2 - 8xy + y^2 = (4x)^2 - 2(4x)(y) + (y)^2 = (4x - y)^2$$

Perhaps you will want to do this step mentally after you feel comfortable with the process.

As we have indicated, factoring is an important algebraic skill. We learned some basic factoring techniques one at a time, but you must be able to apply whichever is (or are) appropriate to the situation. Let's review the techniques and consider a variety of examples that demonstrate their use.

In this chapter, we have discussed

1. Factoring by using the distributive property to factor out a common monomial (or binomial) factor.
2. Factoring by applying the difference-of-two-squares pattern.
3. Factoring by applying the sum-of-two-cubes or the difference-of-two-cubes pattern.
4. Factoring of trinomials into the product of two binomials. (The perfect-square-trinomial pattern is a special case of this technique.)

As a general guideline, always look for a common monomial factor first and then proceed with the other techniques. Study the following examples carefully and be sure that you agree with the indicated factors.

$$2x^2 + 20x + 48 = 2(x^2 + 10x + 24) \qquad 16a^2 - 64 = 16(a^2 - 4)$$
$$= 2(x + 4)(x + 6) \qquad = 16(a + 2)(a - 2)$$

$$3x^3y^3 + 27xy = 3xy(x^2y^2 + 9) \qquad x^2 + 3x - 21 \text{ is not factorable using integers}$$

$$30n^2 - 31n + 5 = (5n - 1)(6n - 5) \qquad t^4 + 3t^2 + 2 = (t^2 + 2)(t^2 + 1)$$

$$2x^3 - 16 = 2(x^3 - 8) = 2(x - 2)(x^2 + 2x + 4)$$

PROBLEM SET 3.6

For Problems 1–56, factor completely each of the polynomials and indicate any that are not factorable using integers.

1. $x^2 + 9x + 20$

2. $x^2 + 11x + 24$

3. $x^2 - 11x + 28$

4. $x^2 - 8x + 12$

5. $a^2 + 5a - 36$

6. $a^2 + 6a - 40$

7. $y^2 + 20y + 84$

8. $y^2 + 21y + 98$

9. $x^2 - 5x - 14$

10. $x^2 - 3x - 54$

11. $x^2 + 9x + 12$

12. $35 - 2x - x^2$

13. $6 + 5x - x^2$

14. $x^2 + 8x - 24$

15. $x^2 + 15xy + 36y^2$

16. $x^2 - 14xy + 40y^2$

17. $a^2 - ab - 56b^2$

18. $a^2 + 2ab - 63b^2$

19. $15x^2 + 23x + 6$

20. $9x^2 + 30x + 16$

21. $12x^2 - x - 6$

22. $20x^2 - 11x - 3$

23. $4a^2 + 3a - 27$

24. $12a^2 + 4a - 5$

25. $3n^2 - 7n - 20$

26. $4n^2 + 7n - 15$

27. $3x^2 + 10x + 4$

28. $4n^2 - 19n + 21$

29. $10n^2 - 29n - 21$

30. $4x^2 - x + 6$

31. $8x^2 + 26x - 45$

32. $6x^2 + 13x - 33$

33. $6 - 35x - 6x^2$

34. $4 - 4x - 15x^2$

35. $20y^2 + 31y - 9$

36. $8y^2 + 22y - 21$

37. $24n^2 - 2n - 5$

38. $3n^2 - 16n - 35$

39. $5n^2 + 33n + 18$

40. $7n^2 + 31n + 12$

41. $x^2 + 25x + 150$

42. $x^2 + 21x + 108$

43. $n^2 - 36n + 320$

44. $n^2 - 26n + 168$

45. $t^2 + 3t - 180$

46. $t^2 - 2t - 143$

47. $t^4 - 5t^2 + 6$

48. $t^4 + 10t^2 + 24$

49. $10x^4 + 3x^2 - 4$

50. $3x^4 + 7x^2 - 6$

51. $x^4 - 9x^2 + 8$

52. $x^4 - x^2 - 12$

53. $18n^4 + 25n^2 - 3$

54. $4n^4 + 3n^2 - 27$

55. $x^4 - 17x^2 + 16$

56. $x^4 - 13x^2 + 36$

Problems 57–94 should help you pull together all of the factoring techniques of this chapter. Factor completely each polynomial, and indicate any that are not factorable using integers.

57. $2t^2 - 8$

58. $14w^2 - 29w - 15$

59. $12x^2 + 7xy - 10y^2$

60. $8x^2 + 2xy - y^2$

61. $18n^3 + 39n^2 - 15n$

62. $n^2 + 18n + 77$

63. $n^2 - 17n + 60$

64. $(x + 5)^2 - y^2$

65. $36a^2 - 12a + 1$

66. $2n^2 - n - 5$

67. $6x^2 + 54$

68. $x^5 - x$

69. $3x^2 + x - 5$

70. $5x^2 + 42x - 27$

71. $x^2 - (y - 7)^2$

72. $2n^3 + 6n^2 + 10n$

73. $1 - 16x^4$

74. $9a^2 - 30a + 25$

75. $4n^2 + 25n + 36$

76. $x^3 - 9x$

77. $n^3 - 49n$

78. $4x^2 + 16$

79. $x^2 - 7x - 8$

80. $x^2 + 3x - 54$

81. $3x^4 - 81x$

82. $x^3 + 125$

83. $x^4 + 6x^2 + 9$

84. $18x^2 - 12x + 2$

85. $x^4 - 5x^2 - 36$

86. $6x^4 - 5x^2 - 21$

87. $6w^2 - 11w - 35$

88. $10x^3 + 15x^2 + 20x$

89. $25n^2 + 64$

90. $4x^2 - 37x + 40$

91. $2n^3 + 14n^2 - 20n$

92. $25t^2 - 100$

93. $2xy + 6x + y + 3$

94. $3xy + 15x - 2y - 10$

■ ■ ■ **Thoughts into words**

95. How can you determine that $x^2 + 5x + 12$ is not factorable using integers?

96. Explain your thought process when factoring $30x^2 + 13x - 56$.

97. Consider the following approach to factoring $12x^2 + 54x + 60$.

$$12x^2 + 54x + 60 = (3x + 6)(4x + 10)$$
$$= 3(x + 2)(2)(2x + 5)$$
$$= 6(x + 2)(2x + 5)$$

Is this a correct factoring process? Do you have any suggestion for the person using this approach?

■ ■ ■ **Further investigations**

For Problems 98–103, factor each trinomial and assume that all variables that appear as exponents represent positive integers.

98. $x^{2a} + 2x^a - 24$

99. $x^{2a} + 10x^a + 21$

100. $6x^{2a} - 7x^a + 2$

101. $4x^{2a} + 20x^a + 25$

102. $12x^{2n} + 7x^n - 12$

103. $20x^{2n} + 21x^n - 5$

Consider the following approach to factoring $(x - 2)^2 + 3(x - 2) - 10$.

$(x - 2)^2 + 3(x - 2) - 10$

$\quad = y^2 + 3y - 10$ Replace $x - 2$ with y.

$\quad = (y + 5)(y - 2)$ Factor.

$\quad = (x - 2 + 5)(x - 2 - 2)$ Replace y with $x - 2$.

$\quad = (x + 3)(x - 4)$

Use this approach to factor Problems 104–109.

104. $(x - 3)^2 + 10(x - 3) + 24$

105. $(x + 1)^2 - 8(x + 1) + 15$

106. $(2x + 1)^2 + 3(2x + 1) - 28$

107. $(3x - 2)^2 - 5(3x - 2) - 36$

108. $6(x - 4)^2 + 7(x - 4) - 3$

109. $15(x + 2)^2 - 13(x + 2) + 2$

3.7 Equations and Problem Solving

The techniques for factoring trinomials that were presented in the previous section provide us with more power to solve equations. That is, the property "$ab = 0$ if and only if $a = 0$ or $b = 0$" continues to play an important role as we solve equations that contain factorable trinomials. Let's consider some examples.

EXAMPLE 1   Solve $x^2 - 11x - 12 = 0$.

Solution

$$x^2 - 11x - 12 = 0$$
$$(x - 12)(x + 1) = 0$$
$$x - 12 = 0 \quad \text{or} \quad x + 1 = 0$$
$$x = 12 \quad \text{or} \quad x = -1$$

The solution set is $\{-1, 12\}$.   ■

EXAMPLE 2   Solve $20x^2 + 7x - 3 = 0$.

Solution

$$20x^2 + 7x - 3 = 0$$
$$(4x - 1)(5x + 3) = 0$$
$$4x - 1 = 0 \quad \text{or} \quad 5x + 3 = 0$$
$$4x = 1 \quad \text{or} \quad 5x = -3$$
$$x = \frac{1}{4} \quad \text{or} \quad x = -\frac{3}{5}$$

The solution set is $\left\{ -\dfrac{3}{5}, \dfrac{1}{4} \right\}$.   ■

EXAMPLE 3   Solve $-2n^2 - 10n + 12 = 0$.

Solution

$$-2n^2 - 10n + 12 = 0$$
$$-2(n^2 + 5n - 6) = 0$$
$$n^2 + 5n - 6 = 0 \qquad \text{Multiply both sides by } -\frac{1}{2}.$$
$$(n + 6)(n - 1) = 0$$
$$n + 6 = 0 \quad \text{or} \quad n - 1 = 0$$
$$n = -6 \quad \text{or} \quad n = 1$$

The solution set is $\{-6, 1\}$.   ■

EXAMPLE 4   Solve $16x^2 - 56x + 49 = 0$.

Solution

$$16x^2 - 56x + 49 = 0$$
$$(4x - 7)^2 = 0$$

$$(4x - 7)(4x - 7) = 0$$

$$4x - 7 = 0 \quad \text{or} \quad 4x - 7 = 0$$

$$4x = 7 \quad \text{or} \quad 4x = 7$$

$$x = \frac{7}{4} \quad \text{or} \quad x = \frac{7}{4}$$

The only solution is $\frac{7}{4}$; thus the solution set is $\left\{ \frac{7}{4} \right\}$.   ■

EXAMPLE 5   Solve $9a(a + 1) = 4$.

Solution

$$9a(a + 1) = 4$$

$$9a^2 + 9a = 4$$

$$9a^2 + 9a - 4 = 0$$

$$(3a + 4)(3a - 1) = 0$$

$$3a + 4 = 0 \quad \text{or} \quad 3a - 1 = 0$$

$$3a = -4 \quad \text{or} \quad 3a = 1$$

$$a = -\frac{4}{3} \quad \text{or} \quad a = \frac{1}{3}$$

The solution set is $\left\{ -\frac{4}{3}, \frac{1}{3} \right\}$.   ■

EXAMPLE 6   Solve $(x - 1)(x + 9) = 11$.

Solution

$$(x - 1)(x + 9) = 11$$

$$x^2 + 8x - 9 = 11$$

$$x^2 + 8x - 20 = 0$$

$$(x + 10)(x - 2) = 0$$

$$x + 10 = 0 \quad \text{or} \quad x - 2 = 0$$

$$x = -10 \quad \text{or} \quad x = 2$$

The solution set is $\{-10, 2\}$.   ■

Problem Solving

As you might expect, the increase in our power to solve equations broadens our base for solving problems. Now we are ready to tackle some problems using equations of the types presented in this section.

P R O B L E M   1

A room contains 78 chairs. The number of chairs per row is one more than twice the number of rows. Find the number of rows and the number of chairs per row.

Solution

Let r represent the number of rows. Then $2r + 1$ represents the number of chairs per row.

$$r(2r + 1) = 78 \qquad \text{The number of rows times the number of chairs}$$
$$\text{per row yields the total number of chairs.}$$
$$2r^2 + r = 78$$
$$2r^2 + r - 78 = 0$$
$$(2r + 13)(r - 6) = 0$$
$$2r + 13 = 0 \qquad \text{or} \qquad r - 6 = 0$$
$$2r = -13 \qquad \text{or} \qquad r = 6$$
$$r = -\frac{13}{2} \qquad \text{or} \qquad r = 6$$

The solution $-\dfrac{13}{2}$ must be disregarded, so there are 6 rows and $2r + 1$ or $2(6) + 1 = 13$ chairs per row. ■

P R O B L E M   2

A strip of uniform width cut from both sides and both ends of an 8-inch by 11-inch sheet of paper reduces the size of the paper to an area of 40 square inches. Find the width of the strip.

Solution

Let x represent the width of the strip, as indicated in Figure 3.19.

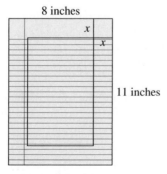

8 inches

x

x

11 inches

Figure 3.19

The length of the paper after the strips of width x are cut from both ends and both sides will be $11 - 2x$, and the width of the newly formed rectangle will be

$8 - 2x$. Because the area ($A = lw$) is to be 40 square inches, we can set up and solve the following equation.

$$(11 - 2x)(8 - 2x) = 40$$

$$88 - 38x + 4x^2 = 40$$

$$4x^2 - 38x + 48 = 0$$

$$2x^2 - 19x + 24 = 0$$

$$(2x - 3)(x - 8) = 0$$

$$2x - 3 = 0 \quad \text{or} \quad x - 8 = 0$$

$$2x = 3 \quad \text{or} \quad x = 8$$

$$x = \frac{3}{2} \quad \text{or} \quad x = 8$$

The solution of 8 must be discarded because the width of the original sheet is only 8 inches. Therefore, the strip to be cut from all four sides must be $1\frac{1}{2}$ inches wide. (Check this answer!) ▮

The Pythagorean theorem, an important theorem pertaining to right triangles, can sometimes serve as a guideline for solving problems that deal with right triangles (see Figure 3.20). The Pythagorean theorem states that "in any right triangle, the square of the longest side (called the hypotenuse) is equal to the sum of the squares of the other two sides (called legs)." Let's use this relationship to help solve a problem.

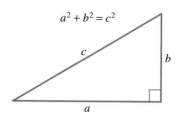

Figure 3.20

PROBLEM 3

One leg of a right triangle is 2 centimeters more than twice as long as the other leg. The hypotenuse is 1 centimeter longer than the longer of the two legs. Find the lengths of the three sides of the right triangle.

Solution

Let l represent the length of the shortest leg. Then $2l + 2$ represents the length of the other leg, and $2l + 3$ represents the length of the hypotenuse. Use the Pythagorean theorem as a guideline to set up and solve the following equation.

$$l^2 + (2l + 2)^2 = (2l + 3)^2$$

$$l^2 + 4l^2 + 8l + 4 = 4l^2 + 12l + 9$$

$$l^2 - 4l - 5 = 0$$

$$(l - 5)(l + 1) = 0$$

$$l - 5 = 0 \quad \text{or} \quad l + 1 = 0$$

$$l = 5 \quad \text{or} \quad l = -1$$

The negative solution must be discarded, so the length of one leg is 5 centimeters; the other leg is $2(5) + 2 = 12$ centimeters long, and the hypotenuse is $2(5) + 3 = 13$ centimeters long.   ∎

PROBLEM SET 3.7

For Problems 1–54, solve each equation. You will need to use factoring techniques that we discussed throughout this chapter.

1. $x^2 + 4x + 3 = 0$

2. $x^2 + 7x + 10 = 0$

3. $x^2 + 18x + 72 = 0$

4. $n^2 + 20n + 91 = 0$

5. $n^2 - 13n + 36 = 0$

6. $n^2 - 10n + 16 = 0$

7. $x^2 + 4x - 12 = 0$

8. $x^2 + 7x - 30 = 0$

9. $w^2 - 4w = 5$

10. $s^2 - 4s = 21$

11. $n^2 + 25n + 156 = 0$

12. $n(n - 24) = -128$

13. $3t^2 + 14t - 5 = 0$

14. $4t^2 - 19t - 30 = 0$

15. $6x^2 + 25x + 14 = 0$

16. $25x^2 + 30x + 8 = 0$

17. $3t(t - 4) = 0$

18. $1 - x^2 = 0$

19. $-6n^2 + 13n - 2 = 0$

20. $(x + 1)^2 - 4 = 0$

21. $2n^3 = 72n$

22. $a(a - 1) = 2$

23. $(x - 5)(x + 3) = 9$

24. $3w^3 - 24w^2 + 36w = 0$

25. $16 - x^2 = 0$

26. $16t^2 - 72t + 81 = 0$

27. $n^2 + 7n - 44 = 0$

28. $2x^3 = 50x$

29. $3x^2 = 75$

30. $x^2 + x - 2 = 0$

31. $15x^2 + 34x + 15 = 0$

32. $20x^2 + 41x + 20 = 0$

33. $8n^2 - 47n - 6 = 0$

34. $7x^2 + 62x - 9 = 0$

35. $28n^2 - 47n + 15 = 0$

36. $24n^2 - 38n + 15 = 0$

37. $35n^2 - 18n - 8 = 0$

38. $8n^2 - 6n - 5 = 0$

39. $-3x^2 - 19x + 14 = 0$

40. $5x^2 = 43x - 24$

41. $n(n + 2) = 360$

42. $n(n + 1) = 182$

43. $9x^4 - 37x^2 + 4 = 0$

44. $4x^4 - 13x^2 + 9 = 0$

45. $3x^2 - 46x - 32 = 0$

46. $x^4 - 9x^2 = 0$

47. $2x^2 + x - 3 = 0$

48. $x^3 + 5x^2 - 36x = 0$

49. $12x^3 + 46x^2 + 40x = 0$

50. $5x(3x - 2) = 0$

51. $(3x - 1)^2 - 16 = 0$

52. $(x + 8)(x - 6) = -24$

53. $4a(a + 1) = 3$

54. $-18n^2 - 15n + 7 = 0$

For Problems 55–70, set up an equation and solve each problem.

55. Find two consecutive integers whose product is 72.

56. Find two consecutive even whole numbers whose product is 224.

57. Find two integers whose product is 105 such that one of the integers is one more than twice the other integer.

58. Find two integers whose product is 104 such that one of the integers is three less than twice the other integer.

59. The perimeter of a rectangle is 32 inches and the area is 60 square inches. Find the length and width of the rectangle.

60. Suppose that the length of a certain rectangle is two centimeters more than three times its width. If the area of the rectangle is 56 square centimeters, find its length and width.

61. The sum of the squares of two consecutive integers is 85. Find the integers.

62. The sum of the areas of two circles is 65π square feet. The length of a radius of the larger circle is 1 foot less than twice the length of a radius of the smaller circle. Find the length of a radius of each circle.

63. The combined area of a square and a rectangle is 64 square centimeters. The width of the rectangle is 2 centimeters more than the length of a side of the square, and the length of the rectangle is 2 centimeters more than its width. Find the dimensions of the square and the rectangle.

64. The Ortegas have an apple orchard that contains 90 trees. The number of trees in each row is 3 more than twice the number of rows. Find the number of rows and the number of trees per row.

65. The lengths of the three sides of a right triangle are represented by consecutive whole numbers. Find the lengths of the three sides.

66. The area of the floor of the rectangular room shown in Figure 3.21 is 175 square feet. The length of the room is $1\frac{1}{2}$ feet longer than the width. Find the length of the room.

67. Suppose that the length of one leg of a right triangle is 3 inches more than the length of the other leg. If the length of the hypotenuse is 15 inches, find the lengths of the two legs.

68. The lengths of the three sides of a right triangle are represented by consecutive even whole numbers. Find the lengths of the three sides.

69. The area of a triangular sheet of paper is 28 square inches. One side of the triangle is 2 inches more than three times the length of the altitude to that side. Find the length of that side and the altitude to the side.

70. A strip of uniform width is shaded along both sides and both ends of a rectangular poster that measures 12 inches by 16 inches (see Figure 3.22). How wide is the shaded strip if one-half of the poster is shaded?

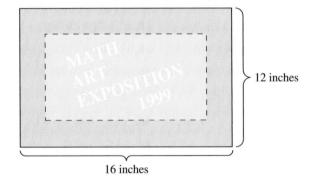

12 inches

16 inches

Figure 3.22

Area = 175 square feet

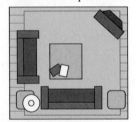

Figure 3.21

■ ■ ■ **Thoughts into words**

71. Discuss the role that factoring plays in solving equations.

72. Explain how you would solve the equation $(x + 6)(x - 4) = 0$ and also how you would solve $(x + 6)(x - 4) = -16$.

73. Explain how you would solve the equation $3(x - 1)(x + 2) = 0$ and also how you would solve the equation $x(x - 1)(x + 2) = 0$.

74. Consider the following two solutions for the equation $(x + 3)(x - 4) = (x + 3)(2x - 1)$.

Solution A

$$(x + 3)(x - 4) = (x + 3)(2x - 1)$$

$$(x + 3)(x - 4) - (x + 3)(2x - 1) = 0$$

$$(x + 3)[x - 4 - (2x - 1)] = 0$$

$$(x + 3)(x - 4 - 2x + 1) = 0$$

$$(x + 3)(-x - 3) = 0$$

$$x + 3 = 0 \quad \text{or} \quad -x - 3 = 0$$

$$x = -3 \quad \text{or} \quad -x = 3$$

$$x = -3 \quad \text{or} \quad x = -3$$

The solution set is $\{-3\}$.

Solution B

$$(x + 3)(x - 4) = (x + 3)(2x - 1)$$

$$x^2 - x - 12 = 2x^2 + 5x - 3$$

$$0 = x^2 + 6x + 9$$

$$0 = (x + 3)^2$$

$$x + 3 = 0$$

$$x = -3$$

The solution set is $\{-3\}$.

Are both approaches correct? Which approach would you use, and why?

SUMMARY

(3.1) A **term** is an indicated product and may contain any number of factors. The variables involved in a term are called **literal factors,** and the numerical factor is called the **numerical coefficient.** Terms that contain variables with only nonnegative integers as exponents are called **monomials.** The **degree** of a monomial is the sum of the exponents of the literal factors.

A **polynomial** is a monomial or a finite sum (or difference) of monomials. We classify polynomials as follows:

Polynomial with one term $\longrightarrow$ Monomial
Polynomial with two terms $\longrightarrow$ Binomial
Polynomial with three terms $\longrightarrow$ Trinomial

Similar terms, or **like terms,** have the same literal factors. The commutative, associative, and distributive properties provide the basis for rearranging, regrouping, and combining similar terms.

(3.2) The following properties provide the basis for multiplying and dividing monomials.

  1. $b^n \cdot b^m = b^{n+m}$

  2. $(b^n)^m = b^{mn}$

  3. $(ab)^n = a^n b^n$

  4. (a) $\dfrac{b^n}{b^m} = b^{n-m}, \quad if \ n > m$

   (b) $\dfrac{b^n}{b^m} = 1, \quad if \ n = m$

(3.3) The commutative and associative properties, the properties of exponents, and the distributive property work together to form a basis for multiplying polynomials. The following can be used as multiplication patterns.

$$(a + b)^2 = a^2 + 2ab + b^2$$
$$(a - b)^2 = a^2 - 2ab + b^2$$
$$(a + b)(a - b) = a^2 - b^2$$
$$(a + b)^3 = a^3 + 3a^2b + 3ab^2 + b^3$$
$$(a - b)^3 = a^3 - 3a^2b + 3ab^2 - b^3$$

(3.4) If a positive integer greater than 1 has no factors that are positive integers other than itself and 1, then it is

called a **prime number.** A positive integer greater than 1 that is not a prime number is called a **composite number.**

The indicated product form that contains only prime factors is called the **prime factorization form** of a number.

An expression such as $ax + bx + ay + by$ can be factored as follows:

$$ax + bx + ay + by = x(a + b) + y(a + b)$$
$$= (a + b)(x + y)$$

This is called **factoring by grouping.**

The distributive property in the form $ab + ac = a(b + c)$ is the basis for **factoring out the highest common monomial factor.**

Expressing polynomials in factored form, and then applying the property $ab = 0$ if and only if $a = 0$ or $b = 0$, provides us with another technique for solving equations.

(3.5) The factoring pattern

$$a^2 - b^2 = (a + b)(a - b)$$

is called the **difference of two squares.**

The difference-of-two-squares factoring pattern, along with the property $ab = 0$ if and only if $a = 0$ or $b = 0$, provides us with another technique for solving equations. The factoring patterns

$$a^3 + b^3 = (a + b)(a^2 - ab + b^2) \qquad and$$
$$a^3 - b^3 = (a - b)(a^2 + ab + b^2)$$

are called the **sum and difference of two cubes.**

(3.6) Expressing a trinomial (for which the coefficient of the squared term is 1) as a product of two binomials is based on the relationship

$$(x + a)(x + b) = x^2 + (a + b)x + ab$$

The coefficient of the middle term is the sum of a and b, and the last term is the product of a and b.

If the coefficient of the squared term of a trinomial does not equal 1, then the following relationship holds.

$$(px + r)(qx + s) = (pq)x^2 + (ps + rq)x + rs$$

The two coefficients of x, ps and rq, must have a sum of $(ps) + (rq)$ and a product of $pqrs$. Thus, to factor something like $6x^2 + 7x - 3$, we need to find two integers whose product is $6(-3) = -18$ and whose sum is 7. The integers are 9 and -2, and we can factor as follows:

$$6x^2 + 7x - 3 = 6x^2 + 9x - 2x - 3$$
$$= 3x(2x + 3) - 1(2x + 3)$$
$$= (2x + 3)(3x - 1)$$

A **perfect-square trinomial** is the result of squaring a binomial. There are two basic perfect-square trinomial factoring patterns.

$$a^2 + 2ab + b^2 = (a + b)^2$$
$$a^2 - 2ab + b^2 = (a - b)^2$$

(3.7) The factoring techniques we discussed in this chapter, along with the property $ab = 0$ if and only if $a = 0$ or $b = 0$, provide the basis for expanding our repertoire of equation-solving processes.

The ability to solve more types of equations increases our capabilities for problem solving.

CHAPTER 3   REVIEW PROBLEM SET

For Problems 1–23, perform the indicated operations and simplify each of the following.

1. $(3x - 2) + (4x - 6) + (-2x + 5)$

2. $(8x^2 + 9x - 3) - (5x^2 - 3x - 1)$

3. $(6x^2 - 2x - 1) + (4x^2 + 2x + 5) - (-2x^2 + x - 1)$

4. $(-5x^2y^3)(4x^3y^4)$ 5. $(-2a^2)(3ab^2)(a^2b^3)$

6. $5a^2(3a^2 - 2a - 1)$ 7. $(4x - 3y)(6x + 5y)$

8. $(x + 4)(3x^2 - 5x - 1)$ 9. $(4x^2y^3)^4$

10. $(3x - 2y)^2$ 11. $(-2x^2y^3z)^3$

12. $\dfrac{-39x^3y^4}{3xy^3}$

13. $[3x - (2x - 3y + 1)] - [2y - (x - 1)]$

14. $(x^2 - 2x - 5)(x^2 + 3x - 7)$

15. $(7 - 3x)(3 + 5x)$ 16. $-(3ab)(2a^2b^3)^2$

17. $\left(\dfrac{1}{2}ab\right)(8a^3b^2)(-2a^3)$ 18. $(7x - 9)(x + 4)$

19. $(3x + 2)(2x^2 - 5x + 1)$ 20. $(3x^{n+1})(2x^{3n-1})$

21. $(2x + 5y)^2$ 22. $(x - 2)^3$

23. $(2x + 5)^3$

For Problems 24–45, factor each polynomial completely. Indicate any that are not factorable using integers.

24. $x^2 + 3x - 28$ 25. $2t^2 - 18$

26. $4n^2 + 9$ 27. $12n^2 - 7n + 1$

28. $x^6 - x^2$ 29. $x^3 - 6x^2 - 72x$

30. $6a^3b + 4a^2b^2 - 2a^2bc$ 31. $x^2 - (y - 1)^2$

32. $8x^2 + 12$ 33. $12x^2 + x - 35$

34. $16n^2 - 40n + 25$ 35. $4n^2 - 8n$

36. $3w^3 + 18w^2 - 24w$ 37. $20x^2 + 3xy - 2y^2$

38. $16a^2 - 64a$ 39. $3x^3 - 15x^2 - 18x$

40. $n^2 - 8n - 128$ 41. $t^4 - 22t^2 - 75$

42. $35x^2 - 11x - 6$ 43. $15 - 14x + 3x^2$

44. $64n^3 - 27$ 45. $16x^3 + 250$

For Problems 46–65, solve each equation.

46. $4x^2 - 36 = 0$ 47. $x^2 + 5x - 6 = 0$

48. $49n^2 - 28n + 4 = 0$ 49. $(3x - 1)(5x + 2) = 0$

50. $(3x - 4)^2 - 25 = 0$ 51. $6a^3 = 54a$

52. $x^5 = x$ 53. $-n^2 + 2n + 63 = 0$

54. $7n(7n + 2) = 8$

55. $30w^2 - w - 20 = 0$

56. $5x^4 - 19x^2 - 4 = 0$

57. $9n^2 - 30n + 25 = 0$

58. $n(2n + 4) = 96$

59. $7x^2 + 33x - 10 = 0$

60. $(x + 1)(x + 2) = 42$

61. $x^2 + 12x - x - 12 = 0$

62. $2x^4 + 9x^2 + 4 = 0$

63. $30 - 19x - 5x^2 = 0$

64. $3t^3 - 27t^2 + 24t = 0$

65. $-4n^2 - 39n + 10 = 0$

For Problems 66–75, set up an equation and solve each problem.

66. Find three consecutive integers such that the product of the smallest and the largest is one less than 9 times the middle integer.

67. Find two integers whose sum is 2 and whose product is −48.

68. Find two consecutive odd whole numbers whose product is 195.

69. Two cars leave an intersection at the same time, one traveling north and the other traveling east. Some time later, they are 20 miles apart, and the car going east has traveled 4 miles farther than the other car. How far has each car traveled?

70. The perimeter of a rectangle is 32 meters and its area is 48 square meters. Find the length and width of the rectangle.

71. A room contains 144 chairs. The number of chairs per row is two less than twice the number of rows. Find the number of rows and the number of chairs per row.

72. The area of a triangle is 39 square feet. The length of one side is 1 foot more than twice the altitude to that side. Find the length of that side and the altitude to the side.

73. A rectangular-shaped pool 20 feet by 30 feet has a sidewalk of uniform width around the pool (see Figure 3.23). The area of the sidewalk is 336 square feet. Find the width of the sidewalk.

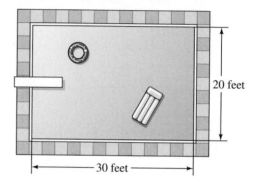

20 feet

30 feet

Figure 3.23

74. The sum of the areas of two squares is 89 square centimeters. The length of a side of the larger square is 3 centimeters more than the length of a side of the smaller square. Find the dimensions of each square.

75. The total surface area of a right circular cylinder is 32π square inches. If the altitude of the cylinder is three times the length of a radius, find the altitude of the cylinder.

CHAPTER 3

TEST

For Problems 1–8, perform the indicated operations and simplify each expression.

1. $(-3x - 1) + (9x - 2) - (4x + 8)$

2. $(-6xy^2)(8x^3y^2)$

3. $(-3x^2y^4)^3$
  4. $(5x - 7)(4x + 9)$

5. $(3n - 2)(2n - 3)$
  6. $(x - 4y)^3$

7. $(x + 6)(2x^2 - x - 5)$
  8. $\dfrac{-70x^4y^3}{5xy^2}$

For Problems 9–14, factor each expression completely.

9. $6x^2 + 19x - 20$
  10. $12x^2 - 3$

11. $64 + t^3$
  12. $30x + 4x^2 - 16x^3$

13. $x^2 - xy + 4x - 4y$
  14. $24n^2 + 55n - 24$

For Problems 15–22, solve each equation.

15. $x^2 + 8x - 48 = 0$
  16. $4n^2 = n$

17. $4x^2 - 12x + 9 = 0$

18. $(n - 2)(n + 7) = -18$

19. $3x^3 + 21x^2 - 54x = 0$

20. $12 + 13x - 35x^2 = 0$

21. $n(3n - 5) = 2$
  22. $9x^2 - 36 = 0$

For Problems 23–25, set up an equation and solve each problem.

23. The perimeter of a rectangle is 30 inches and its area is 54 square inches. Find the length of the longest side of the rectangle.

24. A room contains 105 chairs arranged in rows. The number of rows is one more than twice the number of chairs per row. Find the number of rows.

25. The combined area of a square and a rectangle is 57 square feet. The width of the rectangle is 3 feet more than the length of a side of the square, and the length of the rectangle is 5 feet more than the length of a side of the square. Find the length of the rectangle.

C U M U L A T I V E   R E V I E W   P R O B L E M   S E T   *Chapters 1-3*

For Problems 1–10, evaluate each algebraic expression for the given values of the variables. Don't forget that in some cases it may be helpful to simplify the algebraic expression before evaluating it.

1. $x^2 - 2xy + y^2$ for $x = -2$ and $y = -4$

2. $-n^2 + 2n - 4$ for $n = -3$

3. $2x^2 - 5x + 6$ for $x = 3$

4. $3(2x - 1) - 2(x + 4) - 4(2x - 7)$ for $x = -1$

5. $-(2n - 1) + 5(2n - 3) - 6(3n + 4)$ for $n = 4$

6. $2(a - 4) - (a - 1) + (3a - 6)$ for $a = -5$

7. $(3x^2 - 4x - 7) - (4x^2 - 7x + 8)$ for $x = -4$

8. $-2(3x - 5y) - 4(x + 2y) + 3(-2x - 3y)$ for $x = 2$ and $y = -3$

9. $5(-x^2 - x + 3) - (2x^2 - x + 6) - 2(x^2 + 4x - 6)$ for $x = 2$

10. $3(x^2 - 4xy + 2y^2) - 2(x^2 - 6xy - y^2)$ for $x = -5$ and $y = -2$

For Problems 11–18, perform the indicated operations and express your answers in simplest form.

11. $4(3x - 2) - 2(4x - 1) - (2x + 5)$

12. $(-6ab^2)(2ab)(-3b^3)$

13. $(5x - 7)(6x + 1)$

14. $(-2x - 3)(x + 4)$

15. $(-4a^2b^3)^3$

16. $(x + 2)(5x - 6)(x - 2)$

17. $(x - 3)(x^2 - x - 4)$

18. $(x^2 + x + 4)(2x^2 - 3x - 7)$

For Problems 19–38, factor each of the algebraic expressions completely.

19. $7x^2 - 7$

20. $4a^2 - 4ab + b^2$

21. $3x^2 - 17x - 56$

22. $1 - x^3$

23. $xy - 5x + 2y - 10$

24. $3x^2 - 24x + 48$

25. $4n^4 - n^2 - 3$

26. $32x^4 + 108x$

27. $4x^2 + 36$

28. $6x^2 + 5x - 4$

29. $9x^2 - 30x + 25$

30. $2x^2 + 6xy + x + 3y$

31. $8a^3 + 27b^3$

32. $x^4 - 16$

33. $10m^4n^2 - 2m^3n^3 - 4m^2n^4$

34. $5x(2y + 7z) - 12(2y + 7z)$

35. $3x^2 - x - 10$

36. $25 - 4a^2$

37. $36x^2 + 60x + 25$

38. $64y^3 + 1$

For Problems 39–42, solve each equation for the indicated variable.

39. $5x - 2y = 6$ for x

40. $3x + 4y = 12$ for y

41. $V = 2\pi rh + 2\pi r^2$ for h

42. $\dfrac{1}{R} = \dfrac{1}{R_1} + \dfrac{1}{R_2}$ for R_1

43. Solve $A = P + Prt$ for r, given that $A = \$4997$, $P = \$3800$, and $t = 3$ years.

44. Solve $C = \dfrac{5}{9}(F - 32)$ for C, given that F = 5°.

For Problems 45–62, solve each of the equations.

45. $(x - 2)(x + 5) = 8$

46. $(5n - 2)(3n + 7) = 0$

47. $-2(n - 1) + 3(2n + 1) = -11$

48. $x^2 + 7x - 18 = 0$

49. $8x^2 - 8 = 0$

50. $\dfrac{3}{4}(x - 2) - \dfrac{2}{5}(2x - 3) = \dfrac{1}{5}$

51. $0.1(x - 0.1) - 0.4(x + 2) = -5.31$

52. $\dfrac{2x - 1}{2} - \dfrac{5x + 2}{3} = 3$

53. $|3n - 2| = 7$

54. $|2x - 1| = |x + 4|$

55. $0.08(x + 200) = 0.07x + 20$

56. $2x^2 - 12x - 80 = 0$

57. $x^3 = 16x$

58. $x(x + 2) - 3(x + 2) = 0$

59. $-12n^2 + 5n + 2 = 0$

60. $3y(y + 1) = 90$

61. $2x^3 + 6x^2 - 20x = 0$

62. $(3n - 1)(2n + 3) = (n + 4)(6n - 5)$

For Problems 63–70, solve each of the inequalities.

63. $-5(3n + 4) < -2(7n - 1)$

64. $7(x + 1) - 8(x - 2) < 0$

65. $|2x - 1| > 7$

66. $|3x + 7| < 14$

67. $0.09x + 0.1(x + 200) > 77$

68. $\dfrac{2x - 1}{4} - \dfrac{x - 2}{6} \leq \dfrac{3}{8}$

69. $-(x - 1) + 2(3x - 1) \geq 2(x + 4) - (x - 1)$

70. $\dfrac{1}{4}(x - 2) + \dfrac{3}{7}(2x - 1) < \dfrac{3}{14}$

For Problems 71–84, solve each problem by setting up and solving an appropriate equation or inequality.

71. Find three consecutive odd integers such that three times the first minus the second is one more than the third.

72. Inez has a collection of 48 coins consisting of nickels, dimes, and quarters. The number of dimes is one less than twice the number of nickels, and the number of quarters is ten greater than the number of dimes. How many coins of each denomination are there in the collection?

73. The sum of the present ages of Joey and his mother is 46 years. In 4 years, Joey will be 3 years less than one-half as old as his mother at that time. Find the present ages of Joey and his mother.

74. The difference of the measures of two supplementary angles is 56°. Find the measure of each angle.

75. Norm invested a certain amount of money at 8% interest and $200 more than that amount at 9%. His total yearly interest was $86. How much did he invest at each rate?

76. Sanchez has a collection of pennies, nickels, and dimes worth $9.35. He has five more nickels than pennies and twice as many dimes as pennies. How may coins of each kind does he have?

77. Sandy starts off with her bicycle at 8 miles per hour. Fifty minutes later, Billie starts riding along the same route at 12 miles per hour. How long will it take Billie to overtake Sandy?

78. How many milliliters of pure acid must be added to 150 milliliters of a 30% solution of acid to obtain a 40% solution?

79. A retailer has some carpet that cost him $18.00 a square yard. If he sells it for $30 a square yard, what is his rate of profit based on the selling price?

80. Brad had scores of 88, 92, 93, and 89 on his first four algebra tests. What score must he obtain on the fifth test to have an average better than 90 for the five tests?

81. Suppose that the area of a square is one-half the area of a triangle. One side of the triangle is 16 inches long, and the altitude to that side is the same length as a side of the square. Find the length of a side of the square.

82. A rectangle is twice as long as it is wide, and its area is 98 square meters. Find the length and width of the rectangle.

83. A room contains 96 chairs. The number of chairs per row is four more than the number of rows. Find the number of rows and the number of chairs per row.

84. One leg of a right triangle is 3 feet longer than the other leg. The hypotenuse is 3 feet longer than the longer leg. Find the lengths of the three sides of the right triangle.

Rational Expressions

Computers often work together to compile large processing jobs. Rational numbers are used to express the rate of the processing speed of a computer.

AP/Wide World Photos

It takes Pat 12 hours to complete a task. After he had been working on this task for 3 hours, he was joined by his brother, Liam, and together they finished the job in 5 hours. How long would it take Liam to do the job by himself? We can use the **fractional equation** $\dfrac{5}{12} + \dfrac{5}{h} = \dfrac{3}{4}$ to determine that Liam could do the entire job by himself in 15 hours.

Rational expressions are to algebra what rational numbers are to arithmetic. Most of the work we will do with rational expressions in this chapter parallels the work you have previously done with arithmetic fractions. The same basic properties we use to explain reducing, adding, subtracting, multiplying, and dividing arithmetic fractions will serve as a basis for our work with rational expressions. The techniques of factoring that we studied in Chapter 3 will also play an important role in our discussions. At the end of this chapter, we will work with some fractional equations that contain rational expressions.

InfoTrac Project   Do a subject guide search on pi. Find a periodical article called "Record-Breaking Pi."
Write a short summary of the article. To do the task, the work was divided among several
computers. Suppose three computers worked on the project together for 85 hours. If the first
computer would need 255 hours to complete the job alone and the second computer would need
340 hours to do the job alone, how long would it take the third computer to do the job alone?
(*Hint*: After you write your equation, reduce your fractions, if possible.)

4.1   Simplifying Rational Expressions

We reviewed the basic operations with rational numbers in an informal setting in
Chapter 1. In this review, we relied primarily on your knowledge of arithmetic. At
this time, we want to become a little more formal with our review so that we can
use the work with rational numbers as a basis for operating with rational expressions. We will define a rational expression shortly.

You will recall that any number that can be written in the form $\dfrac{a}{b}$, where a and
b are integers and $b \neq 0$, is called a rational number. The following are examples
of rational numbers.

$$\frac{1}{2} \qquad \frac{3}{4} \qquad \frac{15}{7} \qquad \frac{-5}{6} \qquad \frac{7}{-8} \qquad \frac{-12}{-17}$$

Numbers such as $6, -4, 0, 4\dfrac{1}{2}, 0.7$, and 0.21 are also rational because we can express
them as the indicated quotient of two integers. For example,

$$6 = \frac{6}{1} = \frac{12}{2} = \frac{18}{3} \quad \text{and so on} \qquad\qquad 4\frac{1}{2} = \frac{9}{2}$$

$$-4 = \frac{4}{-1} = \frac{-4}{1} = \frac{8}{-2} \quad \text{and so on} \qquad 0.7 = \frac{7}{10}$$

$$0 = \frac{0}{1} = \frac{0}{2} = \frac{0}{3} \quad \text{and so on} \qquad\qquad 0.21 = \frac{21}{100}$$

Because a rational number is the quotient of two integers, our previous work with
division of integers can help us understand the various forms of rational numbers.
If the signs of the numerator and denominator are different, then the rational number is negative. If the signs of the numerator and denominator are the same, then
the rational number is positive. The next examples and Property 4.1 show
the equivalent forms of rational numbers. Generally, it is preferred to express the
denominator of a rational number as a positive integer.

$$\frac{8}{-2} = \frac{-8}{2} = -\frac{8}{2} = -4 \qquad \frac{12}{3} = \frac{-12}{-3} = 4$$

Observe the following general properties.

> ### PROPERTY 4.1
>
> **1.** $\dfrac{-a}{b} = \dfrac{a}{-b} = -\dfrac{a}{b}$,   where $b \neq 0$
>
> **2.** $\dfrac{-a}{-b} = \dfrac{a}{b}$,   where $b \neq 0$

Therefore, a rational number such as $\dfrac{-2}{5}$ can also be written as $\dfrac{2}{-5}$ or $-\dfrac{2}{5}$.

We use the following property, often referred to as the **fundamental principle of fractions,** to reduce fractions to lowest terms or express fractions in simplest or reduced form.

> ### PROPERTY 4.2
>
> If b and k are nonzero integers and a is any integer, then
>
> $$\frac{a \cdot k}{b \cdot k} = \frac{a}{b}$$

Let's apply Properties 4.1 and 4.2 to the following examples.

E X A M P L E   1   Reduce $\dfrac{18}{24}$ to lowest terms.

Solution

$$\frac{18}{24} = \frac{3 \cdot 6}{4 \cdot 6} = \frac{3}{4}$$

E X A M P L E   2   Change $\dfrac{40}{48}$ to simplest form.

Solution

$$\frac{\overset{5}{\cancel{40}}}{\underset{6}{\cancel{48}}} = \frac{5}{6} \qquad \text{A common factor of 8 was divided out of both numerator and denominator.}$$

E X A M P L E   3   Express $\dfrac{-36}{63}$ in reduced form.

Solution

$$\frac{-36}{63} = -\frac{36}{63} = -\frac{4 \cdot 9}{7 \cdot 9} = -\frac{4}{7}$$

EXAMPLE 4

Reduce $\dfrac{72}{-90}$ to simplest form.

Solution

$$\frac{72}{-90} = -\frac{72}{90} = -\frac{2 \cdot 2 \cdot 2 \cdot 3 \cdot 3}{2 \cdot 3 \cdot 3 \cdot 5} = -\frac{4}{5}$$

Note the different terminology used in Examples 1– 4. Regardless of the terminology, keep in mind that the number is not being changed, but the form of the numeral representing the number is being changed. In Example 1, $\dfrac{18}{24}$ and $\dfrac{3}{4}$ are equivalent fractions; they name the same number. Also note the use of prime factors in Example 4.

Rational Expressions

A **rational expression** is the indicated quotient of two polynomials. The following are examples of rational expressions.

$$\frac{3x^2}{5} \qquad \frac{x - 2}{x + 3} \qquad \frac{x^2 + 5x - 1}{x^2 - 9} \qquad \frac{xy^2 + x^2y}{xy} \qquad \frac{a^3 - 3a^2 - 5a - 1}{a^4 + a^3 + 6}$$

Because we must avoid division by zero, no values that create a denominator of zero can be assigned to variables. Thus the rational expression $\dfrac{x - 2}{x + 3}$ is meaningful for all values of x except $x = -3$. Rather than making restrictions for each individual expression, we will merely assume that all denominators represent nonzero real numbers.

Property 4.2 $\left(\dfrac{a \cdot k}{b \cdot k} = \dfrac{a}{b} \right)$ serves as the basis for simplifying rational expressions, as the next examples illustrate.

EXAMPLE 5

Simplify $\dfrac{15xy}{25y}$.

Solution

$$\frac{15xy}{25y} = \frac{3 \cdot 5 \cdot x \cdot y}{5 \cdot 5 \cdot y} = \frac{3x}{5}$$

EXAMPLE 6

Simplify $\dfrac{-9}{18x^2y}$.

Solution

$$\frac{-9}{18x^2y} = -\frac{\overset{1}{\cancel{9}}}{\underset{2}{\cancel{18}}x^2y} = -\frac{1}{2x^2y}$$

A common factor of 9 was divided out of numerator and denominator.

EXAMPLE 7 Simplify $\dfrac{-28a^2b^2}{-63a^2b^3}$.

Solution

$$\frac{-28a^2b^2}{-63a^2b^3} = \frac{4 \cdot 7 \cdot a^2 \cdot b^2}{9 \cdot 7 \cdot a^2 \cdot b^3} = \frac{4}{9b}$$

The factoring techniques from Chapter 3 can be used to factor numerators and/or denominators so that we can apply the property $\dfrac{a \cdot k}{b \cdot k} = \dfrac{a}{b}$. Examples 8–12 should clarify this process.

EXAMPLE 8 Simplify $\dfrac{x^2 + 4x}{x^2 - 16}$.

Solution

$$\frac{x^2 + 4x}{x^2 - 16} = \frac{x(x + 4)}{(x - 4)(x + 4)} = \frac{x}{x - 4}$$

EXAMPLE 9 Simplify $\dfrac{4a^2 + 12a + 9}{2a + 3}$.

Solution

$$\frac{4a^2 + 12a + 9}{2a + 3} = \frac{(2a + 3)(2a + 3)}{1(2a + 3)} = \frac{2a + 3}{1} = 2a + 3$$

EXAMPLE 10 Simplify $\dfrac{5n^2 + 6n - 8}{10n^2 - 3n - 4}$.

Solution

$$\frac{5n^2 + 6n - 8}{10n^2 - 3n - 4} = \frac{(5n - 4)(n + 2)}{(5n - 4)(2n + 1)} = \frac{n + 2}{2n + 1}$$

EXAMPLE 11 Simplify $\dfrac{6x^3y - 6xy}{x^3 + 5x^2 + 4x}$.

Solution

$$\frac{6x^3y - 6xy}{x^3 + 5x^2 + 4x} = \frac{6xy(x^2 - 1)}{x(x^2 + 5x + 4)} = \frac{6xy(x + 1)(x - 1)}{x(x + 1)(x + 4)} = \frac{6y(x - 1)}{x + 4}$$

Note that in Example 11 we left the numerator of the final fraction in factored form. This is often done if expressions other than monomials are involved. Either $\dfrac{6y(x - 1)}{x + 4}$ or $\dfrac{6xy - 6y}{x + 4}$ is an acceptable answer.

Remember that the quotient of any nonzero real number and its opposite is -1. For example, $\dfrac{6}{-6} = -1$ and $\dfrac{-8}{8} = -1$. Likewise, the indicated quotient of any polynomial and its opposite is equal to -1. For example,

$$\dfrac{a}{-a} = -1 \quad \text{because } a \text{ and } -a \text{ are opposites}$$

$$\dfrac{a - b}{b - a} = -1 \quad \text{because } a - b \text{ and } b - a \text{ are opposites}$$

$$\dfrac{x^2 - 4}{4 - x^2} = -1 \quad \text{because } x^2 - 4 \text{ and } 4 - x^2 \text{ are opposites}$$

Example 12 shows how we use this idea when simplifying rational expressions.

EXAMPLE 12   Simplify $\dfrac{6a^2 - 7a + 2}{10a - 15a^2}$.

Solution

$$\dfrac{6a^2 - 7a + 2}{10a - 15a^2} = \dfrac{(2a - 1)(3a - 2)}{5a(2 - 3a)} \qquad \dfrac{3a - 2}{2 - 3a} = -1$$

$$= (-1)\left(\dfrac{2a - 1}{5a}\right)$$

$$= -\dfrac{2a - 1}{5a} \qquad \text{or} \qquad \dfrac{1 - 2a}{5a} \qquad \blacksquare$$

PROBLEM SET 4.1

For Problems 1–8, express each rational number in reduced form.

1. $\dfrac{27}{36}$ **2.** $\dfrac{14}{21}$ **3.** $\dfrac{45}{54}$

4. $\dfrac{-14}{42}$ **5.** $\dfrac{24}{-60}$ **6.** $\dfrac{45}{-75}$

7. $\dfrac{-16}{-56}$ **8.** $\dfrac{-30}{-42}$

For Problems 9–50, simplify each rational expression.

9. $\dfrac{12xy}{42y}$ **10.** $\dfrac{21xy}{35x}$

11. $\dfrac{18a^2}{45ab}$ **12.** $\dfrac{48ab}{84b^2}$

13. $\dfrac{-14y^3}{56xy^2}$ **14.** $\dfrac{-14x^2y^3}{63xy^2}$

15. $\dfrac{54c^2d}{-78cd^2}$ **16.** $\dfrac{60x^3z}{-64xyz^2}$

17. $\dfrac{-40x^3y}{-24xy^4}$ **18.** $\dfrac{-30x^2y^2z^2}{-35xz^3}$

19. $\dfrac{x^2 - 4}{x^2 + 2x}$ **20.** $\dfrac{xy + y^2}{x^2 - y^2}$

21. $\dfrac{18x + 12}{12x - 6}$ **22.** $\dfrac{20x + 50}{15x - 30}$

23. $\dfrac{a^2 + 7a + 10}{a^2 - 7a - 18}$ **24.** $\dfrac{a^2 + 4a - 32}{3a^2 + 26a + 16}$

25. $\dfrac{2n^2 + n - 21}{10n^2 + 33n - 7}$

26. $\dfrac{4n^2 - 15n - 4}{7n^2 - 30n + 8}$

27. $\dfrac{5x^2 + 7}{10x}$

28. $\dfrac{12x^2 + 11x - 15}{20x^2 - 23x + 6}$

29. $\dfrac{6x^2 + x - 15}{8x^2 - 10x - 3}$

30. $\dfrac{4x^2 + 8x}{x^3 + 8}$

31. $\dfrac{3x^2 - 12x}{x^3 - 64}$

32. $\dfrac{x^2 - 14x + 49}{6x^2 - 37x - 35}$

33. $\dfrac{3x^2 + 17x - 6}{9x^2 - 6x + 1}$

34. $\dfrac{9y^2 - 1}{3y^2 + 11y - 4}$

35. $\dfrac{2x^3 + 3x^2 - 14x}{x^2y + 7xy - 18y}$

36. $\dfrac{3x^3 + 12x}{9x^2 + 18x}$

37. $\dfrac{5y^2 + 22y + 8}{25y^2 - 4}$

38. $\dfrac{16x^3y + 24x^2y^2 - 16xy^3}{24x^2y + 12xy^2 - 12y^3}$

39. $\dfrac{15x^3 - 15x^2}{5x^3 + 5x}$

40. $\dfrac{5n^2 + 18n - 8}{3n^2 + 13n + 4}$

41. $\dfrac{4x^2y + 8xy^2 - 12y^3}{18x^3y - 12x^2y^2 - 6xy^3}$

42. $\dfrac{3 + x - 2x^2}{2 + x - x^2}$

43. $\dfrac{3n^2 + 16n - 12}{7n^2 + 44n + 12}$

44. $\dfrac{x^4 - 2x^2 - 15}{2x^4 + 9x^2 + 9}$

45. $\dfrac{8 + 18x - 5x^2}{10 + 31x + 15x^2}$

46. $\dfrac{6x^4 - 11x^2 + 4}{2x^4 + 17x^2 - 9}$

47. $\dfrac{27x^4 - x}{6x^3 + 10x^2 - 4x}$

48. $\dfrac{64x^4 + 27x}{12x^3 - 27x^2 - 27x}$

49. $\dfrac{-40x^3 + 24x^2 + 16x}{20x^3 + 28x^2 + 8x}$

50. $\dfrac{-6x^3 - 21x^2 + 12x}{-18x^3 - 42x^2 + 120x}$

For Problems 51–58, simplify each rational expression. You will need to use factoring by grouping.

51. $\dfrac{xy + ay + bx + ab}{xy + ay + cx + ac}$

52. $\dfrac{xy + 2y + 3x + 6}{xy + 2y + 4x + 8}$

53. $\dfrac{ax - 3x + 2ay - 6y}{2ax - 6x + ay - 3y}$

54. $\dfrac{x^2 - 2x + ax - 2a}{x^2 - 2x + 3ax - 6a}$

55. $\dfrac{5x^2 + 5x + 3x + 3}{5x^2 + 3x - 30x - 18}$

56. $\dfrac{x^2 + 3x + 4x + 12}{2x^2 + 6x - x - 3}$

57. $\dfrac{2st - 30 - 12s + 5t}{3st - 6 - 18s + t}$

58. $\dfrac{nr - 6 - 3n + 2r}{nr + 10 + 2r + 5n}$

For Problems 59–68, simplify each rational expression. You may want to refer to Example 12 of this section.

59. $\dfrac{5x - 7}{7 - 5x}$

60. $\dfrac{4a - 9}{9 - 4a}$

61. $\dfrac{n^2 - 49}{7 - n}$

62. $\dfrac{9 - y}{y^2 - 81}$

63. $\dfrac{2y - 2xy}{x^2y - y}$

64. $\dfrac{3x - x^2}{x^2 - 9}$

65. $\dfrac{2x^3 - 8x}{4x - x^3}$

66. $\dfrac{x^2 - (y - 1)^2}{(y - 1)^2 - x^2}$

67. $\dfrac{n^2 - 5n - 24}{40 + 3n - n^2}$

68. $\dfrac{x^2 + 2x - 24}{20 - x - x^2}$

■ ■ ■ Thoughts into words

69. Compare the concept of a rational number in arithmetic to the concept of a rational expression in algebra.

70. What role does factoring play in the simplifying of rational expressions?

71. Why is the rational expression $\dfrac{x + 3}{x^2 - 4}$ undefined for $x = 2$ and $x = -2$ but defined for $x = -3$?

72. How would you convince someone that $\dfrac{x - 4}{4 - x} = -1$ for all real numbers except 4?

4.2 Multiplying and Dividing Rational Expressions

We define multiplication of rational numbers in common fraction form as follows:

> ### DEFINITION 4.1
>
> If a, b, c, and d are integers, and b and d are not equal to zero, then
>
> $$\frac{a}{b} \cdot \frac{c}{d} = \frac{a \cdot c}{b \cdot d} = \frac{ac}{bd}$$

To multiply rational numbers in common fraction form, we merely **multiply numerators and multiply denominators,** as the following examples demonstrate. (The steps in the dashed boxes are usually done mentally.)

$$\frac{2}{3} \cdot \frac{4}{5} = \frac{2 \cdot 4}{3 \cdot 5} = \frac{8}{15}$$

$$\frac{-3}{4} \cdot \frac{5}{7} = \frac{-3 \cdot 5}{4 \cdot 7} = \frac{-15}{28} = -\frac{15}{28}$$

$$-\frac{5}{6} \cdot \frac{13}{3} = \frac{-5}{6} \cdot \frac{13}{3} = \frac{-5 \cdot 13}{6 \cdot 3} = \frac{-65}{18} = -\frac{65}{18}$$

We also agree, when multiplying rational numbers, to express the final product in reduced form. The following examples show some different formats used to multiply and simplify rational numbers.

$$\frac{3}{4} \cdot \frac{4}{7} = \frac{3 \cdot 4}{4 \cdot 7} = \frac{3}{7}$$

$$\overset{1}{\underset{1}{\frac{8}{9}}} \cdot \overset{3}{\underset{4}{\frac{27}{32}}} = \frac{3}{4} \qquad \text{A common factor of 9 was divided out of 9 and 27, and a common factor of 8 was divided out of 8 and 32.}$$

$$\left(-\frac{28}{25}\right)\left(-\frac{65}{78}\right) = \frac{2 \cdot 2 \cdot 7 \cdot 5 \cdot 13}{5 \cdot 5 \cdot 2 \cdot 3 \cdot 13} = \frac{14}{15}. \qquad \text{We should recognize that a } \textit{negative times a negative is positive.}\text{ Also, note the use of prime factors to help us recognize common factors.}$$

Multiplication of rational expressions follows the same basic pattern as multiplication of rational numbers in common fraction form. That is to say, *we multiply numerators and multiply denominators and express the final product in simplified or reduced form.* Let's consider some examples.

$$\frac{3x}{4y} \cdot \frac{8y^2}{9x} = \frac{3 \cdot 8 \cdot x \cdot \overset{y}{\cancel{y^2}}}{\underset{3}{\cancel{4}} \cdot 9 \cdot \cancel{x} \cdot \cancel{y}} = \frac{2y}{3}$$

Note that we use the commutative property of multiplication to rearrange the factors in a form that allows us to identify common factors of the numerator and denominator.

$$\frac{-4a}{6a^2b^2} \cdot \frac{9ab}{12a^2} = -\frac{\overset{3}{\cancel{4}} \cdot 9 \cdot \cancel{a^2} \cdot \cancel{b}}{\underset{2}{\cancel{6}} \cdot \underset{3}{\cancel{12}} \cdot \underset{a^2}{\cancel{a^4}} \cdot \underset{b}{\cancel{b^2}}} = -\frac{1}{2a^2b}$$

$$\frac{12x^2y}{-18xy} \cdot \frac{-24xy^2}{56y^3} = \frac{\overset{2}{\cancel{12}} \cdot \overset{3}{\cancel{24}} \cdot \overset{x^2}{\cancel{x^3}} \cdot \cancel{y^3}}{\underset{3}{\cancel{18}} \cdot \underset{7}{\cancel{56}} \cdot \cancel{x} \cdot \underset{y}{\cancel{y^4}}} = \frac{2x^2}{7y}$$

You should recognize that the first fraction is equivalent to $-\dfrac{12x^2y}{18xy}$ and the second to $-\dfrac{24xy^2}{56y^3}$; thus the product is positive.

If the rational expressions contain polynomials (other than monomials) that are factorable, then our work may take on the following format.

EXAMPLE 1   Multiply and simplify $\dfrac{y}{x^2 - 4} \cdot \dfrac{x + 2}{y^2}$.

Solution

$$\frac{y}{x^2 - 4} \cdot \frac{x + 2}{y^2} = \frac{\cancel{y}(\cancel{x + 2})}{\underset{y}{\cancel{y^2}}(\cancel{x + 2})(x - 2)} = \frac{1}{y(x - 2)} \qquad \blacksquare$$

In Example 1, note that we combined the steps of multiplying numerators and denominators and factoring the polynomials. Also note that we left the final answer in factored form. Either $\dfrac{1}{y(x - 2)}$ or $\dfrac{1}{xy - 2y}$ would be an acceptable answer.

EXAMPLE 2   Multiply and simplify $\dfrac{x^2 - x}{x + 5} \cdot \dfrac{x^2 + 5x + 4}{x^4 - x^2}$.

Solution

$$\frac{x^2 - x}{x + 5} \cdot \frac{x^2 + 5x + 4}{x^4 - x^2} = \frac{x(x - 1)}{x + 5} \cdot \frac{(x + 1)(x + 4)}{x^2(x - 1)(x + 1)}$$

$$= \frac{\cancel{x}(\cancel{x - 1})(\cancel{x + 1})(x + 4)}{(x + 5)(\cancel{x^2})(\cancel{x - 1})(\cancel{x + 1})} = \frac{x + 4}{x(x + 5)} \qquad \blacksquare$$

EXAMPLE 3 Multiply and simplify $\dfrac{6n^2 + 7n - 5}{n^2 + 2n - 24} \cdot \dfrac{4n^2 + 21n - 18}{12n^2 + 11n - 15}$.

Solution

$$\dfrac{6n^2 + 7n - 5}{n^2 + 2n - 24} \cdot \dfrac{4n^2 + 21n - 18}{12n^2 + 11n - 15}$$

$$= \dfrac{(3n + 5)(2n - 1)(4n - 3)(n + 6)}{(n + 6)(n - 4)(3n + 5)(4n - 3)} = \dfrac{2n - 1}{n - 4} \qquad \blacksquare$$

Dividing Rational Expressions

We define division of rational numbers in common fraction form as follows:

DEFINITION 4.2

If a, b, c, and d are integers, and b, c, and d are not equal to zero, then

$$\frac{a}{b} \div \frac{c}{d} = \frac{a}{b} \cdot \frac{d}{c} = \frac{ad}{bc}$$

Definition 4.2 states that to divide two rational numbers in fraction form, we **invert the divisor and multiply.** We call the numbers $\dfrac{c}{d}$ and $\dfrac{d}{c}$ **reciprocals** or **multiplicative inverses** of each other because their product is 1. Thus we can describe division by saying **to divide by a fraction, multiply by its reciprocal.** The following examples demonstrate the use of Definition 4.2.

$$\frac{7}{8} \div \frac{5}{6} = \frac{7}{\underset{4}{8}} \cdot \frac{\overset{3}{6}}{5} = \frac{21}{20}, \qquad \frac{-5}{9} \div \frac{15}{18} = -\frac{5}{9} \cdot \frac{\overset{2}{18}}{\underset{3}{15}} = -\frac{2}{3}$$

$$\frac{14}{-19} \div \frac{21}{-38} = \left(-\frac{14}{19}\right) \div \left(-\frac{21}{38}\right) = \left(-\frac{\overset{2}{14}}{19}\right)\left(-\frac{\overset{2}{38}}{\underset{3}{21}}\right) = \frac{4}{3}$$

We define division of algebraic rational expressions in the same way that we define division of rational numbers. That is, the quotient of two rational expressions is the product we obtain when we multiply the first expression by the reciprocal of the second. Consider the following examples.

EXAMPLE 4 Divide and simplify $\dfrac{16x^2y}{24xy^3} \div \dfrac{9xy}{8x^2y^2}$.

Solution

$$\frac{16x^2y}{24xy^3} \div \frac{9xy}{8x^2y^2} = \frac{16x^2y}{24xy^3} \cdot \frac{8x^2y^2}{9xy} = \frac{16 \cdot 8 \cdot \overset{x^2}{\cancel{x^4}} \cdot y^3}{\underset{3}{\cancel{24}} \cdot 9 \cdot x^2 \cdot \underset{y}{\cancel{y^4}}} = \frac{16x^2}{27y}$$

■

EXAMPLE 5   Divide and simplify $\dfrac{3a^2 + 12}{3a^2 - 15a} \div \dfrac{a^4 - 16}{a^2 - 3a - 10}$.

Solution

$$\frac{3a^2 + 12}{3a^2 - 15a} \div \frac{a^4 - 16}{a^2 - 3a - 10} = \frac{3a^2 + 12}{3a^2 - 15a} \cdot \frac{a^2 - 3a - 10}{a^4 - 16}$$

$$= \frac{3(a^2 + 4)}{3a(a - 5)} \cdot \frac{(a - 5)(a + 2)}{(a^2 + 4)(a + 2)(a - 2)}$$

$$= \frac{\overset{1}{\cancel{3}}(\cancel{a^2 + 4})(\cancel{a - 5})(\cancel{a + 2})}{\underset{1}{\cancel{3}}a(\cancel{a - 5})(\cancel{a^2 + 4})(\cancel{a + 2})(a - 2)}$$

$$= \frac{1}{a(a - 2)}$$

■

EXAMPLE 6   Divide and simplify $\dfrac{28t^3 - 51t^2 - 27t}{49t^2 + 42t + 9} \div (4t - 9)$.

Solution

$$\frac{28t^3 - 51t^2 - 27t}{49t^2 + 42t + 9} \div \frac{4t - 9}{1} = \frac{28t^3 - 51t^2 - 27t}{49t^2 + 42t + 9} \cdot \frac{1}{4t - 9}$$

$$= \frac{t(7t + 3)(4t - 9)}{(7t + 3)(7t + 3)} \cdot \frac{1}{(4t - 9)}$$

$$= \frac{t(\cancel{7t + 3})(\cancel{4t - 9})}{(\cancel{7t + 3})(7t + 3)(\cancel{4t - 9})}$$

$$= \frac{t}{7t + 3}$$

■

In a problem such as Example 6, it may be helpful to write the divisor with a denominator of 1. Thus we write $4t - 9$ as $\dfrac{4t - 9}{1}$; its reciprocal is obviously $\dfrac{1}{4t - 9}$.

Let's consider one final example that involves both multiplication and division.

E X A M P L E   7   Perform the indicated operations and simplify.

$$\frac{x^2 + 5x}{3x^2 - 4x - 20} \cdot \frac{x^2y + y}{2x^2 + 11x + 5} \div \frac{xy^2}{6x^2 - 17x - 10}$$

Solution

$$\frac{x^2 + 5x}{3x^2 - 4x - 20} \cdot \frac{x^2y + y}{2x^2 + 11x + 5} \div \frac{xy^2}{6x^2 - 17x - 10}$$

$$= \frac{x^2 + 5x}{3x^2 - 4x - 20} \cdot \frac{x^2y + y}{2x^2 + 11x + 5} \cdot \frac{6x^2 - 17x - 10}{xy^2}$$

$$= \frac{x(x + 5)}{(3x - 10)(x + 2)} \cdot \frac{y(x^2 + 1)}{(2x + 1)(x + 5)} \cdot \frac{(2x + 1)(3x - 10)}{xy^2}$$

$$= \frac{x(x + 5)(\cancel{y})(x^2 + 1)(2x + 1)(3x - 10)}{(3x - 10)(x + 2)(2x + 1)(x + 5)(x)(y^2)} = \frac{x^2 + 1}{y(x + 2)}$$

■

PROBLEM SET 4.2

For Problems 1–12, perform the indicated operations involving rational numbers. Express final answers in reduced form.

For Problems 13–50, perform the indicated operations involving rational expressions. Express final answers in simplest form.

1. $\dfrac{7}{12} \cdot \dfrac{6}{35}$

2. $\dfrac{5}{8} \cdot \dfrac{12}{20}$

3. $\dfrac{-4}{9} \cdot \dfrac{18}{30}$

4. $\dfrac{-6}{9} \cdot \dfrac{36}{48}$

5. $\dfrac{3}{-8} \cdot \dfrac{-6}{12}$

6. $\dfrac{-12}{16} \cdot \dfrac{18}{-32}$

7. $\left(-\dfrac{5}{7}\right) \div \dfrac{6}{7}$

8. $\left(-\dfrac{5}{9}\right) \div \dfrac{10}{3}$

9. $\dfrac{-9}{5} \div \dfrac{27}{10}$

10. $\dfrac{4}{7} \div \dfrac{16}{-21}$

11. $\dfrac{4}{9} \cdot \dfrac{6}{11} \div \dfrac{4}{15}$

12. $\dfrac{2}{3} \cdot \dfrac{6}{7} \div \dfrac{8}{3}$

13. $\dfrac{6xy}{9y^4} \cdot \dfrac{30x^3y}{-48x}$

14. $\dfrac{-14xy^4}{18y^2} \cdot \dfrac{24x^2y^3}{35y^2}$

15. $\dfrac{5a^2b^2}{11ab} \cdot \dfrac{22a^3}{15ab^2}$

16. $\dfrac{10a^2}{5b^2} \cdot \dfrac{15b^3}{2a^4}$

17. $\dfrac{5xy}{8y^2} \cdot \dfrac{18x^2y}{15}$

18. $\dfrac{4x^2}{5y^2} \cdot \dfrac{15xy}{24x^2y^2}$

19. $\dfrac{5x^4}{12x^2y^3} \div \dfrac{9}{5xy}$

20. $\dfrac{7x^2y}{9xy^3} \div \dfrac{3x^4}{2x^2y^2}$

21. $\dfrac{9a^2c}{12bc^2} \div \dfrac{21ab}{14c^3}$

22. $\dfrac{3ab^3}{4c} \div \dfrac{21ac}{12bc^3}$

23. $\dfrac{9x^2y^3}{14x} \cdot \dfrac{21y}{15xy^2} \cdot \dfrac{10x}{12y^3}$

24. $\dfrac{5xy}{7a} \cdot \dfrac{14a^2}{15x} \cdot \dfrac{3a}{8y}$

25. $\dfrac{3x + 6}{5y} \cdot \dfrac{x^2 + 4}{x^2 + 10x + 16}$

26. $\dfrac{5xy}{x + 6} \cdot \dfrac{x^2 - 36}{x^2 - 6x}$

27. $\dfrac{5a^2 + 20a}{a^3 - 2a^2} \cdot \dfrac{a^2 - a - 12}{a^2 - 16}$

28. $\dfrac{2a^2 + 6}{a^2 - a} \cdot \dfrac{a^3 - a^2}{8a - 4}$

29. $\dfrac{3n^2 + 15n - 18}{3n^2 + 10n - 48} \cdot \dfrac{6n^2 - n - 40}{4n^2 + 6n - 10}$

30. $\dfrac{6n^2 + 11n - 10}{3n^2 + 19n - 14} \cdot \dfrac{2n^2 + 6n - 56}{2n^2 - 3n - 20}$

31. $\dfrac{9y^2}{x^2 + 12x + 36} \div \dfrac{12y}{x^2 + 6x}$

32. $\dfrac{7xy}{x^2 - 4x + 4} \div \dfrac{14y}{x^2 - 4}$

33. $\dfrac{x^2 - 4xy + 4y^2}{7xy^2} \div \dfrac{4x^2 - 3xy - 10y^2}{20x^2y + 25xy^2}$

34. $\dfrac{x^2 + 5xy - 6y^2}{xy^2 - y^3} \cdot \dfrac{2x^2 + 15xy + 18y^2}{xy + 4y^2}$

35. $\dfrac{5 - 14n - 3n^2}{1 - 2n - 3n^2} \cdot \dfrac{9 + 7n - 2n^2}{27 - 15n + 2n^2}$

36. $\dfrac{6 - n - 2n^2}{12 - 11n + 2n^2} \cdot \dfrac{24 - 26n + 5n^2}{2 + 3n + n^2}$

37. $\dfrac{3x^4 + 2x^2 - 1}{3x^4 + 14x^2 - 5} \cdot \dfrac{x^4 - 2x^2 - 35}{x^4 - 17x^2 + 70}$

38. $\dfrac{2x^4 + x^2 - 3}{2x^4 + 5x^2 + 2} \cdot \dfrac{3x^4 + 10x^2 + 8}{3x^4 + x^2 - 4}$

39. $\dfrac{3x^2 - 20x + 25}{2x^2 - 7x - 15} \div \dfrac{9x^2 - 3x - 20}{12x^2 + 28x + 15}$

40. $\dfrac{21t^2 + t - 2}{2t^2 - 17t - 9} \div \dfrac{12t^2 - 5t - 3}{8t^2 - 2t - 3}$

41. $\dfrac{10t^3 + 25t}{20t + 10} \cdot \dfrac{2t^2 - t - 1}{t^5 - t}$

42. $\dfrac{t^4 - 81}{t^2 - 6t + 9} \cdot \dfrac{6t^2 - 11t - 21}{5t^2 + 8t - 21}$

43. $\dfrac{4t^2 + t - 5}{t^3 - t^2} \cdot \dfrac{t^4 + 6t^3}{16t^2 + 40t + 25}$

44. $\dfrac{9n^2 - 12n + 4}{n^2 - 4n - 32} \cdot \dfrac{n^2 + 4n}{3n^3 - 2n^2}$

45. $\dfrac{nr + 3n + 2r + 6}{nr + 3n - 3r - 9} \cdot \dfrac{n^2 - 9}{n^3 - 4n}$

46. $\dfrac{xy + xc + ay + ac}{xy - 2xc + ay - 2ac} \cdot \dfrac{2x^3 - 8x}{12x^3 + 20x^2 - 8x}$

47. $\dfrac{x^2 - x}{4y} \cdot \dfrac{10xy^2}{2x - 2} \div \dfrac{3x^2 + 3x}{15x^2y^2}$

48. $\dfrac{4xy^2}{7x} \cdot \dfrac{14x^3y}{12y} \div \dfrac{7y}{9x^3}$

49. $\dfrac{a^2 - 4ab + 4b^2}{6a^2 - 4ab} \cdot \dfrac{3a^2 + 5ab - 2b^2}{6a^2 + ab - b^2} \div \dfrac{a^2 - 4b^2}{8a + 4b}$

50. $\dfrac{2x^2 + 3x}{2x^3 - 10x^2} \cdot \dfrac{x^2 - 8x + 15}{3x^3 - 27x} \div \dfrac{14x + 21}{x^2 - 6x - 27}$

■ ■ ■ **Thoughts into words**

51. Explain in your own words how to divide two rational expressions.

52. Suppose that your friend missed class the day the material in this section was discussed. How could you draw on her background in arithmetic to explain to her how to multiply and divide rational expressions?

53. Give a step-by-step description of how to do the following multiplication problem.

$$\dfrac{x^2 + 5x + 6}{x^2 - 2x - 8} \cdot \dfrac{x^2 - 16}{16 - x^2}$$

4.3   Adding and Subtracting Rational Expressions

We can define addition and subtraction of rational numbers as follows:

DEFINITION 4.3

If a, b, and c are integers and b is not zero, then

$$\frac{a}{b} + \frac{c}{b} = \frac{a+c}{b} \qquad \text{Addition}$$

$$\frac{a}{b} - \frac{c}{b} = \frac{a-c}{b} \qquad \text{Subtraction}$$

We can add or subtract rational numbers with a common denominator by adding or subtracting the numerators and placing the result over the common denominator. The following examples illustrate Definition 4.3.

$$\frac{2}{9} + \frac{3}{9} = \frac{2+3}{9} = \frac{5}{9}$$

$$\frac{7}{8} - \frac{3}{8} = \frac{7-3}{8} = \frac{4}{8} = \frac{1}{2} \qquad \text{Don't forget to reduce!}$$

$$\frac{4}{6} + \frac{-5}{6} = \frac{4+(-5)}{6} = \frac{-1}{6} = -\frac{1}{6}$$

$$\frac{7}{10} + \frac{4}{-10} = \frac{7}{10} + \frac{-4}{10} = \frac{7+(-4)}{10} = \frac{3}{10}$$

We use this same *common denominator* approach when adding or subtracting rational expressions, as in these next examples.

$$\frac{3}{x} + \frac{9}{x} = \frac{3+9}{x} = \frac{12}{x}$$

$$\frac{8}{x-2} - \frac{3}{x-2} = \frac{8-3}{x-2} = \frac{5}{x-2}$$

$$\frac{9}{4y} + \frac{5}{4y} = \frac{9+5}{4y} = \frac{14}{4y} = \frac{7}{2y} \qquad \text{Don't forget to simplify the final answer!}$$

$$\frac{n^2}{n-1} - \frac{1}{n-1} = \frac{n^2-1}{n-1} = \frac{(n+1)(n-1)}{n-1} = n+1$$

$$\frac{6a^2}{2a+1} + \frac{13a+5}{2a+1} = \frac{6a^2+13a+5}{2a+1} = \frac{(2a+1)(3a+5)}{2a+1} = 3a+5$$

In each of the previous examples that involve rational expressions, we should technically restrict the variables to exclude division by zero. For example, $\frac{3}{x} + \frac{9}{x} = \frac{12}{x}$ is true for all real number values for x, except $x = 0$. Likewise, $\frac{8}{x-2} - \frac{3}{x-2} = \frac{5}{x-2}$ as long as x does not equal 2. Rather than taking the time and space to write down restrictions for each problem, we will merely assume that such restrictions exist.

If rational numbers that do not have a common denominator are to be added or subtracted, then we apply the fundamental principle of fractions $\left(\dfrac{a}{b} = \dfrac{ak}{bk}\right)$ to obtain equivalent fractions with a common denominator. Equivalent fractions are fractions such as $\dfrac{1}{2}$ and $\dfrac{2}{4}$ that name the same number. Consider the following example.

$$\frac{1}{2} + \frac{1}{3} = \frac{3}{6} + \frac{2}{6} = \frac{3+2}{6} = \frac{5}{6}$$

$$\left(\begin{array}{c}\dfrac{1}{2} \text{ and } \dfrac{3}{6} \\ \text{are equivalent} \\ \text{fractions.}\end{array}\right)\left(\begin{array}{c}\dfrac{1}{3} \text{ and } \dfrac{2}{6} \\ \text{are equivalent} \\ \text{fractions.}\end{array}\right)$$

Note that we chose 6 as our common denominator and 6 is the **least common multiple** of the original denominators 2 and 3. (The least common multiple of a set of whole numbers is the smallest nonzero whole number divisible by each of the numbers.) In general, we use the least common multiple of the denominators of the fractions to be added or subtracted as a **least common denominator** (LCD).

A least common denominator may be found by inspection or by using the prime-factored forms of the numbers. Let's consider some examples and use each of these techniques.

EXAMPLE 1   Subtract $\dfrac{5}{6} - \dfrac{3}{8}$.

Solution

By inspection, we can see that the LCD is 24. Thus both fractions can be changed to equivalent fractions, each with a denominator of 24.

$$\frac{5}{6} - \frac{3}{8} = \left(\frac{5}{6}\right)\left(\frac{4}{4}\right) - \left(\frac{3}{8}\right)\left(\frac{3}{3}\right) = \frac{20}{24} - \frac{9}{24} = \frac{11}{24}$$

$$\uparrow \qquad\qquad \uparrow$$
$$\text{Form of 1} \quad \text{Form of 1}$$

∎

In Example 1, note that the fundamental principle of fractions, $\dfrac{a}{b} = \dfrac{a \cdot k}{b \cdot k}$, can be written as $\dfrac{a}{b} = \left(\dfrac{a}{b}\right)\left(\dfrac{k}{k}\right)$. This latter form emphasizes the fact that 1 is the multiplication identity element.

EXAMPLE 2 Perform the indicated operations: $\dfrac{3}{5} + \dfrac{1}{6} - \dfrac{13}{15}$

Solution

Again by inspection, we can determine that the LCD is 30. Thus we can proceed as follows:

$$\frac{3}{5} + \frac{1}{6} - \frac{13}{15} = \left(\frac{3}{5}\right)\left(\frac{6}{6}\right) + \left(\frac{1}{6}\right)\left(\frac{5}{5}\right) - \left(\frac{13}{15}\right)\left(\frac{2}{2}\right)$$

$$= \frac{18}{30} + \frac{5}{30} - \frac{26}{30} = \frac{18 + 5 - 26}{30}$$

$$= \frac{-3}{30} = -\frac{1}{10} \qquad \text{Don't forget to reduce!} \quad \blacksquare$$

EXAMPLE 3 Add $\dfrac{7}{18} + \dfrac{11}{24}$.

Solution

Let's use the prime-factored forms of the denominators to help find the LCD.

$$18 = 2 \cdot 3 \cdot 3 \qquad 24 = 2 \cdot 2 \cdot 2 \cdot 3$$

The LCD must contain three factors of 2 because 24 contains three 2s. The LCD must also contain two factors of 3 because 18 has two 3s. Thus the LCD = $2 \cdot 2 \cdot 2 \cdot 3 \cdot 3 = 72$. Now we can proceed as usual.

$$\frac{7}{18} + \frac{11}{24} = \left(\frac{7}{18}\right)\left(\frac{4}{4}\right) + \left(\frac{11}{24}\right)\left(\frac{3}{3}\right) = \frac{28}{72} + \frac{33}{72} = \frac{61}{72} \qquad \blacksquare$$

To add and subtract rational expressions with different denominators, follow the same basic routine that you follow when you add or subtract rational numbers with different denominators. Study the following examples carefully and note the similarity to our previous work with rational numbers.

EXAMPLE 4 Add $\dfrac{x + 2}{4} + \dfrac{3x + 1}{3}$.

Solution

By inspection, we see that the LCD is 12.

$$\frac{x + 2}{4} + \frac{3x + 1}{3} = \left(\frac{x + 2}{4}\right)\left(\frac{3}{3}\right) + \left(\frac{3x + 1}{3}\right)\left(\frac{4}{4}\right)$$

$$= \frac{3(x + 2)}{12} + \frac{4(3x + 1)}{12}$$

$$= \frac{3(x + 2) + 4(3x + 1)}{12}$$

$$= \frac{3x + 6 + 12x + 4}{12}$$

$$= \frac{15x + 10}{12} \qquad \blacksquare$$

Note the final result in Example 4. The numerator, $15x + 10$, could be factored as $5(3x + 2)$. However, because this produces no common factors with the denominator, the fraction cannot be simplified. Thus the final answer can be left as $\frac{15x + 10}{12}$. It would also be acceptable to express it as $\frac{5(3x + 2)}{12}$.

EXAMPLE 5   Subtract $\dfrac{a - 2}{2} - \dfrac{a - 6}{6}$.

Solution

By inspection, we see that the LCD is 6.

$$\frac{a - 2}{2} - \frac{a - 6}{6} = \left(\frac{a - 2}{2}\right)\left(\frac{3}{3}\right) - \frac{a - 6}{6}$$

$$= \frac{3(a - 2)}{6} - \frac{a - 6}{6}$$

$$= \frac{3(a - 2) - (a - 6)}{6} \qquad \text{Be careful with this sign as you move to the next step!}$$

$$= \frac{3a - 6 - a + 6}{6}$$

$$= \frac{2a}{6} = \frac{a}{3} \qquad \text{Don't forget to simplify.} \qquad \blacksquare$$

EXAMPLE 6   Perform the indicated operations: $\dfrac{x + 3}{10} + \dfrac{2x + 1}{15} - \dfrac{x - 2}{18}$.

Solution

If you cannot determine the LCD by inspection, then use the prime-factored forms of the denominators.

$$10 = 2 \cdot 5 \qquad 15 = 3 \cdot 5 \qquad 18 = 2 \cdot 3 \cdot 3$$

The LCD must contain one factor of 2, two factors of 3, and one factor of 5. Thus the LCD is $2 \cdot 3 \cdot 3 \cdot 5 = 90$.

$$\frac{x + 3}{10} + \frac{2x + 1}{15} - \frac{x - 2}{18} = \left(\frac{x + 3}{10}\right)\left(\frac{9}{9}\right) + \left(\frac{2x + 1}{15}\right)\left(\frac{6}{6}\right) - \left(\frac{x - 2}{18}\right)\left(\frac{5}{5}\right)$$

$$= \frac{9(x + 3)}{90} + \frac{6(2x + 1)}{90} - \frac{5(x - 2)}{90}$$

$$= \frac{9(x + 3) + 6(2x + 1) - 5(x - 2)}{90}$$

$$= \frac{9x + 27 + 12x + 6 - 5x + 10}{90}$$

$$= \frac{16x + 43}{90}$$

A denominator that contains variables does not create any serious difficulties; our approach remains basically the same.

EXAMPLE 7 Add $\dfrac{3}{2x} + \dfrac{5}{3y}$.

Solution
Using an LCD of $6xy$, we can proceed as follows:

$$\frac{3}{2x} + \frac{5}{3y} = \left(\frac{3}{2x}\right)\left(\frac{3y}{3y}\right) + \left(\frac{5}{3y}\right)\left(\frac{2x}{2x}\right)$$

$$= \frac{9y}{6xy} + \frac{10x}{6xy}$$

$$= \frac{9y + 10x}{6xy}$$

EXAMPLE 8 Subtract $\dfrac{7}{12ab} - \dfrac{11}{15a^2}$.

Solution
We can prime-factor the numerical coefficients of the denominators to help find the LCD.

$$\left.\begin{array}{l} 12ab = 2 \cdot 2 \cdot 3 \cdot a \cdot b \\ 15a^2 = 3 \cdot 5 \cdot a^2 \end{array}\right\} \longrightarrow \text{LCD} = 2 \cdot 2 \cdot 3 \cdot 5 \cdot a^2 \cdot b = 60a^2b$$

$$\frac{7}{12ab} - \frac{11}{15a^2} = \left(\frac{7}{12ab}\right)\left(\frac{5a}{5a}\right) - \left(\frac{11}{15a^2}\right)\left(\frac{4b}{4b}\right)$$

$$= \frac{35a}{60a^2b} - \frac{44b}{60a^2b}$$

$$= \frac{35a - 44b}{60a^2b}$$

EXAMPLE 9   Add $\dfrac{x}{x-3} + \dfrac{4}{x}$.

Solution

By inspection, the LCD is $x(x-3)$.

$$\frac{x}{x-3} + \frac{4}{x} = \left(\frac{x}{x-3}\right)\left(\frac{x}{x}\right) + \left(\frac{4}{x}\right)\left(\frac{x-3}{x-3}\right)$$

$$= \frac{x^2}{x(x-3)} + \frac{4(x-3)}{x(x-3)}$$

$$= \frac{x^2 + 4(x-3)}{x(x-3)}$$

$$= \frac{x^2 + 4x - 12}{x(x-3)} \qquad \text{or} \qquad \frac{(x+6)(x-2)}{x(x-3)}$$

EXAMPLE 10   Subtract $\dfrac{2x}{x+1} - 3$.

Solution

$$\frac{2x}{x+1} - 3 = \frac{2x}{x+1} - 3\left(\frac{x+1}{x+1}\right)$$

$$= \frac{2x}{x+1} - \frac{3(x+1)}{x+1}$$

$$= \frac{2x - 3(x+1)}{x+1}$$

$$= \frac{2x - 3x - 3}{x+1}$$

$$= \frac{-x - 3}{x+1}$$

PROBLEM SET 4.3

For Problems 1–12, perform the indicated operations involving rational numbers. Be sure to express your answers in reduced form.

1. $\dfrac{1}{4} + \dfrac{5}{6}$

2. $\dfrac{3}{5} + \dfrac{1}{6}$

3. $\dfrac{7}{8} - \dfrac{3}{5}$

4. $\dfrac{7}{9} - \dfrac{1}{6}$

5. $\dfrac{6}{5} + \dfrac{1}{-4}$

6. $\dfrac{7}{8} + \dfrac{5}{-12}$

7. $\dfrac{8}{15} + \dfrac{3}{25}$

8. $\dfrac{5}{9} - \dfrac{11}{12}$

9. $\dfrac{1}{5} + \dfrac{5}{6} - \dfrac{7}{15}$

10. $\dfrac{2}{3} - \dfrac{7}{8} + \dfrac{1}{4}$

11. $\dfrac{1}{3} - \dfrac{1}{4} - \dfrac{3}{14}$

12. $\dfrac{5}{6} - \dfrac{7}{9} - \dfrac{3}{10}$

For Problems 13–66, add or subtract the rational expressions as indicated. Be sure to express your answers in simplest form.

13. $\dfrac{2x}{x-1} + \dfrac{4}{x-1}$

14. $\dfrac{3x}{2x+1} - \dfrac{5}{2x+1}$

15. $\dfrac{4a}{a+2} + \dfrac{8}{a+2}$

16. $\dfrac{6a}{a-3} - \dfrac{18}{a-3}$

17. $\dfrac{3(y-2)}{7y} + \dfrac{4(y-1)}{7y}$

18. $\dfrac{2x-1}{4x^2} + \dfrac{3(x-2)}{4x^2}$

19. $\dfrac{x-1}{2} + \dfrac{x+3}{3}$

20. $\dfrac{x-2}{4} + \dfrac{x+6}{5}$

21. $\dfrac{2a-1}{4} + \dfrac{3a+2}{6}$

22. $\dfrac{a-4}{6} + \dfrac{4a-1}{8}$

23. $\dfrac{n+2}{6} - \dfrac{n-4}{9}$

24. $\dfrac{2n+1}{9} - \dfrac{n+3}{12}$

25. $\dfrac{3x-1}{3} - \dfrac{5x+2}{5}$

26. $\dfrac{4x-3}{6} - \dfrac{8x-2}{12}$

27. $\dfrac{x-2}{5} - \dfrac{x+3}{6} + \dfrac{x+1}{15}$

28. $\dfrac{x+1}{4} + \dfrac{x-3}{6} - \dfrac{x-2}{8}$

29. $\dfrac{3}{8x} + \dfrac{7}{10x}$

30. $\dfrac{5}{6x} - \dfrac{3}{10x}$

31. $\dfrac{5}{7x} - \dfrac{11}{4y}$

32. $\dfrac{5}{12x} - \dfrac{9}{8y}$

33. $\dfrac{4}{3x} + \dfrac{5}{4y} - 1$

34. $\dfrac{7}{3x} - \dfrac{8}{7y} - 2$

35. $\dfrac{7}{10x^2} + \dfrac{11}{15x}$

36. $\dfrac{7}{12a^2} - \dfrac{5}{16a}$

37. $\dfrac{10}{7n} - \dfrac{12}{4n^2}$

38. $\dfrac{6}{8n^2} - \dfrac{3}{5n}$

39. $\dfrac{3}{n^2} - \dfrac{2}{5n} + \dfrac{4}{3}$

40. $\dfrac{1}{n^2} + \dfrac{3}{4n} - \dfrac{5}{6}$

41. $\dfrac{3}{x} - \dfrac{5}{3x^2} - \dfrac{7}{6x}$

42. $\dfrac{7}{3x^2} - \dfrac{9}{4x} - \dfrac{5}{2x}$

43. $\dfrac{6}{5t^2} - \dfrac{4}{7t^3} + \dfrac{9}{5t^3}$

44. $\dfrac{5}{7t} + \dfrac{3}{4t^2} + \dfrac{1}{14t}$

45. $\dfrac{5b}{24a^2} - \dfrac{11a}{32b}$

46. $\dfrac{9}{14x^2y} - \dfrac{4x}{7y^2}$

47. $\dfrac{7}{9xy^3} - \dfrac{4}{3x} + \dfrac{5}{2y^2}$

48. $\dfrac{7}{16a^2b} + \dfrac{3a}{20b^2}$

49. $\dfrac{2x}{x-1} + \dfrac{3}{x}$

50. $\dfrac{3x}{x-4} - \dfrac{2}{x}$

51. $\dfrac{a-2}{a} - \dfrac{3}{a+4}$

52. $\dfrac{a+1}{a} - \dfrac{2}{a+1}$

53. $\dfrac{-3}{4n+5} - \dfrac{8}{3n+5}$

54. $\dfrac{-2}{n-6} - \dfrac{6}{2n+3}$

55. $\dfrac{-1}{x+4} + \dfrac{4}{7x-1}$

56. $\dfrac{-3}{4x+3} + \dfrac{5}{2x-5}$

57. $\dfrac{7}{3x-5} - \dfrac{5}{2x+7}$

58. $\dfrac{5}{x-1} - \dfrac{3}{2x-3}$

59. $\dfrac{5}{3x-2} + \dfrac{6}{4x+5}$

60. $\dfrac{3}{2x+1} + \dfrac{2}{3x+4}$

61. $\dfrac{3x}{2x+5} + 1$

62. $2 + \dfrac{4x}{3x-1}$

63. $\dfrac{4x}{x-5} - 3$

64. $\dfrac{7x}{x+4} - 2$

65. $-1 - \dfrac{3}{2x+1}$

66. $-2 - \dfrac{5}{4x-3}$

67. Recall that the indicated quotient of a polynomial and its opposite is -1. For example, $\dfrac{x-2}{2-x}$ simplifies to -1. Keep this idea in mind as you add or subtract the following rational expressions.

(a) $\dfrac{1}{x-1} - \dfrac{x}{x-1}$ (b) $\dfrac{3}{2x-3} - \dfrac{2x}{2x-3}$

(c) $\dfrac{4}{x-4} - \dfrac{x}{x-4} + 1$ (d) $-1 + \dfrac{2}{x-2} - \dfrac{x}{x-2}$

68. Consider the addition problem $\dfrac{8}{x-2} + \dfrac{5}{2-x}$. Note that the denominators are opposites of each other. If the property $\dfrac{a}{-b} = -\dfrac{a}{b}$ is applied to the second

fraction, we have $\dfrac{5}{2-x} = -\dfrac{5}{x-2}$. Thus we proceed as follows:

$$\dfrac{8}{x-2} + \dfrac{5}{2-x} = \dfrac{8}{x-2} - \dfrac{5}{x-2} = \dfrac{8-5}{x-2} = \dfrac{3}{x-2}$$

Use this approach to do the following problems.

(a) $\dfrac{7}{x-1} + \dfrac{2}{1-x}$ (b) $\dfrac{5}{2x-1} + \dfrac{8}{1-2x}$

(c) $\dfrac{4}{a-3} - \dfrac{1}{3-a}$ (d) $\dfrac{10}{a-9} - \dfrac{5}{9-a}$

(e) $\dfrac{x^2}{x-1} - \dfrac{2x-3}{1-x}$ (f) $\dfrac{x^2}{x-4} - \dfrac{3x-28}{4-x}$

■ ■ ■ **Thoughts into words**

69. What is the difference between the concept of least common multiple and the concept of least common denominator?

70. A classmate tells you that she finds the least common multiple of two counting numbers by listing the multiples of each number and then choosing the smallest number that appears in both lists. Is this a correct procedure? What is the weakness of this procedure?

71. For which real numbers does $\dfrac{x}{x-3} + \dfrac{4}{x}$ equal $\dfrac{(x+6)(x-2)}{x(x-3)}$? Explain your answer.

72. Suppose that your friend does an addition problem as follows:

$$\dfrac{5}{8} + \dfrac{7}{12} = \dfrac{5(12) + 8(7)}{8(12)} = \dfrac{60 + 56}{96} = \dfrac{116}{96} = \dfrac{29}{24}$$

Is this answer correct? If not, what advice would you offer your friend?

4.4 More on Rational Expressions and Complex Fractions

In this section, we expand our work with adding and subtracting rational expressions, and we discuss the process of simplifying complex fractions. Before we begin, however, this seems like an appropriate time to offer a bit of advice regarding your study of algebra. Success in algebra depends on having a good understanding of the concepts as well as on being able to perform the various computations. As for the computational work, you should adopt a carefully organized format that shows as many steps as you need in order to minimize the chances of making careless errors. Don't be eager to find shortcuts for certain computations before you have a thorough understanding of the steps involved in the process. This advice is especially appropriate at the beginning of this section.

Study Examples 1–4 very carefully. Note that the same basic procedure is followed in solving each problem:

STEP 1   Factor the denominators.

STEP 2   Find the LCD.

STEP 3   Change each fraction to an equivalent fraction that has the LCD as its denominator.

STEP 4   Combine the numerators and place over the LCD.

STEP 5   Simplify by performing the addition or subtraction.

STEP 6   Look for ways to reduce the resulting fraction.

EXAMPLE 1   Add $\dfrac{8}{x^2 - 4x} + \dfrac{2}{x}$.

Solution

$$\frac{8}{x^2 - 4x} + \frac{2}{x} = \frac{8}{x(x - 4)} + \frac{2}{x}$$ Factor the denominators

The LCD is $x(x - 4)$.   Find the LCD.

$$= \frac{8}{x(x - 4)} + \left(\frac{2}{x}\right)\left(\frac{x - 4}{x - 4}\right)$$ Change each fraction to an equivalent fraction that has the LCD as its denominator.

$$= \frac{8 + 2(x - 4)}{x(x - 4)}$$ Combine numerators and place over the LCD.

$$= \frac{8 + 2x - 8}{x(x - 4)}$$ Simplify performing the addition or subtraction.

$$= \frac{2x}{x(x - 4)}$$

$$= \frac{2}{x - 4}$$ Reduce.   ∎

E X A M P L E   2   Subtract $\dfrac{a}{a^2 - 4} - \dfrac{3}{a + 2}$.

Solution

$$\frac{a}{a^2 - 4} - \frac{3}{a + 2} = \frac{a}{(a + 2)(a - 2)} - \frac{3}{a + 2}$$ Factor the denominators.

The LCD is $(a + 2)(a - 2)$. Find the LCD.

$$= \frac{a}{(a + 2)(a - 2)} - \left(\frac{3}{a + 2}\right)\left(\frac{a - 2}{a - 2}\right)$$ Change each fraction to an equivalent fraction that has the LCD as its denominator.

$$= \frac{a - 3(a - 2)}{(a + 2)(a - 2)}$$ Combine numerators and place over the LCD.

$$= \frac{a - 3a + 6}{(a + 2)(a - 2)}$$ Simplify performing the addition or subtraction.

$$= \frac{-2a + 6}{(a + 2)(a - 2)} \text{ or } \frac{-2(a - 3)}{(a + 2)(a - 2)}$$ ■

E X A M P L E   3   Add $\dfrac{3n}{n^2 + 6n + 5} + \dfrac{4}{n^2 - 7n - 8}$.

Solution

$$\frac{3n}{n^2 + 6n + 5} + \frac{4}{n^2 - 7n - 8}$$

$$= \frac{3n}{(n + 5)(n + 1)} + \frac{4}{(n - 8)(n + 1)}$$ Factor the denominators.

The LCD is $(n + 5)(n + 1)(n - 8)$. Find the LCD.

$$= \left(\frac{3n}{(n + 5)(n + 1)}\right)\left(\frac{n - 8}{n - 8}\right)$$

$$+ \left(\frac{4}{(n - 8)(n + 1)}\right)\left(\frac{n + 5}{n + 5}\right)$$ Change each fraction to an equivalent fraction that has the LCD as its denominator.

$$= \frac{3n(n - 8) + 4(n + 5)}{(n + 5)(n + 1)(n - 8)}$$ Combine numerators and place over the LCD.

$$= \frac{3n^2 - 24n + 4n + 20}{(n + 5)(n + 1)(n - 8)}$$ Simplify performing the addition or subtraction.

$$= \frac{3n^2 - 20n + 20}{(n + 5)(n + 1)(n - 8)}$$ ■

EXAMPLE 4   Perform the indicated operations.

$$\frac{2x^2}{x^4 - 1} + \frac{x}{x^2 - 1} - \frac{1}{x - 1}$$

Solution

$$\frac{2x^2}{x^4 - 1} + \frac{x}{x^2 - 1} - \frac{1}{x - 1}$$

$$= \frac{2x^2}{(x^2 + 1)(x + 1)(x - 1)} + \frac{x}{(x + 1)(x - 1)} - \frac{1}{x - 1}$$ Factor the denominators

The LCD is $(x^2 + 1)(x + 1)(x - 1)$.   Find the LCD.

$$= \frac{2x^2}{(x^2 + 1)(x + 1)(x - 1)}$$

$$+ \left(\frac{x}{(x + 1)(x - 1)} \right) \left(\frac{x^2 + 1}{x^2 + 1} \right)$$

$$- \left(\frac{1}{x - 1} \right) \frac{(x^2 + 1)(x + 1)}{(x^2 + 1)(x + 1)}$$

Change each fraction to an equivalent fraction that has the LCD as its denominator.

$$= \frac{2x^2 + x(x^2 + 1) - (x^2 + 1)(x + 1)}{(x^2 + 1)(x + 1)(x - 1)}$$ Combine numerators and place over the LCD.

$$= \frac{2x^2 + x^3 + x - x^3 - x^2 - x - 1}{(x^2 + 1)(x + 1)(x - 1)}$$ Simplify performing the addition or subtraction.

$$= \frac{x^2 - 1}{(x^2 + 1)(x + 1)(x - 1)}$$

$$= \frac{(x + 1)(x - 1)}{(x^2 + 1)(x + 1)(x - 1)}$$

$$= \frac{1}{x^2 + 1}$$ Reduce.   ■

Complex Fractions

Complex fractions are fractional forms that contain rational numbers or rational expressions in the numerators and/or denominators. The following are examples of complex fractions.

$$\frac{\dfrac{4}{x}}{\dfrac{2}{xy}} \qquad \frac{\dfrac{1}{2}+\dfrac{3}{4}}{\dfrac{5}{6}-\dfrac{3}{8}} \qquad \frac{\dfrac{3}{x}+\dfrac{2}{y}}{\dfrac{5}{x}-\dfrac{6}{y^2}} \qquad \frac{\dfrac{1}{x}+\dfrac{1}{y}}{2} \qquad \frac{-3}{\dfrac{2}{x}-\dfrac{3}{y}}$$

It is often necessary to **simplify** a complex fraction. We will take each of these five examples and examine some techniques for simplifying complex fractions.

EXAMPLE 5   Simplify $\dfrac{\dfrac{4}{x}}{\dfrac{2}{xy}}$.

Solution

This type of problem is a simple division problem.

$$\frac{\dfrac{4}{x}}{\dfrac{2}{xy}} = \frac{4}{x} \div \frac{2}{xy}$$

$$= \frac{4}{\overset{}{x}} \cdot \frac{xy}{\underset{2}{2}} = 2y \qquad \blacksquare$$

EXAMPLE 6   Simplify $\dfrac{\dfrac{1}{2}+\dfrac{3}{4}}{\dfrac{5}{6}-\dfrac{3}{8}}$.

Let's look at two possible ways to simplify such a problem.

Solution A

Here we will simplify the numerator by performing the addition and simplify the denominator by performing the subtraction. Then the problem is a simple division problem like Example 5.

$$\frac{\dfrac{1}{2}+\dfrac{3}{4}}{\dfrac{5}{6}-\dfrac{3}{8}} = \frac{\dfrac{2}{4}+\dfrac{3}{4}}{\dfrac{20}{24}-\dfrac{9}{24}}$$

$$= \frac{\dfrac{5}{4}}{\dfrac{11}{24}} = \frac{5}{\overset{}{4}} \cdot \frac{\overset{6}{24}}{11}$$

$$= \frac{30}{11}$$

Solution B

Here we find the LCD of all four denominators (2, 4, 6, and 8). The LCD is 24. Use this LCD to multiply the entire complex fraction by a form of 1, specifically $\dfrac{24}{24}$.

$$\frac{\dfrac{1}{2} + \dfrac{3}{4}}{\dfrac{5}{6} - \dfrac{3}{8}} = \left(\frac{24}{24}\right)\left(\frac{\dfrac{1}{2} + \dfrac{3}{4}}{\dfrac{5}{6} - \dfrac{3}{8}}\right)$$

$$= \frac{24\left(\dfrac{1}{2} + \dfrac{3}{4}\right)}{24\left(\dfrac{5}{6} - \dfrac{3}{8}\right)}$$

$$= \frac{24\left(\dfrac{1}{2}\right) + 24\left(\dfrac{3}{4}\right)}{24\left(\dfrac{5}{6}\right) - 24\left(\dfrac{3}{8}\right)}$$

$$= \frac{12 + 18}{20 - 9} = \frac{30}{11}$$

E X A M P L E   7

Simplify $\dfrac{\dfrac{3}{x} + \dfrac{2}{y}}{\dfrac{5}{x} - \dfrac{6}{y^2}}$.

Solution A

Simplify the numerator and the denominator. Then the problem becomes a division problem.

$$\frac{\dfrac{3}{x} + \dfrac{2}{y}}{\dfrac{5}{x} - \dfrac{6}{y^2}} = \frac{\left(\dfrac{3}{x}\right)\left(\dfrac{y}{y}\right) + \left(\dfrac{2}{y}\right)\left(\dfrac{x}{x}\right)}{\left(\dfrac{5}{x}\right)\left(\dfrac{y^2}{y^2}\right) - \left(\dfrac{6}{y^2}\right)\left(\dfrac{x}{x}\right)}$$

$$= \frac{\dfrac{3y}{xy} + \dfrac{2x}{xy}}{\dfrac{5y^2}{xy^2} - \dfrac{6x}{xy^2}}$$

$$= \frac{\dfrac{3y + 2x}{xy}}{\dfrac{5y^2 - 6x}{xy^2}}$$

$$= \frac{3y + 2x}{xy} \div \frac{5y^2 - 6x}{xy^2}$$

$$= \frac{3y + 2x}{\cancel{xy}} \cdot \frac{\overset{y}{\cancel{xy^2}}}{5y^2 - 6x}$$

$$= \frac{y(3y + 2x)}{5y^2 - 6x}$$

Solution B

Here we find the LCD of all four denominators (x, y, x, and y^2). The LCD is xy^2. Use this LCD to multiply the entire complex fraction by a form of 1, specifically $\dfrac{xy^2}{xy^2}$.

$$\frac{\dfrac{3}{x} + \dfrac{2}{y}}{\dfrac{5}{x} - \dfrac{6}{y^2}} = \left(\frac{xy^2}{xy^2}\right)\left(\frac{\dfrac{3}{x} + \dfrac{2}{y}}{\dfrac{5}{x} - \dfrac{6}{y^2}}\right)$$

$$= \frac{xy^2\left(\dfrac{3}{x} + \dfrac{2}{y}\right)}{xy^2\left(\dfrac{5}{x} - \dfrac{6}{y^2}\right)}$$

$$= \frac{xy^2\left(\dfrac{3}{x}\right) + xy^2\left(\dfrac{2}{y}\right)}{xy^2\left(\dfrac{5}{x}\right) - xy^2\left(\dfrac{6}{y^2}\right)}$$

$$= \frac{3y^2 + 2xy}{5y^2 - 6x} \quad \text{or} \quad \frac{y(3y + 2x)}{5y^2 - 6x} \qquad ∎$$

Certainly either approach (Solution A or Solution B) will work with problems such as Examples 6 and 7. Examine Solution B in both examples carefully. This approach works effectively with complex fractions where the LCD of all the denominators is easy to find. (Don't be misled by the length of Solution B for Example 6; we were especially careful to show every step.)

EXAMPLE 8   Simplify $\dfrac{\dfrac{1}{x} + \dfrac{1}{y}}{2}$.

Solution

The number 2 can be written as $\dfrac{2}{1}$; thus the LCD of all three denominators (x, y, and 1) is xy. Therefore, let's multiply the entire complex fraction by a form of 1, specifically $\dfrac{xy}{xy}$.

$$\left(\dfrac{\dfrac{1}{x} + \dfrac{1}{y}}{\dfrac{2}{1}} \right)\left(\dfrac{xy}{xy} \right) = \dfrac{xy\left(\dfrac{1}{x}\right) + xy\left(\dfrac{1}{y}\right)}{2xy}$$

$$= \dfrac{y + x}{2xy}$$ ■

EXAMPLE 9   Simplify $\dfrac{-3}{\dfrac{2}{x} - \dfrac{3}{y}}$.

Solution

$$\left(\dfrac{\dfrac{-3}{1}}{\dfrac{2}{x} - \dfrac{3}{y}} \right)\left(\dfrac{xy}{xy} \right) = \dfrac{-3(xy)}{xy\left(\dfrac{2}{x}\right) - xy\left(\dfrac{3}{y}\right)}$$

$$= \dfrac{-3xy}{2y - 3x}$$ ■

Let's conclude this section with an example that has a complex fraction as part of an algebraic expression.

EXAMPLE 10   Simplify $1 - \dfrac{n}{1 - \dfrac{1}{n}}$.

Solution

First simplify the complex fraction $\dfrac{n}{1 - \dfrac{1}{n}}$ by multiplying by $\dfrac{n}{n}$.

$$\left(\dfrac{n}{1 - \dfrac{1}{n}}\right)\left(\dfrac{n}{n}\right) = \dfrac{n^2}{n - 1}$$

Now we can perform the subtraction.

$$1 - \dfrac{n^2}{n - 1} = \left(\dfrac{n - 1}{n - 1}\right)\left(\dfrac{1}{1}\right) - \dfrac{n^2}{n - 1}$$

$$= \dfrac{n - 1}{n - 1} - \dfrac{n^2}{n - 1}$$

$$= \dfrac{n - 1 - n^2}{n - 1} \quad \text{or} \quad \dfrac{-n^2 + n - 1}{n - 1}$$

PROBLEM SET 4.4

For Problems 1–36, perform the indicated operations and express your answers in simplest form.

1. $\dfrac{2x}{x^2 + 4x} + \dfrac{5}{x}$

2. $\dfrac{3x}{x^2 - 6x} + \dfrac{4}{x}$

3. $\dfrac{4}{x^2 + 7x} - \dfrac{1}{x}$

4. $\dfrac{-10}{x^2 - 9x} - \dfrac{2}{x}$

5. $\dfrac{x}{x^2 - 1} + \dfrac{5}{x + 1}$

6. $\dfrac{2x}{x^2 - 16} + \dfrac{7}{x - 4}$

7. $\dfrac{6a + 4}{a^2 - 1} - \dfrac{5}{a - 1}$

8. $\dfrac{4a - 4}{a^2 - 4} - \dfrac{3}{a + 2}$

9. $\dfrac{2n}{n^2 - 25} - \dfrac{3}{4n + 20}$

10. $\dfrac{3n}{n^2 - 36} - \dfrac{2}{5n + 30}$

11. $\dfrac{5}{x} - \dfrac{5x - 30}{x^2 + 6x} + \dfrac{x}{x + 6}$

12. $\dfrac{3}{x + 1} + \dfrac{x + 5}{x^2 - 1} - \dfrac{3}{x - 1}$

13. $\dfrac{3}{x^2 + 9x + 14} + \dfrac{5}{2x^2 + 15x + 7}$

14. $\dfrac{6}{x^2 + 11x + 24} + \dfrac{4}{3x^2 + 13x + 12}$

15. $\dfrac{1}{a^2 - 3a - 10} - \dfrac{4}{a^2 + 4a - 45}$

16. $\dfrac{6}{a^2 - 3a - 54} - \dfrac{10}{a^2 + 5a - 6}$

17. $\dfrac{3a}{8a^2 - 2a - 3} + \dfrac{1}{4a^2 + 13a - 12}$

18. $\dfrac{2a}{6a^2 + 13a - 5} + \dfrac{a}{2a^2 + a - 10}$

19. $\dfrac{5}{x^2 + 3} - \dfrac{2}{x^2 + 4x - 21}$

20. $\dfrac{7}{x^2 + 1} - \dfrac{3}{x^2 + 7x - 60}$

21. $\dfrac{2}{y^2 + 6y - 16} - \dfrac{4}{y + 8} - \dfrac{3}{y - 2}$

22. $\dfrac{7}{y - 6} - \dfrac{10}{y + 12} + \dfrac{4}{y^2 + 6y - 72}$

23. $x - \dfrac{x^2}{x - 2} + \dfrac{3}{x^2 - 4}$

24. $x + \dfrac{5}{x^2 - 25} - \dfrac{x^2}{x + 5}$

25. $\dfrac{x + 3}{x + 10} + \dfrac{4x - 3}{x^2 + 8x - 20} + \dfrac{x - 1}{x - 2}$

26. $\dfrac{2x - 1}{x + 3} + \dfrac{x + 4}{x - 6} + \dfrac{3x - 1}{x^2 - 3x - 18}$

27. $\dfrac{n}{n - 6} + \dfrac{n + 3}{n + 8} + \dfrac{12n + 26}{n^2 + 2n - 48}$

28. $\dfrac{n - 1}{n + 4} + \dfrac{n}{n + 6} + \dfrac{2n + 18}{n^2 + 10n + 24}$

29. $\dfrac{4x - 3}{2x^2 + x - 1} - \dfrac{2x + 7}{3x^2 + x - 2} - \dfrac{3}{3x - 2}$

30. $\dfrac{2x + 5}{x^2 + 3x - 18} - \dfrac{3x - 1}{x^2 + 4x - 12} + \dfrac{5}{x - 2}$

31. $\dfrac{n}{n^2 + 1} + \dfrac{n^2 + 3n}{n^4 - 1} - \dfrac{1}{n - 1}$

32. $\dfrac{2n^2}{n^4 - 16} - \dfrac{n}{n^2 - 4} + \dfrac{1}{n + 2}$

33. $\dfrac{15x^2 - 10}{5x^2 - 7x + 2} - \dfrac{3x + 4}{x - 1} - \dfrac{2}{5x - 2}$

34. $\dfrac{32x + 9}{12x^2 + x - 6} - \dfrac{3}{4x + 3} - \dfrac{x + 5}{3x - 2}$

35. $\dfrac{t + 3}{3t - 1} + \dfrac{8t^2 + 8t + 2}{3t^2 - 7t + 2} - \dfrac{2t + 3}{t - 2}$

36. $\dfrac{t - 3}{2t + 1} + \dfrac{2t^2 + 19t - 46}{2t^2 - 9t - 5} - \dfrac{t + 4}{t - 5}$

For Problems 37–60, simplify each complex fraction.

37. $\dfrac{\dfrac{1}{2} - \dfrac{1}{4}}{\dfrac{5}{8} + \dfrac{3}{4}}$

38. $\dfrac{\dfrac{3}{8} + \dfrac{3}{4}}{\dfrac{5}{8} - \dfrac{7}{12}}$

39. $\dfrac{\dfrac{3}{28} - \dfrac{5}{14}}{\dfrac{5}{7} + \dfrac{1}{4}}$

40. $\dfrac{\dfrac{5}{9} + \dfrac{7}{36}}{\dfrac{3}{18} - \dfrac{5}{12}}$

41. $\dfrac{\dfrac{5}{6y}}{\dfrac{10}{3xy}}$

42. $\dfrac{\dfrac{9}{8xy^2}}{\dfrac{5}{4x^2}}$

43. $\dfrac{\dfrac{3}{x} - \dfrac{2}{y}}{\dfrac{4}{y} - \dfrac{7}{xy}}$

44. $\dfrac{\dfrac{9}{x} + \dfrac{7}{x^2}}{\dfrac{5}{y} + \dfrac{3}{y^2}}$

45. $\dfrac{\dfrac{6}{a} - \dfrac{5}{b^2}}{\dfrac{12}{a^2} + \dfrac{2}{b}}$

46. $\dfrac{\dfrac{4}{ab} - \dfrac{3}{b^2}}{\dfrac{1}{a} + \dfrac{3}{b}}$

47. $\dfrac{\dfrac{2}{x} - 3}{\dfrac{3}{y} + 4}$

48. $\dfrac{1 + \dfrac{3}{x}}{1 - \dfrac{6}{x}}$

49. $\dfrac{3 + \dfrac{2}{n + 4}}{5 - \dfrac{1}{n + 4}}$

50. $\dfrac{4 + \dfrac{6}{n - 1}}{7 - \dfrac{4}{n - 1}}$

51. $\dfrac{5 - \dfrac{2}{n - 3}}{4 - \dfrac{1}{n - 3}}$

52. $\dfrac{\dfrac{3}{n - 5} - 2}{1 - \dfrac{4}{n - 5}}$

53. $\dfrac{\dfrac{-1}{y - 2} + \dfrac{5}{x}}{\dfrac{3}{x} - \dfrac{4}{xy - 2x}}$

54. $\dfrac{\dfrac{-2}{x} - \dfrac{4}{x + 2}}{\dfrac{3}{x^2 + 2x} + \dfrac{3}{x}}$

55. $\dfrac{\dfrac{2}{x - 3} - \dfrac{3}{x + 3}}{\dfrac{5}{x^2 - 9} - \dfrac{2}{x - 3}}$

56. $\dfrac{\dfrac{2}{x - y} + \dfrac{3}{x + y}}{\dfrac{5}{x + y} - \dfrac{1}{x^2 - y^2}}$

57. $\dfrac{\dfrac{3a}{2 - \dfrac{1}{a}} - 1}$

58. $\dfrac{\dfrac{a}{\dfrac{1}{a} + 4} + 1}$

59. $2 - \dfrac{x}{3 - \dfrac{2}{x}}$

60. $1 + \dfrac{x}{1 + \dfrac{1}{x}}$

---

61. Which of the two techniques presented in the text

would you use to simplify $\dfrac{\dfrac{1}{4} + \dfrac{1}{3}}{\dfrac{3}{4} - \dfrac{1}{6}}$? Which technique

would you use to simplify $\dfrac{\dfrac{3}{8} - \dfrac{5}{7}}{\dfrac{7}{9} + \dfrac{6}{25}}$? Explain your choice for each problem.

62. Give a step-by-step description of how to do the following addition problem.

$$\frac{3x + 4}{8} + \frac{5x - 2}{12}$$

4.5 Dividing Polynomials

In Chapter 3, we saw how the property $\dfrac{b^n}{b^m} = b^{n-m}$, along with our knowledge of dividing integers, is used to divide monomials. For example,

$$\frac{12x^3}{3x} = 4x^2 \qquad \frac{-36x^4y^5}{4xy^2} = -9x^3y^3$$

In Section 4.3, we used $\dfrac{a}{b} + \dfrac{c}{b} = \dfrac{a + c}{b}$ and $\dfrac{a}{b} - \dfrac{c}{b} = \dfrac{a - c}{b}$ as the basis for adding and subtracting rational expressions. These same equalities, viewed as $\dfrac{a + b}{c} = \dfrac{a}{c} + \dfrac{b}{c}$ and $\dfrac{a - c}{b} = \dfrac{a}{b} - \dfrac{c}{b}$, along with our knowledge of dividing monomials, provide the basis for dividing polynomials by monomials. Consider the following examples.

$$\frac{18x^3 + 24x^2}{6x} = \frac{18x^3}{6x} + \frac{24x^2}{6x} = 3x^2 + 4x$$

$$\frac{35x^2y^3 - 55x^3y^4}{5xy^2} = \frac{35x^2y^3}{5xy^2} - \frac{55x^3y^4}{5xy^2} = 7xy - 11x^2y^2$$

To divide a polynomial by a monomial, we divide each term of the polynomial by the monomial. As with many skills, once you feel comfortable with the process, you may then want to perform some of the steps mentally. Your work could take on the following format.

$$\frac{40x^4y^5 + 72x^5y^7}{8x^2y} = 5x^2y^4 + 9x^3y^6 \qquad \frac{36a^3b^4 - 45a^4b^6}{-9a^2b^3} = -4ab + 5a^2b^3$$

In Section 4.1, we saw that a fraction like $\dfrac{3x^2 + 11x - 4}{x + 4}$ can be simplified as follows:

$$\frac{3x^2 + 11x - 4}{x + 4} = \frac{(3x - 1)(x + 4)}{x + 4} = 3x - 1$$

We can obtain the same result by using a dividing process similar to long division in arithmetic.

STEP 1 Use the conventional long-division format, and arrange both the dividend and the divisor in descending powers of the variable.

$$x + 4\overline{)3x^2 + 11x - 4}$$

STEP 2 Find the first term of the quotient by dividing the first term of the dividend by the first term of the divisor.

$$\begin{array}{r} 3x \\ x + 4\overline{)3x^2 + 11x - 4} \end{array}$$

STEP 3 Multiply the entire divisor by the term of the quotient found in Step 2, and position the product to be subtracted from the dividend.

$$\begin{array}{r} 3x \\ x + 4\overline{)3x^2 + 11x - 4} \\ 3x^2 + 12x \end{array}$$

STEP 4 Subtract.

$$\begin{array}{r} 3x \\ x + 4\overline{)3x^2 + 11x - 4} \\ 3x^2 + 12x \\ \hline -x - 4 \end{array}$$

Remember to add the opposite! $\longrightarrow$

$(3x^2 + 11x - 4) - (3x^2 + 12x) = -x - 4$ $\longrightarrow$

STEP 5 Repeat the process beginning with Step 2; use the polynomial that resulted from the subtraction in Step 4 as a new dividend.

$$\begin{array}{r} 3x \;\; -1 \\ x + 4\overline{)3x^2 + 11x - 4} \\ 3x^2 + 12x \\ \hline -x - 4 \\ \underline{-x - 4} \end{array}$$

In the next example, let's *think* in terms of the previous step-by-step procedure but arrange our work in a more compact form.

E X A M P L E 1 Divide $5x^2 + 6x - 8$ by $x + 2$.

Solution

$$\begin{array}{r} 5x \;\; -4 \\ x + 2\overline{)5x^2 + 6x - 8} \\ 5x^2 + 10x \\ \hline -\;4x - 8 \\ \underline{-\;4x - 8} \\ 0 \end{array}$$

Think Steps

1. $\dfrac{5x^2}{x} = 5x$.
2. $5x(x + 2) = 5x^2 + 10x$.
3. $(5x^2 + 6x - 8) - (5x^2 + 10x) = -4x - 8$.
4. $\dfrac{-4x}{x} = -4$.
5. $-4(x + 2) = -4x - 8$. ∎

Recall that to check a division problem, we can multiply the divisor times the quotient and add the remainder. In other words,

Dividend = (Divisor)(Quotient) + (Remainder)

Sometimes the remainder is expressed as a fractional part of the divisor. The relationship then becomes

$$\frac{\text{Dividend}}{\text{Divisor}} = \text{Quotient} + \frac{\text{Remainder}}{\text{Divisor}}$$

EXAMPLE 2   Divide $2x^2 - 3x + 1$ by $x - 5$.

Solution

$$
\require{enclose}
\begin{array}{r}
2x \;+\; 7 \\
x - 5 \enclose{longdiv}{2x^2 \;-\; 3x \;+\; 1} \\
\underline{2x^2 \;-\; 10x} \\
7x \;+\; 1 \\
\underline{7x \;-\; 35} \\
36 \quad \longleftarrow \quad \text{Remainder}
\end{array}
$$

Thus

$$\frac{2x^2 - 3x + 1}{x - 5} = 2x + 7 + \frac{36}{x - 5} \qquad x \neq 5$$

 Check

$$(x - 5)(2x + 7) + 36 \stackrel{?}{=} 2x^2 - 3x + 1$$

$$2x^2 - 3x - 35 + 36 \stackrel{?}{=} 2x^2 - 3x + 1$$

$$2x^2 - 3x + 1 = 2x^2 - 3x + 1 \qquad \blacksquare$$

Each of the next two examples illustrates another point regarding the division process. Study them carefully, and then you should be ready to work the exercises in the next problem set.

EXAMPLE 3   Divide $t^3 - 8$ by $t - 2$.

 Solution

$$
\require{enclose}
\begin{array}{r}
t^2 + 2t \;\;+ 4 \\
t - 2 \enclose{longdiv}{t^3 + 0t^2 + 0t - 8} \\
\underline{t^3 - 2t^2} \\
2t^2 + 0t - 8 \\
\underline{2t^2 - 4t} \\
4t - 8 \\
\underline{4t - 8} \\
0
\end{array}
$$

Note the insertion of a "t-squared" term and a "t term" with zero coefficients.

Check this result!   ◼

EXAMPLE 4

Divide $y^3 + 3y^2 - 2y - 1$ by $y^2 + 2y$.

Solution

$$
\begin{array}{r}
y + 1 \\
y^2 + 2y \overline{\smash{\big)}\ y^3 + 3y^2 - 2y - 1} \\
\underline{y^3 + 2y^2 } \\
y^2 - 2y - 1 \\
\underline{y^2 + 2y } \\
-4y - 1 \quad\longleftarrow\quad \text{Remainder of } -4y - 1
\end{array}
$$

(The division process is complete when the degree of the remainder is less than the degree of the divisor.) Thus

$$
\frac{y^3 + 3y^2 - 2y - 1}{y^2 + 2y} = y + 1 + \frac{-4y - 1}{y^2 + 2y}
$$

∎

If the divisor is of the form $x - k$, where the coefficient of the x term is 1, then the format of the division process described in this section can be simplified by a procedure called **synthetic division.** This procedure is a shortcut for this type of polynomial division. If you are continuing on to study college algebra, then you will want to know synthetic division. If you are not continuing on to college algebra, then you probably will not need a shortcut and the long-division process will be sufficient.

First, let's consider an example and use the usual division process. Then, in a step-by-step fashion, we can observe some shortcuts that will lead us into the synthetic-division procedure. Consider the division problem $(2x^4 + x^3 - 17x^2 + 13x + 2) \div (x - 2)$

$$
\begin{array}{r}
2x^3 + 5x^2 - 7x - 1 \\
x - 2 \overline{\smash{\big)}\ 2x^4 + x^3 - 17x^2 + 13x + 2} \\
\underline{2x^4 - 4x^3 } \\
5x^3 - 17x^2 \\
\underline{5x^3 - 10x^2 } \\
-7x^2 + 13x \\
\underline{-7x^2 + 14x } \\
-x + 2 \\
\underline{-x + 2}
\end{array}
$$

Note that because the dividend $(2x^4 + x^3 - 17x^2 + 13x + 2)$ is written in descending powers of x, the quotient $(2x^3 + 5x^2 - 7x - 1)$ is produced, also in descending powers of x. In other words, the numerical coefficients are the important numbers. Thus let's rewrite this problem in terms of its coefficients.

$$
\begin{array}{r}
2 + 5 \;-\; 7 \;-\; 1 \\[2pt]
1 - 2\,\overline{)2 + 1 - 17 + 13 + 2}\\[2pt]
②- 4 \\[2pt]
\hline
5 \;\ominus\!17 \\[2pt]
⑤- 10 \\[2pt]
\hline
-7 \;+⑬ \\[2pt]
\ominus7 + 14 \\[2pt]
\hline
-1 \;+② \\[2pt]
\ominus1 + 2 \\[2pt]
\hline
\end{array}
$$

Now observe that the numbers that are circled are simply repetitions of the numbers directly above them in the format. Therefore, we can write the process in a more compact form as

$$
\begin{array}{r}
2 \;\; 5 - 7 - 1 \qquad\qquad \textbf{(1)}\\[2pt]
-2\,\overline{)2 \;\; 1 - 17 - 13 \;\; 2}\qquad \textbf{(2)}\\[2pt]
-4 \;-10 \quad 14 \;\; 2 \qquad \textbf{(3)}\\[2pt]
\hline
5 - 7 - 1 \;\; 0 \qquad \textbf{(4)}
\end{array}
$$

where the repetitions are omitted and where 1, the coefficient of x in the divisor, is omitted.

Note that line (4) reveals all of the coefficients of the quotient, line (1), except for the first coefficient of 2. Thus we can begin line (4) with the first coefficient and then use the following form.

$$
\begin{array}{r}
-2\,\overline{)2 \;\; 1 - 17 \quad 13 \quad 2}\qquad \textbf{(5)}\\[2pt]
-4 \;-10 \quad 14 \quad 2 \qquad \textbf{(6)}\\[2pt]
\hline
2 \;\; 5 - 7 - 1 \;\; 0 \qquad \textbf{(7)}
\end{array}
$$

Line (7) contains the coefficients of the quotient, where the 0 indicates the remainder.

Finally, by changing the constant in the divisor to 2 (instead of -2), we can add the corresponding entries in lines (5) and (6) rather than subtract. Thus the final synthetic division form for this problem is

$$
\begin{array}{r}
2\,\overline{)2 \;\; 1 - 17 \quad 13 \quad 2}\\[2pt]
4 \quad 10 \; -14 \; -2 \\[2pt]
\hline
2 \;\; 5 - 7 - 1 \;\; 0
\end{array}
$$

Now let's consider another problem that illustrates a step-by-step procedure for carrying out the synthetic-division process. Suppose that we want to divide $3x^3 - 2x^2 + 6x - 5$ by $x + 4$.

STEP 1   Write the coefficients of the dividend as follows:

$$
\overline{)3 \;\; -2 \;\; 6 \;\; -5}
$$

STEP 2   In the divisor, $(x + 4)$, use -4 instead of 4 so that later we can add rather than subtract.

$$
-4\,\overline{)3 \;\; -2 \;\; 6 \;\; -5}
$$

STEP 3 Bring down the first coeffecient of the dividend (3).

$$-4\overline{)3\quad -2\quad 6\quad -5}$$
$$\overline{3}$$

STEP 4 Multiply$(3)(-4)$, which yields -12; this result is to be added to the second coefficient of the dividend (-2).

$$-4\overline{)3\quad -\ 2\quad 6\quad -5}$$
$$\underline{-12}$$
$$3\quad -14$$

STEP 5 Multiply $(-14)(-4)$, which yields 56; this result is to be added to the third coefficient of the dividend (6).

$$-4\overline{)3\quad -\ 2\quad\ 6\quad -5}$$
$$\underline{-12\quad 56}$$
$$3\quad -14\quad 62$$

STEP 6 Multiply $(62)(-4)$, which yields -248; this result is added to the last term of the dividend (-5).

$$-4\overline{)3\quad -\ 2\quad\ 6\quad -\ 5}$$
$$\underline{-12\quad 56\quad -248}$$
$$3\quad -14\quad 62\quad -253$$

The last row indicates a quotient of $3x^2 - 14x + 62$ and a remainder of -253. Thus we have

$$\frac{3x^3 - 2x^2 + 6x - 5}{x + 4} = 3x^2 - 14x + 62 - \frac{253}{x + 4}$$

We will consider one more example, which shows only the final, compact form for synthetic division.

EXAMPLE 5 Find the quotient and remainder for $(4x^4 - 2x^3 + 6x - 1) \div (x - 1)$.

Solution

$$1\overline{)4\quad -2\quad 0\quad 6\quad -1}$$
$$\underline{\quad\ \ 4\quad 2\quad 2\quad\ \ 8}$$
$$4\quad\ \ 2\quad 2\quad 8\quad\ \ 7$$

Note that a zero has been inserted as the coefficient of the missing x^2 term.

Therefore,

$$\frac{4x^4 - 2x^3 + 6x - 1}{x - 1} = 4x^3 + 2x^2 + 2x + 8 + \frac{7}{x - 1}$$

PROBLEM SET 4.5

For Problems 1–10, perform the indicated divisions of polynomials by monomials.

1. $\dfrac{9x^4 + 18x^3}{3x}$

2. $\dfrac{12x^3 - 24x^2}{6x^2}$

3. $\dfrac{-24x^6 + 36x^8}{4x^2}$

4. $\dfrac{-35x^5 - 42x^3}{-7x^2}$

5. $\dfrac{15a^3 - 25a^2 - 40a}{5a}$

6. $\dfrac{-16a^4 + 32a^3 - 56a^2}{-8a}$

7. $\dfrac{13x^3 - 17x^2 + 28x}{-x}$

8. $\dfrac{14xy - 16x^2y^2 - 20x^3y^4}{-xy}$

9. $\dfrac{-18x^2y^2 + 24x^3y^2 - 48x^2y^3}{6xy}$

10. $\dfrac{-27a^3b^4 - 36a^2b^3 + 72a^2b^5}{9a^2b^2}$

For Problems 11–52, perform the indicated divisions.

11. $\dfrac{x^2 - 7x - 78}{x + 6}$

12. $\dfrac{x^2 + 11x - 60}{x - 4}$

13. $(x^2 + 12x - 160) \div (x - 8)$

14. $(x^2 - 18x - 175) \div (x + 7)$

15. $\dfrac{2x^2 - x - 4}{x - 1}$

16. $\dfrac{3x^2 - 2x - 7}{x + 2}$

17. $\dfrac{15x^2 + 22x - 5}{3x + 5}$

18. $\dfrac{12x^2 - 32x - 35}{2x - 7}$

19. $\dfrac{3x^3 + 7x^2 - 13x - 21}{x + 3}$

20. $\dfrac{4x^3 - 21x^2 + 3x + 10}{x - 5}$

21. $(2x^3 + 9x^2 - 17x + 6) \div (2x - 1)$

22. $(3x^3 - 5x^2 - 23x - 7) \div (3x + 1)$

23. $(4x^3 - x^2 - 2x + 6) \div (x - 2)$

24. $(6x^3 - 2x^2 + 4x - 3) \div (x + 1)$

25. $(x^4 - 10x^3 + 19x^2 + 33x - 18) \div (x - 6)$

26. $(x^4 + 2x^3 - 16x^2 + x + 6) \div (x - 3)$

27. $\dfrac{x^3 - 125}{x - 5}$

28. $\dfrac{x^3 + 64}{x + 4}$

29. $(x^3 + 64) \div (x + 1)$

30. $(x^3 - 8) \div (x - 4)$

31. $(2x^3 - x - 6) \div (x + 2)$

32. $(5x^3 + 2x - 3) \div (x - 2)$

33. $\dfrac{4a^2 - 8ab + 4b^2}{a - b}$

34. $\dfrac{3x^2 - 2xy - 8y^2}{x - 2y}$

35. $\dfrac{4x^3 - 5x^2 + 2x - 6}{x^2 - 3x}$

36. $\dfrac{3x^3 + 2x^2 - 5x - 1}{x^2 + 2x}$

37. $\dfrac{8y^3 - y^2 - y + 5}{y^2 + y}$

38. $\dfrac{5y^3 - 6y^2 - 7y - 2}{y^2 - y}$

39. $(2x^3 + x^2 - 3x + 1) \div (x^2 + x - 1)$

40. $(3x^3 - 4x^2 + 8x + 8) \div (x^2 - 2x + 4)$

41. $(4x^3 - 13x^2 + 8x - 15) \div (4x^2 - x + 5)$

42. $(5x^3 + 8x^2 - 5x - 2) \div (5x^2 - 2x - 1)$

43. $(5a^3 + 7a^2 - 2a - 9) \div (a^2 + 3a - 4)$

44. $(4a^3 - 2a^2 + 7a - 1) \div (a^2 - 2a + 3)$

45. $(2n^4 + 3n^3 - 2n^2 + 3n - 4) \div (n^2 + 1)$

46. $(3n^4 + n^3 - 7n^2 - 2n + 2) \div (n^2 - 2)$

47. $(x^5 - 1) \div (x - 1)$ **48.** $(x^5 + 1) \div (x + 1)$

49. $(x^4 - 1) \div (x + 1)$ **50.** $(x^4 - 1) \div (x - 1)$

51. $(3x^4 + x^3 - 2x^2 - x + 6) \div (x^2 - 1)$

52. $(4x^3 - 2x^2 + 7x - 5) \div (x^2 + 2)$

For problems 53–64, use synthetic division to determine the quotient and remainder.

53. $(x^2 - 8x + 12) \div (x - 2)$

54. $(x^2 + 9x + 18) \div (x + 3)$

55. $(x^2 + 2x - 10) \div (x - 4)$

56. $(x^2 - 10x + 15) \div (x - 8)$

57. $(x^3 - 2x^2 - x + 2) \div (x - 2)$

58. $(x^3 - 5x^2 + 2x + 8) \div (x + 1)$

59. $(x^3 - 7x - 6) \div (x + 2)$

60. $(x^3 + 6x^2 - 5x - 1) \div (x - 1)$

61. $(2x^3 - 5x^2 - 4x + 6) \div (x - 2)$

62. $(3x^4 - x^3 + 2x^2 - 7x - 1) \div (x + 1)$

63. $(x^4 + 4x^3 - 7x - 1) \div (x - 3)$

64. $(2x^4 + 3x^2 + 3) \div (x + 2)$

■ ■ ■ **Thoughts into words**

65. Describe the process of long division of polynomials.

66. Give a step-by-step description of how you would do the following division problem.

$$(4 - 3x - 7x^3) \div (x + 6)$$

67. How do you know by inspection that $3x^2 + 5x + 1$ cannot be the correct answer for the division problem $(3x^3 - 7x^2 - 22x + 8) \div (x - 4)$?

4.6   Fractional Equations

The fractional equations used in this text are of two basic types. One type has only constants as denominators, and the other type contains variables in the denominators.

In Chapter 2, we considered fractional equations that involve only constants in the denominators. Let's briefly review our approach to solving such equations, because we will be using that same basic technique to solve any type of fractional equation.

E X A M P L E   1   Solve $\dfrac{x - 2}{3} + \dfrac{x + 1}{4} = \dfrac{1}{6}$.

Solution

$$\frac{x - 2}{3} + \frac{x + 1}{4} = \frac{1}{6}$$

$$12\left(\frac{x - 2}{3} + \frac{x + 1}{4}\right) = 12\left(\frac{1}{6}\right) \qquad \begin{array}{l}\text{Multiply both sides by 12, which is the} \\ \text{LCD of all of the denominators.}\end{array}$$

$$4(x - 2) + 3(x + 1) = 2$$

$$4x - 8 + 3x + 3 = 2$$

$$7x - 5 = 2$$

$$7x = 7$$

$$x = 1$$

The solution set is $\{1\}$. Check it!   ■

If an equation contains a variable (or variables) in one or more denominators, then we proceed in essentially the same way as in Example 1 **except that we must avoid any value of the variable that makes a denominator zero.** Consider the following examples.

EXAMPLE 2   Solve $\dfrac{5}{n} + \dfrac{1}{2} = \dfrac{9}{n}$.

Solution

First, we need to realize that n cannot equal zero. (Let's indicate this restriction so that it is not forgotten!) Then we can proceed.

$$\frac{5}{n} + \frac{1}{2} = \frac{9}{n}, \qquad n \neq 0$$

$$2n\left(\frac{5}{n} + \frac{1}{2}\right) = 2n\left(\frac{9}{n}\right) \qquad \text{Multiply both sides by the LCD, which is } 2n.$$

$$10 + n = 18$$

$$n = 8$$

The solution set is {8}. Check it!

EXAMPLE 3   Solve $\dfrac{35 - x}{x} = 7 + \dfrac{3}{x}$.

Solution

$$\frac{35 - x}{x} = 7 + \frac{3}{x}, \qquad x \neq 0$$

$$x\left(\frac{35 - x}{x}\right) = x\left(7 + \frac{3}{x}\right) \qquad \text{Multiply both sides by } x.$$

$$35 - x = 7x + 3$$

$$32 = 8x$$

$$4 = x$$

The solution set is {4}.

EXAMPLE 4   Solve $\dfrac{3}{a - 2} = \dfrac{4}{a + 1}$.

Solution

$$\frac{3}{a - 2} = \frac{4}{a + 1}, \qquad a \neq 2 \text{ and } a \neq -1$$

$$(a - 2)(a + 1)\left(\frac{3}{a - 2}\right) = (a - 2)(a + 1)\left(\frac{4}{a + 1}\right) \qquad \begin{array}{l}\text{Multiply both sides} \\ \text{by } (a - 2)(a + 1).\end{array}$$

$$3(a + 1) = 4(a - 2)$$

$$3a + 3 = 4a - 8$$

$$11 = a$$

The solution set is {11}. ∎

Keep in mind that listing the restrictions at the beginning of a problem does not replace checking the potential solutions. In Example 4, the answer 11 needs to be checked in the original equation.

E X A M P L E   5   Solve $\dfrac{a}{a-2} + \dfrac{2}{3} = \dfrac{2}{a-2}$.

Solution

$$\frac{a}{a-2} + \frac{2}{3} = \frac{2}{a-2}, \qquad a \neq 2$$

$$3(a-2)\left(\frac{a}{a-2} + \frac{2}{3}\right) = 3(a-2)\left(\frac{2}{a-2}\right) \qquad \text{Multiply both sides} \\ \text{by } 3(a-2).$$

$$3a + 2(a-2) = 6$$

$$3a + 2a - 4 = 6$$

$$5a = 10$$

$$a = 2$$

Because our initial restriction was $a \neq 2$, we conclude that this equation has no solution. Thus the solution set is $\varnothing$. ∎

Ratio and Proportion

A **ratio** is the comparison of two numbers by division. We often use the fractional form to express ratios. For example, we can write the ratio of a to b as $\dfrac{a}{b}$. A statement of equality between two ratios is called a **proportion.** Thus, if $\dfrac{a}{b}$ and $\dfrac{c}{d}$ are two equal ratios, we can form the proportion $\dfrac{a}{b} = \dfrac{c}{d}$ ($b \neq 0$ and $d \neq 0$). We deduce an important property of proportions as follows:

$$\frac{a}{b} = \frac{c}{d}, \qquad b \neq 0 \text{ and } d \neq 0$$

$$bd\left(\frac{a}{b}\right) = bd\left(\frac{c}{d}\right) \qquad \text{Multiply both sides by } bd.$$

$$ad = bc$$

> ### Cross-Multiplication Property of Proportions
>
> If $\dfrac{a}{b} = \dfrac{c}{d}$ $(b \neq 0$ and $d \neq 0)$, then $ad = bc$.

We can treat some fractional equations as proportions and solve them by using the cross-multiplication idea, as in the next examples.

EXAMPLE 6 Solve $\dfrac{5}{x + 6} = \dfrac{7}{x - 5}$.

Solution

$$\frac{5}{x + 6} = \frac{7}{x - 5}, \qquad x \neq -6 \text{ and } x \neq 5$$

$$5(x - 5) = 7(x + 6) \qquad \text{Apply the cross-multiplication property.}$$

$$5x - 25 = 7x + 42$$

$$-67 = 2x$$

$$-\frac{67}{2} = x$$

The solution set is $\left\{-\dfrac{67}{2}\right\}$. ■

EXAMPLE 7 Solve $\dfrac{x}{7} = \dfrac{4}{x + 3}$.

Solution

$$\frac{x}{7} = \frac{4}{x + 3}, \qquad x \neq -3$$

$$x(x + 3) = 7(4) \qquad \text{Cross-multiplication property}$$

$$x^2 + 3x = 28$$

$$x^2 + 3x - 28 = 0$$

$$(x + 7)(x - 4) = 0$$

$$x + 7 = 0 \quad \text{ or } \quad x - 4 = 0$$

$$x = -7 \quad \text{ or } \quad x = 4$$

The solution set is $\{-7, 4\}$. Check these solutions in the original equation. ■

Problem Solving

The ability to solve fractional equations broadens our base for solving word problems. We are now ready to tackle some word problems that translate into fractional equations.

P R O B L E M   1

The sum of a number and its reciprocal is $\dfrac{10}{3}$. Find the number.

Solution

Let n represent the number. Then $\dfrac{1}{n}$ represents its reciprocal.

$$n + \frac{1}{n} = \frac{10}{3}, \qquad n \neq 0$$

$$3n\left(n + \frac{1}{n}\right) = 3n\left(\frac{10}{3}\right)$$

$$3n^2 + 3 = 10n$$

$$3n^2 - 10n + 3 = 0$$

$$(3n - 1)(n - 3) = 0$$

$$3n - 1 = 0 \qquad \text{or} \qquad n - 3 = 0$$

$$3n = 1 \qquad \text{or} \qquad n = 3$$

$$n = \frac{1}{3} \qquad \text{or} \qquad n = 3$$

If the number is $\dfrac{1}{3}$, then its reciprocal is $\dfrac{1}{\frac{1}{3}} = 3$. If the number is 3, then its reciprocal is $\dfrac{1}{3}$. ∎

Now let's consider a problem where we can use the relationship

$$\frac{\text{Dividend}}{\text{Divisor}} = \text{Quotient} + \frac{\text{Remainder}}{\text{Divisor}}$$

as a guideline.

P R O B L E M   2

The sum of two numbers is 52. If the larger is divided by the smaller, the quotient is 9 and the remainder is 2. Find the numbers.

Solution

Let n represent the smaller number. Then $52 - n$ represents the larger number. Let's use the relationship we discussed previously as a guideline and proceed as follows:

$$\underbrace{\frac{\text{Dividend}}{\text{Divisor}} = \text{Quotient} + \frac{\text{Remainder}}{\text{Divisor}}}$$

$$\frac{52 - n}{n} = 9 + \frac{2}{n}, \qquad n \neq 0$$

$$n\left(\frac{52 - n}{n}\right) = n\left(9 + \frac{2}{n}\right)$$

$$52 - n = 9n + 2$$

$$50 = 10n$$

$$5 = n$$

If $n = 5$, then $52 - n$ equals 47. The numbers are 5 and 47. ∎

We can conveniently set up some problems and solve them using the concepts of ratio and proportion. Let's conclude this section with two such examples.

PROBLEM 3   On a certain map $1\frac{1}{2}$ inches represents 25 miles. If two cities are $5\frac{1}{4}$ inches apart on the map, find the number of miles between the cities (see Figure 4.1).

Solution

Let m represent the number of miles between the two cities. To set up the proportion, we will use a ratio of inches on the map to miles. Be sure to keep the ratio "inches on the map to miles" the same for both sides of the proportion.

$$\frac{1\frac{1}{2}}{25} = \frac{5\frac{1}{4}}{m}, \qquad m \neq 0$$

$$\frac{\frac{3}{2}}{25} = \frac{\frac{21}{4}}{m}$$

$$\frac{3}{2}m = 25\left(\frac{21}{4}\right) \qquad \text{Cross-multiplication property}$$

$$\frac{2}{3}\left(\frac{3}{2}m\right) = \frac{2}{\overset{}{3}}(25)\left(\frac{\overset{7}{21}}{\underset{2}{4}}\right) \qquad \text{Multiply both sides by } \frac{2}{3}.$$

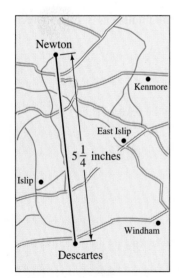

Newton

Kenmore

East Islip

$5\frac{1}{4}$ inches

Islip

Windham

Descartes

Figure 4.1

$$m = \frac{175}{2}$$

$$= 87\frac{1}{2}$$

The distance between the two cities is $87\frac{1}{2}$ miles. ■

PROBLEM 4   A sum of $750 is to be divided between two people in the ratio of 2 to 3. How much does each person receive?

Solution

Let d represent the amount of money that one person receives. Then $750 - d$ represents the amount for the other person.

$$\frac{d}{750 - d} = \frac{2}{3}, \qquad d \neq 750$$

$$3d = 2(750 - d)$$

$$3d = 1500 - 2d$$

$$5d = 1500$$

$$d = 300$$

If $d = 300$, then $750 - d$ equals 450. Therefore, one person receives $300 and the other person receives $450. ■

PROBLEM SET 4.6

For Problems 1–44, solve each equation.

1. $\dfrac{x + 1}{4} + \dfrac{x - 2}{6} = \dfrac{3}{4}$

2. $\dfrac{x + 2}{5} + \dfrac{x - 1}{6} = \dfrac{3}{5}$

3. $\dfrac{x + 3}{2} - \dfrac{x - 4}{7} = 1$

4. $\dfrac{x + 4}{3} - \dfrac{x - 5}{9} = 1$

5. $\dfrac{5}{n} + \dfrac{1}{3} = \dfrac{7}{n}$

6. $\dfrac{3}{n} + \dfrac{1}{6} = \dfrac{11}{3n}$

7. $\dfrac{7}{2x} + \dfrac{3}{5} = \dfrac{2}{3x}$

8. $\dfrac{9}{4x} + \dfrac{1}{3} = \dfrac{5}{2x}$

9. $\dfrac{3}{4x} + \dfrac{5}{6} = \dfrac{4}{3x}$

10. $\dfrac{5}{7x} - \dfrac{5}{6} = \dfrac{1}{6x}$

11. $\dfrac{47 - n}{n} = 8 + \dfrac{2}{n}$

12. $\dfrac{45 - n}{n} = 6 + \dfrac{3}{n}$

13. $\dfrac{n}{65 - n} = 8 + \dfrac{2}{65 - n}$

14. $\dfrac{n}{70 - n} = 7 + \dfrac{6}{70 - n}$

15. $n + \dfrac{1}{n} = \dfrac{17}{4}$

16. $n + \dfrac{1}{n} = \dfrac{37}{6}$

17. $n - \dfrac{2}{n} = \dfrac{23}{5}$

18. $n - \dfrac{3}{n} = \dfrac{26}{3}$

19. $\dfrac{5}{7x - 3} = \dfrac{3}{4x - 5}$

20. $\dfrac{3}{2x - 1} = \dfrac{5}{3x + 2}$

21. $\dfrac{-2}{x - 5} = \dfrac{1}{x + 9}$

22. $\dfrac{5}{2a - 1} = \dfrac{-6}{3a + 2}$

23. $\dfrac{x}{x + 1} - 2 = \dfrac{3}{x - 3}$

24. $\dfrac{x}{x - 2} + 1 = \dfrac{8}{x - 1}$

25. $\dfrac{a}{a+5} - 2 = \dfrac{3a}{a+5}$

26. $\dfrac{a}{a-3} - \dfrac{3}{2} = \dfrac{3}{a-3}$

27. $\dfrac{5}{x+6} = \dfrac{6}{x-3}$

28. $\dfrac{3}{x-1} = \dfrac{4}{x+2}$

29. $\dfrac{3x-7}{10} = \dfrac{2}{x}$

30. $\dfrac{x}{-4} = \dfrac{3}{12x-25}$

31. $\dfrac{x}{x-6} - 3 = \dfrac{6}{x-6}$

32. $\dfrac{x}{x+1} + 3 = \dfrac{4}{x+1}$

33. $\dfrac{3s}{s+2} + 1 = \dfrac{35}{2(3s+1)}$

34. $\dfrac{s}{2s-1} - 3 = \dfrac{-32}{3(s+5)}$

35. $2 - \dfrac{3x}{x-4} = \dfrac{14}{x+7}$

36. $-1 + \dfrac{2x}{x+3} = \dfrac{-4}{x+4}$

37. $\dfrac{n+6}{27} = \dfrac{1}{n}$

38. $\dfrac{n}{5} = \dfrac{10}{n-5}$

39. $\dfrac{3n}{n-1} - \dfrac{1}{3} = \dfrac{-40}{3n-18}$

40. $\dfrac{n}{n+1} + \dfrac{1}{2} = \dfrac{-2}{n+2}$

41. $\dfrac{-3}{4x+5} = \dfrac{2}{5x-7}$

42. $\dfrac{7}{x+4} = \dfrac{3}{x-8}$

43. $\dfrac{2x}{x-2} + \dfrac{15}{x^2-7x+10} = \dfrac{3}{x-5}$

44. $\dfrac{x}{x-4} - \dfrac{2}{x+3} = \dfrac{20}{x^2-x-12}$

For Problems 45–60, set up an algebraic equation and solve each problem.

45. A sum of $1750 is to be divided between two people in the ratio of 3 to 4. How much does each person receive?

46. A blueprint has a scale where 1 inch represents 5 feet. Find the dimensions of a rectangular room that measures $3\dfrac{1}{2}$ inches by $5\dfrac{3}{4}$ inches on the blueprint.

47. One angle of a triangle has a measure of 60° and the measures of the other two angles are in the ratio of 2 to 3. Find the measures of the other two angles.

48. The ratio of the complement of an angle to its supplement is 1 to 4. Find the measure of the angle.

49. The sum of a number and its reciprocal is $\dfrac{53}{14}$. Find the number.

50. The sum of two numbers is 80. If the larger is divided by the smaller, the quotient is 7 and the remainder is 8. Find the numbers.

51. If a home valued at $50,000 is assessed $900 in real estate taxes, then how much, at the same rate, are the taxes on a home valued at $60,000?

52. The ratio of male students to female students at a certain university is 5 to 7. If there is a total of 16,200 students, find the number of male students and the number of female students.

53. Suppose that, together, Laura and Tammy sold $120.75 worth of candy for the annual school fair. If the ratio of Tammy's sales to Laura's sales was 4 to 3, how much did each sell?

54. The total value of a house and a lot is $68,000. If the ratio of the value of the house to the value of the lot is 7 to 1, find the value of the house.

55. The sum of two numbers is 90. If the larger is divided by the smaller, the quotient is 10 and the remainder is 2. Find the numbers.

56. What number must be added to the numerator and denominator of $\dfrac{2}{5}$ to produce a rational number that is equivalent to $\dfrac{7}{8}$?

57. A 20-foot board is to be cut into two pieces whose lengths are in the ratio of 7 to 3. Find the lengths of the two pieces.

58. An inheritance of $300,000 is to be divided between a son and the local heart fund in the ratio of 3 to 1. How much money will the son receive?

59. Suppose that in a certain precinct, 1150 people voted in the last presidential election. If the ratio of female voters to male voters was 3 to 2, how many females and how many males voted?

60. The perimeter of a rectangle is 114 centimeters. If the ratio of its width to its length is 7 to 12, find the dimensions of the rectangle.

■■■ Thoughts into words

61. How could you do Problem 57 without using algebra?

62. Now do Problem 59 using the same approach that you used in Problem 61. What difficulties do you encounter?

63. How can you tell by inspection that the equation

$$\frac{x}{x + 2} = \frac{-2}{x + 2} \text{ has no solution?}$$

64. How would you help someone solve the equation

$$\frac{3}{x} - \frac{4}{x} = \frac{-1}{x}?$$

4.7 More Fractional Equations and Applications

Let's begin this section by considering a few more fractional equations. We will continue to solve them using the same basic techniques as in the previous section. That is, we will multiply both sides of the equation by the least common denominator of all of the denominators in the equation, with the necessary restrictions to avoid division by zero. Some of the denominators in these problems will require factoring before we can determine a least common denominator.

E X A M P L E   1   Solve $\dfrac{x}{2x - 8} + \dfrac{16}{x^2 - 16} = \dfrac{1}{2}$.

Solution

$$\frac{x}{2x - 8} + \frac{16}{x^2 - 16} = \frac{1}{2}$$

$$\frac{x}{2(x - 4)} + \frac{16}{(x + 4)(x - 4)} = \frac{1}{2}, \qquad x \neq 4 \text{ and } x \neq -4$$

$$2(x - 4)(x + 4)\left(\frac{x}{2(x - 4)} + \frac{16}{(x + 4)(x - 4)}\right) = 2(x + 4)(x - 4)\left(\frac{1}{2}\right) \qquad \begin{array}{l}\text{Multiply both} \\ \text{sides by the LCD,} \\ 2(x - 4)(x + 4).\end{array}$$

$$x(x + 4) + 2(16) = (x + 4)(x - 4)$$

$$x^2 + 4x + 32 = x^2 - 16$$

$$4x = -48$$

$$x = -12$$

The solution set is $\{-12\}$. Perhaps you should check it! ■

In Example 1, note that the restrictions were not indicated until the denominators were expressed in factored form. It is usually easier to determine the necessary restrictions at this step.

E X A M P L E   2

Solve $\dfrac{3}{n-5} - \dfrac{2}{2n+1} = \dfrac{n+3}{2n^2 - 9n - 5}$.

Solution

$$\frac{3}{n-5} - \frac{2}{2n+1} = \frac{n+3}{2n^2 - 9n - 5}$$

$$\frac{3}{n-5} - \frac{2}{2n+1} = \frac{n+3}{(2n+1)(n-5)}, \qquad n \neq -\frac{1}{2} \text{ and } n \neq 5$$

$$(2n+1)(n-5)\left(\frac{3}{n-5} - \frac{2}{2n+1}\right) = (2n+1)(n-5)\left(\frac{n+3}{(2n+1)(n-5)}\right) \qquad \begin{array}{l}\text{Multiply both} \\ \text{sides by the LCD,} \\ (2n+1)(n-5).\end{array}$$

$$3(2n+1) - 2(n-5) = n+3$$

$$6n + 3 - 2n + 10 = n + 3$$

$$4n + 13 = n + 3$$

$$3n = -10$$

$$n = -\frac{10}{3}$$

The solution set is $\left\{-\dfrac{10}{3}\right\}$.   ■

E X A M P L E   3

Solve $2 + \dfrac{4}{x-2} = \dfrac{8}{x^2 - 2x}$.

Solution

$$2 + \frac{4}{x-2} = \frac{8}{x^2 - 2x}$$

$$2 + \frac{4}{x-2} = \frac{8}{x(x-2)}, \qquad x \neq 0 \text{ and } x \neq 2$$

$$x(x-2)\left(2 + \frac{4}{x-2}\right) = x(x-2)\left(\frac{8}{x(x-2)}\right) \qquad \begin{array}{l}\text{Multiply both sides} \\ \text{by the LCD, } x(x-2).\end{array}$$

$$2x(x-2) + 4x = 8$$

$$2x^2 - 4x + 4x = 8$$

$$2x^2 = 8$$

$$x^2 = 4$$

$$x^2 - 4 = 0$$

$$(x + 2)(x - 2) = 0$$

$$x + 2 = 0 \quad \text{or} \quad x - 2 = 0$$

$$x = -2 \quad \text{or} \quad x = 2$$

Because our initial restriction indicated that $x \neq 2$, the only solution is -2. Thus the solution set is $\{-2\}$. ■

In Section 2.4, we discussed using the properties of equality to change the form of various formulas. For example, we considered the simple interest formula $A = P + Prt$ and changed its form by solving for P as follows:

$$A = P + Prt$$

$$A = P(1 + rt)$$

$$\frac{A}{1 + rt} = P \qquad \text{Multiply both sides by } \frac{1}{1 + rt}.$$

If the formula is in the form of a fractional equation, then the techniques of these last two sections are applicable. Consider the following example.

E X A M P L E   4

If the original cost of some business property is C dollars and it is depreciated linearly over N years, then its value, V, at the end of T years is given by

$$V = C\left(1 - \frac{T}{N}\right)$$

Solve this formula for N in terms of V, C, and T.

Solution

$$V = C\left(1 - \frac{T}{N}\right)$$

$$V = C - \frac{CT}{N}$$

$$N(V) = N\left(C - \frac{CT}{N}\right) \qquad \text{Multiply both sides by } N.$$

$$NV = NC - CT$$

$$NV - NC = -CT$$

$$N(V - C) = -CT$$

$$N = \frac{-CT}{V - C}$$

$$N = -\frac{CT}{V - C}$$ ■

Problem Solving

In Section 2.4 we solved some uniform motion problems. The formula $d = rt$ was used in the analysis of the problems, and we used guidelines that involve distance relationships. Now let's consider some uniform motion problems where guidelines that involve either times or rates are appropriate. These problems will generate fractional equations to solve.

P R O B L E M   1

An airplane travels 2050 miles in the same time that a car travels 260 miles. If the rate of the plane is 358 miles per hour greater than the rate of the car, find the rate of each.

Solution

Let r represent the rate of the car. Then $r + 358$ represents the rate of the plane. The fact that the times are equal can be a guideline. Remember from the basic formula, $d = rt$, that $t = \dfrac{d}{r}$.

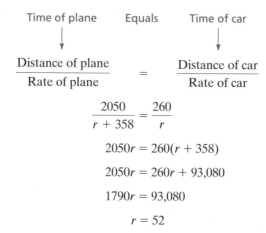

$$\frac{2050}{r + 358} = \frac{260}{r}$$

$$2050r = 260(r + 358)$$

$$2050r = 260r + 93{,}080$$

$$1790r = 93{,}080$$

$$r = 52$$

If $r = 52$, then $r + 358$ equals 410. Thus the rate of the car is 52 miles per hour, and the rate of the plane is 410 miles per hour. ■

P R O B L E M   2

It takes a freight train 2 hours longer to travel 300 miles than it takes an express train to travel 280 miles. The rate of the express train is 20 miles per hour greater than the rate of the freight train. Find the times and rates of both trains.

Solution

Let t represent the time of the express train. Then $t + 2$ represents the time of the freight train. Let's record the information of this problem in a table.

	Distance	Time	Rate = $\dfrac{\text{Distance}}{\text{Time}}$
Express train	280	t	$\dfrac{280}{t}$
Freight train	300	$t + 2$	$\dfrac{300}{t + 2}$

The fact that the rate of the express train is 20 miles per hour greater than the rate of the freight train can be a guideline.

Rate of express	Equals	Rate of freight train plus 20
↓		↓
$\dfrac{280}{t}$	$=$	$\dfrac{300}{t + 2} + 20$

$$t(t + 2)\left(\frac{280}{t}\right) = t(t + 2)\left(\frac{300}{t + 2} + 20\right)$$

$$280(t + 2) = 300t + 20t(t + 2)$$

$$280t + 560 = 300t + 20t^2 + 40t$$

$$280t + 560 = 340t + 20t^2$$

$$0 = 20t^2 + 60t - 560$$

$$0 = t^2 + 3t - 28$$

$$0 = (t + 7)(t - 4)$$

$$t + 7 = 0 \quad \text{or} \quad t - 4 = 0$$

$$t = -7 \quad \text{or} \quad t = 4$$

The negative solution must be discarded, so the time of the express train (t) is 4 hours, and the time of the freight train ($t + 2$) is 6 hours. The rate of the express train $\left(\dfrac{280}{t}\right)$ is $\dfrac{280}{4} = 70$ miles per hour, and the rate of the freight train $\left(\dfrac{300}{t + 2}\right)$ is $\dfrac{300}{6} = 50$ miles per hour.   ■

REMARK: Note that to solve Problem 1 we went directly to a guideline without the use of a table, but for Problem 2 we used a table. Again, remember that this is a personal preference; we are merely acquainting you with a variety of techniques.

Uniform motion problems are a special case of a larger group of problems we refer to as **rate-time problems.** For example, if a certain machine can produce 150 items in 10 minutes, then we say that the machine is producing at a rate of

$\dfrac{150}{10} = 15$ items per minute. Likewise, if a person can do a certain job in 3 hours, then, assuming a constant rate of work, we say that the person is working at a rate of $\dfrac{1}{3}$ of the job per hour. In general, if Q is the quantity of something done in t units of time, then the rate, r, is given by $r = \dfrac{Q}{t}$. We state the rate in terms of *so much quantity per unit of time*. (In uniform motion problems the "quantity" is distance.) Let's consider some examples of rate-time problems.

PROBLEM 3

If Jim can mow a lawn in 50 minutes and his son, Todd, can mow the same lawn in 40 minutes, how long will it take them to mow the lawn if they work together?

Solution

Jim's rate is $\dfrac{1}{50}$ of the lawn per minute and Todd's rate is $\dfrac{1}{40}$ of the lawn per minute.

If we let m represent the number of minutes that they work together, then $\dfrac{1}{m}$ represents their rate when working together. Therefore, because the sum of the individual rates must equal the rate working together, we can set up and solve the following equation.

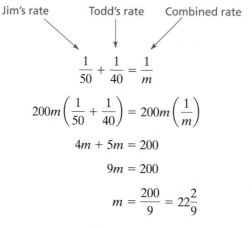

$$\underset{\text{Jim's rate}}{\dfrac{1}{50}} + \underset{\text{Todd's rate}}{\dfrac{1}{40}} = \underset{\text{Combined rate}}{\dfrac{1}{m}}$$

$$200m\left(\dfrac{1}{50} + \dfrac{1}{40}\right) = 200m\left(\dfrac{1}{m}\right)$$

$$4m + 5m = 200$$

$$9m = 200$$

$$m = \dfrac{200}{9} = 22\dfrac{2}{9}$$

It should take them $22\dfrac{2}{9}$ minutes. ∎

PROBLEM 4

Working together, Linda and Kathy can type a term paper in $3\dfrac{3}{5}$ hours. Linda can type the paper by herself in 6 hours. How long would it take Kathy to type the paper by herself?

Solution

Their rate working together is $\dfrac{1}{3\frac{3}{5}} = \dfrac{1}{\frac{18}{5}} = \dfrac{5}{18}$ of the job per hour, and Linda's rate

is $\dfrac{1}{6}$ of the job per hour. If we let h represent the number of hours that it would take

Kathy to do the job by herself, then her rate is $\dfrac{1}{h}$ of the job per hour. Thus, we have

Linda's rate	Kathy's rate	Combined rate
↓	↓	↓
$\dfrac{1}{6}$ $+$	$\dfrac{1}{h}$ $=$	$\dfrac{5}{18}$

Solving this equation yields

$$18h\left(\frac{1}{6} + \frac{1}{h}\right) = 18h\left(\frac{5}{18}\right)$$

$$3h + 18 = 5h$$

$$18 = 2h$$

$$9 = h$$

It would take Kathy 9 hours to type the paper by herself.   ■

Our final example of this section illustrates another approach that some people find meaningful for rate-time problems. For this approach, think in terms of fractional parts of the job. For example, if a person can do a certain job in 5 hours, then at the end of 2 hours, he or she has done $\dfrac{2}{5}$ of the job. (Again, assume a constant rate of work.) At the end of 4 hours, he or she has finished $\dfrac{4}{5}$ of the job; and, in general, at the end of h hours, he or she has done $\dfrac{h}{5}$ of the job. Then, just as in the motion problems where distance equals rate times the time, here the fractional part done equals the working rate times the time. Let's see how this works in a problem.

PROBLEM 5   It takes Pat 12 hours to complete a task. After he had been working for 3 hours, he was joined by his brother Mike, and together they finished the task in 5 hours. How long would it take Mike to do the job by himself?

Solution

Let h represent the number of hours that it would take Mike to do the job by himself.

The fractional part of the job that Pat does equals his working rate times his time. Because it takes Pat 12 hours to do the entire job, his working rate is $\frac{1}{12}$. He works for 8 hours (3 hours before Mike and then 5 hours with Mike). Therefore, Pat's part of the job is $\frac{1}{12}(8) = \frac{8}{12}$. The fractional part of the job that Mike does equals his working rate times his time. Because h represents Mike's time to do the entire job, his working rate is $\frac{1}{h}$. He works for 5 hours. Therefore, Mike's part of the job is $\frac{1}{h}(5) = \frac{5}{h}$. Adding the two fractional parts together results in 1 entire job being done. Let's also show this information in chart form and set up our guideline. Then we can set up and solve the equation.

	Time to do entire job	Working rate	Time working	Fractional part of the job done
Pat	12	$\frac{1}{12}$	8	$\frac{8}{12}$
Mike	h	$\frac{1}{h}$	5	$\frac{5}{h}$

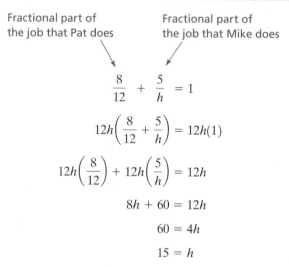

Fractional part of the job that Pat does

Fractional part of the job that Mike does

$$\frac{8}{12} + \frac{5}{h} = 1$$

$$12h\left(\frac{8}{12} + \frac{5}{h}\right) = 12h(1)$$

$$12h\left(\frac{8}{12}\right) + 12h\left(\frac{5}{h}\right) = 12h$$

$$8h + 60 = 12h$$

$$60 = 4h$$

$$15 = h$$

It would take Mike 15 hours to do the entire job by himself.

PROBLEM SET 4.7

For Problems 1–30, solve each equation.

1. $\frac{x}{4x - 4} + \frac{5}{x^2 - 1} = \frac{1}{4}$

2. $\frac{x}{3x - 6} + \frac{4}{x^2 - 4} = \frac{1}{3}$

3. $3 + \frac{6}{t - 3} = \frac{6}{t^2 - 3t}$

4. $2 + \frac{4}{t - 1} = \frac{4}{t^2 - t}$

5. $\dfrac{3}{n-5} + \dfrac{4}{n+7} = \dfrac{2n+11}{n^2+2n-35}$

6. $\dfrac{2}{n+3} + \dfrac{3}{n-4} = \dfrac{2n-1}{n^2-n-12}$

7. $\dfrac{5x}{2x+6} - \dfrac{4}{x^2-9} = \dfrac{5}{2}$ **8.** $\dfrac{3x}{5x+5} - \dfrac{2}{x^2-1} = \dfrac{3}{5}$

9. $1 + \dfrac{1}{n-1} = \dfrac{1}{n^2-n}$ **10.** $3 + \dfrac{9}{n-3} = \dfrac{27}{n^2-3n}$

11. $\dfrac{2}{n-2} - \dfrac{n}{n+5} = \dfrac{10n+15}{n^2+3n-10}$

12. $\dfrac{n}{n+3} + \dfrac{1}{n-4} = \dfrac{11-n}{n^2-n-12}$

13. $\dfrac{2}{2x-3} - \dfrac{2}{10x^2-13x-3} = \dfrac{x}{5x+1}$

14. $\dfrac{1}{3x+4} + \dfrac{6}{6x^2+5x-4} = \dfrac{x}{2x-1}$

15. $\dfrac{2x}{x+3} - \dfrac{3}{x-6} = \dfrac{29}{x^2-3x-18}$

16. $\dfrac{x}{x-4} - \dfrac{2}{x+8} = \dfrac{63}{x^2+4x-32}$

17. $\dfrac{a}{a-5} + \dfrac{2}{a-6} = \dfrac{2}{a^2-11a+30}$

18. $\dfrac{a}{a+2} + \dfrac{3}{a+4} = \dfrac{14}{a^2+6a+8}$

19. $\dfrac{-1}{2x-5} + \dfrac{2x-4}{4x^2-25} = \dfrac{5}{6x+15}$

20. $\dfrac{-2}{3x+2} + \dfrac{x-1}{9x^2-4} = \dfrac{3}{12x-8}$

21. $\dfrac{7y+2}{12y^2+11y-15} - \dfrac{1}{3y+5} = \dfrac{2}{4y-3}$

22. $\dfrac{5y-4}{6y^2+y-12} - \dfrac{2}{2y+3} = \dfrac{5}{3y-4}$

23. $\dfrac{2n}{6n^2+7n-3} - \dfrac{n-3}{3n^2+11n-4} = \dfrac{5}{2n^2+11n+12}$

24. $\dfrac{x+1}{2x^2+7x-4} - \dfrac{x}{2x^2-7x+3} = \dfrac{1}{x^2+x-12}$

25. $\dfrac{1}{2x^2-x-1} + \dfrac{3}{2x^2+x} = \dfrac{2}{x^2-1}$

26. $\dfrac{2}{n^2+4n} + \dfrac{3}{n^2-3n-28} = \dfrac{5}{n^2-6n-7}$

27. $\dfrac{x+1}{x^3-9x} - \dfrac{1}{2x^2+x-21} = \dfrac{1}{2x^2+13x+21}$

28. $\dfrac{x}{2x^2+5x} - \dfrac{x}{2x^2+7x+5} = \dfrac{2}{x^2+x}$

29. $\dfrac{4t}{4t^2-t-3} + \dfrac{2-3t}{3t^2-t-2} = \dfrac{1}{12t^2+17t+6}$

30. $\dfrac{2t}{2t^2+9t+10} + \dfrac{1-3t}{3t^2+4t-4} = \dfrac{4}{6t^2+11t-10}$

For Problems 31–44, solve each equation for the indicated variable.

31. $y = \dfrac{5}{6}x + \dfrac{2}{9}$ for x **32.** $y = \dfrac{3}{4}x - \dfrac{2}{3}$ for x

33. $\dfrac{-2}{x-4} = \dfrac{5}{y-1}$ for y **34.** $\dfrac{7}{y-3} = \dfrac{3}{x+1}$ for y

35. $I = \dfrac{100M}{C}$ for M **36.** $V = C\left(1 - \dfrac{T}{N}\right)$ for T

37. $\dfrac{R}{S} = \dfrac{T}{S+T}$ for R **38.** $\dfrac{1}{R} = \dfrac{1}{S} + \dfrac{1}{T}$ for R

39. $\dfrac{y-1}{x-3} = \dfrac{b-1}{a-3}$ for y **40.** $y = -\dfrac{a}{b}x + \dfrac{c}{d}$ for x

41. $\dfrac{x}{a} + \dfrac{y}{b} = 1$ for y **42.** $\dfrac{y-b}{x} = m$ for y

43. $\dfrac{y-1}{x+6} = \dfrac{-2}{3}$ for y **44.** $\dfrac{y+5}{x-2} = \dfrac{3}{7}$ for y

Set up an equation and solve each of the following problems.

45. Kent drives his Mazda 270 miles in the same time that it takes Dave to drive his Nissan 250 miles. If Kent averages 4 miles per hour faster than Dave, find their rates.

46. Suppose that Wendy rides her bicycle 30 miles in the same time that it takes Kim to ride her bicycle 20 miles. If Wendy rides 5 miles per hour faster than Kim, find the rate of each.

47. An inlet pipe can fill a tank (see Figure 4.2) in 10 minutes. A drain can empty the tank in 12 minutes. If the tank is empty and both the pipe and drain are open, how long will it take before the tank overflows?

Figure 4.2

48. Barry can do a certain job in 3 hours, whereas it takes Sanchez 5 hours to do the same job. How long would it take them to do the job working together?

49. Connie can type 600 words in 5 minutes less than it takes Katie to type 600 words. If Connie types at a rate of 20 words per minute faster than Katie types, find the typing rate of each woman.

50. Walt can mow a lawn in 1 hour, and his son, Malik, can mow the same lawn in 50 minutes. One day Malik started mowing the lawn by himself and worked for 30 minutes. Then Walt joined him and they finished the lawn. How long did it take them to finish mowing the lawn after Walt started to help?

51. Plane A can travel 1400 miles in 1 hour less time than it takes plane B to travel 2000 miles. The rate of plane B is 50 miles per hour greater than the rate of plane A. Find the times and rates of both planes.

52. To travel 60 miles, it takes Sue, riding a moped, 2 hours less time than it takes Doreen to travel 50 miles riding a bicycle. Sue travels 10 miles per hour faster than Doreen. Find the times and rates of both girls.

53. It takes Amy twice as long to deliver papers as it does Nancy. How long would it take each girl to deliver the papers by herself if they can deliver the papers together in 40 minutes?

54. If two inlet pipes are both open, they can fill a pool in 1 hour and 12 minutes. One of the pipes can fill the pool by itself in 2 hours. How long would it take the other pipe to fill the pool by itself?

55. Rod agreed to mow a vacant lot for $12. It took him an hour longer than he had anticipated, so he earned $1 per hour less than he had originally calculated. How long had he anticipated that it would take him to mow the lot?

56. Last week Al bought some golf balls for $20. The next day they were on sale for $.50 per ball less, and he bought $22.50 worth of balls. If he purchased 5 more balls on the second day than on the first day, how many did he buy each day and at what price per ball?

57. Debbie rode her bicycle out into the country for a distance of 24 miles. On the way back, she took a much shorter route of 12 miles and made the return trip in one-half hour less time. If her rate out into the country was 4 miles per hour greater than her rate on the return trip, find both rates.

58. Felipe jogs for 10 miles and then walks another 10 miles. He jogs $2\frac{1}{2}$ miles per hour faster than he walks, and the entire distance of 20 miles takes 6 hours. Find the rate at which he walks and the rate at which he jogs.

■ ■ ■ **Thoughts into words**

59. Why is it important to consider more than one way to do a problem?

60. Write a paragraph or two summarizing the new ideas about problem solving you have acquired thus far in this course.

CHAPTER 4   *SUMMARY*

(4.1) Any number that can be written in the form $\dfrac{a}{b}$, where a and b are integers and $b \neq 0$, is called a **rational number.**

A **rational expression** is defined as the indicated quotient of two polynomials. The following properties pertain to rational numbers and rational expressions.

1. $\dfrac{-a}{b} = \dfrac{a}{-b} = -\dfrac{a}{b}$

2. $\dfrac{-a}{-b} = \dfrac{a}{b}$

3. $\dfrac{a \cdot k}{b \cdot k} = \dfrac{a}{b}$ Fundamental principle of fractions

(4.2) Multiplication and division of rational expressions are based on the following definitions:

1. $\dfrac{a}{b} \cdot \dfrac{c}{d} = \dfrac{ac}{bd}$ Multiplication

2. $\dfrac{a}{b} \div \dfrac{c}{d} = \dfrac{a}{b} \cdot \dfrac{d}{c} = \dfrac{ad}{bc}$ Division

(4.3) Addition and subtraction of rational expressions are based on the following definitions:

1. $\dfrac{a}{b} + \dfrac{c}{b} = \dfrac{a + c}{b}$ Addition

2. $\dfrac{a}{b} - \dfrac{c}{b} = \dfrac{a - c}{b}$ Subtraction

(4.4) The following basic procedure is used to add or subtract rational expressions.

1. Factor the denominators.

2. Find the LCD.

3. Change each fraction to an equivalent fraction that has the LCD as its denominator.

4. Combine the numerators and place over the LCD.

5. Simplify by performing the addition or subtraction.

6. Look for ways to reduce the resulting fraction.

Fractional forms that contain rational numbers or rational expressions in the numerators and/or denominators are called **complex fractions.** The fundamental principle of fractions serves as a basis for simplifying complex fractions.

(4.5) To divide a polynomial by a monomial, we divide each term of the polynomial by the monomial. The procedure for dividing a polynomial by a polynomial, rather than a monomial, resembles the long-division process in arithmetic. (See the examples in Section 4.5.) Synthetic division is a shortcut to the long-division process when the divisor is of the form $x - k$.

(4.6) To solve a fractional equation, it is often easiest to begin by multiplying both sides of the equation by the LCD of all of the denominators in the equation. If an equation contains a variable in one or more denominators, then we must be careful to avoid any value of the variable that makes the denominator zero.

A **ratio** is the comparison of two numbers by division. A statement of equality between two ratios is a **proportion.**

We can treat some fractional equations as proportions, and we can solve them by applying the following property. This property is often called the **cross-multiplication** property:

$$\text{If } \frac{a}{b} = \frac{c}{d}, \text{ then } ad = bc.$$

(4.7) The techniques that we use to solve fractional equations can also be used to change the form of formulas containing rational expressions so that we can use those formulas to solve problems.

CHAPTER 4 REVIEW PROBLEM SET

For Problems 1–6, simplify each rational expression.

1. $\dfrac{26x^2y^3}{39x^4y^2}$

2. $\dfrac{a^2 - 9}{a^2 + 3a}$

3. $\dfrac{n^2 - 3n - 10}{n^2 + n - 2}$

4. $\dfrac{x^4 - 1}{x^3 - x}$

5. $\dfrac{8x^3 - 2x^2 - 3x}{12x^2 - 9x}$

6. $\dfrac{x^4 - 7x^2 - 30}{2x^4 + 7x^2 + 3}$

For Problems 7–10, simplify each complex fraction.

7. $\dfrac{\dfrac{5}{8} - \dfrac{1}{2}}{\dfrac{1}{6} + \dfrac{3}{4}}$

8. $\dfrac{\dfrac{3}{2x} + \dfrac{5}{3y}}{\dfrac{4}{x} - \dfrac{3}{4y}}$

9. $\dfrac{\dfrac{3}{x - 2} - \dfrac{4}{x^2 - 4}}{\dfrac{2}{x + 2} + \dfrac{1}{x - 2}}$

10. $1 - \dfrac{1}{2 - \dfrac{1}{x}}$

For Problems 11–22, perform the indicated operations and express your answers in simplest form.

11. $\dfrac{6xy^2}{7y^3} \div \dfrac{15x^2y}{5x^2}$

12. $\dfrac{9ab}{3a + 6} \cdot \dfrac{a^2 - 4a - 12}{a^2 - 6a}$

13. $\dfrac{n^2 + 10n + 25}{n^2 - n} \cdot \dfrac{5n^3 - 3n^2}{5n^2 + 22n - 15}$

14. $\dfrac{x^2 - 2xy - 3y^2}{x^2 + 9y^2} \div \dfrac{2x^2 + xy - y^2}{2x^2 - xy}$

15. $\dfrac{2x + 1}{5} + \dfrac{3x - 2}{4}$

16. $\dfrac{3}{2n} + \dfrac{5}{3n} - \dfrac{1}{9}$

17. $\dfrac{3x}{x + 7} - \dfrac{2}{x}$

18. $\dfrac{10}{x^2 - 5x} + \dfrac{2}{x}$

19. $\dfrac{3}{n^2 - 5n - 36} + \dfrac{2}{n^2 + 3n - 4}$

20. $\dfrac{3}{2y + 3} + \dfrac{5y - 2}{2y^2 - 9y - 18} - \dfrac{1}{y - 6}$

21. $(18x^2 + 9x - 2) \div (3x + 2)$

22. $(3x^3 + 5x^2 - 6x - 2) \div (x + 4)$

For Problems 23–32, solve each equation.

23. $\dfrac{4x + 5}{3} + \dfrac{2x - 1}{5} = 2$

24. $\dfrac{3}{4x} + \dfrac{4}{5} = \dfrac{9}{10x}$

25. $\dfrac{a}{a - 2} - \dfrac{3}{2} = \dfrac{2}{a - 2}$

26. $\dfrac{4}{5y - 3} = \dfrac{2}{3y + 7}$

27. $n + \dfrac{1}{n} = \dfrac{53}{14}$

28. $\dfrac{1}{2x - 7} + \dfrac{x - 5}{4x^2 - 49} = \dfrac{4}{6x - 21}$

29. $\dfrac{x}{2x + 1} - 1 = \dfrac{-4}{7(x - 2)}$

30. $\dfrac{2x}{-5} = \dfrac{3}{4x - 13}$

31. $\dfrac{2n}{2n^2 + 11n - 21} - \dfrac{n}{n^2 + 5n - 14} = \dfrac{3}{n^2 + 5n - 14}$

32. $\dfrac{2}{t^2 - t - 6} + \dfrac{t + 1}{t^2 + t - 12} = \dfrac{t}{t^2 + 6t + 8}$

33. Solve $\dfrac{y - 6}{x + 1} = \dfrac{3}{4}$ for y.

34. Solve $\dfrac{x}{a} - \dfrac{y}{b} = 1$ for y.

For Problems 35–40, set up an equation and solve the problem.

35. A sum of $1400 is to be divided between two people in the ratio of $\dfrac{3}{5}$. How much does each person receive?

36. Working together, Dan and Julio can mow a lawn in 12 minutes. Julio can mow the lawn by himself in 10 minutes less time than it takes Dan by himself. How long does it take each of them to mow the lawn alone?

37. Suppose that car A can travel 250 miles in 3 hours less time than it takes car B to travel 440 miles. The rate of car B is 5 miles per hour faster than that of car A. Find the rates of both cars.

38. Mark can overhaul an engine in 20 hours, and Phil can do the same job by himself in 30 hours. If they both work together for a time and then Mark finishes the job by himself in 5 hours, how long did they work together?

39. Kelly contracted to paint a house for $640. It took him 20 hours longer than he had anticipated, so he earned $1.60 per hour less than he had calculated. How long had he anticipated that it would take him to paint the house?

40. Nasser rode his bicycle 66 miles in $4\frac{1}{2}$ hours. For the first 40 miles he averaged a certain rate, and then for the last 26 miles he reduced his rate by 3 miles per hour. Find his rate for the last 26 miles.

TEST

For Problems 1–4, simplify each rational expression.

1. $\dfrac{39x^2y^3}{72x^3y}$

2. $\dfrac{3x^2 + 17x - 6}{x^3 - 36x}$

3. $\dfrac{6n^2 - 5n - 6}{3n^2 + 14n + 8}$

4. $\dfrac{2x - 2x^2}{x^2 - 1}$

For Problems 5–13, perform the indicated operations and express your answers in simplest form.

5. $\dfrac{5x^2y}{8x} \cdot \dfrac{12y^2}{20xy}$

6. $\dfrac{5a + 5b}{20a + 10b} \cdot \dfrac{a^2 - ab}{2a^2 + 2ab}$

7. $\dfrac{3x^2 + 10x - 8}{5x^2 + 19x - 4} \div \dfrac{3x^2 - 23x + 14}{x^2 - 3x - 28}$

8. $\dfrac{3x - 1}{4} + \dfrac{2x + 5}{6}$

9. $\dfrac{5x - 6}{3} - \dfrac{x - 12}{6}$

10. $\dfrac{3}{5n} + \dfrac{2}{3} - \dfrac{7}{3n}$

11. $\dfrac{3x}{x - 6} + \dfrac{2}{x}$

12. $\dfrac{9}{x^2 - x} - \dfrac{2}{x}$

13. $\dfrac{3}{2n^2 + n - 10} + \dfrac{5}{n^2 + 5n - 14}$

14. Divide $3x^3 + 10x^2 - 9x - 4$ by $x + 4$.

15. Simplify the complex fraction $\dfrac{\dfrac{3}{2x} - \dfrac{1}{6}}{\dfrac{2}{3x} + \dfrac{3}{4}}$.

16. Solve $\dfrac{x + 2}{y - 4} = \dfrac{3}{4}$ for y.

For Problems 17–22, solve each equation.

17. $\dfrac{x - 1}{2} - \dfrac{x + 2}{5} = -\dfrac{3}{5}$

18. $\dfrac{5}{4x} + \dfrac{3}{2} = \dfrac{7}{5x}$

19. $\dfrac{-3}{4n - 1} = \dfrac{-2}{3n + 11}$

20. $n - \dfrac{5}{n} = 4$

21. $\dfrac{6}{x - 4} - \dfrac{4}{x + 3} = \dfrac{8}{x - 4}$

22. $\dfrac{1}{3x - 1} + \dfrac{x - 2}{9x^2 - 1} = \dfrac{7}{6x - 2}$

For Problems 23–25, set up an equation and solve the problem.

23. The denominator of a rational number is 9 less than three times the numerator. The number in simplest form is $\dfrac{3}{8}$. Find the number.

24. It takes Jodi three times as long to deliver papers as it does Jannie. Together they can deliver the papers in 15 minutes. How long would it take Jodi by herself?

25. René can ride her bike 60 miles in 1 hour less time than it takes Sue to ride 60 miles. René's rate is 3 miles per hour faster than Sue's rate. Find René's rate.

By knowing the time it takes for the pendulum to swing from one side to the other side and back, the formula, $T = 2\pi\sqrt{\dfrac{L}{32}}$, can be solved to find the length of the pendulum.

© Jonathan Nourok/PhotoEdit

Exponents and Radicals

*H*ow long will it take a pendulum that is 1.5 feet long to swing from one side to the other side and back? The formula $T = 2\pi\sqrt{\dfrac{L}{32}}$ can be used to determine that it will take approximately 1.4 seconds.

It is not uncommon in mathematics to find two separately developed concepts that are closely related to each other. In this chapter, we will first develop the concepts of **exponent** and **root** individually and then show how they merge to become even more functional as a unified idea.

InfoTrac Project Do a keyword search on Australian population, and find an article on Australian population growth for September 1999. Find the estimated resident population for September 1999. Round to the nearest thousand and write in scientific notation. Find the number of households and write that number in scientific notation. Using scientific notation, calculate the average number of persons per household in Australia in September 1999. The average consumer debt per household is $10,069 (Australian). Write this number in scientific notation and calculate the total amount of consumer debt for the Australian population in 1999. Express your answer in scientific notation.

5.1 Using Integers as Exponents

Thus far in the text we have used only positive integers as exponents. In Chapter 1 the expression b^n, where b is any real number and n is a positive integer, was defined by

$$b^n = b \cdot b \cdot b \cdot \ldots \cdot b \qquad n \text{ factors of } b$$

Then, in Chapter 3, some of the parts of the following property served as a basis for manipulation with polynomials.

PROPERTY 5.1

If m and n are positive integers and a and b are real numbers (and $b \neq 0$ whenever it appears in a denominator), then

1. $b^n \cdot b^m = b^{n+m}$

2. $(b^n)^m = b^{mn}$

3. $(ab)^n = a^n b^n$

4. $\left(\dfrac{a}{b}\right)^n = \dfrac{a^n}{b^n}$

5. $\dfrac{b^n}{b^m} = b^{n-m}$ when $n > m$

$\dfrac{b^n}{b^m} = 1$ when $n = m$

$\dfrac{b^n}{b^m} = \dfrac{1}{b^{m-n}}$ when $n < m$

We are now ready to extend the concept of an exponent to include the use of zero and the negative integers as exponents.

First, let's consider the use of zero as an exponent. We want to use zero in such a way that the previously listed properties continue to hold. If $b^n \cdot b^m = b^{n+m}$ is to hold, then $x^4 \cdot x^0 = x^{4+0} = x^4$. In other words, x^0 *acts like* 1 because $x^4 \cdot x^0 = x^4$. This line of reasoning suggests the following definition.

DEFINITION 5.1

If b is a nonzero real number, then

$$b^0 = 1$$

According to Definition 5.1, the following statements are all true.

$$5^0 = 1 \qquad\qquad\qquad (-413)^0 = 1$$

$$\left(\frac{3}{11}\right)^0 = 1 \qquad\qquad\qquad n^0 = 1, \quad n \neq 0$$

$$(x^3 y^4)^0 = 1, \quad x \neq 0, y \neq 0$$

We can use a similar line of reasoning to motivate a definition for the use of negative integers as exponents. Consider the example $x^4 \cdot x^{-4}$. If $b^n \cdot b^m = b^{n+m}$ is to hold, then $x^4 \cdot x^{-4} = x^{4+(-4)} = x^0 = 1$. Thus x^{-4} must be the reciprocal of x^4, because their product is 1. That is,

$$x^{-4} = \frac{1}{x^4}$$

This suggests the following general definition.

DEFINITION 5.2

If n is a positive integer and b is a nonzero real number, then

$$b^{-n} = \frac{1}{b^n}$$

According to Definition 5.2, the following statements are all true.

$$x^{-5} = \frac{1}{x^5} \qquad\qquad\qquad 2^{-4} = \frac{1}{2^4} = \frac{1}{16}$$

$$10^{-2} = \frac{1}{10^2} = \frac{1}{100} \text{ or } 0.01 \qquad\qquad \frac{2}{x^{-3}} = \frac{2}{\dfrac{1}{x^3}} = (2)\left(\frac{x^3}{1}\right) = 2x^3$$

$$\left(\frac{3}{4}\right)^{-2} = \frac{1}{\left(\dfrac{3}{4}\right)^2} = \frac{1}{\dfrac{9}{16}} = \frac{16}{9}$$

It can be verified (although it is beyond the scope of this text) that all of the parts of Property 5.1 hold for *all integers*. In fact, the following equality can replace the three separate statements for part (5).

$$\frac{b^n}{b^m} = b^{n-m} \quad \text{for all integers } n \text{ and } m$$

Let's restate Property 5.1 as it holds for all integers and include, at the right, a "name tag" for easy reference.

PROPERTY 5.2

If m and n are integers and a and b are real numbers (and $b \neq 0$ whenever it appears in a denominator), then

1. $b^n \cdot b^m = b^{n+m}$ Product of two powers

2. $(b^n)^m = b^{mn}$ Power of a power

3. $(ab)^n = a^n b^n$ Power of a product

4. $\left(\dfrac{a}{b}\right)^n = \dfrac{a^n}{b^n}$ Power of a quotient

5. $\dfrac{b^n}{b^m} = b^{n-m}$ Quotient of two powers

Having the use of all integers as exponents enables us to work with a large variety of numerical and algebraic expressions. Let's consider some examples that illustrate the use of the various parts of Property 5.2.

EXAMPLE 1   Simplify each of the following numerical expressions.

    (a) $10^{-3} \cdot 10^2$ **(b)** $(2^{-3})^{-2}$ **(c)** $(2^{-1} \cdot 3^2)^{-1}$

    (d) $\left(\dfrac{2^{-3}}{3^{-2}}\right)^{-1}$ **(e)** $\dfrac{10^{-2}}{10^{-4}}$

Solution

    (a) $10^{-3} \cdot 10^2 = 10^{-3+2}$ Product of two powers

$$= 10^{-1}$$

$$= \frac{1}{10^1} = \frac{1}{10}$$

    (b) $(2^{-3})^{-2} = 2^{(-2)(-3)}$ Power of a power

$$= 2^6 = 64$$

    (c) $(2^{-1} \cdot 3^2)^{-1} = (2^{-1})^{-1}(3^2)^{-1}$ Power of a product

$$= 2^1 \cdot 3^{-2}$$

$$= \frac{2^1}{3^2} = \frac{2}{9}$$

(d) $\left(\dfrac{2^{-3}}{3^{-2}}\right)^{-1} = \dfrac{(2^{-3})^{-1}}{(3^{-2})^{-1}}$ Power of a quotient

$= \dfrac{2^3}{3^2} = \dfrac{8}{9}$

(e) $\dfrac{10^{-2}}{10^{-4}} = 10^{-2-(-4)}$ Quotient of two powers

$= 10^2 = 100$ ■

E X A M P L E   2   Simplify each of the following; express final results without using zero or negative integers as exponents.

(a) $x^2 \cdot x^{-5}$ **(b)** $(x^{-2})^4$ **(c)** $(x^2 y^{-3})^{-4}$

(d) $\left(\dfrac{a^3}{b^{-5}}\right)^{-2}$ **(e)** $\dfrac{x^{-4}}{x^{-2}}$

Solution

(a) $x^2 \cdot x^{-5} = x^{2+(-5)}$ Product of two powers

$= x^{-3}$

$= \dfrac{1}{x^3}$

(b) $(x^{-2})^4 = x^{4(-2)}$ Power of a power

$= x^{-8}$

$= \dfrac{1}{x^8}$

(c) $(x^2 y^{-3})^{-4} = (x^2)^{-4}(y^{-3})^{-4}$ Power of a product

$= x^{-4(2)} y^{-4(-3)}$

$= x^{-8} y^{12}$

$= \dfrac{y^{12}}{x^8}$

(d) $\left(\dfrac{a^3}{b^{-5}}\right)^{-2} = \dfrac{(a^3)^{-2}}{(b^{-5})^{-2}}$ Power of a quotient

$= \dfrac{a^{-6}}{b^{10}}$

$= \dfrac{1}{a^6 b^{10}}$

(e) $\dfrac{x^{-4}}{x^{-2}} = x^{-4-(-2)}$ Quotient of two powers

$$= x^{-2}$$

$$= \dfrac{1}{x^2}$$

■

E X A M P L E   3   Find the indicated products and quotients; express your results using positive integral exponents only.

(a) $(3x^2y^{-4})(4x^{-3}y)$ **(b)** $\dfrac{12a^3b^2}{-3a^{-1}b^5}$ **(c)** $\left(\dfrac{15x^{-1}y^2}{5xy^{-4}}\right)^{-1}$

     Solution

(a) $(3x^2y^{-4})(4x^{-3}y) = 12x^{2+(-3)}y^{-4+1}$

$$= 12x^{-1}y^{-3}$$

$$= \dfrac{12}{xy^3}$$

(b) $\dfrac{12a^3b^2}{-3a^{-1}b^5} = -4a^{3-(-1)}b^{2-5}$

$$= -4a^4b^{-3}$$

$$= -\dfrac{4a^4}{b^3}$$

(c) $\left(\dfrac{15x^{-1}y^2}{5xy^{-4}}\right)^{-1} = (3x^{-1-1}y^{2-(-4)})^{-1}$ Note that we are first simplifying inside the parentheses.

$$= (3x^{-2}y^6)^{-1}$$

$$= 3^{-1}x^2y^{-6}$$

$$= \dfrac{x^2}{3y^6}$$

■

The final examples of this section show the simplification of numerical and algebraic expressions that involve sums and differences. In such cases, we use Definition 5.2 to change from negative to positive exponents so that we can proceed in the usual way.

E X A M P L E   4   Simplify $2^{-3} + 3^{-1}$.

Solution

$$2^{-3} + 3^{-1} = \dfrac{1}{2^3} + \dfrac{1}{3^1}$$

$$= \frac{1}{8} + \frac{1}{3}$$

$$= \frac{3}{24} + \frac{8}{24} \qquad \text{Use 24 as the LCD.}$$

$$= \frac{11}{24}$$

EXAMPLE 5   Simplify $(4^{-1} - 3^{-2})^{-1}$.

Solution

$$(4^{-1} - 3^{-2})^{-1} = \left(\frac{1}{4^1} - \frac{1}{3^2} \right)^{-1} \qquad \text{Apply } b^{-n} = \frac{1}{b^n} \text{ to } 4^{-1} \text{ and to } 3^{-2}.$$

$$= \left(\frac{1}{4} - \frac{1}{9} \right)^{-1}$$

$$= \left(\frac{9}{36} - \frac{4}{36} \right)^{-1} \qquad \text{Use 36 as the LCD.}$$

$$= \left(\frac{5}{36} \right)^{-1}$$

$$= \frac{1}{\left(\frac{5}{36} \right)^1} \qquad \text{Apply } b^{-n} = \frac{1}{b^n}.$$

$$= \frac{1}{\frac{5}{36}} = \frac{36}{5}$$

EXAMPLE 6   Express $a^{-1} + b^{-2}$ as a single fraction involving positive exponents only.

Solution

$$a^{-1} + b^{-2} = \frac{1}{a^1} + \frac{1}{b^2} \qquad \text{Use } ab^2 \text{ as the LCD.}$$

$$= \left(\frac{1}{a} \right) \left(\frac{b^2}{b^2} \right) + \left(\frac{1}{b^2} \right) \left(\frac{a}{a} \right) \qquad \text{Change to equivalent fractions with } ab^2 \text{ as the LCD.}$$

$$= \frac{b^2}{ab^2} + \frac{a}{ab^2}$$

$$= \frac{b^2 + a}{ab^2}$$

PROBLEM SET 5.1

For Problems 1–42, simplify each numerical expression.

1. 3^{-3}

2. 2^{-4}

3. -10^{-2}

4. 10^{-3}

5. $\dfrac{1}{3^{-4}}$

6. $\dfrac{1}{2^{-6}}$

7. $-\left(\dfrac{1}{3}\right)^{-3}$

8. $\left(\dfrac{1}{2}\right)^{-3}$

9. $\left(-\dfrac{1}{2}\right)^{-3}$

10. $\left(\dfrac{2}{7}\right)^{-2}$

11. $\left(-\dfrac{3}{4}\right)^{0}$

12. $\dfrac{1}{\left(\dfrac{4}{5}\right)^{-2}}$

13. $\dfrac{1}{\left(\dfrac{3}{7}\right)^{-2}}$

14. $-\left(\dfrac{5}{6}\right)^{0}$

15. $2^7 \cdot 2^{-3}$

16. $3^{-4} \cdot 3^6$

17. $10^{-5} \cdot 10^2$

18. $10^4 \cdot 10^{-6}$

19. $10^{-1} \cdot 10^{-2}$

20. $10^{-2} \cdot 10^{-2}$

21. $(3^{-1})^{-3}$

22. $(2^{-2})^{-4}$

23. $(5^3)^{-1}$

24. $(3^{-1})^3$

25. $(2^3 \cdot 3^{-2})^{-1}$

26. $(2^{-2} \cdot 3^{-1})^{-3}$

27. $(4^2 \cdot 5^{-1})^2$

28. $(2^{-3} \cdot 4^{-1})^{-1}$

29. $\left(\dfrac{2^{-1}}{5^{-2}}\right)^{-1}$

30. $\left(\dfrac{2^{-4}}{3^{-2}}\right)^{-2}$

31. $\left(\dfrac{2^{-1}}{3^{-2}}\right)^{2}$

32. $\left(\dfrac{3^2}{5^{-1}}\right)^{-1}$

33. $\dfrac{3^3}{3^{-1}}$

34. $\dfrac{2^{-2}}{2^3}$

35. $\dfrac{10^{-2}}{10^2}$

36. $\dfrac{10^{-2}}{10^{-5}}$

37. $2^{-2} + 3^{-2}$

38. $2^{-4} + 5^{-1}$

39. $\left(\dfrac{1}{3}\right)^{-1} - \left(\dfrac{2}{5}\right)^{-1}$

40. $\left(\dfrac{3}{2}\right)^{-1} - \left(\dfrac{1}{4}\right)^{-1}$

41. $(2^{-3} + 3^{-2})^{-1}$

42. $(5^{-1} - 2^{-3})^{-1}$

For Problems 43–62, simplify each expression. Express final results without using zero or negative integers as exponents.

43. $x^2 \cdot x^{-8}$

44. $x^{-3} \cdot x^{-4}$

45. $a^3 \cdot a^{-5} \cdot a^{-1}$

46. $b^{-2} \cdot b^3 \cdot b^{-6}$

47. $(a^{-4})^2$

48. $(b^4)^{-3}$

49. $(x^2y^{-6})^{-1}$

50. $(x^5y^{-1})^{-3}$

51. $(ab^3c^{-2})^{-4}$

52. $(a^3b^{-3}c^{-2})^{-5}$

53. $(2x^3y^{-4})^{-3}$

54. $(4x^5y^{-2})^{-2}$

55. $\left(\dfrac{x^{-1}}{y^{-4}}\right)^{-3}$

56. $\left(\dfrac{y^3}{x^{-4}}\right)^{-2}$

57. $\left(\dfrac{3a^{-2}}{2b^{-1}}\right)^{-2}$

58. $\left(\dfrac{2xy^2}{5a^{-1}b^{-2}}\right)^{-1}$

59. $\dfrac{x^{-6}}{x^{-4}}$

60. $\dfrac{a^{-2}}{a^2}$

61. $\dfrac{a^3b^{-2}}{a^{-2}b^{-4}}$

62. $\dfrac{x^{-3}y^{-4}}{x^2y^{-1}}$

For Problems 63–74, find the indicated products and quotients. Express final results using positive integral exponents only.

63. $(2xy^{-1})(3x^{-2}y^4)$

64. $(-4x^{-1}y^2)(6x^3y^{-4})$

65. $(-7a^2b^{-5})(-a^{-2}b^7)$

66. $(-9a^{-3}b^{-6})(-12a^{-1}b^4)$

67. $\dfrac{28x^{-2}y^{-3}}{4x^{-3}y^{-1}}$

68. $\dfrac{63x^2y^{-4}}{7xy^{-4}}$

69. $\dfrac{-72a^2b^{-4}}{6a^3b^{-7}}$

70. $\dfrac{108a^{-5}b^{-4}}{9a^{-2}b}$

71. $\left(\dfrac{35x^{-1}y^{-2}}{7x^4y^3}\right)^{-1}$

72. $\left(\dfrac{-48ab^2}{-6a^3b^5}\right)^{-2}$

73. $\left(\dfrac{-36a^{-1}b^{-6}}{4a^{-1}b^4}\right)^{-2}$

74. $\left(\dfrac{8xy^3}{-4x^4y}\right)^{-3}$

For Problems 75–84, express each of the following as a single fraction involving positive exponents only.

75. $x^{-2} + x^{-3}$

76. $x^{-1} + x^{-5}$

77. $x^{-3} - y^{-1}$

78. $2x^{-1} - 3y^{-2}$

79. $3a^{-2} + 4b^{-1}$

80. $a^{-1} + a^{-1}b^{-3}$

81. $x^{-1}y^{-2} - xy^{-1}$

82. $x^2y^{-2} - x^{-1}y^{-3}$

83. $2x^{-1} - 3x^{-2}$

84. $5x^{-2}y + 6x^{-1}y^{-2}$

■■■ Thoughts into words

85. Is the following simplification process correct?

$$(3^{-2})^{-1} = \left(\frac{1}{3^2}\right)^{-1} = \left(\frac{1}{9}\right)^{-1} = \frac{1}{\left(\frac{1}{9}\right)^1} = 9$$

Could you suggest a better way to do the problem?

86. Explain how to simplify $(2^{-1} \cdot 3^{-2})^{-1}$ and also how to simplify $(2^{-1} + 3^{-2})^{-1}$.

■■■ Further investigations

87. Use a calculator to check your answers for Problems 1–42.

88. Use a calculator to simplify each of the following numerical expressions. Express your answers to the nearest hundredth.

(a) $(2^{-3} + 3^{-3})^{-2}$

(b) $(4^{-3} - 2^{-1})^{-2}$

(c) $(5^{-3} - 3^{-5})^{-1}$

(d) $(6^{-2} + 7^{-4})^{-2}$

(e) $(7^{-3} - 2^{-4})^{-2}$

(f) $(3^{-4} + 2^{-3})^{-3}$

5.2   Roots and Radicals

To **square a number** means to raise it to the second power — that is, to use the number as a factor twice.

$$4^2 = 4 \cdot 4 = 16 \qquad \text{Read "four squared equals sixteen."}$$

$$10^2 = 10 \cdot 10 = 100$$

$$\left(\frac{1}{2}\right)^2 = \frac{1}{2} \cdot \frac{1}{2} = \frac{1}{4}$$

$$(-3)^2 = (-3)(-3) = 9$$

A **square root of a number** is one of its two equal factors. Thus 4 is a square root of 16 because $4 \cdot 4 = 16$. Likewise, -4 is also a square root of 16 because $(-4)(-4) = 16$. In general, a is a square root of b if $a^2 = b$. The following generalizations are a direct consequence of the previous statement.

1. Every positive real number has two square roots; one is positive and the other is negative. They are opposites of each other.
2. Negative real numbers have no real number square roots because any real number except zero is positive when squared.
3. The square root of 0 is 0.

The symbol $\sqrt{}$, called a **radical sign,** is used to designate the nonnegative square root. The number under the radical sign is called the **radicand.** The entire expression, such as $\sqrt{16}$, is called a **radical.**

$\sqrt{16} = 4$ $\sqrt{16}$ indicates the nonnegative or **principal square root** of 16.

$-\sqrt{16} = -4$ $-\sqrt{16}$ indicates the negative square root of 16.

$\sqrt{0} = 0$ Zero has only one square root. Technically, we could write $-\sqrt{0} = -0 = 0$.

$\sqrt{-4}$ is not a real number.

$-\sqrt{-4}$ is not a real number.

In general, the following definition is useful.

DEFINITION 5.3

If $a \geq 0$ and $b \geq 0$, then $\sqrt{b} = a$ if and only if $a^2 = b$; a is called the **principal square root of b.**

To **cube a number** means to raise it to the third power — that is, to use the number as a factor three times.

$2^3 = 2 \cdot 2 \cdot 2 = 8$ Read "two cubed equals eight."

$4^3 = 4 \cdot 4 \cdot 4 = 64$

$\left(\dfrac{2}{3}\right)^3 = \dfrac{2}{3} \cdot \dfrac{2}{3} \cdot \dfrac{2}{3} = \dfrac{8}{27}$

$(-2)^3 = (-2)(-2)(-2) = -8$

A **cube root of a number** is one of its three equal factors. Thus 2 is a cube root of 8 because $2 \cdot 2 \cdot 2 = 8$. (In fact, 2 is the only real number that is a cube root of 8.) Furthermore, -2 is a cube root of -8 because $(-2)(-2)(-2) = -8$. (In fact, -2 is the only real number that is a cube root of -8.)

In general, a is a cube root of b if $a^3 = b$. The following generalizations are a direct consequence of the previous statement.

1. Every positive real number has one positive real number cube root.
2. Every negative real number has one negative real number cube root.
3. The cube root of 0 is 0.

REMARK: Technically, every nonzero real number has three cube roots, but only one of them is a real number. The other two roots are classified as complex numbers. We are restricting our work at this time to the set of real numbers.

The symbol $\sqrt[3]{}$ designates the cube root of a number. Thus we can write

$$\sqrt[3]{8} = 2 \qquad\qquad\qquad \sqrt[3]{\frac{1}{27}} = \frac{1}{3}$$

$$\sqrt[3]{-8} = -2 \qquad\qquad\qquad \sqrt[3]{-\frac{1}{27}} = -\frac{1}{3}$$

In general, the following definition is useful.

DEFINITION 5.4

$\sqrt[3]{b} = a$ if and only if $a^3 = b$.

In Definition 5.4, if b is a positive number, then a, the cube root, is a positive number, whereas if b is a negative number, then a, the cube root, is a negative number. The number a is called the principal cube root of b or simply the cube root of b.

The concept of root can be extended to fourth roots, fifth roots, sixth roots, and, in general, nth roots.

DEFINITION 5.5

The nth root of b is a,   if and only if $a^n = b$.

We can make the following generalizations.

If n is an even positive integer, then the following statements are true.

1. Every positive real number has exactly two real nth roots — one positive and one negative. For example, the real fourth roots of 16 are 2 and -2.
2. Negative real numbers do not have real nth roots. For example, there are no real fourth roots of -16.

If n is an odd positive integer greater than 1, then the following statements are true.

1. Every real number has exactly one real nth root.
2. The real nth root of a positive number is positive. For example, the fifth root of 32 is 2.
3. The real nth root of a negative number is negative. For example, the fifth root of -32 is -2.

The symbol $\sqrt[n]{}$ designates the principal nth root. To complete our terminology, the n in the radical $\sqrt[n]{b}$ is called the index of the radical. If $n = 2$, we commonly write $\sqrt{b}$ instead of $\sqrt[2]{b}$.

The following chart can help summarize this information with respect to $\sqrt[n]{b}$, where n is a positive integer greater than 1.

	If *b* is		
	Positive	**Zero**	**Negative**
n is even	$\sqrt[n]{b}$ is a positive real number	$\sqrt[n]{b} = 0$	$\sqrt[n]{b}$ is not a real number
n is odd	$\sqrt[n]{b}$ is a positive real number	$\sqrt[n]{b} = 0$	$\sqrt[n]{b}$ is a negative real number

Consider the following examples.

$$\sqrt[4]{81} = 3 \qquad \text{because } 3^4 = 81$$
$$\sqrt[5]{32} = 2 \qquad \text{because } 2^5 = 32$$
$$\sqrt[5]{-32} = -2 \qquad \text{because } (-2)^5 = -32$$

$\sqrt[4]{-16}$ is not a real number   because any real number, except zero, is positive when raised to the fourth power

The following property is a direct consequence of Definition 5.5.

PROPERTY 5.3

1. $(\sqrt[n]{b})^n = b$ n is any positive integer greater than 1.
2. $\sqrt[n]{b^n} = b$ n is any positive integer greater than 1 if $b \geq 0$; n is an odd positive integer greater than 1 if $b < 0$.

Because the radical expressions in parts (1) and (2) of Property 5.3 are both equal to b, by the transitive property they are equal to each other. Hence $\sqrt[n]{b^n} = (\sqrt[n]{b})^n$. The arithmetic is usually easier to simplify when we use the form $(\sqrt[n]{b})^n$. The following examples demonstrate the use of Property 5.3.

$$\sqrt{144^2} = (\sqrt{144})^2 = 12^2 = 144$$
$$\sqrt[3]{64^3} = (\sqrt[3]{64})^3 = 4^3 = 64$$

$$\sqrt[3]{(-8)^3} = (\sqrt[3]{-8})^3 = (-2)^3 = -8$$
$$\sqrt[4]{16^4} = (\sqrt[4]{16})^4 = 2^4 = 16$$

Let's use some examples to lead into the next very useful property of radicals.

$$\sqrt{4 \cdot 9} = \sqrt{36} = 6 \quad \text{and} \quad \sqrt{4} \cdot \sqrt{9} = 2 \cdot 3 = 6$$
$$\sqrt{16 \cdot 25} = \sqrt{400} = 20 \quad \text{and} \quad \sqrt{16} \cdot \sqrt{25} = 4 \cdot 5 = 20$$
$$\sqrt[3]{8 \cdot 27} = \sqrt[3]{216} = 6 \quad \text{and} \quad \sqrt[3]{8} \cdot \sqrt[3]{27} = 2 \cdot 3 = 6$$
$$\sqrt[3]{(-8)(27)} = \sqrt[3]{-216} = -6 \quad \text{and} \quad \sqrt[3]{-8} \cdot \sqrt[3]{27} = (-2)(3) = -6$$

In general, we can state the following property.

PROPERTY 5.4

$$\sqrt[n]{bc} = \sqrt[n]{b}\sqrt[n]{c} \qquad \sqrt[n]{b} \text{ and } \sqrt[n]{c} \text{ are real numbers}$$

Property 5.4 states that **the *n*th root of a product is equal to the product of the *n*th roots.**

Simplest Radical Form

The definition of *n*th root, along with Property 5.4, provides the basis for changing radicals to simplest radical form. The concept of **simplest radical form** takes on additional meaning as we encounter more complicated expressions, but for now it simply means that the radicand is not to contain any perfect powers of the index. Let's consider some examples to clarify this idea.

EXAMPLE 1

Express each of the following in simplest radical form.

(a) $\sqrt{8}$ 　　　(b) $\sqrt{45}$ 　　　(c) $\sqrt[3]{24}$ 　　　(d) $\sqrt[3]{54}$

Solution

(a) $\sqrt{8} = \sqrt{4 \cdot 2} = \sqrt{4}\sqrt{2} = 2\sqrt{2}$

　　　　　↑
　　　　4 is a
　　　　perfect
　　　　square.

(b) $\sqrt{45} = \sqrt{9 \cdot 5} = \sqrt{9}\sqrt{5} = 3\sqrt{5}$

　　　　↑
　　　9 is a
　　　perfect
　　　square.

(c) $\sqrt[3]{24} = \sqrt[3]{8 \cdot 3} = \sqrt[3]{8}\sqrt[3]{3} = 2\sqrt[3]{3}$

↑

8 is a
perfect
cube.

(d) $\sqrt[3]{54} = \sqrt[3]{27 \cdot 2} = \sqrt[3]{27}\sqrt[3]{2} = 3\sqrt[3]{2}$

↑

27 is a
perfect
cube.

The first step in each example is to express the radicand of the given radical as the product of two factors, one of which must be a perfect nth power other than 1. Also, observe the radicands of the final radicals. In each case, the radicand cannot have a factor that is a perfect nth power other than 1. We say that the final radicals $2\sqrt{2}, 3\sqrt{5}, 2\sqrt[3]{3}$, and $3\sqrt[3]{2}$ are in **simplest radical form.**

You may vary the steps somewhat in changing to simplest radical form, but the final result should be the same. Consider some different approaches to changing $\sqrt{72}$ to simplest form:

$$\sqrt{72} = \sqrt{9}\sqrt{8} = 3\sqrt{8} = 3\sqrt{4}\sqrt{2} = 3 \cdot 2\sqrt{2} = 6\sqrt{2} \qquad \text{or}$$
$$\sqrt{72} = \sqrt{4}\sqrt{18} = 2\sqrt{18} = 2\sqrt{9}\sqrt{2} = 2 \cdot 3\sqrt{2} = 6\sqrt{2} \qquad \text{or}$$
$$\sqrt{72} = \sqrt{36}\sqrt{2} = 6\sqrt{2}$$

Another variation of the technique for changing radicals to simplest form is to prime-factor the radicand and then to look for perfect nth powers in exponential form. The following example illustrates the use of this technique.

E X A M P L E   2

Express each of the following in simplest radical form.

(a) $\sqrt{50}$ (b) $3\sqrt{80}$ (c) $\sqrt[3]{108}$

Solution

(a) $\sqrt{50} = \sqrt{2 \cdot 5 \cdot 5} = \sqrt{5^2}\sqrt{2} = 5\sqrt{2}$

(b) $3\sqrt{80} = 3\sqrt{2 \cdot 2 \cdot 2 \cdot 2 \cdot 5} = 3\sqrt{2^4}\sqrt{5} = 3 \cdot 2^2\sqrt{5} = 12\sqrt{5}$

(c) $\sqrt[3]{108} = \sqrt[3]{2 \cdot 2 \cdot 3 \cdot 3 \cdot 3} = \sqrt[3]{3^3}\sqrt[3]{4} = 3\sqrt[3]{4}$

Another property of nth roots is demonstrated by the following examples.

$$\sqrt{\frac{36}{9}} = \sqrt{4} = 2 \qquad \text{and} \qquad \frac{\sqrt{36}}{\sqrt{9}} = \frac{6}{3} = 2$$

$$\sqrt[3]{\frac{64}{8}} = \sqrt[3]{8} = 2 \quad \text{and} \quad \frac{\sqrt[3]{64}}{\sqrt[3]{8}} = \frac{4}{2} = 2$$

$$\sqrt[3]{\frac{-8}{64}} = \sqrt[3]{-\frac{1}{8}} = -\frac{1}{2} \quad \text{and} \quad \frac{\sqrt[3]{-8}}{\sqrt[3]{64}} = \frac{-2}{4} = -\frac{1}{2}$$

In general, we can state the following property.

PROPERTY 5.5

$$\sqrt[n]{\frac{b}{c}} = \frac{\sqrt[n]{b}}{\sqrt[n]{c}} \qquad \sqrt[n]{b} \text{ and } \sqrt[n]{c} \text{ are real numbers and } c \neq 0.$$

Property 5.5 states that **the nth root of a quotient is equal to the quotient of the nth roots.**

To evaluate radicals such as $\sqrt{\dfrac{4}{25}}$ and $\sqrt[3]{\dfrac{27}{8}}$, for which the numerator and denominator of the fractional radicand are perfect nth powers, you may use Property 5.5 or merely rely on the definition of nth root.

$$\sqrt{\frac{4}{25}} = \frac{\sqrt{4}}{\sqrt{25}} = \frac{2}{5} \quad \text{or} \quad \sqrt{\frac{4}{25}} = \frac{2}{5} \text{ because } \frac{2}{5} \cdot \frac{2}{5} = \frac{4}{25}$$

$$\uparrow \qquad\qquad\qquad\qquad\qquad \uparrow$$

Property 5.5 $\qquad\qquad\qquad$ Definition of nth root

$$\downarrow \qquad\qquad\qquad\qquad\qquad \downarrow$$

$$\sqrt[3]{\frac{27}{8}} = \frac{\sqrt[3]{27}}{\sqrt[3]{8}} = \frac{3}{2} \quad \text{or} \quad \sqrt[3]{\frac{27}{8}} = \frac{3}{2} \text{ because } \frac{3}{2} \cdot \frac{3}{2} \cdot \frac{3}{2} = \frac{27}{8}$$

Radicals such as $\sqrt{\dfrac{28}{9}}$ and $\sqrt[3]{\dfrac{24}{27}}$, in which only the denominators of the radicand are perfect nth powers, can be simplified as follows:

$$\sqrt{\frac{28}{9}} = \frac{\sqrt{28}}{\sqrt{9}} = \frac{\sqrt{28}}{3} = \frac{\sqrt{4}\sqrt{7}}{3} = \frac{2\sqrt{7}}{3}$$

$$\sqrt[3]{\frac{24}{27}} = \frac{\sqrt[3]{24}}{\sqrt[3]{27}} = \frac{\sqrt[3]{24}}{3} = \frac{\sqrt[3]{8}\sqrt[3]{3}}{3} = \frac{2\sqrt[3]{3}}{3}$$

Before we consider more examples, let's summarize some ideas that pertain to the simplifying of radicals. A radical is said to be in **simplest radical form** if the following conditions are satisfied.

1. No fraction appears with a radical sign. $\sqrt{\dfrac{3}{4}}$ violates this condition.

2. No radical appears in the denominator. $\dfrac{\sqrt{2}}{\sqrt{3}}$ violates this condition.

3. No radicand, when expressed in prime-factored form, contains a factor raised to a power equal to or greater than the index.

$\sqrt{2^3 \cdot 5}$ violates this condition.

Now let's consider an example in which neither the numerator nor the denominator of the radicand is a perfect nth power.

EXAMPLE 3   Simplify $\sqrt{\dfrac{2}{3}}$.

Solution

$$\sqrt{\dfrac{2}{3}} = \dfrac{\sqrt{2}}{\sqrt{3}} = \dfrac{\sqrt{2}}{\sqrt{3}} \cdot \dfrac{\sqrt{3}}{\sqrt{3}} = \dfrac{\sqrt{6}}{3}$$

Form of 1

We refer to the process we used to simplify the radical in Example 3 as **rationalizing the denominator.** Note that the denominator becomes a rational number. The process of rationalizing the denominator can often be accomplished in more than one way, as we will see in the next example.

EXAMPLE 4   Simplify $\dfrac{\sqrt{5}}{\sqrt{8}}$.

Solution A

$$\dfrac{\sqrt{5}}{\sqrt{8}} = \dfrac{\sqrt{5}}{\sqrt{8}} \cdot \dfrac{\sqrt{8}}{\sqrt{8}} = \dfrac{\sqrt{40}}{8} = \dfrac{\sqrt{4}\sqrt{10}}{8} = \dfrac{2\sqrt{10}}{8} = \dfrac{\sqrt{10}}{4}$$

Solution B

$$\dfrac{\sqrt{5}}{\sqrt{8}} = \dfrac{\sqrt{5}}{\sqrt{8}} \cdot \dfrac{\sqrt{2}}{\sqrt{2}} = \dfrac{\sqrt{10}}{\sqrt{16}} = \dfrac{\sqrt{10}}{4}$$

Solution C

$$\dfrac{\sqrt{5}}{\sqrt{8}} = \dfrac{\sqrt{5}}{\sqrt{4}\sqrt{2}} = \dfrac{\sqrt{5}}{2\sqrt{2}} = \dfrac{\sqrt{5}}{2\sqrt{2}} \cdot \dfrac{\sqrt{2}}{\sqrt{2}} = \dfrac{\sqrt{10}}{2\sqrt{4}} = \dfrac{\sqrt{10}}{2(2)} = \dfrac{\sqrt{10}}{4}$$

The three approaches to Example 4 again illustrate the need to think first and only then push the pencil. You may find one approach easier than another. To conclude this section, study the following examples and check the final radicals against the three conditions previously listed for **simplest radical form.**

EXAMPLE 5   Simplify each of the following.

$$\text{(a)}\ \frac{3\sqrt{2}}{5\sqrt{3}} \qquad \text{(b)}\ \frac{3\sqrt{7}}{2\sqrt{18}} \qquad \text{(c)}\ \sqrt[3]{\frac{5}{9}} \qquad \text{(d)}\ \frac{\sqrt[3]{5}}{\sqrt[3]{16}}$$

Solution

$$\text{(a)}\ \frac{3\sqrt{2}}{5\sqrt{3}} = \frac{3\sqrt{2}}{5\sqrt{3}} \cdot \frac{\sqrt{3}}{\sqrt{3}} = \frac{3\sqrt{6}}{5\sqrt{9}} = \frac{3\sqrt{6}}{15} = \frac{\sqrt{6}}{5}$$

$$\uparrow$$
Form of 1

$$\text{(b)}\ \frac{3\sqrt{7}}{2\sqrt{18}} = \frac{3\sqrt{7}}{2\sqrt{18}} \cdot \frac{\sqrt{2}}{\sqrt{2}} = \frac{3\sqrt{14}}{2\sqrt{36}} = \frac{3\sqrt{14}}{12} = \frac{\sqrt{14}}{4}$$

$$\uparrow$$
Form of 1

$$\text{(c)}\ \sqrt[3]{\frac{5}{9}} = \frac{\sqrt[3]{5}}{\sqrt[3]{9}} = \frac{\sqrt[3]{5}}{\sqrt[3]{9}} \cdot \frac{\sqrt[3]{3}}{\sqrt[3]{3}} = \frac{\sqrt[3]{15}}{\sqrt[3]{27}} = \frac{\sqrt[3]{15}}{3}$$

$$\uparrow$$
Form of 1

$$\text{(d)}\ \frac{\sqrt[3]{5}}{\sqrt[3]{16}} = \frac{\sqrt[3]{5}}{\sqrt[3]{16}} \cdot \frac{\sqrt[3]{4}}{\sqrt[3]{4}} = \frac{\sqrt[3]{20}}{\sqrt[3]{64}} = \frac{\sqrt[3]{20}}{4}$$

$$\uparrow$$
Form of 1

Applications of Radicals

Many real-world applications involve radical expressions. For example, police often use the formula $S = \sqrt{30Df}$ to estimate the speed of a car on the basis of the length of the skid marks at the scene of an accident. In this formula, S represents the speed of the car in miles per hour, D represents the length of the skid marks in feet, and f represents a coefficient of friction. For a particular situation, the coefficient of friction is a constant that depends on the type and condition of the road surface.

EXAMPLE 6

Using 0.35 as a coefficient of friction, determine how fast a car was traveling if it skidded 325 feet.

Solution

Substitute 0.35 for f and 325 for D in the formula.

$$S = \sqrt{30Df} = \sqrt{30(325)(0.35)} = 58, \quad \text{to the nearest whole number}$$

The car was traveling at approximately 58 miles per hour. ■

The **period** of a pendulum is the time it takes to swing from one side to the other side and back. The formula

$$T = 2\pi\sqrt{\frac{L}{32}}$$

where T represents the time in seconds and L the length in feet, can be used to determine the period of a pendulum (see Figure 5.1).

Figure 5.1

EXAMPLE 7

Find, to the nearest tenth of a second, the period of a pendulum of length 3.5 feet.

Solution

Let's use 3.14 as an approximation for π and substitute 3.5 for L in the formula.

$$T = 2\pi\sqrt{\frac{L}{32}} = 2(3.14)\sqrt{\frac{3.5}{32}} = 2.1, \quad \text{to the nearest tenth}$$

The period is approximately 2.1 seconds. ■

Radical expressions are also used in some geometric applications. For example, the area of a triangle can be found by using a formula that involves a square root. If a, b, and c represent the lengths of the three sides of a triangle, the formula $K = \sqrt{s(s-a)(s-b)(s-c)}$, known as Heron's formula, can be used to determine the area (K) of the triangle. The letter s represents the semiperimeter of the triangle; that is, $s = \dfrac{a+b+c}{2}$.

EXAMPLE 8

Find the area of a triangular piece of sheet metal that has sides of lengths 17 inches, 19 inches, and 26 inches.

Solution
First, let's find the value of s, the semiperimeter of the triangle.

$$s = \frac{17 + 19 + 26}{2} = 31$$

Now we can use Heron's formula.

$$\begin{aligned} K = \sqrt{s(s-a)(s-b)(s-c)} &= \sqrt{31(31-17)(31-19)(31-26)} \\ &= \sqrt{31(14)(12)(5)} \\ &= \sqrt{20{,}640} \\ &= 161.4, \quad \text{to the nearest tenth} \end{aligned}$$

Thus the area of the piece of sheet metal is approximately 161.4 square inches. ■

REMARK: Note that in Examples 6–8, we did not simplify the radicals. When one is using a calculator to approximate the square roots, there is no need to simplify first.

PROBLEM SET 5.2

For Problems 1–20, evaluate each of the following. For example, $\sqrt{25} = 5$.

1. $\sqrt{64}$

2. $\sqrt{49}$

3. $-\sqrt{100}$

4. $-\sqrt{81}$

5. $\sqrt[3]{27}$

6. $\sqrt[3]{216}$

7. $\sqrt[3]{-64}$

8. $\sqrt[3]{-125}$

9. $\sqrt[4]{81}$

10. $-\sqrt[4]{16}$

11. $\sqrt{\dfrac{16}{25}}$

12. $\sqrt{\dfrac{25}{64}}$

13. $-\sqrt{\dfrac{36}{49}}$

14. $\sqrt{\dfrac{16}{64}}$

15. $\sqrt{\dfrac{9}{36}}$

16. $\sqrt{\dfrac{144}{36}}$

17. $\sqrt[3]{\dfrac{27}{64}}$

18. $\sqrt[3]{-\dfrac{8}{27}}$

19. $\sqrt[3]{8^3}$

20. $\sqrt[4]{16^4}$

For Problems 21–74, change each radical to simplest radical form.

21. $\sqrt{27}$

22. $\sqrt{48}$

23. $\sqrt{32}$

24. $\sqrt{98}$

25. $\sqrt{80}$

26. $\sqrt{125}$

27. $\sqrt{160}$

28. $\sqrt{112}$

29. $4\sqrt{18}$

30. $5\sqrt{32}$

31. $-6\sqrt{20}$

32. $-4\sqrt{54}$

33. $\dfrac{2}{5}\sqrt{75}$

34. $\dfrac{1}{3}\sqrt{90}$

35. $\dfrac{3}{2}\sqrt{24}$

36. $\dfrac{3}{4}\sqrt{45}$

37. $-\dfrac{5}{6}\sqrt{28}$

38. $-\dfrac{2}{3}\sqrt{96}$

39. $\sqrt{\dfrac{19}{4}}$

40. $\sqrt{\dfrac{22}{9}}$

41. $\sqrt{\dfrac{27}{16}}$

42. $\sqrt{\dfrac{8}{25}}$

43. $\sqrt{\dfrac{75}{81}}$

44. $\sqrt{\dfrac{24}{49}}$

45. $\sqrt{\dfrac{2}{7}}$

46. $\sqrt{\dfrac{3}{8}}$

47. $\sqrt{\dfrac{2}{3}}$

48. $\sqrt{\dfrac{7}{12}}$

49. $\dfrac{\sqrt{5}}{\sqrt{12}}$

50. $\dfrac{\sqrt{3}}{\sqrt{7}}$

51. $\dfrac{\sqrt{11}}{\sqrt{24}}$

52. $\dfrac{\sqrt{5}}{\sqrt{48}}$

53. $\dfrac{\sqrt{18}}{\sqrt{27}}$

54. $\dfrac{\sqrt{10}}{\sqrt{20}}$

55. $\dfrac{\sqrt{35}}{\sqrt{7}}$

56. $\dfrac{\sqrt{42}}{\sqrt{6}}$

57. $\dfrac{2\sqrt{3}}{\sqrt{7}}$

58. $\dfrac{3\sqrt{2}}{\sqrt{6}}$

59. $-\dfrac{4\sqrt{12}}{\sqrt{5}}$

60. $\dfrac{-6\sqrt{5}}{\sqrt{18}}$

61. $\dfrac{3\sqrt{2}}{4\sqrt{3}}$

62. $\dfrac{6\sqrt{5}}{5\sqrt{12}}$

63. $\dfrac{-8\sqrt{18}}{10\sqrt{50}}$

64. $\dfrac{4\sqrt{45}}{-6\sqrt{20}}$

65. $\sqrt[3]{16}$

66. $\sqrt[3]{40}$

67. $2\sqrt[3]{81}$

68. $-3\sqrt[3]{54}$

69. $\dfrac{2}{\sqrt[3]{9}}$

70. $\dfrac{3}{\sqrt[3]{3}}$

71. $\dfrac{\sqrt[3]{27}}{\sqrt[3]{4}}$

72. $\dfrac{\sqrt[3]{8}}{\sqrt[3]{16}}$

73. $\dfrac{\sqrt[3]{6}}{\sqrt[3]{4}}$

74. $\dfrac{\sqrt[3]{4}}{\sqrt[3]{2}}$

75. Use a coefficient of friction of 0.4 in the formula from Example 6 and find the speeds of cars that left skid marks of lengths 150 feet, 200 feet, and 350 feet. Express your answers to the nearest mile per hour.

76. Use the formula from Example 7 and find the periods of pendulums of lengths 2 feet, 3 feet, and 4.5 feet. Express your answers to the nearest tenth of a second.

77. Find, to the nearest square centimeter, the area of a triangle that measures 14 centimeters by 16 centimeters by 18 centimeters.

78. Find, to the nearest square yard, the area of a triangular plot of ground that measures 45 yards by 60 yards by 75 yards.

79. Find the area of an equilateral triangle, each of whose sides is 18 inches long. Express the area to the nearest square inch.

80. Find, to the nearest square inch, the area of the quadrilateral in Figure 5.2.

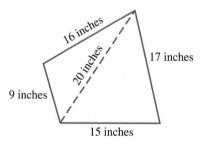

Figure 5.2

■ ■ ■ Thoughts into words

81. Why is $\sqrt{-9}$ not a real number?

82. Why is it that we say 25 has two square roots (5 and −5), but we write $\sqrt{25} = 5$?

83. How is the multiplication property of 1 used when simplifying radicals?

84. How could you find a whole number approximation for $\sqrt{2750}$ if you did not have a calculator or table available?

■ ■ ■ **Further investigations**

85. Use your calculator to find a rational approximation, to the nearest thousandth, for (a) through (i).

    (a) $\sqrt{2}$ **(b)** $\sqrt{75}$ **(c)** $\sqrt{156}$

    (d) $\sqrt{691}$ **(e)** $\sqrt{3249}$ **(f)** $\sqrt{45{,}123}$

    (g) $\sqrt{0.14}$ **(h)** $\sqrt{0.023}$ **(i)** $\sqrt{0.8649}$

86. Sometimes a fairly good estimate can be made of a radical expression by using whole number approximations. For example, $5\sqrt{35} + 7\sqrt{50}$ is approximately $5(6) + 7(7) = 79$. Using a calculator, we find that $5\sqrt{35} + 7\sqrt{50} = 79.1$ to the nearest tenth. In this case our whole number estimate is very good. For (a) through (f), first make a whole number estimate, and then use your calculator to see how well you estimated.

    (a) $3\sqrt{10} - 4\sqrt{24} + 6\sqrt{65}$

    (b) $9\sqrt{27} + 5\sqrt{37} - 3\sqrt{80}$

    (c) $12\sqrt{5} + 13\sqrt{18} + 9\sqrt{47}$

    (d) $3\sqrt{98} - 4\sqrt{83} - 7\sqrt{120}$

    (e) $4\sqrt{170} + 2\sqrt{198} + 5\sqrt{227}$

    (f) $-3\sqrt{256} - 6\sqrt{287} + 11\sqrt{321}$

5.3 Combining Radicals and Simplifying Radicals That Contain Variables

Recall our use of the distributive property as the basis for combining similar terms. For example,

$$3x + 2x = (3 + 2)x = 5x$$

$$8y - 5y = (8 - 5)y = 3y$$

$$\frac{2}{3}a^2 + \frac{3}{4}a^2 = \left(\frac{2}{3} + \frac{3}{4}\right)a^2 = \left(\frac{8}{12} + \frac{9}{12}\right)a^2 = \frac{17}{12}a^2$$

In a like manner, expressions that contain radicals can often be simplified by using the distributive property, as follows:

$$3\sqrt{2} + 5\sqrt{2} = (3 + 5)\sqrt{2} = 8\sqrt{2}$$

$$7\sqrt[3]{5} - 3\sqrt[3]{5} = (7 - 3)\sqrt[3]{5} = 4\sqrt[3]{5}$$

$$4\sqrt{7} + 5\sqrt{7} + 6\sqrt{11} - 2\sqrt{11} = (4 + 5)\sqrt{7} + (6 - 2)\sqrt{11} = 9\sqrt{7} + 4\sqrt{11}$$

Note that *in order to be added or subtracted, radicals must have the same index and the same radicand.* Thus we cannot simplify an expression such as $5\sqrt{2} + 7\sqrt{11}$.

    Simplifying by combining radicals sometimes requires that you first express the given radicals in simplest form and then apply the distributive property. The following examples illustrate this idea.

E X A M P L E   1

Simplify $3\sqrt{8} + 2\sqrt{18} - 4\sqrt{2}$.

Solution

$$\begin{aligned}
3\sqrt{8} + 2\sqrt{18} - 4\sqrt{2} &= 3\sqrt{4}\sqrt{2} + 2\sqrt{9}\sqrt{2} - 4\sqrt{2} \\
&= 3 \cdot 2 \cdot \sqrt{2} + 2 \cdot 3 \cdot \sqrt{2} - 4\sqrt{2} \\
&= 6\sqrt{2} + 6\sqrt{2} - 4\sqrt{2} \\
&= (6 + 6 - 4)\sqrt{2} = 8\sqrt{2}
\end{aligned}$$

■

E X A M P L E   2

Simplify $\dfrac{1}{4}\sqrt{45} + \dfrac{1}{3}\sqrt{20}$.

Solution

$$\begin{aligned}
\frac{1}{4}\sqrt{45} + \frac{1}{3}\sqrt{20} &= \frac{1}{4}\sqrt{9}\sqrt{5} + \frac{1}{3}\sqrt{4}\sqrt{5} \\[2mm]
&= \frac{1}{4} \cdot 3 \cdot \sqrt{5} + \frac{1}{3} \cdot 2 \cdot \sqrt{5} \\[2mm]
&= \frac{3}{4}\sqrt{5} + \frac{2}{3}\sqrt{5} = \left(\frac{3}{4} + \frac{2}{3}\right)\sqrt{5} \\[2mm]
&= \left(\frac{9}{12} + \frac{8}{12}\right)\sqrt{5} = \frac{17}{12}\sqrt{5}
\end{aligned}$$

■

E X A M P L E   3

Simplify $5\sqrt[3]{2} - 2\sqrt[3]{16} - 6\sqrt[3]{54}$.

Solution

$$\begin{aligned}
5\sqrt[3]{2} - 2\sqrt[3]{16} - 6\sqrt[3]{54} &= 5\sqrt[3]{2} - 2\sqrt[3]{8}\sqrt[3]{2} - 6\sqrt[3]{27}\sqrt[3]{2} \\
&= 5\sqrt[3]{2} - 2 \cdot 2 \cdot \sqrt[3]{2} - 6 \cdot 3 \cdot \sqrt[3]{2} \\
&= 5\sqrt[3]{2} - 4\sqrt[3]{2} - 18\sqrt[3]{2} \\
&= (5 - 4 - 18)\sqrt[3]{2} \\
&= -17\sqrt[3]{2}
\end{aligned}$$

■

Radicals That Contain Variables

Before we discuss the process of simplifying radicals that contain variables, there is one technicality that we should call to your attention. Let's look at some examples to clarify the point. Consider the radical $\sqrt{x^2}$.

Let $x = 3$;   then $\sqrt{x^2} = \sqrt{3^2} = \sqrt{9} = 3$.
Let $x = -3$;   then $\sqrt{x^2} = \sqrt{(-3)^2} = \sqrt{9} = 3$.

Thus if $x \geq 0$, then $\sqrt{x^2} = x$, *but* if $x < 0$, then $\sqrt{x^2} = -x$. Using the concept of absolute value, we can state that for all real numbers, $\sqrt{x^2} = |x|$.

Now consider the radical $\sqrt{x^3}$. Because x^3 is negative when x is negative, we need to restrict x to the nonnegative reals when working with $\sqrt{x^3}$. Thus we can write: If $x \geq 0$, then $\sqrt{x^3} = \sqrt{x^2}\sqrt{x} = x\sqrt{x}$, and no absolute-value sign is necessary. Finally, let's consider the radical $\sqrt[3]{x^3}$.

Let $x = 2$;   then $\sqrt[3]{x^3} = \sqrt[3]{2^3} = \sqrt[3]{8} = 2$.
Let $x = -2$; then. $\sqrt[3]{x^3} = \sqrt[3]{(-2)^3} = \sqrt[3]{-8} = -2$.

Thus it is correct to write: $\sqrt[3]{x^3} = x$ for all real numbers, and again no absolute-value sign is necessary.

The previous discussion indicates that technically, every radical expression involving variables in the radicand needs to be analyzed individually in terms of any necessary restrictions imposed on the variables. To help you gain experience with this skill, examples and problems are discussed under Further Investigations in the problem set. For now, however, to avoid considering such restrictions on a problem-to-problem basis, we shall merely assume that all variables represent positive real numbers. Let's consider the process of simplifying radicals that contain variables in the radicand. Study the following examples, and note that the same basic approach we used in Section 5.2 is applied here.

E X A M P L E   4   Simplify each of the following.

      (a) $\sqrt{8x^3}$ **(b)** $\sqrt{45x^3y^7}$ **(c)** $\sqrt{180a^4b^3}$ **(d)** $\sqrt[3]{40x^4y^8}$

Solution

    (a) $\sqrt{8x^3} = \sqrt{4x^2}\sqrt{2x} = 2x\sqrt{2x}$

$4x^2$ is a
perfect square.

    (b) $\sqrt{45x^3y^7} = \sqrt{9x^2y^6}\sqrt{5xy} = 3xy^3\sqrt{5xy}$

$9x^2y^6$ is a
perfect square.

(c) If the numerical coefficient of the radicand is quite large, you may want to look at it in the prime-factored form.

$$\sqrt{180a^4b^3} = \sqrt{2 \cdot 2 \cdot 3 \cdot 3 \cdot 5 \cdot a^4 \cdot b^3}$$
$$= \sqrt{36 \cdot 5 \cdot a^4 \cdot b^3}$$
$$= \sqrt{36a^4b^2}\sqrt{5b}$$
$$= 6a^2b\sqrt{5b}$$

(d) $\sqrt[3]{40x^4y^8} = \sqrt[3]{8x^3y^6}\sqrt[3]{5xy^2} = 2xy^2\sqrt[3]{5xy^2}$

$\uparrow$

$8x^3y^6$ is a
perfect cube.

Before we consider more examples, let's restate (in such a way as to include radicands containing variables) the conditions necessary for a radical to be in simplest radical form.

1. A radicand contains no polynomial factor raised to a power equal to or greater than the index of the radical. $\sqrt{x^3}$ violates this condition.

2. No fraction appears within a radical sign. $\sqrt{\dfrac{2x}{3y}}$ violates this condition.

3. No radical appears in the denominator. $\dfrac{3}{\sqrt[3]{4x}}$ violates this condition.

E X A M P L E   5   Express each of the following in simplest radical form.

(a) $\sqrt{\dfrac{2x}{3y}}$ **(b)** $\dfrac{\sqrt{5}}{\sqrt{12a^3}}$ **(c)** $\dfrac{\sqrt{8x^2}}{\sqrt{27y^5}}$

(d) $\dfrac{3}{\sqrt[3]{4x}}$ **(e)** $\dfrac{\sqrt[3]{16x^2}}{\sqrt[3]{9y^5}}$

Solution

(a) $\sqrt{\dfrac{2x}{3y}} = \dfrac{\sqrt{2x}}{\sqrt{3y}} = \dfrac{\sqrt{2x}}{\sqrt{3y}} \cdot \dfrac{\sqrt{3y}}{\sqrt{3y}} = \dfrac{\sqrt{6xy}}{3y}$

$\uparrow$

Form of 1

(b) $\dfrac{\sqrt{5}}{\sqrt{12a^3}} = \dfrac{\sqrt{5}}{\sqrt{12a^3}} \cdot \dfrac{\sqrt{3a}}{\sqrt{3a}} = \dfrac{\sqrt{15a}}{\sqrt{36a^4}} = \dfrac{\sqrt{15a}}{6a^2}$

$\uparrow$

Form of 1

(c) $\dfrac{\sqrt{8x^2}}{\sqrt{27y^5}} = \dfrac{\sqrt{4x^2}\sqrt{2}}{\sqrt{9y^4}\sqrt{3y}} = \dfrac{2x\sqrt{2}}{3y^2\sqrt{3y}} = \dfrac{2x\sqrt{2}}{3y^2\sqrt{3y}} \cdot \dfrac{\sqrt{3y}}{\sqrt{3y}}$

$= \dfrac{2x\sqrt{6y}}{(3y^2)(3y)} = \dfrac{2x\sqrt{6y}}{9y^3}$

(d) $\dfrac{3}{\sqrt[3]{4x}} = \dfrac{3}{\sqrt[3]{4x}} \cdot \dfrac{\sqrt[3]{2x^2}}{\sqrt[3]{2x^2}} = \dfrac{3\sqrt[3]{2x^2}}{\sqrt[3]{8x^3}} = \dfrac{3\sqrt[3]{2x^2}}{2x}$

(e) $\dfrac{\sqrt[3]{16x^2}}{\sqrt[3]{9y^5}} = \dfrac{\sqrt[3]{16x^2}}{\sqrt[3]{9y^5}} \cdot \dfrac{\sqrt[3]{3y}}{\sqrt[3]{3y}} = \dfrac{\sqrt[3]{48x^2y}}{\sqrt[3]{27y^6}} = \dfrac{\sqrt[3]{8}\sqrt[3]{6x^2y}}{3y^2} = \dfrac{2\sqrt[3]{6x^2y}}{3y^2}$ ∎

Note that in part (c) we did some simplifying first before rationalizing the denominator, whereas in part (b) we proceeded immediately to rationalize the denominator. This is an individual choice, and you should probably do it both ways a few times to decide which you prefer.

PROBLEM SET 5.3

For Problems 1–20, use the distributive property to help simplify each of the following. For example,

$3\sqrt{8} - \sqrt{32} = 3\sqrt{4}\sqrt{2} - \sqrt{16}\sqrt{2}$

$= 3(2)\sqrt{2} - 4\sqrt{2}$

$= 6\sqrt{2} - 4\sqrt{2}$

$= (6 - 4)\sqrt{2} = 2\sqrt{2}$

1. $5\sqrt{18} - 2\sqrt{2}$

2. $7\sqrt{12} + 4\sqrt{3}$

3. $7\sqrt{12} + 10\sqrt{48}$

4. $6\sqrt{8} - 5\sqrt{18}$

5. $-2\sqrt{50} - 5\sqrt{32}$

6. $-2\sqrt{20} - 7\sqrt{45}$

7. $3\sqrt{20} - \sqrt{5} - 2\sqrt{45}$

8. $6\sqrt{12} + \sqrt{3} - 2\sqrt{48}$

9. $-9\sqrt{24} + 3\sqrt{54} - 12\sqrt{6}$

10. $13\sqrt{28} - 2\sqrt{63} - 7\sqrt{7}$

11. $\dfrac{3}{4}\sqrt{7} - \dfrac{2}{3}\sqrt{28}$

12. $\dfrac{3}{5}\sqrt{5} - \dfrac{1}{4}\sqrt{80}$

13. $\dfrac{3}{5}\sqrt{40} + \dfrac{5}{6}\sqrt{90}$

14. $\dfrac{3}{8}\sqrt{96} - \dfrac{2}{3}\sqrt{54}$

15. $\dfrac{3\sqrt{18}}{5} - \dfrac{5\sqrt{72}}{6} + \dfrac{3\sqrt{98}}{4}$

16. $\dfrac{-2\sqrt{20}}{3} + \dfrac{3\sqrt{45}}{4} - \dfrac{5\sqrt{80}}{6}$

17. $5\sqrt[3]{3} + 2\sqrt[3]{24} - 6\sqrt[3]{81}$

18. $-3\sqrt[3]{2} - 2\sqrt[3]{16} + \sqrt[3]{54}$

19. $-\sqrt[3]{16} + 7\sqrt[3]{54} - 9\sqrt[3]{2}$

20. $4\sqrt[3]{24} - 6\sqrt[3]{3} + 13\sqrt[3]{81}$

For Problems 21–64, express each of the following in simplest radical form. All variables represent positive real numbers.

21. $\sqrt{32x}$

22. $\sqrt{50y}$

23. $\sqrt{75x^2}$

24. $\sqrt{108y^2}$

25. $\sqrt{20x^2y}$

26. $\sqrt{80xy^2}$

27. $\sqrt{64x^3y^7}$

28. $\sqrt{36x^5y^6}$

29. $\sqrt{54a^4b^3}$

30. $\sqrt{96a^7b^8}$

31. $\sqrt{63x^6y^8}$

32. $\sqrt{28x^4y^{12}}$

33. $2\sqrt{40a^3}$

34. $4\sqrt{90a^5}$

35. $\dfrac{2}{3}\sqrt{96xy^3}$

36. $\dfrac{4}{5}\sqrt{125x^4y}$

37. $\sqrt{\dfrac{2x}{5y}}$

38. $\sqrt{\dfrac{3x}{2y}}$

39. $\sqrt{\dfrac{5}{12x^4}}$

40. $\sqrt{\dfrac{7}{8x^2}}$

41. $\dfrac{5}{\sqrt{18y}}$

42. $\dfrac{3}{\sqrt{12x}}$

43. $\dfrac{\sqrt{7x}}{\sqrt{8y^5}}$

44. $\dfrac{\sqrt{5y}}{\sqrt{18x^3}}$

45. $\dfrac{\sqrt{18y^3}}{\sqrt{16x}}$

46. $\dfrac{\sqrt{2x^3}}{\sqrt{9y}}$

47. $\dfrac{\sqrt{24a^2b^3}}{\sqrt{7ab^6}}$

48. $\dfrac{\sqrt{12a^2b}}{\sqrt{5a^3b^3}}$

49. $\sqrt[3]{24y}$

50. $\sqrt[3]{16x^2}$

51. $\sqrt[3]{16x^4}$

52. $\sqrt[3]{54x^3}$

53. $\sqrt[3]{56x^6y^8}$

54. $\sqrt[3]{81x^5y^6}$

55. $\sqrt[3]{\dfrac{7}{9x^2}}$

56. $\sqrt[3]{\dfrac{5}{2x}}$

57. $\dfrac{\sqrt[3]{3y}}{\sqrt[3]{16x^4}}$

58. $\dfrac{\sqrt[3]{2y}}{\sqrt[3]{3x}}$

59. $\dfrac{\sqrt[3]{12xy}}{\sqrt[3]{3x^2y^5}}$

60. $\dfrac{5}{\sqrt[3]{9xy^2}}$

61. $\sqrt{8x + 12y}$ [*Hint:* $\sqrt{8x + 12y} = \sqrt{4(2x + 3y)}$]

62. $\sqrt{4x + 4y}$

63. $\sqrt{16x + 48y}$

64. $\sqrt{27x + 18y}$

For Problems 65–74, use the distributive property to help simplify each of the following. All variables represent positive real numbers.

65. $-3\sqrt{4x} + 5\sqrt{9x} + 6\sqrt{16x}$

66. $-2\sqrt{25x} - 4\sqrt{36x} + 7\sqrt{64x}$

67. $2\sqrt{18x} - 3\sqrt{8x} - 6\sqrt{50x}$

68. $4\sqrt{20x} + 5\sqrt{45x} - 10\sqrt{80x}$

69. $5\sqrt{27n} - \sqrt{12n} - 6\sqrt{3n}$

70. $4\sqrt{8n} + 3\sqrt{18n} - 2\sqrt{72n}$

71. $7\sqrt{4ab} - \sqrt{16ab} - 10\sqrt{25ab}$

72. $4\sqrt{ab} - 9\sqrt{36ab} + 6\sqrt{49ab}$

73. $-3\sqrt[3]{2x^3} + 4\sqrt[3]{8x^3} - 3\sqrt[3]{32x^3}$

74. $2\sqrt[3]{40x^5} - 3\sqrt[3]{90x^5} + 5\sqrt[3]{160x^5}$

■ ■ ■ **Thoughts into words**

75. Is the expression $3\sqrt{2} + \sqrt{50}$ in simplest radical form? Defend your answer.

76. Your friend simplified $\dfrac{\sqrt{6}}{\sqrt{8}}$ as follows:

$$\dfrac{\sqrt{6}}{\sqrt{8}} \cdot \dfrac{\sqrt{8}}{\sqrt{8}} = \dfrac{\sqrt{48}}{8} = \dfrac{\sqrt{16}\sqrt{3}}{8} = \dfrac{4\sqrt{3}}{8} = \dfrac{\sqrt{3}}{2}$$

Is this a correct procedure? Can you show her a better way to do this problem?

77. Does $\sqrt{x + y}$ equal $\sqrt{x} + \sqrt{y}$? Defend your answer.

■ ■ ■ **Further investigations**

78. Use your calculator and evaluate each expression in Problems 1–16. Then evaluate the simplified expression that you obtained when doing these problems. Your two results for each problem should be the same.

Consider these problems, where the variables could represent any real number. However, we would still have the restriction that the radical would represent a real number. In other words, the radicand must be nonnegative.

$\sqrt{98x^2} = \sqrt{49x^2}\sqrt{2} = 7|x|\sqrt{2}$ An absolute-value sign is necessary to ensure that the principal root is nonnegative.

$\sqrt{24x^4} = \sqrt{4x^4}\sqrt{6} = 2x^2\sqrt{6}$ Because x^2 is nonnegative, there is no need for an absolute-value sign to ensure that the principal root is nonnegative.

$\sqrt{25x^3} = \sqrt{25x^2}\sqrt{x} = 5x\sqrt{x}$ Because the radicand is defined to be nonnegative, x must be nonnegative, and there is no need for an absolute-value sign to ensure that the principal root is nonnegative.

$\sqrt{18b^5} = \sqrt{9b^4}\sqrt{2b} = 3b^2\sqrt{2b}$ An absolute-value sign is not necessary to ensure that the principal root is nonnegative.

$\sqrt{12y^6} = \sqrt{4y^6}\sqrt{3} = 2|y^3|\sqrt{3}$ An absolute-value sign is necessary to ensure that the principal root is nonnegative.

79. Do the following problems, where the variable could be any real number as long as the radical represents a real number. Use absolute-value signs in the answers as necessary.

(a) $\sqrt{125x^2}$ **(b)** $\sqrt{16x^4}$

(c) $\sqrt{8b^3}$ **(d)** $\sqrt{3y^5}$

(e) $\sqrt{288x^6}$ **(f)** $\sqrt{28m^8}$

(g) $\sqrt{128c^{10}}$ **(h)** $\sqrt{18d^7}$

(i) $\sqrt{49x^2}$ **(j)** $\sqrt{80n^{20}}$

(k) $\sqrt{81h^3}$

5.4 **Products and Quotients Involving Radicals**

As we have seen, Property 5.4 ($\sqrt[n]{bc} = \sqrt[n]{b}\sqrt[n]{c}$) is used to express one radical as the product of two radicals and also to express the product of two radicals as one radical. In fact, we have used the property for both purposes within the framework of simplifying radicals. For example,

$$\frac{\sqrt{3}}{\sqrt{32}} = \frac{\sqrt{3}}{\sqrt{16}\sqrt{2}} = \frac{\sqrt{3}}{4\sqrt{2}} = \frac{\sqrt{3}}{4\sqrt{2}} \cdot \frac{\sqrt{2}}{\sqrt{2}} = \frac{\sqrt{6}}{8}$$

$$\uparrow \qquad \uparrow \qquad \qquad \uparrow \qquad \uparrow$$

$$\sqrt[n]{bc} = \sqrt[n]{b}\sqrt[n]{c} \qquad\qquad \sqrt[n]{b}\sqrt[n]{c} = \sqrt[n]{bc}$$

The following examples demonstrate the use of Property 5.4 to multiply radicals and to express the product in simplest form.

E X A M P L E   1

Multiply and simplify where possible.

(a) $(2\sqrt{3})(3\sqrt{5})$ (b) $(3\sqrt{8})(5\sqrt{2})$

(c) $(7\sqrt{6})(3\sqrt{8})$ (d) $(2\sqrt[3]{6})(5\sqrt[3]{4})$

Solution

(a) $(2\sqrt{3})(3\sqrt{5}) = 2 \cdot 3 \cdot \sqrt{3} \cdot \sqrt{5} = 6\sqrt{15}$

(b) $(3\sqrt{8})(5\sqrt{2}) = 3 \cdot 5 \cdot \sqrt{8} \cdot \sqrt{2} = 15\sqrt{16} = 15 \cdot 4 = 60.$

(c) $(7\sqrt{6})(3\sqrt{8}) = 7 \cdot 3 \cdot \sqrt{6} \cdot \sqrt{8} = 21\sqrt{48} = 21\sqrt{16}\sqrt{3}$

$$= 21 \cdot 4 \cdot \sqrt{3} = 84\sqrt{3}$$

(d) $(2\sqrt[3]{6})(5\sqrt[3]{4}) = 2 \cdot 5 \cdot \sqrt[3]{6} \cdot \sqrt[3]{4} = 10\sqrt[3]{24}$

$$= 10\sqrt[3]{8}\sqrt[3]{3}$$

$$= 10 \cdot 2 \cdot \sqrt[3]{3}$$

$$= 20\sqrt[3]{3}$$

Recall the use of the distributive property when finding the product of a monomial and a polynomial. For example, $3x^2(2x + 7) = 3x^2(2x) + 3x^2(7) = 6x^3 + 21x^2$. In a similar manner, the distributive property and Property 5.4 provide the basis for finding certain special products that involve radicals. The following examples illustrate this idea.

E X A M P L E   2

Multiply and simplify where possible.

(a) $\sqrt{3}(\sqrt{6} + \sqrt{12})$ (b) $2\sqrt{2}(4\sqrt{3} - 5\sqrt{6})$

(c) $\sqrt{6x}(\sqrt{8x} + \sqrt{12xy})$ (d) $\sqrt[3]{2}(5\sqrt[3]{4} - 3\sqrt[3]{16})$

Solution

(a) $\sqrt{3}(\sqrt{6} + \sqrt{12}) = \sqrt{3}\sqrt{6} + \sqrt{3}\sqrt{12}$

$$= \sqrt{18} + \sqrt{36}$$

$$= \sqrt{9}\sqrt{2} + 6$$

$$= 3\sqrt{2} + 6$$

(b) $2\sqrt{2}(4\sqrt{3} - 5\sqrt{6}) = (2\sqrt{2})(4\sqrt{3}) - (2\sqrt{2})(5\sqrt{6})$

$$= 8\sqrt{6} - 10\sqrt{12}$$

$$= 8\sqrt{6} - 10\sqrt{4}\sqrt{3}$$

$$= 8\sqrt{6} - 20\sqrt{3}$$

(c) $\sqrt{6x}(\sqrt{8x} + \sqrt{12xy}) = (\sqrt{6x})(\sqrt{8x}) + (\sqrt{6x})(\sqrt{12xy})$

$$= \sqrt{48x^2} + \sqrt{72x^2y}$$

$$= \sqrt{16x^2}\sqrt{3} + \sqrt{36x^2}\sqrt{2y}$$

$$= 4x\sqrt{3} + 6x\sqrt{2y}$$

(d) $\sqrt[3]{2}(5\sqrt[3]{4} - 3\sqrt[3]{16}) = (\sqrt[3]{2})(5\sqrt[3]{4}) - (\sqrt[3]{2})(3\sqrt[3]{16})$

$$= 5\sqrt[3]{8} - 3\sqrt[3]{32}$$

$$= 5 \cdot 2 - 3\sqrt[3]{8}\sqrt[3]{4}$$

$$= 10 - 6\sqrt[3]{4}$$ ■

The distributive property also plays a central role in determining the product of two binomials. For example, $(x + 2)(x + 3) = x(x + 3) + 2(x + 3) = x^2 + 3x + 2x + 6 = x^2 + 5x + 6$. Finding the product of two binomial expressions that involve radicals can be handled in a similar fashion, as in the next examples.

EXAMPLE 3   Find the following products and simplify.

(a) $(\sqrt{3} + \sqrt{5})(\sqrt{2} + \sqrt{6})$ **(b)** $(2\sqrt{2} - \sqrt{7})(3\sqrt{2} + 5\sqrt{7})$

(c) $(\sqrt{8} + \sqrt{6})(\sqrt{8} - \sqrt{6})$ **(d)** $(\sqrt{x} + \sqrt{y})(\sqrt{x} - \sqrt{y})$

Solution

(a) $(\sqrt{3} + \sqrt{5})(\sqrt{2} + \sqrt{6}) = \sqrt{3}(\sqrt{2} + \sqrt{6}) + \sqrt{5}(\sqrt{2} + \sqrt{6})$

$$= \sqrt{3}\sqrt{2} + \sqrt{3}\sqrt{6} + \sqrt{5}\sqrt{2} + \sqrt{5}\sqrt{6}$$

$$= \sqrt{6} + \sqrt{18} + \sqrt{10} + \sqrt{30}$$

$$= \sqrt{6} + 3\sqrt{2} + \sqrt{10} + \sqrt{30}$$

(b) $(2\sqrt{2} - \sqrt{7})(3\sqrt{2} + 5\sqrt{7}) = 2\sqrt{2}(3\sqrt{2} + 5\sqrt{7})$

$$- \sqrt{7}(3\sqrt{2} + 5\sqrt{7})$$

$$= (2\sqrt{2})(3\sqrt{2}) + (2\sqrt{2})(5\sqrt{7})$$

$$- (\sqrt{7})(3\sqrt{2}) - (\sqrt{7})(5\sqrt{7})$$

$$= 12 + 10\sqrt{14} - 3\sqrt{14} - 35$$

$$= -23 + 7\sqrt{14}$$

(c) $(\sqrt{8} + \sqrt{6})(\sqrt{8} - \sqrt{6}) = \sqrt{8}(\sqrt{8} - \sqrt{6}) + \sqrt{6}(\sqrt{8} - \sqrt{6})$

$$= \sqrt{8}\sqrt{8} - \sqrt{8}\sqrt{6} + \sqrt{6}\sqrt{8} - \sqrt{6}\sqrt{6}$$

$$= 8 - \sqrt{48} + \sqrt{48} - 6$$

$$= 2$$

(d) $(\sqrt{x} + \sqrt{y})(\sqrt{x} - \sqrt{y}) = \sqrt{x}(\sqrt{x} - \sqrt{y}) + \sqrt{y}(\sqrt{x} - \sqrt{y})$

$$= \sqrt{x}\sqrt{x} - \sqrt{x}\sqrt{y} + \sqrt{y}\sqrt{x} - \sqrt{y}\sqrt{y}$$

$$= x - \sqrt{xy} + \sqrt{xy} - y$$

$$= x - y$$

Note parts (c) and (d) of Example 3; they fit the special-product pattern $(a + b)(a - b) = a^2 - b^2$. Furthermore, in each case the final product is in rational form. The factors $a + b$ and $a - b$ are called **conjugates.** This suggests a way of rationalizing the denominator in an expression that contains a binomial denominator with radicals. We will multiply by the conjugate of the binomial denominator. Consider the following example.

E X A M P L E   4

Simplify $\dfrac{4}{\sqrt{5} + \sqrt{2}}$ by rationalizing the denominator.

Solution

$$\frac{4}{\sqrt{5} + \sqrt{2}} = \frac{4}{\sqrt{5} + \sqrt{2}} \cdot \left(\frac{\sqrt{5} - \sqrt{2}}{\sqrt{5} - \sqrt{2}} \right) \quad \text{Form of 1.}$$

$$= \frac{4(\sqrt{5} - \sqrt{2})}{(\sqrt{5} + \sqrt{2})(\sqrt{5} - \sqrt{2})} = \frac{4(\sqrt{5} - \sqrt{2})}{5 - 2}$$

$$= \frac{4(\sqrt{5} - \sqrt{2})}{3} \quad \text{or} \quad \frac{4\sqrt{5} - 4\sqrt{2}}{3}$$

Either answer
is acceptable.

The next examples further illustrate the process of rationalizing and simplifying expressions that contain binomial denominators.

E X A M P L E   5

For each of the following, rationalize the denominator and simplify.

(a) $\dfrac{\sqrt{3}}{\sqrt{6} - 9}$

(b) $\dfrac{7}{3\sqrt{5} + 2\sqrt{3}}$

(c) $\dfrac{\sqrt{x} + 2}{\sqrt{x} - 3}$

(d) $\dfrac{2\sqrt{x} - 3\sqrt{y}}{\sqrt{x} + \sqrt{y}}$

Solution

(a) $\dfrac{\sqrt{3}}{\sqrt{6}-9} = \dfrac{\sqrt{3}}{\sqrt{6}-9} \cdot \dfrac{\sqrt{6}+9}{\sqrt{6}+9}$

$= \dfrac{\sqrt{3}(\sqrt{6}+9)}{(\sqrt{6}-9)(\sqrt{6}+9)}$

$= \dfrac{\sqrt{18}+9\sqrt{3}}{6-81}$

$= \dfrac{3\sqrt{2}+9\sqrt{3}}{-75}$

$= \dfrac{3(\sqrt{2}+3\sqrt{3})}{(-3)(25)}$

$= -\dfrac{\sqrt{2}+3\sqrt{3}}{25} \qquad \text{or} \qquad \dfrac{-\sqrt{2}-3\sqrt{3}}{25}$

(b) $\dfrac{7}{3\sqrt{5}+2\sqrt{3}} = \dfrac{7}{3\sqrt{5}+2\sqrt{3}} \cdot \dfrac{3\sqrt{5}-2\sqrt{3}}{3\sqrt{5}-2\sqrt{3}}$

$= \dfrac{7(3\sqrt{5}-2\sqrt{3})}{(3\sqrt{5}+2\sqrt{3})(3\sqrt{5}-2\sqrt{3})}$

$= \dfrac{7(3\sqrt{5}-2\sqrt{3})}{45-12}$

$= \dfrac{7(3\sqrt{5}-2\sqrt{3})}{33} \qquad \text{or} \qquad \dfrac{21\sqrt{5}-14\sqrt{3}}{33}$

(c) $\dfrac{\sqrt{x}+2}{\sqrt{x}-3} = \dfrac{\sqrt{x}+2}{\sqrt{x}-3} \cdot \dfrac{\sqrt{x}+3}{\sqrt{x}+3} = \dfrac{(\sqrt{x}+2)(\sqrt{x}+3)}{(\sqrt{x}-3)(\sqrt{x}+3)}$

$= \dfrac{x+3\sqrt{x}+2\sqrt{x}+6}{x-9}$

$= \dfrac{x+5\sqrt{x}+6}{x-9}$

(d) $\dfrac{2\sqrt{x}-3\sqrt{y}}{\sqrt{x}+\sqrt{y}} = \dfrac{2\sqrt{x}-3\sqrt{y}}{\sqrt{x}+\sqrt{y}} \cdot \dfrac{\sqrt{x}-\sqrt{y}}{\sqrt{x}-\sqrt{y}}$

$= \dfrac{(2\sqrt{x}-3\sqrt{y})(\sqrt{x}-\sqrt{y})}{(\sqrt{x}+\sqrt{y})(\sqrt{x}-\sqrt{y})}$

$= \dfrac{2x-2\sqrt{xy}-3\sqrt{xy}+3y}{x-y}$

$= \dfrac{2x-5\sqrt{xy}+3y}{x-y}$

■

PROBLEM SET 5.4

For Problems 1–14, multiply and simplify where possible.

1. $\sqrt{6}\sqrt{12}$

2. $\sqrt{8}\sqrt{6}$

3. $(3\sqrt{3})(2\sqrt{6})$

4. $(5\sqrt{2})(3\sqrt{12})$

5. $(4\sqrt{2})(-6\sqrt{5})$

6. $(-7\sqrt{3})(2\sqrt{5})$

7. $(-3\sqrt{3})(-4\sqrt{8})$

8. $(-5\sqrt{8})(-6\sqrt{7})$

9. $(5\sqrt{6})(4\sqrt{6})$

10. $(3\sqrt{7})(2\sqrt{7})$

11. $(2\sqrt[3]{4})(6\sqrt[3]{2})$

12. $(4\sqrt[3]{3})(5\sqrt[3]{9})$

13. $(4\sqrt[3]{6})(7\sqrt[3]{4})$

14. $(9\sqrt[3]{6})(2\sqrt[3]{9})$

For Problems 15–52, find the following products and express answers in simplest radical form. All variables represent nonnegative real numbers.

15. $\sqrt{2}(\sqrt{3} + \sqrt{5})$

16. $\sqrt{3}(\sqrt{7} + \sqrt{10})$

17. $3\sqrt{5}(2\sqrt{2} - \sqrt{7})$

18. $5\sqrt{6}(2\sqrt{5} - 3\sqrt{11})$

19. $2\sqrt{6}(3\sqrt{8} - 5\sqrt{12})$

20. $4\sqrt{2}(3\sqrt{12} + 7\sqrt{6})$

21. $-4\sqrt{5}(2\sqrt{5} + 4\sqrt{12})$

22. $-5\sqrt{3}(3\sqrt{12} - 9\sqrt{8})$

23. $3\sqrt{x}(5\sqrt{2} + \sqrt{y})$

24. $\sqrt{2x}(3\sqrt{y} - 7\sqrt{5})$

25. $\sqrt{xy}(5\sqrt{xy} - 6\sqrt{x})$

26. $4\sqrt{x}(2\sqrt{xy} + 2\sqrt{x})$

27. $\sqrt{5y}(\sqrt{8x} + \sqrt{12y^2})$

28. $\sqrt{2x}(\sqrt{12xy} - \sqrt{8y})$

29. $5\sqrt{3}(2\sqrt{8} - 3\sqrt{18})$

30. $2\sqrt{2}(3\sqrt{12} - \sqrt{27})$

31. $(\sqrt{3} + 4)(\sqrt{3} - 7)$

32. $(\sqrt{2} + 6)(\sqrt{2} - 2)$

33. $(\sqrt{5} - 6)(\sqrt{5} - 3)$

34. $(\sqrt{7} - 2)(\sqrt{7} - 8)$

35. $(3\sqrt{5} - 2\sqrt{3})(2\sqrt{7} + \sqrt{2})$

36. $(\sqrt{2} + \sqrt{3})(\sqrt{5} - \sqrt{7})$

37. $(2\sqrt{6} + 3\sqrt{5})(\sqrt{8} - 3\sqrt{12})$

38. $(5\sqrt{2} - 4\sqrt{6})(2\sqrt{8} + \sqrt{6})$

39. $(2\sqrt{6} + 5\sqrt{5})(3\sqrt{6} - \sqrt{5})$

40. $(7\sqrt{3} - \sqrt{7})(2\sqrt{3} + 4\sqrt{7})$

41. $(3\sqrt{2} - 5\sqrt{3})(6\sqrt{2} - 7\sqrt{3})$

42. $(\sqrt{8} - 3\sqrt{10})(2\sqrt{8} - 6\sqrt{10})$

43. $(\sqrt{6} + 4)(\sqrt{6} - 4)$

44. $(\sqrt{7} - 2)(\sqrt{7} + 2)$

45. $(\sqrt{2} + \sqrt{10})(\sqrt{2} - \sqrt{10})$

46. $(2\sqrt{3} + \sqrt{11})(2\sqrt{3} - \sqrt{11})$

47. $(\sqrt{2x} + \sqrt{3y})(\sqrt{2x} - \sqrt{3y})$

48. $(2\sqrt{x} - 5\sqrt{y})(2\sqrt{x} + 5\sqrt{y})$

49. $2\sqrt[3]{3}(5\sqrt[3]{4} + \sqrt[3]{6})$

50. $2\sqrt[3]{2}(3\sqrt[3]{6} - 4\sqrt[3]{5})$

51. $3\sqrt[3]{4}(2\sqrt[3]{2} - 6\sqrt[3]{4})$

52. $3\sqrt[3]{3}(4\sqrt[3]{9} + 5\sqrt[3]{7})$

For Problems 53–76, rationalize the denominator and simplify. All variables represent positive real numbers.

53. $\dfrac{2}{\sqrt{7} + 1}$

54. $\dfrac{6}{\sqrt{5} + 2}$

55. $\dfrac{3}{\sqrt{2} - 5}$

56. $\dfrac{-4}{\sqrt{6} - 3}$

57. $\dfrac{1}{\sqrt{2} + \sqrt{7}}$

58. $\dfrac{3}{\sqrt{3} + \sqrt{10}}$

59. $\dfrac{\sqrt{2}}{\sqrt{10} - \sqrt{3}}$

60. $\dfrac{\sqrt{3}}{\sqrt{7} - \sqrt{2}}$

61. $\dfrac{\sqrt{3}}{2\sqrt{5} + 4}$

62. $\dfrac{\sqrt{7}}{3\sqrt{2} - 5}$

63. $\dfrac{6}{3\sqrt{7} - 2\sqrt{6}}$

64. $\dfrac{5}{2\sqrt{5} + 3\sqrt{7}}$

65. $\dfrac{\sqrt{6}}{3\sqrt{2} + 2\sqrt{3}}$

66. $\dfrac{3\sqrt{6}}{5\sqrt{3} - 4\sqrt{2}}$

67. $\dfrac{2}{\sqrt{x} + 4}$

68. $\dfrac{3}{\sqrt{x} + 7}$

69. $\dfrac{\sqrt{x}}{\sqrt{x}-5}$

70. $\dfrac{\sqrt{x}}{\sqrt{x}-1}$

73. $\dfrac{\sqrt{x}}{\sqrt{x}+2\sqrt{y}}$

74. $\dfrac{\sqrt{y}}{2\sqrt{x}-\sqrt{y}}$

71. $\dfrac{\sqrt{x}-2}{\sqrt{x}+6}$

72. $\dfrac{\sqrt{x}+1}{\sqrt{x}-10}$

75. $\dfrac{3\sqrt{y}}{2\sqrt{x}-3\sqrt{y}}$

76. $\dfrac{2\sqrt{x}}{3\sqrt{x}+5\sqrt{y}}$

■ ■ ■ **Thoughts into words**

77. How would you help someone rationalize the denominator and simplify $\dfrac{4}{\sqrt{8}+\sqrt{12}}$?

78. Discuss how the distributive property has been used thus far in this chapter.

79. How would you simplify the expression $\dfrac{\sqrt{8}+\sqrt{12}}{\sqrt{2}}$?

■ ■ ■ **Further investigations**

80. Use your calculator to evaluate each expression in Problems 53–66. Then evaluate the results you obtained when you did the problems.

5.5 Equations Involving Radicals

We often refer to equations that contain radicals with variables in a radicand as **radical equations.** In this section we discuss techniques for solving such equations that contain one or more radicals. To solve radical equations, we need the following property of equality.

> ### PROPERTY 5.6
>
> Let a and b be real numbers and n be a positive integer.
>
> If $a = b$, then $a^n = b^n$.

Property 5.6 states that we can raise both sides of an equation to a positive integral power. However, raising both sides of an equation to a positive integral power sometimes produces results that do not satisfy the original equation. Let's consider two examples to illustrate this point.

EXAMPLE 1   Solve $\sqrt{2x-5}=7$.

Solution

$$\sqrt{2x - 5} = 7$$

$$(\sqrt{2x - 5})^2 = 7^2 \qquad \text{Square both sides.}$$

$$2x - 5 = 49$$

$$2x = 54$$

$$x = 27$$

 Check

$$\sqrt{2x - 5} = 7$$

$$\sqrt{2(27) - 5} \overset{?}{=} 7$$

$$\sqrt{49} \overset{?}{=} 7$$

$$7 = 7$$

The solution set for $\sqrt{2x - 5} = 7$ is {27}. ■

EXAMPLE 2   Solve $\sqrt{3a + 4} = -4$.

Solution

$$\sqrt{3a + 4} = -4$$

$$(\sqrt{3a + 4})^2 = (-4)^2 \qquad \text{Square both sides.}$$

$$3a + 4 = 16$$

$$3a = 12$$

$$a = 4$$

 Check

$$\sqrt{3a + 4} = -4$$

$$\sqrt{3(4) + 4} \overset{?}{=} -4$$

$$\sqrt{16} \overset{?}{=} -4$$

$$4 \neq -4$$

Because 4 does not check, the original equation has no real number solution. Thus the solution set is $\varnothing$. ■

In general, raising both sides of an equation to a positive integral power produces an equation that has all of the solutions of the original equation, but it may also have some extra solutions that do not satisfy the original equation. Such extra solutions are called **extraneous solutions.** Therefore, when using Property 5.6, you *must* check each potential solution in the original equation.

Let's consider some examples to illustrate different situations that arise when we are solving radical equations.

E X A M P L E   3

Solve $\sqrt{2t - 4} = t - 2$.

Solution

$$\sqrt{2t - 4} = t - 2$$

$$(\sqrt{2t - 4})^2 = (t - 2)^2 \qquad \text{Square both sides.}$$

$$2t - 4 = t^2 - 4t + 4$$

$$0 = t^2 - 6t + 8$$

$$0 = (t - 2)(t - 4) \qquad \text{Factor the right side.}$$

$$t - 2 = 0 \qquad \text{or} \qquad t - 4 = 0 \qquad \text{Apply: } ab = 0 \text{ if and only if}$$
$$t = 2 \qquad \text{or} \qquad t = 4 \qquad\qquad a = 0 \text{ or } b = 0.$$

 Check

$$\sqrt{2t - 4} = t - 2 \qquad\qquad\qquad \sqrt{2t - 4} = t - 2$$

$$\sqrt{2(2) - 4} \overset{?}{=} 2 - 2, \quad \text{when } t = 2 \quad \text{or} \quad \sqrt{2(4) - 4} \overset{?}{=} 4 - 2, \quad \text{when } t = 4$$

$$\sqrt{0} \overset{?}{=} 0 \qquad\qquad\qquad\qquad \sqrt{4} \overset{?}{=} 2$$

$$0 = 0 \qquad\qquad\qquad\qquad\qquad 2 = 2$$

The solution set is $\{2, 4\}$. ■

E X A M P L E   4

Solve $\sqrt{y} + 6 = y$.

Solution

$$\sqrt{y} + 6 = y$$

$$\sqrt{y} = y - 6$$

$$(\sqrt{y})^2 = (y - 6)^2 \qquad \text{Square both sides.}$$

$$y = y^2 - 12y + 36$$

$$0 = y^2 - 13y + 36$$

$$0 = (y - 4)(y - 9) \qquad \text{Factor the right side.}$$

$$y - 4 = 0 \qquad \text{or} \qquad y - 9 = 0 \qquad \text{Apply: } ab = 0 \text{ if and}$$
$$y = 4 \qquad \text{or} \qquad y = 9 \qquad\qquad \text{only if } a = 0 \text{ or } b = 0.$$

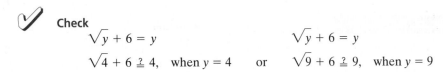 **Check**

$$\sqrt{y} + 6 = y \qquad\qquad\qquad \sqrt{y} + 6 = y$$

$$\sqrt{4} + 6 \overset{?}{=} 4, \quad \text{when } y = 4 \quad \text{or} \quad \sqrt{9} + 6 \overset{?}{=} 9, \quad \text{when } y = 9$$

$$2 + 6 \overset{?}{=} 4 \qquad\qquad 3 + 6 \overset{?}{=} 9$$

$$8 \neq 4 \qquad\qquad 9 = 9$$

The only solution is 9; the solution set is {9}.   ■

In Example 4, note that we changed the form of the original equation $\sqrt{y} + 6 = y$ to $\sqrt{y} = y - 6$ before we squared both sides. Squaring both sides of $\sqrt{y} + 6 = y$ produces $y + 12\sqrt{y} + 36 = y^2$, which is a much more complex equation that still contains a radical. Here again, it pays to think ahead before carrying out all the steps. Now let's consider an example involving a cube root.

EXAMPLE 5   Solve $\sqrt[3]{n^2 - 1} = 2$.

Solution

$$\sqrt[3]{n^2 - 1} = 2$$

$$(\sqrt[3]{n^2 - 1})^3 = 2^3 \qquad \text{Cube both sides.}$$

$$n^2 - 1 = 8$$

$$n^2 - 9 = 0$$

$$(n + 3)(n - 3) = 0$$

$$n + 3 = 0 \qquad \text{or} \qquad n - 3 = 0$$

$$n = -3 \qquad \text{or} \qquad n = 3$$

  Check

$$\sqrt[3]{n^2 - 1} = 2 \qquad\qquad\qquad \sqrt[3]{n^2 - 1} = 2$$

$$\sqrt[3]{(-3)^2 - 1} \overset{?}{=} 2, \quad \text{when } n = -3 \qquad \text{or} \qquad \sqrt[3]{3^2 - 1} \overset{?}{=} 2, \quad \text{when } n = 3$$

$$\sqrt[3]{8} \overset{?}{=} 2 \qquad\qquad\qquad\qquad \sqrt[3]{8} \overset{?}{=} 2$$

$$2 = 2 \qquad\qquad\qquad\qquad\qquad 2 = 2$$

The solution set is {−3, 3}.   ■

It may be necessary to square both sides of an equation, simplify the resulting equation, and then square both sides again. The next example illustrates this type of problem.

EXAMPLE 6   Solve $\sqrt{x + 2} = 7 - \sqrt{x + 9}$.

Solution

$$\sqrt{x + 2} = 7 - \sqrt{x + 9}$$

$$(\sqrt{x + 2})^2 = (7 - \sqrt{x + 9})^2 \qquad \text{Square both sides.}$$

$$x + 2 = 49 - 14\sqrt{x + 9} + x + 9$$

$$x + 2 = x + 58 - 14\sqrt{x + 9}$$

$$-56 = -14\sqrt{x + 9}$$

$$4 = \sqrt{x + 9}$$

$$(4)^2 = (\sqrt{x + 9})^2 \qquad \text{Square both sides.}$$

$$16 = x + 9$$

$$7 = x$$

Check

$$\sqrt{x + 2} = 7 - \sqrt{x + 9}$$

$$\sqrt{7 + 2} \stackrel{?}{=} 7 - \sqrt{7 + 9}$$

$$\sqrt{9} \stackrel{?}{=} 7 - \sqrt{16}$$

$$3 \stackrel{?}{=} 7 - 4$$

$$3 = 3$$

The solution set is {7}. ■

Another Look at Applications

In Section 5.1 we used the formula $S = \sqrt{30Df}$ to approximate how fast a car was traveling on the basis of the length of skid marks. (Remember that S represents the speed of the car in miles per hour, D represents the length of the skid marks in feet, and f represents a coefficient of friction.) This same formula can be used to estimate the length of skid marks that are produced by cars traveling at different rates on various types of road surfaces. To use the formula for this purpose, let's change the form of the equation by solving for D.

$$\sqrt{30Df} = S$$

$$30Df = S^2 \qquad \text{The result of squaring both sides of the original equation}$$

$$D = \frac{S^2}{30f} \qquad \text{D, S, and f are positive numbers, so this final equation and the original one are equivalent.}$$

| EXAMPLE 7 | Suppose that for a particular road surface, the coefficient of friction is 0.35. How far will a car skid when the brakes are applied at 60 miles per hour? |

Solution

We can substitute 0.35 for f and 60 for S in the formula $D = \dfrac{S^2}{30f}$.

$$D = \frac{60^2}{30(0.35)} = 343, \quad \text{to the nearest whole number}$$

The car will skid approximately 343 feet. ∎

REMARK: Pause for a moment and think about the result in Example 7. The coefficient of friction 0.35 refers to a wet concrete road surface. Note that a car traveling at 60 miles per hour on such a surface will skid more than the length of a football field.

PROBLEM SET 5.5

For Problems 1–52, solve each equation. Don't forget to check each of your potential solutions.

1. $\sqrt{5x} = 10$

2. $\sqrt{3x} = 9$

3. $\sqrt{2x} + 4 = 0$

4. $\sqrt{4x} + 5 = 0$

5. $2\sqrt{n} = 5$

6. $5\sqrt{n} = 3$

7. $3\sqrt{n} - 2 = 0$

8. $2\sqrt{n} - 7 = 0$

9. $\sqrt{3y + 1} = 4$

10. $\sqrt{2y - 3} = 5$

11. $\sqrt{4y - 3} - 6 = 0$

12. $\sqrt{3y + 5} - 2 = 0$

13. $\sqrt{2x - 5} = -1$

14. $\sqrt{4x - 3} = -4$

15. $\sqrt{5x + 2} = \sqrt{6x + 1}$

16. $\sqrt{4x + 2} = \sqrt{3x + 4}$

17. $\sqrt{3x + 1} = \sqrt{7x - 5}$

18. $\sqrt{6x + 5} = \sqrt{2x + 10}$

19. $\sqrt{3x - 2} - \sqrt{x + 4} = 0$

20. $\sqrt{7x - 6} - \sqrt{5x + 2} = 0$

21. $5\sqrt{t - 1} = 6$

22. $4\sqrt{t + 3} = 6$

23. $\sqrt{x^2 + 7} = 4$

24. $\sqrt{x^2 + 3} - 2 = 0$

25. $\sqrt{x^2 + 13x + 37} = 1$

26. $\sqrt{x^2 + 5x - 20} = 2$

27. $\sqrt{x^2 - x + 1} = x + 1$

28. $\sqrt{n^2 - 2n - 4} = n$

29. $\sqrt{x^2 + 3x + 7} = x + 2$

30. $\sqrt{x^2 + 2x + 1} = x + 3$

31. $\sqrt{-4x + 17} = x - 3$

32. $\sqrt{2x - 1} = x - 2$

33. $\sqrt{n + 4} = n + 4$

34. $\sqrt{n + 6} = n + 6$

35. $\sqrt{3y} = y - 6$

36. $2\sqrt{n} = n - 3$

37. $4\sqrt{x + 5} = x$

38. $\sqrt{-x} - 6 = x$

39. $\sqrt[3]{x - 2} = 3$

40. $\sqrt[3]{x + 1} = 4$

41. $\sqrt[3]{2x + 3} = -3$

42. $\sqrt[3]{3x - 1} = -4$

43. $\sqrt[3]{2x + 5} = \sqrt[3]{4 - x}$

44. $\sqrt[3]{3x - 1} = \sqrt[3]{2 - 5x}$

45. $\sqrt{x + 19} - \sqrt{x + 28} = -1$

46. $\sqrt{x + 4} = \sqrt{x - 1} + 1$

47. $\sqrt{3x + 1} + \sqrt{2x + 4} = 3$

48. $\sqrt{2x - 1} - \sqrt{x + 3} = 1$

49. $\sqrt{n - 4} + \sqrt{n + 4} = 2\sqrt{n - 1}$

50. $\sqrt{n - 3} + \sqrt{n + 5} = 2\sqrt{n}$

51. $\sqrt{t + 3} - \sqrt{t - 2} = \sqrt{7 - t}$

52. $\sqrt{t + 7} - 2\sqrt{t - 8} = \sqrt{t - 5}$

53. Use the formula given in Example 7 with a coefficient of friction of 0.95. How far will a car skid at 40 miles per hour? at 55 miles per hour? at 65 miles per hour? Express the answers to the nearest foot.

54. Solve the formula $T = 2\pi\sqrt{\dfrac{L}{32}}$ for L. (Remember that in this formula, which was used in Section 5.2, T represents the period of a pendulum expressed in seconds, and L represents the length of the pendulum in feet.)

55. In Problem 54, you should have obtained the equation $L = \dfrac{8T^2}{\pi^2}$. What is the length of a pendulum that has a period of 2 seconds? of 2.5 seconds? of 3 seconds? Express your answers to the nearest tenth of a foot.

■ ■ ■ **Thoughts into words**

56. Explain the concept of extraneous solutions.

57. Explain why possible solutions for radical equations *must* be checked.

58. Your friend makes an effort to solve the equation $3 + 2\sqrt{x} = x$ as follows:

$$(3 + 2\sqrt{x})^2 = x^2$$

$$9 + 12\sqrt{x} + 4x = x^2$$

At this step he stops and doesn't know how to proceed. What help would you give him?

5.6   Merging Exponents and Roots

Recall that the basic properties of positive integral exponents led to a definition for the use of negative integers as exponents. In this section, the properties of integral exponents are used to form definitions for the use of rational numbers as exponents. These definitions will tie together the concepts of exponent and root.

Let's consider the following comparisons.

From our study of radicals, we know that

$$(\sqrt{5})^2 = 5$$

$$(\sqrt[3]{8})^3 = 8$$

$$(\sqrt[4]{21})^4 = 21$$

If $(b^n)^m = b^{mn}$ is to hold when n equals a rational number of the form $\dfrac{1}{p}$, where p is a positive integer greater than 1, then

$$\left(5^{\frac{1}{2}}\right)^2 = 5^{2\left(\frac{1}{2}\right)} = 5^1 = 5$$

$$\left(8^{\frac{1}{3}}\right)^3 = 8^{3\left(\frac{1}{3}\right)} = 8^1 = 8$$

$$\left(21^{\frac{1}{4}}\right)^4 = 21^{4\left(\frac{1}{4}\right)} = 21^1 = 21$$

It would seem reasonable to make the following definition.

DEFINITION 5.6

If b is a real number, n is a positive integer greater than 1, and $\sqrt[n]{b}$ exists, then

$$b^{\frac{1}{n}} = \sqrt[n]{b}$$

Definition 5.6 states that $b^{\frac{1}{n}}$ means the *n*th root of *b*. We shall assume that *b* and *n* are chosen so that $\sqrt[n]{b}$ exists. For example, $(-25)^{\frac{1}{2}}$ is not meaningful at this time because $\sqrt{-25}$ is not a real number. Consider the following examples, which demonstrate the use of Definition 5.6.

$$25^{\frac{1}{2}} = \sqrt{25} = 5 \qquad\qquad 16^{\frac{1}{4}} = \sqrt[4]{16} = 2$$

$$8^{\frac{1}{3}} = \sqrt[3]{8} = 2 \qquad\qquad \left(\frac{36}{49}\right)^{\frac{1}{2}} = \sqrt{\frac{36}{49}} = \frac{6}{7}$$

$$(-27)^{\frac{1}{3}} = \sqrt[3]{-27} = -3$$

The following definition provides the basis for the use of *all* rational numbers as exponents.

DEFINITION 5.7

If $\dfrac{m}{n}$ is a rational number, where *n* is a positive integer greater than 1, and *b* is a real number such that $\sqrt[n]{b}$ exists, then

$$b^{\frac{m}{n}} = \sqrt[n]{b^m} = (\sqrt[n]{b})^m$$

In Definition 5.7, note that the denominator of the exponent is the index of the radical and that the numerator of the exponent is either the exponent of the radicand or the exponent of the root.

Whether we use the form $\sqrt[n]{b^m}$ or the form $(\sqrt[n]{b})^m$ for computational purposes depends somewhat on the magnitude of the problem. Let's use both forms on two problems to illustrate this point.

$$8^{\frac{2}{3}} = \sqrt[3]{8^2} \qquad \text{or} \qquad 8^{\frac{2}{3}} = (\sqrt[3]{8})^2$$
$$= \sqrt[3]{64} \qquad\qquad\qquad = 2^2$$
$$= 4 \qquad\qquad\qquad\quad = 4$$
$$27^{\frac{2}{3}} = \sqrt[3]{27^2} \qquad \text{or} \qquad 27^{\frac{2}{3}} = (\sqrt[3]{27})^2$$
$$= \sqrt[3]{729} \qquad\qquad\qquad = 3^2$$
$$= 9 \qquad\qquad\qquad\quad = 9$$

To compute $8^{\frac{2}{3}}$, either form seems to work about as well as the other one. However, to compute $27^{\frac{2}{3}}$, it should be obvious that $(\sqrt[3]{27})^2$ is much easier to handle than $\sqrt[3]{27^2}$.

EXAMPLE 1   Simplify each of the following numerical expressions.

(a) $25^{\frac{3}{2}}$ **(b)** $16^{\frac{3}{4}}$ **(c)** $(32)^{-\frac{2}{5}}$

(d) $(-64)^{\frac{2}{3}}$ **(e)** $-8^{\frac{1}{3}}$

Solution

(a) $25^{\frac{3}{2}} = (\sqrt{25})^3 = 5^3 = 125$

(b) $16^{\frac{3}{4}} = (\sqrt[4]{16})^3 = 2^3 = 8$

(c) $(32)^{-\frac{2}{5}} = \dfrac{1}{(32)^{\frac{2}{5}}} = \dfrac{1}{(\sqrt[5]{32})^2} = \dfrac{1}{2^2} = \dfrac{1}{4}$

(d) $(-64)^{\frac{2}{3}} = (\sqrt[3]{-64})^2 = (-4)^2 = 16$

(e) $-8^{\frac{1}{3}} = -\sqrt[3]{8} = -2$ ■

The basic laws of exponents that we stated in Property 5.2 are true for all rational exponents. Therefore, from now on we will use Property 5.2 for rational as well as integral exponents.

Some problems can be handled better in exponential form and others in radical form. Thus we must be able to switch forms with a certain amount of ease. Let's consider some examples where we switch from one form to the other.

EXAMPLE 2   Write each of the following expressions in radical form.

(a) $x^{\frac{3}{4}}$ (b) $3y^{\frac{2}{5}}$ (c) $x^{\frac{1}{4}}y^{\frac{3}{4}}$ (d) $(x + y)^{\frac{2}{3}}$

Solution

(a) $x^{\frac{3}{4}} = \sqrt[4]{x^3}$ (b) $3y^{\frac{2}{5}} = 3\sqrt[5]{y^2}$

(c) $x^{\frac{1}{4}}y^{\frac{3}{4}} = (xy^3)^{\frac{1}{4}} = \sqrt[4]{xy^3}$ (d) $(x + y)^{\frac{2}{3}} = \sqrt[3]{(x + y)^2}$ ■

EXAMPLE 3   Write each of the following using positive rational exponents.

(a) $\sqrt{xy}$ (b) $\sqrt[4]{a^3b}$ (c) $4\sqrt[3]{x^2}$ (d) $\sqrt[5]{(x + y)^4}$

Solution

(a) $\sqrt{xy} = (xy)^{\frac{1}{2}} = x^{\frac{1}{2}}y^{\frac{1}{2}}$ (b) $\sqrt[4]{a^3b} = (a^3b)^{\frac{1}{4}} = a^{\frac{3}{4}}b^{\frac{1}{4}}$

(c) $4\sqrt[3]{x^2} = 4x^{\frac{2}{3}}$ (d) $\sqrt[5]{(x + y)^4} = (x + y)^{\frac{4}{5}}$ ■

The properties of exponents provide the basis for simplifying algebraic expressions that contain rational exponents, as these next examples illustrate.

EXAMPLE 4   Simplify each of the following. Express final results using positive exponents only.

(a) $\left(3x^{\frac{1}{2}}\right)\left(4x^{\frac{2}{3}}\right)$ (b) $\left(5a^{\frac{1}{3}}b^{\frac{1}{2}}\right)^2$ (c) $\dfrac{12y^{\frac{1}{3}}}{6y^{\frac{1}{2}}}$ (d) $\left(\dfrac{3x^{\frac{2}{5}}}{2y^{\frac{2}{3}}}\right)^4$

Solution

(a) $\left(3x^{\frac{1}{2}}\right)\left(4x^{\frac{2}{3}}\right) = 3 \cdot 4 \cdot x^{\frac{1}{2}} \cdot x^{\frac{2}{3}}$

$\qquad\qquad = 12x^{\frac{1}{2}+\frac{2}{3}}$ $\qquad\qquad b^n \cdot b^m = b^{n+m}$

$\qquad\qquad = 12x^{\frac{3}{6}+\frac{4}{6}}$ $\qquad\qquad$ Use 6 as LCD.

$\qquad\qquad = 12x^{\frac{7}{6}}$

(b) $\left(5a^{\frac{1}{3}}b^{\frac{1}{2}}\right)^2 = 5^2 \cdot \left(a^{\frac{1}{3}}\right)^2 \cdot \left(b^{\frac{1}{2}}\right)^2$ $\qquad (ab)^n = a^n b^n$

$\qquad\qquad = 25a^{\frac{2}{3}}b$ $\qquad\qquad\qquad (b^n)^m = b^{mn}$

(c) $\dfrac{12y^{\frac{1}{3}}}{6y^{\frac{1}{2}}} = 2y^{\frac{1}{3}-\frac{1}{2}}$ $\qquad\qquad \dfrac{b^n}{b^m} = b^{n-m}$

$\qquad\qquad = 2y^{\frac{2}{6}-\frac{3}{6}}$

$\qquad\qquad = 2y^{-\frac{1}{6}}$

$\qquad\qquad = \dfrac{2}{y^{\frac{1}{6}}}$

(d) $\left(\dfrac{3x^{\frac{2}{5}}}{2y^{\frac{2}{3}}}\right)^4 = \dfrac{\left(3x^{\frac{2}{5}}\right)^4}{\left(2y^{\frac{2}{3}}\right)^4}$ $\qquad\qquad \left(\dfrac{a}{b}\right)^n = \dfrac{a^n}{b^n}$

$\qquad\qquad = \dfrac{3^4 \cdot \left(x^{\frac{2}{5}}\right)^4}{2^4 \cdot \left(y^{\frac{2}{3}}\right)^4}$ $\qquad\qquad (ab)^n = a^n b^n$

$\qquad\qquad = \dfrac{81x^{\frac{8}{5}}}{16y^{\frac{8}{3}}}$ $\qquad\qquad (b^n)^m = b^{mn}$ ■

The link between exponents and roots also provides a basis for multiplying and dividing some radicals even if they have different indexes. The general procedure is as follows:

1. Change from radical form to exponential form.
2. Apply the properties of exponents.
3. Then change back to radical form.

The three parts of Example 5 illustrate this process.

E X A M P L E   5   Perform the indicated operations and express the answers in simplest radical form.

(a) $\sqrt{2}\sqrt[3]{2}$ $\qquad\qquad$ (b) $\dfrac{\sqrt{5}}{\sqrt[3]{5}}$ $\qquad\qquad$ (c) $\dfrac{\sqrt{4}}{\sqrt[3]{2}}$

Solution

(a) $\sqrt{2}\sqrt[3]{2} = 2^{\frac{1}{2}} \cdot 2^{\frac{1}{3}}$

$= 2^{\frac{1}{2}+\frac{1}{3}}$

$= 2^{\frac{3}{6}+\frac{2}{6}}$ Use 6 as LCD.

$= 2^{\frac{5}{6}}$

$= \sqrt[6]{2^5} = \sqrt[6]{32}$

(b) $\dfrac{\sqrt{5}}{\sqrt[3]{5}} = \dfrac{5^{\frac{1}{2}}}{5^{\frac{1}{3}}}$

$= 5^{\frac{1}{2}-\frac{1}{3}}$

$= 5^{\frac{3}{6}-\frac{2}{6}}$ Use 6 as LCD.

$= 5^{\frac{1}{6}} = \sqrt[6]{5}$

(c) $\dfrac{\sqrt{4}}{\sqrt[3]{2}} = \dfrac{4^{\frac{1}{2}}}{2^{\frac{1}{3}}}$

$= \dfrac{(2^2)^{\frac{1}{2}}}{2^{\frac{1}{3}}}$

$= \dfrac{2^1}{2^{\frac{1}{3}}}$

$= 2^{1-\frac{1}{3}}$

$= 2^{\frac{2}{3}} = \sqrt[3]{2^2} = \sqrt[3]{4}$

PROBLEM SET 5.6

For Problems 1–30, evaluate each numerical expression.

1. $81^{\frac{1}{2}}$

2. $64^{\frac{1}{2}}$

3. $27^{\frac{1}{3}}$

4. $(-32)^{\frac{1}{5}}$

5. $(-8)^{\frac{1}{3}}$

6. $\left(-\dfrac{27}{8}\right)^{\frac{1}{3}}$

7. $-25^{\frac{1}{2}}$

8. $-64^{\frac{1}{3}}$

9. $36^{-\frac{1}{2}}$

10. $81^{-\frac{1}{2}}$

11. $\left(\dfrac{1}{27}\right)^{-\frac{1}{3}}$

12. $\left(-\dfrac{8}{27}\right)^{-\frac{1}{3}}$

13. $4^{\frac{3}{2}}$

14. $64^{\frac{2}{3}}$

15. $27^{\frac{4}{3}}$

16. $4^{\frac{7}{2}}$

17. $(-1)^{\frac{7}{3}}$

18. $(-8)^{\frac{4}{3}}$

19. $-4^{\frac{5}{2}}$

20. $-16^{\frac{3}{2}}$

21. $\left(\dfrac{27}{8}\right)^{\frac{4}{3}}$

22. $\left(\dfrac{8}{125}\right)^{\frac{2}{3}}$

23. $\left(\dfrac{1}{8}\right)^{-\frac{2}{3}}$

24. $\left(-\dfrac{1}{27}\right)^{-\frac{2}{3}}$

25. $64^{-\frac{7}{6}}$

26. $32^{-\frac{4}{5}}$

27. $-25^{\frac{3}{2}}$

28. $-16^{\frac{3}{4}}$

29. $125^{\frac{4}{3}}$

30. $81^{\frac{5}{4}}$

For Problems 31–44, write each of the following in radical form. For example,

$$3x^{\frac{2}{3}} = 3\sqrt[3]{x^2}$$

31. $x^{\frac{4}{3}}$

32. $x^{\frac{2}{5}}$

33. $3x^{\frac{1}{2}}$

34. $5x^{\frac{1}{4}}$

35. $(2y)^{\frac{1}{3}}$

36. $(3xy)^{\frac{1}{2}}$

37. $(2x - 3y)^{\frac{1}{2}}$

38. $(5x + y)^{\frac{1}{3}}$

39. $(2a - 3b)^{\frac{2}{3}}$

40. $(5a + 7b)^{\frac{3}{5}}$

41. $x^{\frac{2}{3}}y^{\frac{1}{3}}$

42. $x^{\frac{3}{7}}y^{\frac{5}{7}}$

43. $-3x^{\frac{1}{5}}y^{\frac{2}{5}}$

44. $-4x^{\frac{3}{4}}y^{\frac{1}{4}}$

For Problems 45–58, write each of the following using positive rational exponents. For example,

$$\sqrt{ab} = (ab)^{\frac{1}{2}} = a^{\frac{1}{2}}b^{\frac{1}{2}}$$

45. $\sqrt{5y}$

46. $\sqrt{2xy}$

47. $3\sqrt{y}$

48. $5\sqrt{ab}$

49. $\sqrt[3]{xy^2}$

50. $\sqrt[5]{x^2y^4}$

51. $\sqrt[4]{a^2b^3}$

52. $\sqrt[6]{ab^5}$

53. $\sqrt[5]{(2x - y)^3}$

54. $\sqrt[7]{(3x - y)^4}$

55. $5x\sqrt{y}$

56. $4y\sqrt[3]{x}$

57. $-\sqrt[3]{x + y}$

58. $-\sqrt[5]{(x - y)^2}$

For Problems 59–80, simplify each of the following. Express final results using positive exponents only. For example,

$$\left(2x^{\frac{1}{2}}\right)\left(3x^{\frac{1}{3}}\right) = 6x^{\frac{5}{6}}$$

59. $\left(2x^{\frac{2}{5}}\right)\left(6x^{\frac{1}{4}}\right)$

60. $\left(3x^{\frac{1}{4}}\right)\left(5x^{\frac{1}{3}}\right)$

61. $\left(y^{\frac{2}{3}}\right)\left(y^{-\frac{1}{4}}\right)$

62. $\left(y^{\frac{1}{4}}\right)\left(y^{-\frac{1}{2}}\right)$

63. $\left(x^{\frac{2}{5}}\right)\left(4x^{-\frac{1}{2}}\right)$

64. $\left(2x^{\frac{1}{3}}\right)\left(x^{-\frac{1}{2}}\right)$

65. $\left(4x^{\frac{1}{2}}y\right)^2$

66. $\left(3x^{\frac{1}{4}}y^{\frac{1}{5}}\right)^3$

67. $(8x^6y^3)^{\frac{1}{3}}$

68. $(9x^2y^4)^{\frac{1}{2}}$

69. $\dfrac{24x^{\frac{3}{5}}}{6x^{\frac{1}{3}}}$

70. $\dfrac{18x^{\frac{1}{2}}}{9x^{\frac{1}{3}}}$

71. $\dfrac{48b^{\frac{1}{3}}}{12b^{\frac{3}{4}}}$

72. $\dfrac{56a^{\frac{1}{6}}}{8a^{\frac{1}{4}}}$

73. $\left(\dfrac{6x^{\frac{2}{5}}}{7y^{\frac{2}{3}}}\right)^2$

74. $\left(\dfrac{2x^{\frac{1}{3}}}{3y^{\frac{1}{4}}}\right)^4$

75. $\left(\dfrac{x^2}{y^3}\right)^{-\frac{1}{2}}$

76. $\left(\dfrac{a^3}{b^{-2}}\right)^{-\frac{1}{3}}$

77. $\left(\dfrac{18x^{\frac{1}{3}}}{9x^{\frac{1}{4}}}\right)^2$

78. $\left(\dfrac{72x^{\frac{3}{4}}}{6x^{\frac{1}{2}}}\right)^2$

79. $\left(\dfrac{60a^{\frac{1}{5}}}{15a^{\frac{3}{4}}}\right)^2$

80. $\left(\dfrac{64a^{\frac{1}{3}}}{16a^{\frac{5}{9}}}\right)^3$

For Problems 81–90, perform the indicated operations and express answers in simplest radical form. (See Example 5.)

81. $\sqrt[3]{3}\sqrt{3}$

82. $\sqrt{2}\sqrt[4]{2}$

83. $\sqrt[4]{6}\sqrt{6}$

84. $\sqrt[3]{5}\sqrt{5}$

85. $\dfrac{\sqrt[3]{3}}{\sqrt[4]{3}}$

86. $\dfrac{\sqrt{2}}{\sqrt[3]{2}}$

87. $\dfrac{\sqrt[3]{8}}{\sqrt[4]{4}}$

88. $\dfrac{\sqrt{9}}{\sqrt[3]{3}}$

89. $\dfrac{\sqrt[4]{27}}{\sqrt{3}}$

90. $\dfrac{\sqrt[3]{16}}{\sqrt[6]{4}}$

■ ■ ■ **Thoughts into words**

91. Your friend keeps getting an error message when evaluating $-4^{\frac{5}{2}}$ on his calculator. What error is he probably making?

92. Explain how you would evaluate $27^{\frac{2}{3}}$ without a calculator.

■ ■ ■ **Further investigations**

93. Use your calculator to evaluate each of the following.

   (a) $\sqrt[3]{1728}$

   (b) $\sqrt[3]{5832}$

   (c) $\sqrt[4]{2401}$

   (d) $\sqrt[4]{65,536}$

   (e) $\sqrt[5]{161,051}$

   (f) $\sqrt[5]{6,436,343}$

94. Definition 5.7 states that

$$b^{\frac{m}{n}} = \sqrt[n]{b^m} = (\sqrt[n]{b})^m$$

Use your calculator to verify each of the following.

   (a) $\sqrt[3]{27^2} = (\sqrt[3]{27})^2$ **(b)** $\sqrt[3]{8^5} = (\sqrt[3]{8})^5$

(c) $\sqrt[4]{16^3} = (\sqrt[4]{16})^3$ **(d)** $\sqrt[3]{16^2} = (\sqrt[3]{16})^2$

(e) $\sqrt[5]{9^4} = (\sqrt[5]{9})^4$ **(f)** $\sqrt[3]{12^4} = (\sqrt[3]{12})^4$

95. Use your calculator to evaluate each of the following.

(a) $16^{\frac{5}{2}}$ **(b)** $25^{\frac{7}{2}}$

(c) $16^{\frac{9}{4}}$ **(d)** $27^{\frac{5}{3}}$

(e) $343^{\frac{2}{3}}$ **(f)** $512^{\frac{4}{3}}$

96. Use your calculator to estimate each of the following to the nearest one-thousandth.

(a) $7^{\frac{4}{3}}$ **(b)** $10^{\frac{4}{5}}$

(c) $12^{\frac{3}{5}}$ **(d)** $19^{\frac{2}{5}}$

(e) $7^{\frac{3}{4}}$ **(f)** $10^{\frac{5}{4}}$

97. (a) Because $\frac{4}{5} = 0.8$, we can evaluate $10^{\frac{4}{5}}$ by evaluating $10^{0.8}$, which involves a shorter sequence of "calculator steps." Evaluate parts (b), (c), (d), (e), and (f) of Problem 96 and take advantage of decimal exponents.
(b) What problem is created when we try to evaluate $7^{\frac{4}{3}}$ by changing the exponent to decimal form?

5.7   Scientific Notation

Many applications of mathematics involve the use of very large or very small numbers.

1. The speed of light is approximately 29,979,200,000 centimeters per second.
2. A light year—the distance that light travels in 1 year—is approximately 5,865,696,000,000 miles.
3. A millimicron equals 0.000000001 of a meter.

Working with numbers of this type in standard decimal form is quite cumbersome. It is much more convenient to represent very small and very large numbers in **scientific notation.** The expression $(N)(10)^k$, where N is a number greater than or equal to 1 and less than 10, written in decimal form, and k is any integer, is commonly called scientific notation or the scientific form of a number. Consider the following examples, which show a comparison between ordinary decimal notation and scientific notation.

Ordinary notation	Scientific notation
2.14	$(2.14)(10)^0$
31.78	$(3.178)(10)^1$
412.9	$(4.129)(10)^2$
8,000,000	$(8)(10)^6$
0.14	$(1.4)(10)^{-1}$
0.0379	$(3.79)(10)^{-2}$
0.00000049	$(4.9)(10)^{-7}$

To switch from ordinary notation to scientific notation, you can use the following procedure.

> Write the given number as the product of a number greater than or equal to 1 and less than 10, and a power of 10. The exponent of 10 is determined by counting the number of places that the decimal point was moved when going from the original number to the number greater than or equal to 1 and less than 10. This exponent is (a) negative if the original number is less than 1, (b) positive if the original number is greater than 10, and (c) 0 if the original number itself is between 1 and 10.

Thus we can write

$$0.00467 = (4.67)(10)^{-3}$$
$$87,000 \ = (8.7)(10)^{4}$$
$$3.1416 \ = (3.1416)(10)^{0}$$

We can express the applications given earlier in scientific notation as follows:

Speed of light $29,979,200,000 = (2.99792)(10)^{10}$ centimeters per second.

Light year $5,865,696,000,000 = (5.865696)(10)^{12}$ miles.

Metric units   A millimicron is $0.000000001 = (1)(10)^{-9}$ meter.

To switch from scientific notation to ordinary decimal notation, you can use the following procedure.

> Move the decimal point the number of places indicated by the exponent of 10. The decimal point is moved to the right if the exponent is positive and to the left if the exponent is negative.

Thus we can write

$$(4.78)(10)^{4} = 47,800$$
$$(8.4)(10)^{-3} = 0.0084$$

Scientific notation can frequently be used to simplify numerical calculations. We merely change the numbers to scientific notation and use the appropriate properties of exponents. Consider the following examples.

EXAMPLE 1   Perform the indicated operations.

(a) $(0.00024)(20,000)$

(b) $\dfrac{7,800,000}{0.0039}$

(c) $\dfrac{(0.00069)(0.0034)}{(0.0000017)(0.023)}$

(d) $\sqrt{0.000004}$

Solution

(a) $(0.00024)(20{,}000) = (2.4)(10)^{-4}(2)(10)^4$

$= (2.4)(2)(10)^{-4}(10)^4$

$= (4.8)(10)^0$

$= (4.8)(1)$

$= 4.8$

(b) $\dfrac{7{,}800{,}000}{0.0039} = \dfrac{(7.8)(10)^6}{(3.9)(10)^{-3}}$

$= (2)(10)^9$

$= 2{,}000{,}000{,}000$

(c) $\dfrac{(0.00069)(0.0034)}{(0.0000017)(0.023)} = \dfrac{(6.9)(10)^{-4}(3.4)(10)^{-3}}{(1.7)(10)^{-6}(2.3)(10)^{-2}}$

$= \dfrac{\overset{3}{(\cancel{6.9})}\,\overset{2}{(\cancel{3.4})}(10)^{-7}}{(\cancel{1.7})(\cancel{2.3})(10)^{-8}}$

$= (6)(10)^1$

$= 60$

(d) $\sqrt{0.00004} = \sqrt{(4)(10)^{-6}}$

$= \left((4)(10)^{-6}\right)^{\frac{1}{2}}$

$= 4^{\frac{1}{2}}\left((10)^{-6}\right)^{\frac{1}{2}}$

$= (2)(10)^{-3}$

$= 0.002$

■

EXAMPLE   2   The speed of light is approximately $(1.86)(10^5)$ miles per second. When the earth is $(9.3)(10^7)$ miles away from the sun, how long does it take light from the sun to reach the earth?

Solution

We will use the formula $t = \dfrac{d}{r}$.

$t = \dfrac{(9.3)(10^7)}{(1.86)(10^5)}$

$t = \dfrac{(9.3)}{(1.86)}(10^2)$ Subtract exponents.

$t = (5)(10^2) = 500$ seconds

At this distance it takes light about 500 seconds to travel from the sun to the earth. To find the answer in minutes, divide 500 seconds by 60 seconds/minute. That gives a result of approximately 8.33 minutes. ∎

Many calculators are equipped to display numbers in scientific notation. The display panel shows the number between 1 and 10 and the appropriate exponent of 10. For example, evaluating $(3,800,000)^2$ yields

| 1.444E13 |

Thus $(3,800,000)^2 = (1.444)(10)^{13} = 14,440,000,000,000$.

Similarly, the answer for $(0.000168)^2$ is displayed as

| 2.8224E-8 |

Thus $(0.000168)^2 = (2.8224)(10)^{-8} = 0.000000028224$.

Calculators vary as to the number of digits displayed in the number between 1 and 10 when scientific notation is used. For example, we used two different calculators to estimate $(6729)^6$ and obtained the following results.

| 9.2833E22 |

| 9.283316768E22 |

Obviously, you need to know the capabilities of your calculator when working with problems in scientific notation. Many calculators also allow the entry of a number in scientific notation. Such calculators are equipped with an enter-the-exponent key (often labeled as $\boxed{\text{EE}}$ or $\boxed{\text{EEX}}$). Thus a number such as $(3.14)(10)^8$ might be entered as follows:

Enter	Press	Display
3.14	$\boxed{\text{EE}}$	3.14E
8		3.14E8

or

Enter	Press	Display
3.14	$\boxed{\text{EE}}$	3.14 ^{00}
8		3.14 ^{08}

A $\boxed{\text{MODE}}$ key is often used on calculators to let you choose normal decimal notation, scientific notation, or engineering notation. (The abbreviations Norm, Sci, and Eng are commonly used.) If the calculator is in scientific mode, then a number can be entered and changed to scientific form by pressing the $\boxed{\text{ENTER}}$ key. For example, when we enter 589 and press the $\boxed{\text{ENTER}}$ key, the display will show 5.89E2. Likewise, when the calculator is in scientific mode, the answers to computational problems are given in scientific form. For example, the answer for $(76)(533)$ is given as 4.0508E4.

It should be evident from this brief discussion that even when you are using a calculator, you need to have a thorough understanding of scientific notation.

PROBLEM SET 5.7

For Problems 1–18, write each of the following in scientific notation. For example

$$27800 = (2.78)(10)^4$$

1. 89

2. 117

3. 4290

4. 812,000

5. 6,120,000

6. 72,400,000

7. 40,000,000

8. 500,000,000

9. 376.4

10. 9126.21

11. 0.347

12. 0.2165

13. 0.0214

14. 0.0037

15. 0.00005

16. 0.00000082

17. 0.00000000194

18. 0.00000000003

For Problems 19–32, write each of the following in ordinary decimal notation. For example,

$$(3.18)(10)^2 = 318$$

19. $(2.3)(10)^1$

20. $(1.62)(10)^2$

21. $(4.19)(10)^3$

22. $(7.631)(10)^4$

23. $(5)(10)^8$

24. $(7)(10)^9$

25. $(3.14)(10)^{10}$

26. $(2.04)(10)^{12}$

27. $(4.3)(10)^{-1}$

28. $(5.2)(10)^{-2}$

29. $(9.14)(10)^{-4}$

30. $(8.76)(10)^{-5}$

31. $(5.123)(10)^{-8}$

32. $(6)(10)^{-9}$

For Problems 33–50, use scientific notation and the properties of exponents to help you perform the following operations.

33. $(0.0037)(0.00002)$

34. $(0.00003)(0.00025)$

35. $(0.00007)(11,000)$

36. $(0.000004)(120,000)$

37. $\dfrac{360,000,000}{0.0012}$

38. $\dfrac{66,000,000,000}{0.022}$

39. $\dfrac{0.000064}{16,000}$

40. $\dfrac{0.00072}{0.0000024}$

41. $\dfrac{(60,000)(0.006)}{(0.0009)(400)}$

42. $\dfrac{(0.00063)(960,000)}{(3,200)(0.0000021)}$

43. $\dfrac{(0.0045)(60,000)}{(1800)(0.00015)}$

44. $\dfrac{(0.00016)(300)(0.028)}{0.064}$

45. $\sqrt{9,000,000}$

46. $\sqrt{0.00000009}$

47. $\sqrt[3]{8000}$

48. $\sqrt[3]{0.001}$

49. $(90,000)^{\frac{3}{2}}$

50. $(8000)^{\frac{2}{3}}$

51. Avogadro's number, 602,000,000,000,000,000,000,000, is the number of atoms in 1 mole of a substance. Express this number in scientific notation.

52. The Social Security program paid out approximately $33,200,000,000 in benefits in May 2000. Express this number in scientific notation.

53. Carlos's first computer had a processing speed of $(1.6)(10^6)$ hertz. He recently purchased a laptop computer with a processing speed of $(1.33)(10^9)$ hertz. Approximately how many times faster is the processing speed of his laptop than that of his first computer? Express the result in decimal form.

54. Alaska has an area of approximately $(6.15)(10^5)$ square miles. In 1999 the state had a population of approximately 619,000 people. Compute the population density to the nearest hundredth. Population density is the number of people per square mile. Express the result in decimal form.

55. In the year 2000 the public debt of the United States was approximately $5,700,000,000,000. For July 2000, the census reported that 275,000,000 people lived in the United States. Convert these figures to scientific notation and compute the average debt per person. Express the result in scientific notation.

56. The space shuttle can travel at approximately 410,000 miles per day. If the shuttle could travel to Mars and Mars was 140,000,000 miles away, how many days would it take the shuttle to travel to Mars? Express the result in decimal form.

57. Atomic masses are measured in atomic mass units (amu). The amu, $(1.66)(10^{-27})$ kilograms, is defined as $\dfrac{1}{12}$ the mass of a common carbon atom. Find the mass of a carbon atom in kilograms. Express the result in scientific notation.

58. The field of view of a microscope is $(4)(10^{-4})$ meters. If a single cell organism occupies $\frac{1}{5}$ of the field of view, find the length of the organism in meters. Express the result in scientific notation.

59. The mass of an electron is $(9.11)(10^{-31})$ kilogram, and the mass of a proton is $(1.67)(10^{-27})$ kilogram. Approx-

imately how many times more is the weight of a proton than the weight of an electron? Express the result in decimal form.

60. A square pixel on a computer screen has a side of length $(1.17)(10^{-2})$ inches. Find the approximate area of the pixel in inches. Express the result in decimal form.

■ ■ ■ Thoughts into words

61. Explain the importance of scientific notation.

62. Why do we need scientific notation even when using calculators and computers?

■ ■ ■ Further investigations

63. Sometimes it is more convenient to express a number as a product of a power of 10 and a number that is not between 1 and 10. For example, suppose that we want to calculate $\sqrt{640,000}$. We can proceed as follows:

$$\sqrt{640,000} = \sqrt{(64)(10)^4}$$
$$= ((64)(10)^4)^{\frac{1}{2}}$$
$$= (64)^{\frac{1}{2}}(10^4)^{\frac{1}{2}}$$
$$= (8)(10)^2$$
$$= 8(100) = 800$$

Compute each of the following without a calculator, and then use a calculator to check your answers.

(a) $\sqrt{49,000,000}$ **(b)** $\sqrt{0.0025}$

(c) $\sqrt{14,400}$ **(d)** $\sqrt{0.000121}$

(e) $\sqrt[3]{27,000}$ **(f)** $\sqrt[3]{0.000064}$

64. Use your calculator to evaluate each of the following. Express final answers in ordinary notation.

(a) $(27,000)^2$ **(b)** $(450,000)^2$

(c) $(14,800)^2$ **(d)** $(1700)^3$

(e) $(900)^4$ **(f)** $(60)^5$

(g) $(0.0213)^2$ **(h)** $(0.000213)^2$

(i) $(0.000198)^2$ **(j)** $(0.000009)^3$

65. Use your calculator to estimate each of the following. Express final answers in scientific notation with the number between 1 and 10 rounded to the nearest one-thousandth.

(a) $(4576)^4$ **(b)** $(719)^{10}$

(c) $(28)^{12}$ **(d)** $(8619)^6$

(e) $(314)^5$ **(f)** $(145,723)^2$

66. Use your calculator to estimate each of the following. Express final answers in ordinary notation rounded to the nearest one-thousandth.

(a) $(1.09)^5$ **(b)** $(1.08)^{10}$

(c) $(1.14)^7$ **(d)** $(1.12)^{20}$

(e) $(0.785)^4$ **(f)** $(0.492)^5$

CHAPTER 5

SUMMARY

(5.1) The following properties form the basis for manipulating with exponents.

1. $b^n \cdot b^m = b^{n+m}$ Product of two powers

2. $(b^n)^m = b^{mn}$ Power of a power

3. $(ab)^n = a^n b^n$ Power of a product

4. $\left(\dfrac{a}{b}\right)^n = \dfrac{a^n}{b^n}$ Power of a quotient

5. $\dfrac{b^n}{b^m} = b^{n-m}$ Quotient of two powers

(5.2) and (5.3) The **principal nth root of b** is designated by $\sqrt[n]{b}$, where n is the **index** and b is the **radicand.**

A radical expression is in **simplest radical form** if

1. A radicand contains no polynomial factor raised to a power equal to or greater than the index of the radical,
2. No fraction appears within a radical sign, and
3. No radical appears in the denominator.

The following properties are used to express radicals in simplest form.

$$\sqrt[n]{bc} = \sqrt[n]{b}\sqrt[n]{c} \qquad \sqrt[n]{\frac{b}{c}} = \frac{\sqrt[n]{b}}{\sqrt[n]{c}}$$

Simplifying by combining radicals sometimes requires that we first express the given radicals in simplest form and then apply the distributive property.

(5.4) The distributive property and the property $\sqrt[n]{b}\sqrt[n]{c} = \sqrt[n]{bc}$ are used to find products of expressions that involve radicals.

The special-product pattern $(a + b)(a - b) = a^2 - b^2$ suggests a procedure for **rationalizing the denominator** of an expression that contains a binomial denominator with radicals.

(5.5) Equations that contain radicals with variables in a radicand are called **radical equations.** The property "if

$a = b$, then $a^n = b^n$" forms the basis for solving radical equations. Raising both sides of an equation to a positive integral power may produce **extraneous solutions**—that is, solutions that do not satisfy the original equation. Therefore, you must check each potential solution.

(5.6) If b is a real number, n is a positive integer greater than 1, and $\sqrt[n]{b}$ exists, then

$$b^{\frac{1}{n}} = \sqrt[n]{b}$$

Thus $b^{\frac{1}{n}}$ means **the nth root of b.**

If $\dfrac{m}{n}$ is a rational number, n is a positive integer greater than 1, and b is a real number such that $\sqrt[n]{b}$ exists, then

$$b^{\frac{m}{n}} = \sqrt[n]{b^m} = (\sqrt[n]{b})^m$$

Both $\sqrt[n]{b^m}$ and $(\sqrt[n]{b})^m$ can be used for computational purposes.

We need to be able to switch back and forth between **exponential form** and **radical form.** The link between exponents and roots provides a basis for multiplying and dividing some radicals even if they have different indexes.

(5.7) The **scientific form** of a number is expressed as

$$(N)(10)^k$$

where N is a number greater than or equal to 1 and less than 10, written in decimal form, and k is an integer. Scientific notation is often convenient to use with very small and very large numbers. For example, 0.000046 can be expressed as $(4.6)(10^{-5})$, and 92,000,000 can be written as $(9.2)(10)^7$.

Scientific notation can often be used to simplify numerical calculations. For example,

$$(0.000016)(30,000) = (1.6)(10)^{-5}(3)(10)^4$$
$$= (4.8)(10)^{-1} = 0.48$$

CHAPTER 5 REVIEW PROBLEM SET

For Problems 1–12, evaluate each of the following numerical expressions.

1. 4^{-3}

2. $\left(\dfrac{2}{3}\right)^{-2}$

3. $(3^2 \cdot 3^{-3})^{-1}$

4. $\sqrt[3]{-8}$

5. $\sqrt[4]{\dfrac{16}{81}}$

6. $4^{\frac{5}{2}}$

7. $(-1)^{\frac{2}{3}}$

8. $\left(\dfrac{8}{27}\right)^{\frac{2}{3}}$

9. $-16^{\frac{3}{2}}$

10. $\dfrac{2^3}{2^{-2}}$

11. $(4^{-2} \cdot 4^2)^{-1}$

12. $\left(\dfrac{3^{-1}}{3^2}\right)^{-1}$

For Problems 13–24, express each of the following radicals in simplest radical form. Assume the variables represent positive real numbers.

13. $\sqrt{54}$

14. $\sqrt{48x^3y}$

15. $\dfrac{4\sqrt{3}}{\sqrt{6}}$

16. $\sqrt{\dfrac{5}{12x^3}}$

17. $\sqrt[3]{56}$

18. $\dfrac{\sqrt[3]{2}}{\sqrt[3]{9}}$

19. $\sqrt{\dfrac{9}{5}}$

20. $\sqrt{\dfrac{3x^3}{7}}$

21. $\sqrt[3]{108x^4y^8}$

22. $\dfrac{3}{4}\sqrt{150}$

23. $\dfrac{2}{3}\sqrt{45xy^3}$

24. $\dfrac{\sqrt{8x^2}}{\sqrt{2x}}$

For Problems 25–32, multiply and simplify. Assume the variables represent nonnegative real numbers.

25. $(3\sqrt{8})(4\sqrt{5})$

26. $(5\sqrt[3]{2})(6\sqrt[3]{4})$

27. $3\sqrt{2}(4\sqrt{6} - 2\sqrt{7})$

28. $(\sqrt{x} + 3)(\sqrt{x} - 5)$

29. $(2\sqrt{5} - \sqrt{3})(2\sqrt{5} + \sqrt{3})$

30. $(3\sqrt{2} + \sqrt{6})(5\sqrt{2} - 3\sqrt{6})$

31. $(2\sqrt{a} + \sqrt{b})(3\sqrt{a} - 4\sqrt{b})$

32. $(4\sqrt{8} - \sqrt{2})(\sqrt{8} + 3\sqrt{2})$

For Problems 33–36, rationalize the denominator and simplify.

33. $\dfrac{4}{\sqrt{7} - 1}$

34. $\dfrac{\sqrt{3}}{\sqrt{8} + \sqrt{5}}$

35. $\dfrac{3}{2\sqrt{3} + 3\sqrt{5}}$

36. $\dfrac{3\sqrt{2}}{2\sqrt{6} - \sqrt{10}}$

For Problems 37–42, simplify each of the following and express the final results using positive exponents.

37. $(x^{-3}y^4)^{-2}$

38. $\left(\dfrac{2a^{-1}}{3b^4}\right)^{-3}$

39. $(4x^{\frac{1}{2}})(5x^{\frac{1}{5}})$

40. $\dfrac{42a^{\frac{3}{4}}}{6a^{\frac{1}{3}}}$

41. $\left(\dfrac{x^3}{y^4}\right)^{-\frac{1}{3}}$

42. $\left(\dfrac{6x^{-2}}{2x^4}\right)^{-2}$

For Problems 43–46, use the distributive property to help simplify each of the following.

43. $3\sqrt{45} - 2\sqrt{20} - \sqrt{80}$

44. $4\sqrt[3]{24} + 3\sqrt[3]{3} - 2\sqrt[3]{81}$

45. $3\sqrt{24} - \dfrac{2\sqrt{54}}{5} + \dfrac{\sqrt{96}}{4}$

46. $-2\sqrt{12x} + 3\sqrt{27x} - 5\sqrt{48x}$

For Problems 47 and 48, express each as a single fraction involving positive exponents only.

47. $x^{-2} + y^{-1}$

48. $a^{-2} - 2a^{-1}b^{-1}$

For Problems 49–56, solve each equation.

49. $\sqrt{7x - 3} = 4$

50. $\sqrt{2y + 1} = \sqrt{5y - 11}$

51. $\sqrt{2x} = x - 4$

52. $\sqrt{n^2 - 4n - 4} = n$

53. $\sqrt[3]{2x - 1} = 3$

54. $\sqrt{t^2 + 9t - 1} = 3$

55. $\sqrt{x^2 + 3x - 6} = x$

56. $\sqrt{x + 1} - \sqrt{2x} = -1$

For Problems 57–64, use scientific notation and the properties of exponents to help perform the following calculations.

57. $(0.00002)(0.0003)$

58. $(120,000)(300,000)$

59. $(0.000015)(400,000)$

60. $\dfrac{0.000045}{0.0003}$

61. $\dfrac{(0.00042)(0.0004)}{0.006}$

62. $\sqrt{0.000004}$

63. $\sqrt[3]{0.000000008}$

64. $(4,000,000)^{\frac{3}{2}}$

CHAPTER 5 *TEST*

For Problems 1–4, simplify each of the numerical expressions.

1. $(4)^{-\frac{5}{2}}$

2. $-16^{\frac{5}{4}}$

3. $\left(\dfrac{2}{3}\right)^{-4}$

4. $\left(\dfrac{2^{-1}}{2^{-2}}\right)^{-2}$

For Problems 5–9, express each radical expression in simplest radical form. Assume the variables represent positive real numbers.

5. $\sqrt{63}$

6. $\sqrt[3]{108}$

7. $\sqrt{52x^4y^3}$

8. $\dfrac{5\sqrt{18}}{3\sqrt{12}}$

9. $\sqrt{\dfrac{7}{24x^3}}$

10. Multiply and simplify: $(4\sqrt{6})(3\sqrt{12})$

11. Multiply and simplify: $(3\sqrt{2} + \sqrt{3})(\sqrt{2} - 2\sqrt{3})$

12. Simplify by combining similar radicals:
$2\sqrt{50} - 4\sqrt{18} - 9\sqrt{32}$

13. Rationalize the denominator and simplify:
$\dfrac{3\sqrt{2}}{4\sqrt{3} - \sqrt{8}}$

14. Simplify and express the answer using positive exponents: $\left(\dfrac{2x^{-1}}{3y}\right)^{-2}$

15. Simplify and express the answer using positive exponents: $\dfrac{-84a^{\frac{1}{2}}}{7a^{\frac{4}{5}}}$

16. Express $x^{-1} + y^{-3}$ as a single fraction involving positive exponents.

17. Multiply and express the answer using positive exponents: $\left(3x^{-\frac{1}{2}}\right)\left(4x^{\frac{3}{4}}\right)$

18. Multiply and simplify:
$(3\sqrt{5} - 2\sqrt{3})(3\sqrt{5} + 2\sqrt{3})$

For Problems 19 and 20, use scientific notation and the properties of exponents to help with the calculations.

19. $\dfrac{(0.00004)(300)}{0.00002}$

20. $\sqrt{0.000009}$

For Problems 21–25, solve each equation.

21. $\sqrt{3x + 1} = 3$

22. $\sqrt[3]{3x + 2} = 2$

23. $\sqrt{x} = x - 2$

24. $\sqrt{5x - 2} = \sqrt{3x + 8}$

25. $\sqrt{x^2 - 10x + 28} = 2$

The Pythagorean theorem is applied throughout the construction industry when right angles are involved.

© Jeff Greenberg/PhotoEdit

Quadratic Equations and Inequalities

A page for a magazine contains 70 square inches of type. The height of the page is twice the width. If the margin around the type is 2 inches uniformly, what are the dimensions of a page? We can use the quadratic equation $(x - 4)(2x - 4) = 70$ to determine that the page measures 9 inches by 18 inches.

Solving equations is one of the central themes of this text. Let's pause for a moment and reflect on the different types of equations that we have solved in the last six chapters.

As this chart shows, we have solved second-degree equations in one variable, but only those for which the polynomial is factorable. In this chapter we will expand our work to include more general types of second-degree equations, as well as inequalities in one variable.

Type of equation	Examples
First-degree equations in one variable	$3x + 2x = x - 4$; $5(x + 4) = 12$; $\dfrac{x + 2}{3} + \dfrac{x - 1}{4} = 2$

(continues)

Type of equation	Examples
Second-degree equations in one variable *that are factorable*	$x^2 + 5x = 0$; $x^2 + 5x + 6 = 0$; $x^2 - 9 = 0$; $x^2 - 10x + 25 = 0$
Fractional equations	$\dfrac{2}{x} + \dfrac{3}{x} = 4$; $\dfrac{5}{a-1} = \dfrac{6}{a-2}$; $\dfrac{2}{x^2 - 9} + \dfrac{3}{x+3} = \dfrac{4}{x-3}$
Radical equations	$\sqrt{x} = 2$; $\sqrt{3x - 2} = 5$; $\sqrt{5y + 1} = \sqrt{3y + 4}$

InfoTrac Project   Do a subject guide search on area measurement and find a periodical article on measuring land. Write a brief summary of the article. How did we get the measurement called *mile*? A piece of land is in the shape of a right triangle. One leg of the right triangle is 3 kilometers longer than the other. If the hypotenuse is 6 kilometers longer than the shorter leg, how long is each side of the piece of land?

6.1 Complex Numbers

Because the square of any real number is nonnegative, a simple equation such as $x^2 = -4$ has no solutions in the set of real numbers. To handle this situation, we can expand the set of real numbers into a larger set called the **complex numbers.** In this section we will instruct you on how to manipulate complex numbers.

To provide a solution for the equation $x^2 + 1 = 0$, we use the number i, such that

$$i^2 = -1$$

The number i is not a real number and is often called the **imaginary unit,** but the number i^2 is the real number -1. The imaginary unit i is used to define a complex number as follows:

DEFINITION 6.1

A **complex number** is any number that can be expressed in the form

$$a + bi$$

where a and b are real numbers.

The form $a + bi$ is called the **standard form** of a complex number. The real number a is called the **real part** of the complex number, and b is called the **imaginary part.** (Note that b is a real number even though it is called the imaginary part.) The following list exemplifies this terminology.

1. The number $7 + 5i$ is a complex number that has a real part of 7 and an imaginary part of 5.
2. The number $\frac{2}{3} + i\sqrt{2}$ is a complex number that has a real part of $\frac{2}{3}$ and an imaginary part of $\sqrt{2}$. (It is easy to mistake $\sqrt{2}i$ for $\sqrt{2i}$. Thus we commonly write $i\sqrt{2}$ instead of $\sqrt{2}i$ to avoid any difficulties with the radical sign.)
3. The number $-4 - 3i$ can be written in the standard form $-4 + (-3i)$ and therefore is a complex number that has a real part of -4 and an imaginary part of -3. [The form $-4 - 3i$ is often used, but we know that it means $-4 + (-3i)$.]
4. The number $-9i$ can be written as $0 + (-9i)$; thus it is a complex number that has a real part of 0 and an imaginary part of -9. (Complex numbers, such as $-9i$, for which $a = 0$ and $b \neq 0$ are called **pure imaginary numbers.**)
5. The real number 4 can be written as $4 + 0i$ and is thus a complex number that has a real part of 4 and an imaginary part of 0.

Look at item 5 in this list. We see that the set of real numbers is a subset of the set of complex numbers. The following diagram indicates the organizational format of the complex numbers.

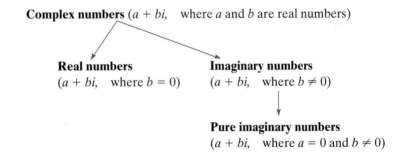

Complex numbers $(a + bi,$ where a and b are real numbers)

Real numbers
$(a + bi,$ where $b = 0)$

Imaginary numbers
$(a + bi,$ where $b \neq 0)$

Pure imaginary numbers
$(a + bi,$ where $a = 0$ and $b \neq 0)$

Two complex numbers $a + bi$ and $c + di$ are said to be **equal** if and only if $a = c$ and $b = d$.

Adding and Subtracting Complex Numbers

To **add complex numbers,** we simply add their real parts and add their imaginary parts. Thus

$$(a + bi) + (c + di) = (a + c) + (b + d)i$$

The following examples show addition of two complex numbers.

1. $(4 + 3i) + (5 + 9i) = (4 + 5) + (3 + 9)i = 9 + 12i$

2. $(-6 + 4i) + (8 - 7i) = (-6 + 8) + (4 - 7)i$

$$= 2 - 3i$$

3. $\left(\dfrac{1}{2} + \dfrac{3}{4}i\right) + \left(\dfrac{2}{3} + \dfrac{1}{5}i\right) = \left(\dfrac{1}{2} + \dfrac{2}{3}\right) + \left(\dfrac{3}{4} + \dfrac{1}{5}\right)i$

$$= \left(\dfrac{3}{6} + \dfrac{4}{6}\right) + \left(\dfrac{15}{20} + \dfrac{4}{20}\right)i$$

$$= \dfrac{7}{6} + \dfrac{19}{20}i$$

The set of complex numbers is closed with respect to addition; that is, the sum of two complex numbers is a complex number. Furthermore, the commutative and associative properties of addition hold for all complex numbers. The addition identity element is $0 + 0i$ (or simply the real number 0). The additive inverse of $a + bi$ is $-a - bi$, because

$$(a + bi) + (-a - bi) = 0$$

To **subtract complex numbers,** $c + di$ from $a + bi$, add the additive inverse of $c + di$. Thus

$$(a + bi) - (c + di) = (a + bi) + (-c - di)$$

$$= (a - c) + (b - d)i$$

In other words, we subtract the real parts and subtract the imaginary parts, as in the next examples.

1. $(9 + 8i) - (5 + 3i) = (9 - 5) + (8 - 3)i$

$$= 4 + 5i$$

2. $(3 - 2i) - (4 + 10i) = (3 - 4) + (-2 - (-10))i$

$$= -1 + 8i$$

Products and Quotients of Complex Numbers

Because $i^2 = -1$, i is a square root of -1, so we let $i = \sqrt{-1}$. It should also be evident that $-i$ is a square root of -1, because

$$(-i)^2 = (-i)(-i) = i^2 = -1$$

Thus, in the set of complex numbers, -1 has two square roots, i and $-i$. We express these symbolically as

$$\sqrt{-1} = i \qquad \text{and} \qquad -\sqrt{-1} = -i$$

Let us extend our definition so that in the set of complex numbers every negative real number has two square roots. We simply define $\sqrt{-b}$, where b is a positive real number, to be the number whose square is $-b$. Thus

$$(\sqrt{-b})^2 = -b, \quad \text{for } b > 0$$

Furthermore, because $(i\sqrt{b})(i\sqrt{b}) = i^2(b) = -1(b) = -b$, we see that

$$\sqrt{-b} = i\sqrt{b}$$

In other words, a square root of any negative real number can be represented as the product of a real number and the imaginary unit i. Consider the following examples.

$$\sqrt{-4} = i\sqrt{4} = 2i$$

$$\sqrt{-17} = i\sqrt{17}$$

$$\sqrt{-24} = i\sqrt{24} = i\sqrt{4}\sqrt{6} = 2i\sqrt{6} \qquad \text{Note that we simplified the radical } \sqrt{24} \text{ to } 2\sqrt{6}.$$

We should also observe that $-\sqrt{-b}$, where $b > 0$, is a square root of $-b$ because

$$(-\sqrt{-b})^2 = (-i\sqrt{b})^2 = i^2(b) = -1(b) = -b$$

Thus, in the set of complex numbers, $-b$ (where $b > 0$) has two square roots, $i\sqrt{b}$ and $-i\sqrt{b}$. We express these symbolically as

$$\sqrt{-b} = i\sqrt{b} \qquad \text{and} \qquad -\sqrt{-b} = -i\sqrt{b}$$

We must be very careful with the use of the symbol $\sqrt{-b}$, where $b > 0$. Some real number properties that involve the square root symbol do not hold if the square root symbol does not represent a real number. For example, $\sqrt{a}\sqrt{b} = \sqrt{ab}$ does not hold if a and b are both negative numbers.

Correct $\sqrt{-4}\sqrt{-9} = (2i)(3i) = 6i^2 = 6(-1) = -6$

Incorrect $\sqrt{-4}\sqrt{-9} = \sqrt{(-4)(-9)} = \sqrt{36} = 6$

To avoid difficulty with this idea, you should rewrite all expressions of the form $\sqrt{-b}$, where $b > 0$, in the form $i\sqrt{b}$ before doing any computations. The following examples further demonstrate this point.

1. $\sqrt{-5}\sqrt{-7} = (i\sqrt{5})(i\sqrt{7}) = i^2\sqrt{35} = (-1)\sqrt{35} = -\sqrt{35}$

2. $\sqrt{-2}\sqrt{-8} = (i\sqrt{2})(i\sqrt{8}) = i^2\sqrt{16} = (-1)(4) = -4$

3. $\sqrt{-6}\sqrt{-8} = (i\sqrt{6})(i\sqrt{8}) = i^2\sqrt{48} = (-1)\sqrt{16}\sqrt{3} = -4\sqrt{3}$

4. $\dfrac{\sqrt{-75}}{\sqrt{-3}} = \dfrac{i\sqrt{75}}{i\sqrt{3}} = \dfrac{\sqrt{75}}{\sqrt{3}} = \sqrt{\dfrac{75}{3}} = \sqrt{25} = 5$

5. $\dfrac{\sqrt{-48}}{\sqrt{12}} = \dfrac{i\sqrt{48}}{\sqrt{12}} = i\sqrt{\dfrac{48}{12}} = i\sqrt{4} = 2i$

Complex numbers have a binomial form, so we find the product of two complex numbers in the same way that we find the product of two binomials. Then, by replacing i^2 with -1, we are able to simplify and express the final result in standard form. Consider the following examples.

6. $(2 + 3i)(4 + 5i) = 2(4 + 5i) + 3i(4 + 5i)$

$$= 8 + 10i + 12i + 15i^2$$

$$= 8 + 22i + 15i^2$$

$$= 8 + 22i + 15(-1) = -7 + 22i$$

7. $(-3 + 6i)(2 - 4i) = -3(2 - 4i) + 6i(2 - 4i)$

$$= -6 + 12i + 12i - 24i^2$$

$$= -6 + 24i - 24(-1)$$

$$= -6 + 24i + 24 = 18 + 24i$$

8. $(1 - 7i)^2 = (1 - 7i)(1 - 7i)$

$$= 1(1 - 7i) - 7i(1 - 7i)$$

$$= 1 - 7i - 7i + 49i^2$$

$$= 1 - 14i + 49(-1)$$

$$= 1 - 14i - 49$$

$$= -48 - 14i$$

9. $(2 + 3i)(2 - 3i) = 2(2 - 3i) + 3i(2 - 3i)$

$$= 4 - 6i + 6i - 9i^2$$

$$= 4 - 9(-1)$$

$$= 4 + 9$$

$$= 13$$

Example 9 illustrates an important situation: The complex numbers $2 + 3i$ and $2 - 3i$ are conjugates of each other. In general, two complex numbers $a + bi$ and $a - bi$ are called **conjugates** of each other. *The product of a complex number and its conjugate is always a real number,* which can be shown as follows:

$$(a + bi)(a - bi) = a(a - bi) + bi(a - bi)$$

$$= a^2 - abi + abi - b^2i^2$$

$$= a^2 - b^2(-1)$$

$$= a^2 + b^2$$

We use conjugates to simplify expressions such as $\dfrac{3i}{5 + 2i}$ that indicate the quotient of two complex numbers. To eliminate i in the denominator and change the indicated quotient to the standard form of a complex number, we can multiply both the numerator and the denominator by the conjugate of the denominator as follows:

$$\frac{3i}{5 + 2i} = \frac{3i(5 - 2i)}{(5 + 2i)(5 - 2i)}$$

$$= \frac{15i - 6i^2}{25 - 4i^2}$$

$$= \frac{15i - 6(-1)}{25 - 4(-1)}$$

$$= \frac{15i + 6}{29}$$

$$= \frac{6}{29} + \frac{15}{29}i$$

The following examples further clarify the process of dividing complex numbers.

10. $\dfrac{2 - 3i}{4 - 7i} = \dfrac{(2 - 3i)(4 + 7i)}{(4 - 7i)(4 + 7i)}$ $4 + 7i$ is the conjugate of $4 - 7i$.

$$= \frac{8 + 14i - 12i - 21i^2}{16 - 49i^2}$$

$$= \frac{8 + 2i - 21(-1)}{16 - 49(-1)}$$

$$= \frac{8 + 2i + 21}{16 + 49}$$

$$= \frac{29 + 2i}{65}$$

$$= \frac{29}{65} + \frac{2}{65}i$$

11. $\dfrac{4 - 5i}{2i} = \dfrac{(4 - 5i)(-2i)}{(2i)(-2i)}$ $-2i$ is the conjugate of $2i$.

$$= \frac{-8i + 10i^2}{-4i^2}$$

$$= \frac{-8i + 10(-1)}{-4(-1)}$$

$$= \frac{-8i - 10}{4}$$

$$= -\frac{5}{2} - 2i$$

In Example 11, where the denominator is a pure imaginary number, we can change to standard form by choosing a multiplier other than the conjugate. Consider the following alternative approach for Example 11.

$$\frac{4 - 5i}{2i} = \frac{(4 - 5i)(i)}{(2i)(i)}$$

$$= \frac{4i - 5i^2}{2i^2}$$

$$= \frac{4i - 5(-1)}{2(-1)}$$

$$= \frac{4i + 5}{-2}$$

$$= -\frac{5}{2} - 2i$$

PROBLEM SET 6.1

For Problems 1–8, label each statement true or false.

1. Every complex number is a real number.

2. Every real number is a complex number.

3. The real part of the complex number $6i$ is 0.

4. Every complex number is a pure imaginary number.

5. The sum of two complex numbers is always a complex number.

6. The imaginary part of the complex number 7 is 0.

7. The sum of two complex numbers is sometimes a real number.

8. The sum of two pure imaginary numbers is always a pure imaginary number.

For Problems 9–26, add or subtract as indicated.

9. $(6 + 3i) + (4 + 5i)$

10. $(5 + 2i) + (7 + 10i)$

11. $(-8 + 4i) + (2 + 6i)$

12. $(5 - 8i) + (-7 + 2i)$

13. $(3 + 2i) - (5 + 7i)$

14. $(1 + 3i) - (4 + 9i)$

15. $(-7 + 3i) - (5 - 2i)$

16. $(-8 + 4i) - (9 - 4i)$

17. $(-3 - 10i) + (2 - 13i)$

18. $(-4 - 12i) + (-3 + 16i)$

19. $(4 - 8i) - (8 - 3i)$

20. $(12 - 9i) - (14 - 6i)$

21. $(-1 - i) - (-2 - 4i)$

22. $(-2 - 3i) - (-4 - 14i)$

23. $\left(\frac{3}{2} + \frac{1}{3}i\right) + \left(\frac{1}{6} - \frac{3}{4}i\right)$

24. $\left(\frac{2}{3} - \frac{1}{5}i\right) + \left(\frac{3}{5} - \frac{3}{4}i\right)$

25. $\left(-\frac{5}{9} + \frac{3}{5}i\right) - \left(\frac{4}{3} - \frac{1}{6}i\right)$

26. $\left(\frac{3}{8} - \frac{5}{2}i\right) - \left(\frac{5}{6} + \frac{1}{7}i\right)$

For Problems 27–42, write each of the following in terms of i and simplify. For example,

$$\sqrt{-20} = i\sqrt{20} = i\sqrt{4}\sqrt{5} = 2i\sqrt{5}$$

27. $\sqrt{-81}$

28. $\sqrt{-49}$

29. $\sqrt{-14}$

30. $\sqrt{-33}$

31. $\sqrt{-\frac{16}{25}}$

32. $\sqrt{-\frac{64}{36}}$

33. $\sqrt{-18}$

34. $\sqrt{-84}$

35. $\sqrt{-75}$

36. $\sqrt{-63}$

37. $3\sqrt{-28}$

38. $5\sqrt{-72}$

39. $-2\sqrt{-80}$

40. $-6\sqrt{-27}$

41. $12\sqrt{-90}$

42. $9\sqrt{-40}$

For Problems 43–60, write each of the following in terms of i, perform the indicated operations, and simplify. For example,

$$\sqrt{-3}\sqrt{-8} = (i\sqrt{3})(i\sqrt{8})$$

$$= i^2\sqrt{24}$$

$$= (-1)\sqrt{4}\sqrt{6}$$

$$= -2\sqrt{6}$$

43. $\sqrt{-4}\sqrt{-16}$

44. $\sqrt{-81}\sqrt{-25}$

45. $\sqrt{-3}\sqrt{-5}$

46. $\sqrt{-7}\sqrt{-10}$

47. $\sqrt{-9}\sqrt{-6}$

48. $\sqrt{-8}\sqrt{-16}$

49. $\sqrt{-15}\sqrt{-5}$

50. $\sqrt{-2}\sqrt{-20}$

51. $\sqrt{-2}\sqrt{-27}$

52. $\sqrt{-3}\sqrt{-15}$

53. $\sqrt{6}\sqrt{-8}$

54. $\sqrt{-75}\sqrt{3}$

55. $\dfrac{\sqrt{-25}}{\sqrt{-4}}$

56. $\dfrac{\sqrt{-81}}{\sqrt{-9}}$

57. $\dfrac{\sqrt{-56}}{\sqrt{-7}}$

58. $\dfrac{\sqrt{-72}}{\sqrt{-6}}$

59. $\dfrac{\sqrt{-24}}{\sqrt{6}}$

60. $\dfrac{\sqrt{-96}}{\sqrt{2}}$

For Problems 61–84, find each of the products and express the answers in the standard form of a complex number.

61. $(5i)(4i)$

62. $(-6i)(9i)$

63. $(7i)(-6i)$

64. $(-5i)(-12i)$

65. $(3i)(2-5i)$

66. $(7i)(-9+3i)$

67. $(-6i)(-2-7i)$

68. $(-9i)(-4-5i)$

69. $(3+2i)(5+4i)$

70. $(4+3i)(6+i)$

71. $(6-2i)(7-i)$

72. $(8-4i)(7-2i)$

73. $(-3-2i)(5+6i)$

74. $(-5-3i)(2-4i)$

75. $(9+6i)(-1-i)$

76. $(10+2i)(-2-i)$

77. $(4+5i)^2$

78. $(5-3i)^2$

79. $(-2-4i)^2$

80. $(-3-6i)^2$

81. $(6+7i)(6-7i)$

82. $(5-7i)(5+7i)$

83. $(-1+2i)(-1-2i)$

84. $(-2-4i)(-2+4i)$

For Problems 85–100, find each of the following quotients and express the answers in the standard form of a complex number.

85. $\dfrac{3i}{2+4i}$

86. $\dfrac{4i}{5+2i}$

87. $\dfrac{-2i}{3-5i}$

88. $\dfrac{-5i}{2-4i}$

89. $\dfrac{-2+6i}{3i}$

90. $\dfrac{-4-7i}{6i}$

91. $\dfrac{2}{7i}$

92. $\dfrac{3}{10i}$

93. $\dfrac{2+6i}{1+7i}$

94. $\dfrac{5+i}{2+9i}$

95. $\dfrac{3+6i}{4-5i}$

96. $\dfrac{7-3i}{4-3i}$

97. $\dfrac{-2+7i}{-1+i}$

98. $\dfrac{-3+8i}{-2+i}$

99. $\dfrac{-1-3i}{-2-10i}$

100. $\dfrac{-3-4i}{-4-11i}$

■ ■ ■ Thoughts into words

101. Why is the set of real numbers a subset of the set of complex numbers?

102. Can the sum of two nonreal complex numbers be a real number? Defend your answer.

103. Can the product of two nonreal complex numbers be a real number? Defend your answer.

6.2 Quadratic Equations

A second-degree equation in one variable contains the variable with an exponent of 2, but no higher power. Such equations are also called **quadratic equations.** The following are examples of quadratic equations.

$$x^2 = 36 \qquad y^2 + 4y = 0 \qquad x^2 + 5x - 2 = 0$$

$$3n^2 + 2n - 1 = 0 \qquad 5x^2 + x + 2 = 3x^2 - 2x - 1$$

A quadratic equation in the variable x can also be defined as any equation that can be written in the form

$$ax^2 + bx + c = 0$$

where a, b, and c are real numbers and $a \neq 0$. The form $ax^2 + bx + c = 0$ is called the **standard form** of a quadratic equation.

In previous chapters you solved quadratic equations (the term *quadratic* was not used at that time) by factoring and applying the property "$ab = 0$ if and only if $a = 0$ or $b = 0$." Let's review a few such examples.

E X A M P L E   1

Solve $3n^2 + 14n - 5 = 0$.

Solution

$$3n^2 + 14n - 5 = 0$$

$$(3n - 1)(n + 5) = 0 \qquad \text{Factor the left side.}$$

$$3n - 1 = 0 \qquad \text{or} \qquad n + 5 = 0 \qquad \text{Apply "}ab = 0\text{ if and}$$
$$3n = 1 \qquad \text{or} \qquad n = -5 \qquad \text{only if } a = 0 \text{ or } b = 0."$$

$$n = \frac{1}{3} \qquad \text{or} \qquad n = -5$$

The solution set is $\left\{-5, \dfrac{1}{3}\right\}$. ∎

E X A M P L E   2

Solve $x^2 + 3kx - 10k^2 = 0$ for x.

Solution

$$x^2 + 3kx - 10k^2 = 0$$

$$(x + 5k)(x - 2k) = 0 \qquad \text{Factor the left side.}$$

$$x + 5k = 0 \qquad \text{or} \qquad x - 2k = 0 \qquad \text{Apply "}ab = 0\text{ if and}$$
$$x = -5k \qquad \text{or} \qquad x = 2k \qquad \text{only if } a = 0 \text{ or } b = 0."$$

The solution set is $\{-5k, 2k\}$. ∎

E X A M P L E   3   Solve $2\sqrt{x} = x - 8$.

Solution

$$2\sqrt{x} = x - 8$$

$$(2\sqrt{x})^2 = (x - 8)^2 \qquad \text{Square both sides.}$$

$$4x = x^2 - 16x + 64$$

$$0 = x^2 - 20x + 64$$

$$0 = (x - 16)(x - 4) \qquad \text{Factor the right side.}$$

$$x - 16 = 0 \quad \text{or} \quad x - 4 = 0 \qquad \text{Apply "}ab = 0 \text{ if and only if } a = 0 \text{ or } b = 0.\text{"}$$

$$x = 16 \quad \text{or} \quad x = 4$$

 Check

$$2\sqrt{x} = x - 8 \qquad\qquad 2\sqrt{x} = x - 8$$

$$2\sqrt{16} \stackrel{?}{=} 16 - 8 \quad \text{or} \quad 2\sqrt{4} \stackrel{?}{=} 4 - 8$$

$$2(4) \stackrel{?}{=} 8 \qquad\qquad 2(2) \stackrel{?}{=} -4$$

$$8 = 8 \qquad\qquad 4 \neq -4$$

The solution set is $\{16\}$. ■

We should make two comments about Example 3. First, remember that applying the property "if $a = b$, then $a^n = b^n$" might produce extraneous solutions. Therefore, we *must* check all potential solutions. Second, the equation $2\sqrt{x} = x - 8$ is said to be of **quadratic form** because it can be written as $2x^{\frac{1}{2}} = \left(x^{\frac{1}{2}}\right)^2 - 8$. More will be said about the phrase *quadratic form* later.

Let's consider quadratic equations of the form $x^2 = a$, where x is the variable and a is any real number. We can solve $x^2 = a$ as follows:

$$x^2 = a$$

$$x^2 - a = 0$$

$$x^2 - (\sqrt{a})^2 = 0 \qquad\qquad a = (\sqrt{a})^2$$

$$(x - \sqrt{a})(x + \sqrt{a}) = 0 \qquad\qquad \text{Factor the left side.}$$

$$x - \sqrt{a} = 0 \quad \text{or} \quad x + \sqrt{a} = 0 \qquad \text{Apply "}ab = 0 \text{ if and only if } a = 0 \text{ or } b = 0.\text{"}$$

$$x = \sqrt{a} \quad \text{or} \quad x = -\sqrt{a}.$$

The solutions are $\sqrt{a}$ and $-\sqrt{a}$. We can state this result as a general property and use it to solve certain types of quadratic equations.

PROPERTY 6.1

For any real number a,

$$x^2 = a \quad \text{if and only if } x = \sqrt{a} \text{ or } x = -\sqrt{a}$$

(The statement $x = \sqrt{a}$ or $x = -\sqrt{a}$ can be written as $x = \pm\sqrt{a}$.)

Property 6.1, along with our knowledge of square roots, makes it very easy to solve quadratic equations of the form $x^2 = a$.

EXAMPLE 4   Solve $x^2 = 45$.

Solution
$$x^2 = 45$$
$$x = \pm\sqrt{45}$$
$$x = \pm3\sqrt{5} \qquad \sqrt{45} = \sqrt{9}\sqrt{5} = 3\sqrt{5}$$

The solution set is $\{\pm3\sqrt{5}\}$.   ■

EXAMPLE 5   Solve $x^2 = -9$.

Solution
$$x^2 = -9$$
$$x = \pm\sqrt{-9}$$
$$x = \pm3i \qquad \sqrt{-9} = i\sqrt{9} = 3i$$

Thus the solution set is $\{\pm3i\}$.   ■

EXAMPLE 6   Solve $7n^2 = 12$.

Solution
$$7n^2 = 12$$
$$n^2 = \frac{12}{7}$$
$$n = \pm\sqrt{\frac{12}{7}}$$
$$n = \pm\frac{2\sqrt{21}}{7} \qquad \sqrt{\frac{12}{7}} = \frac{\sqrt{12}}{\sqrt{7}} \cdot \frac{\sqrt{7}}{\sqrt{7}} = \frac{\sqrt{84}}{7} = \frac{\sqrt{4}\sqrt{21}}{7} = \frac{2\sqrt{21}}{7}$$

The solution set is $\left\{\pm\dfrac{2\sqrt{21}}{7}\right\}$.   ■

EXAMPLE 7   Solve $(3n + 1)^2 = 25$.

Solution

$$(3n + 1)^2 = 25$$

$$(3n + 1) = \pm\sqrt{25}$$

$$3n + 1 = \pm 5$$

$$3n + 1 = 5 \quad \text{or} \quad 3n + 1 = -5$$

$$3n = 4 \quad \text{or} \quad 3n = -6$$

$$n = \frac{4}{3} \quad \text{or} \quad n = -2$$

The solution set is $\left\{-2, \dfrac{4}{3}\right\}$. ∎

EXAMPLE 8   Solve $(x - 3)^2 = -10$.

Solution

$$(x - 3)^2 = -10$$

$$x - 3 = \pm\sqrt{-10}$$

$$x - 3 = \pm i\sqrt{10}$$

$$x = 3 \pm i\sqrt{10}$$

Thus the solution set is $\{3 \pm i\sqrt{10}\}$. ∎

REMARK: Take another look at the equations in Examples 5 and 8. We should immediately realize that the solution sets will consist only of nonreal complex numbers, because any nonzero real number squared is positive.

Sometimes it may be necessary to change the form before we can apply Property 6.1. Let's consider one example to illustrate this idea.

EXAMPLE 9   Solve $3(2x - 3)^2 + 8 = 44$.

Solution

$$3(2x - 3)^2 + 8 = 44$$

$$3(2x - 3)^2 = 36$$

$$(2x - 3)^2 = 12$$

$$2x - 3 = \pm\sqrt{12}$$

$$2x - 3 = \pm 2\sqrt{3}$$

$$2x = 3 \pm 2\sqrt{3}$$

$$x = \frac{3 \pm 2\sqrt{3}}{2}$$

The solution set is $\left\{\dfrac{3 \pm 2\sqrt{3}}{2}\right\}$.

■

Back to the Pythagorean Theorem

Our work with radicals, Property 6.1, and the Pythagorean theorem form a basis for solving a variety of problems that pertain to right triangles.

E X A M P L E   1 0   A 50-foot rope hangs from the top of a flagpole. When pulled taut to its full length, the rope reaches a point on the ground 18 feet from the base of the pole. Find the height of the pole to the nearest tenth of a foot.

Solution

Let's make a sketch (Figure 6.1) and record the given information.

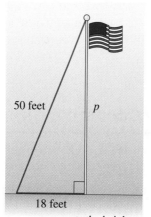

Use the Pythagorean theorem to solve for p as follows:

$$p^2 + 18^2 = 50^2$$

$$p^2 + 324 = 2500$$

$$p^2 = 2176$$

$$p = \sqrt{2176} = 46.6, \quad \text{to the nearest tenth}$$

The height of the flagpole is approximately 46.6 feet.

50 feet / p

18 feet

p represents the height of the flagpole.

Figure 6.1

■

There are two special kinds of right triangles that we use extensively in later mathematics courses. The first is the **isosceles right triangle,** which is a right triangle that has both legs of the same length. Let's consider a problem that involves an isosceles right triangle.

E X A M P L E   1 1   Find the length of each leg of an isosceles right triangle that has a hypotenuse of length 5 meters.

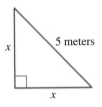

Figure 6.2

Solution

Let's sketch an isosceles right triangle and let x represent the length of each leg (Figure 6.2). Then we can apply the Pythagorean theorem.

$$x^2 + x^2 = 5^2$$

$$2x^2 = 25$$

$$x^2 = \frac{25}{2}$$

$$x = \pm\sqrt{\frac{25}{2}} = \pm\frac{5}{\sqrt{2}} = \pm\frac{5\sqrt{2}}{2}$$

Each leg is $\dfrac{5\sqrt{2}}{2}$ meters long. ■

REMARK: In Example 10 we made no attempt to express $\sqrt{2176}$ in simplest radical form because the answer was to be given as a rational approximation to the nearest tenth. However, in Example 11 we left the final answer in radical form and therefore expressed it in simplest radical form.

The second special kind of right triangle that we use frequently is one that contains acute angles of 30° and 60°. In such a right triangle, which we refer to as a **30°– 60° right triangle,** the side opposite the 30° angle is equal in length to one-half of the length of the hypotenuse. This relationship, along with the Pythagorean theorem, provides us with another problem-solving technique.

E X A M P L E   1 2

Suppose that a 20-foot ladder is leaning against a building and makes an angle of 60° with the ground. How far up the building does the top of the ladder reach? Express your answer to the nearest tenth of a foot.

Solution

Figure 6.3 depicts this situation. The side opposite the 30° angle equals one-half of the hypotenuse, so it is of length $\frac{1}{2}(20) = 10$ feet. Now we can apply the Pythagorean theorem.

$$h^2 + 10^2 = 20^2$$

$$h^2 + 100 = 400$$

$$h^2 = 300$$

$$h = \sqrt{300} = 17.3, \quad \text{to the nearest tenth}$$

The top of the ladder touches the building at a point approximately 17.3 feet from the ground. ■

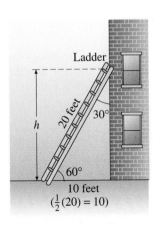

Figure 6.3

PROBLEM SET 6.2

For Problems 1–20, solve each of the quadratic equations by factoring and applying the property "$ab = 0$ if and only if $a = 0$ or $b = 0$." If necessary, return to Chapter 3 and review the factoring techniques presented there.

1. $x^2 - 9x = 0$

2. $x^2 + 5x = 0$

3. $x^2 = -3x$

4. $x^2 = 15x$

5. $3y^2 + 12y = 0$

6. $6y^2 - 24y = 0$

7. $5n^2 - 9n = 0$

8. $4n^2 + 13n = 0$

9. $x^2 + x - 30 = 0$

10. $x^2 - 8x - 48 = 0$

11. $x^2 - 19x + 84 = 0$

12. $x^2 - 21x + 104 = 0$

13. $2x^2 + 19x + 24 = 0$

14. $4x^2 + 29x + 30 = 0$

15. $15x^2 + 29x - 14 = 0$

16. $24x^2 + x - 10 = 0$

17. $25x^2 - 30x + 9 = 0$

18. $16x^2 - 8x + 1 = 0$

19. $6x^2 - 5x - 21 = 0$

20. $12x^2 - 4x - 5 = 0$

For Problems 21–26, solve each radical equation. Don't forget, you *must* check potential solutions.

21. $3\sqrt{x} = x + 2$

22. $3\sqrt{2x} = x + 4$

23. $\sqrt{2x} = x - 4$

24. $\sqrt{x} = x - 2$

25. $\sqrt{3x + 6} = x$

26. $\sqrt{5x + 10} = x$

For Problems 27–34, solve each equation for x by factoring and applying the property "$ab = 0$ if and only if $a = 0$ or $b = 0$."

27. $x^2 - 5kx = 0$

28. $x^2 + 7kx = 0$

29. $x^2 = 16k^2x$

30. $x^2 = 25k^2x$

31. $x^2 - 12kx + 35k^2 = 0$

32. $x^2 - 3kx - 18k^2 = 0$

33. $2x^2 + 5kx - 3k^2 = 0$

34. $3x^2 - 20kx - 7k^2 = 0$

For Problems 35–70, use Property 6.1 to help solve each quadratic equation.

35. $x^2 = 1$

36. $x^2 = 81$

37. $x^2 = -36$

38. $x^2 = -49$

39. $x^2 = 14$

40. $x^2 = 22$

41. $n^2 - 28 = 0$

42. $n^2 - 54 = 0$

43. $3t^2 = 54$

44. $4t^2 = 108$

45. $2t^2 = 7$

46. $3t^2 = 8$

47. $15y^2 = 20$

48. $14y^2 = 80$

49. $10x^2 + 48 = 0$

50. $12x^2 + 50 = 0$

51. $24x^2 = 36$

52. $12x^2 = 49$

53. $(x - 2)^2 = 9$

54. $(x + 1)^2 = 16$

55. $(x + 3)^2 = 25$

56. $(x - 2)^2 = 49$

57. $(x + 6)^2 = -4$

58. $(3x + 1)^2 = 9$

59. $(2x - 3)^2 = 1$

60. $(2x + 5)^2 = -4$

61. $(n - 4)^2 = 5$

62. $(n - 7)^2 = 6$

63. $(t + 5)^2 = 12$

64. $(t - 1)^2 = 18$

65. $(3y - 2)^2 = -27$

66. $(4y + 5)^2 = 80$

67. $3(x + 7)^2 + 4 = 79$

68. $2(x + 6)^2 - 9 = 63$

69. $2(5x - 2)^2 + 5 = 25$

70. $3(4x - 1)^2 + 1 = -17$

For Problems 71–76, a and b represent the lengths of the legs of a right triangle, and c represents the length of the hypotenuse. Express answers in simplest radical form.

71. Find c if $a = 4$ centimeters and $b = 6$ centimeters.

72. Find c if $a = 3$ meters and $b = 7$ meters.

73. Find a if $c = 12$ inches and $b = 8$ inches.

74. Find a if $c = 8$ feet and $b = 6$ feet.

75. Find b if $c = 17$ yards and $a = 15$ yards.

76. Find b if $c = 14$ meters and $a = 12$ meters.

For Problems 77–80, use the isosceles right triangle in Figure 6.4. Express your answers in simplest radical form.

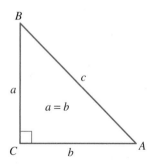

Figure 6.4

77. If $b = 6$ inches, find c.

78. If $a = 7$ centimeters, find c.

79. If $c = 8$ meters, find a and b.

80. If $c = 9$ feet, find a and b.

For Problems 81–86, use the triangle in Figure 6.5. Express your answers in simplest radical form.

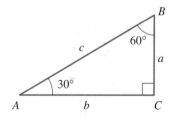

Figure 6.5

81. If $a = 3$ inches, find b and c.

82. If $a = 6$ feet, find b and c.

83. If $c = 14$ centimeters, find a and b.

84. If $c = 9$ centimeters, find a and b.

85. If $b = 10$ feet, find a and c.

86. If $b = 8$ meters, find a and c.

87. A 24-foot ladder resting against a house reaches a windowsill 16 feet above the ground. How far is the foot of the ladder from the foundation of the house? Express your answer to the nearest tenth of a foot.

88. A 62-foot guy-wire makes an angle of 60° with the ground and is attached to a telephone pole (see Figure 6.6). Find the distance from the base of the pole to the point on the pole where the wire is attached. Express your answer to the nearest tenth of a foot.

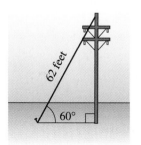

Figure 6.6

89. A rectangular plot measures 16 meters by 34 meters. Find, to the nearest meter, the distance from one corner of the plot to the corner diagonally opposite.

90. Consecutive bases of a square-shaped baseball diamond are 90 feet apart (see Figure 6.7). Find, to the nearest tenth of a foot, the distance from first base diagonally across the diamond to third base.

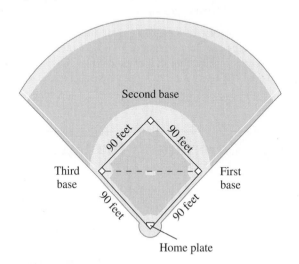

Figure 6.7

91. A diagonal of a square parking lot is 75 meters. Find, to the nearest meter, the length of a side of the lot.

▪▪▪ Thoughts into words

92. Explain why the equation $(x + 2)^2 + 5 = 1$ has no real number solutions.

93. Suppose that your friend solved the equation $(x + 3)^2 = 25$ as follows:

$$(x + 3)^2 = 25$$

$$x^2 + 6x + 9 = 25$$

$$x^2 + 6x - 16 = 0$$

$$(x + 8)(x - 2) = 0$$

$$x + 8 = 0 \quad \text{or} \quad x - 2 = 0$$

$$x = -8 \quad \text{or} \quad x = 2$$

Is this a correct approach to the problem? Would you offer any suggestion about an easier approach to the problem?

▪▪▪ Further investigations

94. Suppose that we are given a cube with edges 12 centimeters in length. Find the length of a diagonal from a lower corner to the diagonally opposite upper corner. Express your answer to the nearest tenth of a centimeter.

95. Suppose that we are given a rectangular box with a length of 8 centimeters, a width of 6 centimeters, and a height of 4 centimeters. Find the length of a diagonal from a lower corner to the upper corner diagonally opposite. Express your answer to the nearest tenth of a centimeter.

96. The converse of the Pythagorean theorem is also true. It states, "If the measures a, b, and c of the sides of a triangle are such that $a^2 + b^2 = c^2$, then the triangle is a right triangle with a and b the measures of the legs and c the measure of the hypotenuse." Use the converse of the Pythagorean theorem to determine which of the triangles with sides of the following measures are right triangles.

    (a) 9, 40, 41   **(b)** 20, 48, 52
    (c) 19, 21, 26   **(d)** 32, 37, 49
    (e) 65, 156, 169   **(f)** 21, 72, 75

97. Find the length of the hypotenuse (h) of an isosceles right triangle if each leg is s units long. Then use this relationship to redo Problems 77–80.

98. Suppose that the side opposite the 30° angle in a 30°–60° right triangle is s units long. Express the length of the hypotenuse and the length of the other leg in terms of s. Then use these relationships and redo Problems 81–86.

6.3   Completing the Square

Thus far we have solved quadratic equations by factoring and applying the property "$ab = 0$ if and only if $a = 0$ or $b = 0$" or by applying the property "$x^2 = a$ if and only if $x = \pm\sqrt{a}$." In this section we examine another method called **completing the square,** which will give us the power to solve any quadratic equation.

A factoring technique we studied in Chapter 3 relied on recognizing **perfect-square trinomials.** In each of the following, the perfect-square trinomial on the right side is the result of squaring the binomial on the left side.

$$(x + 4)^2 = x^2 + 8x + 16 \qquad (x - 6)^2 = x^2 - 12x + 36$$

$$(x + 7)^2 = x^2 + 14x + 49 \qquad (x - 9)^2 = x^2 - 18x + 81$$

$$(x + a)^2 = x^2 + 2ax + a^2$$

Note that in each of the square trinomials, the constant term is equal to the square of one-half of the coefficient of the x term. This relationship enables us to form a perfect-square trinomial by adding a proper constant term. To find the constant term, take one-half of the coefficient of the x term and then square the result. For example, suppose that we want to form a perfect-square trinomial from $x^2 + 10x$. The coefficient of the x term is 10. Because $\frac{1}{2}(10) = 5$ and $5^2 = 25$, the constant term should be 25. The perfect-square trinomial that can be formed is $x^2 + 10x + 25$. This perfect-square trinomial can be factored and expressed as $(x + 5)^2$. Let's use the previous ideas to help solve some quadratic equations.

E X A M P L E   1   Solve $x^2 + 10x - 2 = 0$.

Solution

$$x^2 + 10x - 2 = 0$$

$$x^2 + 10x = 2 \qquad \text{Isolate the } x^2 \text{ and } x \text{ terms.}$$

$$\frac{1}{2}(10) = 5 \text{ and } 5^2 = 25 \qquad \text{Take } \frac{1}{2} \text{ of the coefficient of the } x \text{ term and then square the result.}$$

$$x^2 + 10x + 25 = 2 + 25 \qquad \text{Add 25 to } both \text{ sides of the equation.}$$

$$(x + 5)^2 = 27 \qquad \text{Factor the perfect-square trinomial.}$$

$$x + 5 = \pm\sqrt{27} \qquad \text{Now solve by applying Property 6.1.}$$

$$x + 5 = \pm 3\sqrt{3}$$

$$x = -5 \pm 3\sqrt{3}$$

The solution set is $\{-5 \pm 3\sqrt{3}\}$. ■

Note from Example 1 that the method of completing the square to solve a quadratic equation is merely what the name implies. A perfect-square trinomial is formed, then the equation can be changed to the necessary form for applying the property "$x^2 = a$ if and only if $x = \pm\sqrt{a}$." Let's consider another example.

E X A M P L E   2   Solve $x(x + 8) = -23$.

Solution

$$x(x + 8) = -23$$

$$x^2 + 8x = -23 \qquad \text{Apply the distributive property.}$$

$$\frac{1}{2}(8) = 4 \text{ and } 4^2 = 16 \qquad \text{Take } \frac{1}{2} \text{ of the coefficient of the } x \text{ term and then square the result.}$$

$$x^2 + 8x + 16 = -23 + 16 \qquad \text{Add 16 to \textit{both} sides of the equation.}$$

$$(x + 4)^2 = -7 \qquad \text{Factor the perfect-square trinomial.}$$

$$x + 4 = \pm\sqrt{-7} \qquad \text{Now solve by applying Property 6.1.}$$

$$x + 4 = \pm i\sqrt{7}$$

$$x = -4 \pm i\sqrt{7}$$

The solution set is $\{-4 \pm i\sqrt{7}\}$.

EXAMPLE 3   Solve $x^2 - 3x + 1 = 0$.

Solution

$$x^2 - 3x + 1 = 0$$

$$x^2 - 3x = -1$$

$$x^2 - 3x + \frac{9}{4} = -1 + \frac{9}{4} \qquad \frac{1}{2}(3) = \frac{3}{2} \text{ and } \left(\frac{3}{2}\right)^2 = \frac{9}{4}$$

$$\left(x - \frac{3}{2}\right)^2 = \frac{5}{4}$$

$$x - \frac{3}{2} = \pm\sqrt{\frac{5}{4}}$$

$$x - \frac{3}{2} = \pm\frac{\sqrt{5}}{2}$$

$$x = \frac{3}{2} \pm \frac{\sqrt{5}}{2}$$

$$x = \frac{3 \pm \sqrt{5}}{2}$$

The solution set is $\left\{\dfrac{3 \pm \sqrt{5}}{2}\right\}$.

In Example 3 note that because the coefficient of the x term is odd, we are forced into the realm of fractions. Using common fractions rather than decimals enables us to apply our previous work with radicals.

The relationship for a perfect-square trinomial that states that the constant term is equal to the square of one-half of the coefficient of the x term holds only if the coefficient of x^2 is 1. Thus we must make an adjustment when solving quadratic equations that have a coefficient of x^2 other than 1. We will need to apply the multiplication property of equality so that the coefficient of the x^2 term becomes 1. The next example shows how to make this adjustment.

EXAMPLE 4

Solve $2x^2 + 12x - 5 = 0$.

Solution

$$2x^2 + 12x - 5 = 0$$

$$2x^2 + 12x = 5$$

$$x^2 + 6x = \frac{5}{2} \qquad \text{Multiply both sides by } \frac{1}{2}.$$

$$x^2 + 6x + 9 = \frac{5}{2} + 9 \qquad \frac{1}{2}(6) \text{ and } 3^2 = 9$$

$$x^2 + 6x + 9 = \frac{23}{2}$$

$$(x + 3)^2 = \frac{23}{2}$$

$$x + 3 = \pm\sqrt{\frac{23}{2}}$$

$$x + 3 = \pm\frac{\sqrt{46}}{2} \qquad \sqrt{\frac{23}{2}} = \frac{\sqrt{23}}{\sqrt{2}} \cdot \frac{\sqrt{2}}{\sqrt{2}} = \frac{\sqrt{46}}{2}$$

$$x = -3 \pm \frac{\sqrt{46}}{2}$$

$$x = \frac{-6}{2} \pm \frac{\sqrt{46}}{2} \qquad \text{Common denominator of 2}$$

$$x = \frac{-6 \pm \sqrt{46}}{2}$$

The solution set is $\left\{ \dfrac{-6 \pm \sqrt{46}}{2} \right\}$. ■

As we mentioned earlier, we can use the method of completing the square to solve *any* quadratic equation. To illustrate, let's use it to solve an equation that could also be solved by factoring.

EXAMPLE 5

Solve $x^2 - 2x - 8 = 0$ by completing the square.

Solution

$$x^2 - 2x - 8 = 0$$

$$x^2 - 2x = 8$$

$$x^2 - 2x + 1 = 8 + 1 \qquad \frac{1}{2}(-2) = -1 \text{ and } (-1)^2 = 1$$

$$(x - 1)^2 = 9$$

$$x - 1 = \pm 3$$

$$x - 1 = 3 \quad \text{or} \quad x - 1 = -3$$

$$x = 4 \quad \text{or} \quad x = -2$$

The solution set is $\{-2, 4\}$.   ∎

Solving the equation in Example 5 by factoring would be easier than completing the square. Remember, however, that the method of completing the square will work with any quadratic equation.

PROBLEM SET 6.3

For Problems 1–14, solve each quadratic equation by using (a) the factoring method and (b) the method of completing the square.

1. $x^2 - 4x - 60 = 0$

2. $x^2 + 6x - 16 = 0$

3. $x^2 - 14x = -40$

4. $x^2 - 18x = -72$

5. $x^2 - 5x - 50 = 0$

6. $x^2 + 3x - 18 = 0$

7. $x(x + 7) = 8$

8. $x(x - 1) = 30$

9. $2n^2 - n - 15 = 0$

10. $3n^2 + n - 14 = 0$

11. $3n^2 + 7n - 6 = 0$

12. $2n^2 + 7n - 4 = 0$

13. $n(n + 6) = 160$

14. $n(n - 6) = 216$

For Problems 15–38, use the method of completing the square to solve each quadratic equation.

15. $x^2 + 4x - 2 = 0$

16. $x^2 + 2x - 1 = 0$

17. $x^2 + 6x - 3 = 0$

18. $x^2 + 8x - 4 = 0$

19. $y^2 - 10y = 1$

20. $y^2 - 6y = -10$

21. $n^2 - 8n + 17 = 0$

22. $n^2 - 4n + 2 = 0$

23. $n(n + 12) = -9$

24. $n(n + 14) = -4$

25. $n^2 + 2n + 6 = 0$

26. $n^2 + n - 1 = 0$

27. $x^2 + 3x - 2 = 0$

28. $x^2 + 5x - 3 = 0$

29. $x^2 + 5x + 1 = 0$

30. $x^2 + 7x + 2 = 0$

31. $y^2 - 7y + 3 = 0$

32. $y^2 - 9y + 30 = 0$

33. $2x^2 + 4x - 3 = 0$

34. $2t^2 - 4t + 1 = 0$

35. $3n^2 - 6n + 5 = 0$

36. $3x^2 + 12x - 2 = 0$

37. $3x^2 + 5x - 1 = 0$

38. $2x^2 + 7x - 3 = 0$

For Problems 39–60, solve each quadratic equation using the method that seems most appropriate.

39. $x^2 + 8x - 48 = 0$

40. $x^2 + 5x - 14 = 0$

41. $2n^2 - 8n = -3$

42. $3x^2 + 6x = 1$

43. $(3x - 1)(2x + 9) = 0$

44. $(5x + 2)(x - 4) = 0$

45. $(x + 2)(x - 7) = 10$

46. $(x - 3)(x + 5) = -7$

47. $(x - 3)^2 = 12$

48. $x^2 = 16x$

49. $3n^2 - 6n + 4 = 0$

50. $2n^2 - 2n - 1 = 0$

51. $n(n + 8) = 240$

52. $t(t - 26) = -160$

53. $3x^2 + 5x = -2$

54. $2x^2 - 7x = -5$

55. $4x^2 - 8x + 3 = 0$

56. $9x^2 + 18x + 5 = 0$

57. $x^2 + 12x = 4$

58. $x^2 + 6x = -11$

59. $4(2x + 1)^2 - 1 = 11$

60. $5(x + 2)^2 + 1 = 16$

61. Use the method of completing the square to solve $ax^2 + bx + c = 0$ for x, where a, b, and c are real numbers and $a \neq 0$.

■ ■ ■ **Thoughts into words**

62. Explain the process of completing the square to solve a quadratic equation.

63. Give a step-by-step description of how to solve $3x^2 + 9x - 4 = 0$ by completing the square.

■ ■ ■ **Further investigations**

Solve Problems 64–67 for the indicated variable. Assume that all letters represent positive numbers.

64. $\dfrac{x^2}{a^2} - \dfrac{y^2}{b^2} = 1$ for y

65. $\dfrac{x^2}{a^2} + \dfrac{y^2}{b^2} = 1$ for x

66. $s = \dfrac{1}{2}gt^2$ for t

67. $A = \pi r^2$ for r

Solve each of the following equations for x.

68. $x^2 + 8ax + 15a^2 = 0$

69. $x^2 - 5ax + 6a^2 = 0$

70. $10x^2 - 31ax - 14a^2 = 0$

71. $6x^2 + ax - 2a^2 = 0$

72. $4x^2 + 4bx + b^2 = 0$

73. $9x^2 - 12bx + 4b^2 = 0$

6.4   Quadratic Formula

As we saw in the last section, the method of completing the square can be used to solve any quadratic equation. Thus, if we apply the method of completing the square to the equation $ax^2 + bx + c = 0$, where a, b, and c are real numbers and $a \neq 0$, we can produce a formula for solving quadratic equations. This formula can then be used to solve any quadratic equation. Let's solve $ax^2 + bx + c = 0$ by completing the square.

$$ax^2 + bx + c = 0$$

$$ax^2 + bx = -c \qquad \text{Isolate the } x^2 \text{ and } x \text{ terms.}$$

$$x^2 + \frac{b}{a}x = -\frac{c}{a} \qquad \text{Multiply both sides by } \frac{1}{a}.$$

$$x^2 + \frac{b}{a}x + \frac{b^2}{4a^2} = -\frac{c}{a} + \frac{b^2}{4a^2} \qquad \frac{1}{2}\left(\frac{b}{a}\right) = \frac{b}{2a} \text{ and } \left(\frac{b}{2a}\right)^2 = \frac{b^2}{4a^2}$$

Complete the square by adding $\dfrac{b^2}{4a^2}$ to both sides.

$$x^2 + \frac{b}{a}x + \frac{b^2}{4a^2} = -\frac{4ac}{4a^2} + \frac{b^2}{4a^2} \qquad \text{Common denominator of } 4a^2 \text{ on right side}$$

$$x^2 + \frac{b}{a}x + \frac{b^2}{4a^2} = \frac{b^2}{4a^2} - \frac{4ac}{4a^2} \qquad \text{Commutative property}$$

$$\left(x + \frac{b}{2a}\right)^2 = \frac{b^2 - 4ac}{4a^2}$$

The right side is combined into a single fraction.

$$x + \frac{b}{2a} = \pm\sqrt{\frac{b^2 - 4ac}{4a^2}}$$

$$x + \frac{b}{2a} = \pm\frac{\sqrt{b^2 - 4ac}}{\sqrt{4a^2}}$$

$$x + \frac{b}{2a} = \pm\frac{\sqrt{b^2 - 4ac}}{2a}$$

$\sqrt{4a^2} = |2a|$ but $2a$ can be used because of the use of $\pm$.

$$x + \frac{b}{2a} = \frac{\sqrt{b^2 - 4ac}}{2a} \qquad \text{or} \qquad x + \frac{b}{2a} = -\frac{\sqrt{b^2 - 4ac}}{2a}$$

$$x = -\frac{b}{2a} + \frac{\sqrt{b^2 - 4ac}}{2a} \qquad \text{or} \qquad x = -\frac{b}{2a} - \frac{\sqrt{b^2 - 4ac}}{2a}$$

$$x = \frac{-b + \sqrt{b^2 - 4ac}}{2a} \qquad \text{or} \qquad x = \frac{-b - \sqrt{b^2 - 4ac}}{2a}$$

The quadratic formula is usually stated as follows:

Quadratic Formula

$$x = \frac{-b \pm \sqrt{b^2 - 4ac}}{2a}, \qquad a \neq 0$$

We can use the quadratic formula to solve *any* quadratic equation by expressing the equation in the standard form $ax^2 + bx + c = 0$ and substituting the values for a, b, and c into the formula. Let's consider some examples.

EXAMPLE 1   Solve $x^2 + 5x + 2 = 0$.

Solution

$$x^2 + 5x + 2 = 0$$

The given equation is in standard form with $a = 1$, $b = 5$, and $c = 2$. Let's substitute these values into the formula and simplify.

$$x = \frac{-b \pm \sqrt{b^2 - 4ac}}{2a}$$

$$x = \frac{-5 \pm \sqrt{5^2 - 4(1)(2)}}{2(1)}$$

$$x = \frac{-5 \pm \sqrt{25 - 8}}{2}$$

$$x = \frac{-5 \pm \sqrt{17}}{2}$$

The solution set is $\left\{ \dfrac{-5 \pm \sqrt{17}}{2} \right\}$.

■

EXAMPLE 2   Solve $x^2 - 2x - 4 = 0$.

Solution

$$x^2 - 2x - 4 = 0$$

We need to think of $x^2 - 2x - 4 = 0$ as $x^2 + (-2)x + (-4) = 0$ to determine the values $a = 1$, $b = -2$, and $c = -4$. Let's substitute these values into the quadratic formula and simplify.

$$x = \frac{-b \pm \sqrt{b^2 - 4ac}}{2a}$$

$$x = \frac{-(-2) \pm \sqrt{(-2)^2 - 4(1)(-4)}}{2(1)}$$

$$x = \frac{2 \pm \sqrt{4 + 16}}{2}$$

$$x = \frac{2 \pm \sqrt{20}}{2}$$

$$x = \frac{2 \pm 2\sqrt{5}}{2}$$

$$x = \frac{2(1 \pm \sqrt{5})}{2}$$

The solution set is $\{1 \pm \sqrt{5}\}$.

■

EXAMPLE 3   Solve $x^2 - 2x + 19 = 0$.

Solution

$$x^2 - 2x + 19 = 0.$$

We can substitute $a = 1$, $b = -2$, and $c = 19$.

$$x = \frac{-b \pm \sqrt{b^2 - 4ac}}{2a}$$

$$x = \frac{-(-2) \pm \sqrt{(-2)^2 - 4(1)(19)}}{2(1)}$$

$$x = \frac{2 \pm \sqrt{4 - 76}}{2}$$

$$x = \frac{2 \pm \sqrt{-72}}{2}$$

$$x = \frac{2 \pm 6i\sqrt{2}}{2} \qquad \sqrt{-72} = i\sqrt{72} = i\sqrt{36}\sqrt{2} = 6i\sqrt{2}$$

$$x = \frac{2(1 \pm 3i\sqrt{2})}{2}$$

The solution set is $\{1 \pm 3i\sqrt{2}\}$. ■

EXAMPLE 4

Solve $2x^2 + 4x - 3 = 0$.

Solution

$$2x^2 + 4x - 3 = 0$$

Here $a = 2$, $b = 4$, and $c = -3$. Solving by using the quadratic formula is unlike solving by completing the square in that there is no need to make the coefficient of the x^2 equal to 1.

$$x = \frac{-b \pm \sqrt{b^2 - 4ac}}{2a}$$

$$x = \frac{-4 \pm \sqrt{4^2 - 4(2)(-3)}}{2(2)}$$

$$x = \frac{-4 \pm \sqrt{16 + 24}}{4}$$

$$x = \frac{-4 \pm \sqrt{40}}{4}$$

$$x = \frac{-4 \pm 2\sqrt{10}}{4}$$

$$x = \frac{2(-2 \pm \sqrt{10})}{4}$$

$$x = \frac{-2 \pm \sqrt{10}}{2}$$

The solution set is $\left\{ \dfrac{-2 \pm \sqrt{10}}{2} \right\}$. ■

EXAMPLE 5

Solve $n(3n - 10) = 25$.

Solution

$$n(3n - 10) = 25$$

First, we need to change the equation to the standard form $an^2 + bn + c = 0$.

$$n(3n - 10) = 25$$

$$3n^2 - 10n = 25$$

$$3n^2 - 10n - 25 = 0$$

Now we can substitute $a = 3$, $b = -10$, and $c = -25$ into the quadratic formula.

$$n = \frac{-b \pm \sqrt{b^2 - 4ac}}{2a}$$

$$n = \frac{-(-10) \pm \sqrt{(-10)^2 - 4(3)(-25)}}{2(3)}$$

$$n = \frac{10 \pm \sqrt{100 + 300}}{2(3)}$$

$$n = \frac{10 \pm \sqrt{400}}{6}$$

$$n = \frac{10 \pm 20}{6}$$

$$n = \frac{10 + 20}{6} \quad \text{or} \quad n = \frac{10 - 20}{6}$$

$$n = 5 \quad \text{or} \quad n = -\frac{5}{3}$$

The solution set is $\left\{ -\frac{5}{3}, 5 \right\}$.

In Example 5, note that we used the variable n. The quadratic formula is usually stated in terms of x, but it certainly can be applied to quadratic equations in other variables. Also note in Example 5 that the polynomial $3n^2 - 10n - 25$ can be factored as $(3n + 5)(n - 5)$. Therefore, we could also solve the equation $3n^2 - 10n - 25 = 0$ by using the factoring approach. Section 6.5 will offer some guidance in deciding which approach to use for a particular equation.

Nature of Roots

The quadratic formula makes it easy to determine the nature of the roots of a quadratic equation without completely solving the equation. The number

$$b^2 - 4ac$$

which appears under the radical sign in the quadratic formula, is called the **discriminant** of the quadratic equation. The discriminant is the indicator of the kind of roots the equation has. For example, suppose that you start to solve the equation $x^2 - 4x + 7 = 0$ as follows:

$$x = \frac{-b \pm \sqrt{b^2 - 4ac}}{2a}$$

$$x = \frac{-(-4) \pm \sqrt{(-4)^2 - 4(1)(7)}}{2(1)}$$

$$x = \frac{4 \pm \sqrt{16 - 28}}{2}$$

$$x = \frac{4 \pm \sqrt{-12}}{2}$$

At this stage you should be able to look ahead and realize that you will obtain two complex solutions for the equation. (Note, by the way, that these solutions are complex conjugates.) In other words, the discriminant, -12, indicates what type of roots you will obtain.

We make the following general statements relative to the roots of a quadratic equation of the form $ax^2 + bx + c = 0$.

1. If $b^2 - 4ac < 0$, then the equation has two nonreal complex solutions.
2. If $b^2 - 4ac = 0$, then the equation has one real solution.
3. If $b^2 - 4ac > 0$, then the equation has two real solutions.

The following examples illustrate each of these situations. (You may want to solve the equations completely to verify the conclusions.)

Equation	Discriminant	Nature of roots
$x^2 - 3x + 7 = 0$	$b^2 - 4ac = (-3)^2 - 4(1)(7)$ $= 9 - 28$ $= -19$	Two nonreal complex solutions
$9x^2 - 12x + 4 = 0$	$b^2 - 4ac = (-12)^2 - 4(9)(4)$ $= 144 - 144$ $= 0$	One real solution
$2x^2 + 5x - 3 = 0$	$b^2 - 4ac = (5)^2 - 4(2)(-3)$ $= 25 + 24$ $= 49$	Two real solutions

There is another very useful relationship that involves the roots of a quadratic equation and the numbers a, b, and c of the general form $ax^2 + bx + c = 0$. Suppose that we let x_1 and x_2 be the two roots generated by the quadratic formula. Thus we have

$$x_1 = \frac{-b + \sqrt{b^2 - 4ac}}{2a} \quad \text{and} \quad x_2 = \frac{-b - \sqrt{b^2 - 4ac}}{2a}$$

REMARK: A clarification is called for at this time. Previously, we made the statement that if $b^2 - 4ac = 0$, then the equation has one real solution. Technically, such an equation has two solutions, but they are equal. For example, each factor of $(x - 2)(x - 2) = 0$ produces a solution, but both solutions are the number 2. We sometimes refer to this as one real solution with a *multiplicity of two*. Using the idea of multiplicity of roots, we can say that every quadratic equation has two roots.

Now let's consider the sum and product of the two roots.

$$\textbf{Sum} \quad x_1 + x_2 = \frac{-b + \sqrt{b^2 - 4ac}}{2a} + \frac{-b - \sqrt{b^2 - 4ac}}{2a} = \frac{-2b}{2a} = \boxed{-\frac{b}{a}}$$

$$\textbf{Product} \quad (x_1)(x_2) = \left(\frac{-b + \sqrt{b^2 - 4ac}}{2a} \right)\left(\frac{-b - \sqrt{b^2 - 4ac}}{2a} \right)$$

$$= \frac{b^2 - (b^2 - 4ac)}{4a^2}$$

$$= \frac{b^2 - b^2 + 4ac}{4a^2}$$

$$= \frac{4ac}{4a^2} = \boxed{\frac{c}{a}}$$

These relationships provide another way of checking potential solutions when solving quadratic equations. For instance, back in Example 3 we solved the equation $x^2 - 2x + 19 = 0$ and obtained solutions of $1 + 3i\sqrt{2}$ and $1 - 3i\sqrt{2}$. Let's check these solutions by using the sum and product relationships.

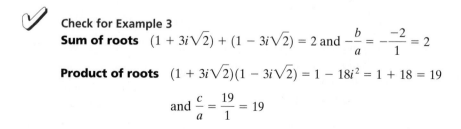

Check for Example 3

Sum of roots $(1 + 3i\sqrt{2}) + (1 - 3i\sqrt{2}) = 2$ and $-\dfrac{b}{a} = -\dfrac{-2}{1} = 2$

Product of roots $(1 + 3i\sqrt{2})(1 - 3i\sqrt{2}) = 1 - 18i^2 = 1 + 18 = 19$

$$\text{and } \frac{c}{a} = \frac{19}{1} = 19$$

Likewise, a check for Example 4 is as follows:

Check for Example 4

Sum of roots $\left(\dfrac{-2 + \sqrt{10}}{2}\right) + \left(\dfrac{-2 - \sqrt{10}}{2}\right) = -\dfrac{4}{2} = -2$

and $-\dfrac{b}{a} = -\dfrac{4}{2} = -2$

Product of roots $\left(\dfrac{-2 + \sqrt{10}}{2}\right)\left(\dfrac{-2 - \sqrt{10}}{2}\right) = -\dfrac{6}{4} = -\dfrac{3}{2}$

and $\dfrac{c}{a} = \dfrac{-3}{2} = -\dfrac{3}{2}$

Note that for both Examples 3 and 4, it was much easier to check by using the sum and product relationships than it would have been to check by substituting back into the original equation. Don't forget that the values for *a*, *b*, and *c* come from a quadratic equation of the form $ax^2 + bx + c = 0$. In Example 5, if we are going to check the potential solutions by using the sum and product relationships, we must be certain that we made no errors when changing the given equation $n(3n - 10) = 25$ to the form $3n^2 - 10n - 25 = 0$.

PROBLEM SET 6.4

For each quadratic equation in Problems 1–10, first use the discriminant to determine whether the equation has two nonreal complex solutions, one real solution with a multiplicity of two, or two real solutions. Then solve the equation.

1. $x^2 + 4x - 21 = 0$

2. $x^2 - 3x - 54 = 0$

3. $9x^2 - 6x + 1 = 0$

4. $4x^2 + 20x + 25 = 0$

5. $x^2 - 7x + 13 = 0$

6. $2x^2 - x + 5 = 0$

7. $15x^2 + 17x - 4 = 0$

8. $8x^2 + 18x - 5 = 0$

9. $3x^2 + 4x = 2$

10. $2x^2 - 6x = -1$

For Problems 11–50, use the quadratic formula to solve each of the quadratic equations. Check your solutions by using the sum and product relationships.

11. $x^2 + 2x - 1 = 0$

12. $x^2 + 4x - 1 = 0$

13. $n^2 + 5n - 3 = 0$

14. $n^2 + 3n - 2 = 0$

15. $a^2 - 8a = 4$

16. $a^2 - 6a = 2$

17. $n^2 + 5n + 8 = 0$

18. $2n^2 - 3n + 5 = 0$

19. $x^2 - 18x + 80 = 0$

20. $x^2 + 19x + 70 = 0$

21. $-y^2 = -9y + 5$

22. $-y^2 + 7y = 4$

23. $2x^2 + x - 4 = 0$

24. $2x^2 + 5x - 2 = 0$

25. $4x^2 + 2x + 1 = 0$

26. $3x^2 - 2x + 5 = 0$

27. $3a^2 - 8a + 2 = 0$

28. $2a^2 - 6a + 1 = 0$

29. $-2n^2 + 3n + 5 = 0$

30. $-3n^2 - 11n + 4 = 0$

31. $3x^2 + 19x + 20 = 0$

32. $2x^2 - 17x + 30 = 0$

33. $36n^2 - 60n + 25 = 0$

34. $9n^2 + 42n + 49 = 0$

35. $4x^2 - 2x = 3$

36. $6x^2 - 4x = 3$

37. $5x^2 - 13x = 0$

38. $7x^2 + 12x = 0$

39. $3x^2 = 5$

40. $4x^2 = 3$

41. $6t^2 + t - 3 = 0$

42. $2t^2 + 6t - 3 = 0$

43. $n^2 + 32n + 252 = 0$ **44.** $n^2 - 4n - 192 = 0$

45. $12x^2 - 73x + 110 = 0$ **46.** $6x^2 + 11x - 255 = 0$

47. $-2x^2 + 4x - 3 = 0$ **48.** $-2x^2 + 6x - 5 = 0$

49. $-6x^2 + 2x + 1 = 0$ **50.** $-2x^2 + 4x + 1 = 0$

■ ■ ■ Thoughts into words

51. Your friend states that the equation $-2x^2 + 4x - 1 = 0$ must be changed to $2x^2 - 4x + 1 = 0$ (by multiplying both sides by -1) before the quadratic formula can be applied. Is she right about this? If not, how would you convince her she is wrong?

52. Another of your friends claims that the quadratic formula can be used to solve the equation $x^2 - 9 = 0$. How would you react to this claim?

53. Why must we change the equation $3x^2 - 2x = 4$ to $3x^2 - 2x - 4 = 0$ before applying the quadratic formula?

■ ■ ■ Further investigations

The solution set for $x^2 - 4x - 37 = 0$ is $\{2 \pm \sqrt{41}\}$. With a calculator, we found a rational approximation, to the nearest one-thousandth, for each of these solutions.

$$2 - \sqrt{41} = -4.403 \quad \text{and} \quad 2 + \sqrt{41} = 8.403$$

Thus the solution set is $\{-4.403, 8.403\}$, with the answers rounded to the nearest one-thousandth.

Solve each of the equations in Problems 54–63, expressing solutions to the nearest one-thousandth.

54. $x^2 - 6x - 10 = 0$ **55.** $x^2 - 16x - 24 = 0$

56. $x^2 + 6x - 44 = 0$ **57.** $x^2 + 10x - 46 = 0$

58. $x^2 + 8x + 2 = 0$ **59.** $x^2 + 9x + 3 = 0$

60. $4x^2 - 6x + 1 = 0$ **61.** $5x^2 - 9x + 1 = 0$

62. $2x^2 - 11x - 5 = 0$ **63.** $3x^2 - 12x - 10 = 0$

For Problems 64–66, use the discriminant to help solve each problem.

64. Determine k so that the solutions of $x^2 - 2x + k = 0$ are complex but nonreal.

65. Determine k so that $4x^2 - kx + 1 = 0$ has two equal real solutions.

66. Determine k so that $3x^2 - kx - 2 = 0$ has real solutions.

6.5 More Quadratic Equations and Applications

Which method should be used to solve a particular quadratic equation? There is no hard and fast answer to that question; it depends on the type of equation and on your personal preference. In the following examples we will state reasons for choosing a specific technique. However, keep in mind that usually this is a decision you must make as the need arises. That's why you need to be familiar with the strengths and weaknesses of each method.

EXAMPLE 1   Solve $2x^2 - 3x - 1 = 0$.

Solution

Because of the leading coefficient of 2 and the constant term of -1, there are very few factoring possibilities to consider. Therefore, with such problems, first try the factoring approach. Unfortunately, this particular polynomial is not factorable using integers. Let's use the quadratic formula to solve the equation.

$$x = \frac{-b \pm \sqrt{b^2 - 4ac}}{2a}$$

$$x = \frac{-(-3) \pm \sqrt{(-3)^2 - 4(2)(-1)}}{2(2)}$$

$$x = \frac{3 \pm \sqrt{9 + 8}}{4}$$

$$x = \frac{3 \pm \sqrt{17}}{4}$$

Check

We can use the sum-of-roots and the product-of-roots relationships for our checking purposes.

Sum of roots $\dfrac{3 + \sqrt{17}}{4} + \dfrac{3 - \sqrt{17}}{4} = \dfrac{6}{4} = \dfrac{3}{2}$ and $-\dfrac{b}{a} = -\dfrac{-3}{2} = \dfrac{3}{2}$

Product of roots $\left(\dfrac{3 + \sqrt{17}}{4}\right)\left(\dfrac{3 - \sqrt{17}}{4}\right) = \dfrac{9 - 17}{16} = -\dfrac{8}{16} = -\dfrac{1}{2}$ and

$$\frac{c}{a} = \frac{-1}{2} = -\frac{1}{2}$$

The solution set is $\left\{\dfrac{3 \pm \sqrt{17}}{4}\right\}$.   ■

EXAMPLE 2   Solve $\dfrac{3}{n} + \dfrac{10}{n + 6} = 1$.

Solution

$$\frac{3}{n} + \frac{10}{n + 6} = 1, \qquad n \neq 0 \text{ and } n \neq -6$$

$$n(n + 6)\left(\frac{3}{n} + \frac{10}{n + 6}\right) = 1(n)(n + 6) \qquad \text{Multiply both sides by } n(n + 6),\text{ which is the LCD.}$$

$$3(n + 6) + 10n = n(n + 6)$$

$$3n + 18 + 10n = n^2 + 6n$$

$$13n + 18 = n^2 + 6n$$
$$0 = n^2 - 7n - 18$$

This equation is an easy one to consider for possible factoring, and it factors as follows:

$$0 = (n - 9)(n + 2)$$

$$n - 9 = 0 \quad \text{or} \quad n + 2 = 0$$

$$n = 9 \quad \text{or} \quad n = -2$$

 Check

Substituting 9 and −2 back into the original equation, we obtain

$$\frac{3}{n} + \frac{10}{n + 6} = 1 \qquad\qquad \frac{3}{n} + \frac{10}{n + 6} = 1$$

$$\frac{3}{9} + \frac{10}{9 + 6} \overset{?}{=} 1 \qquad\qquad \frac{3}{-2} + \frac{10}{-2 + 6} \overset{?}{=} 1$$

$$\frac{1}{3} + \frac{10}{15} \overset{?}{=} 1 \qquad \text{or} \qquad -\frac{3}{2} + \frac{10}{4} \overset{?}{=} 1$$

$$\frac{1}{3} + \frac{2}{3} \overset{?}{=} 1 \qquad\qquad -\frac{3}{2} + \frac{5}{2} \overset{?}{=} 1$$

$$1 = 1 \qquad\qquad \frac{2}{2} = 1$$

The solution set is $\{-2, 9\}$. ■

We should make two comments about Example 2. First, note the indication of the initial restrictions $n \neq 0$ and $n \neq -6$. Remember that we need to do this when solving fractional equations. Second, the sum-of-roots and product-of-roots relationships were not used for checking purposes in this problem. Those relationships would check the validity of our work only from the step $0 = n^2 - 7n - 18$ to the finish. In other words, an error made in changing the original equation to quadratic form would not be detected by checking the sum and product of potential roots. With such a problem, the only *absolute check* is to substitute the potential solutions back into the original equation.

E X A M P L E   3   Solve $x^2 + 22x + 112 = 0$.

Solution

The size of the constant term makes the factoring approach a little cumbersome for this problem. Furthermore, because the leading coefficient is 1 and the coefficient of the x term is even, the method of completing the square will work effectively.

$$x^2 + 22x + 112 = 0$$

$$x^2 + 22x = -112$$

$$x^2 + 22x + 121 = -112 + 121$$

$$(x + 11)^2 = 9$$

$$x + 11 = \pm\sqrt{9}$$

$$x + 11 = \pm 3$$

$$x + 11 = 3 \quad \text{or} \quad x + 11 = -3$$

$$x = -8 \quad \text{or} \quad x = -14$$

Check

Sum of roots $-8 + (-14) = -22$ and $-\dfrac{b}{a} = -22$

Product of roots $(-8)(-14) = 112$ and $\dfrac{c}{a} = 112$

The solution set is $\{-14, -8\}$.

E X A M P L E   4   Solve $x^4 - 4x^2 - 96 = 0$.

Solution

An equation such as $x^4 - 4x^2 - 96 = 0$ is not a quadratic equation, but we can solve it using the techniques that we use on quadratic equations. That is, we can factor the polynomial and apply the property "$ab = 0$ if and only if $a = 0$ or $b = 0$" as follows:

$$x^4 - 4x^2 - 96 = 0$$

$$(x^2 - 12)(x^2 + 8) = 0$$

$$x^2 - 12 = 0 \quad \text{or} \quad x^2 + 8 = 0$$

$$x^2 = 12 \quad \text{or} \quad x^2 = -8$$

$$x = \pm\sqrt{12} \quad \text{or} \quad x = \pm\sqrt{-8}$$

$$x = \pm 2\sqrt{3} \quad \text{or} \quad x = \pm 2i\sqrt{2}$$

The solution set is $\{\pm 2\sqrt{3}, \pm 2i\sqrt{2}\}$. (We will leave the check for this problem for you to do!)

REMARK: Another approach to Example 4 would be to substitute y for x^2 and y^2 for x^4. The equation $x^4 - 4x^2 - 96 = 0$ becomes the quadratic equation $y^2 - 4y - 96 = 0$. Thus we say that $x^4 - 4x^2 - 96 = 0$ is of *quadratic form*. Then we could solve the quadratic equation $y^2 - 4y - 96 = 0$ and use the equation $y = x^2$ to determine the solutions for x.

Applications

Before we conclude this section with some word problems that can be solved using quadratic equations, let's restate the suggestions we made, in an earlier chapter, for solving word problems.

Suggestions for Solving Word Problems

1. Read the problem carefully and make certain that you understand the meanings of all the words. Be especially alert for any technical terms used in the statement of the problem.
2. Read the problem a second time (perhaps even a third time) to get an overview of the situation being described and to determine the known facts, as well as what is to be found.
3. Sketch any figure, diagram, or chart that might be helpful in analyzing the problem.
4. Choose a meaningful variable to represent an unknown quantity in the problem (perhaps l, if the length of a rectangle is an unknown quantity) and represent any other unknowns in terms of that variable.
5. Look for a guideline that you can use to set up an equation. A guideline might be a formula such as $A = lw$ or a relationship such as "the fractional part of a job done by Bill plus the fractional part of the job done by Mary equals the total job."
6. Form an equation that contains the variable and translates the conditions of the guideline from English to algebra.
7. Solve the equation and use the solutions to determine all facts requested in the problem.
8. **Check all answers back into the original statement of the problem.**

Keep these suggestions in mind as we now consider some word problems.

PROBLEM 1

A page for a magazine contains 70 square inches of type. The height of a page is twice the width. If the margin around the type is to be 2 inches uniformly, what are the dimensions of a page?

Solution

Let x represent the width of a page. Then $2x$ represents the height of a page. Now let's draw and label a model of a page (Figure 6.8).

Width of   Height of   Area of
typed   typed   typed
material   material   material

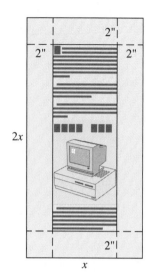

$$(x - 4)(2x - 4) = 70$$

$$2x^2 - 12x + 16 = 70$$

$$2x^2 - 12x - 54 = 0$$

$$x^2 - 6x - 27 = 0$$

$$(x - 9)(x + 3) = 0$$

$$x - 9 = 0 \quad \text{or} \quad x + 3 = 0$$

$$x = 9 \quad \text{or} \quad x = -3$$

Figure 6.8

Disregard the negative solution; the page must be 9 inches wide, and its height is $2(9) = 18$ inches. ◼

   Let's use our knowledge of quadratic equations to analyze some applications of the business world. For example, if P dollars is invested at r rate of interest compounded annually for t years, then the amount of money, A, accumulated at the end of t years is given by the formula

$$A = P(1 + r)^t$$

This compound interest formula serves as a guideline for the next problem.

PROBLEM 2

Suppose that $100 is invested at a certain rate of interest compounded annually for 2 years. If the accumulated value at the end of 2 years is $121, find the rate of interest.

Solution

Let r represent the rate of interest. Substitute the known values into the compound interest formula to yield

$$A = P(1 + r)^t$$

$$121 = 100(1 + r)^2$$

Solving this equation, we obtain

$$\frac{121}{100} = (1 + r)^2$$

$$\pm\sqrt{\frac{121}{100}} = (1 + r)$$

$$\pm\frac{11}{10} = 1 + r$$

$$1 + r = \frac{11}{10} \qquad \text{or} \qquad 1 + r = -\frac{11}{10}$$

$$r = -1 + \frac{11}{10} \qquad \text{or} \qquad r = -1 - \frac{11}{10}$$

$$r = \frac{1}{10} \qquad \text{or} \qquad r = -\frac{21}{10}$$

We must disregard the negative solution, so $r = \dfrac{1}{10}$ is the only solution. Change $\dfrac{1}{10}$ to a percent and the rate of interest is 10%. ■

PROBLEM 3

On a 130-mile trip from Orlando to Sarasota, Roberto encountered a heavy thunderstorm for the last 40 miles of the trip. During the thunderstorm he averaged 20 miles per hour slower than before the storm. The entire trip took $2\frac{1}{2}$ hours. How fast did he travel before the storm?

Solution

Let x represent Roberto's rate before the thunderstorm. Then $x - 20$ represents his speed during the thunderstorm. Because $t = \dfrac{d}{r}$, then $\dfrac{90}{x}$ represents the time traveling before the storm and $\dfrac{40}{x - 20}$ represents the time traveling during the storm. The following guideline sums up the situation.

Time traveling before the storm	Plus	Time traveling after the storm	Equals	Total time
↓		↓		↓
$\dfrac{90}{x}$	$+$	$\dfrac{40}{x - 20}$	$=$	$\dfrac{5}{2}$

Solving this equation, we obtain

$$2x(x - 20)\left(\frac{90}{x} + \frac{40}{x - 20}\right) = 2x(x - 20)\left(\frac{5}{2}\right)$$

$$2x(x - 20)\left(\frac{90}{x}\right) + 2x(x - 20)\left(\frac{40}{x - 20}\right) = 2x(x - 20)\left(\frac{5}{2}\right)$$

$$180(x - 20) + 2x(40) = 5x(x - 20)$$

$$180x - 3600 + 80x = 5x^2 - 100x$$

$$0 = 5x^2 - 360x + 3600$$

$$0 = 5(x^2 - 72x + 720)$$

$$0 = 5(x - 60)(x - 12)$$

$$x - 60 = 0 \quad \text{or} \quad x - 12 = 0$$

$$x = 60 \quad \text{or} \quad x = 12$$

We discard the solution of 12 because it would be impossible to drive 20 miles per hour slower than 12 miles per hour; thus Roberto's rate before the thunderstorm was 60 miles per hour. ▪

PROBLEM 4

A businesswoman bought a parcel of land on speculation for $120,000. She subdivided the land into lots, and when she had sold all but 18 lots at a profit of $6000 per lot, she had regained the entire cost of the land. How many lots were sold and at what price per lot?

Solution

Let x represent the number of lots sold. Then $x + 18$ represents the total number of lots. Therefore, $\dfrac{120,000}{x}$ represents the selling price per lot, and $\dfrac{120,000}{x + 18}$ represents the cost per lot. The following equation sums up the situation.

Selling price per lot	Equals	Cost per lot	Plus	$6000
$\dfrac{120,000}{x}$	$=$	$\dfrac{120,000}{x + 18}$	$+$	6000

Solving this equation, we obtain

$$x(x + 18)\left(\frac{120,000}{x}\right) = \left(\frac{120,000}{x + 18} + 6000\right)(x)(x + 18)$$

$$120,000(x + 18) = 120,000x + 6000x(x + 18)$$

$$120,000x + 2,160,000 = 120,000x + 6000x^2 + 108,000x$$

$$0 = 6000x^2 + 108,000x - 2,160,000$$

$$0 = x^2 + 18x - 360$$

The method of completing the square works very well with this equation.

$$x^2 + 18x = 360$$

$$x^2 + 18x + 81 = 441$$

$$(x + 9)^2 = 441$$

$$x + 9 = \pm\sqrt{441}$$

$$x + 9 = \pm 21$$

$$x + 9 = 21 \quad \text{or} \quad x + 9 = -21$$

$$x = 12 \quad \text{or} \quad x = -30$$

We discard the negative solution; thus 12 lots were sold at $\dfrac{120{,}000}{x} = \dfrac{120{,}000}{12} =$ $10,000 per lot.

◼

PROBLEM 5

Barry bought a number of shares of stock for $600. A week later the value of the stock had increased $3 per share, and he sold all but 10 shares and regained his original investment of $600. How many shares did he sell and at what price per share?

Solution

Let s represent the number of shares Barry sold. Then $s + 10$ represents the number of shares purchased. Therefore, $\dfrac{600}{s}$ represents the selling price per share, and $\dfrac{600}{s + 10}$ represents the cost per share.

Selling price per share ↓   Cost per share ↓

$$\frac{600}{s} \quad = \quad \frac{600}{s + 10} + 3$$

Solving this equation yields

$$s(s + 10)\left(\frac{600}{s}\right) = \left(\frac{600}{s + 10} + 3\right)(s)(s + 10)$$

$$600(s + 10) = 600s + 3s(s + 10)$$

$$600s + 6000 = 600s + 3s^2 + 30s$$

$$0 = 3s^2 + 30s - 6000$$

$$0 = s^2 + 10s - 2000$$

Use the quadratic formula to obtain

$$s = \frac{-10 \pm \sqrt{10^2 - 4(1)(-2000)}}{2(1)}$$

$$s = \frac{-10 \pm \sqrt{100 + 8000}}{2}$$

$$s = \frac{-10 \pm \sqrt{8100}}{2}$$

$$s = \frac{-10 \pm 90}{2}$$

$$s = \frac{-10 + 90}{2} \quad \text{or} \quad s = \frac{-10 - 90}{2}$$

$$s = 40 \quad \text{or} \quad s = -50$$

We discard the negative solution, and we know that 40 shares were sold at $\frac{600}{s} = \frac{600}{40} = \$15$ per share. ∎

This next problem set contains a large variety of word problems. Not only are there some business applications similar to those we discussed in this section, but there are also more problems of the types we discussed in Chapters 3 and 4. Try to give them your best shot without referring to the examples in earlier chapters.

PROBLEM SET 6.5

For Problems 1–20, solve each quadratic equation using the method that seems most appropriate to you.

1. $x^2 - 4x - 6 = 0$

2. $x^2 - 8x - 4 = 0$

3. $3x^2 + 23x - 36 = 0$

4. $n^2 + 22n + 105 = 0$

5. $x^2 - 18x = 9$

6. $x^2 + 20x = 25$

7. $2x^2 - 3x + 4 = 0$

8. $3y^2 - 2y + 1 = 0$

9. $135 + 24n + n^2 = 0$

10. $28 - x - 2x^2 = 0$

11. $(x - 2)(x + 9) = -10$

12. $(x + 3)(2x + 1) = -3$

13. $2x^2 - 4x + 7 = 0$

14. $3x^2 - 2x + 8 = 0$

15. $x^2 - 18x + 15 = 0$

16. $x^2 - 16x + 14 = 0$

17. $20y^2 + 17y - 10 = 0$

18. $12x^2 + 23x - 9 = 0$

19. $4t^2 + 4t - 1 = 0$

20. $5t^2 + 5t - 1 = 0$

For Problems 21–40, solve each equation.

21. $n + \frac{3}{n} = \frac{19}{4}$

22. $n - \frac{2}{n} = -\frac{7}{3}$

23. $\frac{3}{x} + \frac{7}{x - 1} = 1$

24. $\frac{2}{x} + \frac{5}{x + 2} = 1$

25. $\frac{12}{x - 3} + \frac{8}{x} = 14$

26. $\frac{16}{x + 5} - \frac{12}{x} = -2$

27. $\frac{3}{x - 1} - \frac{2}{x} = \frac{5}{2}$

28. $\frac{4}{x + 1} + \frac{2}{x} = \frac{5}{3}$

29. $\frac{6}{x} + \frac{40}{x + 5} = 7$

30. $\frac{12}{t} + \frac{18}{t + 8} = \frac{9}{2}$

31. $\frac{5}{n - 3} - \frac{3}{n + 3} = 1$

32. $\frac{3}{t + 2} + \frac{4}{t - 2} = 2$

33. $x^4 - 18x^2 + 72 = 0$

34. $x^4 - 21x^2 + 54 = 0$

35. $3x^4 - 35x^2 + 72 = 0$

36. $5x^4 - 32x^2 + 48 = 0$

37. $3x^4 + 17x^2 + 20 = 0$

38. $4x^4 + 11x^2 - 45 = 0$

39. $6x^4 - 29x^2 + 28 = 0$

40. $6x^4 - 31x^2 + 18 = 0$

For Problems 41–70, set up an equation and solve each problem.

41. Find two consecutive whole numbers such that the sum of their squares is 145.

42. Find two consecutive odd whole numbers such that the sum of their squares is 74.

43. Two positive integers differ by 3, and their product is 108. Find the numbers.

44. Suppose that the sum of two numbers is 20 and the sum of their squares is 232. Find the numbers.

45. Find two numbers such that their sum is 10 and their product is 22.

46. Find two numbers such that their sum is 6 and their product is 7.

47. Suppose that the sum of two whole numbers is 9 and the sum of their reciprocals is $\frac{1}{2}$. Find the numbers.

48. The difference between two whole numbers is 8 and the difference between their reciprocals is $\frac{1}{6}$. Find the two numbers.

49. The sum of the lengths of the two legs of a right triangle is 21 inches. If the length of the hypotenuse is 15 inches, find the length of each leg.

50. The length of a rectangular floor is 1 meter less than twice its width. If a diagonal of the rectangle is 17 meters, find the length and width of the floor.

51. A rectangular plot of ground measuring 12 meters by 20 meters is surrounded by a sidewalk of a uniform width (see Figure 6.9). The area of the sidewalk is 68 square meters. Find the width of the walk.

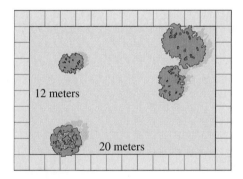

Figure 6.9

52. A 5-inch by 7-inch picture is surrounded by a frame of uniform width. The area of the picture and frame together is 80 square inches. Find the width of the frame.

53. The perimeter of a rectangle is 44 inches and its area is 112 square inches. Find the length and width of the rectangle.

54. A rectangular piece of cardboard is 2 units longer than it is wide. From each of its corners a square piece 2 units on a side is cut out. The flaps are then turned up to form an open box that has a volume of 70 cubic units. Find the length and width of the original piece of cardboard.

55. Charlotte's time to travel 250 miles is 1 hour more than Lorraine's time to travel 180 miles. Charlotte drove 5 miles per hour faster than Lorraine. How fast did each one travel?

56. Larry's time to travel 156 miles is 1 hour more than Terrell's time to travel 108 miles. Terrell drove 2 miles per hour faster than Larry. How fast did each one travel?

57. On a 570-mile trip, Andy averaged 5 miles per hour faster for the last 240 miles than he did for the first 330 miles. The entire trip took 10 hours. How fast did he travel for the first 330 miles?

58. On a 135-mile bicycle excursion, Maria averaged 5 miles per hour faster for the first 60 miles than she did for the last 75 miles. The entire trip took 8 hours. Find her rate for the first 60 miles.

59. It takes Terry 2 hours longer to do a certain job than it takes Tom. They worked together for 3 hours; then Tom left and Terry finished the job in 1 hour. How long would it take each of them to do the job alone?

60. Suppose that Arlene can mow the entire lawn in 40 minutes less time with the power mower than she can with the push mower. One day the power mower broke down after she had been mowing for 30 minutes. She finished the lawn with the push mower in 20 minutes. How long does it take Arlene to mow the entire lawn with the power mower?

61. A student did a word processing job for $24. It took him 1 hour longer than he expected, and therefore he earned $4 per hour less than he anticipated. How long did he expect that it would take to do the job?

62. A group of students agreed that each would chip in the same amount to pay for a party that would cost $100. Then they found 5 more students interested in the party and in sharing the expenses. This decreased the amount each had to pay by $1. How many students were involved in the party and how much did each student have to pay?

63. A group of students agreed that each would contribute the same amount to buy their favorite teacher an $80 birthday gift. At the last minute, 2 of the students decided not to chip in. This increased the amount that the remaining students had to pay by $2 per student. How many students actually contributed to the gift?

64. A retailer bought a number of special mugs for $48. She decided to keep two of the mugs for herself but then had to change the price to $3 a mug above the original cost per mug. If she sells the remaining mugs for $70, how many mugs did she buy and at what price per mug did she sell them?

65. Tony bought a number of shares of stock for $720. A month later the value of the stock increased by $8 per share, and he sold all but 20 shares and received $800. How many shares did he sell and at what price per share?

66. The formula $D = \dfrac{n(n - 3)}{2}$ yields the number of diagonals, D, in a polygon of n sides. Find the number of sides of a polygon that has 54 diagonals.

67. The formula $S = \dfrac{n(n + 1)}{2}$ yields the sum, S, of the first n natural numbers $1, 2, 3, 4, \ldots$. How many consecutive natural numbers starting with 1 will give a sum of 1275?

68. At a point 16 yards from the base of a tower, the distance to the top of the tower is 4 yards more than the height of the tower (see Figure 6.10). Find the height of the tower.

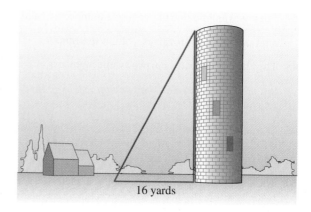

16 yards

Figure 6.10

69. Suppose that $500 is invested at a certain rate of interest compounded annually for 2 years. If the accumulated value at the end of 2 years is $594.05, find the rate of interest.

70. Suppose that $10,000 is invested at a certain rate of interest compounded annually for 2 years. If the accumulated value at the end of 2 years is $12,544, find the rate of interest.

■ ■ ■ **Thoughts into words**

71. How would you solve the equation $x^2 - 4x = 252$? Explain your choice of the method that you would use.

72. Explain how you would solve $(x - 2)(x - 7) = 0$ and also how you would solve $(x - 2)(x - 7) = 4$.

73. One of our problem-solving suggestions is to look for a guideline that can be used to help determine an equation. What does this suggestion mean to you?

74. Can a quadratic equation with integral coefficients have exactly one nonreal complex solution? Explain your answer.

■ ■ ■ **Further investigations**

For Problems 75–81, solve each equation.

75. $x - 9\sqrt{x} + 18 = 0$ [*Hint:* Let $y = \sqrt{x}$.]

76. $x - 4\sqrt{x} + 3 = 0$

77. $x + \sqrt{x} - 2 = 0$

78. $x^{\frac{2}{3}} + x^{\frac{1}{3}} - 6 = 0$ [*Hint:* Let $y = x^{\frac{1}{3}}$.]

79. $6x^{\frac{2}{3}} - 5x^{\frac{1}{3}} - 6 = 0$

80. $x^{-2} + 4x^{-1} - 12 = 0$

81. $12x^{-2} - 17x^{-1} - 5 = 0$

The following equations are also quadratic in form. To solve, begin by raising each side of the equation to the appropriate power so that the exponent will become an integer. Then, to solve the resulting quadratic equation, you may use the square-root property, factoring, or the quadratic formula, as is most appropriate. Be aware that raising each side of the equation to a power may introduce extraneous roots; therefore, be sure to check your solutions. Study the following example before you begin the problems.

Solve

$$(x + 3)^{\frac{2}{3}} = 1$$

$$\left[(x + 3)^{\frac{2}{3}}\right]^{3} = 1^{3}$$ Raise both sides to the third power.

$$(x + 3)^{2} = 1$$

$$x^{2} + 6x + 9 = 1$$

$$x^{2} + 6x + 8 = 0$$

$$(x + 4)(x + 2) = 0$$

$$x + 4 = 0 \quad \text{or} \quad x + 2 = 0$$

$$x = -4 \quad \text{or} \quad x = -2$$

Both solutions do check. The solution set is $\{-4, -2\}$.

For problems 82–90, solve each equation.

82. $(5x + 6)^{\frac{1}{2}} = x$

83. $(3x + 4)^{\frac{1}{2}} = x$

84. $x^{\frac{2}{3}} = 2$

85. $x^{\frac{2}{5}} = 2$

86. $(2x + 6)^{\frac{1}{2}} = x$

87. $(2x - 4)^{\frac{2}{3}} = 1$

88. $(4x + 5)^{\frac{2}{3}} = 2$

89. $(6x + 7)^{\frac{1}{2}} = x + 2$

90. $(5x + 21)^{\frac{1}{2}} = x + 3$

6.6   Quadratic and Other Nonlinear Inequalities

We refer to the equation $ax^2 + bx + c = 0$ as the standard form of a quadratic equation in one variable. Similarly, the following forms express **quadratic inequalities** in one variable.

$$ax^2 + bx + c > 0 \qquad\qquad ax^2 + bx + c < 0$$

$$ax^2 + bx + c \geq 0 \qquad\qquad ax^2 + bx + c \leq 0$$

We can use the number line very effectively to help solve quadratic inequalities where the quadratic polynomial is factorable. Let's consider some examples to illustrate the procedure.

E X A M P L E   1

Solve and graph the solutions for $x^2 + 2x - 8 > 0$.

Solution

First, let's factor the polynomial.

$$x^2 + 2x - 8 > 0$$

$$(x + 4)(x - 2) > 0$$

On a number line (Figure 6.11), we indicate that at $x = 2$ and $x = -4$, the product $(x + 4)(x - 2)$ equals zero. The numbers -4 and 2 divide the number line into three intervals: (1) the numbers less than -4, (2) the numbers between -4 and 2, and (3) the numbers greater than 2. We can choose a **test number** from each of these intervals and see how it affects the signs of the factors $x + 4$ and $x - 2$ and, consequently, the

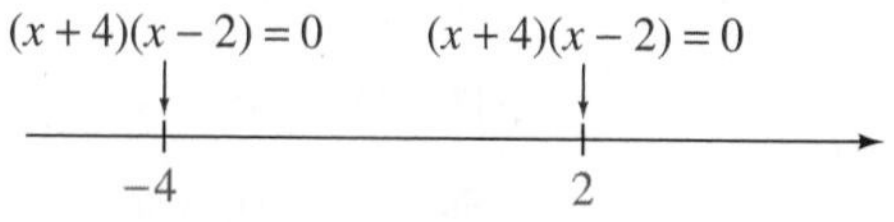

$(x + 4)(x - 2) = 0$ $(x + 4)(x - 2) = 0$

$$\begin{array}{ccc} & \downarrow & & \downarrow \\ \hline & -4 & & 2 \end{array}$$

Figure 6.11

sign of the product of these factors. For example, if $x < -4$ (try $x = -5$), then $x + 4$ is negative and $x - 2$ is negative, so their product is positive. If $-4 < x < 2$ (try $x = 0$), then $x + 4$ is positive and $x - 2$ is negative, so their product is negative. If $x > 2$ (try $x = 3$), then $x + 4$ is positive and $x - 2$ is positive, so their product is positive. This information can be conveniently arranged using a number line, as shown in Figure 6.12. Note the open circles at -4 and 2 to indicate that they are not included in the solution set.

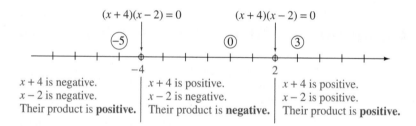

$(x + 4)(x - 2) = 0$ $(x + 4)(x - 2) = 0$

$x + 4$ is negative.	$x + 4$ is positive.	$x + 4$ is positive.
$x - 2$ is negative.	$x - 2$ is negative.	$x - 2$ is positive.
Their product is **positive**.	Their product is **negative**.	Their product is **positive**.

Figure 6.12

Thus the given inequality, $x^2 + 2x - 8 > 0$, is satisfied by numbers less than -4 along with numbers greater than 2. Using interval notation, the solution set is $(-\infty, -4) \cup (2, \infty)$. These solutions can be shown on a number line (Figure 6.13).

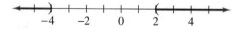

Figure 6.13

We refer to numbers such as -4 and 2 in the preceding example (where the given polynomial or algebraic expression equals zero or is undefined) as **critical numbers.** Let's consider some additional examples that make use of critical numbers and test numbers.

EXAMPLE 2

Solve and graph the solutions for $x^2 + 2x - 3 \leq 0$.

Solution

First, factor the polynomial.

$$x^2 + 2x - 3 \leq 0$$

$$(x + 3)(x - 1) \leq 0$$

Second, locate the values for which $(x + 3)(x - 1)$ equals zero. We put dots at -3 and 1 to remind ourselves that these two numbers are to be included in the solution set because the given statement includes equality. Now let's choose a test number from each of the three intervals, and record the sign behavior of the factors $(x + 3)$ and $(x - 1)$ (Figure 6.14).

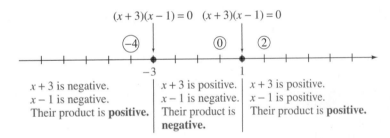

Figure 6.14

Therefore, the solution set is $[-3, 1]$, and it can be graphed as in Figure 6.15.

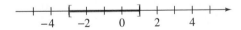

Figure 6.15

Examples 1 and 2 have indicated a systematic approach for solving quadratic inequalities where the polynomial is factorable. This same type of number line analysis can also be used to solve indicated quotients such as $\dfrac{x + 1}{x - 5} > 0$.

EXAMPLE 3

Solve and graph the solutions for $\dfrac{x+1}{x-5} > 0$.

Solution

First, indicate that at $x = -1$ the given quotient equals zero, and at $x = 5$ the quotient is undefined. Second, choose test numbers from each of the three intervals, and record the sign behavior of $(x + 1)$ and $(x - 5)$ as in Figure 6.16.

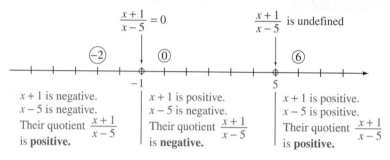

Figure 6.16

Therefore, the solution set is $(-\infty, -1) \cup (5, \infty)$, and its graph is shown in Figure 6.17.

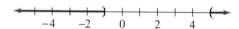

Figure 6.17

EXAMPLE 4

Solve $\dfrac{x+2}{x+4} \leq 0$.

Solution

The indicated quotient equals zero at $x = -2$ and is undefined at $x = -4$. (Note that -2 is to be included in the solution set, but -4 is not to be included.) Now let's choose some test numbers and record the sign behavior of $(x + 2)$ and $(x + 4)$ as in Figure 6.18.

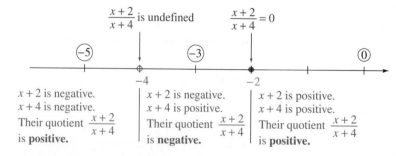

Figure 6.18

Therefore, the solution set is $(-4, -2]$.

The final example illustrates that sometimes we need to change the form of the given inequality before we use the number line analysis.

EXAMPLE 5   Solve $\dfrac{x}{x + 2} \geq 3$.

Solution

First, let's change the form of the given inequality as follows:

$$\frac{x}{x + 2} \geq 3$$

$$\frac{x}{x + 2} - 3 \geq 0 \qquad \text{Add } -3 \text{ to both sides.}$$

$$\frac{x - 3(x + 2)}{x + 2} \geq 0 \qquad \text{Express the left side over a common denominator.}$$

$$\frac{x - 3x - 6}{x + 2} \geq 0$$

$$\frac{-2x - 6}{x + 2} \geq 0$$

Now we can proceed as we did with the previous examples. If $x = -3$, then $\dfrac{-2x - 6}{x + 2}$ equals zero; and if $x = -2$, then $\dfrac{-2x - 6}{x + 2}$ is undefined. Then, choosing test numbers, we can record the sign behavior of $(-2x - 6)$ and $(x + 2)$ as in Figure 6.19.

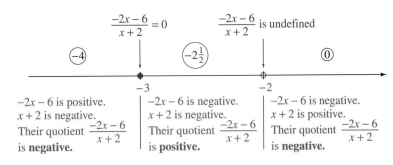

Figure 6.19

Therefore, the solution set is $[-3, -2)$. Perhaps you should check a few numbers from this solution set back into the original inequality!

PROBLEM SET 6.6

For Problems 1–20, solve each inequality and graph its solution set on a number line.

1. $(x + 2)(x - 1) > 0$

2. $(x - 2)(x + 3) > 0$

3. $(x + 1)(x + 4) < 0$

4. $(x - 3)(x - 1) < 0$

5. $(2x - 1)(3x + 7) \geq 0$

6. $(3x + 2)(2x - 3) \geq 0$

7. $(x + 2)(4x - 3) \leq 0$

8. $(x - 1)(2x - 7) \leq 0$

9. $(x + 1)(x - 1)(x - 3) > 0$

10. $(x + 2)(x + 1)(x - 2) > 0$

11. $x(x + 2)(x - 4) \leq 0$

12. $x(x + 3)(x - 3) \leq 0$

13. $\dfrac{x + 1}{x - 2} > 0$

14. $\dfrac{x - 1}{x + 2} > 0$

15. $\dfrac{x - 3}{x + 2} < 0$

16. $\dfrac{x + 2}{x - 4} < 0$

17. $\dfrac{2x - 1}{x} \geq 0$

18. $\dfrac{x}{3x + 7} \geq 0$

19. $\dfrac{-x + 2}{x - 1} \leq 0$

20. $\dfrac{3 - x}{x + 4} \leq 0$

For Problems 21–46, solve each inequality.

21. $x^2 + 2x - 35 < 0$

22. $x^2 + 3x - 54 < 0$

23. $x^2 - 11x + 28 > 0$

24. $x^2 + 11x + 18 > 0$

25. $3x^2 + 13x - 10 \leq 0$

26. $4x^2 - x - 14 \leq 0$

27. $8x^2 + 22x + 5 \geq 0$

28. $12x^2 - 20x + 3 \geq 0$

29. $x(5x - 36) > 32$

30. $x(7x + 40) < 12$

31. $x^2 - 14x + 49 \geq 0$

32. $(x + 9)^2 \geq 0$

33. $4x^2 + 20x + 25 \leq 0$

34. $9x^2 - 6x + 1 \leq 0$

35. $(x + 1)(x - 3)^2 > 0$

36. $(x - 4)^2(x - 1) \leq 0$

37. $\dfrac{2x}{x + 3} > 4$

38. $\dfrac{x}{x - 1} > 2$

39. $\dfrac{x - 1}{x - 5} \leq 2$

40. $\dfrac{x + 2}{x + 4} \leq 3$

41. $\dfrac{x + 2}{x - 3} > -2$

42. $\dfrac{x - 1}{x - 2} < -1$

43. $\dfrac{3x + 2}{x + 4} \leq 2$

44. $\dfrac{2x - 1}{x + 2} \geq -1$

45. $\dfrac{x + 1}{x - 2} < 1$

46. $\dfrac{x + 3}{x - 4} \geq 1$

■ ■ ■ Thoughts into words

47. Explain how to solve the inequality $(x + 1)(x - 2)(x - 3) > 0$.

48. Explain how to solve the inequality $(x - 2)^2 > 0$ by inspection.

49. Your friend looks at the inequality $1 + \dfrac{1}{x} > 2$ and without any computation states that the solution set is all real numbers between 0 and 1. How can she do that?

50. Why is the solution set for $(x - 2)^2 \geq 0$ the set of all real numbers?

51. Why is the solution set for $(x - 2)^2 \leq 0$ the set $\{2\}$?

■ ■ ■ Further investigations

52. The product $(x - 2)(x + 3)$ is positive if both factors are negative *or* if both factors are positive. Therefore, we can solve $(x - 2)(x + 3) > 0$ as follows:

$(x - 2 < 0 \text{ and } x + 3 < 0) \text{ or } (x - 2 > 0 \text{ and } x + 3 > 0)$
$(x < 2 \text{ and } x < -3) \text{ or } (x > 2 \text{ and } x > -3)$
$x < -3 \text{ or } x > 2$

The solution set is $(-\infty, -3) \cup (2, \infty)$. Use this type of analysis to solve each of the following.

(a) $(x - 2)(x + 7) > 0$

(b) $(x - 3)(x + 9) \geq 0$

(c) $(x + 1)(x - 6) \leq 0$

(d) $(x + 4)(x - 8) < 0$

(e) $\dfrac{x + 4}{x - 7} > 0$

(f) $\dfrac{x - 5}{x + 8} \leq 0$

CHAPTER 6

SUMMARY

(6.1) A number of the form $a + bi$, where a and b are real numbers and i is the imaginary unit defined by $i = \sqrt{-1}$, is a **complex number.**

Two complex numbers $a + bi$ and $c + di$ are said to be equal if and only if $a = c$ and $b = d$.

We describe addition and subtraction of complex numbers as follows:

$$(a + bi) + (c + di) = (a + c) + (b + d)i$$

$$(a + bi) - (c + di) = (a - c) + (b - d)i$$

We can represent a square root of any negative real number as the product of a real number and the imaginary unit i. That is,

$$\sqrt{-b} = i\sqrt{b}, \quad \text{where } b \text{ is a positive real number}$$

The product of two complex numbers conforms with the product of two binomials. The **conjugate** of $a + bi$ is $a - bi$. The product of a complex number and its conjugate is a real number. Therefore, conjugates are used to simplify expressions such as $\dfrac{4 + 3i}{5 - 2i}$, which indicate the quotient of two complex numbers.

(6.2) The **standard form for a quadratic equation** in one variable is

$$ax^2 + bx + c = 0$$

where a, b, and c are real numbers and $a \neq 0$.

Some quadratic equations can be solved by factoring and applying the property "$ab = 0$ if and only if $a = 0$ or $b = 0$."

Don't forget that applying the property "if $a = b$, then $a^n = b^n$" might produce extraneous solutions. Therefore, we *must* check all potential solutions.

We can solve some quadratic equations by applying the property "$x^2 = a$ if and only if $x = \pm\sqrt{a}$."

(6.3) To solve a quadratic equation of the form $x^2 + bx = k$ by **completing the square,** we (1) add $\left(\dfrac{b}{2}\right)^2$ to both sides, (2) factor the left side, and (3) apply the property "$x^2 = a$ if and only if $x = \pm\sqrt{a}$."

(6.4) We can solve any quadratic equation of the form $ax^2 + bx + c = 0$ by the **quadratic formula,** which we usually state as

$$x = \frac{-b \pm \sqrt{b^2 - 4ac}}{2a}$$

The **discriminant,** $b^2 - 4ac$, can be used to determine the nature of the roots of a quadratic equation as follows:

1. If $b^2 - 4ac < 0$, then the equation has two nonreal complex solutions.
2. If $b^2 - 4ac = 0$, then the equation has two equal real solutions.
3. If $b^2 - 4ac > 0$, then the equation has two unequal real solutions.

If x_1 and x_2 are roots of a quadratic equation, then the following relationships exist.

$$x_1 + x_2 = -\frac{b}{a} \quad \text{and} \quad (x_1)(x_2) = \frac{c}{a}$$

These **sum-of-roots and product-of-roots relationships** can be used to check potential solutions of quadratic equations.

(6.5) To review the strengths and weaknesses of the three basic methods for solving a quadratic equation (factoring, completing the square, and the quadratic formula), go back over the examples in this section.

Keep the following suggestions in mind as you solve word problems.

1. Read the problem carefully.
2. Sketch any figure, diagram, or chart that might help you organize and analyze the problem.
3. Choose a meaningful variable.

4. Look for a guideline that can be used to set up an equation.

5. Form an equation that translates the guideline from English to algebra.

6. Solve the equation and use the solutions to determine all facts requested in the problem.

7. Check all answers back into the original statement of the problem.

(6.6) The number line, along with **critical numbers** and **test numbers,** provides a good basis for solving **quadratic inequalities** where the polynomial is factorable. We can use this same basic approach to solve inequalities, such as $\dfrac{3x + 1}{x - 4} > 0$, that indicate quotients.

CHAPTER 6 REVIEW PROBLEM SET

For Problems 1–8, perform the indicated operations and express the answers in the standard form of a complex number.

1. $(-7 + 3i) + (9 - 5i)$

2. $(4 - 10i) - (7 - 9i)$

3. $5i(3 - 6i)$

4. $(5 - 7i)(6 + 8i)$

5. $(-2 - 3i)(4 - 8i)$

6. $(4 - 3i)(4 + 3i)$

7. $\dfrac{4 + 3i}{6 - 2i}$

8. $\dfrac{-1 - i}{-2 + 5i}$

For Problems 9–12, find the discriminant of each equation and determine whether the equation has (1) two nonreal complex solutions, (2) one real solution with a multiplicity of two, or (3) two real solutions. Do not solve the equations.

9. $4x^2 - 20x + 25 = 0$

10. $5x^2 - 7x + 31 = 0$

11. $7x^2 - 2x - 14 = 0$

12. $5x^2 - 2x = 4$

For Problems 13–31, solve each equation.

13. $x^2 - 17x = 0$

14. $(x - 2)^2 = 36$

15. $(2x - 1)^2 = -64$

16. $x^2 - 4x - 21 = 0$

17. $x^2 + 2x - 9 = 0$

18. $x^2 - 6x = -34$

19. $4\sqrt{x} = x - 5$

20. $3n^2 + 10n - 8 = 0$

21. $n^2 - 10n = 200$

22. $3a^2 + a - 5 = 0$

23. $x^2 - x + 3 = 0$

24. $2x^2 - 5x + 6 = 0$

25. $2a^2 + 4a - 5 = 0$

26. $t(t + 5) = 36$

27. $x^2 + 4x + 9 = 0$

28. $(x - 4)(x - 2) = 80$

29. $\dfrac{3}{x} + \dfrac{2}{x + 3} = 1$

30. $2x^4 - 23x^2 + 56 = 0$

31. $\dfrac{3}{n - 2} = \dfrac{n + 5}{4}$

For Problems 32–35, solve each inequality and indicate the solution set on a number line graph.

32. $x^2 + 3x - 10 > 0$

33. $2x^2 + x - 21 \le 0$

34. $\dfrac{x - 4}{x + 6} \ge 0$

35. $\dfrac{2x - 1}{x + 1} > 4$

For Problems 36–43, set up an equation and solve each problem.

36. Find two numbers whose sum is 6 and whose product is 2.

37. Sherry bought a number of shares of stock for $250. Six months later the value of the stock had increased by $5 per share, and she sold all but 5 shares and regained her original investment plus a profit of $50. How many shares did she sell and at what price per share?

38. Andre traveled 270 miles in 1 hour more time than it took Sandy to travel 260 miles. Sandy drove 7 miles per hour faster than Andre. How fast did each one travel?

39. The area of a square is numerically equal to twice its perimeter. Find the length of a side of the square.

40. Find two consecutive even whole numbers such that the sum of their squares is 164.

41. The perimeter of a rectangle is 38 inches and its area is 84 square inches. Find the length and width of the rectangle.

42. It takes Billy 2 hours longer to do a certain job than it takes Reena. They worked together for 2 hours; then Reena left and Billy finished the job in 1 hour. How long would it take each of them to do the job alone?

43. A company has a rectangular parking lot 40 meters wide and 60 meters long. The company plans to increase the area of the lot by 1100 square meters by adding a strip of equal width to one side and one end. Find the width of the strip to be added.

TEST

1. Find the product $(3 - 4i)(5 + 6i)$ and express the result in the standard form of a complex number.

2. Find the quotient $\dfrac{2 - 3i}{3 + 4i}$ and express the result in the standard form of a complex number.

For Problems 3–15, solve each equation.

3. $x^2 = 7x$

4. $(x - 3)^2 = 16$

5. $x^2 + 3x - 18 = 0$

6. $x^2 - 2x - 1 = 0$

7. $5x^2 - 2x + 1 = 0$

8. $x^2 + 30x = -224$

9. $(3x - 1)^2 + 36 = 0$

10. $(5x - 6)(4x + 7) = 0$

11. $(2x + 1)(3x - 2) = 55$

12. $n(3n - 2) = 40$

13. $x^4 + 12x^2 - 64 = 0$

14. $\dfrac{3}{x} + \dfrac{2}{x + 1} = 4$

15. $3x^2 - 2x - 3 = 0$

16. Does the equation $4x^2 + 20x + 25 = 0$ have (a) two nonreal complex solutions, (b) two equal real solutions, or (c) two unequal real solutions?

17. Does the equation $4x^2 - 3x = -5$ have (a) two nonreal complex solutions, (b) two equal real solutions, or (c) two unequal real solutions?

For Problems 18–20, solve each inequality and express the solution set using interval notation.

18. $x^2 - 3x - 54 \leq 0$

19. $\dfrac{3x - 1}{x + 2} > 0$

20. $\dfrac{x - 2}{x + 6} \geq 3$

For Problems 21–25, set up an equation and solve each problem.

21. A 24-foot ladder leans against a building and makes an angle of $60°$ with the ground. How far up on the building does the top of the ladder reach? Express your answer to the nearest tenth of a foot.

22. A rectangular plot of ground measures 16 meters by 24 meters. Find, to the nearest meter, the distance from one corner of the plot to the diagonally opposite corner.

23. Dana bought a number of shares of stock for a total of $3000. Three months later the stock had increased in value by $5 per share, and she sold all but 50 shares and regained her original investment of $3000. How many shares did she sell?

24. The perimeter of a rectangle is 41 inches and its area is 91 square inches. Find the length of its shortest side.

25. The sum of two numbers is 6 and their product is 4. Find the larger of the two numbers.

CUMULATIVE REVIEW PROBLEM SET *Chapters 1-6*

For Problems 1–5, evaluate each algebraic expression for the given values of the variables.

1. $\dfrac{4a^2b^3}{12a^3b}$ for $a = 5$ and $b = -8$

2. $\dfrac{\frac{1}{x} + \frac{1}{y}}{\frac{1}{x} - \frac{1}{y}}$ for $x = 4$ and $y = 7$

3. $\dfrac{3}{n} + \dfrac{5}{2n} - \dfrac{4}{3n}$ for $n = 25$

4. $\dfrac{4}{x - 1} - \dfrac{2}{x + 2}$ for $x = \dfrac{1}{2}$

5. $2\sqrt{2x + y} - 5\sqrt{3x - y}$ for $x = 5$ and $y = 6$

For Problems 6–17, perform the indicated operations and express the answers in simplified form.

6. $(3a^2b)(-2ab)(4ab^3)$

7. $(x + 3)(2x^2 - x - 4)$

8. $\dfrac{6xy^2}{14y} \cdot \dfrac{7x^2y}{8x}$

9. $\dfrac{a^2 + 6a - 40}{a^2 - 4a} \div \dfrac{2a^2 + 19a - 10}{a^3 + a^2}$

10. $\dfrac{3x + 4}{6} - \dfrac{5x - 1}{9}$

11. $\dfrac{4}{x^2 + 3x} + \dfrac{5}{x}$

12. $\dfrac{3n^2 + n}{n^2 + 10n + 16} \cdot \dfrac{2n^2 - 8}{3n^3 - 5n^2 - 2n}$

13. $\dfrac{3}{5x^2 + 3x - 2} - \dfrac{2}{5x^2 - 22x + 8}$

14. $\dfrac{y^3 - 7y^2 + 16y - 12}{y - 2}$

15. $(4x^3 - 17x^2 + 7x + 10) \div (4x - 5)$

16. $(3\sqrt{2} + 2\sqrt{5})(5\sqrt{2} - \sqrt{5})$

17. $(\sqrt{x} - 3\sqrt{y})(2\sqrt{x} + 4\sqrt{y})$

For Problems 18–25, evaluate each of the numerical expressions.

18. $-\sqrt{\dfrac{9}{64}}$

19. $\sqrt[3]{-\dfrac{8}{27}}$

20. $\sqrt[3]{0.008}$

21. $32^{-\frac{1}{5}}$

22. $3^0 + 3^{-1} + 3^{-2}$

23. $-9^{\frac{3}{2}}$

24. $\left(\dfrac{3}{4}\right)^{-2}$

25. $\dfrac{1}{\left(\dfrac{2}{3}\right)^{-3}}$

For Problems 26–31, factor each of the algebraic expressions completely.

26. $3x^4 + 81x$

27. $6x^2 + 19x - 20$

28. $12 + 13x - 14x^2$

29. $9x^4 + 68x^2 - 32$

30. $2ax - ay - 2bx + by$

31. $27x^3 - 8y^3$

For Problems 32–55, solve each of the equations.

32. $3(x - 2) - 2(3x + 5) = 4(x - 1)$

33. $0.06n + 0.08(n + 50) = 25$

34. $4\sqrt{x} + 5 = x$

35. $\sqrt[3]{n^2 - 1} = -1$

36. $6x^2 - 24 = 0$

37. $a^2 + 14a + 49 = 0$

38. $3n^2 + 14n - 24 = 0$

39. $\dfrac{2}{5x - 2} = \dfrac{4}{6x + 1}$

40. $\sqrt{2x - 1} - \sqrt{x + 2} = 0$

41. $5x - 4 = \sqrt{5x - 4}$

42. $|3x - 1| = 11$

43. $(3x - 2)(4x - 1) = 0$

44. $(2x + 1)(x - 2) = 7$

45. $\dfrac{5}{6x} - \dfrac{2}{3} = \dfrac{7}{10x}$

46. $\dfrac{3}{y + 4} + \dfrac{2y - 1}{y^2 - 16} = \dfrac{-2}{y - 4}$

47. $6x^4 - 23x^2 - 4 = 0$

48. $3n^3 + 3n = 0$

49. $n^2 - 13n - 114 = 0$

50. $12x^2 + x - 6 = 0$

51. $x^2 - 2x + 26 = 0$

52. $(x + 2)(x - 6) = -15$

53. $(3x - 1)(x + 4) = 0$

54. $x^2 + 4x + 20 = 0$

55. $2x^2 - x - 4 = 0$

For Problems 56 – 65, solve each inequality and express the solution set using interval notation.

56. $6 - 2x \geq 10$

57. $4(2x - 1) < 3(x + 5)$

58. $\dfrac{n + 1}{4} + \dfrac{n - 2}{12} > \dfrac{1}{6}$

59. $|2x - 1| < 5$

60. $|3x + 2| > 11$

61. $\dfrac{1}{2}(3x - 1) - \dfrac{2}{3}(x + 4) \leq \dfrac{3}{4}(x - 1)$

62. $x^2 - 2x - 8 \leq 0$

63. $3x^2 + 14x - 5 > 0$

64. $\dfrac{x + 2}{x - 7} \geq 0$

65. $\dfrac{2x - 1}{x + 3} < 1$

For Problems 66 – 74, solve each problem by setting up and solving an appropriate equation.

66. How many liters of a 60%-acid solution must be added to 14 liters of a 10%-acid solution to produce a 25%-acid solution?

67. A sum of $2250 is to be divided between two people in the ratio of 2 to 3. How much does each person receive?

68. The length of a picture without its border is 7 inches less than twice its width. If the border is 1 inch wide and its area is 62 square inches, what are the dimensions of the picture alone?

69. Working together, Lolita and Doug can paint a shed in 3 hours and 20 minutes. If Doug can paint the shed by himself in 10 hours, how long would it take Lolita to paint the shed by herself?

70. Angie bought some golf balls for $14. If each ball had cost $.25 less, she could have purchased one more ball for the same amount of money. How many golf balls did Angie buy?

71. A jogger who can run an 8-minute mile starts half a mile ahead of a jogger who can run a 6-minute mile. How long will it take the faster jogger to catch the slower jogger?

72. Suppose that $100 is invested at a certain rate of interest compounded annually for 2 years. If the accumulated value at the end of 2 years is $114.49, find the rate of interest.

73. A room contains 120 chairs arranged in rows. The number of chairs per row is one less than twice the number of rows. Find the number of chairs per row.

74. Bjorn bought a number of shares of stock for $2800. A month later the value of the stock had increased $6 per share, and he sold all but 60 shares and regained his original investment of $2800. How many shares did he sell?

Rene Descartes, a philosopher and mathematician, developed a system for locating a point on a plane. This system is our current rectangular coordinate grid used for graphing and is named the Cartesian coordinate system.

© Leonard de Selva/CORBIS

Linear Equations and Inequalities in Two Variables

René Descartes, a French mathematician of the 17th century, was able to transform geometric problems into an algebraic setting so that he could use the tools of algebra to solve the problems. This connecting of algebraic and geometric ideas is the foundation of a branch of mathematics called **analytic geometry,** today more commonly called **coordinate geometry.** Basically, there are two kinds of problems in coordinate geometry: Given an algebraic equation, find its geometric graph; and given a set of conditions pertaining to a geometric graph, find its algebraic equation. We discuss problems of both types in this chapter.

InfoTrac Project

Do a keyword search on circumference. Look for articles discussing how the circumference of a patient's leg, waist, or neck is being used in studying health problems. Write a short paragraph telling what you found. Using 12 other people, choose one of those body parts, (neck, leg, or waist) and measure the height of the person and the circumference of the body part you have chosen. Make a table of values, using height as your independent variable. Make a conjecture about whether your graph will be increasing or decreasing. Graph your values. What conclusions might you draw from your graph?

Now, using weight and the circumference measurement, make a table of values and create a graph. What are your conclusions this time? Choose two data points from your data and develop the equation of the line passing through these two points. Connect the points on the graph. Does the line seem to fit the data? Choosing the weight measurement from another data point, determine what the circumference would have been if your equation had fit the data. What is the difference between the actual measurement and the predicted value for your equation?

7.1   Rectangular Coordinate System

Consider two number lines, one vertical and one horizontal, perpendicular to each other at the point we associate with zero on both lines (Figure 7.1). We refer to these number lines as the **horizontal and vertical axes** or, together, as the **coordinate axes.** They partition the plane into four regions called **quadrants.** The quadrants are numbered counterclockwise from I through IV as indicated in Figure 7.1. The point of intersection of the two axes is called the **origin.**

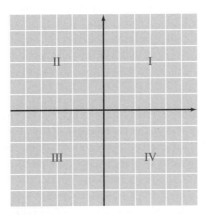

Figure 7.1

It is now possible to set up a one-to-one correspondence between **ordered pairs** of real numbers and the points in a plane. To each ordered pair of real numbers there corresponds a unique point in the plane, and to each point in the plane there corresponds a unique ordered pair of real numbers. A part of this correspondence is illustrated in Figure 7.2. The ordered pair $(3, 2)$ means that the point A is located three units to the right of and two units up from the origin. (The ordered pair $(0, 0)$ is associated with the origin O.) The ordered pair $(-3, -5)$ means that the point D is located three units to the left and five units down from the origin.

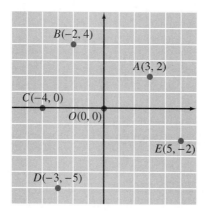

Figure 7.2

REMARK: The notation $(-2, 4)$ was used earlier in this text to indicate an interval of the real number line. Now we are using the same notation to indicate an ordered pair of real numbers. This double meaning should not be confusing because the context of the material will always indicate which meaning of the notation is being used. Throughout this chapter, we will be using the ordered-pair interpretation.

In general we refer to the real numbers a and b in an ordered pair (a, b) associated with a point as the **coordinates of the point.** The first number, a, called the **abscissa,** is the directed distance of the point from the vertical axis measured parallel to the horizontal axis. The second number, b, called the **ordinate,** is the directed distance of the point from the horizontal axis measured parallel to the vertical axis (Figure 7.3a). Thus, in the first quadrant all points have a positive abscissa and a positive ordinate. In the second quadrant all points have a negative abscissa and a positive ordinate. We have indicated the sign situations for all four quadrants in Figure 7.3(b). This system of associating points in a plane with pairs of real numbers is called the **rectangular coordinate system** or the **Cartesian coordinate system.**

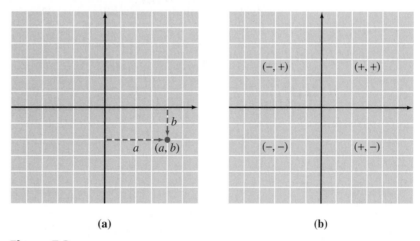

(a)   (b)

Figure 7.3

Historically, the rectangular coordinate system provided the basis for the development of the branch of mathematics called **analytic geometry,** or what we presently refer to as **coordinate geometry.** In this discipline, René Descartes, a French 17th-century mathematician, was able to transform geometric problems into an algebraic setting and then use the tools of algebra to solve the problems.

Basically, there are two kinds of problems to solve in coordinate geometry:

1. Given an algebraic equation, find its geometric graph.
2. Given a set of conditions pertaining to a geometric figure, find its algebraic equation.

In this chapter we will discuss problems of both types.

Let's begin by considering the solutions for the equation $y = x + 2$. A **solution** of an equation in two variables is an ordered pair of real numbers that satisfies the equation. When using the variables x and y, we agree that the first number of an ordered pair is a value of x and the second number is a value of y. We see that $(1, 3)$ is a solution for $y = x + 2$ because if x is replaced by 1 and y by 3, the true numerical statement $3 = 1 + 2$ is obtained. Likewise, $(-2, 0)$ is a solution because $0 = -2 + 2$ is a true statement. We can find infinitely many pairs of real numbers that satisfy $y = x + 2$ by arbitrarily choosing values for x, and for each value of x

we choose, we can determine a corresponding value for y. Let's use a table to record some of the solutions for $y = x + 2$.

Choose x	Determine y from $y = x + 2$	Solutions for $y = x + 2$
0	2	$(0, 2)$
1	3	$(1, 3)$
3	5	$(3, 5)$
5	7	$(5, 7)$
-2	0	$(-2, 0)$
-4	-2	$(-4, -2)$
-6	-4	$(-6, -4)$

We can plot the ordered pairs as points in a coordinate plane and use the horizontal axis as the x axis and the vertical axis as the y axis, as in Figure 7.4(a). The straight line that contains the points in Figure 7.4(b) is called the **graph of the equation** $y = x + 2$.

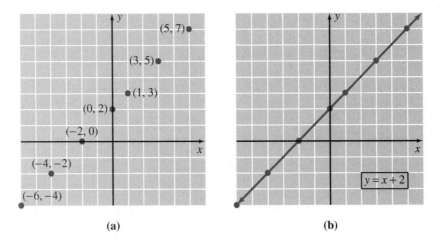

(a)   (b)

Figure 7.4

REMARK: It is important to recognize that all points on the x axis have ordered pairs of the form $(a, 0)$ associated with them. That is, the second number in the ordered pair is 0. Likewise, all points on the y axis have ordered pairs of the form $(0, b)$ associated with them.

Graphing Techniques

For the remainder of this section we will use examples to introduce some basic graphing techniques. Then, toward the end of the section, we will summarize these techniques for you.

E X A M P L E   1

Graph $2x + 3y = 6$.

Solution

First, let's find the points of this graph that fall on the coordinate axes. Let $x = 0$; then

$$2(0) + 3y = 6$$

$$3y = 6$$

$$y = 2$$

Thus $(0, 2)$ is a solution and locates a point of the graph on the y axis. Let $y = 0$; then

$$2x + 3(0) = 6$$

$$2x = 6$$

$$x = 3$$

Thus $(3, 0)$ is a solution and locates a point of the graph on the x axis.

Second, let's change the form of the equation to make it easier to find some additional solutions. We can either solve for x in terms of y or solve for y in terms of x. Let's solve for y in terms of x.

$$2x + 3y = 6$$

$$3y = 6 - 2x$$

$$y = \frac{6 - 2x}{3}$$

Third, a table of values can be formed that includes the two points we found previously.

x	y
0	2
3	0
6	-2
-3	4
-6	6

Plotting these points, we see that they lie in a straight line, and we obtain Figure 7.5.

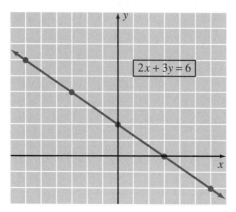

Figure 7.5

REMARK: Look again at the table of values in Example 1. Note that values of x were chosen such that integers were obtained for y. That is not necessary, but it does make things easier from a computational standpoint.

The points (3, 0) and (0, 2) in Figure 7.5 are special points. They are the points of the graph that are on the coordinate axes. That is, they yield the x intercept and the y intercept of the graph. Let's define in general the *intercepts* of a graph.

The x coordinates of the points that a graph has in common with the x axis are called the x **intercepts** of the graph. (To compute the x intercepts, let $y = 0$ and solve for x.)

The y coordinates of the points that a graph has in common with the y axis are called the y **intercepts** of the graph. (To compute the y intercepts, let $x = 0$ and solve for y.)

E X A M P L E   2   Graph $y = x^2 - 4$

Solution

Let's begin by finding the intercepts. If $x = 0$, then

$$y = 0^2 - 4 = -4$$

The point $(0, -4)$ is on the graph. If $y = 0$, then

$$0 = x^2 - 4$$

$$0 = (x + 2)(x - 2)$$

$$x + 2 = 0 \quad \text{or} \quad x - 2 = 0$$

$$x = -2 \quad \text{or} \quad x = 2$$

The points $(-2, 0)$ and $(2, 0)$ are on the graph. The given equation is in a convenient form for setting up a table of values.

x	y	
0	−4	
−2	0	Intercepts
2	0	
1	−3	
−1	−3	
3	5	Other points
−3	5	

Plotting these points and connecting them with a smooth curve produces Figure 7.6.

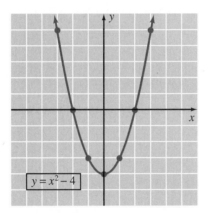

$$y = x^2 - 4$$

Figure 7.6

The curve in Figure 7.6 is called a parabola; we will study parabolas in more detail in a later chapter. However, at this time we want to emphasize that the parabola in Figure 7.6 is said to be *symmetric with respect to the y axis*. In other words, the *y* axis is a line of symmetry. Each half of the curve is a mirror image of the other half through the *y* axis. Note, in the table of values, that for each ordered pair (x, y), the ordered pair $(-x, y)$ is also a solution. A general test for y axis symmetry can be stated as follows:

y Axis Symmetry
The graph of an equation is symmetric with respect to the y axis if replacing x with $-x$ results in an equivalent equation.

The equation $y = x^2 - 4$ exhibits symmetry with respect to the y axis because replacing x with $-x$ produces $y = (-x)^2 - 4 = x^2 - 4$. Let's test some equations for such symmetry. We will replace x with $-x$ and check for an equivalent equation.

Equation	Test for symmetry with respect to the y axis	Equivalent equation	Symmetric with respect to the y axis?
$y = -x^2 + 2$	$y = -(-x)^2 + 2 = -x^2 + 2$	Yes	Yes
$y = 2x^2 + 5$	$y = 2(-x)^2 + 5 = 2x^2 + 5$	Yes	Yes
$y = x^4 + x^2$	$y = (-x)^4 + (-x)^2$ $= x^4 + x^2$	Yes	Yes
$y = x^3 + x^2$	$y = (-x)^3 + (-x)^2$ $= -x^3 + x^2$	No	No
$y = x^2 + 4x + 2$	$y = (-x)^2 + 4(-x) + 2$ $= x^2 - 4x + 2$	No	No

Some equations yield graphs that have x axis symmetry. In the next example we will see the graph of a parabola that is symmetric with respect to the x axis.

E X A M P L E   3

Graph $x = y^2$.

Solution

First, we see that $(0, 0)$ is on the graph and determines both intercepts. Second, the given equation is in a convenient form for setting up a table of values.

x	y	
0	0	Intercepts
1	1	
1	-1	Other points
4	2	
4	-2	

Plotting these points and connecting them with a smooth curve produces Figure 7.7.

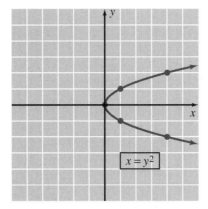

Figure 7.7

The parabola in Figure 7.7 is said to be *symmetric with respect to the x axis*. Each half of the curve is a mirror image of the other half through the x axis. Also note, in the table of values, that for each ordered pair (x, y), the ordered pair $(x, -y)$ is a solution. A general test for x axis symmetry can be stated as follows:

x Axis Symmetry

The graph of an equation is symmetric with respect to the x axis if replacing y with $-y$ results in an equivalent equation.

The equation $x = y^2$ exhibits x axis symmetry because replacing y with $-y$ produces $x = (-y)^2 = y^2$. Let's test some equations for x axis symmetry. We will replace y with $-y$ and check for an equivalent equation.

Equation	Test for symmetry with respect to the x axis	Equivalent equation	Symmetric with respect to the x axis?
$x = y^2 + 5$	$x = (-y)^2 + 5 = y^2 + 5$	Yes	Yes
$x = -3y^2$	$x = -3(-y)^2 = -3y^2$	Yes	Yes
$x = y^3 + 2$	$x = (-y)^3 + 2 = -y^3 + 2$	No	No
$x = y^2 - 5y + 6$	$x = (-y)^2 - 5(-y) + 6$		
	$= y^2 + 5y + 6$	No	No

In addition to y axis and x axis symmetry, some equations yield graphs that have symmetry with respect to the origin. In the next example we will see a graph that is symmetric with respect to the origin.

EXAMPLE 4 Graph $y = \dfrac{1}{x}$.

Solution

First, let's find the intercepts. Let $x = 0$; then $y = \dfrac{1}{x}$ becomes $y = \dfrac{1}{0}$, and $\dfrac{1}{0}$ is undefined. Thus there is no y intercept. Let $y = 0$; then $y = \dfrac{1}{x}$ becomes $0 = \dfrac{1}{x}$, and there are no values of x that will satisfy this equation. In other words, this graph has no points on either the x axis or the y axis. Second, let's set up a table of values and keep in mind that neither x nor y can equal zero.

x	y
$\dfrac{1}{2}$	2
1	1
2	$\dfrac{1}{2}$
3	$\dfrac{1}{3}$
$-\dfrac{1}{2}$	-2
-1	-1
-2	$-\dfrac{1}{2}$
-3	$-\dfrac{1}{3}$

In Figure 7.8(a) we plotted the points associated with the solutions from the table. Because the graph does not intersect either axis, it must consist of two branches. Thus, connecting the points in the first quadrant with a smooth curve and then connecting the points in the third quadrant with a smooth curve, we obtain the graph shown in Figure 7.8(b).

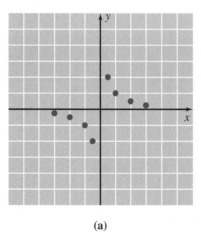

(a)

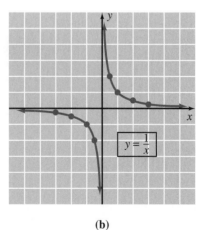

(b)

Figure 7.8

The curve in Figure 7.8 is said to be *symmetric with respect to the origin.* Each half of the curve is a mirror image of the other half through the origin. Note, in the table of values, that for each ordered pair (x, y), the ordered pair $(-x, -y)$ is also a solution. A general test for origin symmetry can be stated as follows:

Origin Symmetry

The graph of an equation is symmetric with respect to the origin if replacing x with $-x$ and y with $-y$ results in an equivalent equation.

The equation $y = \dfrac{1}{x}$ exhibits symmetry with respect to the origin because replacing y with $-y$ and x with $-x$ produces $-y = \dfrac{1}{-x}$, which is equivalent to $y = \dfrac{1}{x}$.

Let's test some equations for symmetry with respect to the origin. We will replace y with $-y$, replace x with $-x$, and then check for an equivalent equation.

Equation	Test for symmetry with respect to the origin	Equivalent equation	Symmetric with respect to the origin?
$y = x^3$	$(-y) = (-x)^3$ $-y = -x^3$ $y = x^3$	Yes	Yes
$x^2 + y^2 = 4$	$(-x)^2 + (-y)^2 = 4$ $x^2 + y^2 = 4$	Yes	Yes
$y = x^2 - 3x + 4$	$(-y) = (-x)^2 - 3(-x) + 4$ $-y = x^2 + 3x + 4$ $y = -x^2 - 3x - 4$	No	No

Let's pause for a moment and pull together the graphing techniques that we have introduced thus far. Following is a list of graphing suggestions. The order of the suggestions indicates the order in which we usually attack a new graphing problem.

1. Determine what type of symmetry the equation exhibits.
2. Find the intercepts.
3. Solve the equation for y in terms of x or for x in terms of y if it is not already in such a form.
4. Set up a table of ordered pairs that satisfy the equation. The type of symmetry will affect your choice of values in the table. (We will illustrate this in a moment.)
5. Plot the points associated with the ordered pairs from the table, and connect them with a smooth curve. Then, if appropriate, reflect this part of the curve according to the symmetry shown by the equation.

EXAMPLE 5   Graph $x^2y = -2$.

Solution

Because replacing x with $-x$ produces $(-x)^2y = -2$ or, equivalently, $x^2y = -2$, the equation exhibits y axis symmetry. There are no intercepts because neither x nor y can equal 0. Solving the equation for y produces $y = \dfrac{-2}{x^2}$. The equation exhibits y axis symmetry, so let's use only positive values for x and then reflect the curve across the y axis.

x	y
1	-2
2	$-\dfrac{1}{2}$
3	$-\dfrac{2}{9}$
4	$-\dfrac{1}{8}$
$\dfrac{1}{2}$	-8

Let's plot the points determined by the table, connect them with a smooth curve, and reflect this portion of the curve across the y axis. Figure 7.9 is the result of this process.

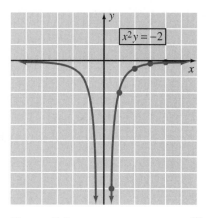

Figure 7.9

EXAMPLE 6   Graph $x = y^3$.

Solution

Because replacing x with $-x$ and y with $-y$ produces $-x = (-y)^3 = -y^3$, which is equivalent to $x = y^3$, the given equation exhibits origin symmetry. If $x = 0$, then $y = 0$, so the origin is a point of the graph. The given equation is in an easy form for deriving a table of values.

x	y
0	0
8	2
$\dfrac{1}{8}$	$\dfrac{1}{2}$
$\dfrac{27}{64}$	$\dfrac{3}{4}$

Let's plot the points determined by the table, connect them with a smooth curve, and reflect this portion of the curve through the origin to produce Figure 7.10.

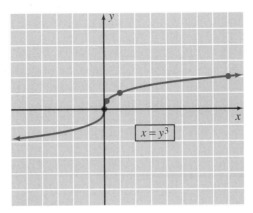

Figure 7.10

Graphing Utilities

The term **graphing utility** is used in current literature to refer to either a graphing calculator (see Figure 7.11) or a computer with a graphing software package. (We will frequently use the phrase *use a graphing calculator* to mean "use a graphing calculator or a computer with the appropriate software.")

These devices have a large range of capabilities that enable the user not only to obtain a quick sketch of a graph but also to study various characteristics of it, such as the x intercepts, y intercepts, and turning points of a curve. We will introduce some of these features of graphing utilities as we need them in the text. Because there are so many different types of graphing utilities available, we will use mostly generic terminology and let you consult your user's manual for specific key-punching instructions. We urge you to study the graphing utility examples in this text even if you do not have access to a graphing calculator or a computer. The examples were chosen to reinforce concepts under discussion.

Courtesy Texas Instruments

Figure 7.11

E X A M P L E   7   Use a graphing utility to obtain a graph of the equation $x = y^3$.

Solution

First, we may need to solve the equation for y in terms of x. (We say we "may need to" because some graphing utilities are capable of graphing two-variable equations without solving for y in terms of x.)

$$y = \sqrt[3]{x} = x^{1/3}$$

Now we can enter the expression $x^{1/3}$ for Y_1 and obtain the graph shown in Figure 7.12.

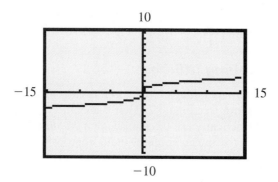

Figure 7.12

As indicated in Figure 7.12, the **viewing rectangle** of a graphing utility is a portion of the xy plane shown on the display of the utility. In this display, the boundaries were set so that $-15 \le x \le 15$ and $-10 \le y \le 10$. These boundaries were set automatically; however, boundaries can be reassigned as necessary, which is an important feature of graphing utilities.

PROBLEM SET 7.1

1. Indicate the quadrant that contains each of the points represented by the following ordered pairs.
   (a) $(-3, -4)$ **(b)** $(5, -6)$
   (c) $(-6, 8)$ **(d)** $(-1, -6)$
   (e) $(4, 10)$ **(f)** $(9, -1)$

2. (a) What is the first coordinate of any point on the vertical axis in a rectangular coordinate system?
   (b) What is the second coordinate of any point on the horizontal axis in a rectangular coordinate system?

3. (a) In which quadrants do the coordinates of a point have the same sign?
   (b) In which quadrants do the coordinates of a point have opposite signs?
   (c) In which quadrants is the abscissa negative?
   (d) In which quadrants is the ordinate negative?

For each of the points in Problems 4–9, determine the points that are symmetric with respect to (a) the x axis, (b) the y axis, and (c) the origin.

4. $(2, 6)$

5. $(-3, 1)$

6. $(-2, -4)$

7. $(7, -2)$

8. $(0, -4)$

9. $(5, 0)$

For Problems 10–23, determine the type(s) of symmetry (symmetry with respect to the x axis, y axis, and/or origin) exhibited by the graph of each of the following equations. Do not sketch the graph.

10. $x^2 + 2y = 4$

11. $-3x + 2y^2 = -4$

12. $x = -y^2 + 5$

13. $y = 4x^2 + 13$

14. $xy = -6$

15. $2x^2y^2 = 5$

16. $2x^2 + 3y^2 = 9$

17. $x^2 - 2x - y^2 = 4$

18. $y = x^2 - 6x - 4$

19. $y = 2x^2 - 7x - 3$

20. $y = x$

21. $y = 2x$

22. $y = x^4 + 4$

23. $y = x^4 - x^2 + 2$

For Problems 24–57, graph each of the equations.

24. $y = x + 1$

25. $y = x - 4$

26. $y = 3x - 6$

27. $y = 2x + 4$

28. $y = -2x + 1$

29. $y = -3x - 1$

30. $y = \dfrac{2}{3}x - 1$

31. $y = -\dfrac{1}{3}x + 2$

32. $y = \dfrac{1}{3}x$

33. $y = \dfrac{1}{2}x$

34. $2x + y = 6$

35. $2x - y = 4$

36. $x + 3y = -3$

37. $x - 2y = 2$

38. $y = x^2 - 1$

39. $y = x^2 + 2$

40. $y = -x^3$

41. $y = x^3$

42. $y = \dfrac{2}{x^2}$

43. $y = \dfrac{-1}{x^2}$

44. $y = 2x^2$

45. $y = -3x^2$

46. $xy = -3$

47. $xy = 2$

48. $x^2y = 4$

49. $xy^2 = -4$

50. $y^3 = x^2$

51. $y^2 = x^3$

52. $y = \dfrac{-2}{x^2 + 1}$

53. $y = \dfrac{4}{x^2 + 1}$

54. $x = -y^3$

55. $y = x^4$

56. $y = -x^4$

57. $x = -y^3 + 2$

■ ■ ■ Thoughts into words

58. How would you convince someone that there are infinitely many ordered pairs of real numbers that satisfy $x + y = 7$?

59. What is the graph of $x = 0$? What is the graph of $y = 0$? Explain your answers.

60. Is a graph symmetric with respect to the origin if it is symmetric with respect to both axes? Defend your answer.

61. Is a graph symmetric with respect to both axes if it is symmetric with respect to the origin? Defend your answer.

 ### Graphing calculator activities

This is the first of many appearances of a group of problems called graphing calculator activities. These problems are specifically designed for those of you who have access to a graphing calculator or a computer with an appropriate software package. Within the framework of these problems, you will be given the opportunity to reinforce concepts we discussed in the text; lay groundwork for concepts we will introduce later in the text; predict shapes and locations of graphs on the basis of your previous graphing experiences; solve problems that are unreasonable or perhaps impossible to solve without a graphing utility; and in general become familiar with the capabilities and limitations of your graphing utility.

This first set of activities is designed to help you get started with your graphing utility by setting different boundaries for the viewing rectangle; you will notice the effect on the graphs produced. These boundaries are usually set by using a menu displayed by a key marked either WINDOW or RANGE. You may need to consult the user's manual for specific key-punching instructions.

62. Graph the equation $y = \dfrac{1}{x}$ (Example 4) using the following boundaries.

(a) $-15 \leq x \leq 15$ and $-10 \leq y \leq 10$
(b) $-10 \leq x \leq 10$ and $-10 \leq y \leq 10$
(c) $-5 \leq x \leq 5$ and $-5 \leq y \leq 5$

63. Graph the equation $y = \dfrac{-2}{x^2}$ (Example 5), using the following boundaries.

(a) $-15 \leq x \leq 15$ and $-10 \leq y \leq 10$
(b) $-5 \leq x \leq 5$ and $-10 \leq y \leq 10$
(c) $-5 \leq x \leq 5$ and $-10 \leq y \leq 1$

64. Graph the two equations $y = \pm\sqrt{x}$ (Example 3) on the same set of axes, using the following boundaries. (Let $Y_1 = \sqrt{x}$ and $Y_2 = -\sqrt{x}$)

(a) $-15 \leq x \leq 15$ and $-10 \leq y \leq 10$
(b) $-1 \leq x \leq 15$ and $-10 \leq y \leq 10$
(c) $-1 \leq x \leq 15$ and $-5 \leq y \leq 5$

65. Graph $y = \dfrac{1}{x}$, $y = \dfrac{5}{x}$, $y = \dfrac{10}{x}$, and $y = \dfrac{20}{x}$ on the same set of axes. (Choose your own boundaries.) What effect does increasing the constant seem to have on the graph?

66. Graph $y = \dfrac{10}{x}$ and $y = \dfrac{-10}{x}$ on the same set of axes. What relationship exists between the two graphs?

67. Graph $y = \dfrac{10}{x^2}$ and $y = \dfrac{-10}{x^2}$ on the same set of axes. What relationship exists between the two graphs?

7.2   Linear Equations in Two Variables

It is advantageous to be able to recognize the kind of graph that a certain type of equation produces. For example, if we recognize that the graph of $3x + 2y = 12$ is a straight line, then it becomes a simple matter to find two points and sketch the line. Let's pursue the graphing of straight lines in a little more detail.

In general, any equation of the form $Ax + By = C$, where A, B, and C are constants (A and B not both zero) and x and y are variables, is a **linear equation,** and its graph is a straight line. Two points of clarification about this description of a linear equation should be made. First, the choice of x and y for variables is arbitrary. Any two letters could be used to represent the variables. For example, an equation such as $3r + 2s = 9$ can be considered a linear equation in two variables. So that we are not constantly changing the labeling of the coordinate axes when graphing equations, however, it is much easier to use the same two variables in all equations. Thus we will go along with convention and use x and y as variables. Second, the phrase "any equation of the form $Ax + By = C$" technically means "any equation of the form $Ax + By = C$ or equivalent to that form." For example, the equation $y = 2x - 1$ is equivalent to $-2x + y = -1$ and thus is linear and produces a straight-line graph.

The knowledge that any equation of the form $Ax + By = C$ produces a straight-line graph, along with the fact that two points determine a straight line, makes graphing linear equations a simple process. We merely find two solutions (such as the intercepts), plot the corresponding points, and connect the points with a straight line. It is usually wise to find a third point as a check point. Let's consider an example.

E X A M P L E   1

Graph $3x - 2y = 12$.

Solution

First, let's find the intercepts. Let $x = 0$; then

$$3(0) - 2y = 12$$

$$-2y = 12$$

$$y = -6$$

Thus $(0, -6)$ is a solution. Let $y = 0$; then

$$3x - 2(0) = 12$$

$$3x = 12$$

$$x = 4$$

Thus $(4, 0)$ is a solution. Now let's find a third point to serve as a check point. Let $x = 2$; then

$$3(2) - 2y = 12$$

$$6 - 2y = 12$$

$$-2y = 6$$

$$y = -3$$

Thus $(2, -3)$ is a solution. Plot the points associated with these three solutions and connect them with a straight line to produce the graph of $3x - 2y = 12$ in Figure 7.13.

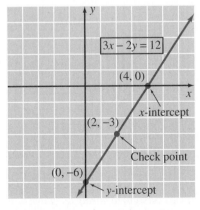

Figure 7.13

Let's review our approach to Example 1. Note that we did not solve the equation for y in terms of x or for x in terms of y. Because we know the graph is a straight line, there is no need for any extensive table of values; thus there is no need to change the form of the original equation. Furthermore, the solution $(2, -3)$ served as a check point. If it had not been on the line determined by the two intercepts, then we would have known that an error had been made.

EXAMPLE 2   Graph $2x + 3y = 7$.

Solution

Without showing all of our work, the following table indicates the intercepts and a check point.

x	y	
0	$\dfrac{7}{3}$	
$\dfrac{7}{2}$	0	Intercepts
2	1	Checkpoint

The points from the table are plotted and the graph of $2x + 3y = 7$ is shown in Figure 7.14.

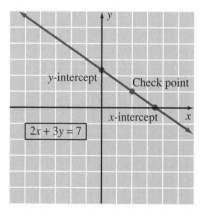

Figure 7.14

It is helpful to recognize some *special* straight lines. For example, the graph of any equation of the form $Ax + By = C$, where $C = 0$ (the constant term is zero), is a straight line that contains the origin. Let's consider an example.

EXAMPLE 3   Graph $y = 2x$.

Solution

Obviously $(0, 0)$ is a solution. (Also, notice that $y = 2x$ is equivalent to $-2x + y = 0$; thus it fits the condition $Ax + By = C$, where $C = 0$.) Because both the x intercept and the y intercept are determined by the point $(0, 0)$, another point is necessary to determine the line. Then a third point should be found as a check point.

The graph of $y = 2x$ is shown in Figure 7.15.

x	y	
0	0	Intercepts
2	4	Additional point
−1	−2	Check point

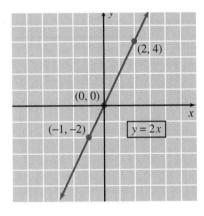

Figure 7.15

EXAMPLE 4   Graph $x = 2$.

Solution

Because we are considering linear equations in *two variables,* the equation $x = 2$ is equivalent to $x + 0(y) = 2$. Now we can see that any value of y can be used, but the x value must always be 2. Therefore, some of the solutions are $(2, 0)$, $(2, 1)$, $(2, 2)$, $(2, -1)$, and $(2, -2)$. The graph of all solutions of $x = 2$ is the vertical line in Figure 7.16.

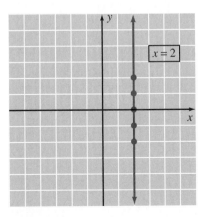

Figure 7.16

EXAMPLE 5

Graph $y = -3$.

Solution

The equation $y = -3$ is equivalent to $0(x) + y = -3$. Thus any value of x can be used, but the value of y must be -3. Some solutions are $(0, -3)$, $(1, -3)$, $(2, -3)$, $(-1, -3)$, and $(-2, -3)$. The graph of $y = -3$ is the horizontal line in Figure 7.17.

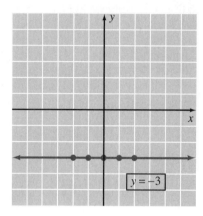

Figure 7.17

In general, the graph of any equation of the form $Ax + By = C$, where $A = 0$ or $B = 0$ (not both), is a line parallel to one of the axes. More specifically, any equation of the form $x = a$, where a is a constant, is a line parallel to the y axis that has an x intercept of a. Any equation of the form $y = b$, where b is a constant, is a line parallel to the x axis that has a y intercept of b.

REMARK: Even though it is not particularly helpful for graphing straight lines, you should recognize that the equations in Examples 3, 4, and 5 do exhibit symmetry. In Example 3, the equation $y = 2x$ exhibits origin symmetry. Likewise, in Example 4 the equation $x = 2$ exhibits x axis symmetry, and in Example 5 the equation $y = -3$ exhibits y axis symmetry.

EXAMPLE 6

Use a graphing utility to obtain a graph of the line $2.1x + 5.3y = 7.9$.

Solution

First, let's solve the equation for y in terms of x.

$$2.1x + 5.3y = 7.9$$

$$5.3y = 7.9 - 2.1x$$

$$y = \frac{7.9 - 2.1x}{5.3}$$

Now we can enter the expression $\dfrac{7.9 - 2.1x}{5.3}$ for Y_1 and obtain the graph as shown in Figure 7.18.

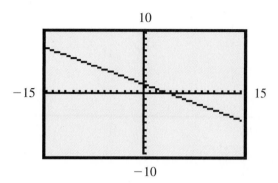

Figure 7.18

Linear Relationships

There are numerous applications of linear relationships. For example, suppose that a retailer has a number of items that she wants to sell at a profit of 30% of the cost of each item. If we let s represent the selling price and c the cost of each item, then the equation

$$s = c + 0.3c = 1.3c$$

can be used to determine the selling price of each item based on the cost of the item. In other words, if the cost of an item is \$4.50, then it should be sold for $s = (1.3)(4.5) = \$5.85$.

The equation $s = 1.3c$ can be used to determine the following table of values. Reading from the table, we see that if the cost of an item is \$15, then it should be sold for \$19.50 in order to yield a profit of 30% of the cost. Furthermore, because this is a linear relationship, we can obtain exact values between values given in the table.

c	1	5	10	15	20
s	1.3	6.5	13	19.5	26

For example, a c value of 12.5 is halfway between c values of 10 and 15, so the corresponding s value is halfway between the s values of 13 and 19.5. Therefore, a c value of 12.5 produces an s value of

$$s = 13 + \frac{1}{2}(19.5 - 13) = 16.25$$

Thus, if the cost of an item is \$12.50, it should be sold for \$16.25.

Now let's graph this linear relationship. We can label the horizontal axis *c*, label the vertical axis *s*, and use the origin along with one ordered pair from the table to produce the straight-line graph in Figure 7.19. (Because of the type of application, we use only nonnegative values for *c* and *s*.)

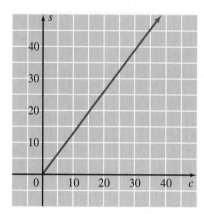

Figure 7.19

From the graph we can approximate *s* values on the basis of given *c* values. For example, if *c* = 30, then by reading up from 30 on the *c* axis to the line and then across to the *s* axis, we see that *s* is a little less than 40. (An exact *s* value of 39 is obtained by using the equation *s* = 1.3*c*.)

Many formulas that are used in various applications are linear equations in two variables. For example, the formula $C = \dfrac{5}{9}(F - 32)$, which is used to convert temperatures from the Fahrenheit scale to the Celsius scale, is a linear relationship. Using this equation, we can determine that 14°F is equivalent to $C = \dfrac{5}{9}(14 - 32) = \dfrac{5}{9}(-18) = -10°C$. Let's use the equation $C = \dfrac{5}{9}(F - 32)$ to complete the following table.

F	−22	−13	5	32	50	68	86
C	−30	−25	−15	0	10	20	30

Reading from the table, we see for example that −13°F = −25°C and 68°F = 20°C.

To graph the equation $C = \dfrac{5}{9}(F - 32)$ we can label the horizontal axis F, label the vertical axis C, and plot two ordered pairs (F, C) from the table. Figure 7.20 shows the graph of the equation.

From the graph we can approximate C values on the basis of given F values. For example, if F = 80°, then by reading up from 80 on the F axis to the line and then across to the C axis, we see that C is approximately 25°. Likewise, we can ob-

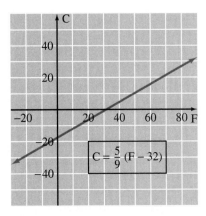

Figure 7.20

tain approximate F values on the basis of given C values. For example, if C = −25°, then by reading across from −25 on the C axis to the line and then up to the F axis, we see that F is approximately −15°.

PROBLEM SET 7.2

For Problems 1–33, graph each of the linear equations.

1. $x + 2y = 4$

2. $2x + y = 6$

3. $2x − y = 2$

4. $3x − y = 3$

5. $3x + 2y = 6$

6. $2x + 3y = 6$

7. $5x − 4y = 20$

8. $4x − 3y = −12$

9. $x + 4y = −6$

10. $5x + y = −2$

11. $−x − 2y = 3$

12. $−3x − 2y = 12$

13. $y = x + 3$

14. $y = x − 1$

15. $y = −2x − 1$

16. $y = 4x + 3$

17. $y = \dfrac{1}{2}x + \dfrac{2}{3}$

18. $y = \dfrac{2}{3}x − \dfrac{3}{4}$

19. $y = −x$

20. $y = x$

21. $y = 3x$

22. $y = −4x$

23. $x = 2y − 1$

24. $x = −3y + 2$

25. $y = −\dfrac{1}{4}x + \dfrac{1}{6}$

26. $y = −\dfrac{1}{2}x − \dfrac{1}{2}$

27. $2x − 3y = 0$

28. $3x + 4y = 0$

29. $x = 0$

30. $y = 0$

31. $y = 2$

32. $x = −3$

33. $−3y = −x + 3$

34. Suppose that the daily profit from an ice cream stand is given by the equation $p = 2n − 4$, where n represents the number of gallons of ice cream mix used in a day and p represents the number of dollars of profit. Label the horizontal axis n and the vertical axis p, and graph the equation $p = 2n − 4$ for nonnegative values of n.

35. The cost (c) of playing an online computer game for a time (t) in hours is given by the equation $c = 3t + 5$. Label the horizontal axis t and the vertical axis c, and graph the equation for nonnegative values of t.

36. The area of a sidewalk whose width is fixed at 3 feet can be given by the equation $A = 3l$, where A represents the area in square feet and l represents the length in feet. Label the horizontal axis l and the vertical axis A, and graph the equation $A = 3l$ for nonnegative values of l.

37. An online grocery store charges for delivery based on the equation $C = 0.30p$, where C represents the cost in dollars and p represents the weight of the groceries in pounds. Label the horizontal axis p and the vertical axis C, and graph the equation $C = 0.30p$ for non-negative values of p.

38. (a) The equation $F = \dfrac{9}{5}C + 32$ can be used to convert from degrees Celsius to degrees Fahrenheit. Complete the following table.

C	0	5	10	15	20	−5	−10	−15	−20	−25
F										

(b) Graph the equation $F = \dfrac{9}{5}C + 32$.

(c) Use your graph from part (b) to approximate values for F when C = 25°, 30°, −30°, and −40°.

(d) Check the accuracy of your readings from the graph in part (c) by using the equation $F = \dfrac{9}{5}C + 32$.

39. (a) Digital Solutions charges for help-desk services according to the equation $c = 0.25m + 10$, where c represents the cost in dollars and m represents the minutes of service. Complete the following table.

m	5	10	15	20	30	60
c						

(b) Label the horizontal axis m and the vertical axis c, and graph the equation $c = 0.25m + 10$ for non-negative values of m.

(c) Use the graph from part (b) to approximate values for c when m = 25, 40, and 45.

(d) Check the accuracy of your readings from the graph in part (c) by using the equation $c = 0.25m + 10$.

■ ■ ▨ **Thoughts into words**

40. How do we know that the graph of $y = -3x$ is a straight line that contains the origin?

41. How do we know that the graphs of $2x - 3y = 6$ and $-2x + 3y = -6$ are the same line?

42. What is the graph of the conjunction $x = 2$ and $y = 4$? What is the graph of the disjunction $x = 2$ or $y = 4$? Explain your answers.

43. Your friend claims that the graph of the equation $x = 2$ is the point $(2, 0)$. How do you react to this claim?

■ ■ ▨ **Further investigations**

From our work with absolute value, we know that $|x + y| = 1$ is equivalent to $x + y = 1$ or $x + y = -1$. Therefore, the graph of $|x + y| = 1$ consists of the two lines $x + y = 1$ and $x + y = -1$. Graph each of the following.

44. $|x + y| = 1$

45. $|x - y| = 4$

46. $|2x - y| = 4$

47. $|3x + 2y| = 6$

▦ **Graphing calculator activities**

48. (a) Graph $y = 3x + 4$, $y = 2x + 4$, $y = -4x + 4$, and $y = -2x + 4$ on the same set of axes.

(b) Graph $y = \dfrac{1}{2}x - 3$, $y = 5x - 3$, $y = 0.1x - 3$, and $y = -7x - 3$ on the same set of axes.

(c) What characteristic do all lines of the form $y = ax + 2$ (where a is any real number) share?

49. (a) Graph $y = 2x - 3$, $y = 2x + 3$, $y = 2x - 6$, and $y = 2x + 5$ on the same set of axes.

(b) Graph $y = -3x + 1$, $y = -3x + 4$, $y = -3x - 2$, and $y = -3x - 5$ on the same set of axes.

(c) Graph $y = \frac{1}{2}x + 3$, $y = \frac{1}{2}x - 4$, $y = \frac{1}{2}x + 5$, and $y = \frac{1}{2}x - 2$ on the same set of axes.

(d) What relationship exists among all lines of the form $y = 3x + b$, where b is any real number?

50. (a) Graph $2x + 3y = 4$, $2x + 3y = -6$, $4x - 6y = 7$, and $8x + 12y = -1$ on the same set of axes.

(b) Graph $5x - 2y = 4$, $5x - 2y = -3$, $10x - 4y = 3$, and $15x - 6y = 30$ on the same set of axes.

(c) Graph $x + 4y = 8$, $2x + 8y = 3$, $x - 4y = 6$, and $3x + 12y = 10$ on the same set of axes.

(d) Graph $3x - 4y = 6$, $3x + 4y = 10$, $6x - 8y = 20$, and $6x - 8y = 24$ on the same set of axes.

(e) For each of the following pairs of lines, (a) predict whether they are parallel lines, and (b) graph each pair of lines to check your prediction.

(1) $5x - 2y = 10$ and $5x - 2y = -4$
(2) $x + y = 6$ and $x - y = 4$
(3) $2x + y = 8$ and $4x + 2y = 2$
(4) $y = 0.2x + 1$ and $y = 0.2x - 4$
(5) $3x - 2y = 4$ and $3x + 2y = 4$
(6) $4x - 3y = 8$ and $8x - 6y = 3$
(7) $2x - y = 10$ and $6x - 3y = 6$
(8) $x + 2y = 6$ and $3x - 6y = 6$

51. Now let's use a graphing calculator to get a graph of $C = \frac{5}{9}(F - 32)$. By letting $F = x$ and $C = y$, we obtain Figure 7.21.

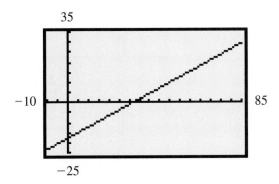

35

-10 85

-25

Figure 7.21

Pay special attention to the boundaries on x. These values were chosen so that the fraction

$$\frac{(\text{Maximum value of } x) \text{ minus } (\text{Minimum value of } x)}{95}$$

would be equal to 1. The viewing window of the graphing calculator used to produce Figure 7.21 is 95 pixels (dots) wide. Therefore, we use 95 as the denominator of the fraction. We chose the boundaries for y to make sure that the cursor would be visible on the screen when we looked for certain values.

Now let's use the TRACE feature of the graphing calculator to complete the following table. Note that the cursor moves in increments of 1 as we trace along the graph.

F	-5	5	9	11	12	20	30	45	60
C									

(This was accomplished by setting the aforementioned fraction equal to 1.) By moving the cursor to each of the F values, we can complete the table as follows.

F	-5	5	9	11	12	20	30	45	60
C	-21	-15	-13	-12	-11	-7	-1	7	16

The C values are expressed to the nearest degree. Use your calculator and check the values in the table by using the equation $C = \frac{5}{9}(F - 32)$.

52. (a) Use your graphing calculator to graph $F = \frac{9}{5}C + 32$. Be sure to set boundaries on the horizontal axis so that when you are using the trace feature, the cursor will move in increments of 1.

(b) Use the TRACE feature and check your answers for part (a) of Problem 38.

7.3 Linear Inequalities in Two Variables

Linear inequalities in two variables are of the form $Ax + By > C$ or $Ax + By < C$, where A, B, and C are real numbers. (Combined linear equality and inequality statements are of the form $Ax + By \geq C$ or $Ax + By \leq C$.)

Graphing linear inequalities is almost as easy as graphing linear equations. The following discussion leads into a simple, step-by-step process. Let's consider the following equation and related inequalities.

$$x + y = 2 \qquad x + y > 2 \qquad x + y < 2$$

The graph of $x + y = 2$ is shown in Figure 7.22. The line divides the plane into two half planes, one above the line and one below the line. In Figure 7.23(a) we indi-

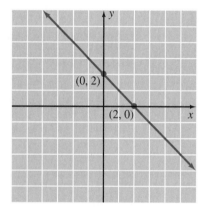

Figure 7.22

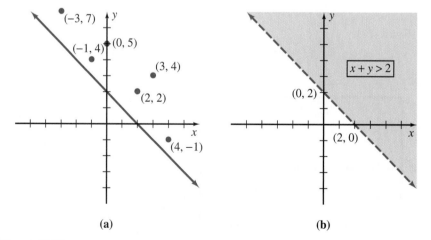

(a)   (b)

Figure 7.23

cated several points in the half plane above the line. Note that for each point, the ordered pair of real numbers satisfies the inequality $x + y > 2$. This is true for *all points* in the half plane above the line. Therefore, the graph of $x + y > 2$ is the half plane above the line, as indicated by the shaded portion in Figure 7.23(b). We use a dashed line to indicate that points on the line do *not* satisfy $x + y > 2$. We would use a solid line if we were graphing $x + y \geq 2$.

In Figure 7.24(a) several points were indicated in the half plane below the line $x + y = 2$. Note that for each point, the ordered pair of real numbers satisfies the inequality $x + y < 2$. This is true for *all points* in the half plane below the line. Thus the graph of $x + y < 2$ is the half plane below the line, as indicated in Figure 7.24(b).

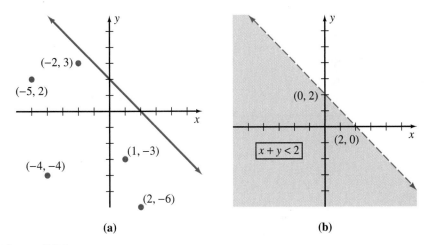

(a)   (b)

Figure 7.24

To graph a linear inequality, we suggest the following steps.

1. First, graph the corresponding equality. Use a solid line if equality is included in the original statement. Use a dashed line if equality is not included.
2. Choose a "test point" not on the line and substitute its coordinates into the inequality. (The origin is a convenient point to use if it is not on the line.)
3. The graph of the original inequality is
   (a) the half plane that contains the test point if the inequality is satisfied by that point, or
   (b) the half plane that does not contain the test point if the inequality is not satisfied by the point.

Let's apply these steps to some examples.

E X A M P L E   1   Graph $x - 2y > 4$.

Solution

STEP 1   Graph $x - 2y = 4$ as a dashed line because equality is not included in $x - 2y > 4$ (Figure 7.25).

STEP 2   Choose the origin as a test point and substitute its coordinates into the inequality.

$$x - 2y > 4 \quad \text{becomes } 0 - 2(0) > 4, \text{ which is false.}$$

STEP 3   Because the test point did not satisfy the given inequality, the graph is the half plane that does not contain the test point. Thus the graph of $x - 2y > 4$ is the half plane below the line, as indicated in Figure 7.25.

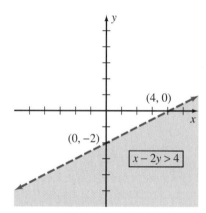

Figure 7.25

E X A M P L E   2   Graph $3x + 2y \leq 6$.

Solution

STEP 1   Graph $3x + 2y = 6$ as a solid line because equality is included in $3x + 2y \leq 6$ (Figure 7.26).

STEP 2   Choose the origin as a test point and substitute its coordinates into the given statement.

$$3x + 2y \leq 6 \quad \text{becomes } 3(0) + 2(0) \leq 6, \text{ which is true.}$$

STEP 3   Because the test point satisfies the given statement, all points in the same half plane as the test point satisfy the statement. Thus the graph of $3x + 2y \leq 6$ consists of the line and the half plane below the line (Figure 7.26).

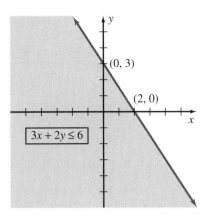

Figure 7.26

| **E X A M P L E   3** | Graph $y \leq 3x$. |

Solution

> **STEP 1**   Graph $y = 3x$ as a solid line because equality is included in the statement $y \leq 3x$ (Figure 7.27).

> **STEP 2**   The origin is on the line, so we must choose some other point as a test point. Let's try $(2, 1)$.
>
> $$y \leq 3x \quad \text{becomes } 1 \leq 3(2), \text{ which is a true statement.}$$

> **STEP 3**   Because the test point satisfies the given inequality, the graph is the half plane that contains the test point. Thus the graph of $y \leq 3x$ consists of the line and the half plane below the line, as indicated in Figure 7.27.

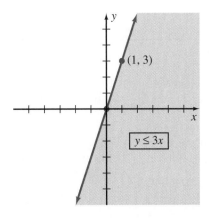

Figure 7.27

PROBLEM SET 7.3

For Problems 1–24, graph each of the inequalities.

1. $x - y > 2$

2. $x + y > 4$

3. $x + 3y < 3$

4. $2x - y > 6$

5. $2x + 5y \geq 10$

6. $3x + 2y \leq 4$

7. $y \leq -x + 2$

8. $y \geq -2x - 1$

9. $y > -x$

10. $y < x$

11. $2x - y \geq 0$

12. $x + 2y \geq 0$

13. $-x + 4y - 4 \leq 0$

14. $-2x + y - 3 \leq 0$

15. $y > -\dfrac{3}{2}x - 3$

16. $2x + 5y > -4$

17. $y < -\dfrac{1}{2}x + 2$

18. $y < -\dfrac{1}{3}x + 1$

19. $x \leq 3$

20. $y \geq -2$

21. $x > 1$ and $y < 3$

22. $x > -2$ and $y > -1$

23. $x \leq -1$ and $y < 1$

24. $x < 2$ and $y \geq -2$

■ ■ ■ **Thoughts into words**

25. Why is the point $(-4, 1)$ not a good test point to use when graphing $5x - 2y > -22$?

26. Explain how you would graph the inequality $-3 > x - 3y$.

■ ■ ■ **Further investigations**

27. Graph $|x| < 2$. [*Hint:* Remember that $|x| < 2$ is equivalent to $-2 < x < 2$.]

28. Graph $|y| > 1$.

29. Graph $|x + y| < 1$.

30. Graph $|x - y| > 2$.

▦ **Graphing calculator activities**

31. This is a good time for you to become acquainted with the DRAW features of your graphing calculator. Again, you may need to consult your user's manual for specific key-punching instructions. Return to Examples 1, 2, and 3 of this section and use your graphing calculator to graph the inequalities.

32. Use a graphing calculator to check your graphs for Problems 1–24.

33. Use the DRAW feature of your graphing calculator to draw each of the following.
   (a) A line segment between $(-2, -4)$ and $(-2, 5)$
   (b) A line segment between $(2, 2)$ and $(5, 2)$
   (c) A line segment between $(2, 3)$ and $(5, 7)$
   (d) A triangle with vertices at $(1, -2)$, $(3, 4)$, and $(-3, 6)$

7.4   **Distance and Slope**

As we work with the rectangular coordinate system, it is sometimes necessary to express the length of certain line segments. In other words, we need to be able to find the distance between two points. Let's first consider two specific examples and then develop the general distance formula.

E X A M P L E   1

Find the distance between the points $A(2, 2)$ and $B(5, 2)$ and also between the points $C(-2, 5)$ and $D(-2, -4)$.

Solution

Let's plot the points and draw $\overline{AB}$ as in Figure 7.28. Because $\overline{AB}$ is parallel to the x axis, its length can be expressed as $|5 - 2|$ or $|2 - 5|$. (The absolute-value symbol is used to ensure a nonnegative value.) Thus the length of $\overline{AB}$ is 3 units. Likewise, the length of $\overline{CD}$ is $|5 - (-4)| = |-4 - 5| = 9$ units.

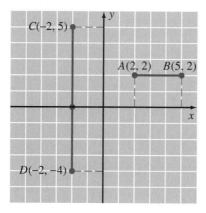

Figure 7.28

E X A M P L E   2

Find the distance between the points $A(2, 3)$ and $B(5, 7)$.

Solution

Let's plot the points and form a right triangle as indicated in Figure 7.29. Note that the coordinates of point C are $(5, 3)$. Because $\overline{AC}$ is parallel to the horizontal axis, its length is easily determined to be 3 units. Likewise, $\overline{CB}$ is parallel to the vertical axis and its length is 4 units. Let d represent the length of $\overline{AB}$ and apply the Pythagorean theorem to obtain

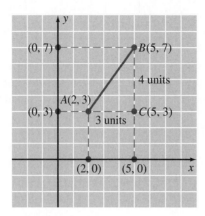

$d^2 = 3^2 + 4^2$

$d^2 = 9 + 16$

$d^2 = 25$

$d = \pm\sqrt{25} = \pm5$

Distance between is a nonnegative value, so the length of $\overline{AB}$ is 5 units.

Figure 7.29

We can use the approach we used in Example 2 to develop a general distance formula for finding the distance between any two points in a coordinate plane. The development proceeds as follows:

1. Let $P_1(x_1, y_1)$ and $P_2(x_2, y_2)$ represent any two points in a coordinate plane.
2. Form a right triangle as indicated in Figure 7.30. The coordinates of the vertex of the right angle, point R, are (x_2, y_1).

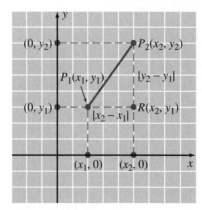

Figure 7.30

The length of $\overline{P_1R}$ is $|x_2 - x_1|$ and the length of $\overline{RP_2}$ is $|y_2 - y_1|$. (The absolute-value symbol is used to ensure a nonnegative value.) Let d represent the length of P_1P_2 and apply the Pythagorean theorem to obtain

$$d^2 = |x_2 - x_1|^2 + |y_2 - y_1|^2$$

Because $|a|^2 = a^2$, the **distance formula** can be stated as

$$d = \sqrt{(x_2 - x_1)^2 + (y_2 - y_1)^2}$$

It makes no difference which point you call P_1 or P_2 when using the distance formula. If you forget the formula, don't panic. Just form a right triangle and apply the Pythagorean theorem as we did in Example 2. Let's consider an example that demonstrates the use of the distance formula.

E X A M P L E   3

Find the distance between $(-1, 4)$ and $(1, 2)$.

Solution

Let $(-1, 4)$ be P_1 and $(1, 2)$ be P_2. Using the distance formula, we obtain

$$d = \sqrt{[(1 - (-1))]^2 + (2 - 4)^2}$$
$$= \sqrt{2^2 + (-2)^2}$$
$$= \sqrt{4 + 4}$$
$$= \sqrt{8} = 2\sqrt{2} \qquad \text{Express the answer in simplest radical form.}$$

The distance between the two points is $2\sqrt{2}$ units. ∎

In Example 3, we did not sketch a figure because of the simplicity of the problem. However, sometimes it is helpful to use a figure to organize the given information and aid in the analysis of the problem, as we see in the next example.

E X A M P L E   4

Verify that the points $(-3, 6)$, $(3, 4)$, and $(1, -2)$ are vertices of an isosceles triangle. (An isosceles triangle has two sides of the same length.)

Solution

Let's plot the points and draw the triangle (Figure 7.31). Use the distance formula to find the lengths d_1, d_2, and d_3, as follows:

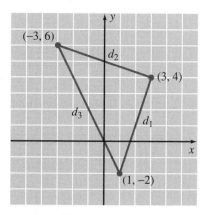

$$d_1 = \sqrt{(3 - 1)^2 + (4 - (-2))^2}$$
$$= \sqrt{2^2 + 6^2} = \sqrt{40} = 2\sqrt{10}$$
$$d_2 = \sqrt{(-3 - 3)^2 + (6 - 4)^2}$$
$$= \sqrt{(-6)^2 + 2^2} = \sqrt{40}$$
$$= 2\sqrt{10}$$
$$d_3 = \sqrt{(-3 - 1)^2 + (6 - (-2))^2}$$
$$= \sqrt{(-4)^2 + 8^2} = \sqrt{80} = 4\sqrt{5}$$

Figure 7.31

Because $d_1 = d_2$, we know that it is an isosceles triangle. ∎

Slope of a Line

In coordinate geometry, the concept of **slope** is used to describe the "steepness" of lines. The slope of a line is the ratio of the vertical change to the horizontal change as we move from one point on a line to another point. This is illustrated in Figure 7.32 with points P_1 and P_2.

A precise definition for slope can be given by considering the coordinates of the points P_1, P_2, and R as indicated in Figure 7.33. The horizontal change as we move from P_1 to P_2 is $x_2 - x_1$ and the vertical change is $y_2 - y_1$. Thus the following definition for slope is given.

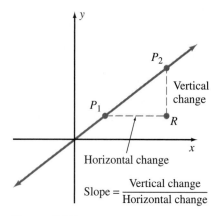

Figure 7.32

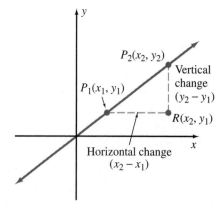

Figure 7.33

DEFINITION 7.1

If points P_1 and P_2 with coordinates (x_1, y_1) and (x_2, y_2), respectively, are any two different points on a line, then the slope of the line (denoted by m) is

$$m = \frac{y_2 - y_1}{x_2 - x_1}, \qquad x_2 \neq x_1$$

Because $\dfrac{y_2 - y_1}{x_2 - x_1} = \dfrac{y_1 - y_2}{x_1 - x_2}$, how we designate P_1 and P_2 is not important. Let's use Definition 7.1 to find the slopes of some lines.

EXAMPLE 5   Find the slope of the line determined by each of the following pairs of points, and graph the lines.

     (a) $(-1, 1)$ and $(3, 2)$ **(b)** $(4, -2)$ and $(-1, 5)$
     (c) $(2, -3)$ and $(-3, -3)$

Solution

     (a) Let $(-1, 1)$ be P_1 and $(3, 2)$ be P_2 (Figure 7.34).

$$m = \frac{y_2 - y_1}{x_2 - x_1} = \frac{2 - 1}{3 - (-1)} = \frac{1}{4}$$

     (b) Let $(4, -2)$ be P_1 and $(-1, 5)$ be P_2 (Figure 7.35).

$$m = \frac{y_2 - y_1}{x_2 - x_1} = \frac{5 - (-2)}{-1 - 4} = \frac{7}{-5} = -\frac{7}{5}$$

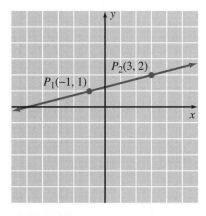

Figure 7.34

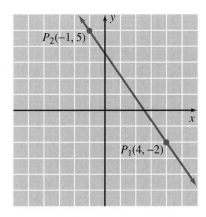

Figure 7.35

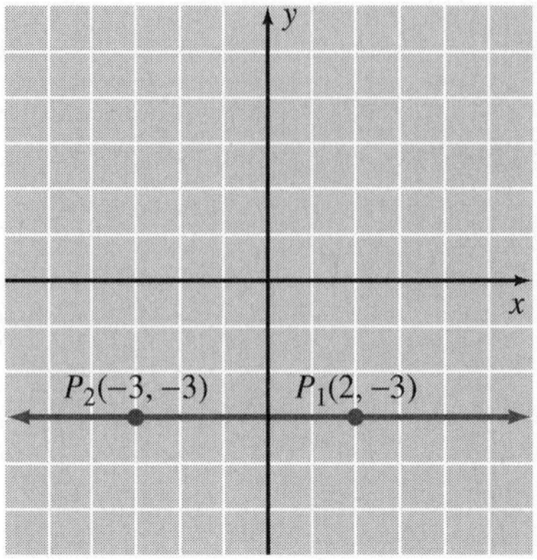

Figure 7.36

(c) Let $(2, -3)$ be P_1 and $(-3, -3)$ be P_2 (Figure 7.36).

$$m = \frac{y_2 - y_1}{x_2 - x_1}$$

$$= \frac{-3 - (-3)}{-3 - 2}$$

$$= \frac{0}{-5} = 0$$

The three parts of Example 5 represent the three basic possibilities for slope; that is, the slope of a line can be positive, negative, or zero. A line that has a positive slope rises as we move from left to right, as in Figure 7.34. A line that has a negative slope falls as we move from left to right, as in Figure 7.35. A horizontal line, as in Figure 7.36, has a slope of zero. Finally, we need to realize that *the concept of slope is undefined for vertical lines.* This is due to the fact that for any vertical line, the horizontal change as we move from one point on the line to another is zero. Thus the ratio $\dfrac{y_2 - y_1}{x_2 - x_1}$ will have a denominator of zero and be undefined. Accordingly, the restriction $x_2 \neq x_1$ is imposed in Definition 7.1.

One final idea pertaining to the concept of slope needs to be emphasized. The slope of a line is a **ratio,** the ratio of vertical change to horizontal change. A slope of $\dfrac{2}{3}$ means that for every 2 units of vertical change there must be a corresponding 3 units of horizontal change. Thus, starting at some point on a line that has a slope of $\dfrac{2}{3}$, we could locate other points on the line as follows:

$$\frac{2}{3} = \frac{4}{6} \qquad \longrightarrow \text{ by moving 4 units } up \text{ and 6 units to the } right$$

$$\frac{2}{3} = \frac{8}{12} \qquad \longrightarrow \text{ by moving 8 units } up \text{ and 12 units to the } right$$

$$\frac{2}{3} = \frac{-2}{-3} \qquad \longrightarrow \text{ by moving 2 units } down \text{ and 3 units to the } left$$

Likewise, if a line has a slope of $-\dfrac{3}{4}$, then by starting at some point on the line we could locate other points on the line as follows:

$$-\frac{3}{4} = \frac{-3}{4} \qquad \longrightarrow \text{ by moving 3 units } \textit{down} \text{ and 4 units to the } \textit{right}$$

$$-\frac{3}{4} = \frac{3}{-4} \qquad \longrightarrow \text{ by moving 3 units } \textit{up} \text{ and 4 units to the } \textit{left}$$

$$-\frac{3}{4} = \frac{-9}{12} \qquad \longrightarrow \text{ by moving 9 units } \textit{down} \text{ and 12 units to the } \textit{right}$$

$$-\frac{3}{4} = \frac{15}{-20} \qquad \longrightarrow \text{ by moving 15 units } \textit{up} \text{ and 20 units to the } \textit{left}$$

EXAMPLE 6   Graph the line that passes through the point $(0, -2)$ and has a slope of $\dfrac{1}{3}$.

Solution

To graph, plot the point $(0, -2)$. Furthermore, because the slope $= \dfrac{\text{vertical change}}{\text{horizontal change}} = \dfrac{1}{3}$, we can locate another point on the line by starting from the point $(0, -2)$ and moving 1 unit up and 3 units to the right to obtain the point $(3, -1)$. Because two points determine a line, we can draw the line (Figure 7.37).

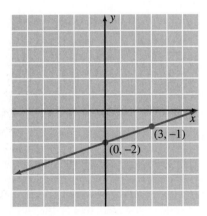

Figure 7.37

REMARK: Because $m = \dfrac{1}{3} = \dfrac{-1}{-3}$, we can locate another point by moving 1 unit down and 3 units to the left from the point $(0, -2)$.

EXAMPLE 7   Graph the line that passes through the point $(1, 3)$ and has a slope of -2.

Solution

To graph the line, plot the point $(1, 3)$. We know that $m = -2 = \dfrac{-2}{1}$. Furthermore, because the slope $= \dfrac{\text{vertical change}}{\text{horizontal change}} = \dfrac{-2}{1}$, we can locate another point on the line by starting from the point $(1, 3)$ and moving 2 units down and 1 unit to the right to obtain the point $(2, 1)$. Because two points determine a line, we can draw the line (Figure 7.38).

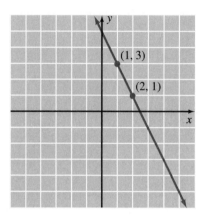

Figure 7.38

REMARK: Because $m = -2 = \dfrac{-2}{1} = \dfrac{2}{-1}$ we can locate another point by moving 2 units up and 1 unit to the left from the point $(1, 3)$. ∎

Applications of Slope

The concept of slope has many real-world applications even though the word *slope* is often not used. The concept of slope is used in most situations where an incline is involved. Hospital beds are hinged in the middle so that both the head end and the foot end can be raised or lowered; that is, the slope of either end of the bed can be changed. Likewise, treadmills are designed so that the incline (slope) of the platform can be adjusted. A roofer, when making an estimate to replace a roof, is concerned not only about the total area to be covered but also about the pitch of the roof. (Contractors do not define *pitch* as identical with the mathematical definition of slope, but both concepts refer to "steepness.") In Figure 7.39, the two roofs might require the same amount of shingles, but the roof on the left will take longer to complete because the pitch is so great that scaffolding will be required.

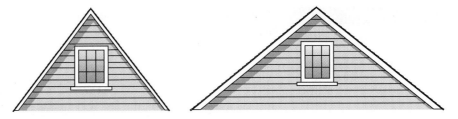

Figure 7.39

The concept of slope is also used in the construction of flights of stairs (Figure 7.40). The terms *rise* and *run* are commonly used, and the steepness (slope) of the stairs can be expressed as the ratio of rise to run. In Figure 7.40, the stairs on the left, where the ratio of rise to run is $\frac{10}{11}$, are steeper than the stairs on the right, which have a ratio of $\frac{7}{11}$.

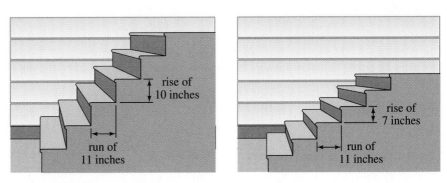

rise of
10 inches

run of
11 inches

rise of
7 inches

run of
11 inches

Figure 7.40

In highway construction, the word *grade* is used for the concept of slope. For example, in Figure 7.41 the highway is said to have a grade of 17%. This means that for every horizontal distance of 100 feet, the highway rises or drops 17 feet. In other words, the slope of the highway is $\frac{17}{100}$.

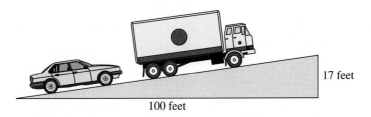

17 feet

100 feet

Figure 7.41

EXAMPLE 8   A certain highway has a 3% grade. How many feet does it rise in a horizontal distance of 1 mile?

Solution

A 3% grade means a slope of $\dfrac{3}{100}$. Therefore, if we let y represent the unknown vertical distance and use the fact that 1 mile = 5280 feet, we can set up and solve the following proportion.

$$\frac{3}{100} = \frac{y}{5280}$$

$$100y = 3(5280) = 15{,}840$$

$$y = 158.4$$

The highway rises 158.4 feet in a horizontal distance of 1 mile. ▪

PROBLEM SET 7.4

For Problems 1–12, find the distance between each of the pairs of points. Express answers in simplest radical form.

1. $(-2, -1), (7, 11)$ **2.** $(2, 1), (10, 7)$

3. $(1, -1), (3, -4)$ **4.** $(-1, 3), (2, -2)$

5. $(6, -4), (9, -7)$ **6.** $(-5, 2), (-1, 6)$

7. $(-3, 3), (0, -3)$ **8.** $(-2, -4), (4, 0)$

9. $(1, -6), (-5, -6)$ **10.** $(-2, 3), (-2, -7)$

11. $(1, 7), (4, -2)$ **12.** $(6, 4), (-4, -8)$

13. Verify that the points $(-3, 1)$, $(5, 7)$, and $(8, 3)$ are vertices of a right triangle. [*Hint:* If $a^2 + b^2 = c^2$, then it is a right triangle with the right angle opposite side c.]

14. Verify that the points $(0, 3)$, $(2, -3)$, and $(-4, -5)$ are vertices of an isosceles triangle.

15. Verify that the points $(7, 12)$ and $(11, 18)$ divide the line segment joining $(3, 6)$ and $(15, 24)$ into three segments of equal length.

16. Verify that $(3, 1)$ is the midpoint of the line segment joining $(-2, 6)$ and $(8, -4)$.

For Problems 17–28, graph the line determined by the two points and find the slope of the line.

17. $(1, 2), (4, 6)$ **18.** $(3, 1), (-2, -2)$

19. $(-4, 5), (-1, -2)$ **20.** $(-2, 5), (3, -1)$

21. $(2, 6), (6, -2)$ **22.** $(-2, -1), (2, -5)$

23. $(-6, 1), (-1, 4)$ **24.** $(-3, 3), (2, 3)$

25. $(-2, -4), (2, -4)$ **26.** $(1, -5), (4, -1)$

27. $(0, -2), (4, 0)$ **28.** $(-4, 0), (0, -6)$

29. Find x if the line through $(-2, 4)$ and $(x, 6)$ has a slope of $\dfrac{2}{9}$.

30. Find y if the line through $(1, y)$ and $(4, 2)$ has a slope of $\dfrac{5}{3}$.

31. Find x if the line through $(x, 4)$ and $(2, -5)$ has a slope of $-\dfrac{9}{4}$.

32. Find y if the line through $(5, 2)$ and $(-3, y)$ has a slope of $-\dfrac{7}{8}$.

For Problems 33–40, you are given one point on a line and the slope of the line. Find the coordinates of three other points on the line.

33. $(2, 5)$, $m = \dfrac{1}{2}$　　**34.** $(3, 4)$, $m = \dfrac{5}{6}$

35. $(-3, 4)$, $m = 3$　　**36.** $(-3, -6)$, $m = 1$

37. $(5, -2)$, $m = -\dfrac{2}{3}$　　**38.** $(4, -1)$, $m = -\dfrac{3}{4}$

39. $(-2, -4)$, $m = -2$　　**40.** $(-5, 3)$, $m = -3$

For Problems 41–48, graph the line that passes through the given point and has the given slope.

41. $(3, 1)$　$m = \dfrac{2}{3}$　　**42.** $(-1, 0)$　$m = \dfrac{3}{4}$

43. $(-2, 3)$　$m = -1$　　**44.** $(1, -4)$　$m = -3$

45. $(0, 5)$　$m = \dfrac{-1}{4}$　　**46.** $(-3, 4)$　$m = \dfrac{-3}{2}$

47. $(2, -2)$　$m = \dfrac{3}{2}$　　**48.** $(3, -4)$　$m = \dfrac{5}{2}$

For Problems 49–58, find the coordinates of two points on the given line, and then use those coordinates to find the slope of the line.

49. $2x + 3y = 6$　　**50.** $4x + 5y = 20$

51. $x - 2y = 4$　　**52.** $3x - y = 12$

53. $4x - 7y = 12$　　**54.** $2x + 7y = 11$

55. $y = 4$　　**56.** $x = 3$

57. $y = -5x$　　**58.** $y - 6x = 0$

59. A certain highway has a 2% grade. How many feet does it rise in a horizontal distance of 1 mile? (1 mile = 5280 feet)

60. The grade of a highway up a hill is 30%. How much change in horizontal distance is there if the vertical height of the hill is 75 feet?

61. Suppose that a highway rises a distance of 215 feet in a horizontal distance of 2640 feet. Express the grade of the highway to the nearest tenth of a percent.

62. If the ratio of rise to run is to be $\dfrac{3}{5}$ for some steps and the rise is 19 centimeters, find the run to the nearest centimeter.

63. If the ratio of rise to run is to be $\dfrac{2}{3}$ for some steps and the run is 28 centimeters, find the rise to the nearest centimeter.

64. Suppose that a county ordinance requires a $2\dfrac{1}{4}$% "fall" for a sewage pipe from the house to the main pipe at the street. How much vertical drop must there be for a horizontal distance of 45 feet? Express the answer to the nearest tenth of a foot.

■ ■ ■ Thoughts into words

65. How would you explain the concept of slope to someone who was absent from class the day it was discussed?

66. If one line has a slope of $\dfrac{2}{5}$ and another line has a slope of $\dfrac{3}{7}$, which line is steeper? Explain your answer.

67. Suppose that a line has a slope of $\dfrac{2}{3}$ and contains the point $(4, 7)$. Are the points $(7, 9)$ and $(1, 3)$ also on the line? Explain your answer.

■ ■ ■ Further investigations

68. Sometimes it is necessary to find the coordinate of a point on a number line that is located somewhere between two given points. For example, suppose that we want to find the coordinate (x) of the point located two-thirds of the distance from 2 to 8. Because the total distance from 2 to 8 is $8 - 2 = 6$ units, we can start at 2 and move $\frac{2}{3}(6) = 4$ units toward 8. Thus $x = 2 + \frac{2}{3}(6) = 2 + 4 = 6$.

For each of the following, find the coordinate of the indicated point on a number line.

(a) Two-thirds of the distance from 1 to 10
(b) Three-fourths of the distance from -2 to 14
(c) One-third of the distance from -3 to 7
(d) Two-fifths of the distance from -5 to 6
(e) Three-fifths of the distance from -1 to -11
(f) Five-sixths of the distance from 3 to -7

69. Now suppose that we want to find the coordinates of point P, which is located two-thirds of the distance from $A(1, 2)$ to $B(7, 5)$ in a coordinate plane. We have plotted the given points A and B in Figure 7.42 to help with the analysis of this problem. Point D is two-thirds of the distance from A to C because parallel lines cut off proportional segments on every transversal that intersects the lines. Thus $\overline{AC}$ can be treated as a segment of a number line, as shown in Figure 7.43.

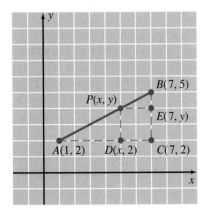

Figure 7.42

Figure 7.43

Therefore,

$$x = 1 + \frac{2}{3}(7 - 1) = 1 + \frac{2}{3}(6) = 5$$

Similarly, $\overline{CB}$ can be treated as a segment of a number line, as shown in Figure 7.44. Therefore,

$$y = 2 + \frac{2}{3}(5 - 2) = 2 + \frac{2}{3}(3) = 4$$

The coordinates of point P are $(5, 4)$.

Figure 7.44

For each of the following, find the coordinates of the indicated point in the xy plane.

(a) One-third of the distance from $(2, 3)$ to $(5, 9)$
(b) Two-thirds of the distance from $(1, 4)$ to $(7, 13)$
(c) Two-fifths of the distance from $(-2, 1)$ to $(8, 11)$
(d) Three-fifths of the distance from $(2, -3)$ to $(-3, 8)$
(e) Five-eighths of the distance from $(-1, -2)$ to $(4, -10)$
(f) Seven-eighths of the distance from $(-2, 3)$ to $(-1, -9)$

70. Suppose we want to find the coordinates of the midpoint of a line segment. Let $P(x, y)$ represent the midpoint of the line segment from $A(x_1, y_1)$ to $B(x_2, y_2)$. Using the method in Problem 68, the formula for the x coordinate of the midpoint is $x = x_1 + \frac{1}{2}(x_2 - x_1)$. This formula can be simplified algebraically to produce a simpler formula.

$$x = x_1 + \frac{1}{2}(x_2 - x_1)$$

$$x = x_1 + \frac{1}{2}x_2 - \frac{1}{2}x_1$$

$$x = \frac{1}{2}x_1 + \frac{1}{2}x_2$$

$$x = \frac{x_1 + x_2}{2}$$

Hence the x coordinate of the midpoint can be interpreted as the average of the x coordinates of the endpoints of the line segment. A similar argument for the y coordinate of the midpoint gives the following formula.

$$y = \frac{y_1 + y_2}{2}$$

For each of the pairs of points, use the formula to find the midpoint of the line segment between the points.
(a) $(3, 1)$ and $(7, 5)$
(b) $(-2, 8)$ and $(6, 4)$
(c) $(-3, 2)$ and $(5, 8)$

(d) $(4, 10)$ and $(9, 25)$
(e) $(-4, -1)$ and $(-10, 5)$
(f) $(5, 8)$ and $(-1, 7)$

Graphing calculator activities

71. Remember that we did some work with parallel lines back in the graphing calculator activities in Problem Set 7.2. Now let's do some work with perpendicular lines. Be sure to set your boundaries so that the distance between tick marks is the same on both axes.

(a) Graph $y = 4x$ and $y = -\dfrac{1}{4}x$ on the same set of axes. Do they appear to be perpendicular lines?

(b) Graph $y = 3x$ and $y = \dfrac{1}{3}x$ on the same set of axes. Do they appear to be perpendicular lines?

(c) Graph $y = \dfrac{2}{5}x - 1$ and $y = -\dfrac{5}{2}x + 2$ on the same set of axes. Do they appear to be perpendicular lines?

(d) Graph $y = \dfrac{3}{4}x - 3$, $y = \dfrac{4}{3}x + 2$, and $y = -\dfrac{4}{3}x + 2$ on the same set of axes. Does there appear to be a pair of perpendicular lines?

(e) On the basis of your results in parts (a) through (d), make a statement about how we can recognize perpendicular lines from their equations.

72. For each of the following pairs of equations, (1) predict whether they represent parallel lines, perpendicular lines, or lines that intersect but are not perpendicular, and (2) graph each pair of lines to check your prediction.
(a) $5.2x + 3.3y = 9.4$ and $5.2x + 3.3y = 12.6$
(b) $1.3x - 4.7y = 3.4$ and $1.3x - 4.7y = 11.6$
(c) $2.7x + 3.9y = 1.4$ and $2.7x - 3.9y = 8.2$
(d) $5x - 7y = 17$ and $7x + 5y = 19$
(e) $9x + 2y = 14$ and $2x + 9y = 17$
(f) $2.1x + 3.4y = 11.7$ and $3.4x - 2.1y = 17.3$

7.5 Determining the Equation of a Line

To review, there are basically two types of problems to solve in coordinate geometry:

1. Given an algebraic equation, find its geometric graph.
2. Given a set of conditions pertaining to a geometric figure, find its algebraic equation.

Problems of type 1 have been our primary concern thus far in this chapter. Now let's analyze some problems of type 2 that deal specifically with straight lines. Given certain facts about a line, we need to be able to determine its algebraic equation. Let's consider some examples.

EXAMPLE 1 Find the equation of the line that has a slope of $\frac{2}{3}$ and contains the point $(1, 2)$.

Solution

First, let's draw the line and record the given information. Then choose a point (x, y) that represents any point on the line other than the given point $(1, 2)$. (See Figure 7.45.)

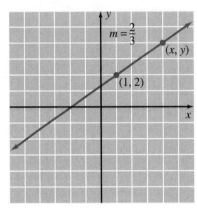

The slope determined by $(1, 2)$ and (x, y) is $\frac{2}{3}$. Thus

$$\frac{y - 2}{x - 1} = \frac{2}{3}$$

$$2(x - 1) = 3(y - 2)$$

$$2x - 2 = 3y - 6$$

$$2x - 3y = -4$$

Figure 7.45

EXAMPLE 2 Find the equation of the line that contains $(3, 2)$ and $(-2, 5)$.

Solution

First, let's draw the line determined by the given points (Figure 7.46); if we know two points, we can find the slope.

$$m = \frac{y_2 - y_1}{x_2 - x_1} = \frac{3}{-5} = -\frac{3}{5}$$

Now we can use the same approach as in Example 1.

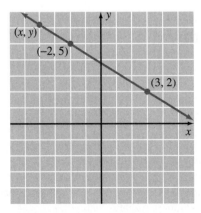

Figure 7.46

Form an equation using a variable point (x, y), one of the two given points, and the slope of $-\dfrac{3}{5}$.

$$\frac{y - 5}{x + 2} = \frac{3}{-5} \qquad \left(-\frac{3}{5} = \frac{3}{-5}\right)$$

$$3(x + 2) = -5(y - 5)$$

$$3x + 6 = -5y + 25$$

$$3x + 5y = 19$$

E X A M P L E   3   Find the equation of the line that has a slope of $\dfrac{1}{4}$ and a y intercept of 2.

Solution

A y intercept of 2 means that the point $(0, 2)$ is on the line (Figure 7.47).

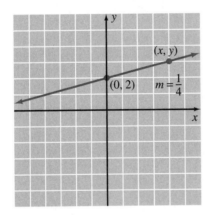

Figure 7.47

Choose a variable point (x, y) and proceed as in the previous examples.

$$\frac{y - 2}{x - 0} = \frac{1}{4}$$

$$1(x - 0) = 4(y - 2)$$

$$x = 4y - 8$$

$$x - 4y = -8$$

Perhaps it would be helpful to pause a moment and look back over Examples 1, 2, and 3. Note that we used the same basic approach in all three situations. We chose a variable point (x, y) and used it to determine the equation that satisfies the conditions given in the problem. The approach we took in the previous examples can be generalized to produce some special forms of equations of straight lines.

Point-Slope Form

E X A M P L E   4

Find the equation of the line that has a slope of m and contains the point (x_1, y_1).

Solution

Choose (x, y) to represent any other point on the line (Figure 7.48), and the slope of the line is therefore given by

$$m = \frac{y - y_1}{x - x_1}, \qquad x \neq x_1$$

from which

$$y - y_1 = m(x - x_1)$$

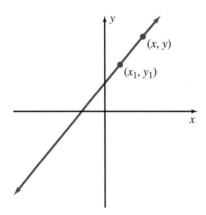

Figure 7.48

We refer to the equation

$$y - y_1 = m(x - x_1)$$

as the **point-slope form** of the equation of a straight line. Instead of the approach we used in Example 1, we could use the point-slope form to write the equation of a line with a given slope that contains a given point. For example, we can determine the equation of the line that has a slope of $\frac{3}{5}$ and contains the point $(2, 4)$ as follows:

$$y - y_1 = m(x - x_1)$$

Substitute $(2, 4)$ for (x_1, y_1) and $\dfrac{3}{5}$ for m.

$$y - 4 = \frac{3}{5}(x - 2)$$

$$5(y - 4) = 3(x - 2)$$

$$5y - 20 = 3x - 6$$

$$-14 = 3x - 5y$$

Slope-Intercept Form

E X A M P L E   5   Find the equation of the line that has a slope of m and a y intercept of b.

Solution

A y intercept of b means that the line contains the point $(0, b)$, as in Figure 7.49. Therefore, we can use the point-slope form as follows:

$$y - y_1 = m(x - x_1)$$

$$y - b = m(x - 0)$$

$$y - b = mx$$

$$y = mx + b$$

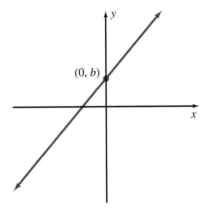

Figure 7.49

We refer to the equation

$$y = mx + b$$

as the **slope-intercept form** of the equation of a straight line. We use it for three primary purposes, as the next three examples illustrate.

EXAMPLE 6 Find the equation of the line that has a slope of $\frac{1}{4}$ and a y intercept of 2.

Solution

This is a restatement of Example 3, but this time we will use the slope-intercept form ($y = mx + b$) of a line to write its equation. Because $m = \frac{1}{4}$ and $b = 2$, we can substitute these values into $y = mx + b$.

$$y = mx + b$$

$$y = \frac{1}{4}x + 2$$

$$4y = x + 8 \qquad \text{Multiply both sides by 4.}$$

$$x - 4y = -8 \qquad \text{Same result as in Example 3.} \qquad ■$$

EXAMPLE 7 Find the slope of the line when the equation is $3x + 2y = 6$.

Solution

We can solve the equation for y in terms of x and then compare it to the slope-intercept form to determine its slope. Thus

$$3x + 2y = 6$$

$$2y = -3x + 6$$

$$y = -\frac{3}{2}x + 3$$

$$y = -\frac{3}{2}x + 3 \qquad y = mx + b$$

The slope of the line is $-\frac{3}{2}$. Furthermore, the y intercept is 3. ■

EXAMPLE 8 Graph the line determined by the equation $y = \frac{2}{3}x - 1$.

Solution

Comparing the given equation to the general slope-intercept form, we see that the slope of the line is $\frac{2}{3}$ and the y intercept is -1. Because the y intercept is -1, we can plot the point $(0, -1)$. Then, because the slope is $\frac{2}{3}$, let's move 3 units to the right and 2 units up from $(0, -1)$ to locate the point $(3, 1)$. The two points $(0, -1)$ and

(3, 1) determine the line in Figure 7.50. (Again, you should determine a third point as a check point.)

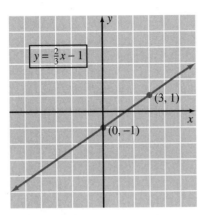

Figure 7.50

In general, if the equation of a nonvertical line is written in slope-intercept form ($y = mx + b$), the coefficient of x is the slope of the line, and the constant term is the y intercept. (Remember that the concept of slope is not defined for a vertical line.)

Parallel and Perpendicular Lines

We can use two important relationships between lines and their slopes to solve certain kinds of problems. It can be shown that nonvertical parallel lines have the same slope and that two nonvertical lines are perpendicular if the product of their slopes is −1. (Details for verifying these facts are left to another course.) In other words, if two lines have slopes m_1 and m_2, respectively, then

1. The two lines are parallel if and only if $m_1 = m_2$.
2. The two lines are perpendicular if and only if $(m_1)(m_2) = -1$.

The following examples demonstrate the use of these properties.

EXAMPLE 9

(a) Verify that the graphs of $2x + 3y = 7$ and $4x + 6y = 11$ are parallel lines.
(b) Verify that the graphs of $8x - 12y = 3$ and $3x + 2y = 2$ are perpendicular lines.

Solution

(a) Let's change each equation to slope-intercept form.

$$2x + 3y = 7 \quad \longrightarrow \quad 3y = -2x + 7$$

$$y = -\frac{2}{3}x + \frac{7}{3}$$

$$4x + 6y = 11 \qquad \longrightarrow \qquad 6y = -4x + 11$$

$$y = -\frac{4}{6}x + \frac{11}{6}$$

$$y = -\frac{2}{3}x + \frac{11}{6}$$

Both lines have a slope of $-\frac{2}{3}$, but they have different y intercepts. Therefore, the two lines are parallel.

(b) Solving each equation for y in terms of x, we obtain

$$8x - 12y = 3 \qquad \longrightarrow \qquad -12y = -8x + 3$$

$$y = \frac{8}{12}x - \frac{3}{12}$$

$$y = \frac{2}{3}x - \frac{1}{4}$$

$$3x + 2y = 2 \qquad \longrightarrow \qquad 2y = -3x + 2$$

$$y = -\frac{3}{2}x + 1$$

Because $\left(\dfrac{2}{3}\right)\left(-\dfrac{3}{2}\right) = -1$ (the product of the two slopes is -1), the lines are perpendicular. ■

REMARK: The statement "the product of two slopes is -1" is the same as saying that the two slopes are negative reciprocals of each other; that is, $m_1 = -\dfrac{1}{m_2}$.

E X A M P L E   1 0

Find the equation of the line that contains the point $(1, 4)$ and is parallel to the line determined by $x + 2y = 5$.

Solution

First, let's draw a figure to help in our analysis of the problem (Figure 7.51). Because the line through $(1, 4)$ is to be parallel to the line determined by $x + 2y = 5$, it must have the same slope. Let's find the slope by changing $x + 2y = 5$ to the slope-intercept form.

$$x + 2y = 5$$

$$2y = -x + 5$$

$$y = -\frac{1}{2}x + \frac{5}{2}$$

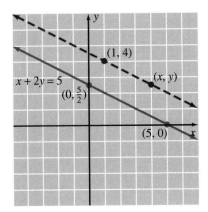

Figure 7.51

The slope of both lines is $-\dfrac{1}{2}$. Now we can choose a variable point (x, y) on the line through $(1, 4)$ and proceed as we did in earlier examples.

$$\frac{y - 4}{x - 1} = \frac{1}{-2}$$

$$1(x - 1) = -2(y - 4)$$

$$x - 1 = -2y + 8$$

$$x + 2y = 9$$

■

E X A M P L E   1 1

Find the equation of the line that contains the point $(-1, -2)$ and is perpendicular to the line determined by $2x - y = 6$.

Solution

First, let's draw a figure to help in our analysis of the problem (Figure 7.52). Because the line through $(-1, -2)$ is to be perpendicular to the line determined by $2x - y = 6$, its slope must be the negative reciprocal of the slope of $2x - y = 6$. Let's find the slope of $2x - y = 6$ by changing it to the slope-intercept form.

$$2x - y = 6$$

$$-y = -2x + 6$$

$$y = 2x - 6 \qquad \text{The slope is 2.}$$

The slope of the desired line is $-\dfrac{1}{2}$ (the negative reciprocal of 2), and we can proceed as before by using a variable point (x, y).

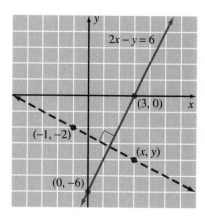

Figure 7.52

$$\frac{y + 2}{x + 1} = \frac{1}{-2}$$

$$1(x + 1) = -2(y + 2)$$

$$x + 1 = -2y - 4$$

$$x + 2y = -5$$

We use two forms of equations of straight lines extensively. They are the **standard form** and the **slope-intercept form,** and we describe them as follows.

Standard Form $Ax + By = C$, where B and C are integers and A is a non-negative integer (A and B not both zero).

Slope-Intercept Form $y = mx + b$, where m is a real number representing the slope and b is a real number representing the y intercept.

PROBLEM SET 7.5

For Problems 1–8, write the equation of the line that has the indicated slope and contains the indicated point. Express final equations in standard form.

1. $m = \dfrac{1}{2}$, $(3, 5)$

2. $m = \dfrac{1}{3}$, $(2, 3)$

3. $m = 3$, $(-2, 4)$

4. $m = -2$, $(-1, 6)$

5. $m = -\dfrac{3}{4}$, $(-1, -3)$

6. $m = -\dfrac{3}{5}$, $(-2, -4)$

7. $m = \dfrac{5}{4}$, $(4, -2)$

8. $m = \dfrac{3}{2}$, $(8, -2)$

For Problems 9–18, write the equation of the line that contains the indicated pair of points. Express final equations in standard form.

9. $(2, 1), (6, 5)$

10. $(-1, 2), (2, 5)$

11. $(-2, -3), (2, 7)$

12. $(-3, -4), (1, 2)$

13. $(-3, 2), (4, 1)$

14. $(-2, 5), (3, -3)$

15. $(-1, -4), (3, -6)$

16. $(3, 8), (7, 2)$

17. $(0, 0), (5, 7)$

18. $(0, 0), (-5, 9)$

For Problems 19–26, write the equation of the line that has the indicated slope (m) and y intercept (b). Express final equations in slope-intercept form.

19. $m = \dfrac{3}{7}, \quad b = 4$

20. $m = \dfrac{2}{9}, \quad b = 6$

21. $m = 2, \quad b = -3$

22. $m = -3, \quad b = -1$

23. $m = -\dfrac{2}{5}, \quad b = 1$

24. $m = -\dfrac{3}{7}, \quad b = 4$

25. $m = 0, \quad b = -4$

26. $m = \dfrac{1}{5}, \quad b = 0$

For Problems 27–42, write the equation of the line that satisfies the given conditions. Express final equations in standard form.

27. x intercept of 2 and y intercept of -4

28. x intercept of -1 and y intercept of -3

29. x intercept of -3 and slope of $-\dfrac{5}{8}$

30. x intercept of 5 and slope of $-\dfrac{3}{10}$

31. Contains the point $(2, -4)$ and is parallel to the y axis

32. Contains the point $(-3, -7)$ and is parallel to the x axis

33. Contains the point $(5, 6)$ and is perpendicular to the y axis

34. Contains the point $(-4, 7)$ and is perpendicular to the x axis

35. Contains the point $(1, 3)$ and is parallel to the line $x + 5y = 9$

36. Contains the point $(-1, 4)$ and is parallel to the line $x - 2y = 6$

37. Contains the origin and is parallel to the line $4x - 7y = 3$

38. Contains the origin and is parallel to the line $-2x - 9y = 4$

39. Contains the point $(-1, 3)$ and is perpendicular to the line $2x - y = 4$

40. Contains the point $(-2, -3)$ and is perpendicular to the line $x + 4y = 6$

41. Contains the origin and is perpendicular to the line $-2x + 3y = 8$

42. Contains the origin and is perpendicular to the line $y = -5x$

For Problems 43–48, change the equation to slope-intercept form and determine the slope and y intercept of the line.

43. $3x + y = 7$

44. $5x - y = 9$

45. $3x + 2y = 9$

46. $x - 4y = 3$

47. $x = 5y + 12$

48. $-4x - 7y = 14$

For Problems 49–56, use the slope-intercept form to graph the following lines.

49. $y = \dfrac{2}{3}x - 4$

50. $y = \dfrac{1}{4}x + 2$

51. $y = 2x + 1$

52. $y = 3x - 1$

53. $y = -\dfrac{3}{2}x + 4$

54. $y = -\dfrac{5}{3}x + 3$

55. $y = -x + 2$

56. $y = -2x + 4$

For Problems 57–66, graph the following lines using the technique that seems most appropriate.

57. $y = -\dfrac{2}{5}x - 1$

58. $y = -\dfrac{1}{2}x + 3$

59. $x + 2y = 5$

60. $2x - y = 7$

61. $-y = -4x + 7$

62. $3x = 2y$

63. $7y = -2x$

64. $y = -3$

65. $x = 2$

66. $y = -x$

For Problems 67–70, the situations can be described by the use of linear equations in two variables. If two pairs of values are known, then we can determine the equation by using the approach we used in Example 2 of this section. For each of the following, assume that the relationship can be expressed as a linear equation in two variables, and use the given information to determine the equation. Express the equation in slope-intercept form.

67. A company uses 7 pounds of fertilizer for a lawn that measures 5000 square feet and 12 pounds for a lawn that measures 10,000 square feet. Let y represent the pounds of fertilizer and x the square footage of the lawn.

68. A new diet fad claims that a person weighing 140 pounds should consume 1490 daily calories and that a 200-pound person should consume 1700 calories. Let y represent the calories and x the weight of the person in pounds.

69. Two banks on opposite corners of a town square had signs that displayed the current temperature. One bank displayed the temperature in degrees Celsius and the other in degrees Fahrenheit. A temperature of 10°C was displayed at the same time as a temperature of 50°F. On another day, a temperature of −5°C was displayed at the same time as a temperature of 23°F. Let y represent the temperature in degrees Fahrenheit and x the temperature in degrees Celsius.

70. An accountant has a schedule of depreciation for some business equipment. The schedule shows that after 12 months the equipment is worth $7600 and that after 20 months it is worth $6000. Let y represent the worth and x represent the time in months.

■ ■ ■ Thoughts into words

71. What does it mean to say that two points determine a line?

72. How would you help a friend determine the equation of the line that is perpendicular to $x - 5y = 7$ and contains the point $(5, 4)$?

73. Explain how you would find the slope of the line $y = 4$.

■ ■ ■ Further investigations

74. The equation of a line that contains the two points (x_1, y_1) and (x_2, y_2) is $\dfrac{y - y_1}{x - x_1} = \dfrac{y_2 - y_1}{x_2 - x_1}$. We often refer to this as the **two-point form** of the equation of a straight line. Use the two-point form and write the equation of the line that contains each of the indicated pairs of points. Express final equations in standard form.
   (a) $(1, 1)$ and $(5, 2)$
   (b) $(2, 4)$ and $(-2, -1)$
   (c) $(-3, 5)$ and $(3, 1)$
   (d) $(-5, 1)$ and $(2, -7)$

75. Let $Ax + By = C$ and $A'x + B'y = C'$ represent two lines. Change both of these equations to slope-intercept form, and then verify each of the following properties.
   (a) If $\dfrac{A}{A'} = \dfrac{B}{B'} \neq \dfrac{C}{C'}$, then the lines are parallel.
   (b) If $AA' = -BB'$, then the lines are perpendicular.

76. The properties in Problem 75 provide us with another way to write the equation of a line parallel or perpendicular to a given line that contains a given point not on the line. For example, suppose that we want the equation of the line perpendicular to $3x + 4y = 6$ that contains the point $(1, 2)$. The form $4x - 3y = k$, where k is a constant, represents a family of lines perpendicular to $3x + 4y = 6$ because we have satisfied the condition $AA' = -BB'$. Therefore, to find what specific line of the family contains $(1, 2)$, we substitute 1 for x and 2 for y to determine k.

$$4x - 3y = k$$

$$4(1) - 3(2) = k$$

$$-2 = k$$

Thus the equation of the desired line is $4x - 3y = -2$.

Use the properties from Problem 75 to help write the equation of each of the following lines.
(a) Contains $(1, 8)$ and is parallel to $2x + 3y = 6$

(b) Contains $(-1, 4)$ and is parallel to $x - 2y = 4$

(c) Contains $(2, -7)$ and is perpendicular to $3x - 5y = 10$

(d) Contains $(-1, -4)$ and is perpendicular to $2x + 5y = 12$

77. The problem of finding the perpendicular bisector of a line segment presents itself often in the study of analytic geometry. As with any problem of writing the equation of a line, you must determine the slope of the line and a point that the line passes through. A perpendicular bisector passes through the midpoint of the line segment and has a slope that is the negative reciprocal of the slope of the line segment. The problem can be solved as follows:

Find the perpendicular bisector of the line segment between the points $(1, -2)$ and $(7, 8)$.

The midpoint of the line segment is $\left(\dfrac{1 + 7}{2}, \dfrac{-2 + 8}{2} \right)$
$= (4, 3)$.

The slope of the line segment is $m = \dfrac{8 - (-2)}{7 - 1}$

$$= \frac{10}{6} = \frac{5}{3}.$$

Hence the perpendicular bisector will pass through the point $(4, 3)$ and have a slope of $m = -\dfrac{3}{5}$.

$$y - 3 = -\frac{3}{5}(x - 4)$$

$$5(y - 3) = -3(x - 4)$$

$$5y - 15 = -3x + 12$$

$$3x + 5y = 27$$

Thus the equation of the perpendicular bisector of the line segment between the points $(1, -2)$ and $(7, 8)$ is $3x + 5y = 27$.

Find the perpendicular bisector of the line segment between the points for the following. Write the equation in standard form.
(a) $(-1, 2)$ and $(3, 0)$
(b) $(6, -10)$ and $(-4, 2)$
(c) $(-7, -3)$ and $(5, 9)$
(d) $(0, 4)$ and $(12, -4)$

📟 Graphing calculator activities

78. Predict whether each of the following pairs of equations represents parallel lines, perpendicular lines, or lines that intersect but are not perpendicular. Then graph each pair of lines to check your predictions. (The properties presented in Problem 75 should be very helpful.)
(a) $5.2x + 3.3y = 9.4$ and $5.2x + 3.3y = 12.6$
(b) $1.3x - 4.7y = 3.4$ and $1.3x - 4.7y = 11.6$
(c) $2.7x + 3.9y = 1.4$ and $2.7x - 3.9y = 8.2$
(d) $5x - 7y = 17$ and $7x + 5y = 19$
(e) $9x + 2y = 14$ and $2x + 9y = 17$
(f) $2.1x + 3.4y = 11.7$ and $3.4x - 2.1y = 17.3$
(g) $7.1x - 2.3y = 6.2$ and $2.3x + 7.1y = 9.9$
(h) $-3x + 9y = 12$ and $9x - 3y = 14$
(i) $2.6x - 5.3y = 3.4$ and $5.2x - 10.6y = 19.2$
(j) $4.8x - 5.6y = 3.4$ and $6.1x + 7.6y = 12.3$

SUMMARY

(7.1) The **Cartesian** (or **rectangular**) **coordinate system** is used to graph ordered pairs of real numbers. The first number, a, of the ordered pair (a, b) is called the **abscissa,** and the second number, b, is called the **ordinate;** together they are referred to as the **coordinates** of a point.

Two basic kinds of problems exist in coordinate geometry:

1. Given an algebraic equation, find its geometric graph.
2. Given a set of conditions that pertains to a geometric figure, find its algebraic equation.

A **solution** of an equation in two variables is an ordered pair of real numbers that satisfies the equation.

The following suggestions are offered for **graphing an equation** in two variables.

1. Determine what type of symmetry the equation exhibits.
2. Find the intercepts.
3. Solve the equation for y in terms of x or for x in terms of y if it is not already in such a form.
4. Set up a table of ordered pairs that satisfy the equation. The type of symmetry will affect your choice of values in the table.
5. Plot the points associated with the ordered pairs from the table, and connect them with a smooth curve. Then, if appropriate, reflect this part of the curve according to the symmetry shown by the equation.

(7.2) Any equation of the form $Ax + By = C$, where A, B, and C are constants (A and B not both zero) and x and y are variables, is a **linear equation** and its graph is a **straight line.**

Any equation of the form $Ax + By = C$, where $C = 0$, is a straight line that contains the origin.

Any equation of the form $x = a$, where a is a constant, is a line parallel to the y axis that has an x intercept of a.

Any equation of the form $y = b$, where b is a constant, is a line parallel to the x axis that has a y intercept of b.

(7.3) **Linear inequalities** in two variables are of the form $Ax + By > C$ or $Ax + By < C$. To **graph a linear inequality,** we suggest the following steps.

1. First, graph the corresponding equality. Use a solid line if equality is included in the original statement. Use a dashed line if equality is not included.
2. Choose a test point not on the line and substitute its coordinates into the inequality.
3. The graph of the original inequality is
   (a) the half plane that contains the test point if the inequality is satisfied by that point, or
   (b) the half plane that does not contain the test point if the inequality is not satisfied by the point.

(7.4) The distance between any two points (x_1, y_1) and (x_2, y_2) is given by the **distance formula,**

$$d = \sqrt{(x_2 - x_1)^2 + (y_2 - y_1)^2}$$

The **slope** (denoted by m) of a line determined by the points (x_1, y_1) and (x_2, y_2) is given by the slope formula,

$$m = \frac{y_2 - y_1}{x_2 - x_1}, \qquad x_2 \neq x_1$$

(7.5) The equation $y = mx + b$ is referred to as the **slope-intercept form** of the equation of a straight line. If the equation of a nonvertical line is written in this y form, the coefficient of x is the slope of the line and the constant term is the y intercept.

If two lines have slopes m_1 and m_2, respectively, then

1. The two lines are parallel if and only if $m_1 = m_2$.
2. The two lines are perpendicular if and only if $(m_1)(m_2) = -1$.

To determine the equation of a straight line given a set of conditions, we can use the point-slope form, $y - y_1 = m(x - x_1)$, or $\dfrac{y - y_1}{x - x_1} = m$. The conditions generally fall into one of the following four categories.

1. Given the slope and a point contained in the line
2. Given two points contained in the line
3. Given a point contained in the line and that the line is parallel to another line
4. Given a point contained in the line and that the line is perpendicular to another line

The result can then be expressed in standard form or slope-intercept form.

CHAPTER 7   REVIEW PROBLEM SET

1. Find the slope of the line determined by each pair of points.

   (a) $(3, 4), (-2, -2)$ (b) $(-2, 3), (4, -1)$

2. Find y if the line through $(-4, 3)$ and $(12, y)$ has a slope of $\dfrac{1}{8}$.

3. Find x if the line through $(x, 5)$ and $(3, -1)$ has a slope of $-\dfrac{3}{2}$.

4. Find the slope of each of the following lines.

   (a) $4x + y = 7$ (b) $2x - 7y = 3$

5. Find the lengths of the sides of a triangle whose vertices are at $(2, 3), (5, -1)$, and $(-4, -5)$.

6. Find the distance between each of the pairs of points.

   (a) $(-1, 4), (1, -2)$ (b) $(5, 0), (2, 7)$

7. Verify that $(1, 6)$ is the midpoint of the line segment joining $(3, 2)$ and $(-1, 10)$.

For Problems 8–15, write the equation of the line that satisfies the stated conditions. Express final equations in standard form.

8. Containing the points $(-1, 2)$ and $(3, -5)$

9. Having a slope of $-\dfrac{3}{7}$ and a y intercept of 4

10. Containing the point $(-1, -6)$ and having a slope of $\dfrac{2}{3}$

11. Containing the point $(2, 5)$ and parallel to the line $x - 2y = 4$

12. Containing the point $(-2, -6)$ and perpendicular to the line $3x + 2y = 12$

13. Containing the points $(0, 4)$ and $(2, 6)$

14. Containing the point $(3, -5)$ and having a slope of -1

15. Containing the point $(-8, 3)$ and parallel to the line $4x + y = 7$

For Problems 16–25, graph each equation.

16. $2x - y = 6$

17. $y = 2x - 5$

18. $y = -2x - 1$

19. $y = -4x$

20. $-3x - 2y = 6$

21. $x = 2y + 4$

22. $5x - y = -5$

23. $y = -\dfrac{1}{2}x + 3$

24. $y = \dfrac{3x - 4}{2}$

25. $y = 4$

26. $2x + 3y = 0$

27. $y = \dfrac{3}{5}x - 4$

28. $x = 1$

29. $x = -3$

30. $y = -2$

31. $2x - 3y = 3$

32. $y = x^3 + 2$

33. $y = -x^3$

34. $y = x^2 + 3$

35. $y = -2x^2 - 1$

For Problems 36–41, graph each inequality.

36. $-x + 3y < -6$

37. $x + 2y \geq 4$

38. $2x - 3y \leq 6$

39. $y > -\dfrac{1}{2}x + 3$

40. $y < 2x - 5$

41. $y \geq \dfrac{2}{3}x$

42. A certain highway has a 6% grade. How many feet does it rise in a horizontal distance of 1 mile?

43. If the ratio of rise to run is to be $\dfrac{2}{3}$ for the steps of a staircase, and the run is 12 inches, find the rise.

44. Find the slope of any line that is perpendicular to the line $-3x + 5y = 7$.

45. Find the slope of any line that is parallel to the line $4x + 5y = 10$.

46. The taxes for a primary residence can be described by a linear relationship. Find the equation for the relationship if the taxes for a home valued at $200,000 are $2400 and the taxes are $3150 when the home is valued

at \$250,000. Let y be the taxes and x the value of the home. Write the equation in slope-intercept form.

47. The freight charged by a trucking firm for a parcel under 200 pounds depends on the miles it is being shipped. To ship a 150-pound parcel 300 miles, it costs \$40. If the same parcel is shipped 1000 miles, the cost is \$180. Assume the relationship between the cost and miles is linear. Find the equation for the relationship. Let y be the cost and x be the miles. Write the equation in slope-intercept form.

48. On a final exam in math class, the number of points earned has a linear relationship with the number of correct answers. John got 96 points when he answered 12 questions correctly. Kimberly got 144 points when she answered 18 questions correctly. Find the equation for the relationship. Let y be the number of points and x be the number of correct answers. Write the equation in slope-intercept form.

49. The time needed to install computer cabling has a linear relationship with the number of feet of cable being installed. It takes $1\frac{1}{2}$ hours to install 300 feet, and 1050 feet can be installed in 4 hours. Find the equation for the relationship. Let y be the feet of cable installed and x be the time in hours. Write the equation in slope-intercept form.

50. Determine the type(s) of symmetry (symmetry with respect to the x axis, y axis, and/or origin) exhibited by the graph of each of the following equations. Do not sketch the graph.

(a) $y = x^2 + 4$ **(b)** $xy = -4$
(c) $y = -x^3$ **(d)** $x = y^4 + 2y^2$

TEST

1. Find the slope of the line determined by the points $(-2, 4)$ and $(3, -2)$.

2. Find the slope of the line determined by the equation $3x - 7y = 12$.

3. Find the length of the line segment whose endpoints are $(4, 2)$ and $(-3, -1)$. Express the answer in simplest radical form.

4. Find the equation of the line that has a slope of $-\dfrac{3}{2}$ and contains the point $(4, -5)$. Express the equation in standard form.

5. Find the equation of the line that contains the points $(-4, 2)$ and $(2, 1)$. Express the equation in slope-intercept form.

6. Find the equation of the line that is parallel to the line $5x + 2y = 7$ and contains the point $(-2, -4)$. Express the equation in standard form.

7. Find the equation of the line that is perpendicular to the line $x - 6y = 9$ and contains the point $(4, 7)$. Express the equation in standard form.

8. What kind(s) of symmetry does the graph of $y = 9x$ exhibit?

9. What kind(s) of symmetry does the graph of $y^2 = x^2 + 6$ exhibit?

10. What kind(s) of symmetry does the graph of $x^2 + 6x + 2y^2 - 8 = 0$ exhibit?

11. What is the slope of all lines that are parallel to the line $7x - 2y = 9$?

12. What is the slope of all lines that are perpendicular to the line $4x + 9y = -6$?

13. Find the x intercept of the line $y = \dfrac{3}{5}x - \dfrac{2}{3}$.

14. Find the y intercept of the line $\dfrac{3}{4}x - \dfrac{2}{5}y = \dfrac{1}{4}$.

15. The grade of a highway up a hill is 25%. How much change in horizontal distance is there if the vertical height of the hill is 120 feet?

16. Suppose that a highway rises 200 feet in a horizontal distance of 3000 feet. Express the grade of the highway to the nearest tenth of a percent.

17. If the ratio of rise to run is to be $\dfrac{3}{4}$ for the steps of a staircase, and the rise is 32 centimeters, find the run to the nearest centimeter.

For Problems 18–23, graph each equation.

18. $y = -x^2 - 3$

19. $y = -x - 3$

20. $-3x + y = 5$

21. $3y = 2x$

22. $\dfrac{1}{3}x + \dfrac{1}{2}y = 2$

23. $y = \dfrac{-x - 1}{4}$

For Problems 24 and 25, graph each inequality.

24. $2x - y < 4$

25. $3x + 2y \geq 6$

Examples of conic sections, in particular parabolas and ellipses, can be found in corporate logos throughout the world.

© AFP/CORBIS

Conic Sections

P arabolas, circles, ellipses, and hyperbolas can be formed when a plane intersects a conical surface as shown in Figure 8.1; we often refer to these curves as the **conic sections.** A flashlight produces a "cone of light" that can be cut by the plane of a wall to illustrate the conic sections. Try shining a flashlight against a wall at different angles to produce a circle, an ellipse, a parabola, and one branch of a hyperbola. (You may find it difficult to distinguish between a parabola and a branch of a hyperbola.)

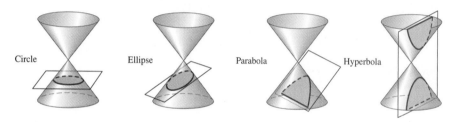

Figure 8.1

InfoTrac Project

Do a keyword search on parabolas and find an article titled "Fast Food and Fine Architecture." Write a summary of the article, mentioning some of the examples of parabolic arches found in Budapest. What is the name of the person who introduced parabolic arches into Hungarian architecture? In the article a world-renowned American trademark that uses parabolic arches is mentioned. Find the equation of one of these arches if it opens downward, has its vertex at (2.5, 9) and passes through (5, 0). Graph your equation and check your results on a graphing utility.

8.1 Graphing Parabolas

In general the graph of any equation of the form $y = ax^2 + bx + c$, where a, b, and c are real numbers and $a \neq 0$, is a parabola. At this time we want to develop a very easy and systematic way of graphing parabolas without the use of a graphing calculator. As we work with parabolas, we will use the vocabulary indicated in Figure 8.2.

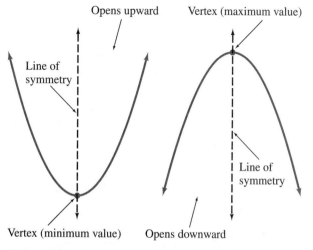

Figure 8.2

Let's begin by using the concepts of intercepts and symmetry to help us sketch the graph of the equation $y = x^2$.

E X A M P L E   1   Graph $y = x^2$.

Solution

If we replace x with $-x$, the given equation becomes $y = (-x)^2 = x^2$; therefore, we have y axis symmetry. The origin, $(0, 0)$, is a point of the graph. We can recognize from the equation that 0 is the minimum value of y; hence the point $(0, 0)$ is the vertex of the parabola. Now we can set up a table of values that uses nonnegative values for x. Plot the points determined by the table, connect them with a smooth curve, and reflect that portion of the curve across the y axis to produce Figure 8.3.

x	y
0	0
$\frac{1}{2}$	$\frac{1}{4}$
1	1
2	4
3	9

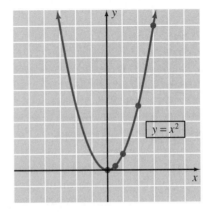

Figure 8.3

To graph parabolas, we need to be able to:

1. Find the vertex.
2. Determine whether the parabola opens upward or downward.
3. Locate two points on opposite sides of the line of symmetry.
4. Compare the parabola to the basic parabola $y = x^2$.

To graph parabolas produced by the various types of equations such as $y = x^2 + k$, $y = ax^2$, $y = (x - h)^2$, and $y = a(x - h)^2 + k$, we can compare these equations to that of the basic parabola, $y = x^2$. First, let's consider some equations of the form $y = x^2 + k$, where k is a constant.

E X A M P L E   2   Graph $y = x^2 + 1$.

Solution

Let's set up a table of values to compare y values for $y = x^2 + 1$ to corresponding y values for $y = x^2$.

x	$y = x^2$	$y = x^2 + 1$
0	0	1
1	1	2
2	4	5
−1	1	2
−2	4	5

It should be evident that y values for $y = x^2 + 1$ are *1 greater than* corresponding y values for $y = x^2$. For example, if $x = 2$ then $y = 4$ for the equation $y = x^2$, but if $x = 2$ then $y = 5$ for the equation $y = x^2 + 1$. Thus the graph of $y = x^2 + 1$ is the same as the graph of $y = x^2$, but moved up 1 unit (Figure 8.4). The vertex will move from $(0, 0)$ to $(0, 1)$.

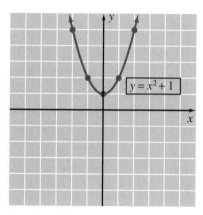

Figure 8.4

EXAMPLE 3

Graph $y = x^2 - 2$.

Solution

The y values for $y = x^2 - 2$ are *2 less than* the corresponding y values for $y = x^2$, as indicated in the following table.

x	$y = x^2$	$y = x^2 - 2$
0	0	−2
1	1	−1
2	4	2
−1	1	−1
−2	4	2

Thus the graph of $y = x^2 - 2$ is the same as the graph of $y = x^2$ but moved down 2 units (Figure 8.5). The vertex will move from $(0, 0)$ to $(0, -2)$.

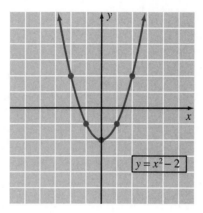

Figure 8.5

In general, the graph of a quadratic equation of the form $y = x^2 + k$ is the same as the graph of $y = x^2$ but moved up or down $|k|$ units, depending on whether k is positive or negative.

Now, let's consider some quadratic equations of the form $y = ax^2$, where a is a nonzero constant.

E X A M P L E   4   Graph $y = 2x^2$.

Solution

Again, let's use a table to make some comparisons of y values.

x	$y = x^2$	$y = 2x^2$
0	0	0
1	1	2
2	4	8
−1	1	2
−2	4	8

Obviously, the y values for $y = 2x^2$ are *twice* the corresponding y values for $y = x^2$. Thus the parabola associated with $y = 2x^2$ has the same vertex (the origin) as the graph of $y = x^2$, but it is narrower (Figure 8.6).

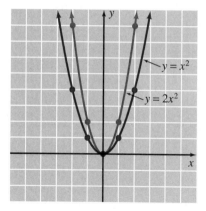

Figure 8.6

EXAMPLE 5   Graph $y = \dfrac{1}{2}x^2$.

Solution

The following table indicates some comparisons of y values.

x	$y = x^2$	$y = \dfrac{1}{2}x^2$
0	0	0
1	1	$\dfrac{1}{2}$
2	4	2
−1	1	$\dfrac{1}{2}$
−2	4	2

The y values for $y = \dfrac{1}{2}x^2$ are one-half of the corresponding y values for $y = x^2$. Therefore, the graph of $y = \dfrac{1}{2}x^2$ has the same vertex (the origin) as the graph of $y = x^2$, but it is wider (Figure 8.7).

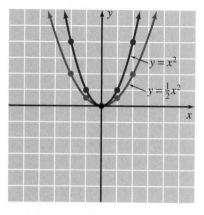

Figure 8.7

EXAMPLE 6

Graph $y = -x^2$.

Solution

x	$y = x^2$	$y = -x^2$
0	0	0
1	1	−1
2	4	−4
−1	1	−1
−2	4	−4

The y values for $y = -x^2$ are the opposites of the corresponding y values for $y = x^2$. Thus the graph of $y = -x^2$ has the same vertex (the origin) as the graph of $y = x^2$, but it is a reflection across the x axis of the basic parabola (Figure 8.8).

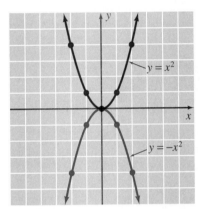

Figure 8.8

> In general, the graph of a quadratic equation of the form $y = ax^2$ has its vertex at the origin and opens upward if a is positive and downward if a is negative. The parabola is narrower than the basic parabola if $|a| > 1$ and wider if $|a| < 1$.

Let's continue our investigation of quadratic equations by considering those of the form $y = (x - h)^2$, where h is a nonzero constant.

EXAMPLE 7

Graph $y = (x - 2)^2$.

Solution

A fairly extensive table of values reveals a pattern.

x	$y = x^2$	$y = (x - 2)^2$
−2	4	16
−1	1	9
0	0	4
1	1	1
2	4	0
3	9	1
4	16	4
5	25	9

Note that $y = (x - 2)^2$ and $y = x^2$ take on the same y values, but for different values of x. More specifically, if $y = x^2$ achieves a certain y value at x equals a constant, then $y = (x - 2)^2$ achieves the same y value at x equals the *constant plus two*. In other words, the graph of $y = (x - 2)^2$ is the same as the graph of $y = x^2$ but moved 2 units to the right (Figure 8.9). The vertex will move from $(0, 0)$ to $(2, 0)$.

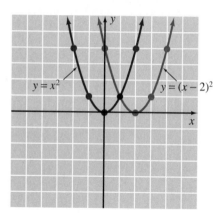

$y = x^2$ $y = (x - 2)^2$

Figure 8.9

E X A M P L E   8

Graph $y = (x + 3)^2$.

Solution

x	$y = x^2$	$y = (x + 3)^2$
−3	9	0
−2	4	1
−1	1	4
0	0	9
1	1	16
2	4	25
3	9	36

If $y = x^2$ achieves a certain y value at x equals a constant, then $y = (x + 3)^2$ achieves that same y value at x equals that *constant minus three*. Therefore, the graph of $y = (x + 3)^2$ is the same as the graph of $y = x^2$ but moved 3 units to the left (Figure 8.10). The vertex will move from $(0, 0)$ to $(-3, 0)$.

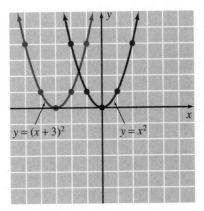

Figure 8.10

In general, the graph of a quadratic equation of the form $y = (x - h)^2$ is the same as the graph of $y = x^2$ but moved to the right h units if h is positive or moved to the left $|h|$ units if h is negative.

$y = (x - 4)^2$ ⟶ Moved to the right 4 units

$y = (x + 2)^2 = (x - (-2))^2$ ⟶ Moved to the left 2 units

The following diagram summarizes our work with graphing quadratic equations.

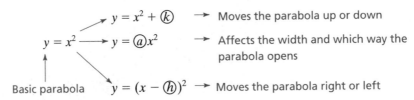

Equations of the form $y = x^2 + k$ and $y = ax^2$ are symmetric about the y axis. The next two examples of this section show how we can combine these ideas to graph a quadratic equation of the form $y = a(x - h)^2 + k$.

E X A M P L E   9

Graph $y = 2(x - 3)^2 + 1$.

Solution

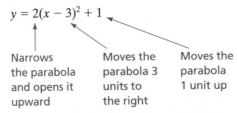

The vertex will be located at the point (3, 1). In addition to the vertex, two points are located to determine the parabola. The parabola is drawn in Figure 8.11.

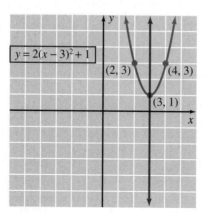

Figure 8.11

   E X A M P L E   1 0   Graph $y = -\dfrac{1}{2}(x + 1)^2 - 2$.

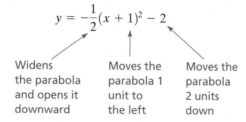   **Solution**

$$y = -\frac{1}{2}(x + 1)^2 - 2$$

Widens
the parabola
and opens it
downward

Moves the
parabola 1
unit to
the left

Moves the
parabola
2 units
down

The parabola is drawn in Figure 8.12.

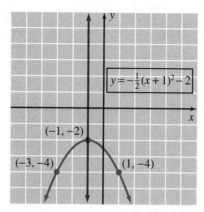

Figure 8.12

Finally, we can use a graphing utility to demonstrate some of the ideas of this section. Let's graph $y = x^2$, $y = -3(x - 7)^2 - 1$, $y = 2(x + 9)^2 + 5$, and $y = -0.2(x + 8)^2 - 3.5$ on the same set of axes, as shown in Figure 8.13. Certainly, Figure 8.13 is consistent with the ideas we presented in this section.

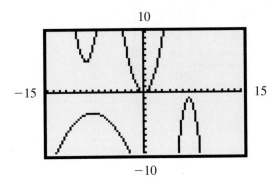

Figure 8.13

PROBLEM SET 8.1

For Problems 1–30, graph each parabola.

1. $y = x^2 + 2$

2. $y = x^2 + 3$

3. $y = x^2 - 1$

4. $y = x^2 - 5$

5. $y = 4x^2$

6. $y = 3x^2$

7. $y = -3x^2$

8. $y = -4x^2$

9. $y = \dfrac{1}{3}x^2$

10. $y = \dfrac{1}{4}x^2$

11. $y = -\dfrac{1}{2}x^2$

12. $y = -\dfrac{2}{3}x^2$

13. $y = (x - 1)^2$

14. $y = (x - 3)^2$

15. $y = (x + 4)^2$

16. $y = (x + 2)^2$

17. $y = 3x^2 + 2$

18. $y = 2x^2 + 3$

19. $y = -2x^2 - 2$

20. $y = \dfrac{1}{2}x^2 - 2$

21. $y = (x - 1)^2 - 2$

22. $y = (x - 2)^2 + 3$

23. $y = (x + 2)^2 + 1$

24. $y = (x + 1)^2 - 4$

25. $y = 3(x - 2)^2 - 4$

26. $y = 2(x + 3)^2 - 1$

27. $y = -(x + 4)^2 + 1$

28. $y = -(x - 1)^2 + 1$

29. $y = -\dfrac{1}{2}(x + 1)^2 - 2$

30. $y = -3(x - 4)^2 - 2$

■ ■ ■ **Thoughts into words**

31. Write a few paragraphs that summarize the ideas we presented in this section for someone who was absent from class that day.

32. How would you convince someone that $y = (x + 3)^2$ is the basic parabola moved 3 units to the left but that $y = (x - 3)^2$ is the basic parabola moved 3 units to the right?

33. How does the graph of $-y = x^2$ compare to the graph of $y = x^2$? Explain your answer.

34. How does the graph of $y = 4x^2$ compare to the graph of $y = 2x^2$? Explain your answer.

Graphing calculator activities

35. Use a graphing calculator to check your graphs for Problems 21–30.

36. **(a)** Graph $y = x^2$, $y = 2x^2$, $y = 3x^2$, and $y = 4x^2$ on the same set of axes.

(b) Graph $y = x^2$, $y = \dfrac{3}{4}x^2$, $y = \dfrac{1}{2}x^2$, and $y = \dfrac{1}{5}x^2$ on the same set of axes.

(c) Graph $y = x^2$, $y = -x^2$, $y = -3x^2$, and $y = -\dfrac{1}{4}x^2$ on the same set of axes.

37. **(a)** Graph $y = x^2$, $y = (x - 2)^2$, $y = (x - 3)^2$, and $y = (x - 5)^2$ on the same set of axes.

(b) Graph $y = x^2$, $y = (x + 1)^2$, $y = (x + 3)^2$, and $y = (x + 6)^2$ on the same set of axes.

38. **(a)** Graph $y = x^2$, $y = (x - 2)^2 + 3$, $y = (x + 4)^2 - 2$, and $y = (x - 6)^2 - 4$ on the same set of axes.

(b) Graph $y = x^2$, $y = 2(x + 1)^2 + 4$, $y = 3(x - 1)^2 - 3$, and $y = \dfrac{1}{2}(x - 5)^2 + 2$ on the same set of axes.

(c) Graph $y = x^2$, $y = -(x - 4)^2 - 3$, $y = -2(x + 3)^2 - 1$, and $y = -\dfrac{1}{2}(x - 2)^2 + 6$ on the same set of axes.

39. **(a)** Graph $y = x^2 - 12x + 41$ and $y = x^2 + 12x + 41$ on the same set of axes. What relationship seems to exist between the two graphs?

(b) Graph $y = x^2 - 8x + 22$ and $y = -x^2 + 8x - 22$ on the same set of axes. What relationship seems to exist between the two graphs?

(c) Graph $y = x^2 + 10x + 29$ and $y = -x^2 + 10x - 29$ on the same set of axes. What relationship seems to exist between the two graphs?

(d) Summarize your findings for parts (a) through (c).

8.2   More Parabolas and Some Circles

We are now ready to graph quadratic equations of the form $y = ax^2 + bx + c$, where a, b, and c are real numbers and $a \neq 0$. The general approach is one of changing equations of the form $y = ax^2 + bx + c$ to the form $y = a(x - h)^2 + k$. Then we can proceed to graph them as we did in the previous section. The process of *completing the square* is used to make the necessary change in the form of the equations. Let's consider some examples to illustrate the details.

E X A M P L E   1   Graph $y = x^2 + 6x + 8$.

Solution

$$y = x^2 + 6x + 8$$

$$y = (x^2 + 6x + \underline{}) - (\underline{}) + 8 \qquad \text{Complete the square.}$$

$$y = (x^2 + 6x + 9) - (9) + 8 \qquad \frac{1}{2}(6) = 3 \text{ and } 3^2 = 9. \text{ Add 9 and also}$$

$$y = (x + 3)^2 - 1 \qquad \qquad \text{subtract 9 to compensate for the 9 that was added.}$$

The graph of $y = (x + 3)^2 - 1$ is the basic parabola moved 3 units to the left and 1 unit down (Figure 8.14).

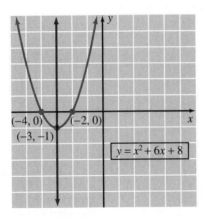

Figure 8.14

EXAMPLE 2

Graph $y = x^2 - 3x - 1$.

Solution

$$y = x^2 - 3x - 1$$

$$y = (x^2 - 3x + \_\_) - (\_\_) - 1 \qquad \text{Complete the square.}$$

$$y = \left(x^2 - 3x + \frac{9}{4} \right) - \frac{9}{4} - 1 \qquad \frac{1}{2}(-3) = -\frac{3}{2} \text{ and } \left(-\frac{3}{2} \right)^2 = \frac{9}{4}. \text{ Add and}$$

$$\text{subtract } \frac{9}{4}.$$

$$y = \left(x - \frac{3}{2} \right)^2 - \frac{13}{4}$$

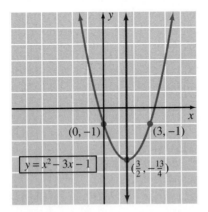

The graph of $y = \left(x - \frac{3}{2} \right)^2 - \frac{13}{4}$ is the basic parabola moved $1\frac{1}{2}$ units to the right and $3\frac{1}{4}$ units down (Figure 8.15).

Figure 8.15

If the coefficient of x^2 is not 1, then a slight adjustment has to be made before we apply the process of completing the square. The next two examples illustrate this situation.

E X A M P L E   3

Graph $y = 2x^2 + 8x + 9$.

Solution

$$y = 2x^2 + 8x + 9$$

$$y = 2(x^2 + 4x) + 9$$ Factor a 2 from the *x*-variable terms.

$$y = 2(x^2 + 4x + \underline{\ \ }) - (2)(\underline{\ \ }) + 9$$ Complete the square. Note that the number being subtracted will be multiplied by a factor of 2.

$$y = 2(x^2 + 4x + 4) - 2(4) + 9$$ $\frac{1}{2}(4) = 2$ and $2^2 = 4$

$$y = 2(x^2 + 4x + 4) - 8 + 9$$

$$y = 2(x + 2)^2 + 1$$

See Figure 8.16 for the graph of $y = 2(x + 2)^2 + 1$.

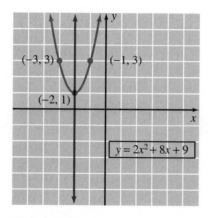

(−3, 3)   (−1, 3)

(−2, 1)

$y = 2x^2 + 8x + 9$

Figure 8.16

E X A M P L E   4

Graph $y = -3x^2 + 6x - 5$.

Solution

$$y = -3x^2 + 6x - 5$$

$$y = -3(x^2 - 2x) - 5$$ Factor a −3 from the *x*-variable terms.

$$y = -3(x^2 - 2x + \underline{\ \ }) - (-3)(\underline{\ \ }) - 5$$ Complete the square. Note that the number being subtracted will be multiplied by a factor of −3.

$$y = -3(x^2 - 2x + 1) - (-3)(1) - 5$$ $\frac{1}{2}(-2) = -1$ and $(-1)^2 = 1$

$$y = -3(x^2 - 2x + 1) + 3 - 5$$

$$y = -3(x - 1)^2 - 2$$

The graph of $y = -3(x - 1)^2 - 2$ is shown in Figure 8.17.

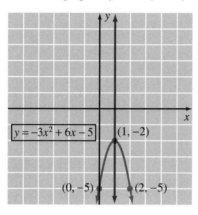

Figure 8.17

Circles

The distance formula, $d = \sqrt{(x_2 - x_1)^2 + (y_2 - y_1)^2}$ (developed in Section 7.4), when it applies to the definition of a circle produces what is known as the **standard equation of a circle.** We start with a precise definition of a circle.

DEFINITION 8.1

A **circle** is the set of all points in a plane equidistant from a given fixed point called the **center.** A line segment determined by the center and any point on the circle is called a **radius.**

Let's consider a circle that has a radius of length r and a center at (h, k) on a coordinate system (Figure 8.18).

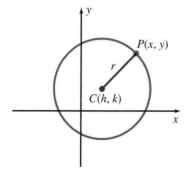

By using the distance formula, we can express the length of a radius (denoted by r) for any point $P(x, y)$ on the circle, as

$$r = \sqrt{(x - h)^2 + (y - k)^2}$$

Figure 8.18

Thus, squaring both sides of the equation, we obtain the **standard form of the equation of a circle:**

$$(x - h)^2 + (y - k)^2 = r^2$$

We can use the standard form of the equation of a circle to solve two basic kinds of circle problems:

1. Given the coordinates of the center and the length of a radius of a circle, find its equation.
2. Given the equation of a circle, find its center and the length of a radius.

Let's look at some examples of such problems.

E X A M P L E   5   Write the equation of a circle that has its center at $(3, -5)$ and a radius of length 6 units.

Solution

Let's substitute 3 for h, -5 for k, and 6 for r into the standard form $(x - h)^2 + (y - k)^2 = r^2$ that becomes $(x - 3)^2 + (y + 5)^2 = 6^2$, which we can simplify as follows:

$$(x - 3)^2 + (y + 5)^2 = 6^2$$

$$x^2 - 6x + 9 + y^2 + 10y + 25 = 36$$

$$x^2 + y^2 - 6x + 10y - 2 = 0$$

∎

Note in Example 5 that we simplified the equation to the form $x^2 + y^2 + Dx + Ey + F = 0$, where D, E, and F are integers. This is another form that we commonly use when working with circles.

E X A M P L E   6   Graph $x^2 + y^2 + 4x - 6y + 9 = 0$.

Solution

This equation is of the form $x^2 + y^2 + Dx + Ey + F = 0$, so its graph is a circle. We can change the given equation into the form $(x - h)^2 + (y - k)^2 = r^2$ by completing the square on x and on y as follows:

$$x^2 + y^2 + 4x - 6y + 9 = 0$$

$$(x^2 + 4x + \underline{\quad}) + (y^2 - 6y + \underline{\quad}) = -9$$

$$(x^2 + 4x + 4) + (y^2 - 6y + 9) = -9 + 4 + 9$$

Added 4 to complete the square on x Added 9 to complete the square on y Added 4 and 9 to compensate for the 4 and 9 added on the left side

$$(x + 2)^2 + (y - 3)^2 = 4$$
$$(x - (-2))^2 + (y - 3)^2 = 2^2$$

$$\uparrow \qquad\qquad \uparrow \qquad \uparrow$$
$$h \qquad\qquad k \qquad r$$

The center of the circle is at $(-2, 3)$ and the length of a radius is 2 (Figure 8.19).

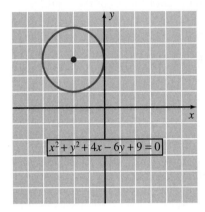

$$x^2 + y^2 + 4x - 6y + 9 = 0$$

Figure 8.19

As demonstrated by Examples 5 and 6, both forms, $(x - h)^2 + (y - k)^2 = r^2$ and $x^2 + y^2 + Dx + Ey + F = 0$, play an important role when we are solving problems that deal with circles.

Finally, we need to recognize that the standard form of a circle that has its center at the origin is $x^2 + y^2 = r^2$. This is merely the result of letting $h = 0$ and $k = 0$ in the general standard form.

$$(x - h)^2 + (y - k)^2 = r^2$$
$$(x - 0)^2 + (y - 0)^2 = r^2$$
$$x^2 + y^2 = r^2$$

Thus, by inspection we can recognize that $x^2 + y^2 = 9$ is a circle with its center at the origin; the length of a radius is 3 units. Likewise, the equation of a circle that has its center at the origin and a radius of length 6 units is $x^2 + y^2 = 36$.

When using a graphing utility to graph a circle, we need to solve the equation for y in terms of x. This will produce two equations that can be graphed on the same set of axes. Furthermore, as with any graph, it may be necessary to change the boundaries on x or y (or both) to obtain a complete graph. If the circle appears oblong, you may want to use a zoom square option so that the graph will appear as a circle. Let's consider an example.

EXAMPLE 7

Use a graphing utility to graph $x^2 - 40x + y^2 + 351 = 0$.

Solution

First, we need to solve for y in terms of x.

$$x^2 - 40x + y^2 + 351 = 0$$

$$y^2 = -x^2 + 40x - 351$$

$$y = \pm\sqrt{-x^2 + 40x - 351}$$

Now we can make the following assignments.

$$Y_1 = \sqrt{-x^2 + 40x - 351}$$

$$Y_2 = -Y_1$$

(Note that we assigned Y_2 in terms of Y_1. By doing this we avoid repetitive key strokes and thus reduce the chance for errors. You may need to consult your user's manual for instructions on how to keystroke $-Y_1$.) Figure 8.20 shows the graph.

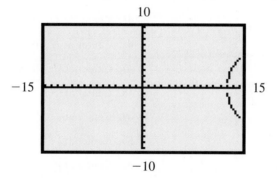

Figure 8.20

Because we know from the original equation that this graph should be a circle, we need to make some adjustments on the boundaries in order to get a complete graph. This can be done by completing the square on the original equation to change its form to $(x - 20)^2 + y^2 = 49$ or simply by a trial-and-error process. By changing the boundaries on x such that $-15 \le x \le 30$, we obtain Figure 8.21.

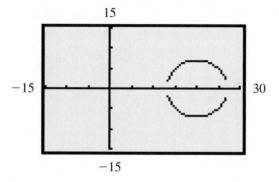

Figure 8.21

PROBLEM SET 8.2

For Problems 1–22, graph each parabola.

1. $y = x^2 - 6x + 13$

2. $y = x^2 - 4x + 7$

3. $y = x^2 + 2x + 6$

4. $y = x^2 + 8x + 14$

5. $y = x^2 - 5x + 3$

6. $y = x^2 + 3x + 1$

7. $y = x^2 + 7x + 14$

8. $y = x^2 - x - 1$

9. $y = 3x^2 - 6x + 5$

10. $y = 2x^2 + 4x + 7$

11. $y = 4x^2 - 24x + 32$

12. $y = 3x^2 + 24x + 49$

13. $y = -2x^2 - 4x - 5$

14. $y = -2x^2 + 8x - 5$

15. $y = -x^2 + 8x - 21$

16. $y = -x^2 - 6x - 7$

17. $y = 2x^2 - x + 2$

18. $y = 2x^2 + 3x + 1$

19. $y = 3x^2 + 2x + 1$

20. $y = 3x^2 - x - 1$

21. $y = -3x^2 - 7x - 2$

22. $y = -2x^2 + x - 2$

For Problems 23–34, find the center and the length of a radius of each circle.

23. $x^2 + y^2 - 2x - 6y - 6 = 0$

24. $x^2 + y^2 + 4x - 12y + 39 = 0$

25. $x^2 + y^2 + 6x + 10y + 18 = 0$

26. $x^2 + y^2 - 10x + 2y + 1 = 0$

27. $x^2 + y^2 = 10$

28. $x^2 + y^2 + 4x + 14y + 50 = 0$

29. $x^2 + y^2 - 16x + 6y + 71 = 0$

30. $x^2 + y^2 = 12$

31. $x^2 + y^2 + 6x - 8y = 0$

32. $x^2 + y^2 - 16x + 30y = 0$

33. $4x^2 + 4y^2 + 4x - 32y + 33 = 0$

34. $9x^2 + 9y^2 - 6x - 12y - 40 = 0$

For Problems 35–44, graph each circle.

35. $x^2 + y^2 = 25$

36. $x^2 + y^2 = 36$

37. $(x - 1)^2 + (y + 2)^2 = 9$

38. $(x + 3)^2 + (y - 2)^2 = 1$

39. $x^2 + y^2 + 6x - 2y + 6 = 0$

40. $x^2 + y^2 - 4x - 6y - 12 = 0$

41. $x^2 + y^2 + 4y - 5 = 0$

42. $x^2 + y^2 - 4x + 3 = 0$

43. $x^2 + y^2 + 4x + 4y - 8 = 0$

44. $x^2 + y^2 - 6x + 6y + 2 = 0$

For problems 45–54, write the equation of each circle. Express the final equation in the form $x^2 + y^2 + Dx + E_y + F = 0$.

45. Center at $(3, 5)$ and $r = 5$

46. Center at $(2, 6)$ and $r = 7$

47. Center at $(-4, 1)$ and $r = 8$

48. Center at $(-3, 7)$ and $r = 6$

49. Center at $(-2, -6)$ and $r = 3\sqrt{2}$

50. Center at $(-4, -5)$ and $r = 2\sqrt{3}$

51. Center at $(0, 0)$ and $r = 2\sqrt{5}$

52. Center at $(0, 0)$ and $r = \sqrt{7}$

53. Center at $(5, -8)$ and $r = 4\sqrt{6}$

54. Center at $(4, -10)$ and $r = 8\sqrt{2}$

55. Find the equation of the circle that passes through the origin and has its center at $(0, 4)$.

56. Find the equation of the circle that passes through the origin and has its center at $(-6, 0)$.

57. Find the equation of the circle that passes through the origin and has its center at $(-4, 3)$.

58. Find the equation of the circle that passes through the origin and has its center at $(8, -15)$.

■ ■ ■ **Thoughts into words**

59. What is the graph of $x^2 + y^2 = -4$? Explain your answer.

60. On which axis does the center of the circle $x^2 + y^2 - 8y + 7 = 0$ lie? Defend your answer.

61. Give a step-by-step description of how you would help someone graph the parabola $y = 2x^2 - 12x + 9$.

■ ■ ■ **Further investigations**

62. The points (x, y) and (y, x) are mirror images of each other across the line $y = x$. Therefore, by interchanging x and y in the equation $y = ax^2 + bx + c$, we obtain the equation of its mirror image across the line $y = x$ — namely, $x = ay^2 + by + c$. Thus to graph $x = y^2 + 2$, we can first graph $y = x^2 + 2$ and then reflect it across the line $y = x$, as indicated in Figure 8.22.

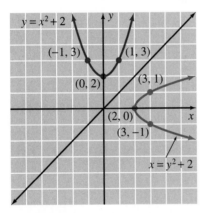

Figure 8.22

Graph each of the following parabolas.
(a) $x = y^2$
(b) $x = -y^2$
(c) $x = y^2 - 1$
(d) $x = -y^2 + 3$
(e) $x = -2y^2$
(f) $x = 3y^2$
(g) $x = y^2 + 4y + 7$
(h) $x = y^2 - 2y - 3$

63. By expanding $(x - h)^2 + (y - k)^2 = r^2$, we obtain $x^2 - 2hx + h^2 + y^2 - 2ky + k^2 - r^2 = 0$. When we compare this result to the form $x^2 + y^2 + Dx + Ey + F = 0$, we see that $D = -2h$, $E = -2k$, and $F = h^2 + k^2 - r^2$. Therefore, the center and length of a radius of a circle can be found by using $h = \dfrac{D}{-2}$, $k = \dfrac{E}{-2}$, and $r = \sqrt{h^2 + k^2 - F}$. Use these relationships to find the center and the length of a radius of each of the following circles.

(a) $x^2 + y^2 - 2x - 8y + 8 = 0$
(b) $x^2 + y^2 + 4x - 14y + 49 = 0$
(c) $x^2 + y^2 + 12x + 8y - 12 = 0$
(d) $x^2 + y^2 - 16x + 20y + 115 = 0$
(e) $x^2 + y^2 - 12y - 45 = 0$
(f) $x^2 + y^2 + 14x = 0$

▦ **Graphing calculator activities**

64. Use a graphing calculator to check your graphs for Problems 1–22.

65 Use a graphing calculator to graph the circles in Problems 23–26. Be sure that your graphs are consistent with the center and the length of a radius that you found when you did the problems.

66. Graph each of the following parabolas and circles. Be sure to set your boundaries so that you get a complete graph.

(a) $x^2 + 24x + y^2 + 135 = 0$
(b) $y = x^2 - 4x + 18$
(c) $x^2 + y^2 - 18y + 56 = 0$
(d) $x^2 + y^2 + 24x + 28y + 336 = 0$
(e) $y = -3x^2 - 24x - 58$
(f) $y = x^2 - 10x + 3$

8.3 Graphing Ellipses

In the previous section, we found that the graph of the equation $x^2 + y^2 = 36$ is a circle of radius 6 units with its center at the origin. More generally, it is true that any equation of the form $Ax^2 + By^2 = C$, where $A = B$ and where A, B, and C are nonzero constants that have the same sign, is a circle with the center at the origin. For example, $3x^2 + 3y^2 = 12$ is equivalent to $x^2 + y^2 = 4$ (divide both sides of the given equation by 3), and thus it is a circle of radius 2 units with its center at the origin.

The general equation $Ax^2 + By^2 = C$ can be used to describe other geometric figures by changing the restrictions on A and B. For example, if A, B, and C are of the same sign, but $A \neq B$, then the graph of the equation $Ax^2 + By^2 = C$ is an **ellipse.** Let's consider two examples.

E X A M P L E   1

Graph $4x^2 + 25y^2 = 100$.

Solution

Let's find the x and y intercepts. Let $x = 0$; then

$$4(0)^2 + 25y^2 = 100$$

$$25y^2 = 100$$

$$y^2 = 4$$

$$y = \pm 2$$

Thus the points $(0, 2)$ and $(0, -2)$ are on the graph. Let $y = 0$; then

$$4x^2 + 25(0)^2 = 100$$

$$4x^2 = 100$$

$$x^2 = 25$$

$$x = \pm 5$$

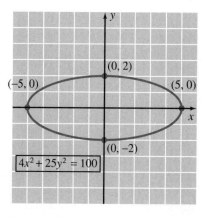

Thus the points $(5, 0)$ and $(-5, 0)$ are also on the graph. We know that this figure is an ellipse, so we plot the four points and we get a pretty good sketch of the figure (Figure 8.23).

Figure 8.23

In Figure 8.23, the line segment with endpoints at $(-5, 0)$ and $(5, 0)$ is called the **major axis** of the ellipse. The shorter line segment with endpoints at $(0, -2)$ and $(0, 2)$ is called the **minor axis.** Establishing the endpoints of the major and minor axes provides a basis for sketching an ellipse. The point of intersection of the major and minor axes is called the **center** of the ellipse.

E X A M P L E   2

Graph $9x^2 + 4y^2 = 36$.

Solution

Again, let's find the x and y intercepts. Let $x = 0$; then

$$9(0)^2 + 4y^2 = 36$$
$$4y^2 = 36$$
$$y^2 = 9$$
$$y = \pm 3$$

Thus the points $(0, 3)$ and $(0, -3)$ are on the graph. Let $y = 0$; then

$$9x^2 + 4(0)^2 = 36$$
$$9x^2 = 36$$
$$x^2 = 4$$
$$x = \pm 2$$

Thus the points $(2, 0)$ and $(-2, 0)$ are also on the graph. The ellipse is sketched in Figure 8.24.

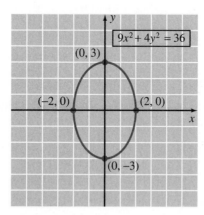

Figure 8.24

In Figure 8.24, the major axis has endpoints at $(0, -3)$ and $(0, 3)$ and the minor axis has endpoints at $(-2, 0)$ and $(2, 0)$. The ellipses in Figures 8.23 and 8.24 are symmetric about the x axis and about the y axis. In other words, both the x axis and the y axis serve as **axes of symmetry.**

Now we turn to some ellipses whose centers are not at the origin but whose major and minor axes are parallel to the x axis and the y axis. We can graph such ellipses in much the same way as we handled circles in Section 8.2. Let's consider two examples to illustrate the procedure.

E X A M P L E   3

Graph $4x^2 + 24x + 9y^2 - 36y + 36 = 0$.

Solution

Let's complete the square on x and y as follows:

$$4x^2 + 24x + 9y^2 - 36y + 36 = 0$$

$$4(x^2 + 6x + \underline{}) + 9(y^2 - 4y + \underline{}) = -36$$

$$4(x^2 + 6x + 9) + 9(y^2 - 4y + 4) = -36 + 36 + 36$$

$$4(x + 3)^2 + 9(y - 2)^2 = 36$$

$$4(x - (-3))^2 + 9(y - 2)^2 = 36$$

Because 4, 9, and 36 are of the same sign and $4 \neq 9$, the graph is an ellipse. The center of the ellipse is at $(-3, 2)$. We can find the endpoints of the major and minor axes as follows: Use the equation $4(x + 3)^2 + 9(y - 2)^2 = 36$ and let $y = 2$ (the y coordinate of the center).

$$4(x + 3)^2 + 9(2 - 2)^2 = 36$$

$$4(x + 3)^2 = 36$$

$$(x + 3)^2 = 9$$

$$x + 3 = \pm 3$$

$$x + 3 = 3 \quad \text{or} \quad x + 3 = -3$$

$$x = 0 \quad \text{or} \quad x = -6$$

This gives the points $(0, 2)$ and $(-6, 2)$. These are the coordinates of the endpoints of the major axis. Now let $x = -3$ (the x coordinate of the center).

$$4(-3 + 3)^2 + 9(y - 2)^2 = 36$$

$$9(y - 2)^2 = 36$$

$$(y - 2)^2 = 4$$

$$y - 2 = \pm 2$$

$$y - 2 = 2 \quad \text{or} \quad y - 2 = -2$$

$$y = 4 \quad \text{or} \quad y = 0$$

This gives the points $(-3, 4)$ and $(-3, 0)$. These are the coordinates of the endpoints of the minor axis. The ellipse is shown in Figure 8.25.

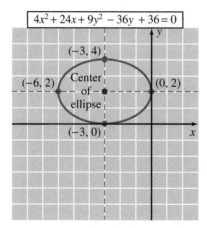

$$4x^2 + 24x + 9y^2 - 36y + 36 = 0$$

Figure 8.25

E X A M P L E   4

Graph $4x^2 - 16x + y^2 + 6y + 9 = 0$.

Solution

First let's complete the square on x and on y.

$$4x^2 - 16x + y^2 + 6y + 9 = 0$$

$$4(x^2 - 4x + \underline{\ \ }) + (y^2 + 6y + \underline{\ \ }) = -9$$

$$4(x^2 - 4x + 4) + (y^2 + 6y + 9) = -9 + 16 + 9$$

$$4(x - 2)^2 + (y + 3)^2 = 16$$

The center of the ellipse is at $(2, -3)$. Now let $x = 2$ (the x coordinate of the center).

$$4(2 - 2)^2 + (y + 3)^2 = 16$$

$$(y + 3)^2 = 16$$

$$y + 3 = \pm 4$$

$$y + 3 = -4 \quad \text{or} \quad y + 3 = 4$$

$$y = -7 \quad \text{or} \quad y = 1$$

This gives the points $(2, -7)$ and $(2, 1)$. These are the coordinates of the endpoints of the major axis. Now let $y = -3$ (the y coordinate of the center).

$$4(x - 2)^2 + (-3 + 3)^2 = 16$$

$$4(x - 2)^2 = 16$$

$$(x - 2)^2 = 4$$

$$x - 2 = \pm 2$$

$$x - 2 = -2 \quad \text{or} \quad x - 2 = 2$$

$$x = 0 \quad \text{or} \quad x = 4$$

This gives the points $(0, -3)$ and $(4, -3)$. These are the coordinates of the endpoints of the minor axis. The ellipse is shown in Figure 8.26.

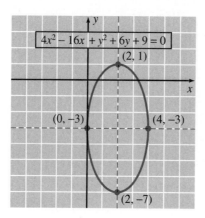

Figure 8.26

PROBLEM SET 8.3

For Problems 1–16, graph each ellipse.

1. $x^2 + 4y^2 = 36$

2. $x^2 + 4y^2 = 16$

3. $9x^2 + y^2 = 36$

4. $16x^2 + 9y^2 = 144$

5. $4x^2 + 3y^2 = 12$

6. $5x^2 + 4y^2 = 20$

7. $16x^2 + y^2 = 16$

8. $9x^2 + 2y^2 = 18$

9. $25x^2 + 2y^2 = 50$

10. $12x^2 + y^2 = 36$

11. $4x^2 + 8x + 16y^2 - 64y + 4 = 0$

12. $9x^2 - 36x + 4y^2 - 24y + 36 = 0$

13. $x^2 + 8x + 9y^2 + 36y + 16 = 0$

14. $4x^2 - 24x + y^2 + 4y + 24 = 0$

15. $4x^2 + 9y^2 - 54y + 45 = 0$

16. $x^2 + 2x + 4y^2 - 15 = 0$

■ ■ ▨ **Thoughts into words**

17. Is the graph of $x^2 + y^2 = 4$ the same as the graph of $y^2 + x^2 = 4$? Explain your answer.

18. Is the graph of $x^2 + y^2 = 0$ a circle? If so, what is the length of a radius?

19. Is the graph of $4x^2 + 9y^2 = 36$ the same as the graph of $9x^2 + 4y^2 = 36$? Explain your answer.

20. What is the graph of $x^2 + 2y^2 = -16$? Explain your answer.

Graphing calculator activities

21. Use a graphing calculator to graph the ellipses in Examples 1– 4 of this section.

22. Use a graphing calculator to check your graphs for Problems 11–16.

8.4 Graphing Hyperbolas

The graph of an equation of the form $Ax^2 + By^2 = C$, where A, B, and C are nonzero real numbers and A and B are of unlike signs, is a **hyperbola.** Let's use some examples to illustrate a procedure for graphing hyperbolas.

EXAMPLE 1

Graph $x^2 - y^2 = 9$.

Solution

If we let $y = 0$, we obtain

$$x^2 - 0^2 = 0$$

$$x^2 = 9$$

$$x = \pm 3$$

Thus the points $(3, 0)$ and $(-3, 0)$ are on the graph. If we let $x = 0$, we obtain

$$0^2 - y^2 = 9$$

$$-y^2 = 9$$

$$y^2 = -9$$

Because $y^2 = -9$ has no real number solutions, there are no points of the y axis on this graph. That is, the graph does not intersect the y axis. Now let's solve the given equation for y so that we have a more convenient form for finding other solutions.

$$x^2 - y^2 = 9$$

$$-y^2 = 9 - x^2$$

$$y^2 = x^2 - 9$$

$$y = \pm\sqrt{x^2 - 9}$$

The radicand, $x^2 - 9$, must be nonnegative, so the values we choose for x must be greater than or equal to 3 or less than or equal to -3. With this in mind, we can form the following table of values.

x	y	
3	0	Intercepts
-3	0	
4	$\pm\sqrt{7}$	
-4	$\pm\sqrt{7}$	Other points
5	± 4	
-5	± 4	

We plot these points and draw the hyperbola as in Figure 8.27. (This graph is also symmetric about both axes.)

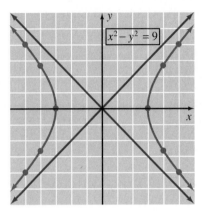

Figure 8.27

Note the blue lines in Figure 8.27; they are called **asymptotes.** Each branch of the hyperbola approaches one of these lines but does not intersect it. Therefore, the ability to sketch the asymptotes of a hyperbola is very helpful when we are graphing the hyperbola. Fortunately, the equations of the asymptotes are easy to determine. They can be found by replacing the constant term in the given equation of the hyperbola with 0 and solving for y. (The reason why this works will become evident in a later course.) Thus, for the hyperbola in Example 3, we obtain

$$x^2 - y^2 = 0$$
$$y^2 = x^2$$
$$y = \pm x$$

Thus the two lines $y = x$ and $y = -x$ are the asymptotes indicated by the blue lines in Figure 8.27.

EXAMPLE 2

Graph $y^2 - 5x^2 = 4$.

Solution

If we let $x = 0$, we obtain

$$y^2 - 5(0)^2 = 4$$
$$y^2 = 4$$
$$y = \pm 2$$

The points $(0, 2)$ and $(0, -2)$ are on the graph. If we let $y = 0$, we obtain

$$0^2 - 5x^2 = 4$$

$$-5x^2 = 4$$

$$x^2 = -\frac{4}{5}$$

Because $x^2 = -\frac{4}{5}$ has no real number solutions, we know that this hyperbola does not intersect the x axis. Solving the given equation for y yields

$$y^2 - 5x^2 = 4$$

$$y^2 = 5x^2 + 4$$

$$y = \sqrt{5x^2 + 4}$$

The following table shows some additional solutions for the equation.

x	y	
0	2	Intercepts
0	−2	
1	±3	Other points
−1	±3	
2	$\pm\sqrt{24}$	
−2	$\pm\sqrt{24}$	

The equations of the asymptotes are determined as follows:

$$y^2 - 5x^2 = 0$$

$$y^2 = 5x^2$$

$$y = \pm\sqrt{5}x$$

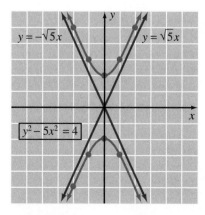

Sketch the asymptotes and plot the points determined by the table of values to determine the hyperbola in Figure 8.28. (Note that this hyperbola is also symmetric about the x axis and the y axis.)

Figure 8.28

EXAMPLE 3   Graph $4x^2 - 9y^2 = 36$.

Solution

If we let $x = 0$, we obtain

$$4(0)^2 - 9y^2 = 36$$

$$-9y^2 = 36$$

$$y^2 = -4$$

Because $y^2 = -4$ has no real number solutions, we know that this hyperbola does not intersect the y axis. If we let $y = 0$, we obtain

$$4x^2 - 9(0)^2 = 36$$

$$4x^2 = 36$$

$$x^2 = 9$$

$$x = \pm 3$$

Thus the points $(3, 0)$ and $(-3, 0)$ are on the graph. Now let's solve the equation for y in terms of x and set up a table of values.

$$4x^2 - 9y^2 = 36$$

$$-9y^2 = 36 - 4x^2$$

$$9y^2 = 4x^2 - 36$$

$$y^2 = \frac{4x^2 - 36}{9}$$

$$y = \pm\frac{\sqrt{4x^2 - 36}}{3}$$

x	y	
3	0	Intercepts
−3	0	
4	$\pm\dfrac{2\sqrt{7}}{3}$	
−4	$\pm\dfrac{2\sqrt{7}}{3}$	Other points
5	$\pm\dfrac{8}{3}$	
−5	$\pm\dfrac{8}{3}$	

The equations of the asymptotes are found as follows:

$$4x^2 - 9y^2 = 0$$

$$-9y^2 = -4x^2$$

$$9y^2 = 4x^2$$

$$y^2 = \frac{4x^2}{9}$$

$$y = \pm\frac{2}{3}x$$

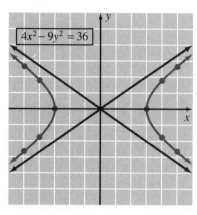

$4x^2 - 9y^2 = 36$

Sketch the asymptotes and plot the points determined by the table to determine the hyperbola as shown in Figure 8.29.

Figure 8.29

Now let's consider hyperbolas that are not symmetric with respect to the origin but are symmetric with respect to lines parallel to one of the axes — that is, vertical and horizontal lines. Again, let's use examples to illustrate a procedure for graphing such hyperbolas.

EXAMPLE 4

Graph $4x^2 - 8x - y^2 - 4y - 16 = 0$.

Solution

Completing the square on x and y, we obtain

$$4x^2 - 8x - y^2 - 4y - 16 = 0$$

$$4(x^2 - 2x + \underline{}) - (y^2 + 4y + \underline{}) = 16$$

$$4(x^2 - 2x + 1) - (y^2 + 4y + 4) = 16 + 4 - 4$$

$$4(x - 1)^2 - (y + 2)^2 = 16$$

$$4(x - 1)^2 - 1(y - (-2))^2 = 16$$

Because 4 and −1 are of opposite signs, the graph is a hyperbola. The center of the hyperbola is at $(1, -2)$.

Now, using the equation $4(x - 1)^2 - (y + 2)^2 = 16$, we can proceed as follows: Let $y = -2$; then

$$4(x - 1)^2 - (-2 + 2)^2 = 16$$
$$4(x - 1)^2 = 16$$
$$(x - 1)^2 = 4$$
$$x - 1 = \pm 2$$

$$x - 1 = 2 \quad \text{or} \quad x - 1 = -2$$
$$x = 3 \quad \text{or} \quad x = -1$$

Thus the hyperbola intersects the horizontal line $y = -2$ at $(3, -2)$ and at $(-1, -2)$. Let $x = 1$; then

$$4(1 - 1)^2 - (y + 2)^2 = 16$$
$$-(y + 2)^2 = 16$$
$$(y + 2)^2 = -16$$

Because $(y + 2)^2 = -16$ has no real number solutions, we know that the hyperbola does not intersect the vertical line $x = 1$. We replace the constant term of $4(x - 1)^2 - (y + 2)^2 = 16$ with 0 and solve for y to produce the equations of the asymptotes as follows:

$$4(x - 1)^2 - (y + 2)^2 = 0$$

The left side can be factored using the pattern of the difference of squares.

$$(2(x - 1) + (y + 2))(2(x - 1) - (y + 2)) = 0$$
$$(2x - 2 + y + 2)(2x - 2 - y - 2) = 0$$
$$(2x + y)(2x - y - 4) = 0$$

$$2x + y = 0 \quad \text{or} \quad 2x - y - 4 = 0$$
$$y = -2x \quad \text{or} \quad 2x - 4 = y$$

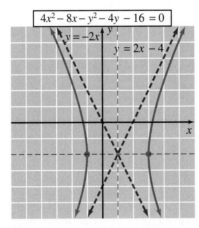

Thus the equations of the asymptotes are $y = -2x$ and $y = 2x - 4$. Sketching the asymptotes and plotting the two points $(3, -2)$ and $(-1, -2)$, we can draw the hyperbola as shown in Figure 8.30.

Figure 8.30

E X A M P L E   5

Graph $y^2 - 4y - 4x^2 - 24x - 36 = 0$.

Solution

First let's complete the square on x and on y.

$$y^2 - 4y - 4x^2 - 24x - 36 = 0$$

$$(y^2 - 4y + \underline{}) - 4(x^2 + 6x + \underline{}) = 36$$

$$(y^2 - 4y + 4) - 4(x^2 + 6x + 9) = 36 + 4 - 36$$

$$(y - 2)^2 - 4(x + 3)^2 = 4$$

The center of the hyperbola is at $(-3, 2)$. Now let $y = 2$.

$$(2 - 2)^2 - 4(x + 3)^2 = 4$$

$$-4(x + 3)^2 = 4$$

$$(x + 3)^2 = -1$$

Because $(x + 3)^2 = -1$ has no real number solutions, the graph does not intersect the line $y = 2$. Now let $x = -3$.

$$(y - 2)^2 - 4(-3 + 3)^2 = 4$$

$$(y - 2)^2 = 4$$

$$y - 2 = \pm 2$$

$$y - 2 = -2 \quad \text{or} \quad y - 2 = 2$$

$$y = 0 \quad \text{or} \quad y = 4$$

Therefore, the hyperbola intersects the line $x = -3$ at $(-3, 0)$ and $(-3, 4)$. Now, to find the equations of the asymptotes, let's replace the constant term of $(y - 2)^2 - 4(x + 3)^2 = 4$ with 0 and solve for y.

$$(y - 2)^2 - 4(x + 3)^2 = 0$$

$$[(y - 2) + 2(x + 3)][(y - 2) - 2(x + 3)] = 0$$

$$(y - 2 + 2x + 6)(y - 2 - 2x - 6) = 0$$

$$(y + 2x + 4)(y - 2x - 8) = 0$$

$$y + 2x + 4 = 0 \quad \text{or} \quad y - 2x - 8 = 0$$

$$y = -2x - 4 \quad \text{or} \quad y = 2x + 8$$

Therefore, the equations of the asymptotes are $y = -2x - 4$ and $y = 2x + 8$. Drawing the asymptotes and plotting the points $(-3, 0)$ and $(-3, 4)$, we can graph the hyperbola as shown in Figure 8.31.

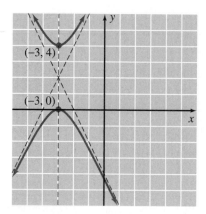

Figure 8.31

As a way of summarizing our work with conic sections, let's focus our attention on the continuity pattern used in this chapter. In Sections 8.1 and 8.2, we studied parabolas by considering variations of the basic quadratic equation $y = ax^2 + bx + c$. Also in Section 8.2, we used the definition of a circle to generate a standard form for the equation of a circle. Then, in Sections 8.3 and 8.4, we discussed ellipses and hyperbolas, not from a definition viewpoint, but by considering variations of the equations $Ax^2 + By^2 = C$ and $A(x - h)^2 + B(y - k)^2 = C$. In a subsequent mathematics course, parabolas, ellipses, and hyperbolas will be developed from a definition viewpoint. That is, first each concept will be defined, and then the definition will be used to generate a standard form of its equation.

PROBLEM SET 8.4

For Problems 1–6, find the intercepts and the equations for the asymptotes.

1. $x^2 - 9y^2 = 16$

2. $16x^2 - y^2 = 25$

3. $y^2 - 9x^2 = 36$

4. $4x^2 - 9y^2 = 16$

5. $25x^2 - 9y^2 = 4$

6. $y^2 - x^2 = 16$

For Problems 7–10, find the equations for the asymptotes.

7. $x^2 + 4x - y^2 - 6y - 30 = 0$

8. $y^2 - 8y - x^2 - 4x + 3 = 0$

9. $9x^2 - 18x - 4y^2 - 24y - 63 = 0$

10. $4x^2 + 24x - y^2 + 4y + 28 = 0$

For Problems 11–28, graph each hyperbola.

11. $x^2 - y^2 = 1$

12. $x^2 - y^2 = 4$

13. $y^2 - 4x^2 = 9$

14. $4y^2 - x^2 = 16$

15. $5x^2 - 2y^2 = 20$

16. $9x^2 - 4y^2 = 9$

17. $y^2 - 16x^2 = 4$

18. $y^2 - 9x^2 = 16$

19. $-4x^2 + y^2 = -4$

20. $-9x^2 + y^2 = -36$

21. $25y^2 - 3x^2 = 75$

22. $16y^2 - 5x^2 = 80$

23. $-4x^2 + 32x + 9y^2 - 18y - 91 = 0$

24. $x^2 - 4x - y^2 + 6y - 14 = 0$

25. $-4x^2 + 24x + 16y^2 + 64y - 36 = 0$

26. $x^2 + 4x - 9y^2 + 54y - 113 = 0$

27. $4x^2 - 24x - 9y^2 = 0$ **28.** $16y^2 + 64y - x^2 = 0$

29. The graphs of equations of the form $xy = k$, where k is a nonzero constant, are also hyperbolas, sometimes referred to as rectangular hyperbolas. Graph each of the following.

(a) $xy = 3$ **(b)** $xy = 5$
(c) $xy = -2$ **(d)** $xy = -4$

30. What is the graph of $xy = 0$? Defend your answer.

31. We have graphed various equations of the form $Ax^2 + By^2 = C$, where C is a nonzero constant. Now graph each of the following.

(a) $x^2 + y^2 = 0$ **(b)** $2x^2 + 3y^2 = 0$
(c) $x^2 - y^2 = 0$ **(d)** $4y^2 - x^2 = 0$

■ ■ ■ Thoughts into words

32. Explain the concept of an asymptote.

33. Explain how asymptotes can be used to help graph hyperbolas.

34. Are the graphs of $x^2 - y^2 = 0$ and $y^2 - x^2 = 0$ identical? Are the graphs of $x^2 - y^2 = 4$ and $y^2 - x^2 = 4$ identical? Explain your answers.

 ### Graphing calculator activities

35. To graph the hyperbola in Example 1 of this section, we can make the following assignments for the graphing calculator.

$Y_1 = \sqrt{x^2 - 9}$ $Y_2 = -Y_1$
$Y_3 = x$ $Y_4 = -Y_3$

Do this and see if your graph agrees with Figure 8.27. Also graph the asymptotes and hyperbolas for Examples 2 and 3.

36. Use a graphing calculator to check your graphs for Problems 11–16.

37. Use a graphing calculator to check your graphs for Problems 23–28.

38. For each of the following equations, (1) predict the type and location of the graph, and (2) use your graphing calculator to check your predictions.

(a) $x^2 + y^2 = 100$ **(b)** $x^2 - y^2 = 100$
(c) $y^2 - x^2 = 100$ **(d)** $y = -x^2 + 9$
(e) $2x^2 + y^2 = 14$ **(f)** $x^2 + 2y^2 = 14$
(g) $x^2 + 2x + y^2 - 4 = 0$ **(h)** $x^2 + y^2 - 4y - 2 = 0$
(i) $y = x^2 + 16$ **(j)** $y^2 = x^2 + 16$
(k) $9x^2 - 4y^2 = 72$ **(l)** $4x^2 - 9y^2 = 72$
(m) $y^2 = -x^2 - 4x + 6$

SUMMARY

(8.1) and (8.2) The graph of any quadratic equation of the form $y = ax^2 + bx + c$, where a, b, and c are real numbers and $a \neq 0$, is a **parabola.**

The following diagram summarizes the graphing of parabolas.

$$y = x^2 + \boxed{k} \longrightarrow \text{Moves the parabola up or down}$$

$$y = x^2 \longrightarrow y = \boxed{a}x^2 \longrightarrow \text{Affects the width and which way the parabola opens}$$

$$\text{Basic} \quad y = (x - \boxed{h})^2 \longrightarrow \text{Moves the parabola right or left}$$
$$\text{parabola}$$

The **standard form of the equation of a circle** with its center at (h, k) and a radius of length r is

$$(x - h)^2 + (y - k)^2 = r^2$$

The standard form of the equation of a circle with its center at the origin and a radius of length r is

$$x^2 + y^2 = r^2$$

(8.3) The graph of an equation of the form $Ax^2 + By^2 = C$ or of the form $A(x - h)^2 + B(y - k)^2 = C$, where A, B, and C are nonzero real numbers of the same sign and $A \neq B$, is an **ellipse.**

(8.4) The graph of an equation of the form $Ax^2 + By^2 = C$ or of the form $A(x - h)^2 + B(y - k)^2 = C$, where A, B, and C are nonzero real numbers with A and B of unlike signs, is a **hyperbola.**

The equations of the asymptotes of a hyperbola can be found by replacing the constant term of the equation of the hyperbola with zero and solving the resulting equation for y.

Circles, ellipses, parabolas, and hyperbolas are often referred to as **conic sections.**

CHAPTER 8 REVIEW PROBLEM SET

For Problems 1–6, find the vertex of each parabola.

1. $y = x^2 + 6$

2. $y = -x^2 - 8$

3. $y = (x + 3)^2 - 1$

4. $y = x^2 - 14x + 54$

5. $y = -x^2 + 12x - 44$

6. $y = 3x^2 + 24x + 39$

For Problems 7–9, write the equation of the circle satisfying the given conditions. Express your answers in the form $x^2 + y^2 + Dx + Ey + F = 0$.

7. Center at $(2, -6)$ and $r = 5$

8. Center at $(-4, -8)$ and $r = 2\sqrt{3}$

9. Center at $(0, 5)$ and passes through the origin.

For Problems 10–13, find the center and the length of a radius for each circle.

10. $x^2 + 14x + y^2 - 8y + 16 = 0$

11. $x^2 + 16x + y^2 + 39 = 0$

12. $x^2 - 12x + y^2 + 16y = 0$

13. $x^2 + y^2 = 24$

For Problems 14–19, find the length of the major axis and the length of the minor axis of each ellipse.

14. $16x^2 + y^2 = 64$

15. $16x^2 + 9y^2 = 144$

16. $4x^2 + 25y^2 = 100$

17. $2x^2 + 7y^2 = 28$

18. $x^2 - 4x + 9y^2 + 54y + 76 = 0$

19. $9x^2 + 72x + 4y^2 - 8y + 112 = 0$

For Problems 20–25, find the equations of the asymptotes of each hyperbola.

20. $x^2 - 9y^2 = 25$

21. $4x^2 - y^2 = 16$

22. $9y^2 - 25x^2 = 36$

23. $16y^2 - 4x^2 = 17$

24. $25x^2 + 100x - 4y^2 + 24y - 36 = 0$

25. $36y^2 - 288y - x^2 + 2x + 539 = 0$

For Problems 26–39, graph each equation.

26. $9x^2 + y^2 = 81$

27. $9x^2 - y^2 = 81$

28. $y = -2x^2 + 3$

29. $y = 4x^2 - 16x + 19$

30. $x^2 + 4x + y^2 + 8y + 11 = 0$

31. $4x^2 - 8x + y^2 + 8y + 4 = 0$

32. $y^2 + 6y - 4x^2 - 24x - 63 = 0$

33. $y = -2x^2 - 4x - 3$

34. $x^2 - y^2 = -9$

35. $4x^2 + 16y^2 + 96y = 0$

36. $(x - 3)^2 + (y + 1)^2 = 4$

37. $(x + 1)^2 + (y - 2)^2 = 4$

38. $x^2 + y^2 - 6x - 2y + 4 = 0$

39. $x^2 + y^2 - 2y - 8 = 0$

TEST

For Problems 1–4, find the vertex of each parabola.

1. $y = -2x^2 + 9$

2. $y = -x^2 + 2x + 6$

3. $y = 4x^2 + 32x + 62$

4. $y = x^2 - 6x + 9$

For Problems 5–7, write the equation of the circle that satisfies the given conditions. Express your answers in the form $x^2 + y^2 + Dx + Ey + F = 0$.

5. Center at $(-4, 0)$ and $r = 3\sqrt{5}$

6. Center at $(2, 8)$ and $r = 3$

7. Center at $(-3, -4)$ and $r = 5$

For Problems 8–10, find the center and the length of a radius of each circle.

8. $x^2 + y^2 = 32$

9. $x^2 - 12x + y^2 + 8y + 3 = 0$

10. $x^2 + 10x + y^2 + 2y - 38 = 0$

11. Find the length of the major axis of the ellipse $9x^2 + 2y^2 = 32$.

12. Find the length of the minor axis of the ellipse $8x^2 + 3y^2 = 72$.

13. Find the length of the major axis of the ellipse $3x^2 - 12x + 5y^2 + 10y - 10 = 0$.

14. Find the length of the minor axis of the ellipse $8x^2 - 32x + 5y^2 + 30y + 45 = 0$.

For Problems 15–17, find the equations of the asymptotes for each hyperbola.

15. $y^2 - 16x^2 = 36$

16. $25x^2 - 16y^2 = 50$

17. $x^2 - 2x - 25y^2 - 50y - 54 = 0$

For Problems 18–25, graph each equation.

18. $x^2 - 4y^2 = -16$

19. $y = x^2 + 4x$

20. $x^2 + 2x + y^2 + 8y + 8 = 0$

21. $2x^2 + 3y^2 = 12$

22. $y = 2x^2 + 12x + 22$

23. $9x^2 - y^2 = 9$

24. $3x^2 - 12x + 5y^2 + 10y - 10 = 0$

25. $x^2 - 4x - y^2 + 4y - 9 = 0$

The price of goods may be decided by using a function to describe the relationship between the price and the demand. Such a function gives us a means of studying the demand when the price is varied.

CHAPTER 9

© Bill Aron /PhotoEdit

Functions

9.1   *Relations and Functions*

9.2   *Functions: Their Graphs and Applications*

9.3   *Graphing Made Easy Via Transformations*

9.4   *Composition of Functions*

9.5   *Inverse Functions*

9.6   *Direct and Inverse Variations*

A golf pro-shop operator finds that she can sell 30 sets of golf clubs at $500 per set in a year. Furthermore, she predicts that for each $25 decrease in price, 3 additional sets of golf clubs could be sold. At what price should she sell the clubs to maximize gross income? We can use the quadratic function $f(x) = (30 + 3x)(500 - 25x)$ to determine that the clubs should be sold at $375 per set.

One of the fundamental concepts of mathematics is the function. Functions are used to unify mathematics and also to apply mathematics to many real-world problems. Functions provide a means of studying quantities that vary with one another — that is, change in one quantity causes a corresponding change in the other.

In this chapter we will (1) introduce the basic ideas that pertain to the function concept, (2) review and extend some concepts from Chapter 8, and (3) discuss some applications of functions.

431

InfoTrac Project

Do a keyword search on linear relationships. Find an article concerning language skills in cocaine-exposed infants. Write a brief paragraph on the general conclusion of the article. Suppose the relationship between the level (x) of benzoylecgonine (a cocaine by-product) in the newborn's meconium and the age level of auditory comprehension ($f(x)$) at age 1 can be expressed by the following equation: $f(x) = -0.057x + 1.0$. What is the age level of auditory comprehension for a child who was not exposed to cocaine during fetal development? As the level of benzoylecgonine rises, what happens to the age level of auditory comprehension? What would the level of benzoylecgonine need to be for a child to have the age level of auditory comprehension of a 6-month-old child? Graph this relationship on a graphing utility.

9.1 Relations and Functions

Mathematically, a function is a special kind of **relation,** so we will begin our discussion with a simple definition of a relation.

DEFINITION 9.1

A **relation** is a set of ordered pairs.

Thus a set of ordered pairs such as $\{(1, 2), (3, 7), (8, 14)\}$ is a relation. The set of all first components of the ordered pairs is the **domain** of the relation, and the set of all second components is the **range** of the relation. The relation $\{(1, 2), (3, 7), (8, 14)\}$ has a domain of $\{1, 3, 8\}$ and a range of $\{2, 7, 14\}$.

The ordered pairs we refer to in Definition 9.1 may be generated by various means, such as a graph or a chart. However, one of the most common ways of generating ordered pairs is by using equations. Because the solution set of an equation in two variables is a set of ordered pairs, such an equation describes a relation. Each of the following equations describes a relation between the variables x and y. We have listed some of the infinitely many ordered pairs (x, y) of each relation.

1. $x^2 + y^2 = 4$: $(1, \sqrt{3}), (1, -\sqrt{3}), (0, 2), (0, -2)$
2. $y^2 = x^3$: $(0, 0), (1, 1), (1, -1), (4, 8), (4, -8)$
3. $y = x + 2$: $(0, 2), (1, 3), (2, 4), (-1, 1), (5, 7)$
4. $y = \dfrac{1}{x - 1}$: $\left(0, -1\right), (2, 1), \left(3, \dfrac{1}{2}\right), \left(-1, -\dfrac{1}{2}\right), \left(-2, -\dfrac{1}{3}\right)$
5. $y = x^2$: $(0, 0), (1, 1), (2, 4), (-1, 1), (-2, 4)$

Now we direct your attention to the ordered pairs associated with equations 3, 4, and 5. Note that in each case, no two ordered pairs have the same first component. Such a set of ordered pairs is called a **function.**

DEFINITION 9.2

A **function** is a relation in which no two ordered pairs have the same first component.

Stated another way, Definition 9.2 means that a function is a relation wherein each member of the domain is assigned *one and only one* member of the range. Thus it is easy to determine that each of the following sets of ordered pairs is a function.

$$f = \{(x, y) | y = x + 2\}$$

$$g = \left\{ (x, y) \middle| y = \frac{1}{x - 1} \right\}$$

$$h = \{(x, y) | y = x^2\}$$

In each case there is one and only one value of y (an element of the range) associated with each value of x (an element of the domain).

Note that we named the previous functions f, g, and h. It is customary to name functions by means of a single letter, and the letters f, g, and h are often used. We would suggest more meaningful choices when functions are used to portray real-world situations. For example, if a problem involves a profit function, then naming the function p or even P would seem natural.

The symbol for a function can be used along with a variable that represents an element in the domain to represent the associated element in the range. For example, suppose that we have a function f specified in terms of the variable x. The symbol $f(x)$, which is read "f of x" or "the value of f at x," represents the element in the range associated with the element x from the domain. The function $f = \{(x, y) | y = x + 2\}$ can be written as $f = \{(x, f(x)) | f(x) = x + 2\}$ and is usually shortened to read "f is the function determined by the equation $f(x) = x + 2$."

REMARK: Be careful with the notation $f(x)$. As we stated above, it means the value of the function f at x. It does not mean f times x.

This **function notation** is very convenient for computing and expressing various values of the function. For example, the value of the function $f(x) = 3x - 5$ at $x = 1$ is

$$f(1) = 3(1) - 5 = -2$$

Likewise, the functional values for $x = 2$, $x = -1$, and $x = 5$ are

$$f(2) = 3(2) - 5 = 1$$

$$f(-1) = 3(-1) - 5 = -8$$

$$f(5) = 3(5) - 5 = 10$$

Thus this function f contains the ordered pairs $(1, -2)$, $(2, 1)$, $(-1, -8)$, $(5, 10)$, and in general all ordered pairs of the form $(x, f(x))$, where $f(x) = 3x - 5$ and x is any real number.

It may be helpful for you to picture the concept of a function in terms of a *function machine,* as in Figure 9.1. Each time that a value of x is put into the machine, the equation $f(x) = x + 2$ is used to generate one and only one value for $f(x)$ to be ejected from the machine. For example, if 3 is put into this machine, then $f(3) = 3 + 2 = 5$, and 5 is ejected. Thus the ordered pair (3, 5) is one element of the function. Now let's look at some examples to illustrate these ideas about functions.

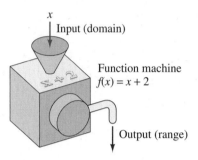

x

Input (domain)

Function machine
$f(x) = x + 2$

Output (range)

Figure 9.1

E X A M P L E   1

Determine whether the relation $\{(x, y) \mid y^2 = x\}$ is a function and specify its domain and range.

Solution

Because $y^2 = x$ is equivalent to $y = \pm \sqrt{x}$, to each value of x there are assigned *two* values for y. Therefore, this relation is not a function. The expression $\sqrt{x}$ requires that x be nonnegative, so the domain (D) is

$$D = \{x \mid x \geq 0\}$$

The domain can also be stated in interval notation as $D: [0, \infty)$. In the next few examples, domain and range will be shown with both set builder notation and interval notation. Consider the advantages and disadvantages of each format so that you can begin to decide, in each case, which format is the most appropriate. The answers given in the back of the book are shown in both set builder notation and interval notation.

To each nonnegative real number, the relation assigns two real numbers, $\sqrt{x}$ and $-\sqrt{x}$. Thus the range (R) is

$$R = \{y \mid y \text{ is a real number}\} \qquad \text{or} \qquad R: (-\infty, \infty)$$

$\blacksquare$

E X A M P L E   2

Consider the function $f(x) = x^2$.

    (a) Specify its domain.
    (b) Determine its range.
    (c) Evaluate $f(-2)$, $f(0)$, and $f(4)$.

Solution

(a) Any real number can be squared; therefore, the domain (D) is

$$D = \{x \mid x \text{ is a real number}\} \quad \text{or} \quad D: (-\infty, \infty)$$

(b) Squaring a real number always produces a nonnegative result. Thus the range (R) is

$$R = \{f(x) \mid f(x) \geq 0\} \quad \text{or} \quad R: [0, \infty)$$

(c) $f(-2) = (-2)^2 = 4$

$f(0) = (0)^2 = 0$

$f(4) = (4)^2 = 16$ ■

For our purposes in this text, if the domain of a function is not specifically indicated or determined by a real-world application, then we assume the domain to be all **real number** replacements for the variable, which represents an element in the domain that will produce **real number** functional values. Consider the following examples.

E X A M P L E   3   Specify the domain for each of the following:

(a) $f(x) = \dfrac{1}{x - 1}$ **(b)** $f(t) = \dfrac{1}{t^2 - 4}$ **(c)** $f(s) = \sqrt{s - 3}$

Solution

(a) We can replace x with any real number except 1, because 1 makes the denominator zero. Thus the domain is given by

$$D = \{x \mid x \neq 1\} \quad \text{or} \quad D: (-\infty, 1) \cup (1, \infty)$$

Here you may consider set builder notation to be easier than interval notation for expressing the domain.

(b) We need to eliminate any value of t that will make the denominator zero, so let's solve the equation $t^2 - 4 = 0$.

$t^2 - 4 = 0$

$t^2 = 4$

$t = \pm 2$

The domain is the set

$$D = \{t \mid t \neq -2 \text{ and } t \neq 2\} \quad \text{or} \quad D: (-\infty, -2) \cup (-2, 2) \cup (2, \infty)$$

When the domain is all real numbers except a few numbers, set builder notation is the more compact notation.

(c) The radicand, $s - 3$, must be nonnegative.

$s - 3 \geq 0$

$s \geq 3$

The domain is the set

$$D = \{s \mid s \geq 3\} \quad \text{or} \quad D: [3, \infty)$$ ■

E X A M P L E   4   If $f(x) = -2x + 7$ and $g(x) = x^2 - 5x + 6$, find $f(3)$, $f(-4)$, $f(b)$, $f(3c)$, $g(2)$, $g(-1)$, $g(a)$, and $g(a + 4)$.

Solution

$$f(x) = -2x + 7 \qquad\qquad g(x) = x^2 - 5x + 6$$

$$f(3) = -2(3) + 7 = 1 \qquad\qquad g(2) = (2)^2 - 5(2) + 6 = 0$$

$$f(-4) = -2(-4) + 7 = 15 \qquad\qquad g(-1) = (-1)^2 - 5(-1) + 6 = 12$$

$$f(b) = -2(b) + 7 = -2b + 7 \qquad\qquad g(a) = (a)^2 - 5(a) + 6 = a^2 - 5a + 6$$

$$f(3c) = -2(3c) + 7 = -6c + 7 \qquad g(a + 4) = (a + 4)^2 - 5(a + 4) + 6$$

$$= a^2 + 8a + 16 - 5a - 20 + 6$$

$$= a^2 + 3a + 2 \qquad\blacksquare$$

In Example 4, note that we are working with two different functions in the same problem. Thus different names, f and g, are used.

The quotient $\dfrac{f(a + h) - f(a)}{h}$ is often called a **difference quotient,** and we use it extensively with functions when studying the limit concept in calculus. The next two examples show how we found the difference quotient for two specific functions.

E X A M P L E   5   If $f(x) = 3x - 5$, find $\dfrac{f(a + h) - f(a)}{h}$.

Solution

$$f(a + h) = 3(a + h) - 5$$

$$= 3a + 3h - 5$$

and

$$f(a) = 3a - 5$$

Therefore,

$$f(a + h) - f(a) = (3a + 3h - 5) - (3a - 5)$$

$$= 3a + 3h - 5 - 3a + 5$$

$$= 3h$$

and

$$\frac{f(a + h) - f(a)}{h} = \frac{3h}{h} = 3 \qquad\blacksquare$$

EXAMPLE 6   If $f(x) = x^2 + 2x - 3$, find $\dfrac{f(a + h) - f(a)}{h}$.

Solution

$$f(a + h) = (a + h)^2 + 2(a + h) - 3$$
$$= a^2 + 2ah + h^2 + 2a + 2h - 3$$

and

$$f(a) = a^2 + 2a - 3$$

Therefore,

$$f(a + h) - f(a) = (a^2 + 2ah + h^2 + 2a + 2h - 3) - (a^2 + 2a - 3)$$
$$= a^2 + 2ah + h^2 + 2a + 2h - 3 - a^2 - 2a + 3$$
$$= 2ah + h^2 + 2h$$

and

$$\frac{f(a + h) - f(a)}{h} = \frac{2ah + h^2 + 2h}{h}$$
$$= \frac{h(2a + h + 2)}{h}$$
$$= 2a + h + 2 \qquad\blacksquare$$

Functions and functional notation provide the basis for describing many real-world relationships. The next example illustrates this point.

EXAMPLE 7   Suppose a factory determines that the overhead for producing a quantity of a certain item is $500 and that the cost for each item is $25. Express the total expenses as a function of the number of items produced, and compute the expenses for producing 12, 25, 50, 75, and 100 items.

Solution

Let n represent the number of items produced. Then $25n + 500$ represents the total expenses. Let's use E to represent the expense function, so that we have

$$E(n) = 25n + 500, \quad \text{where } n \text{ is a whole number}$$

from which we obtain

$$E(12) = 25(12) + 500 = 800$$
$$E(25) = 25(25) + 500 = 1125$$
$$E(50) = 25(50) + 500 = 1750$$
$$E(75) = 25(75) + 500 = 2375$$
$$E(100) = 25(100) + 500 = 3000$$

Thus the total expenses for producing 12, 25, 50, 75, and 100 items are $800, $1125, $1750, $2375, and $3000, respectively. $\blacksquare$

PROBLEM SET 9.1

For Problems 1–10, specify the domain and the range for each relation. Also state whether or not the relation is a function.

1. $\{(1, 5), (2, 8), (3, 11), (4, 14)\}$

2. $\{(0, 0), (2, 10), (4, 20), (6, 30), (8, 40)\}$

3. $\{(0, 5), (0, -5), (1, 2\sqrt{6}), (1, -2\sqrt{6})\}$

4. $\{(1, 1), (1, 2), (1, -1), (1, -2), (1, 3)\}$

5. $\{(1, 2), (2, 5), (3, 10), (4, 17), (5, 26)\}$

6. $\{(-1, 5), (0, 1), (1, -3), (2, -7)\}$

7. $\{(x, y)| 5x - 2y = 6\}$

8. $\{(x, y)| y = -3x\}$

9. $\{(x, y)| x^2 = y^3\}$

10. $\{(x, y)| x^2 - y^2 = 16\}$

For Problems 11–36, specify the domain for each of the functions.

11. $f(x) = 7x - 2$

12. $f(x) = x^2 + 1$

13. $f(x) = \dfrac{1}{x - 1}$

14. $f(x) = \dfrac{-3}{x + 4}$

15. $g(x) = \dfrac{3x}{4x - 3}$

16. $g(x) = \dfrac{5x}{2x + 7}$

17. $h(x) = \dfrac{2}{(x + 1)(x - 4)}$

18. $h(x) = \dfrac{-3}{(x - 6)(2x + 1)}$

19. $f(x) = \dfrac{14}{x^2 + 3x - 40}$

20. $f(x) = \dfrac{7}{x^2 - 8x - 20}$

21. $f(x) = \dfrac{-4}{x^2 + 6x}$

22. $f(x) = \dfrac{9}{x^2 - 12x}$

23. $f(t) = \dfrac{4}{t^2 + 9}$

24. $f(t) = \dfrac{8}{t^2 + 1}$

25. $f(t) = \dfrac{3t}{t^2 - 4}$

26. $f(t) = \dfrac{-2t}{t^2 - 25}$

27. $h(x) = \sqrt{x + 4}$

28. $h(x) = \sqrt{5x - 3}$

29. $f(s) = \sqrt{4s - 5}$

30. $f(s) = \sqrt{s - 2} + 5$

31. $f(x) = \sqrt{x^2 - 16}$

32. $f(x) = \sqrt{x^2 - 49}$

33. $f(x) = \sqrt{x^2 - 3x - 18}$

34. $f(x) = \sqrt{x^2 + 4x - 32}$

35. $f(x) = \sqrt{1 - x^2}$

36. $f(x) = \sqrt{9 - x^2}$

37. If $f(x) = 5x - 2$, find $f(0)$, $f(2)$, $f(-1)$, and $f(-4)$.

38. If $f(x) = -3x - 4$, find $f(-2)$, $f(-1)$, $f(3)$, and $f(5)$.

39. If $f(x) = \dfrac{1}{2}x - \dfrac{3}{4}$, find $f(-2)$, $f(0)$, $f\left(\dfrac{1}{2}\right)$, $f\left(\dfrac{2}{3}\right)$

40. If $g(x) = x^2 + 3x - 1$, find $g(1)$, $g(-1)$, $g(3)$, and $g(-4)$.

41. If $g(x) = 2x^2 - 5x - 7$, find $g(-1)$, $g(2)$, $g(-3)$, and $g(4)$.

42. If $h(x) = -x^2 - 3$, find $h(1)$, $h(-1)$, $h(-3)$, and $h(5)$.

43. If $h(x) = -2x^2 - x + 4$, find $h(-2)$, $h(-3)$, $h(4)$, and $h(5)$.

44. If $f(x) = \sqrt{x - 1}$, find $f(1)$, $f(5)$, $f(13)$, and $f(26)$.

45. If $f(x) = \sqrt{2x + 1}$, find $f(3)$, $f(4)$, $f(10)$, and $f(12)$.

46. If $f(x) = \dfrac{3}{x - 2}$, find $f(3)$, $f(0)$, $f(-1)$, and $f(-5)$.

47. If $f(x) = \dfrac{-4}{x + 3}$, find $f(1)$, $f(-1)$, $f(3)$, and $f(-6)$.

48. If $f(x) = 2x^2 - 7$ and $g(x) = x^2 + x - 1$, find $f(-2)$, $f(3)$, $g(-4)$, and $g(5)$.

49. If $f(x) = 5x^2 - 2x + 3$ and $g(x) = -x^2 + 4x - 5$, find $f(-2)$, $f(3)$, $g(-4)$, and $g(6)$.

50. If $f(x) = |3x - 2|$ and $g(x) = |x| + 2$, find $f(1)$, $f(-1)$, $g(2)$, and $g(-3)$.

51. If $f(x) = 3|x| - 1$ and $g(x) = -|x| + 1$, find $f(-2)$, $f(3)$, $g(-4)$, and $g(5)$.

For Problems 52–59, find $\dfrac{f(a + h) - f(a)}{h}$ for each of the given functions.

52. $f(x) = 5x - 4$

53. $f(x) = -3x + 6$

54. $f(x) = x^2 + 5$

55. $f(x) = -x^2 - 1$

56. $f(x) = x^2 - 3x + 7$ **57.** $f(x) = 2x^2 - x + 8$

58. $f(x) = -3x^2 + 4x - 1$ **59.** $f(x) = -4x^2 - 7x - 9$

60. Suppose that the cost function for producing a certain item is given by $C(n) = 3n + 5$, where n represents the number of items produced. Compute $C(150)$, $C(500)$, $C(750)$, and $C(1500)$.

61. The height of a projectile fired vertically into the air (neglecting air resistance) at an initial velocity of 64 feet per second is a function of the time (t) and is given by the equation

$$h(t) = 64t - 16t^2$$

Compute $h(1)$, $h(2)$, $h(3)$, and $h(4)$.

62. The profit function for selling n items is given by $P(n) = -n^2 + 500n - 61{,}500$. Compute $P(200)$, $P(230)$, $P(250)$, and $P(260)$.

63. A car rental agency charges $50 per day plus $.32 a mile. Therefore, the daily charge for renting a car is a function of the number of miles traveled (m) and can be expressed as $C(m) = 50 + 0.32m$. Compute $C(75)$, $C(150)$, $C(225)$, and $C(650)$.

64. The equation $A(r) = \pi r^2$ expresses the area of a circular region as a function of the length of a radius (r). Use 3.14 as an approximation for π and compute $A(2)$, $A(3)$, $A(12)$, and $A(17)$.

65. The equation $I(r) = 500r$ expresses the amount of simple interest earned by an investment of $500 for 1 year as a function of the rate of interest (r). Compute $I(0.11)$, $I(0.12)$, $I(0.135)$, and $I(0.15)$.

■ ■ ■ Thoughts into words

66. Are all functions also relations? Are all relations also functions? Defend your answers.

67. What does it mean to say that the domain of a function may be restricted if the function represents a real-world situation? Give two or three examples of such situations.

68. Does $f(a + b) = f(a) + f(b)$ for all functions? Defend your answer.

69. Are there any functions for which $f(a + b) = f(a) + f(b)$? Defend your answer.

9.2 Functions: Their Graphs and Applications

In Section 7.2, we made statements such as "The graph of the solution set of the equation $y = x - 1$ (or simply the graph of the equation $y = x - 1$) is a line that contains the points $(0, -1)$ and $(1, 0)$." Because the equation $y = x - 1$ (which can be written as $f(x) = x - 1$) can be used to specify a function, that line we previously referred to is also called the **graph of the function specified by the equation** or simply the **graph of the function.** Generally speaking, the graph of any equation that determines a function is also called the graph of the function. Thus the graphing techniques we discussed earlier will continue to play an important role as we graph functions.

As we use the function concept in our study of mathematics, it is helpful to classify certain types of functions and become familiar with their equations, characteristics, and graphs. In this section we will discuss two special types of functions — **linear** and **quadratic functions.** These functions are merely an outgrowth of our earlier study of linear and quadratic equations.

Linear Functions

Any function that can be written in the form

$$f(x) = ax + b$$

where a and b are real numbers, is called a **linear function.** The following equations are examples of linear functions.

$$f(x) = -3x + 6 \qquad f(x) = 2x + 4 \qquad f(x) = -\frac{1}{2}x - \frac{3}{4}$$

Graphing linear functions is quite easy because the graph of every linear function is a straight line. Therefore, all we need to do is determine two points of the graph and draw the line determined by those two points. You may want to continue using a third point as a check point.

EXAMPLE 1   Graph the function $f(x) = -3x + 6$.

Solution

Because $f(0) = 6$, the point $(0, 6)$ is on the graph. Likewise, because $f(1) = 3$, the point $(1, 3)$ is on the graph. Plot these two points and draw the line determined by the two points to produce Figure 9.2.

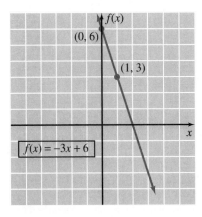

Figure 9.2

REMARK: Note that in Figure 9.2, we labeled the vertical axis $f(x)$. We could also label it y, because $f(x) = -3x + 6$ and $y = -3x + 6$ mean the same thing. We will continue to use the label $f(x)$ in this chapter to help you adjust to the function notation.

E X A M P L E   2   Graph the function $f(x) = x$.

Solution

The equation $f(x) = x$ can be written as $f(x) = 1x + 0$; thus it is a linear function. Because $f(0) = 0$ and $f(2) = 2$, the points $(0, 0)$ and $(2, 2)$ determine the line in Figure 9.3. The function $f(x) = x$ is often called the **identity function.**

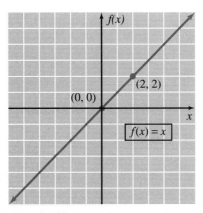

Figure 9.3

   As we use function notation to graph functions, it is often helpful to think of the ordinate of every point on the graph as the value of the function at a specific value of x. Geometrically, this functional value is the directed distance of the point from the x axis, as illustrated in Figure 9.4 with the function $f(x) = 2x - 4$. For example, consider the graph of the function $f(x) = 2$. The function $f(x) = 2$ means that every functional value is 2, or, geometrically, that every point on the graph is 2 units above the x axis. Thus the graph is the horizontal line shown in Figure 9.5.

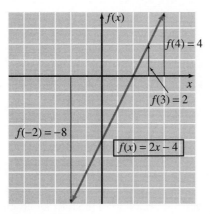

Figure 9.4

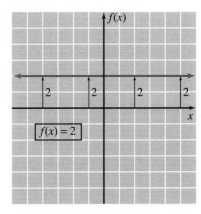

Figure 9.5

Any linear function of the form $f(x) = ax + b$, where $a = 0$, is called a **constant function,** and its graph is a horizontal line.

Applications of Linear Functions

We worked with some applications of linear equations in Section 7.2. Let's consider some additional applications that use the concept of a linear function to connect mathematics to the real world.

EXAMPLE 3

The cost for burning a 60-watt light bulb is given by the function $c(h) = 0.0036h$, where h represents the number of hours that the bulb is burning.

(a) How much does it cost to burn a 60-watt bulb for 3 hours per night for a 30-day month?
(b) Graph the function $c(h) = 0.0036h$.
(c) Suppose that a 60-watt light bulb is left burning in a closet for a week before it is discovered and turned off. Use the graph from part (b) to approximate the cost of allowing the bulb to burn for a week. Then use the function to find the exact cost.

Solution

(a) $c(90) = 0.0036(90) = 0.324$. The cost, to the nearest cent, is \$.32.
(b) Because $c(0) = 0$ and $c(100) = 0.36$, we can use the points $(0, 0)$ and $(100, 0.36)$ to graph the linear function $c(h) = 0.0036h$ (Figure 9.6).

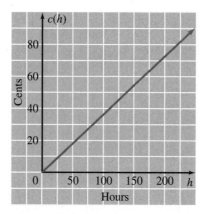

Figure 9.6

(c) If the bulb burns for 24 hours per day for a week, it burns for $24(7) = 168$ hours. Reading from the graph, we can approximate 168 on the horizontal axis, read up to the line, and then read across to the vertical axis. It looks as if it will cost approximately 60 cents. Using $c(h) = 0.0036h$, we obtain exactly $c(168) = 0.0036(168) = 0.6048$.

EXAMPLE 4

The EZ Car Rental charges a fixed amount per day plus an amount per mile for renting a car. For two different day trips, Ed has rented a car from EZ. He paid $70 for 100 miles on one day and $120 for 350 miles on another day. Determine the linear function that the EZ Car Rental uses to determine its daily rental charges.

Solution

The linear function $f(x) = ax + b$, where x represents the number of miles, models this situation. Ed's two day trips can be represented by the ordered pairs (100, 70) and (350, 120). From these two ordered pairs we can determine a, which is the slope of the line.

$$a = \frac{120 - 70}{350 - 100} = \frac{50}{250} = \frac{1}{5} = 0.2$$

Thus $f(x) = ax + b$ becomes $f(x) = 0.2x + b$. Now either ordered pair can be used to determine the value of b. Using (100, 70), we have $f(100) = 70$; therefore,

$$f(100) = 0.2(100) + b = 70$$

$$b = 50$$

The linear function is $f(x) = 0.2x + 50$. In other words, the EZ Car Rental charges a daily fee of $50 plus $.20 per mile. ∎

EXAMPLE 5

Suppose that Ed (Example 4) also has access to the A-OK Car Rental agency, which charges a daily fee of $25 plus $.30 per mile. Should Ed use the EZ Car Rental from Example 4 or A-OK Car Rental?

Solution

The linear function $g(x) = 0.3x + 25$, where x represents the number of miles, can be used to determine the daily charges of A-OK Car Rental. Let's graph this function and $f(x) = 0.2x + 50$ from Example 4 on the same set of axes (Figure 9.7).

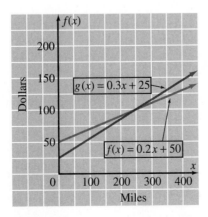

Figure 9.7

Now we see that the two functions have equal values at the point of intersection of the two lines. To find the coordinates of this point, we can set $0.3x + 25$ equal to $0.2x + 50$ and solve for x.

$$0.3x + 25 = 0.2x + 50$$

$$0.1x = 25$$

$$x = 250$$

If $x = 250$, then $0.3(250) + 25 = 100$, and the point of intersection is $(250, 100)$. Again looking at the lines in Figure 9.7, we see that Ed should use A-OK Car Rental for day trips of less than 250 miles, but he should use EZ Car Rental for day trips of more than 250 miles. ◼

Quadratic Functions

Any function that can be written in the form

$$f(x) = ax^2 + bx + c$$

where a, b, and c are real numbers with $a \neq 0$, is called a **quadratic function.** The following equations are examples of quadratic functions.

$$f(x) = 3x^2 \qquad f(x) = -2x^2 + 5x \qquad f(x) = 4x^2 - 7x + 1$$

The techniques discussed in Chapter 8 relative to graphing quadratic equations of the form $y = ax^2 + bx + c$ provide the basis for graphing quadratic functions. Let's review some work from Chapter 8 with an example.

E X A M P L E   6   Graph the function $f(x) = 2x^2 - 4x + 5$.

Solution

$$f(x) = 2x^2 - 4x + 5$$

$$= 2(x^2 - 2x + \underline{}) + 5 \qquad \text{Recall the process of completing the}$$

$$= 2(x^2 - 2x + 1) + 5 - 2 \qquad \text{square!}$$

$$= 2(x - 1)^2 + 3$$

From this form we can obtain the following information about the parabola.

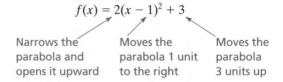

$$f(x) = 2(x - 1)^2 + 3$$

Narrows the parabola and opens it upward   Moves the parabola 1 unit to the right   Moves the parabola 3 units up

Thus the parabola can be drawn as shown in Figure 9.8.

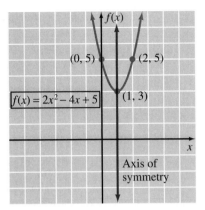

$(0, 5)$ $(2, 5)$

$f(x) = 2x^2 - 4x + 5$ $(1, 3)$

Axis of symmetry

Figure 9.8

In general, if we complete the square on

$$f(x) = ax^2 + bx + c$$

we obtain

$$f(x) = a\left(x^2 + \frac{b}{a}x + \underline{\quad}\right) + c$$

$$= a\left(x^2 + \frac{b}{a}x + \frac{b^2}{4a^2}\right) + c - \frac{b^2}{4a}$$

$$= a\left(x + \frac{b}{2a}\right)^2 + \frac{4ac - b^2}{4a}$$

Therefore, the parabola associated with $f(x) = ax^2 + bx + c$ has its vertex at $\left(-\dfrac{b}{2a}, \dfrac{4ac - b^2}{4a}\right)$ and the equation of its axis of symmetry is $x = -\dfrac{b}{2a}$. These facts are illustrated in Figure 9.9.

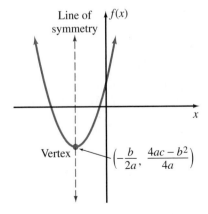

Line of symmetry

Vertex $\left(-\dfrac{b}{2a}, \dfrac{4ac - b^2}{4a}\right)$

Figure 9.9

By using the information from Figure 9.9, we now have another way of graphing quadratic functions of the form $f(x) = ax^2 + bx + c$, shown by the following steps.

1. Determine whether the parabola opens upward (if $a > 0$) or downward (if $a < 0$).

2. Find $-\dfrac{b}{2a}$, which is the x coordinate of the vertex.

3. Find $f\left(-\dfrac{b}{2a}\right)$, which is the y coordinate of the vertex. $\left(\text{You could also find the } y \text{ coordinate by evaluating } \dfrac{4ac - b^2}{4a}.\right)$

4. Locate another point on the parabola, and also locate its image across the line of symmetry, $x = -\dfrac{b}{2a}$.

The three points in Steps 2, 3, and 4 should determine the general shape of the parabola. Let's use these steps in the following two examples.

EXAMPLE 7 Graph $f(x) = 3x^2 - 6x + 5$.

Solution

STEP 1   Because a = 3, the parabola opens upward.

STEP 2 $-\dfrac{b}{2a} = -\dfrac{-6}{6} = 1$

STEP 3 $f\left(-\dfrac{b}{2a}\right) = f(1) = 3 - 6 + 5 = 2$. Thus the vertex is at $(1, 2)$.

STEP 4   Letting $x = 2$, we obtain $f(2) = 12 - 12 + 5 = 5$. Thus $(2, 5)$ is on the graph and so is its reflection $(0, 5)$ across the line of symmetry $x = 1$.

The three points $(1, 2), (2, 5)$, and $(0, 5)$ are used to graph the parabola in Figure 9.10.

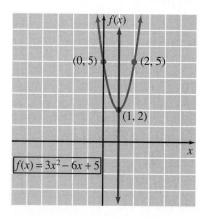

Figure 9.10

EXAMPLE 8   Graph $f(x) = -x^2 - 4x - 7$.

Solution

STEP 1   Because $a = -1$, the parabola opens downward.

STEP 2 $-\dfrac{b}{2a} = -\dfrac{-4}{-2} = -2.$

STEP 3 $f\left(-\dfrac{b}{2a}\right) = f(-2) = -(-2)^2 - 4(-2) - 7 = -3.$ So the vertex is at $(-2, -3)$.

STEP 4   Letting $x = 0$, we obtain $f(0) = -7$. Thus $(0, -7)$ is on the graph and so is its reflection $(-4, -7)$ across the line of symmetry $x = -2$.

The three points $(-2, -3)$, $(0, -7)$ and $(-4, -7)$ are used to draw the parabola in Figure 9.11.

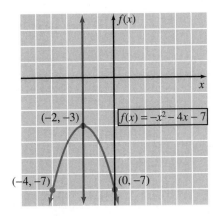

Figure 9.11

In summary, to graph a quadratic function, we have two methods.

1. We can express the function in the form $f(x) = a(x - h)^2 + k$ and use the values of a, h, and k to determine the parabola.
2. We can express the function in the form $f(x) = ax^2 + bx + c$ and use the approach demonstrated in Examples 7 and 8.

Problem Solving Using Quadratic Functions

As we have seen, the vertex of the graph of a quadratic function is either the lowest or the highest point on the graph. Thus the term *minimum value* or *maximum value* of a function is often used in applications of the parabola. The x value of the vertex indicates where the minimum or maximum occurs, and $f(x)$ yields the minimum or maximum value of the function. Let's consider some examples that illustrate these ideas.

E X A M P L E   9

A farmer has 120 rods of fencing and wants to enclose a rectangular plot of land that requires fencing on only three sides because it is bounded by a river on one side. Find the length and width of the plot that will maximize the area.

Solution

Let x represent the width; then $120 - 2x$ represents the length, as indicated in Figure 9.12.

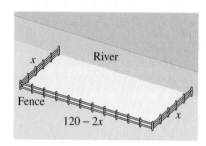

Figure 9.12

The function $A(x) = x(120 - 2x)$ represents the area of the plot in terms of the width x. Because

$$A(x) = x(120 - 2x)$$
$$= 120x - 2x^2$$
$$= -2x^2 + 120x$$

we have a quadratic function with $a = -2, b = 120$, and $c = 0$. Therefore, the x value where the maximum value of the function is obtained is

$$-\frac{b}{2a} = -\frac{120}{2(-2)} = 30$$

If $x = 30$, then $120 - 2x = 120 - 2(30) = 60$. Thus the farmer should make the plot 30 rods wide and 60 rods long to maximize the area at $(30)(60) = 1800$ square rods. ■

E X A M P L E   1 0

Find two numbers whose sum is 30, such that the sum of their squares is a minimum.

Solution

Let x represent one of the numbers; then $30 - x$ represents the other number. By expressing the sum of the squares as a function of x, we obtain

$$f(x) = x^2 + (30 - x)^2$$

which can be simplified to

$$f(x) = x^2 + 900 - 60x + x^2$$

$$= 2x^2 - 60x + 900$$

This is a quadratic function with $a = 2$, $b = -60$, and $c = 900$. Therefore, the x value where the minimum occurs is

$$-\frac{b}{2a} = -\frac{-60}{4} = 15$$

If $x = 15$, then $30 - x = 30 - (15) = 15$. Thus the two numbers should both be 15. ■

EXAMPLE 11

A golf pro-shop operator finds that she can sell 30 sets of golf clubs at $500 per set in a year. Furthermore, she predicts that for each $25 decrease in price, three more sets of golf clubs could be sold. At what price should she sell the clubs to maximize gross income?

Solution

Sometimes, when we are analyzing such a problem, it helps to set up a table.

	Number of sets	×	Price per set	=	Income
3 additional sets can be sold for a $25 decrease in price	30	×	$500	=	$15,000
	33	×	$475	=	$15,675
	36	×	$450	=	$16,200

Let x represent the number of $25 decreases in price. Then we can express the income as a function of x as follows:

$$f(x) = (30 + 3x)(500 - 25x)$$

Number of sets   Price per set

When we simplify, we obtain

$$f(x) = 15,000 - 750x + 1500x - 75x^2$$

$$= -75x^2 + 750x + 15,000$$

Completing the square yields

$$f(x) = -75x^2 + 750x + 15,000$$

$$= -75(x^2 - 10x + \_\_) + 15,000$$

$$= -75(x^2 - 10x + 25) + 15,000 + 1875$$

$$= -75(x - 5)^2 + 16,875$$

From this form we know that the vertex of the parabola is at (5, 16875). Thus 5 decreases of $25 each—that is, a $125 reduction in price—will give a maximum income of $16,875. The golf clubs should be sold at $375 per set.   ■

What we know about parabolas and the process of completing the square can be helpful when we are using a graphing utility to graph a quadratic function. Consider the following example.

E X A M P L E   1 2   Use a graphing utility to obtain the graph of the quadratic function

$$f(x) = -x^2 + 37x - 311$$

 Solution

First, we know that the parabola opens downward and that its width is the same as that of the basic parabola $f(x) = x^2$. Then we can start the process of completing the square to determine an approximate location of the vertex.

$$f(x) = -x^2 + 37x - 311$$
$$= -(x^2 - 37x + \underline{\quad}) - 311$$
$$= -\left[x^2 - 37x + \left(\frac{37}{2}\right)^2\right] - 311 + \left(\frac{37}{2}\right)^2$$
$$= -[(x^2 - 37x + (18.5)^2] - 311 + 342.25$$
$$= -(x - 18.5)^2 + 31.25$$

Thus the vertex is near $x = 18$ and $y = 31$. Therefore, setting the boundaries of the viewing rectangle so that $-2 \le x \le 25$ and $-10 \le y \le 35$, we obtain the graph shown in Figure 9.13.

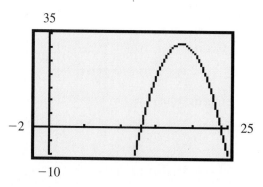

Figure 9.13   ■

REMARK: The graph in Figure 9.13 is sufficient for most purposes, because it shows the vertex and the x intercepts of the parabola. Certainly other boundaries could be used that would also give this information.

PROBLEM SET 9.2

Graph each of the following linear and quadratic functions (Problems 1–30).

1. $f(x) = 2x - 4$

2. $f(x) = 3x + 3$

3. $f(x) = -2x^2$

4. $f(x) = -4x^2$

5. $f(x) = -3x$

6. $f(x) = -4x$

7. $f(x) = -(x + 1)^2 - 2$

8. $f(x) = -(x - 2)^2 + 4$

9. $f(x) = -x + 3$

10. $f(x) = -2x - 4$

11. $f(x) = x^2 + 2x - 2$

12. $f(x) = x^2 - 4x - 1$

13. $f(x) = -x^2 + 6x - 8$

14. $f(x) = -x^2 - 8x - 15$

15. $f(x) = -3$

16. $f(x) = 1$

17. $f(x) = 2x^2 - 20x + 52$

18. $f(x) = 2x^2 + 12x + 14$

19. $f(x) = -3x^2 + 6x$

20. $f(x) = -4x^2 - 8x$

21. $f(x) = x^2 - x + 2$

22. $f(x) = x^2 + 3x + 2$

23. $f(x) = 2x^2 + 10x + 11$

24. $f(x) = 2x^2 - 10x + 15$

25. $f(x) = -2x^2 - 1$

26. $f(x) = -3x^2 + 2$

27. $f(x) = -3x^2 + 12x - 7$

28. $f(x) = -3x^2 - 18x - 23$

29. $f(x) = -2x^2 + 14x - 25$

30. $f(x) = -2x^2 - 10x - 14$

31. The cost for burning a 75-watt bulb is given by the function $c(h) = 0.0045h$, where h represents the number of hours that the bulb burns.
(a) How much does it cost to burn a 75-watt bulb for 3 hours per night for a 31-day month? Express your answer to the nearest cent.
(b) Graph the function $c(h) = 0.0045h$.
(c) Use the graph in part (b) to approximate the cost of burning a 75-watt bulb for 225 hours.
(d) Use $c(h) = 0.0045h$ to find the exact cost, to the nearest cent, of burning a 75-watt bulb for 225 hours.

32. The Rent-Me Car Rental charges \$15 per day plus \$.22 per mile to rent a car. Determine a linear function that can be used to calculate daily car rentals. Then use that function to determine the cost of renting a car for a day and driving 175 miles; 220 miles; 300 miles; 460 miles.

33. The ABC Car Rental uses the function $f(x) = 26$ for any daily use of a car up to and including 200 miles. For driving more than 200 miles per day, ABC uses the function $g(x) = 26 + 0.15(x - 200)$ to determine the charges. How much would ABC charge for daily driving of 150 miles? of 230 miles? of 360 miles? of 430 miles?

34. Suppose that a car rental agency charges a fixed amount per day plus an amount per mile for renting a car. Heidi rented a car one day and paid \$80 for 200 miles. On another day she rented a car from the same agency and paid \$117.50 for 350 miles. Find the linear function that the agency could use to determine its daily rental charges.

35. A retailer has a number of items that she wants to sell and make a profit of 40% of the cost of each item. The function $s(c) = c + 0.4c = 1.4c$, where c represents the cost of an item, can be used to determine the selling price. Find the selling price of items that cost \$1.50, \$3.25, \$14.80, \$21, and \$24.20.

36. Zack wants to sell five items that cost him \$1.20, \$2.30, \$6.50, \$12, and \$15.60. He wants to make a profit of 60% of the cost. Create a function that you can use to determine the selling price of each item, and then use the function to calculate each selling price.

37. "All Items 20% Off Marked Price" is a sign at a local golf course. Create a function and then use it to determine how much one has to pay for each of the following marked items: a \$9.50 hat, a \$15 umbrella, a \$75 pair of golf shoes, a \$12.50 golf glove, a \$750 set of golf clubs.

38. The linear depreciation method assumes that an item depreciates the same amount each year. Suppose a new piece of machinery costs $32,500 and it depreciates $1950 each year for t years.
  (a) Set up a linear function that yields the value of the machinery after t years.
  (b) Find the value of the machinery after 5 years.
  (c) Find the value of the machinery after 8 years.
  (d) Graph the function from part (a).
  (e) Use the graph from part (d) to approximate how many years it takes for the value of the machinery to become zero.
  (f) Use the function to determine how long it takes for the value of the machinery to become zero.

39. Suppose that the cost function for a particular item is given by the equation $C(x) = 2x^2 - 320x + 12{,}920$, where x represents the number of items. How many items should be produced to minimize the cost?

40. Suppose that the equation $p(x) = -2x^2 + 280x - 1000$, where x represents the number of items sold, describes the profit function for a certain business. How many items should be sold to maximize the profit?

41. Find two numbers whose sum is 30, such that the sum of the square of one number plus ten times the other number is a minimum.

42. The height of a projectile fired vertically into the air (neglecting air resistance) at an initial velocity of 96 feet per second is a function of the time and is given by the equation $f(x) = 96x - 16x^2$, where x represents the time. Find the highest point reached by the projectile.

43. Two hundred and forty meters of fencing is available to enclose a rectangular playground. What should be the dimensions of the playground to maximize the area?

44. Find two numbers whose sum is 50 and whose product is a maximum.

45. A cable TV company has 1000 subscribers, and each pays $15 per month. On the basis of a survey, company managers feel that for each decrease of $.25 on the monthly rate, they could obtain 20 additional subscribers. At what rate will maximum revenue be obtained and how many subscribers will it take at that rate?

46. A motel advertises that it will provide dinner, dancing, and drinks for $50 per couple at a New Year's Eve party. It must have a guarantee of 30 couples. Furthermore, it will agree that for each couple in excess of 30, it will reduce the price per couple for all attending by $.50. How many couples will it take to maximize the motel's revenue?

■ ■ ■ **Thoughts into words**

47. Give a step-by-step description of how you would use the ideas of this section to graph $f(x) = -4x^2 + 16x - 13$.

48. Is $f(x) = (3x - 2) - (2x + 1)$ a linear function? Explain your answer.

49. Suppose that Bianca walks at a constant rate of 3 miles per hour. Explain what it means that the distance Bianca walks is a linear function of the time that she walks.

 Graphing calculator activities

50. Use a graphing calculator to check your graphs for Problems 17–30.

51. Graph each of the following parabolas, and keep in mind that you may need to change the dimensions of the viewing window to obtain a good picture.
  (a) $f(x) = x^2 - 2x + 12$ **(b)** $f(x) = -x^2 - 4x - 16$
  (c) $f(x) = x^2 + 12x + 44$ **(d)** $f(x) = x^2 - 30x + 229$
  (e) $f(x) = -2x^2 + 8x - 19$

52. Graph each of the following parabolas, and use the TRACE feature to find whole number estimates of the vertex. Then either complete the square or use $\left(-\dfrac{b}{2a}, \dfrac{4ac - b^2}{4a}\right)$ to find the vertex.
  (a) $f(x) = x^2 - 6x + 3$
  (b) $f(x) = x^2 - 18x + 66$
  (c) $f(x) = -x^2 + 8x - 3$
  (d) $f(x) = -x^2 + 24x - 129$

(e) $f(x) = 14x^2 - 7x + 1$
(f) $f(x) = -0.5x^2 + 5x - 8.5$

53. (a) Graph $f(x) = |x|$, $f(x) = 2|x|$, $f(x) = 4|x|$, and
$f(x) = \frac{1}{2}|x|$ on the same set of axes.

(b) Graph $f(x) = |x|$, $f(x) = -|x|$, $f(x) = -3|x|$, and
$f(x) = -\frac{1}{2}|x|$ on the same set of axes.

(c) Use your results from parts (a) and (b) to make a conjecture about the graphs of $f(x) = a|x|$, where a is a nonzero real number.

(d) Graph $f(x) = |x|$, $f(x) = |x| + 3$, $f(x) = |x| - 4$, and $f(x) = |x| + 1$ on the same set of axes. Make a con-

jecture about the graphs of $f(x) = |x| + k$, where k is a nonzero real number.

(e) Graph $f(x) = |x|$, $f(x) = |x - 3|$, $f(x) = |x - 1|$, and $f(x) = |x + 4|$ on the same set of axes. Make a conjecture about the graphs of $f(x) = |x - h|$, where h is a nonzero real number.

(f) On the basis of your results from parts (a) through (e), sketch each of the following graphs. Then use a graphing calculator to check your sketches.

(1) $f(x) = |x - 2| + 3$ **(2)** $f(x) = |x + 1| - 4$

(3) $f(x) = 2|x - 4| - 1$ **(4)** $f(x) = -3|x + 2| + 4$

(5) $f(x) = \frac{1}{2}|x - 3| - 2$

9.3   Graphing Made Easy Via Transformations

Figures 9.14–9.16 show the graphs of the functions $f(x) = x^2$, $f(x) = x^3$, and $f(x) = \dfrac{1}{x}$, respectively.

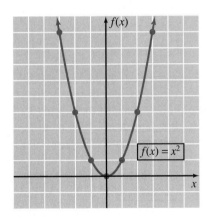

Figure 9.14

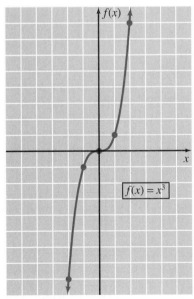

Figure 9.15

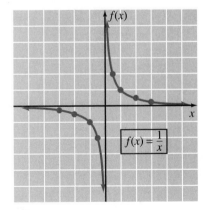

Figure 9.16

To graph a new function—that is, one you are not familiar with — use some of the graphing suggestions we offered in Chapter 7. We will restate those suggestions in terms of function vocabulary and notation. Pay special attention to suggestions 2 and 3, where we have restated the concepts of intercepts and symmetry using function notation.

1. Determine the domain of the function.

2. Determine any types of symmetry that the equation possesses. If $f(-x) = f(x)$, then the function exhibits y axis symmetry. If $f(-x) = -f(x)$, then the function exhibits origin symmetry. (Note that the definition of a function rules out the possibility that the graph of a function has x axis symmetry.)

3. Find the y intercept (we are labeling the y axis with $f(x)$) by evaluating $f(0)$. Find the x intercept by finding the value(s) of x such that $f(x) = 0$.

4. Set up a table of ordered pairs that satisfy the equation. The type of symmetry and the domain will affect your choice of values of x in the table.

5. Plot the points associated with the ordered pairs and connect them with a smooth curve. Then, if appropriate, reflect this part of the curve according to any symmetries the graph exhibits.

Let's consider some examples where we can use some of these suggestions.

EXAMPLE 1   Graph $f(x) = \sqrt{x}$.

Solution

The radicand must be nonnegative, so the domain is the set of nonnegative real numbers. Because $x \geq 0$, $f(-x)$ is not a real number; thus there is no symmetry for this graph. We see that $f(0) = 0$, so both intercepts are 0. That is, the origin $(0, 0)$ is a point of the graph. Now let's set up a table of values, keeping in mind that $x \geq 0$. Plotting these points and connecting them with a smooth curve produces Figure 9.17.

x	$f(x)$
0	0
1	1
4	2
9	3

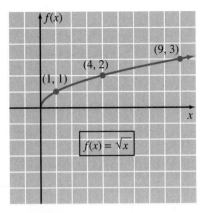

Figure 9.17

Sometimes a new function is defined in terms of old functions. In such cases, the definition plays an important role in the study of the new function. Consider the following example.

EXAMPLE 2

Graph the function $f(x) = |x|$.

Solution

The concept of absolute value is defined for all real numbers as

$$|x| = x \quad \text{if } x \geq 0$$

$$|x| = -x \quad \text{if } x < 0$$

Therefore, we can express the absolute-value function as

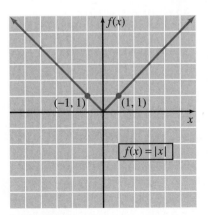

$$f(x) = |x| = \begin{cases} x & \text{if } x \geq 0 \\ -x & \text{if } x < 0 \end{cases}$$

The graph of $f(x) = x$ for $x \geq 0$ is the ray in the first quadrant, and the graph of $f(x) = -x$ for $x < 0$ is the half-line in the second quadrant, as indicated in Figure 9.18.

Figure 9.18

REMARK: Note in Example 2 that the equation $f(x) = |x|$ does exhibit y axis symmetry because $f(-x) = |-x| = |x|$. Even though we did not use the symmetry idea to sketch the curve, you should recognize that the symmetry does exist.

Translations of the Basic Curves

From our work in Chapter 8, we know that the graph of $f(x) = x^2 + 3$ is the graph of $f(x) = x^2$ moved up 3 units. Likewise, the graph of $f(x) = x^2 - 2$ is the graph of $f(x) = x^2$ moved down 2 units. Now we will describe in general the concept of **vertical translation.**

> ### Vertical Translation
>
> The graph of $y = f(x) + k$ is the graph of $y = f(x)$ shifted k units upward if $k > 0$ or shifted $|k|$ units downward if $k < 0$.

In Figure 9.19, we obtain the graph of $f(x) = |x| + 2$ by shifting the graph of $f(x) = |x|$ upward 2 units, and we obtain the graph of $f(x) = |x| - 3$ by shifting the graph of $f(x) = |x|$ downward 3 units. (Remember that we can write $f(x) = |x| - 3$ as $f(x) = |x| + (-3)$.)

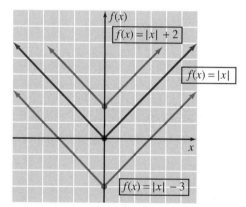

Figure 9.19

We also graphed horizontal translations of the basic parabola in Chapter 8. For example, the graph of $f(x) = (x - 4)^2$ is the graph of $f(x) = x^2$ shifted 4 units to the right, and the graph of $f(x) = (x + 5)^2$ is the graph of $f(x) = x^2$ shifted 5 units to the left. We describe the general concept of a **horizontal translation** as follows:

> ### Horizontal Translation
>
> The graph of $y = f(x - h)$ is the graph of $y = f(x)$ shifted h units to the right if $h > 0$ or shifted $|h|$ units to the left if $h < 0$.

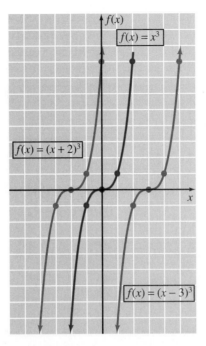

Figure 9.20

In Figure 9.20, we obtain the graph of $f(x) = (x - 3)^3$ by shifting the graph of $f(x) = x^3$ to the right 3 units. Likewise, we obtain the graph of $f(x) = (x + 2)^3$ by shifting the graph of $f(x) = x^3$ to the left 2 units.

Reflections of the Basic Curves

From our work in Chapter 8, we know that the graph of $f(x) = -x^2$ is the graph of $f(x) = x^2$ reflected through the x axis. We describe the general concept of an **x axis reflection** as follows:

x Axis Reflection

The graph of $y = -f(x)$ is the graph of $y = f(x)$ reflected through the x axis.

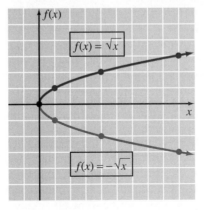

Figure 9.21

In Figure 9.21, we obtain the graph of $f(x) = -\sqrt{x}$ by reflecting the graph of $f(x) = \sqrt{x}$ through the x axis. Reflections are sometimes referred to as **mirror images.** Thus, in Figure 9.21, if we think of the x axis as a mirror, the graphs of $f(x) = \sqrt{x}$ and $f(x) = -\sqrt{x}$ are mirror images of each other.

In Chapter 8, we did not consider a y axis reflection of the basic parabola $f(x) = x^2$ because it is symmetric with respect to the y axis. In other words, a y axis reflection of $f(x) = x^2$ produces the same figure in the same location. At this time we will describe the general concept of a y axis reflection.

---

y Axis Reflection

The graph of $y = f(-x)$ is the graph of $y = f(x)$ reflected through the y axis.

Now suppose that we want to do a y axis reflection of $f(x) = \sqrt{x}$. Because $f(x) = \sqrt{x}$ is defined for $x \geq 0$, the y axis reflection $f(x) = \sqrt{-x}$ is defined for $-x \geq 0$, which is equivalent to $x \leq 0$. Figure 9.22 shows the y axis reflection of $f(x) = \sqrt{x}$.

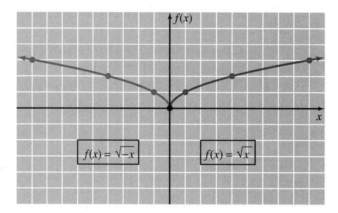

Figure 9.22

Vertical Stretching and Shrinking

Translations and reflections are called **rigid transformations** because the basic shape of the curve being transformed is not changed. In other words, only the positions of the graphs are changed. Now we want to consider some transformations that distort the shape of the original figure somewhat.

In Chapter 8, we graphed the equation $y = 2x^2$ by doubling the y coordinates of the ordered pairs that satisfy the equation $y = x^2$. We obtained a parabola with its vertex at the origin, symmetric with respect to the y axis, but narrower than the basic parabola. Likewise, we graphed the equation $y = \frac{1}{2}x^2$ by halving the y coordinates of the ordered pairs that satisfy $y = x^2$. In this case, we obtained a parabola with its vertex at the origin, symmetric with respect to the y axis, but wider than the basic parabola.

We can use the concepts of narrower and wider to describe parabolas, but they cannot be used to describe some other curves accurately. Instead, we use the more general concepts of vertical stretching and shrinking.

> **Vertical Stretching and Shrinking**
>
> The graph of $y = cf(x)$ is obtained from the graph of $y = f(x)$ by multiplying the y coordinates of $y = f(x)$ by c. If $c > 1$, the graph is said to be *stretched* by a factor of c, and if $0 < c < 1$, the graph is said to be *shrunk* by a factor of c.

In Figure 9.23, the graph of $f(x) = 2\sqrt{x}$ is obtained by doubling the y coordinates of points on the graph of $f(x) = \sqrt{x}$. Likewise, in Figure 9.23, the graph of $f(x) = \dfrac{1}{2}\sqrt{x}$ is obtained by halving the y coordinates of points on the graph of $f(x) = \sqrt{x}$.

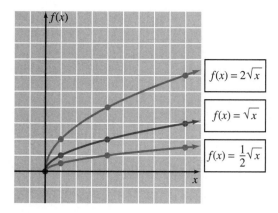

Figure 9.23

Successive Transformations

Some curves are the result of performing more than one transformation on a basic curve. Let's consider the graph of a function that involves a stretching, a reflection, a horizontal translation, and a vertical translation of the basic absolute-value function.

EXAMPLE 3

Graph $f(x) = -2|x - 3| + 1$.

Solution

This is the basic absolute-value curve stretched by a factor of two, reflected through the x axis, shifted 3 units to the right, and shifted 1 unit upward. To sketch the graph, we locate the point $(3, 1)$ and then determine a point on each of the rays. The graph is shown in Figure 9.24.

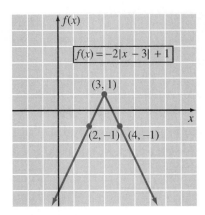

Figure 9.24

REMARK: Note in Example 3 that we did not sketch the original basic curve $f(x) = |x|$ or any of the intermediate transformations. However, it is helpful to mentally picture each transformation. This locates the point $(3, 1)$ and establishes the fact that the two rays point downward. Then a point on each ray determines the final graph.

You also need to realize that changing the order of doing the transformations may produce an incorrect graph. In Example 3, performing the translations first, followed by the stretching and x axis reflection, would produce an incorrect graph that has its vertex at $(3, -1)$ instead of $(3, 1)$. Unless parentheses indicate otherwise, stretchings, shrinkings, and x axis reflections should be performed before translations.

E X A M P L E   4

Graph the function $f(x) = \dfrac{1}{x + 2} + 3$.

Solution

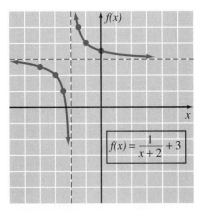

Figure 9.25

This is the basic curve $f(x) = \dfrac{1}{x}$ moved 2 units to the left and 3 units upward. Remember that the x axis is a horizontal asymptote and the y axis a vertical asymptote for the curve $f(x) = \dfrac{1}{x}$. Thus, for this curve, the vertical asymptote is shifted 2 units to the left, and its equation is $x = -2$. Likewise, the horizontal asymptote is shifted 3 units upward, and its equation is $y = 3$. Therefore, in Figure 9.25 we have drawn the asymptotes as dashed lines and then located a few points to help determine each branch of the curve.

Finally, let's use a graphing utility to give another illustration of the concepts of stretching and shrinking a curve.

E X A M P L E   5   If $f(x) = \sqrt{25 - x^2}$, sketch a graph of $y = 2(f(x))$ and $y = \dfrac{1}{2}(f(x))$.

Solution

If $y = f(x) = \sqrt{25 - x^2}$, then

$$y = 2(f(x)) = 2\sqrt{25 - x^2}, \quad \text{and} \quad y = \frac{1}{2}(f(x)) = \frac{1}{2}\sqrt{25 - x^2}$$

Graphing all three of these functions on the same set of axes produces Figure 9.26.

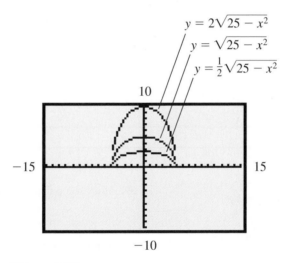

Figure 9.26

PROBLEM SET 9.3

For Problems 1–34, graph each of the functions.

1. $f(x) = -x^3$

2. $f(x) = x^3 - 2$

3. $f(x) = -(x - 4)^2 + 2$

4. $f(x) = -2(x + 3)^2 - 4$

5. $f(x) = \dfrac{1}{x} - 2$

6. $f(x) = \dfrac{1}{x - 2}$

7. $f(x) = |x - 1| + 2$

8. $f(x) = -|x + 2|$

9. $f(x) = \dfrac{1}{2}|x|$

10. $f(x) = -2|x|$

11. $f(x) = -2\sqrt{x}$

12. $f(x) = 2\sqrt{x - 1}$

13. $f(x) = \sqrt{x + 2} - 3$

14. $f(x) = -\sqrt{x + 2} + 2$

15. $f(x) = \dfrac{2}{x - 1} + 3$

16. $f(x) = \dfrac{3}{x + 3} - 4$

17. $f(x) = \sqrt{2 - x}$

18. $f(x) = \sqrt{-1 - x}$

19. $f(x) = -3(x - 2)^2 - 1$

20. $f(x) = (x + 5)^2 - 2$

21. $f(x) = 3(x - 2)^3 - 1$

22. $f(x) = -2(x + 1)^3 + 2$

23. $f(x) = 2x^3 + 3$

24. $f(x) = -2x^3 - 1$

25. $f(x) = -2\sqrt{x + 3} + 4$ **26.** $f(x) = -3\sqrt{x - 1} + 2$

27. $f(x) = \dfrac{-2}{x+2} + 2$ **28.** $f(x) = \dfrac{-1}{x-1} - 1$

29. $f(x) = \dfrac{x-1}{x}$ **30.** $f(x) = \dfrac{x+2}{x}$

31. $f(x) = -3|x+4| + 3$ **32.** $f(x) = -2|x-3| - 4$

33. $f(x) = 4|x| + 2$ **34.** $f(x) = -3|x| - 4$

35. Suppose that the graph of $y = f(x)$ with a domain of $-2 \le x \le 2$ is shown in Figure 9.27.

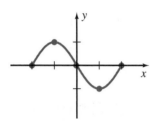

Figure 9.27

Sketch the graph of each of the following transformations of $y = f(x)$.
(a) $y = f(x) + 3$ **(b)** $y = f(x-2)$
(c) $y = -f(x)$ **(d)** $y = f(x+3) - 4$

36. Use the definition of absolute value to help you sketch the following graphs.
(a) $f(x) = x + |x|$ **(b)** $f(x) = x - |x|$
(c) $f(x) = |x| - x$ **(d)** $f(x) = \dfrac{x}{|x|}$
(e) $f(x) = \dfrac{|x|}{x}$

■■■ Thoughts into words

37. Is the graph of $f(x) = x^2 + 2x + 4$ a y axis reflection of $f(x) = x^2 - 2x + 4$? Defend your answer.

38. Is the graph of $f(x) = x^2 - 4x - 7$ an x axis reflection of $f(x) = x^2 + 4x + 7$? Defend your answer.

39. Your friend claims that the graph of $f(x) = \dfrac{2x+1}{x}$ is the graph of $f(x) = \dfrac{1}{x}$ shifted 2 units upward. How could you decide whether she is correct?

    ### Graphing calculator activities

40. Use a graphing calculator to check your graphs for Problems 12–27.

41. Use a graphing calculator to check your graphs for Problem 36.

42. For each of the following, answer the question on the basis of your knowledge of transformations, and then use a graphing calculator to check your answer.
(a) Is the graph of $f(x) = 2x^2 + 8x + 13$ a y axis reflection of $f(x) = 2x^2 - 8x + 13$?
(b) Is the graph of $f(x) = 3x^2 - 12x + 16$ an x axis reflection of $f(x) = -3x^2 + 12x - 16$?
(c) Is the graph of $f(x) = \sqrt{4 - x}$ a y axis reflection of $f(x) = \sqrt{x + 4}$?

(d) Is the graph of $f(x) = \sqrt{3 - x}$ a y axis reflection of $f(x) = \sqrt{x - 3}$?
(e) Is the graph of $f(x) = -x^3 + x + 1$ a y axis reflection of $f(x) = x^3 - x + 1$?
(f) Is the graph of $f(x) = -(x - 2)^3$ an x axis reflection of $f(x) = (x - 2)^3$?
(g) Is the graph of $f(x) = -x^3 - x^2 - x + 1$ an x axis reflection of $f(x) = x^3 + x^2 + x - 1$?
(h) Is the graph of $f(x) = \dfrac{3x + 1}{x}$ a vertical translation of $f(x) = \dfrac{1}{x}$ upward 3 units?

(i) Is the graph of $f(x) = 2 + \dfrac{1}{x}$ a y axis reflection of

$f(x) = \dfrac{2x - 1}{x}$?

43. Are the graphs of $f(x) = 2\sqrt{x}$ and $g(x) = \sqrt{2x}$ identical? Defend your answer.

44. Are the graphs of $f(x) = \sqrt{x + 4}$ and $g(x) = \sqrt{-x + 4}$ y axis reflections of each other? Defend your answer.

9.4 Composition of Functions

The basic operations of addition, subtraction, multiplication, and division can be performed on functions. However, there is an additional operation, called **composition,** that we will use in the next section. Let's start with the definition and an illustration of this operation.

> ### DEFINITION 9.3
>
> The **composition** of functions f and g is defined by
>
> $$(f \circ g)(x) = f(g(x))$$
>
> for all x in the domain of g such that $g(x)$ is in the domain of f.

The left side, $(f \circ g)(x)$, of the equation in Definition 9.3 can be read as "the composition of f and g," and the right side, $f(g(x))$, can be read as "f of g of x." It may also be helpful for you to picture Definition 9.3 as two function machines hooked together to produce another function (often called a **composite function**) as illustrated in Figure 9.28. Note that what comes out of the function g is substituted into the function f. Thus composition is sometimes called the substitution of functions.

Figure 9.28 also vividly illustrates the fact that $f \circ g$ is defined for all x in the domain of g such that $g(x)$ is in the domain of f. In other words, what comes out of g must be capable of being fed into f. Let's consider some examples.

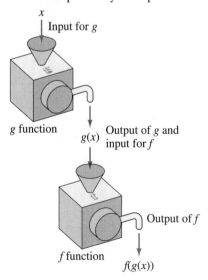

Figure 9.28

EXAMPLE 1

If $f(x) = x^2$ and $g(x) = x - 3$, find $(f \circ g)(x)$ and determine its domain.

Solution

Applying Definition 9.3, we obtain

$$(f \circ g)(x) = f(g(x))$$
$$= f(x - 3)$$
$$= (x - 3)^2$$

Because g and f are both defined for all real numbers, so is $f \circ g$.

EXAMPLE 2

If $f(x) = \sqrt{x}$ and $g(x) = x - 4$, find $(f \circ g)(x)$ and determine its domain.

Solution

Applying Definition 9.3, we obtain

$$(f \circ g)(x) = f(g(x))$$
$$= f(x - 4)$$
$$= \sqrt{x - 4}$$

The domain of g is all real numbers, but the domain of f is only the nonnegative real numbers. Thus $g(x)$, which is $x - 4$, has to be nonnegative. Therefore,

$$x - 4 \geq 0$$
$$x \geq 4$$

and the domain of $f \circ g$ is D $= \{x | x \geq 4\}$ or D: $[4, \infty)$.

Definition 9.3, with f and g interchanged, defines the composition of g and f as $(g \circ f)(x) = g(f(x))$.

EXAMPLE 3

If $f(x) = x^2$ and $g(x) = x - 3$, find $(g \circ f)(x)$ and determine its domain.

Solution

$$(g \circ f)(x) = g(f(x))$$
$$= g(x^2)$$
$$= x^2 - 3$$

Because f and g are both defined for all real numbers, the domain of $g \circ f$ is the set of all real numbers.

The results of Examples 1 and 3 demonstrate an important idea: that the composition of functions is *not a commutative operation*. In other words, it is not true that $f \circ g = g \circ f$ for all functions f and g. However, as we will see in the next section, there is a special class of functions where $f \circ g = g \circ f$.

EXAMPLE 4

If $f(x) = 2x + 3$ and $g(x) = \sqrt{x - 1}$, determine each of the following.

(a) $(f \circ g)(x)$ (b) $(g \circ f)(x)$ (c) $(f \circ g)(5)$ (d) $(g \circ f)(7)$

Solution

(a) $(f \circ g)(x) = f(g(x))$

$$= f(\sqrt{x - 1})$$

$$= 2\sqrt{x - 1} + 3 \qquad \text{D} = \{x | x \geq 1\} \text{ or D: } [1, \infty)$$

(b) $(g \circ f)(x) = g(f(x))$

$$= g(2x + 3)$$

$$= \sqrt{2x + 3 - 1}$$

$$= \sqrt{2x + 2} \qquad \text{D} = \{x | x \geq -1\} \text{ or D: } [-1, \infty)$$

(c) $(f \circ g)(5) = 2\sqrt{5 - 1} + 3 = 7$

(d) $(g \circ f)(7) = \sqrt{2(7) + 2} = 4$

EXAMPLE 5

If $f(x) = \dfrac{2}{x - 1}$ and $g(x) = \dfrac{1}{x}$, find $(f \circ g)(x)$ and $(g \circ f)(x)$. Determine the domain for each composite function.

Solution

$$(f \circ g)(x) = f(g(x))$$

$$= f\left(\frac{1}{x}\right)$$

$$= \frac{2}{\frac{1}{x} - 1} = \frac{2}{\frac{1 - x}{x}}$$

$$= \frac{2x}{1 - x}$$

The domain of g is all real numbers except 0, and the domain of f is all real numbers except 1. Because $g(x)$, which is $\dfrac{1}{x}$, cannot equal 1,

$$\frac{1}{x} \neq 1$$

$$x \neq 1$$

Therefore, the domain of $f \circ g$ is $D = \{x | x \neq 0 \text{ and } x \neq 1\}$ or D: $(-\infty, 0) \cup (0, 1) \cup (1, \infty)$.

$$(g \circ f)(x) = g(f(x))$$

$$= g\left(\frac{2}{x - 1}\right)$$

$$= \frac{1}{\dfrac{2}{x - 1}}$$

$$= \frac{x - 1}{2}$$

The domain of f is all real numbers except 1, and the domain of g is all real numbers except 0. Because $f(x)$, which is $\dfrac{2}{x - 1}$, will never equal 0, the domain of $g \circ f$ is $D = \{x | x \neq 1\}$ or D: $(-\infty, 1) \cup (1, \infty)$. ∎

A graphing utility can be used to find the graph of a composite function without actually forming the function algebraically. Let's see how this works.

E X A M P L E   6

If $f(x) = x^3$ and $g(x) = x - 4$, use a graphing utility to obtain the graph of $y = (f \circ g)(x)$ and the graph of $y = (g \circ f)(x)$.

Solution

To find the graph of $y = (f \circ g)(x)$, we can make the following assignments.

$$Y_1 = x - 4$$

$$Y_2 = (Y_1)^3$$

(Note that we have substituted Y_1 for x in $f(x)$ and assigned this expression to Y_2, much the same way as we would algebraically.) Now, by showing only the graph of Y_2, we obtain Figure 9.29.

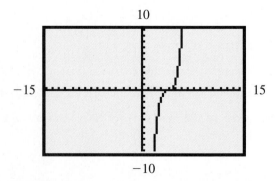

Figure 9.29

To find the graph of $y = (g \circ f)(x)$, we can make the following assignments.

$$Y_1 = x^3$$

$$Y_2 = Y_1 - 4$$

The graph of $y = (g \circ f)(x)$ is the graph of Y_2, as shown in Figure 9.30.

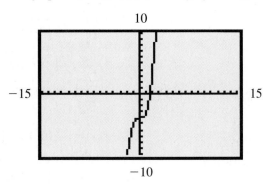

Figure 9.30

Take another look at Figures 9.29 and 9.30. Note that in Figure 9.29 the graph of $y = (f \circ g)(x)$ is the basic cubic curve $f(x) = x^3$ shifted 4 units to the right. Likewise, in Figure 9.30 the graph of $y = (g \circ f)(x)$ is the basic cubic curve shifted 4 units downward. These are examples of a more general concept of using composite functions to represent various geometric transformations.

PROBLEM SET 9.4

For Problems 1–12, determine the indicated functional values.

1. If $f(x) = 9x - 2$ and $g(x) = -4x + 6$, find $(f \circ g)(-2)$ and $(g \circ f)(4)$.

2. If $f(x) = -2x - 6$ and $g(x) = 3x + 10$, find $(f \circ g)(5)$ and $(g \circ f)(-3)$.

3. If $f(x) = 4x^2 - 1$ and $g(x) = 4x + 5$, find $(f \circ g)(1)$ and $(g \circ f)(4)$.

4. If $f(x) = -5x + 2$ and $g(x) = -3x^2 + 4$, find $(f \circ g)(-2)$ and $(g \circ f)(-1)$.

5. If $f(x) = \dfrac{1}{x}$ and $g(x) = \dfrac{2}{x - 1}$, find $(f \circ g)(2)$ and $(g \circ f)(-1)$.

6. If $f(x) = \dfrac{2}{x - 1}$ and $g(x) = -\dfrac{3}{x}$, find $(f \circ g)(1)$ and $(g \circ f)(-1)$.

7. If $f(x) = \dfrac{1}{x - 2}$ and $g(x) = \dfrac{4}{x - 1}$, find $(f \circ g)(3)$ and $(g \circ f)(2)$.

8. If $f(x) = \sqrt{x + 6}$ and $g(x) = 3x - 1$, find $(f \circ g)(-2)$ and $(g \circ f)(-2)$.

9. If $f(x) = \sqrt{3x - 2}$ and $g(x) = -x + 4$, find $(f \circ g)(1)$ and $(g \circ f)(6)$.

10. If $f(x) = -5x + 1$ and $g(x) = \sqrt{4x + 1}$, find $(f \circ g)(6)$ and $(g \circ f)(-1)$.

11. If $f(x) = |4x - 5|$ and $g(x) = x^3$, find $(f \circ g)(-2)$ and $(g \circ f)(2)$.

12. If $f(x) = -x^3$ and $g(x) = |2x + 4|$, find $(f \circ g)(-1)$ and $(g \circ f)(-3)$.

For Problems 13–30, determine $(f \circ g)(x)$ and $(g \circ f)(x)$ for each pair of functions. Also specify the domain of $(f \circ g)(x)$ and $(g \circ f)(x)$.

13. $f(x) = 3x$ and $g(x) = 5x - 1$

14. $f(x) = 4x - 3$ and $g(x) = -2x$

15. $f(x) = -2x + 1$ and $g(x) = 7x + 4$

16. $f(x) = 6x - 5$ and $g(x) = -x + 6$

17. $f(x) = 3x + 2$ and $g(x) = x^2 + 3$

18. $f(x) = -2x + 4$ and $g(x) = 2x^2 - 1$

19. $f(x) = 2x^2 - x + 2$ and $g(x) = -x + 3$

20. $f(x) = 3x^2 - 2x - 4$ and $g(x) = -2x + 1$

21. $f(x) = \dfrac{3}{x}$ and $g(x) = 4x - 9$

22. $f(x) = -\dfrac{2}{x}$ and $g(x) = -3x + 6$

23. $f(x) = \sqrt{x + 1}$ and $g(x) = 5x + 3$

24. $f(x) = 7x - 2$ and $g(x) = \sqrt{2x - 1}$

25. $f(x) = \dfrac{1}{x}$ and $g(x) = \dfrac{1}{x - 4}$

26. $f(x) = \dfrac{2}{x + 3}$ and $g(x) = -\dfrac{3}{x}$

27. $f(x) = \sqrt{x}$ and $g(x) = \dfrac{4}{x}$

28. $f(x) = \dfrac{2}{x}$ and $g(x) = |x|$

29. $f(x) = \dfrac{3}{2x}$ and $g(x) = \dfrac{1}{x + 1}$

30. $f(x) = \dfrac{4}{x - 2}$ and $g(x) = \dfrac{3}{4x}$

For Problems 31–38, show that $(f \circ g)(x) = x$ and $(g \circ f)(x) = x$ for each pair of functions.

31. $f(x) = 3x$ and $g(x) = \dfrac{1}{3}x$

32. $f(x) = -2x$ and $g(x) = -\dfrac{1}{2}x$

33. $f(x) = 4x + 2$ and $g(x) = \dfrac{x - 2}{4}$

34. $f(x) = 3x - 7$ and $g(x) = \dfrac{x + 7}{3}$

35. $f(x) = \dfrac{1}{2}x + \dfrac{3}{4}$ and $g(x) = \dfrac{4x - 3}{2}$

36. $f(x) = \dfrac{2}{3}x - \dfrac{1}{5}$ and $g(x) = \dfrac{3}{2}x + \dfrac{3}{10}$

37. $f(x) = -\dfrac{1}{4}x - \dfrac{1}{2}$ and $g(x) = -4x - 2$

38. $f(x) = -\dfrac{3}{4}x + \dfrac{1}{3}$ and $g(x) = -\dfrac{4}{3}x + \dfrac{4}{9}$

■ ■ ■ **Thoughts into words**

39. How would you explain the concept of composition of functions to a friend who missed class the day it was discussed?

40. Explain why the composition of functions is not a commutative operation.

 Graphing calculator activities

41. For each of the following, (1) predict the general shape and location of the graph and then (2) use your graph-   ing calculator to graph the function and thus check your prediction. (Your knowledge of the graphs of the basic functions that are being added or subtracted should be helpful when you make your predictions.)

(a) $f(x) = x^3 + x^2$ 　　　**(b)** $f(x) = x^3 - x^2$
(c) $f(x) = x^2 - x^3$ 　　　**(d)** $f(x) = |x| + \sqrt{x}$
(e) $f(x) = |x| - \sqrt{x}$ 　　**(f)** $f(x) = \sqrt{x} - |x|$

42. For each of the following, use your graphing calculator to find the graph of $y = (f \circ g)(x)$ and $y = (g \circ f)(x)$.

Then algebraically find $(f \circ g)(x)$ and $(g \circ f)(x)$ to see whether your results agree.

(a) $f(x) = x^2$ and $g(x) = x - 3$
(b) $f(x) = x^3$ and $g(x) = x + 4$
(c) $f(x) = x - 2$ and $g(x) = -x^3$
(d) $f(x) = x + 6$ and $g(x) = \sqrt{x}$
(e) $f(x) = \sqrt{x}$ and $g(x) = x - 5$

9.5 Inverse Functions

Graphically, the distinction between a relation and a function can be easily recognized. In Figure 9.31, we sketched four graphs. Which of these are graphs of functions and which are graphs of relations that are not functions? Think in terms of the principle that to each member of the domain there is assigned one and only one member of the range; this is the basis for what is known as the **vertical-line test** for functions. Because each value of x produces only one value of $f(x)$, any vertical line drawn through a graph of a function must not intersect the graph in more than one point. Therefore, parts (a) and (c) of Figure 9.31 are graphs of functions, and parts (b) and (d) are graphs of relations that are not functions.

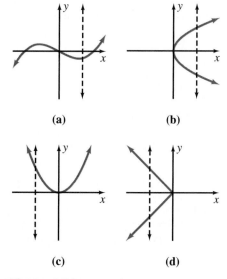

(a) 　　　　(b)

(c) 　　　　(d)

Figure 9.31

We can also make a useful distinction between two basic types of functions. Consider the graphs of the two functions $f(x) = 2x - 1$ and $f(x) = x^2$ in Figure 9.32. In part (a), any *horizontal line* will intersect the graph in no more than one point. Therefore, every value of $f(x)$ has only one value of x associated with it. Any function that has the additional property of having only one value of x associated with each value of $f(x)$ is called a **one-to-one function.** The function $f(x) = x^2$ is not a one-to-one function because the horizontal line in part (b) of Figure 9.32 intersects the parabola in two points.

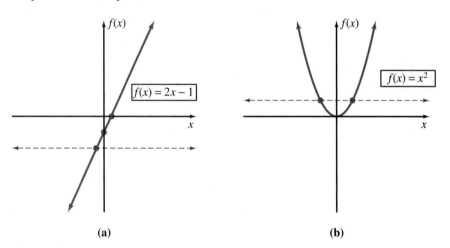

(a)   (b)

Figure 9.32

In terms of ordered pairs, a one-to-one function does not contain any ordered pairs that have the same second component. For example,

$$f = \{(1, 3), (2, 6), (4, 12)\}$$

is a one-to-one function, but

$$g = \{(1, 2), (2, 5), (-2, 5)\}$$

is not a one-to-one function.

If the components of each ordered pair of a given one-to-one function are interchanged, the resulting function and the given function are called **inverses** of each other. Thus

$$\{(1, 3), (2, 6), (4, 12)\} \quad \text{and} \quad \{(3, 1), (6, 2), (12, 4)\}$$

are **inverse functions.** The inverse of a function f is denoted by f^{-1} (which is read "f inverse" or "the inverse of f"). If (a, b) is an ordered pair of f, then (b, a) is an ordered pair of f^{-1}. The domain and range of f^{-1} are the range and domain, respectively, of f.

REMARK: Do not confuse the -1 in f^{-1} with a negative exponent. The symbol f^{-1} does not mean $\dfrac{1}{f^1}$ but, rather, refers to the inverse function of function f.

Graphically, two functions that are inverses of each other are mirror images with reference to the line $y = x$. This is due to the fact that ordered pairs (a, b) and (b, a) are mirror images with respect to the line $y = x$, as illustrated in Figure 9.33. Therefore, if we know the graph of a function f, as in Figure 9.34(a), then we can determine the graph of f^{-1} by reflecting f across the line $y = x$ (Figure 9.34b).

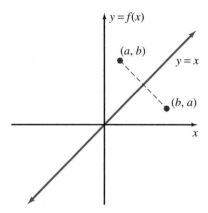

Figure 9.33

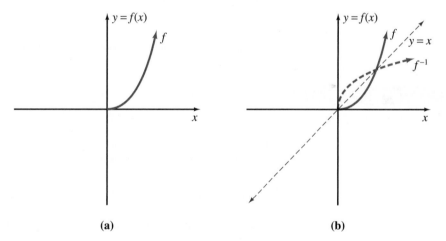

(a)   **(b)**

Figure 9.34

Another useful way of viewing inverse functions is in terms of composition. Basically, inverse functions *undo* each other, and this can be more formally stated as follows: If f and g are inverses of each other, then

1. $(f \circ g)(x) = f(g(x)) = x$ for all x in domain of g
2. $(g \circ f)(x) = g(f(x)) = x$ for all x in domain of f

As we will see in a moment, this relationship of inverse functions can be used to verify whether two functions are indeed inverses of each other.

Finding Inverse Functions

The idea of inverse functions undoing each other provides the basis for a rather informal approach to finding the inverse of a function. Consider the function

$$f(x) = 2x + 1$$

To each x this function assigns *twice x plus 1*. To undo this function, we could *subtract 1 and divide by 2*. Thus the inverse should be

$$f^{-1}(x) = \frac{x-1}{2}$$

Now let's verify that f and f^{-1} are inverses of each other.

$$(f \circ f^{-1})(x) = f(f^{-1}(x)) = f\left(\frac{x-1}{2}\right) = 2\left(\frac{x-1}{2}\right) + 1 = x$$

and

$$(f^{-1} \circ f)(x) = f^{-1}(f(x)) = f^{-1}(2x+1) = \frac{2x+1-1}{2} = x$$

Thus the inverse of f is given by

$$f^{-1}(x) = \frac{x-1}{2}$$

Let's consider another example of finding an inverse function by the undoing process.

EXAMPLE 1   Find the inverse of $f(x) = 3x - 5$.

Solution

To each x, the function f assigns *three times x minus 5*. To undo this, we can *add 5 and then divide by 3*, so the inverse should be

$$f^{-1}(x) = \frac{x+5}{3}$$

To verify that f and f^{-1} are inverses, we can show that

$$(f \circ f^{-1})(x) = f(f^{-1}(x)) = f\left(\frac{x+5}{3}\right)$$

$$= 3\left(\frac{x+5}{3}\right) - 5 = x$$

and

$$(f^{-1} \circ f)(x) = f^{-1}(f(x)) = f^{-1}(3x - 5)$$

$$= \frac{3x - 5 + 5}{3} = x$$

Thus f and f^{-1} are inverses, and we can write

$$f^{-1}(x) = \frac{x + 5}{3}$$

This informal approach may not work very well with more complex functions, but it does emphasize how inverse functions are related to each other. A more formal and systematic technique for finding the inverse of a function can be described as follows:

1. Replace the symbol $f(x)$ by y.
2. Interchange x and y.
3. Solve the equation for y in terms of x.
4. Replace y by the symbol $f^{-1}(x)$.

Now let's use two examples to illustrate this technique.

E X A M P L E   2   Find the inverse of $f(x) = -3x + 11$.

Solution

When we replace $f(x)$ by y, the given equation becomes

$$y = -3x + 11$$

Interchanging x and y produces

$$x = -3y + 11$$

Now, solving for y yields

$$x = -3y + 11$$

$$3y = -x + 11$$

$$y = \frac{-x + 11}{3}$$

Finally, replacing y by $f^{-1}(x)$, we can express the inverse function as

$$f^{-1}(x) = \frac{-x + 11}{3}$$

E X A M P L E   3   Find the inverse of $f(x) = \frac{3}{2}x - \frac{1}{4}$.

Solution

When we replace $f(x)$ by y, the given equation becomes

$$y = \frac{3}{2}x - \frac{1}{4}$$

Interchanging x and y produces

$$x = \frac{3}{2}y - \frac{1}{4}$$

Now, solving for y yields

$$x = \frac{3}{2}y - \frac{1}{4}$$

$$4(x) = 4\left(\frac{3}{2}y - \frac{1}{4}\right) \qquad \text{Multiply both sides by the LCD.}$$

$$4x = 4\left(\frac{3}{2}y\right) - 4\left(\frac{1}{4}\right)$$

$$4x = 6y - 1$$

$$4x + 1 = 6y$$

$$\frac{4x + 1}{6} = y$$

$$\frac{2}{3}x + \frac{1}{6} = y$$

Finally, replacing y by $f^{-1}(x)$, we can express the inverse function as

$$f^{-1}(x) = \frac{2}{3}x + \frac{1}{6}$$

For both Examples 2 and 3, you should be able to show that $(f \circ f^{-1})(x) = x$ and $(f^{-1} \circ f)(x) = x$.

PROBLEM SET 9.5

For Problems 1–8 (Figures 9.35–9.42), identify each graph as (a) the graph of a function or (b) the graph of a relation that is not a function. Use the vertical-line test.

1.

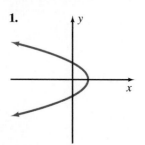

Figure 9.35

2.

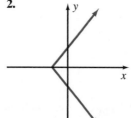

Figure 9.36

3.

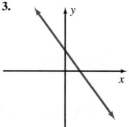

Figure 9.37

4.

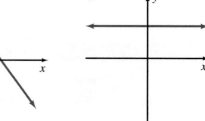

Figure 9.38

5.

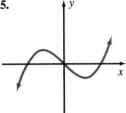

Figure 9.39

6.

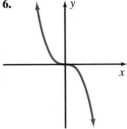

Figure 9.40

13.

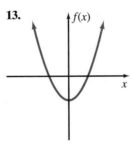

Figure 9.47

14.

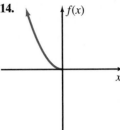

Figure 9.48

7.

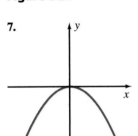

Figure 9.41

8.

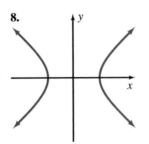

Figure 9.42

15.

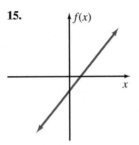

Figure 9.49

16.

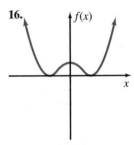

Figure 9.50

For Problems 9–16 (Figures 9.43–9.50), identify each graph as (a) the graph of a one-to-one function or (b) the graph of a function that is not one-to-one. Use the horizontal-line test.

9.

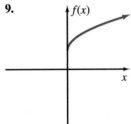

Figure 9.43

10.

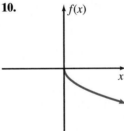

Figure 9.44

11.

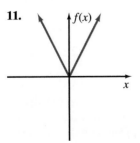

Figure 9.45

12.

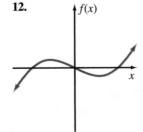

Figure 9.46

For Problems 17–20, (a) list the domain and range of the given function, (b) form the inverse function, and (c) list the domain and range of the inverse function.

17. $f = \{(1, 3), (2, 6), (3, 11), (4, 18)\}$

18. $f = \{(0, -4), (1, -3), (4, -2)\}$

19. $f = \{(-2, -1), (-1, 1), (0, 5), (5, 10)\}$

20. $f = \{(-1, 1), (-2, 4), (1, 9), (2, 12)\}$

For Problems 21–30, find the inverse of the given function by using the "undoing process," and then verify that $(f \circ f^{-1})(x) = x$ and $(f^{-1} \circ f)(x) = x$.

21. $f(x) = 5x - 4$

22. $f(x) = 7x + 9$

23. $f(x) = -2x + 1$

24. $f(x) = -4x - 3$

25. $f(x) = \dfrac{4}{5}x$

26. $f(x) = -\dfrac{2}{3}x$

27. $f(x) = \dfrac{1}{2}x + 4$

28. $f(x) = \dfrac{3}{4}x - 2$

29. $f(x) = \dfrac{1}{3}x - \dfrac{2}{5}$

30. $f(x) = \dfrac{2}{5}x + \dfrac{1}{3}$

For Problems 31–40, find the inverse of the given function by using the process illustrated in Examples 2 and 3 of this section, and then verify that $(f \circ f^{-1})(x) = x$ and $(f^{-1} \circ f)(x) = x$.

31. $f(x) = 9x + 4$

32. $f(x) = 8x - 5$

33. $f(x) = -5x - 4$

34. $f(x) = -6x + 2$

35. $f(x) = -\dfrac{2}{3}x + 7$

36. $f(x) = -\dfrac{3}{5}x + 1$

37. $f(x) = \dfrac{4}{3}x - \dfrac{1}{4}$

38. $f(x) = \dfrac{5}{2}x + \dfrac{2}{7}$

39. $f(x) = -\dfrac{3}{7}x - \dfrac{2}{3}$

40. $f(x) = -\dfrac{3}{5}x + \dfrac{3}{4}$

For Problems 41–50, (a) find the inverse of the given function, and (b) graph the given function and its inverse on the same set of axes.

41. $f(x) = 4x$

42. $f(x) = \dfrac{2}{5}x$

43. $f(x) = -\dfrac{1}{3}x$

44. $f(x) = -6x$

45. $f(x) = 3x - 3$

46. $f(x) = 2x + 2$

47. $f(x) = -2x - 4$

48. $f(x) = -3x + 9$

49. $f(x) = x^2, x \geq 0$

50. $f(x) = x^2 + 2, x \geq 0$

■ ■ ■ **Thoughts into words**

51. Does the function $f(x) = 4$ have an inverse? Explain your answer.

52. Explain why every nonconstant linear function has an inverse.

■ ■ ■ **Further investigations**

53. The composition idea can also be used to find the inverse of a function. For example, to find the inverse of $f(x) = 5x + 3$, we could proceed as follows:

$$f(f^{-1}(x)) = 5(f^{-1}(x)) + 3 \quad \text{and} \quad f(f^{-1}(x)) = x$$

Therefore, equating the two expressions for $f(f^{-1}(x))$, we obtain

$$5(f^{-1}(x)) + 3 = x$$
$$5(f^{-1}(x)) = x - 3$$
$$f^{-1}(x) = \dfrac{x - 3}{5}$$

Use this approach to find the inverse of each of the following functions.

(a) $f(x) = 2x + 1$

(b) $f(x) = 3x - 2$

(c) $f(x) = -4x + 5$

(d) $f(x) = -x + 1$

(e) $f(x) = 2x$

(f) $f(x) = -5x$

■ **Graphing calculator activities**

54. For Problems 31–40, graph the given function, the inverse function that you found, and $f(x) = x$ on the same set of axes. In each case the given function and its inverse should produce graphs that are reflections of each other through the line $f(x) = x$.

55. Let's use a graphing calculator to show that $(f \circ g)(x) = x$ and $(g \circ f)(x) = x$ for two functions that we think are inverses of each other. Consider the functions

$f(x) = 3x + 4$ and $g(x) = \dfrac{x-4}{3}$. We can make the following assignments.

f: $Y_1 = 3x + 4$

g: $Y_2 = \dfrac{x-4}{3}$

$f \circ g$: $Y_3 = 3Y_2 + 4$

$g \circ f$: $Y_4 = \dfrac{Y_1 - 4}{3}$

Now we can graph Y_3 and Y_4 and show that they both produce the line $f(x) = x$.

Use this approach to check your answers for Problems 41–50.

56. Use the approach demonstrated in Problem 55 to show that $f(x) = x^2 - 2$ (for $x \geq 0$) and $g(x) = \sqrt{x+2}$ (for $x \geq -2$) are inverses of each other.

9.6 Direct and Inverse Variations

"The distance a car travels at a fixed rate varies directly as the time." "At a constant temperature, the volume of an enclosed gas varies inversely as the pressure." Such statements illustrate two basic types of functional relationships, called **direct** and **inverse variation,** that are widely used, especially in the physical sciences. These relationships can be expressed by equations that specify functions. The purpose of this section is to investigate these special functions.

The statement "y varies directly as x" means

$$y = kx$$

where k is a nonzero constant called the **constant of variation.** The phrase "y is directly proportional to x" is also used to indicate direct variation; k is then referred to as the **constant of proportionality.**

REMARK: Note that the equation $y = kx$ defines a function and could be written as $f(x) = kx$ by using function notation. However, in this section it is more convenient to avoid the function notation and use variables that are meaningful in terms of the physical entities involved in the problem.

Statements that indicate direct variation may also involve powers of x. For example, "y varies directly as the square of x" can be written as

$$y = kx^2$$

In general, "y varies directly as the nth power of x $(n > 0)$" means

$$y = kx^n$$

The three types of problems that deal with direct variation are

1. To translate an English statement into an equation that expresses the direct variation
2. To find the constant of variation from given values of the variables
3. To find additional values of the variables once the constant of variation has been determined

Let's consider an example of each of these types of problems.

E X A M P L E   1

Translate the statement "the tension on a spring varies directly as the distance it is stretched" into an equation, and use k as the constant of variation.

Solution

If we let t represent the tension and d the distance, the equation becomes

$$t = kd$$ ■

E X A M P L E   2

If A varies directly as the square of s, and $A = 28$ when $s = 2$, find the constant of variation.

Solution

Because A varies directly as the square of s, we have

$$A = ks^2$$

Substituting $A = 28$ and $s = 2$, we obtain

$$28 = k(2)^2$$

Solving this equation for k yields

$$28 = 4k$$

$$7 = k$$

The constant of variation is 7. ■

E X A M P L E   3

If y is directly proportional to x and if $y = 6$ when $x = 9$, find the value of y when $x = 24$.

Solution

The statement "y is directly proportional to x" translates into

$$y = kx$$

If we let $y = 6$ and $x = 9$, the constant of variation becomes

$$6 = k(9)$$

$$6 = 9k$$

$$\frac{6}{9} = k$$

$$\frac{2}{3} = k$$

Thus the specific equation is $y = \dfrac{2}{3}x$. Now, letting $x = 24$, we obtain

$$y = \frac{2}{3}(24) = 16$$

The required value of y is 16. ■

Inverse Variation

We define the second basic type of variation, called **inverse variation,** as follows: The statement "y varies inversely as x" means

$$y = \frac{k}{x}$$

where k is a nonzero constant; again we refer to it as the constant of variation. The phrase "y is inversely proportional to x" is also used to express inverse variation. As with direct variation, statements that indicate inverse variation may involve powers of x. For example, "y varies inversely as the square of x" can be written as

$$y = \frac{k}{x^2}$$

In general, "y varies inversely as the nth power of x ($n > 0$)" means

$$y = \frac{k}{x^n}$$

The following examples illustrate the three basic kinds of problems we run across that involve inverse variation.

E X A M P L E   4

Translate the statement "the length of a rectangle of a fixed area varies inversely as the width" into an equation that uses k as the constant of variation.

Solution
Let l represent the length and w the width, and the equation is

$$l = \frac{k}{w}$$

∎

E X A M P L E   5

If y is inversely proportional to x, and $y = 4$ when $x = 12$, find the constant of variation.

Solution
Because y is inversely proportional to x, we have

$$y = \frac{k}{x}$$

Substituting $y = 4$ and $x = 12$, we obtain

$$4 = \frac{k}{12}$$

Solving this equation for k by multiplying both sides of the equation by 12 yields

$$k = 48$$

The constant of variation is 48.   ■

EXAMPLE 6

Suppose the number of days it takes to complete a construction job varies inversely as the number of people assigned to the job. If it takes 7 people 8 days to do the job, how long would it take 14 people to complete the job?

Solution

Let d represent the number of days and p the number of people. The phrase "number of days . . . varies inversely as the number of people" translates into

$$d = \frac{k}{p}$$

Let $d = 8$ when $p = 7$, and the constant of variation becomes

$$8 = \frac{k}{7}$$

$$k = 56$$

Thus the specific equation is

$$d = \frac{56}{p}$$

Now, let $p = 14$ to obtain

$$d = \frac{56}{14}$$

$$d = 4$$

It should take 14 people 4 days to complete the job.   ■

The terms *direct* and *inverse,* as applied to variation, refer to the relative behavior of the variables involved in the equation. That is, in *direct* variation ($y = kx$), an assignment of *increasing* absolute values for x produces increasing absolute values for y, whereas in *inverse* variation $\left(y = \frac{k}{x} \right)$, an assignment of increasing absolute values for x produces *decreasing* absolute values for y.

Joint Variation

Variation may involve more than two variables. The following table illustrates some variation statements and their equivalent algebraic equations that use k as the constant of variation.

Variation statement	Algebraic equation
1. y varies jointly as x and z.	$y = kxz$
2. y varies jointly as x, z, and w.	$y = kxzw$
3. V varies jointly as h and the square of r.	$V = khr^2$
4. h varies directly as V and inversely as w.	$h = \dfrac{kV}{w}$
5. y is directly proportional to x and inversely proportional to the square of z.	$y = \dfrac{kx}{z^2}$
6. y varies jointly as w and z and inversely as x.	$y = \dfrac{kwz}{x}$

Statements 1, 2, and 3 illustrate the concept of **joint variation.** Statements 4 and 5 show that both direct and inverse variation may occur in the same problem. Statement 6 combines joint variation with inverse variation.

The two final examples of this section illustrate some of these variation situations.

EXAMPLE 7

The length of a rectangular box with a fixed height varies directly as the volume and inversely as the width. If the length is 12 centimeters when the volume is 960 cubic centimeters and the width is 8 centimeters, find the length when the volume is 700 centimeters and the width is 5 centimeters.

Solution

Use l for length, V for volume, and w for width, and the phrase "length varies directly as the volume and inversely as the width" translates into

$$l = \frac{kV}{w}$$

Substitute $l = 12$, $V = 960$, and $w = 8$, and the constant of variation becomes

$$12 = \frac{k(960)}{8}$$

$$12 = 120k$$

$$\frac{1}{10} = k$$

Thus the specific equation is

$$l = \frac{\dfrac{1}{10}V}{w} = \frac{V}{10w}$$

Now let $V = 700$ and $w = 5$ to obtain

$$l = \frac{700}{10(5)} = \frac{700}{50} = 14$$

The length is 14 centimeters.   ■

EXAMPLE 8   Suppose that y varies jointly as x and z and inversely as w. If $y = 154$ when $x = 6$, $z = 11$, and $w = 3$, find the constant of variation.

Solution

The statement "y varies jointly as x and z and inversely as w" translates into

$$y = \frac{kxz}{w}$$

Substitute $y = 154$, $x = 6$, $z = 11$, and $w = 3$ to obtain

$$154 = \frac{k(6)(11)}{3}$$

$$154 = 22k$$

$$7 = k$$

The constant of variation is 7.   ■

PROBLEM SET 9.6

For Problems 1–10, translate each statement of variation into an equation, and use k as the constant of variation.

1. y varies inversely as the square of x.

2. y varies directly as the cube of x.

3. C varies directly as g and inversely as the cube of t.

4. V varies jointly as l and w.

5. The volume (V) of a sphere is directly proportional to the cube of its radius (r).

6. At a constant temperature, the volume (V) of a gas varies inversely as the pressure (P).

7. The surface area (S) of a cube varies directly as the square of the length of an edge (e).

8. The intensity of illumination (I) received from a source of light is inversely proportional to the square of the distance (d) from the source.

9. The volume (V) of a cone varies jointly as its height and the square of its radius.

10. The volume (V) of a gas varies directly as the absolute temperature (T) and inversely as the pressure (P).

For Problems 11–24, find the constant of variation for each of the stated conditions.

11. y varies directly as x, and $y = 8$ when $x = 12$.

12. y varies directly as x, and $y = 60$ when $x = 24$.

13. y varies directly as the square of x, and $y = -144$ when $x = 6$.

14. y varies directly as the cube of x, and $y = 48$ when $x = -2$.

15. V varies jointly as B and h, and $V = 96$ when $B = 24$ and $h = 12$.

16. A varies jointly as b and h, and $A = 72$ when $b = 16$ and $h = 9$.

17. y varies inversely as x, and $y = -4$ when $x = \dfrac{1}{2}$.

18. y varies inversely as x, and $y = -6$ when $x = \dfrac{4}{3}$.

19. r varies inversely as the square of t, and $r = \dfrac{1}{8}$ when $t = 4$.

20. r varies inversely as the cube of t, and $r = \dfrac{1}{16}$ when $t = 4$.

21. y varies directly as x and inversely as z, and $y = 45$ when $x = 18$ and $z = 2$.

22. y varies directly as x and inversely as z, and $y = 24$ when $x = 36$ and $z = 18$.

23. y is directly proportional to x and inversely proportional to the square of z, and $y = 81$ when $x = 36$ and $z = 2$.

24. y is directly proportional to the square of x and inversely proportional to the cube of z, and $y = 4\dfrac{1}{2}$ when $x = 6$ and $z = 4$.

Solve each of the following problems.

25. If y is directly proportional to x, and $y = 36$ when $x = 48$, find the value of y when $x = 12$.

26. If y is directly proportional to x, and $y = 42$ when $x = 28$, find the value of y when $x = 38$.

27. If y is inversely proportional to x, and $y = \dfrac{1}{9}$ when $x = 12$, find the value of y when $x = 8$.

28. If y is inversely proportional to x, and $y = \dfrac{1}{35}$ when $x = 14$, find the value of y when $x = 16$.

29. If A varies jointly as b and h, and $A = 60$ when $b = 12$ and $h = 10$, find A when $b = 16$ and $h = 14$.

30. If V varies jointly as B and h, and $V = 51$ when $B = 17$ and $h = 9$, find V when $B = 19$ and $h = 12$.

31. The volume of a gas at a constant temperature varies inversely as the pressure. What is the volume of a gas under pressure of 25 pounds if the gas occupies 15 cubic centimeters under a pressure of 20 pounds?

32. The time required for a car to travel a certain distance varies inversely as the rate at which it travels. If it takes 4 hours at 50 miles per hour to travel the distance, how long will it take at 40 miles per hour?

33. The volume (V) of a gas varies directly as the temperature (T) and inversely as the pressure (P). If $V = 48$ when $T = 320$ and $P = 20$, find V when $T = 280$ and $P = 30$.

34. The distance that a freely falling body falls varies directly as the square of the time it falls. If a body falls 144 feet in 3 seconds, how far will it fall in 5 seconds?

35. The period (the time required for one complete oscillation) of a simple pendulum varies directly as the square root of its length. If a pendulum 12 feet long has a period of 4 seconds, find the period of a pendulum 3 feet long.

36. The simple interest earned by a certain amount of money varies jointly as the rate of interest and the time (in years) that the money is invested. If the money is invested at 12% for 2 years, $120 is earned. How much is earned if the money is invested at 14% for 3 years?

37. The electrical resistance of a wire varies directly as its length and inversely as the square of its diameter. If the resistance of 200 meters of wire that has a diameter of $\dfrac{1}{2}$ centimeter is 1.5 ohms, find the resistance of 400 meters of wire with a diameter of $\dfrac{1}{4}$ centimeter.

38. The volume of a cylinder varies jointly as its altitude and the square of the radius of its base. If the volume of a cylinder is 1386 cubic centimeters when the radius of the base is 7 centimeters and its altitude is 9 centimeters, find the volume of a cylinder that has a base of radius 14 centimeters. The altitude of the cylinder is 5 centimeters.

39. The simple interest earned by a certain amount of money varies jointly as the rate of interest and the time (in years) that the money is invested.
   (a) If some money invested at 11% for 2 years earns $385, how much would the same amount earn at 12% for 1 year?
   (b) If some money invested at 12% for 3 years earns $819, how much would the same amount earn at 14% for 2 years?
   (c) If some money invested at 14% for 4 years earns $1960, how much would the same amount earn at 15% for 2 years?

40. The period (the time required for one complete oscillation) of a simple pendulum varies directly as the square root of its length. If a pendulum 9 inches long has a period of 2.4 seconds, find the period of a pendulum 12 inches long. Express your answer to the nearest tenth of a second.

41. The volume of a cylinder varies jointly as its altitude and the square of the radius of its base. If a cylinder that has a base with a radius of 5 meters and has an altitude of 7 meters has a volume of 549.5 cubic meters, find the volume of a cylinder that has a base with a radius of 9 meters and has an altitude of 14 meters.

42. If y is directly proportional to x and inversely proportional to the square of z, and if $y = 0.336$ when $x = 6$ and $z = 5$, find the constant of variation.

43. If y is inversely proportional to the square root of x, and if $y = 0.08$ when $x = 225$, find y when $x = 625$.

■ ■ ▨ Thoughts into words

44. How would you explain the difference between direct variation and inverse variation?

45. Suppose that y varies directly as the square of x. Does doubling the value of x also double the value of y? Explain your answer.

46. Suppose that y varies inversely as x. Does doubling the value of x also double the value of y? Explain your answer.

SUMMARY

(9.1) A **relation** is a set of ordered pairs; a **function** is a relation in which no two ordered pairs have the same first component. The **domain** of a relation (or function) is the set of all first components, and the **range** is the set of all second components.

Single symbols such as *f, g,* and *h* are commonly used to name functions. The symbol $f(x)$ represents the element in the range associated with x from the domain. Thus if $f(x) = 3x + 7$, then $f(1) = 3(1) + 7 = 10$.

(9.2) Any function that can be written in the form

$$f(x) = ax + b$$

where a and b are real numbers, is a **linear function.** The graph of a linear function is a straight line.

Any function that can be written in the form

$$f(x) = ax^2 + bx + c$$

where a, b, and c are real numbers and $a \neq 0$, is a **quadratic function.** The graph of any quadratic function is a **parabola,** which can be drawn using either of the following methods.

1. Express the function in the form $f(x) = a(x - h)^2 + k$ and use the values of a, h, and k to determine the parabola.
2. Express the function in the form $f(x) = ax^2 + bx + c$ and use the fact that the vertex is at

$$\left(-\frac{b}{2a}, f\left(-\frac{b}{2a}\right)\right)$$

and the axis of symmetry is

$$x = -\frac{b}{2a}$$

We can solve some applications that involve maximum and minimum values with our knowledge of parabolas that are generated by quadratic functions.

(9.3) Another important graphing technique is to be able to recognize equations of the transformations of basic curves. We have worked with the following transformations in this chapter.

Vertical Translation The graph of $y = f(x) + k$ is the graph of $y = f(x)$ shifted k units upward if $k > 0$ or shifted $|k|$ units downward if $k < 0$.

Horizontal Translation The graph of $y = f(x - h)$ is the graph of $y = f(x)$ shifted h units to the right if $h > 0$ or shifted $|h|$ units to the left if $h < 0$.

x Axis Reflection The graph of $y = -f(x)$ is the graph of $y = f(x)$ reflected through the x axis.

y Axis Reflection The graph of $y = f(-x)$ is the graph of $y = f(x)$ reflected through the y axis.

Vertical Stretching and Shrinking The graph of $y = cf(x)$ is obtained from the graph of $y = f(x)$ by multiplying the y coordinates of $y = f(x)$ by c. If $c > 1$, the graph is said to be **stretched** by a factor of c, and if $0 < c < 1$, the graph is said to be **shrunk** by a factor of c.

We list the following suggestions for graphing functions you are not familiar with.

1. Determine the domain of the function.
2. Determine any type of symmetry exhibited by the equation.
3. Find the intercepts.
4. Set up a table of values that satisfy the equation.
5. Plot the points associated with the ordered pairs and connect them with a smooth curve. Then, if appropriate, reflect this part of the curve according to any symmetry the graph exhibits.

(9.4) The **composition** of two functions f and g is defined by

$$(f \circ g)(x) = f(g(x))$$

for all x in the domain of g such that $g(x)$ is in the domain of f. Remember that the composition of functions is not a commutative operation.

(9.5) A **one-to-one function** is a function such that no two ordered pairs have the same second component.

If the components of each ordered pair of a given one-to-one function are interchanged, the resulting function and the given function are **inverses** of each other. The inverse of a function f is denoted by f^{-1}.

Graphically, two functions that are inverses of each other are mirror images with reference to the line $y = x$.

We can show that two functions f and f^{-1} are inverses of each other by verifying that

1. $(f^{-1} \circ f)(x) = x$ for all x in the domain of f.
2. $(f \circ f^{-1})(x) = x$ for all x in the domain of f^{-1}.

A technique for finding the inverse of a function is as follows:

1. Let $y = f(x)$.
2. Interchange x and y.
3. Solve the equation for y in terms of x.
4. $f^{-1}(x)$ is determined by the final equation.

(9.6) The equation $y = kx$ (k is a nonzero constant) defines a function called **direct variation**. The equation $y = \dfrac{k}{x}$ defines a function called **inverse variation**. In both cases, k is called the **constant of variation.**

CHAPTER 9 REVIEW PROBLEM SET

For Problems 1–4, specify the domain of each function.

1. $f = \{(1, 3), (2, 5), (4, 9)\}$

2. $f(x) = \dfrac{4}{x - 5}$

3. $f(x) = \dfrac{3}{x^2 + 4x}$

4. $f(x) = \sqrt{x^2 - 25}$

5. If $f(x) = x^2 - 2x - 1$, find $f(2), f(-3)$, and $f(a)$.

6. If $f(x) = 2x^2 + x - 7$, find $\dfrac{f(a + h) - f(a)}{h}$.

For Problems 7–16, graph each of the functions.

7. $f(x) = 4$ **8.** $f(x) = -3x + 2$

9. $f(x) = x^2 + 2x + 2$ **10.** $f(x) = |x| + 4$

11. $f(x) = -|x - 2|$ **12.** $f(x) = \sqrt{x - 2} - 3$

13. $f(x) = \dfrac{1}{x^2}$ **14.** $f(x) = -\dfrac{1}{2}x^2$

15. $f(x) = -3x^2 + 6x - 2$ **16.** $f(x) = -\sqrt{x + 1} - 2$

17. Find the coordinates of the vertex and the equation of the line of symmetry for each of the following parabolas.
(a) $f(x) = x^2 + 10x - 3$
(b) $f(x) = -2x^2 - 14x + 9$

For Problems 18–20, determine $(f \circ g)(x)$ and $(g \circ f)(x)$ for each pair of functions.

18. $f(x) = 2x - 3$ and $g(x) = 3x - 4$

19. $f(x) = x - 4$ and $g(x) = x^2 - 2x + 3$

20. $f(x) = x^2 - 5$ and $g(x) = -2x + 5$

For Problems 21–23, find the inverse (f^{-1}) of the given function.

21. $f(x) = 6x - 1$ **22.** $f(x) = \dfrac{2}{3}x + 7$

23. $f(x) = -\dfrac{3}{5}x - \dfrac{2}{7}$

24. If y varies directly as x and inversely as z, and if $y = 21$ when $x = 14$ and $z = 6$, find the constant of variation.

25. If y varies jointly as x and the square root of z, and if $y = 60$ when $x = 2$ and $z = 9$, find y when $x = 3$ and $z = 16$.

26. The weight of a body above the surface of the earth varies inversely as the square of its distance from the center of the earth. Assume that the radius of the earth is 4000 miles. How much would a man weigh 1000 miles above the earth's surface if he weighs 200 pounds on the surface?

27. Find two numbers whose sum is 40 and whose product is a maximum.

28. Find two numbers whose sum is 50 such that the square of one number plus six times the other number is a minimum.

29. Suppose that 50 students are able to raise $250 for a party when each one contributes $5. Furthermore, they figure that for each additional student they can find to contribute, the cost per student will decrease by a nickel. How many additional students will they need to maximize the amount of money they will have for a party?

30. The surface area of a cube varies directly as the square of the length of an edge. If the surface area of a cube that has edges 8 inches long is 384 square inches, find the surface area of a cube that has edges 10 inches long.

31. The cost for burning a 100-watt bulb is given by the function $c(h) = 0.006h$, where h represents the number of hours that the bulb burns. How much, to the nearest cent, does it cost to burn a 100-watt bulb for 4 hours per night for a 30-day month?

32. "All Items 30% Off Marked Price" is a sign in a local department store. Form a function and then use it to determine how much one has to pay for each of the following marked items: a $65 pair of shoes, a $48 pair of slacks, a $15.50 belt.

TEST

1. Determine the domain of the function $f(x) = \dfrac{-3}{2x^2 + 7x - 4}$.

2. Determine the domain of the function $f(x) = \sqrt{5 - 3x}$.

3. If $f(x) = -\dfrac{1}{2}x + \dfrac{1}{3}$, find $f(-3)$.

4. If $f(x) = -x^2 - 6x + 3$, find $f(-2)$.

5. Find the vertex of the parabola $f(x) = -2x^2 - 24x - 69$.

6. If $f(x) = 3x^2 + 2x - 5$, find $\dfrac{f(a + h) - f(a)}{h}$.

7. If $f(x) = -3x + 4$ and $g(x) = 7x + 2$, find $(f \circ g)(x)$.

8. If $f(x) = 2x + 5$ and $g(x) = 2x^2 - x + 3$, find $(g \circ f)(x)$.

9. If $f(x) = \dfrac{3}{x - 2}$ and $g(x) = \dfrac{2}{x}$, find $(f \circ g)(x)$.

For Problems 10–12, find the inverse of the given function.

10. $f(x) = 5x - 9$

11. $f(x) = -3x - 6$

12. $f(x) = \dfrac{2}{3}x - \dfrac{3}{5}$

13. If y varies inversely as x, and if $y = \dfrac{1}{2}$ when $x = -8$, find the constant of variation.

14. If y varies jointly as x and z, and if $y = 18$ when $x = 8$ and $z = 9$, find y when $x = 5$ and $z = 12$.

15. Find two numbers whose sum is 60, such that the sum of the square of one number plus twelve times the other number is a minimum.

16. The simple interest earned by a certain amount of money varies jointly as the rate of interest and the time (in years) that the money is invested. If $140 is earned for a certain amount of money invested at 7% for 5 years, how much is earned if the same amount is invested at 8% for 3 years?

For Problems 17–19, use the concepts of translation and/or reflection to describe how the second curve can be obtained from the first curve.

17. $f(x) = x^3$, $f(x) = (x - 6)^3 - 4$

18. $f(x) = |x|$, $f(x) = -|x| + 8$

19. $f(x) = \sqrt{x}$, $f(x) = -\sqrt{x + 5} + 7$

For Problems 20–25, graph each function.

20. $f(x) = -x - 1$

21. $f(x) = -2x^2 - 12x - 14$

22. $f(x) = 2\sqrt{x} - 2$

23. $f(x) = 3|x - 2| - 1$

24. $f(x) = -\dfrac{1}{x} + 3$

25. $f(x) = \sqrt{-x + 2}$

AP/Wide World Photos

When mixing different solutions, a chemist could use a system of equations to determine how much of each solution is needed to produce a specific concentration.

Systems of Equations

A 10%-salt solution is to be mixed with a 20%-salt solution to produce 20 gallons of a 17.5%-salt solution. How many gallons of the 10% solution and how many gallons of the 20% solution will be needed? The two equations $x + y = 20$ and $0.10x + 0.20y = 0.175(20)$, where x represents the number of gallons of the 10% solution and y represents the number of gallons of the 20% solution, algebraically represent the conditions of the problem. The two equations considered together form a **system of linear equations,** and the problem can be solved by solving this system of equations.

Throughout most of this chapter, we will consider systems of linear equations and their applications. We will discuss various techniques for solving systems of linear equations.

489

InfoTrac Project   Do a keyword search on per capita debt and find an article on municipal debt and Arkansas cities. Write a brief summary of the article. The article mentions that one source of debt in cities is revenue bonds. Suppose the equation relating the net revenue (y) (in hundreds of thousands of dollars) received by the cities to the face value (x) of the bonds that are sold could be expressed by $y = x - 600{,}000$. Explain what you think the 600,000 represents in the equation. At some point in time, the bonds will mature and become an expense to the cities. Suppose the equation representing the relationship of the cost (y) (in hundreds of thousands of dollars) to the face value (x) of the bonds could be expressed by $y = 0.12x$. Explain what the 0.12 represents. Using a graphing utility, graph the two equations and find the coordinates of the point where net revenue and cost are the same. For what face value of the bonds does this occur? Now solve this system of equations using algebra. Were your results the same?

10.1   Systems of Two Linear Equations in Two Variables

In Chapter 7, we stated that any equation of the form $Ax + By = C$, where A, B, and C are real numbers (A and B not both zero) is a *linear equation* in the two variables x and y, and its graph is a straight line. Two linear equations in two variables considered together form a **system of two linear equations in two variables.** Here are a few examples:

$$\begin{pmatrix} x + y = 6 \\ x - y = 2 \end{pmatrix} \qquad \begin{pmatrix} 3x + 2y = 1 \\ 5x - 2y = 23 \end{pmatrix} \qquad \begin{pmatrix} 4x - 5y = 21 \\ 3x + 7y = -38 \end{pmatrix}$$

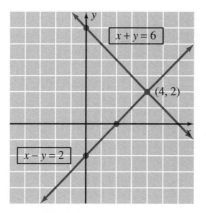

To **solve a system,** such as one of the above, means to find all of the ordered pairs that satisfy both equations in the system. For example, if we graph the two equations $x + y = 6$ and $x - y = 2$ on the same set of axes, as in Figure 10.1, then the ordered pair associated with the point of intersection of the two lines is the solution of the system. Thus we say that $\{(4, 2)\}$ is the solution set of the system

$$\begin{pmatrix} x + y = 6 \\ x - y = 2 \end{pmatrix}$$

Figure 10.1

To check, substitute 4 for x and 2 for y in the two equations, which yields

$x + y$ becomes $4 + 2 = 6$ A true statement

$x - y$ becomes $4 - 2 = 2$ A true statement

Because the graph of a linear equation in two variables is a straight line, there are three possible situations that can occur when we solve a system of two linear equations in two variables. We illustrate these cases in Figure 10.2.

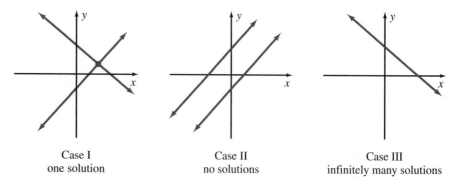

Case I
one solution

Case II
no solutions

Case III
infinitely many solutions

Figure 10.2

CASE I   The graphs of the two equations are two lines intersecting in *one* point. There is *one solution,* and the system is called a **consistent system.**

CASE II   The graphs of the two equations are parallel lines. There is *no solution,* and the system is called an **inconsistent system.**

CASE III   The graphs of the two equations are the same line, and there are *infinitely many solutions* to the system. Any pair of real numbers that satisfies one of the equations will also satisfy the other equation, and we say that the equations are **dependent.**

Thus, as we solve a system of two linear equations in two variables, we know what to expect. The system will have no solutions, one ordered pair as a solution, or infinitely many ordered pairs as solutions.

Substitution Method

It should be evident that solving systems of equations by graphing requires accurate graphs. In fact, unless the solutions are integers, it is quite difficult to obtain exact solutions from a graph. Thus we will consider some other methods for solving systems of equations.

We describe the **substitution method,** which works quite well with systems of two linear equations in two unknowns, as follows:

STEP 1 Solve one of the equations for one variable in terms of the other variable if neither equation is in such a form. (If possible, make a choice that will avoid fractions.)

STEP 2 Substitute the expression obtained in Step 1 into the other equation to produce an equation with one variable.

STEP 3 Solve the equation obtained in Step 2.

STEP 4 Use the solution obtained in Step 3, along with the expression obtained in Step 1, to determine the solution of the system.

Now let's look at some examples that illustrate the substitution method.

EXAMPLE 1 Solve the system $\begin{pmatrix} x + y = 16 \\ y = x + 2 \end{pmatrix}$.

Solution

Because the second equation states that y equals $x + 2$, we can substitute $x + 2$ for y in the first equation.

$$x + y = 16 \xrightarrow{\text{Substitute } x + 2 \text{ for } y} x + (x + 2) = 16$$

Now we have an equation with one variable that we can solve in the usual way.

$$x + (x + 2) = 16$$
$$2x + 2 = 16$$
$$2x = 14$$
$$x = 7$$

Substituting 7 for x in one of the two original equations (let's use the second one) yields

$$y = 7 + 2 = 9$$

To check, we can substitute 7 for x and 9 for y in both of the original equations.

$$7 + 9 = 16 \qquad \text{A true statement}$$
$$9 = 7 + 2 \qquad \text{A true statement}$$

The solution set is $\{(7, 9)\}$. ∎

E X A M P L E   2

Solve the system $\left(\begin{array}{l} x = 3y - 25 \\ 4x + 5y = 19 \end{array}\right)$.

Solution

In this case the first equation states that x equals $3y - 25$. Therefore, we can substitute $3y - 25$ for x in the second equation.

$$\text{Substitute } 3y - 25 \text{ for } x$$
$$4x + 5y = 19 \xrightarrow{\hspace{3cm}} 4(3y - 25) + 5y = 19$$

Solving this equation yields

$$4(3y - 25) + 5y = 19$$
$$12y - 100 + 5y = 19$$
$$17y = 119$$
$$y = 7$$

Substituting 7 for y in the first equation produces

$$x = 3(7) - 25$$
$$= 21 - 25 = -4$$

The solution set is $\{(-4, 7)\}$; check it.   ■

E X A M P L E   3

Solve the system $\left(\begin{array}{l} 3x - 7y = 2 \\ x + 4y = 1 \end{array}\right)$.

Solution

Let's solve the second equation for x in terms of y.

$$x + 4y = 1$$
$$x = 1 - 4y$$

Now we can substitute $1 - 4y$ for x in the first equation.

$$\text{Substitute } 1 - 4y \text{ for } x$$
$$3x - 7y = 2 \xrightarrow{\hspace{3cm}} 3(1 - 4y) - 7y = 2$$

Let's solve this equation for y.

$$3(1 - 4y) - 7y = 2$$
$$3 - 12y - 7y = 2$$
$$-19y = -1$$
$$y = \frac{1}{19}$$

Finally, we can substitute $\dfrac{1}{19}$ for y in the equation $x = 1 - 4y$.

$$x = 1 - 4\left(\dfrac{1}{19}\right)$$

$$= 1 - \dfrac{4}{19}$$

$$= \dfrac{15}{19}$$

The solution set is $\left\{\left(\dfrac{15}{19}, \dfrac{1}{19}\right)\right\}$. ∎

EXAMPLE 4 Solve the system $\begin{pmatrix} 5x - 6y = -4 \\ 3x + 2y = -8 \end{pmatrix}$.

Solution

Note that solving either equation for either variable will produce a fractional form. Let's solve the second equation for y in terms of x.

$$3x + 2y = -8$$

$$2y = -8 - 3x$$

$$y = \dfrac{-8 - 3x}{2}$$

Now we can substitute $\dfrac{-8 - 3x}{2}$ for y in the first equation.

$$5x - 6y = -4 \xrightarrow{\text{Substitute } \dfrac{-8 - 3x}{2} \text{ for } y} 5x - 6\left(\dfrac{-8 - 3x}{2}\right) = -4$$

Solving this equation yields

$$5x - 6\left(\dfrac{-8 - 3x}{2}\right) = -4$$

$$5x - 3(-8 - 3x) = -4$$

$$5x + 24 + 9x = -4$$

$$14x = -28$$

$$x = -2$$

Substituting -2 for x in $y = \dfrac{-8 - 3x}{2}$ yields

$$y = \frac{-8 - 3(-2)}{2}$$

$$= \frac{-8 + 6}{2}$$

$$= \frac{-2}{2}$$

$$= -1$$

The solution set is $\{(-2, -1)\}$. ◼

Problem Solving

Many word problems that we solved earlier in this text using one variable and one equation can also be solved using a system of two linear equations in two variables. In fact, in many of these problems you may find it much more natural to use two variables. Let's consider some examples.

EXAMPLE 5   Anita invested some money at 8% and $400 more than that amount at 9%. The yearly interest from the two investments was $87. How much did Anita invest at each rate?

Solution

Let x represent the amount invested at 8% and let y represent the amount invested at 9%. The problem translates into the following system.

Amount invested at 9% was $400 more than at 8%. ⟶ $\left(\begin{array}{l} y = x + 400 \\ 0.08x + 0.09y = 87 \end{array} \right)$
Yearly interest from the two investments was $87. ⟶

From the first equation we can substitute $x + 400$ for y in the second equation and solve for x.

$$0.08x + 0.09(x + 400) = 87$$

$$0.08x + 0.09x + 36 = 87$$

$$0.17x = 51$$

$$x = 300$$

Therefore, $300 is invested at 8% and $300 + $400 = $700 is invested at 9%. ◼

The two-variable expression $10t + u$ can be used to represent any two-digit number. The t represents the tens digit and the u represents the units digit. For example, if $t = 5$ and $u = 2$, then $10t + u$ becomes $10(5) + 2 = 52$. We use this general representation for a two-digit number in the next problem.

E X A M P L E   6

The tens digit of a two-digit number is 2 more than twice the units digit. The number with the digits reversed is 45 less than the original number. Find the original number.

Solution

Let u represent the units digit of the original number. Let t represent the tens digit of the original number. Then $10t + u$ represents the original number and $10u + t$ represents the number with the digits reversed. The problem translates into the following system.

$$\begin{pmatrix} t = 2u + 2 \\ 10u + t = 10t + u - 45 \end{pmatrix}$$

The tens digit is 2 more than twice the units digit.
The number with the digits reversed is 45 less than the original number.

When we simplify the second equation, the system becomes

$$\begin{pmatrix} t = 2u + 2 \\ t - u = 5 \end{pmatrix}$$

From the first equation, we can substitute $2u + 2$ for t in the second equation and solve.

$$t - u = 5$$
$$2u + 2 - u = 5$$
$$u = 3$$

Substitute 3 for u in $t = 2u + 2$ to obtain

$$t = 2u + 2$$
$$= 2(3) + 2 = 8$$

The tens digit is 8 and the units digit is 3, so the number is 83. (You should check to see whether 83 satisfies the original conditions stated in the problem.) ∎

In our final example of this section, we will use a graphing utility to help solve a system of equations.

E X A M P L E   7

Solve the system $\begin{pmatrix} 1.14x + 2.35y = -7.12 \\ 3.26x - 5.05y = 26.72 \end{pmatrix}$.

Solution

First, we need to solve each equation for y in terms of x. Thus the system becomes

$$\begin{pmatrix} y = \dfrac{-7.12 - 1.14x}{2.35} \\ y = \dfrac{3.26x - 26.72}{5.05} \end{pmatrix}$$

Now we can enter both of these equations into a graphing utility and obtain Figure 10.3. In this figure it appears that the point of intersection is at approximately $x = 2$ and $y = -4$. By direct substitution into the given equations, we can verify that the point of intersection is exactly $(2, -4)$.

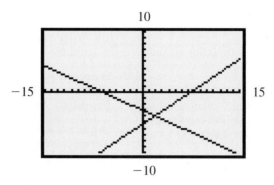

Figure 10.3

PROBLEM SET 10.1

For Problems 1–10, use the graphing approach to determine whether the system is consistent, the system is inconsistent, or the equations are dependent. If the system is consistent, find the solution set from the graph and check it.

1. $\begin{pmatrix} x - y = 1 \\ 2x + y = 8 \end{pmatrix}$

2. $\begin{pmatrix} 3x + y = 0 \\ x - 2y = -7 \end{pmatrix}$

3. $\begin{pmatrix} 4x + 3y = -5 \\ 2x - 3y = -7 \end{pmatrix}$

4. $\begin{pmatrix} 2x - y = 9 \\ 4x - 2y = 11 \end{pmatrix}$

5. $\begin{pmatrix} \dfrac{1}{2}x + \dfrac{1}{4}y = 9 \\ 4x + 2y = 72 \end{pmatrix}$

6. $\begin{pmatrix} 5x + 2y = -9 \\ 4x - 3y = 2 \end{pmatrix}$

7. $\begin{pmatrix} \dfrac{1}{2}x - \dfrac{1}{3}y = 3 \\ x + 4y = -8 \end{pmatrix}$

8. $\begin{pmatrix} 4x - 9y = -60 \\ \dfrac{1}{3}x - \dfrac{3}{4}y = -5 \end{pmatrix}$

9. $\begin{pmatrix} x - \dfrac{y}{2} = -4 \\ 8x - 4y = -1 \end{pmatrix}$

10. $\begin{pmatrix} 3x - 2y = 7 \\ 6x + 5y = -4 \end{pmatrix}$

For Problems 11–36, solve each system by using the substitution method.

11. $\begin{pmatrix} x + y = 20 \\ x = y - 4 \end{pmatrix}$

12. $\begin{pmatrix} x + y = 23 \\ y = x - 5 \end{pmatrix}$

13. $\begin{pmatrix} y = -3x - 18 \\ 5x - 2y = -8 \end{pmatrix}$

14. $\begin{pmatrix} 4x - 3y = 33 \\ x = -4y - 25 \end{pmatrix}$

15. $\begin{pmatrix} x = -3y \\ 7x - 2y = -69 \end{pmatrix}$

16. $\begin{pmatrix} 9x - 2y = -38 \\ y = -5x \end{pmatrix}$

17. $\begin{pmatrix} 2x + 3y = 11 \\ 3x - 2y = -3 \end{pmatrix}$

18. $\begin{pmatrix} 3x - 4y = -14 \\ 4x + 3y = 23 \end{pmatrix}$

19. $\begin{pmatrix} 3x - 4y = 9 \\ x = 4y - 1 \end{pmatrix}$

20. $\begin{pmatrix} y = 3x - 5 \\ 2x + 3y = 6 \end{pmatrix}$

21. $\begin{pmatrix} y = \dfrac{2}{5}x - 1 \\ 3x + 5y = 4 \end{pmatrix}$

22. $\begin{pmatrix} y = \dfrac{3}{4}x - 5 \\ 5x - 4y = 9 \end{pmatrix}$

23. $\begin{pmatrix} 7x - 3y = -2 \\ x = \dfrac{3}{4}y + 1 \end{pmatrix}$

24. $\begin{pmatrix} 5x - y = 9 \\ x = \dfrac{1}{2}y - 3 \end{pmatrix}$

25. $\begin{pmatrix} 2x + y = 12 \\ 3x - y = 13 \end{pmatrix}$

26. $\begin{pmatrix} -x + 4y = -22 \\ x - 7y = 34 \end{pmatrix}$

27. $\begin{pmatrix} 4x + 3y = -40 \\ 5x - y = -12 \end{pmatrix}$

28. $\begin{pmatrix} x - 5y = 33 \\ -4x + 7y = -41 \end{pmatrix}$

29. $\begin{pmatrix} 3x + y = 2 \\ 11x - 3y = 5 \end{pmatrix}$

30. $\begin{pmatrix} 2x - y = 9 \\ 7x + 4y = 1 \end{pmatrix}$

31. $\begin{pmatrix} 3x + 5y = 22 \\ 4x - 7y = -39 \end{pmatrix}$

32. $\begin{pmatrix} 2x - 3y = -16 \\ 6x + 7y = 16 \end{pmatrix}$

33. $\begin{pmatrix} 4x - 5y = 3 \\ 8x + 15y = -24 \end{pmatrix}$ **34.** $\begin{pmatrix} 2x + 3y = 3 \\ 4x - 9y = -4 \end{pmatrix}$

35. $\begin{pmatrix} 6x - 3y = 4 \\ 5x + 2y = -1 \end{pmatrix}$ **36.** $\begin{pmatrix} 7x - 2y = 1 \\ 4x + 5y = 2 \end{pmatrix}$

For Problems 37–50, solve each problem by setting up and solving an appropriate system of equations.

37. Doris invested some money at 7% and some money at 8%. She invested $6000 more at 8% than she did at 7%. Her total yearly interest from the two investments was $780. How much did Doris invest at each rate?

38. Suppose that Gus invested a total of $8000, part of it at 8% and the remainder at 9%. His yearly income from the two investments was $690. How much did he invest at each rate?

39. The sum of the digits of a two-digit number is 11. The tens digit is one more than four times the units digit. Find the number.

40. The units digit of a two-digit number is one less than three times the tens digit. If the sum of the digits is 11, find the number.

41. Find two numbers whose sum is 131 such that one number is five less than three times the other.

42. The difference of two numbers is 75. The larger number is three less than four times the smaller number. Find the numbers.

43. In a class of 50 students, the number of females is two more than five times the number of males. How many females are there in the class?

44. In a recent survey, one thousand registered voters were asked about their political preferences. The number of males in the survey was five less than one-half of the number of females. Find the number of males in the survey.

45. The perimeter of a rectangle is 94 inches. The length of the rectangle is 7 inches more than the width. Find the dimensions of the rectangle.

46. Two angles are supplementary, and the measure of one of them is 20° less than three times the measure of the other angle. Find the measure of each angle.

47. A deposit slip listed $700 in cash to be deposited. There were 100 bills, some of them five-dollar bills and the remainder ten-dollar bills. How many bills of each denomination were deposited?

48. Cindy has 30 coins, consisting of dimes and quarters, that total $5.10. How many coins of each kind does she have?

49. The income from a student production was $10,000. The price of a student ticket was $3, and nonstudent tickets were sold at $5 each. Three thousand tickets were sold. How many tickets of each kind were sold?

50. Sue bought 3 packages of cookies and 2 sacks of potato chips for $3.65. Later she bought 2 more packages of cookies and 5 additional sacks of potato chips for $4.23. Find the price of a package of cookies.

■ ■ ■ Thoughts into words

51. Give a general description of how to use the substitution method to solve a system of two linear equations in two variables.

52. Is it possible for a system of two linear equations in two variables to have exactly two solutions? Defend your answer.

53. Explain how you would solve the system $\begin{pmatrix} 2x + 5y = 5 \\ 5x - y = 9 \end{pmatrix}$ using the substitution method.

Graphing calculator activities

54. Use your graphing calculator to help determine whether, in Problems 1–10, the system is consistent, the system is inconsistent, or the equations are dependent.

55. Use your graphing calculator to help determine the solution set for each of the following systems. Be sure to check your answers.

(a) $\begin{pmatrix} 3x - y = 30 \\ 5x - y = 46 \end{pmatrix}$

(b) $\begin{pmatrix} 1.2x + 3.4y = 25.4 \\ 3.7x - 2.3y = 14.4 \end{pmatrix}$

(c) $\begin{pmatrix} 1.98x + 2.49y = 13.92 \\ 1.19x + 3.45y = 16.18 \end{pmatrix}$

(d) $\begin{pmatrix} 2x - 3y = 10 \\ 3x + 5y = 53 \end{pmatrix}$

(e) $\begin{pmatrix} 4x - 7y = -49 \\ 6x + 9y = 219 \end{pmatrix}$

(f) $\begin{pmatrix} 3.7x - 2.9y = -14.3 \\ 1.6x + 4.7y = -30 \end{pmatrix}$

10.2 Elimination-by-Addition Method

We found in the previous section that the substitution method for solving a system of two equations and two unknowns works rather well. However, as the number of equations and unknowns increases, the substitution method becomes quite unwieldy. In this section we are going to introduce another method, called the **elimination-by-addition** method. We shall introduce it here using systems of two linear equations in two unknowns and then, in the next section, extend its use to three linear equations in three unknowns.

The elimination-by-addition method involves replacing systems of equations with simpler, equivalent systems until we obtain a system whereby we can easily extract the solutions. **Equivalent systems of equations are systems that have exactly the same solution set.** We can apply the following operations or transformations to a system of equations to produce an equivalent system.

1. Any two equations of the system can be interchanged.
2. Both sides of an equation of the system can be multiplied by any nonzero real number.
3. Any equation of the system can be replaced by the *sum* of that equation and a nonzero multiple of another equation.

Now let's see how to apply these operations to solve a system of two linear equations in two unknowns.

EXAMPLE 1

Solve the system $\begin{pmatrix} 3x + 2y = 1 \\ 5x - 2y = 23 \end{pmatrix}$.   **(1)** **(2)**

Solution

Let's replace equation (2) with an equation we form by multiplying equation (1) by 1 and then adding that result to equation (2).

$$\begin{pmatrix} 3x + 2y = 1 \\ 8x \quad\quad = 24 \end{pmatrix}$$ **(3)** **(4)**

From equation (4) we can easily obtain the value of x.

$$8x = 24$$

$$x = 3$$

Then we can substitute 3 for x in equation (3).

$$3x + 2y = 1$$

$$3(3) + 2y = 1$$

$$2y = -8$$

$$y = -4$$

The solution set is $\{(3, -4)\}$. Check it!

E X A M P L E   2

Solve the system $\begin{pmatrix} x + 5y = & -2 \\ 3x - 4y = & -25 \end{pmatrix}$.

(1)
(2)

Solution

Let's replace equation (2) with an equation we form by multiplying equation (1) by -3 and then adding that result to equation (2).

$$\begin{pmatrix} x + 5y = & -2 \\ -19y = & -19 \end{pmatrix}$$

(3)
(4)

From equation (4) we can obtain the value of y.

$$-19y = -19$$

$$y = 1$$

Now we can substitute 1 for y in equation (3).

$$x + 5y = -2$$

$$x + 5(1) = -2$$

$$x = -7$$

The solution set is $\{(-7, 1)\}$.

Note that our objective has been to produce an equivalent system of equations whereby one of the variables can be *eliminated* from one equation. We accomplish this by multiplying one equation of the system by an appropriate number and then *adding* that result to the other equation. Thus the method is called **elimination by addition.** Let's look at another example.

E X A M P L E   3

Solve the system $\begin{pmatrix} 2x + 5y = & 4 \\ 5x - 7y = & -29 \end{pmatrix}$.

(1)
(2)

Solution

Let's form an equivalent system where the second equation has no x term. First, we can multiply equation (2) by -2.

$$\begin{pmatrix} 2x + 5y = & 4 \\ -10x + 14y = & 58 \end{pmatrix}$$

(3)
(4)

Now we can replace equation (4) with an equation that we form by multiplying equation (3) by 5 and then adding that result to equation (4).

$$\begin{pmatrix} 2x + 5y = 4 \\ 39y = 78 \end{pmatrix}$$

(5)
(6)

From equation (6) we can find the value of y.

$$39y = 78$$

$$y = 2$$

Now we can substitute 2 for y in equation (5).

$$2x + 5y = 4$$

$$2x + 5(2) = 4$$

$$2x = -6$$

$$x = -3$$

The solution set is $\{(-3, 2)\}$.   ■

EXAMPLE 4   Solve the system $\begin{pmatrix} 3x - 2y = 5 \\ 2x + 7y = 9 \end{pmatrix}$.

(1)
(2)

Solution

We can start by multiplying equation (2) by -3.

$$\begin{pmatrix} 3x - 2y = 5 \\ -6x - 21y = -27 \end{pmatrix}$$

(3)
(4)

Now we can replace equation (4) with an equation we form by multiplying equation (3) by 2 and then adding that result to equation (4).

$$\begin{pmatrix} 3x - 2y = 5 \\ -25y = -17 \end{pmatrix}$$

(5)
(6)

From equation (6) we can find the value of y.

$$-25y = -17$$

$$y = \frac{17}{25}$$

Now we can substitute $\frac{17}{25}$ for y in equation (5).

$$3x - 2y = 5$$

$$3x - 2\left(\frac{17}{25}\right) = 5$$

$$3x - \frac{34}{25} = 5$$

$$3x = 5 + \frac{34}{25}$$

$$3x = \frac{125}{25} + \frac{34}{25}$$

$$3x = \frac{159}{25}$$

$$x = \left(\frac{159}{25}\right)\left(\frac{1}{3}\right) = \frac{53}{25}$$

The solution set is $\left\{\left(\frac{53}{25}, \frac{17}{25}\right)\right\}$. (Perhaps you should check this result!)   ■

Which Method To Use

Both the elimination-by-addition and the substitution methods can be used to obtain exact solutions for any system of two linear equations in two unknowns. Sometimes the issue is that of deciding which method to use on a particular system. As we have seen with the examples thus far in this section and those of the previous section, many systems lend themselves to one or the other method by the original format of the equations. Let's emphasize that point with some more examples.

E X A M P L E   5   Solve the system $\begin{pmatrix} 4x - 3y = & 4 \\ 10x + 9y = & -1 \end{pmatrix}$.   **(1)** **(2)**

Solution

Because changing the form of either equation in preparation for the substitution method would produce a fractional form, we are probably better off using the elimination-by-addition method. Let's replace equation (2) with an equation we form by multiplying equation (1) by 3 and then adding that result to equation (2).

$$\begin{pmatrix} 4x - 3y = & 4 \\ 22x = & 11 \end{pmatrix}$$ **(3)** **(4)**

From equation (4) we can determine the value of x.

$$22x = 11$$

$$x = \frac{11}{22} = \frac{1}{2}$$

Now we can substitute $\frac{1}{2}$ for x in equation (3).

$$4x - 3y = 4$$

$$4\left(\frac{1}{2}\right) - 3y = 4$$

$$2 - 3y = 4$$

$$-3y = 2$$

$$y = -\frac{2}{3}$$

The solution set is $\left\{\left(\frac{1}{2}, -\frac{2}{3}\right)\right\}$.   ■

EXAMPLE 6   Solve the system $\begin{pmatrix} 6x + 5y = -3 \\ y = -2x - 7 \end{pmatrix}$.   **(1)**
(2)

Solution

Because the second equation is of the form "y equals," let's use the substitution method. From the second equation we can substitute $-2x - 7$ for y in the first equation.

$$6x + 5y = -3 \xrightarrow{\text{Substitute } -2x - 7 \text{ for } y} 6x + 5(-2x - 7) = -3$$

Solving this equation yields

$$6x + 5(-2x - 7) = -3$$

$$6x - 10x - 35 = -3$$

$$-4x - 35 = -3$$

$$-4x = 32$$

$$x = -8$$

Substitute -8 for x in the second equation to obtain

$$y = -2(-8) - 7 = 16 - 7 = 9$$

The solution set is $\{(-8, 9)\}$.   ■

Sometimes we need to simplify the equations of a system before we can decide which method to use for solving the system. Let's consider an example of that type.

EXAMPLE 7   Solve the system $\begin{pmatrix} \dfrac{x - 2}{4} + \dfrac{y + 1}{3} = 2 \\ \dfrac{x + 1}{7} + \dfrac{y - 3}{2} = \dfrac{1}{2} \end{pmatrix}$.   **(1)**
(2)

Solution

First, we need to simplify the two equations. Let's multiply both sides of equation (1) by 12 and simplify.

$$12\left(\frac{x - 2}{4} + \frac{y + 1}{3}\right) = 12(2)$$

$$3(x - 2) + 4(y + 1) = 24$$

$$3x - 6 + 4y + 4 = 24$$

$$3x + 4y - 2 = 24$$

$$3x + 4y = 26$$

Let's multiply both sides of equation (2) by 14.

$$14\left(\frac{x + 1}{7} + \frac{y - 3}{2}\right) = 14\left(\frac{1}{2}\right)$$

$$2(x + 1) + 7(y - 3) = 7$$

$$2x + 2 + 7y - 21 = 7$$

$$2x + 7y - 19 = 7$$

$$2x + 7y = 26$$

Now we have the following system to solve.

$$\begin{pmatrix} 3x + 4y = 26 \\ 2x + 7y = 26 \end{pmatrix} \qquad \begin{matrix} \textbf{(3)} \\ \textbf{(4)} \end{matrix}$$

Probably the easiest approach is to use the elimination-by-addition method. We can start by multiplying equation (4) by -3.

$$\begin{pmatrix} 3x + 4y = 26 \\ -6x - 21y = -78 \end{pmatrix} \qquad \begin{matrix} \textbf{(5)} \\ \textbf{(6)} \end{matrix}$$

Now we can replace equation (6) with an equation we form by multiplying equation (5) by 2 and then adding that result to equation (6).

$$\begin{pmatrix} 3x + 4y = 26 \\ -13y = -26 \end{pmatrix} \qquad \begin{matrix} \textbf{(7)} \\ \textbf{(8)} \end{matrix}$$

From equation (8) we can find the value of y.

$$-13y = -26$$

$$y = 2$$

Now we can substitute 2 for y in equation (7).

$$3x + 4y = 26$$

$$3x + 4(2) = 26$$

$$3x = 18$$

$$x = 6$$

The solution set is $\{(6, 2)\}$.   ■

REMARK: Don't forget that to check a problem like Example 7 you must check the potential solutions back in the original equations.

In Section 10.1, we discussed the fact that you can tell whether a system of two linear equations in two unknowns has no solution, one solution, or infinitely many solutions by graphing the equations of the system. That is, the two lines may be parallel (no solution), or they may intersect in one point (one solution), or they may coincide (infinitely many solutions).

From a practical viewpoint, the systems that have one solution deserve most of our attention. However, we need to be able to deal with the other situations; they do occur occasionally. Let's use two examples to illustrate the type of thing that happens when we encounter *no solution* or *infinitely many solutions* when using either the elimination-by-addition method or the substitution method.

EXAMPLE 8

Solve the system $\left(\begin{array}{l} y = 3x - 1 \\ -9x + 3y = 4 \end{array} \right)$.

(1)
(2)

Solution

Using the substitution method, we can proceed as follows:

$$-9x + 3y = 4 \xrightarrow{\text{Substitute } 3x - 1 \text{ for } y} -9x + 3(3x - 1) = 4$$

Solving this equation yields

$$-9x + 3(3x - 1) = 4$$
$$-9x + 9x - 3 = 4$$
$$-3 = 4$$

The *false numerical statement*, $-3 = 4$, implies that the system has *no solution*. (You may want to graph the two lines to verify this conclusion!)   ∎

EXAMPLE 9

Solve the system $\left(\begin{array}{l} 5x + y = 2 \\ 10x + 2y = 4 \end{array} \right)$.

(1)
(2)

Solution

Use the elimination-by-addition method and proceed as follows: Let's replace equation (2) with an equation we form by multiplying equation (1) by -2 and then adding that result to equation (2).

$$\left(\begin{array}{l} 5x + y = 2 \\ 0 + 0 = 0 \end{array} \right)$$

(3)
(4)

The *true numerical statement*, $0 + 0 = 0$, implies that the system has *infinitely many solutions*. Any ordered pair that satisfies one of the equations will also satisfy the other equation. Thus the solution set can be expressed as

$$\{(x, y) | 5x + y = 2\}$$

∎

PROBLEM SET 10.2

For Problems 1–16, use the elimination-by-addition method to solve each system.

1. $\begin{pmatrix} 2x + 3y = -1 \\ 5x - 3y = 29 \end{pmatrix}$ **2.** $\begin{pmatrix} 3x - 4y = -30 \\ 7x + 4y = 10 \end{pmatrix}$

3. $\begin{pmatrix} 6x - 7y = 15 \\ 6x + 5y = -21 \end{pmatrix}$ **4.** $\begin{pmatrix} 5x + 2y = -4 \\ 5x - 3y = 6 \end{pmatrix}$

5. $\begin{pmatrix} x - 2y = -12 \\ 2x + 9y = 2 \end{pmatrix}$ **6.** $\begin{pmatrix} x - 4y = 29 \\ 3x + 2y = -11 \end{pmatrix}$

7. $\begin{pmatrix} 4x + 7y = -16 \\ 6x - y = -24 \end{pmatrix}$ **8.** $\begin{pmatrix} 6x + 7y = 17 \\ 3x + y = -4 \end{pmatrix}$

9. $\begin{pmatrix} 3x - 2y = 5 \\ 2x + 5y = -3 \end{pmatrix}$ **10.** $\begin{pmatrix} 4x + 3y = -4 \\ 3x - 7y = 34 \end{pmatrix}$

11. $\begin{pmatrix} 7x - 2y = 4 \\ 7x - 2y = 9 \end{pmatrix}$ **12.** $\begin{pmatrix} 5x - y = 6 \\ 10x - 2y = 12 \end{pmatrix}$

13. $\begin{pmatrix} 5x + 4y = 1 \\ 3x - 2y = -1 \end{pmatrix}$ **14.** $\begin{pmatrix} 2x - 7y = -2 \\ 3x + y = 1 \end{pmatrix}$

15. $\begin{pmatrix} 8x - 3y = 13 \\ 4x + 9y = 3 \end{pmatrix}$ **16.** $\begin{pmatrix} 10x - 8y = -11 \\ 8x + 4y = -1 \end{pmatrix}$

For Problems 17– 44, solve each system by using either the substitution or the elimination-by-addition method, whichever seems more appropriate.

17. $\begin{pmatrix} 5x + 3y = -7 \\ 7x - 3y = 55 \end{pmatrix}$ **18.** $\begin{pmatrix} 4x - 7y = 21 \\ -4x + 3y = -9 \end{pmatrix}$

19. $\begin{pmatrix} x = 5y + 7 \\ 4x + 9y = 28 \end{pmatrix}$ **20.** $\begin{pmatrix} 11x - 3y = -60 \\ y = -38 - 6x \end{pmatrix}$

21. $\begin{pmatrix} x = -6y + 79 \\ x = 4y - 41 \end{pmatrix}$ **22.** $\begin{pmatrix} y = 3x + 34 \\ y = -8x - 54 \end{pmatrix}$

23. $\begin{pmatrix} 4x - 3y = 2 \\ 5x - y = 3 \end{pmatrix}$ **24.** $\begin{pmatrix} 3x - y = 9 \\ 5x + 7y = 1 \end{pmatrix}$

25. $\begin{pmatrix} 5x - 2y = 1 \\ 10x - 4y = 7 \end{pmatrix}$ **26.** $\begin{pmatrix} 4x + 7y = 2 \\ 9x - 2y = 1 \end{pmatrix}$

27. $\begin{pmatrix} 3x - 2y = 7 \\ 5x + 7y = 1 \end{pmatrix}$ **28.** $\begin{pmatrix} 2x - 3y = 4 \\ y = \dfrac{2}{3}x - \dfrac{4}{3} \end{pmatrix}$

29. $\begin{pmatrix} -2x + 5y = -16 \\ x = \dfrac{3}{4}y + 1 \end{pmatrix}$ **30.** $\begin{pmatrix} y = \dfrac{2}{3}x - \dfrac{3}{4} \\ 2x + 3y = 11 \end{pmatrix}$

31. $\begin{pmatrix} y = \dfrac{2}{3}x - 4 \\ 5x - 3y = 9 \end{pmatrix}$ **32.** $\begin{pmatrix} 5x - 3y = 7 \\ x = \dfrac{3y}{4} - \dfrac{1}{3} \end{pmatrix}$

33. $\begin{pmatrix} \dfrac{x}{6} + \dfrac{y}{3} = 3 \\ \dfrac{5x}{2} - \dfrac{y}{6} = -17 \end{pmatrix}$ **34.** $\begin{pmatrix} \dfrac{3x}{4} - \dfrac{2y}{3} = 31 \\ \dfrac{7x}{5} + \dfrac{y}{4} = 22 \end{pmatrix}$

35. $\begin{pmatrix} -(x - 6) + 6(y + 1) = 58 \\ 3(x + 1) - 4(y - 2) = -15 \end{pmatrix}$

36. $\begin{pmatrix} -2(x + 2) + 4(y - 3) = -34 \\ 3(x + 4) - 5(y + 2) = 23 \end{pmatrix}$

37. $\begin{pmatrix} 5(x + 1) - (y + 3) = -6 \\ 2(x - 2) + 3(y - 1) = 0 \end{pmatrix}$

38. $\begin{pmatrix} 2(x - 1) - 3(y + 2) = 30 \\ 3(x + 2) + 2(y - 1) = -4 \end{pmatrix}$

39. $\begin{pmatrix} \dfrac{1}{2}x - \dfrac{1}{3}y = 12 \\ \dfrac{3}{4}x + \dfrac{2}{3}y = 4 \end{pmatrix}$ **40.** $\begin{pmatrix} \dfrac{2}{3}x + \dfrac{1}{5}y = 0 \\ \dfrac{3}{2}x - \dfrac{3}{10}y = -15 \end{pmatrix}$

41. $\begin{pmatrix} \dfrac{2x}{3} - \dfrac{y}{2} = -\dfrac{5}{4} \\ \dfrac{x}{4} + \dfrac{5y}{6} = \dfrac{17}{16} \end{pmatrix}$ **42.** $\begin{pmatrix} \dfrac{x}{2} + \dfrac{y}{3} = \dfrac{5}{72} \\ \dfrac{x}{4} + \dfrac{5y}{2} = -\dfrac{17}{48} \end{pmatrix}$

43. $\begin{pmatrix} \dfrac{3x + y}{2} + \dfrac{x - 2y}{5} = 8 \\ \dfrac{x - y}{3} - \dfrac{x + y}{6} = \dfrac{10}{3} \end{pmatrix}$

44. $\begin{pmatrix} \dfrac{x - y}{4} - \dfrac{2x - y}{3} = -\dfrac{1}{4} \\ \dfrac{2x + y}{3} + \dfrac{x + y}{2} = \dfrac{17}{6} \end{pmatrix}$

For Problems 45 –57, solve each problem by setting up and solving an appropriate system of equations.

45. A 10%-salt solution is to be mixed with a 20%-salt solution to produce 20 gallons of a 17.5%-salt solution. How many gallons of the 10% solution and how many gallons of the 20% solution will be needed?

46. A small-town library buys a total of 35 books that cost $462. Some of the books cost $12 each, and the remainder cost $14 each. How many books of each price did the library buy?

47. Suppose that on a particular day the cost of 3 tennis balls and 2 golf balls is $7. The cost of 6 tennis balls and 3 golf balls is $12. Find the cost of 1 tennis ball and the cost of 1 golf ball.

48. For moving purposes, the Hendersons bought 25 cardboard boxes for $97.50. There were two kinds of boxes; the large ones cost $7.50 per box and the small ones were $3 per box. How many boxes of each kind did they buy?

49. A motel in a suburb of Chicago rents double rooms for $42 per day and single rooms for $22 per day. If a total of 55 rooms were rented for $2010, how many of each kind were rented?

50. Suppose that one solution contains 50% alcohol and another solution contains 80% alcohol. How many liters of each solution should be mixed to make 10.5 liters of a 70%-alcohol solution?

51. Suppose that a fulcrum is placed so that weights of 40 pounds and 80 pounds are in balance. Furthermore, suppose that when 20 pounds is added to the 40-pound weight, the 80-pound weight must be moved $1\frac{1}{2}$ feet farther from the fulcrum to preserve the balance. Find the original distance between the two weights.

52. If a certain two-digit number is divided by the sum of its digits, the quotient is 2. If the digits are reversed, the new number is nine less than five times the original number. Find the original number.

53. If the numerator of a certain fraction is increased by 5 and the denominator is decreased by 1, the resulting fraction is $\frac{8}{3}$. However, if the numerator of the original fraction is doubled and the denominator is increased by 7, then the resulting fraction is $\frac{6}{11}$. Find the original fraction.

54. A man bought 2 pounds of coffee and 1 pound of butter for a total of $7.75. A month later, the prices had not changed (this makes it a fictitious problem), and he bought 3 pounds of coffee and 2 pounds of butter for $12.50. Find the price per pound of both the coffee and the butter.

55. Suppose that we have a rectangular-shaped book cover. If the width is increased by 2 centimeters and the length is decreased by 1 centimeter, the area is increased by 28 square centimeters. However, if the width is decreased by 1 centimeter and the length is increased by 2 centimeters, then the area is increased by 10 square centimeters. Find the dimensions of the book cover.

56. A blueprint indicates a master bedroom in the shape of a rectangle. If the width is increased by 2 feet and the length remains the same, then the area is increased by 36 square feet. However, if the width is increased by 1 foot and the length is increased by 2 feet, then the area is increased by 48 square feet. Find the dimensions of the room as indicated on the blueprint.

57. A fulcrum is placed so that weights of 60 pounds and 100 pounds are in balance. If 20 pounds is subtracted from the 100-pound weight, then the 60-pound weight must be moved 1 foot closer to the fulcrum to preserve the balance. Find the original distance between the 60-pound and 100-pound weights.

■ ■ ■ **Thoughts into words**

58. Give a general description of how to use the elimination-by-addition method to solve a system of two linear equations in two variables.

59. Explain how you would solve the system

$$\begin{pmatrix} 3x - 4y = -1 \\ 2x - 5y = 9 \end{pmatrix}$$

using the elimination-by-addition method.

60. How do you decide whether to solve a system of linear equations in two variables by using the substitution method or by using the elimination-by-addition method?

■ ■ ■ **Further investigations**

61. There is another way of telling whether a system of two linear equations in two unknowns is consistent or inconsistent or whether the equations are dependent without taking the time to graph each equation. It can be shown that any system of the form

$$a_1 x + b_1 y = c_1$$

$$a_2 x + b_2 y = c_2$$

has one and only one solution if

$$\frac{a_1}{a_2} \neq \frac{b_1}{b_2} \qquad \text{Consistent}$$

that it has no solution if

$$\frac{a_1}{a_2} = \frac{b_1}{b_2} \neq \frac{c_1}{c_2} \qquad \text{Inconsistent}$$

and that it has infinitely many solutions if

$$\frac{a_1}{a_2} = \frac{b_1}{b_2} = \frac{c_1}{c_2} \qquad \text{Dependent}$$

For each of the following systems, determine whether the system is consistent, the system is inconsistent, or the equations are dependent.

(a) $\begin{pmatrix} 4x - 3y = 7 \\ 9x + 2y = 5 \end{pmatrix}$ **(b)** $\begin{pmatrix} 5x - y = 6 \\ 10x - 2y = 19 \end{pmatrix}$

(c) $\begin{pmatrix} 5x - 4y = 11 \\ 4x + 5y = 12 \end{pmatrix}$ **(d)** $\begin{pmatrix} x + 2y = 5 \\ x - 2y = 9 \end{pmatrix}$

(e) $\begin{pmatrix} x - 3y = 5 \\ 3x - 9y = 15 \end{pmatrix}$ **(f)** $\begin{pmatrix} 4x + 3y = 7 \\ 2x - y = 10 \end{pmatrix}$

(g) $\begin{pmatrix} 3x + 2y = 4 \\ y = -\dfrac{3}{2}x - 1 \end{pmatrix}$ **(h)** $\begin{pmatrix} y = \dfrac{4}{3}x - 2 \\ 4x - 3y = 6 \end{pmatrix}$

62. A system such as

$$\begin{pmatrix} \dfrac{3}{x} + \dfrac{2}{y} = 2 \\ \dfrac{2}{x} - \dfrac{3}{y} = \dfrac{1}{4} \end{pmatrix}$$

is not a system of linear equations but can be transformed into a linear system by changing variables. For

example, when we substitute u for $\dfrac{1}{x}$ and v for $\dfrac{1}{y}$, the system cited becomes

$$\begin{pmatrix} 3u + 2v = 2 \\ 2u - 3v = \dfrac{1}{4} \end{pmatrix}$$

We can solve this "new" system either by elimination by addition or by substitution (we will leave the details for you) to produce $u = \dfrac{1}{2}$ and $v = \dfrac{1}{4}$. Therefore, because $u = \dfrac{1}{x}$ and $v = \dfrac{1}{y}$, we have

$$\frac{1}{x} = \frac{1}{2} \qquad \text{and} \qquad \frac{1}{y} = \frac{1}{4}$$

Solving these equations yields

$$x = 2 \qquad \text{and} \qquad y = 4$$

The solution set of the original system is $\{(2, 4)\}$. Solve each of the following systems.

(a) $\begin{pmatrix} \dfrac{1}{x} + \dfrac{2}{y} = \dfrac{7}{12} \\ \dfrac{3}{x} - \dfrac{2}{y} = \dfrac{5}{12} \end{pmatrix}$ **(b)** $\begin{pmatrix} \dfrac{2}{x} + \dfrac{3}{y} = \dfrac{19}{15} \\ -\dfrac{2}{x} + \dfrac{1}{y} = -\dfrac{7}{15} \end{pmatrix}$

(c) $\begin{pmatrix} \dfrac{3}{x} - \dfrac{2}{y} = \dfrac{13}{6} \\ \dfrac{2}{x} + \dfrac{3}{y} = 0 \end{pmatrix}$ **(d)** $\begin{pmatrix} \dfrac{4}{x} + \dfrac{1}{y} = 11 \\ \dfrac{3}{x} - \dfrac{5}{y} = -9 \end{pmatrix}$

(e) $\begin{pmatrix} \dfrac{5}{x} - \dfrac{2}{y} = 23 \\ \dfrac{4}{x} + \dfrac{3}{y} = \dfrac{23}{2} \end{pmatrix}$ **(f)** $\begin{pmatrix} \dfrac{2}{x} - \dfrac{7}{y} = \dfrac{9}{10} \\ \dfrac{5}{x} + \dfrac{4}{y} = -\dfrac{41}{20} \end{pmatrix}$

63. Solve the following system for x and y.

$$\begin{pmatrix} a_1 x + b_1 y = c_1 \\ a_2 x + b_2 y = c_2 \end{pmatrix}$$

Graphing calculator activities

64. Use a graphing calculator to check your answers for Problem 61.

65. Use a graphing calculator to check your answers for Problem 62.

10.3 Systems of Three Linear Equations in Three Variables

Consider a linear equation in three variables x, y, and z, such as $3x - 2y + z = 7$. Any **ordered triple** (x, y, z) that makes the equation a true numerical statement is said to be a solution of the equation. For example, the ordered triple $(2, 1, 3)$ is a solution because $3(2) - 2(1) + 3 = 7$. However, the ordered triple $(5, 2, 4)$ is not a solution because $3(5) - 2(2) + 4 \neq 7$. There are infinitely many solutions in the solution set.

REMARK: The concept of a *linear* equation is generalized to include equations of more than two variables. Thus an equation such as $5x - 2y + 9z = 8$ is called a linear equation in three variables; the equation $5x - 7y + 2z - 11w = 1$ is called a linear equation in four variables; and so on.

To *solve* a system of three linear equations in three variables, such as

$$\begin{pmatrix} 3x - y + 2z = 13 \\ 4x + 2y + 5z = 30 \\ 5x - 3y - z = 3 \end{pmatrix}$$

means to find all of the ordered triples that satisfy all three equations. In other words, the solution set of the system is the intersection of the solution sets of all three equations in the system.

The graph of a linear equation in three variables is a **plane,** not a line. In fact, graphing equations in three variables requires the use of a three-dimensional coordinate system. Thus using a graphing approach to solve systems of three linear equations in three variables is not at all practical. However, a simple graphical analysis does give us some idea of what we can expect as we begin solving such systems.

In general, because each linear equation in three variables produces a plane, a system of three such equations produces three planes. There are various ways in which three planes can be related. For example, they may be mutually parallel, or two of the planes may be parallel and the third one intersect each of the two. (You may want to analyze all of the other possibilities for the three planes!) However, for our purposes at this time, we need to realize that from a solution set viewpoint, a system of three linear equations in three variables produces one of the following possibilities.

1. There is *one ordered triple* that satisfies all three equations. The three planes have a common point of intersection, as indicated in Figure 10.4.

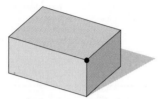

Figure 10.4

2. There are *infinitely many* ordered triples in the solution set, all of which are coordinates of points on a line common to the planes. This can happen if three planes have a common line of intersection (Figure 10.5a) or if two of the planes coincide and the third plane intersects them (Figure 10.5b).

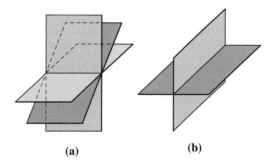

(a)   (b)

Figure 10.5

3. There are *infinitely many* ordered triples in the solution set, all of which are coordinates of points on a plane. This happens if the three planes coincide, as illustrated in Figure 10.6.

Figure 10.6

4. The solution set is *empty;* it is $\emptyset$. This can happen in various ways, as we see in Figure 10.7. Note that in each situation there are no points common to all three planes.

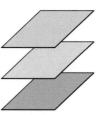

(a) Three parallel planes

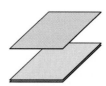

(b) Two planes coincide and the third one is parallel to the coinciding planes.

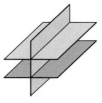

(c) Two planes are parallel and the third intersects them in parallel lines.

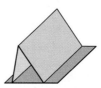

(d) No two planes are parallel, but two of them intersect in a line that is parallel to the third plane.

Figure 10.7

Now that we know what possibilities exist, let's consider finding the solution sets for some systems. Our approach will be the elimination-by-addition method, whereby we replace systems with equivalent systems until we obtain a system where we can easily determine the solution set. Let's start with an example that allows us to determine the solution set without changing to another, equivalent system.

EXAMPLE 1   Solve the system $\begin{pmatrix} 2x - 3y + 5z = -5 \\ 2y - 3z = 4 \\ 4z = -8 \end{pmatrix}.$

          (1)
          (2)
          (3)

Solution

From equation (3) we can find the value of z.

$$4z = -8$$

$$z = -2$$

Now we can substitute -2 for z in equation (2).

$$2y - 3z = 4$$

$$2y - 3(-2) = 4$$

$$2y + 6 = 4$$

$$2y = -2$$

$$y = -1$$

Finally, we can substitute -2 for z and -1 for y in equation (1).

$$2x - 3y + 5z = -5$$

$$2x - 3(-1) + 5(-2) = -5$$

$$2x + 3 - 10 = -5$$

$$2x - 7 = -5$$

$$2x = 2$$

$$x = 1$$

The solution set is $\{(1, -1, -2)\}$.   ■

Note the format of the equations in the system of Example 1. The first equation contains all three variables, the second equation has only two variables, and the third equation has only one variable. This allowed us to solve the third equation and then to use "back-substitution" to find the values of the other variables. Now let's consider an example where we have to make one replacement of an equivalent system.

E X A M P L E   2   Solve the system $\begin{pmatrix} 3x + 2y - 7z = -34 \\ y + 5z = 21 \\ 3y - 2z = -22 \end{pmatrix}$.

    (1)
    (2)
    (3)

Solution

Let's replace equation (3) with an equation we form by multiplying equation (2) by -3 and then adding that result to equation (3).

$$\begin{pmatrix} 3x + 2y - 7z = -34 \\ y + 5z = 21 \\ -17z = -85 \end{pmatrix}.$$

    (4)
    (5)
    (6)

From equation (6), we can find the value of z.

$$-17z = -85$$

$$z = 5$$

Now we can substitute 5 for z in equation (5).

$$y + 5z = 21$$

$$y + 5(5) = 21$$

$$y = -4$$

Finally, we can substitute 5 for z and -4 for y in equation (4).

$$3x + 2y - 7z = -34$$

$$3x + 2(-4) - 7(5) = -34$$

$$3x - 8 - 35 = -34$$

$$3x - 43 = -34$$

$$3x = 9$$

$$x = 3$$

The solution set is $\{(3, -4, 5)\}$. ∎

Now let's consider some examples where we have to make more than one re-placement of equivalent systems.

E X A M P L E   3   Solve the system $\begin{pmatrix} x - y + 4z = -29 \\ 3x - 2y - z = -6 \\ 2x - 5y + 6z = -55 \end{pmatrix}$.   **(1)**
 (2)
 (3)

Solution

Let's replace equation (2) with an equation we form by multiplying equation (1) by -3 and then adding that result to equation (2). Let's also replace equation (3) with an equation we form by multiplying equation (1) by -2 and then adding that result to equation (3).

$$\begin{pmatrix} x - y + 4z = -29 \\ y - 13z = 81 \\ -3y - 2z = 3 \end{pmatrix}$$ **(4)**
 (5)
 (6)

Now let's replace equation (6) with an equation we form by multiplying equation (5) by 3 and then adding that result to equation (6).

$$\begin{pmatrix} x - y + 4z = -29 \\ y - 13z = 81 \\ -41z = 246 \end{pmatrix}$$ **(7)**
 (8)
 (9)

From equation (9) we can determine the value of z.

$$-41z = 246$$

$$z = -6$$

Now we can substitute -6 for z in equation (8).

$$y - 13z = 81$$

$$y - 13(-6) = 81$$

$$y + 78 = 81$$

$$y = 3$$

Finally, we can substitute -6 for z and 3 for y in equation (7).

$$x - y + 4z = -29$$

$$x - 3 + 4(-6) = -29$$

$$x - 3 - 24 = -29$$

$$x - 27 = -29$$

$$x = -2$$

The solution set is $\{(-2, 3, -6)\}$. ∎

EXAMPLE 4 Solve the system $\begin{pmatrix} 3x - 4y + z = 14 \\ 5x + 3y - 2z = 27 \\ 7x - 9y + 4z = 31 \end{pmatrix}$.

(1)
(2)
(3)

Solution

A glance at the coefficients in the system indicates that eliminating the z terms from equations (2) and (3) would be easy. Let's replace equation (2) with an equation we form by multiplying equation (1) by 2 and then adding that result to equation (2). Let's also replace equation (3) with an equation we form by multiplying equation (1) by -4 and then adding that result to equation (3).

$$\begin{pmatrix} 3x - 4y + z = 14 \\ 11x - 5y = 55 \\ -5x + 7y = -25 \end{pmatrix}$$

(4)
(5)
(6)

Now let's eliminate the y terms from equations (5) and (6). First let's multiply equation (6) by 5.

$$\begin{pmatrix} 3x - 4y + z = 14 \\ 11x - 5y = 55 \\ -25x + 35y = -125 \end{pmatrix}$$

(7)
(8)
(9)

Now we can replace equation (9) with an equation we form by multiplying equation (8) by 7 and then adding that result to equation (9).

$$\begin{pmatrix} 3x - 4y + z = 14 \\ 11x - 5y = 55 \\ 52x = 260 \end{pmatrix}$$

(10)
(11)
(12)

From equation (12), we can determine the value of x.

$$52x = 260$$

$$x = 5$$

Now we can substitute 5 for x in equation (11).

$$11x - 5y = 55$$

$$11(5) - 5y = 55$$

$$-5y = 0$$

$$y = 0$$

Finally, we can substitute 5 for x and 0 for y in equation (10).

$$3x - 4y + z = 14$$

$$3(5) - 4(0) + z = 14$$

$$15 - 0 + z = 14$$

$$z = -1$$

The solution set is $\{(5, 0, -1)\}$. ■

E X A M P L E   5

Solve the system $\begin{pmatrix} x - 2y + 3z = 1 \\ 3x - 5y - 2z = 4 \\ 2x - 4y + 6z = 7 \end{pmatrix}$.

    (1)
    (2)
    (3)

Solution

A glance at the coefficients indicates that it should be easy to eliminate the x terms from equations (2) and (3). We can replace equation (2) with an equation we form by multiplying equation (1) by -3 and then adding that result to equation (2). Likewise, we can replace equation (3) with an equation we form by multiplying equation (1) by -2 and then adding that result to equation (3).

$$\begin{pmatrix} x - 2y + 3z = 1 \\ y - 11z = 1 \\ 0 + 0 + 0 = 5 \end{pmatrix}$$

    (4)
    (5)
    (6)

The false statement, $0 = 5$, indicates that the system is inconsistent and that the solution set is therefore $\varnothing$. (If you were to graph this system, equations (1) and (3) would produce parallel planes, which is the situation depicted back in Figure 10.7c.) ■

E X A M P L E   6

Solve the system $\begin{pmatrix} 2x - y + 4z = 1 \\ 3x + 2y - z = 5 \\ 5x - 6y + 17z = -1 \end{pmatrix}$.

    (1)
    (2)
    (3)

Solution

A glance at the coefficients indicates that it is easy to eliminate the y terms from equations (2) and (3). We can replace equation (2) with an equation we form by multiplying equation (1) by 2 and then adding that result to equation (2). Likewise, we can replace equation (3) with an equation we form by multiplying equation (1) by -6 and then adding that result to equation (3).

$$\begin{pmatrix} 2x - y + 4z = & 1 \\ 7x \quad\;\; + 7z = & 7 \\ -7x \quad\;\; - 7z = & -7 \end{pmatrix}$$

(4)
(5)
(6)

Now let's replace equation (6) with an equation we form by multiplying equation (5) by 1 and then adding that result to equation (6).

$$\begin{pmatrix} 2x - y + 4z = 1 \\ 7x \quad\;\; + 7z = 7 \\ 0 + 0 = 0 \end{pmatrix}$$

(7)
(8)
(9)

The true numerical statement, $0 + 0 = 0$, indicates that the system has infinitely many solutions. (The graph of this system is shown in Figure 10.5a). ■

REMARK: It can be shown that the solutions for the system in Example 6 are of the form $(t, 3 - 2t, 1 - t)$, where t is any real number. For example, if we let $t = 2$, then we get the ordered triple $(2, -1, -1)$ and this triple will satisfy all three of the original equations. For our purposes in this text, we shall simply indicate that such a system has infinitely many solutions.

PROBLEM SET 10.3

Solve each of the following systems. If the solution set is $\varnothing$ or if it contains infinitely many solutions, then so indicate.

1. $\begin{pmatrix} x + 2y - 3z = 2 \\ 3y - z = 13 \\ 3y + 5z = 25 \end{pmatrix}$ **2.** $\begin{pmatrix} 2x + 3y - 4z = -10 \\ 2y + 3z = 16 \\ 2y - 5z = -16 \end{pmatrix}$

3. $\begin{pmatrix} 3x + 2y - 2z = 14 \\ x \quad\;\; - 6z = 16 \\ 2x \quad\;\; + 5z = -2 \end{pmatrix}$ **4.** $\begin{pmatrix} 3x + 2y - z = -11 \\ 2x - 3y = -1 \\ 4x + 5y = -13 \end{pmatrix}$

5. $\begin{pmatrix} 2x - y + z = 0 \\ 3x - 2y + 4z = 11 \\ 5x + y - 6z = -32 \end{pmatrix}$ **6.** $\begin{pmatrix} x - 2y + 3z = 7 \\ 2x + y + 5z = 17 \\ 3x - 4y - 2z = 1 \end{pmatrix}$

7. $\begin{pmatrix} 4x - y + z = 5 \\ 3x + y + 2z = 4 \\ x - 2y - z = 1 \end{pmatrix}$ **8.** $\begin{pmatrix} 2x - y + 3z = -14 \\ 4x + 2y - z = 12 \\ 6x - 3y + 4z = -22 \end{pmatrix}$

9. $\begin{pmatrix} x - y + 2z = 4 \\ 2x - 2y + 4z = 7 \\ 3x - 3y + 6z = 1 \end{pmatrix}$ **10.** $\begin{pmatrix} x + y - z = 2 \\ 3x - 4y + 2z = 5 \\ 2x + 2y - 2z = 7 \end{pmatrix}$

11. $\begin{pmatrix} x - 2y + z = -4 \\ 2x + 4y - 3z = -1 \\ -3x - 6y + 7z = 4 \end{pmatrix}$ **12.** $\begin{pmatrix} 2x - y + 3z = 1 \\ 4x + 7y - z = 7 \\ x + 4y - 2z = 3 \end{pmatrix}$

13. $\begin{pmatrix} 3x - 2y + 4z = 6 \\ 9x + 4y - z = 0 \\ 6x - 8y - 3z = 3 \end{pmatrix}$ **14.** $\begin{pmatrix} 2x - y + 3z = 0 \\ 3x + 2y - 4z = 0 \\ 5x - 3y + 2z = 0 \end{pmatrix}$

15. $\begin{pmatrix} 3x - y + 4z = 9 \\ 3x + 2y - 8z = -12 \\ 9x + 5y - 12z = -23 \end{pmatrix}$ **16.** $\begin{pmatrix} 5x - 3y + z = 1 \\ 2x - 5y = -2 \\ 3x - 2y - 4z = -27 \end{pmatrix}$

17. $\begin{pmatrix} 4x - y + 3z = -12 \\ 2x + 3y - z = 8 \\ 6x + y + 2z = -8 \end{pmatrix}$ **18.** $\begin{pmatrix} x + 3y - 2z = 19 \\ 3x - y - z = 7 \\ -2x + 5y + z = 2 \end{pmatrix}$

19. $\begin{pmatrix} x + y + z = 1 \\ 2x - 3y + 6z = 1 \\ -x + y + z = 0 \end{pmatrix}$ **20.** $\begin{pmatrix} 3x + 2y - 2z = -2 \\ x - 3y + 4z = -13 \\ -2x + 5y + 6z = 29 \end{pmatrix}$

Solve each of the following problems by setting up and solving a system of three linear equations in three variables.

21. The sum of the digits of a three-digit number is 14. The number is 14 larger than 20 times the tens digit. The sum of the tens digit and the units digit is 12 larger than the hundreds digit. Find the number.

22. The sum of the digits of a three-digit number is 13. The sum of the hundreds digit and the tens digit is 1 less than the units digit. The sum of three times the hundreds digit and four times the units digit is 26 more than twice the tens digit. Find the number.

23. Two bottles of catsup, 2 jars of peanut butter, and 1 jar of pickles cost $4.20. Three bottles of catsup, 4 jars of peanut butter, and 2 jars of pickles cost $7.70. Four bottles of catsup, 3 jars of peanut butter, and 5 jars of pickles cost $9.80. Find the cost per bottle of catsup and the cost per jar of peanut butter, and the cost per jar of pickles.

24. Five pounds of potatoes, 1 pound of onions, and 2 pounds of apples cost $1.26. Two pounds of potatoes, 3 pounds of onions, and 4 pounds of apples cost $1.88. Three pounds of potatoes, 4 pounds of onions, and 1 pound of apples cost $1.24. Find the price per pound for each item.

25. The sum of three numbers is 20. The sum of the first and third numbers is 2 more than twice the second number. The third number minus the first yields three times the second number. Find the numbers.

26. The sum of three numbers is 40. The third number is 10 less than the sum of the first two numbers. The second number is 1 larger than the first. Find the numbers.

27. The sum of the measures of the angles of a triangle is $180°$. The largest angle is twice the smallest angle. The sum of the smallest and the largest angle is twice the other angle. Find the measure of each angle.

28. A box contains $2 in nickels, dimes, and quarters. There are 19 coins in all, and there are twice as many nickels as dimes. How many coins of each kind are there?

29. Part of $3000 is invested at 12%, another part at 13%, and the remainder at 14%. The total yearly income from the three investments is $400. The sum of the amounts invested at 12% and 13% equals the amount invested at 14%. Determine how much is invested at each rate.

30. The perimeter of a triangle is 45 centimeters. The longest side is 4 centimeters less than twice the shortest side. The sum of the lengths of the shortest and longest sides is 7 centimeters less than three times the length of the remaining side. Find the lengths of all three sides of the triangle.

■ ■ ■ Thoughts into words

31. Give a step-by-step description of how to solve the system

$$\begin{pmatrix} x - 2y + 3z = -23 \\ 5y - 2z = 32 \\ 4z = -24 \end{pmatrix}$$

32. Describe how you would solve the system

$$\begin{pmatrix} x & - 3z = & 4 \\ 3x - 2y + 7z = -1 \\ 2x & + z = & 9 \end{pmatrix}$$

10.4 Matrix Approach to Solving Systems

The primary objective of this chapter is to introduce a variety of techniques for solving systems of linear equations. The techniques we have discussed thus far lend themselves to "small" systems. As the number of equations and variables increases, the systems become more difficult to solve and require other techniques. In these next three sections we will continue to work with small systems for the sake of convenience, but you will learn some techniques that can be extended to larger systems. This section introduces a matrix approach to solving systems.

A **matrix** is simply an array of numbers arranged in horizontal rows and vertical columns. For example, the matrix

$$2 \text{ rows} \longrightarrow \begin{bmatrix} 2 & 1 & -4 \\ 5 & -7 & 6 \end{bmatrix}$$

$$\uparrow \quad \uparrow \quad \uparrow$$

3 columns

has 2 rows and 3 columns, which we refer to as a 2 × 3 (read "two-by-three") matrix. Some additional examples of matrices (*matrices* is the plural of *matrix*) are as follows:

$$3 \times 2 \qquad 2 \times 2 \qquad 1 \times 4 \qquad 5 \times 1$$

$$\begin{bmatrix} 3 & 2 \\ -1 & 4 \\ 5 & 7 \end{bmatrix} \quad \begin{bmatrix} 4 & 1 \\ 0 & -5 \end{bmatrix} \quad [1 \quad 2 \quad 6 \quad 8] \quad \begin{bmatrix} 3 \\ 7 \\ 10 \\ 2 \\ -4 \end{bmatrix}$$

In general, a matrix of m rows and n columns is called a matrix of dimension $m \times n$. With every system of linear equations we can associate a matrix that consists of the coefficients and constant terms. For example, with the system

$$\begin{pmatrix} x - 3y = -17 \\ 2x + 7y = 31 \end{pmatrix}$$

we can associate the matrix

$$\begin{bmatrix} 1 & -3 & \vdots & -17 \\ 2 & 7 & \vdots & 31 \end{bmatrix}$$

which is called the **augmented matrix** of the system. The dashed line separates the coefficients from the constant terms; technically it is not necessary.

Because augmented matrices represent systems of equations, we can operate with them as we do with systems of equations. Our previous work with systems of equations was based on the following properties.

1. Any two equations of a system may be interchanged.

Example When we interchange the two equations, the system $\begin{pmatrix} 2x - 5y = 9 \\ x + 3y = 4 \end{pmatrix}$ is equivalent to the system $\begin{pmatrix} x + 3y = 4 \\ 2x - 5y = 9 \end{pmatrix}$.

2. Any equation of the system may be multiplied by a nonzero constant.

Example When we multiply the top equation by -2, the system $\begin{pmatrix} x + 3y = 4 \\ 2x - 5y = 9 \end{pmatrix}$ is equivalent to the system $\begin{pmatrix} -2x - 6y = -8 \\ 2x - 5y = 9 \end{pmatrix}$.

3. Any equation of the system can be replaced by adding a nonzero multiple of another equation to that equation.

Example   When we add -2 times the first equation to the second equation, the system $\begin{pmatrix} x + 3y = 4 \\ 2x - 5y = 9 \end{pmatrix}$ is equivalent to the system $\begin{pmatrix} x + 3y = 4 \\ -11y = 1 \end{pmatrix}$.

Each of the properties geared to solving a system of equations produces a corresponding property of the augmented matrix of the system. For example, exchanging two equations of a system corresponds to exchanging two rows of the augmented matrix that represents the system. We usually refer to these properties as elementary row operations, and we can state them as follows:

> ### Elementary Row Operations
>
> **1.** Any two rows of an augmented matrix can be interchanged.
> **2.** Any row can be multiplied by a nonzero constant.
> **3.** Any row of the augmented matrix can be replaced by adding a nonzero multiple of another row to that row.

Using the elementary row operations on an augmented matrix provides a basis for solving systems of linear equations. Study the following examples very carefully; keep in mind that the general scheme, called **Gaussian elimination,** is one of using elementary row operations on a matrix to continue replacing a system of equations with an equivalent system until a system is obtained where the solutions are easily determined. We will use a format similar to the one we used in the previous section, except that we will represent systems of equations by matrices.

EXAMPLE 1   Solve the system $\begin{pmatrix} x - 3y = -17 \\ 2x + 7y = 31 \end{pmatrix}$.

Solution

The augmented matrix of the system is

$$\begin{bmatrix} 1 & -3 & \vdots & -17 \\ 2 & 7 & \vdots & 31 \end{bmatrix}$$

We can multiply row one by -2 and add this result to row two to produce a new row two.

$$\begin{bmatrix} 1 & -3 & \vdots & -17 \\ 0 & 13 & \vdots & 65 \end{bmatrix}$$

This matrix represents the system

$$\begin{pmatrix} x - 3y = -17 \\ 13y = 65 \end{pmatrix}$$

From the last equation we can determine the value of y.

$$13y = 65$$

$$y = 5$$

Now we can substitute 5 for y in the equation $x - 3y = -17$.

$$x - 3(5) = -17$$

$$x - 15 = -17$$

$$x = -2$$

The solution set is $\{(-2, 5)\}$. ■

EXAMPLE 2 Solve the system $\begin{pmatrix} 3x + 2y = \ \ 3 \\ 30x - 6y = 17 \end{pmatrix}$.

Solution

The augmented matrix of the system is

$$\begin{bmatrix} 3 & 2 & \vdots & 3 \\ 30 & -6 & \vdots & 17 \end{bmatrix}$$

We can multiply row one by -10 and add this result to row two to produce a new row two.

$$\begin{bmatrix} 3 & 2 & \vdots & 3 \\ 0 & -26 & \vdots & -13 \end{bmatrix}$$

This matrix represents the system

$$\begin{pmatrix} 3x + 2y = \ \ \ 3 \\ -26y = \ -13 \end{pmatrix}$$

From the last equation we can determine the value of y.

$$-26y = -13$$

$$y = \frac{-13}{-26} = \frac{1}{2}$$

Now we can substitute $\frac{1}{2}$ for y in the equation $3x + 2y = 3$.

$$3x + 2\left(\frac{1}{2}\right) = 3$$

$$3x + 1 = 3$$

$$3x = 2$$

$$x = \frac{2}{3}$$

The solution set is $\left\{\left(\frac{2}{3}, \frac{1}{2}\right)\right\}$. ■

EXAMPLE 3   Solve the system $\begin{pmatrix} 2x - 3y - z = -2 \\ x - 2y + 3z = 9 \\ 3x + y - 5z = -8 \end{pmatrix}$.

Solution

The augmented matrix of the system is

$$\begin{bmatrix} 2 & -3 & -1 & \vdots & -2 \\ 1 & -2 & 3 & \vdots & 9 \\ 3 & 1 & -5 & \vdots & -8 \end{bmatrix}$$

Let's begin by interchanging the top two rows.

$$\begin{bmatrix} 1 & -2 & 3 & \vdots & 9 \\ 2 & -3 & -1 & \vdots & -2 \\ 3 & 1 & -5 & \vdots & -8 \end{bmatrix}$$

Now we can multiply row one by -2 and add this result to row two to produce a new row two. Also, we can multiply row one by -3 and add this result to row three to produce a new row three.

$$\begin{bmatrix} 1 & -2 & 3 & \vdots & 9 \\ 0 & 1 & -7 & \vdots & -20 \\ 0 & 7 & -14 & \vdots & -35 \end{bmatrix}$$

Now we can multiply row two by -7 and add this result to row three to produce a new row three.

$$\begin{bmatrix} 1 & -2 & 3 & \vdots & 9 \\ 0 & 1 & -7 & \vdots & -20 \\ 0 & 0 & 35 & \vdots & 105 \end{bmatrix}$$

This last matrix represents the system $\begin{pmatrix} x - 2y + 3z = 9 \\ y - 7z = -20 \\ 35z = 105 \end{pmatrix}$, which is said to be in **triangular form.** We can use the third equation to determine the value of z.

$$35z = 105$$

$$z = 3$$

Now we can substitute 3 for z in the second equation.

$$y - 7z = -20$$

$$y - 7(3) = -20$$

$$y - 21 = -20$$

$$y = 1$$

Finally, we can substitute 3 for z and 1 for y in the first equation.

$$x - 2y + 3z = 9$$

$$x - 2(1) + 3(3) = 9$$

$$x - 2 + 9 = 9$$

$$x + 7 = 9$$

$$x = 2$$

The solution set is $\{(2, 1, 3)\}$. ∎

At this time it might be very helpful for you to look back at Example 3 of Section 10.3 and then to take another look at Example 3 of this section. Note that our approach to both problems is basically the same, except that in this section we are using matrices to represent the systems of equations.

PROBLEM SET 10.4

Solve each of the following systems and use matrices as we did in the examples of this section.

1. $\begin{pmatrix} x - 2y = 14 \\ 4x + 5y = 4 \end{pmatrix}$

2. $\begin{pmatrix} x + 5y = -3 \\ 3x - 2y = -26 \end{pmatrix}$

3. $\begin{pmatrix} 3x + 7y = -40 \\ x + 4y = -20 \end{pmatrix}$

4. $\begin{pmatrix} 7x - 9y = 53 \\ x - 3y = 11 \end{pmatrix}$

5. $\begin{pmatrix} x - 3y = 4 \\ 4x - 5y = 3 \end{pmatrix}$

6. $\begin{pmatrix} x + 3y = 7 \\ 2x - 4y = 9 \end{pmatrix}$

7. $\begin{pmatrix} 6x + 7y = -15 \\ 4x - 9y = 31 \end{pmatrix}$

8. $\begin{pmatrix} 5x - 3y = -16 \\ 6x + 5y = -2 \end{pmatrix}$

9. $\begin{pmatrix} x + 3y - 4z = 5 \\ -2x - 5y + z = 9 \\ 7x - y - z = -2 \end{pmatrix}$

10. $\begin{pmatrix} x - y + 5z = -2 \\ -3x + 2y + z = 17 \\ 4x - 5y - 3z = -36 \end{pmatrix}$

11. $\begin{pmatrix} x - 2y - 3z = -11 \\ 2x - 3y + z = 7 \\ -3x - 5y + 7z = 14 \end{pmatrix}$

12. $\begin{pmatrix} x + y + 3z = -8 \\ 3x + 2y - 5z = 19 \\ 5x - y - 4z = 23 \end{pmatrix}$

13. $\begin{pmatrix} y + 3z = -3 \\ 2x - 5z = 18 \\ 3x - y + 2z = 5 \end{pmatrix}$

14. $\begin{pmatrix} x - z = -1 \\ -2x + y + 3z = 4 \\ 3x - 4y = 31 \end{pmatrix}$

15. $\begin{pmatrix} -x - 5y + 2z = -5 \\ 3x + 14y - z = 13 \\ 4x - 3y + 5z = -26 \end{pmatrix}$

16. $\begin{pmatrix} -x - 3y + 4z = -3 \\ 3x + 8y - z = 27 \\ 5x - y + 2z = -5 \end{pmatrix}$

17. $\begin{pmatrix} x + 2y - z = -5 \\ 3x + 4y + 2z = -8 \\ -2x - y + 5z = 10 \end{pmatrix}$

18. $\begin{pmatrix} x - 3y + 2z = 0 \\ 2x - 4y - 3z = 19 \\ -3x - y + z = -11 \end{pmatrix}$

19. $\begin{pmatrix} -3x + 2y - z = 12 \\ 5x + 2y - 3z = 6 \\ x - y + 5z = -10 \end{pmatrix}$

20. $\begin{pmatrix} -2x - 3y + 5z = 15 \\ 4x - y + 2z = -4 \\ x + y - 3z = -7 \end{pmatrix}$

21. $\begin{pmatrix} -2x + 5y - z = -1 \\ 4x + y - 5z = 23 \\ x - 2y + 3z = -7 \end{pmatrix}$

22. $\begin{pmatrix} 2x + 5y + z = 1 \\ x + 2y - 3z = -13 \\ 3x - y - 2z = -4 \end{pmatrix}$

■ ■ ■ Thoughts into words

23. What is a matrix? What is an augmented matrix of a system of linear equations?

24. Describe how to use matrices to solve the system $\begin{pmatrix} x - 2y = 5 \\ 2x + 7y = 9 \end{pmatrix}$.

■ ■ ■ Further investigations

25. Solve the system $\begin{pmatrix} x - 3y - 2z + w = -3 \\ -2x + 7y + z - 2w = -1 \\ 3x - 7y - 3z + 3w = -5 \\ 5x + y + 4z - 2w = 18 \end{pmatrix}$.

26. Solve the system $\begin{pmatrix} x - 2y + 2z - w = -2 \\ -3x + 5y - z - 3w = 2 \\ 2x + 3y + 3z + 5w = -9 \\ 4x - y - z - 2w = 8 \end{pmatrix}$.

27. Suppose that the augmented matrix of a system of three equations in three variables can be changed to the following matrix.

$$\begin{bmatrix} 1 & 1 & -2 & \vdots & 4 \\ 0 & -5 & 11 & \vdots & -13 \\ 0 & 0 & 0 & \vdots & -9 \end{bmatrix}$$

What can be said about the solution set of the system?

28. Suppose that the augmented matrix of a system of three linear equations in three variables can be changed to the following matrix.

$$\begin{bmatrix} 1 & 0 & 1 & \vdots & 1 \\ 0 & 1 & -1 & \vdots & 0 \\ 0 & 0 & 0 & \vdots & 0 \end{bmatrix}$$

What can be said about the solution set of the system?

Graphing calculator activities

29. If your graphing calculator has the capability to manipulate matrices, this is a good time to become familiar with those operations. You may need to refer to your calculator manual for the specific instructions. To begin the familiarization process, load your calculator with the three augmented matrices in Examples 1, 2, and 3 of this section. Then, for each one, carry out the row operations as described in the text.

10.5 Determinants

A **square matrix** is one that has the same number of rows as columns. Associated with each square matrix that has real number entries is a real number called the **determinant** of the matrix. For a 2×2 matrix

$$\begin{bmatrix} a_1 & b_1 \\ a_2 & b_2 \end{bmatrix}$$

the determinant is written as

$$\begin{vmatrix} a_1 & b_1 \\ a_2 & b_2 \end{vmatrix}$$

and defined by

$$\begin{vmatrix} a_1 & b_1 \\ a_2 & b_2 \end{vmatrix} = a_1b_2 - a_2b_1 \tag{1}$$

Note that a determinant is simply a number and that the determinant notation used on the left side of equation (1) is a way of expressing the number on the right side.

EXAMPLE 1   Find the determinant of the matrix $\begin{bmatrix} 3 & -2 \\ 5 & 8 \end{bmatrix}$.

Solution

In this case, $a_1 = 3$, $b_1 = -2$, $a_2 = 5$, and $b_2 = 8$. Thus we have

$$\begin{vmatrix} 3 & -2 \\ 5 & 8 \end{vmatrix} = 3(8) - 5(-2) = 24 + 10 = 34 \qquad \blacksquare$$

Finding the determinant of a square matrix is commonly called *evaluating the determinant,* and the matrix notation is sometimes omitted.

EXAMPLE 2   Evaluate $\begin{vmatrix} -3 & 5 \\ 1 & 2 \end{vmatrix}$.

Solution

$$\begin{vmatrix} -3 & 5 \\ 1 & 2 \end{vmatrix} = -3(2) - 1(5) = -11 \qquad \blacksquare$$

Cramer's Rule

Determinants provide the basis for another method of solving linear systems. Consider the system

$$\begin{pmatrix} a_1x + b_1y = c_1 \\ a_2x + b_2y = c_2 \end{pmatrix} \qquad\qquad \begin{matrix} (1) \\ (2) \end{matrix}$$

We shall solve this system by using the elimination method; observe that our solutions can be conveniently written in determinant form. To solve for x, we can multiply equation (1) by b_2 and equation (2) by $-b_1$ and then add.

$$
\begin{aligned}
a_1b_2x + b_1b_2y &= c_1b_2 \\
-a_2b_1x - b_1b_2y &= -c_2b_1 \\
\hline
a_1b_2x - a_2b_1x &= c_1b_2 - c_2b_1 \\
(a_1b_2 - a_2b_1)x &= c_1b_2 - c_2b_1 \\
x &= \frac{c_1b_2 - c_2b_1}{a_1b_2 - a_2b_1} \qquad \text{If } a_1b_2 - a_2b_1 \neq 0
\end{aligned}
$$

To solve for y, we can multiply equation (1) by $-a_2$ and equation (2) by a_1 and add.

$$-a_1a_2x - a_2b_1y = -a_2c_1$$
$$\underline{a_1a_2x + a_1b_2y = a_1c_2}$$
$$a_1b_2y - a_2b_1y = a_1c_2 - a_2c_1$$
$$(a_1b_2 - a_2b_1)y = a_1c_2 - a_2c_1$$
$$y = \frac{a_1c_2 - a_2c_1}{a_1b_2 - a_2b_1} \qquad \text{If } a_1b_2 - a_2b_1 \neq 0$$

We can express the solutions for x and y in determinant form as follows:

$$x = \frac{c_1b_2 - c_2b_1}{a_1b_2 - a_2b_1} = \frac{\begin{vmatrix} c_1 & b_1 \\ c_2 & b_2 \end{vmatrix}}{\begin{vmatrix} a_1 & b_1 \\ a_2 & b_2 \end{vmatrix}} \qquad y = \frac{a_1c_2 - a_2c_1}{a_1b_2 - a_2b_1} = \frac{\begin{vmatrix} a_1 & c_1 \\ a_2 & c_2 \end{vmatrix}}{\begin{vmatrix} a_1 & b_1 \\ a_2 & b_2 \end{vmatrix}}$$

For convenience, we shall denote the three determinants in the solution as

$$\begin{vmatrix} a_1 & b_1 \\ a_2 & b_2 \end{vmatrix} = D \qquad \begin{vmatrix} c_1 & b_1 \\ c_2 & b_2 \end{vmatrix} = D_x \qquad \begin{vmatrix} a_1 & c_1 \\ a_2 & c_2 \end{vmatrix} = D_y.$$

Note that the elements of D are the coefficients of the variables in the given system. In D_x, we obtain the elements by replacing the coefficients of x with the respective constants. In D_y, we replace the coefficients of y with the respective constants. This method of using determinants to solve a system of two linear equations in two variables is called **Cramer's rule.** We state it as follows:

Cramer's Rule

Given the system

$$\begin{pmatrix} a_1x + b_1y = c_1 \\ a_2x + b_2y = c_2 \end{pmatrix} \quad \text{with } a_1b_2 - a_2b_1 \neq 0$$

then

$$x = \frac{\begin{vmatrix} c_1 & b_1 \\ c_2 & b_2 \end{vmatrix}}{\begin{vmatrix} a_1 & b_1 \\ a_2 & b_2 \end{vmatrix}} = \frac{D_x}{D} \quad \text{and} \quad y = \frac{\begin{vmatrix} a_1 & c_1 \\ a_2 & c_2 \end{vmatrix}}{\begin{vmatrix} a_1 & b_1 \\ a_2 & b_2 \end{vmatrix}} = \frac{D_y}{D}$$

Let's use Cramer's rule to solve some systems.

EXAMPLE 3

Solve the system $\begin{pmatrix} x + 2y = 11 \\ 2x - y = 2 \end{pmatrix}$.

Solution

Let's find D, D_x, and D_y.

$$D = \begin{vmatrix} 1 & 2 \\ 2 & -1 \end{vmatrix} = -1 - 4 = -5$$

$$D_x = \begin{vmatrix} 11 & 2 \\ 2 & -1 \end{vmatrix} = -11 - 4 = -15$$

$$D_y = \begin{vmatrix} 1 & 11 \\ 2 & 2 \end{vmatrix} = 2 - 22 = -20$$

Thus we have

$$x = \frac{D_x}{D} = \frac{-15}{-5} = 3$$

$$y = \frac{D_y}{D} = \frac{-20}{-5} = 4$$

The solution set is $\{(3, 4)\}$, which we can verify, as always, by substituting back into the original equations.

REMARK: Note that Cramer's rule has a restriction, $a_1b_2 - a_2b_1 \neq 0$; that is, $D \neq 0$. Thus it is a good idea to find D first. Then if $D = 0$, Cramer's rule does not apply, and you must use one of the other methods to determine whether the solution set is empty or has infinitely many solutions.

EXAMPLE 4

Solve the system $\begin{pmatrix} 2x - 3y = -8 \\ 3x + 5y = 7 \end{pmatrix}$.

Solution

$$D = \begin{vmatrix} 2 & -3 \\ 3 & 5 \end{vmatrix} = 10 - (-9) = 19$$

$$D_x = \begin{vmatrix} -8 & -3 \\ 7 & 5 \end{vmatrix} = -40 - (-21) = -19$$

$$D_y = \begin{vmatrix} 2 & -8 \\ 3 & 7 \end{vmatrix} = 14 - (-24) = 38$$

Thus we obtain

$$x = \frac{D_x}{D} = \frac{-19}{19} = -1 \quad \text{and} \quad y = \frac{D_y}{D} = \frac{38}{19} = 2$$

The solution set is $\{(-1, 2)\}$.

E X A M P L E   5 Solve the system $\left(\begin{matrix} y = -2x - 2 \\ 4x - 5y = 17 \end{matrix} \right)$.

Solution

First, we must change the form of the first equation so that the system fits the form given in Cramer's rule. The equation $y = -2x - 2$ can be written as $2x + y = -2$. The system now becomes

$$\left(\begin{matrix} 2x + y = -2 \\ 4x - 5y = 17 \end{matrix} \right)$$

and we can proceed as before.

$$D = \begin{vmatrix} 2 & 1 \\ 4 & -5 \end{vmatrix} = -10 - 4 = -14$$

$$D_x = \begin{vmatrix} -2 & 1 \\ 17 & -5 \end{vmatrix} = 10 - 17 = -7$$

$$D_y = \begin{vmatrix} 2 & -2 \\ 4 & 17 \end{vmatrix} = 34 - (-8) = 42$$

Thus the solutions are

$$x = \frac{D_x}{D} = \frac{-7}{-14} = \frac{1}{2} \quad \text{and} \quad y = \frac{D_y}{D} = \frac{42}{-14} = -3.$$

The solution set is $\left\{ \left(\frac{1}{2}, -3 \right) \right\}$. ■

PROBLEM SET 10.5

Evaluate each of the following determinants.

1. $\begin{vmatrix} 6 & 2 \\ 4 & 3 \end{vmatrix}$

2. $\begin{vmatrix} 7 & 6 \\ 2 & 5 \end{vmatrix}$

3. $\begin{vmatrix} 4 & 7 \\ 8 & 2 \end{vmatrix}$

4. $\begin{vmatrix} 3 & 9 \\ 6 & 4 \end{vmatrix}$

5. $\begin{vmatrix} -3 & 2 \\ 7 & 5 \end{vmatrix}$

6. $\begin{vmatrix} 5 & 1 \\ 8 & -4 \end{vmatrix}$

7. $\begin{vmatrix} 8 & -3 \\ 6 & 4 \end{vmatrix}$

8. $\begin{vmatrix} 5 & 9 \\ -3 & 6 \end{vmatrix}$

9. $\begin{vmatrix} -3 & 2 \\ 5 & -6 \end{vmatrix}$

10. $\begin{vmatrix} -2 & 4 \\ 9 & -7 \end{vmatrix}$

11. $\begin{vmatrix} 3 & -3 \\ -6 & 8 \end{vmatrix}$

12. $\begin{vmatrix} 6 & -5 \\ -8 & 12 \end{vmatrix}$

13. $\begin{vmatrix} -7 & -2 \\ -2 & 4 \end{vmatrix}$

14. $\begin{vmatrix} 6 & -1 \\ -8 & -3 \end{vmatrix}$

15. $\begin{vmatrix} -2 & -3 \\ -4 & -5 \end{vmatrix}$

16. $\begin{vmatrix} -9 & -7 \\ -6 & -4 \end{vmatrix}$

17. $\begin{vmatrix} \frac{1}{4} & -2 \\ \frac{3}{2} & 8 \end{vmatrix}$

18. $\begin{vmatrix} -\frac{2}{3} & 10 \\ -\frac{1}{2} & 6 \end{vmatrix}$

19. $\begin{vmatrix} \frac{3}{2} & -\frac{1}{2} \\ \frac{1}{2} & -\frac{2}{5} \end{vmatrix}$

20. $\begin{vmatrix} -\frac{1}{4} & \frac{1}{3} \\ \frac{3}{2} & \frac{2}{3} \end{vmatrix}$

Use Cramer's rule to find the solution set for each of the following systems.

21. $\begin{pmatrix} 2x + y = 14 \\ 3x - y = 1 \end{pmatrix}$

22. $\begin{pmatrix} 4x - y = 11 \\ 2x + 3y = 23 \end{pmatrix}$

23. $\begin{pmatrix} -x + 3y = 17 \\ 4x - 5y = -33 \end{pmatrix}$

24. $\begin{pmatrix} 5x + 2y = -15 \\ 7x - 3y = 37 \end{pmatrix}$

25. $\begin{pmatrix} 9x + 5y = -8 \\ 7x - 4y = -22 \end{pmatrix}$

26. $\begin{pmatrix} 8x - 11y = 3 \\ -x + 4y = -3 \end{pmatrix}$

27. $\begin{pmatrix} x + 5y = 4 \\ 3x + 15y = -1 \end{pmatrix}$

28. $\begin{pmatrix} 4x - 7y = 0 \\ 7x + 2y = 0 \end{pmatrix}$

29. $\begin{pmatrix} 6x - y = 0 \\ 5x + 4y = 29 \end{pmatrix}$

30. $\begin{pmatrix} 3x - 4y = 2 \\ 9x - 12y = 6 \end{pmatrix}$

31. $\begin{pmatrix} -4x + 3y = 3 \\ 4x - 6y = -5 \end{pmatrix}$

32. $\begin{pmatrix} x - 2y = -1 \\ x = -6y + 5 \end{pmatrix}$

33. $\begin{pmatrix} 6x - 5y = 1 \\ 4x + 7y = 2 \end{pmatrix}$

34. $\begin{pmatrix} y = 3x + 5 \\ y = 6x + 6 \end{pmatrix}$

35. $\begin{pmatrix} 7x + 2y = -1 \\ y = -x + 2 \end{pmatrix}$

36. $\begin{pmatrix} 9x - y = -2 \\ y = 4 - 8x \end{pmatrix}$

37. $\begin{pmatrix} -\dfrac{2}{3}x + \dfrac{1}{2}y = -7 \\ \dfrac{1}{3}x - \dfrac{3}{2}y = 6 \end{pmatrix}$

38. $\begin{pmatrix} \dfrac{1}{2}x + \dfrac{2}{3}y = -6 \\ \dfrac{1}{4}x - \dfrac{1}{3}y = -1 \end{pmatrix}$

39. $\begin{pmatrix} x + \dfrac{2}{3}y = -6 \\ -\dfrac{1}{4}x + 3y = -8 \end{pmatrix}$

40. $\begin{pmatrix} 3x - \dfrac{1}{2}y = 6 \\ -2x + \dfrac{1}{3}y = -4 \end{pmatrix}$

■ ■ ■ Thoughts into words

41. Explain the difference between a matrix and a determinant.

42. Give a step-by-step description of how you would solve the system $\begin{pmatrix} 3x - 2y = 7 \\ 5x + 9y = 14 \end{pmatrix}$ using determinants.

■ ■ ■ Further investigations

43. Verify each of the following. The variables represent real numbers.

(a) $\begin{vmatrix} a & b \\ a & b \end{vmatrix} = 0$

(b) $\begin{vmatrix} a & a \\ b & b \end{vmatrix} = 0$

(c) $\begin{vmatrix} a & b \\ c & d \end{vmatrix} = -\begin{vmatrix} b & a \\ d & c \end{vmatrix}$

(d) $\begin{vmatrix} a & b \\ c & d \end{vmatrix} = -\begin{vmatrix} c & d \\ a & b \end{vmatrix}$

(e) $k\begin{vmatrix} a & b \\ c & d \end{vmatrix} = \begin{vmatrix} ka & b \\ kc & d \end{vmatrix}$

(f) $k\begin{vmatrix} a & b \\ c & d \end{vmatrix} = \begin{vmatrix} ka & kb \\ c & d \end{vmatrix}$

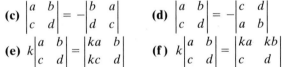 **Graphing calculator activities**

44. Use the determinant function of your graphing calculator to check your answers for Problems 1–16.

45. Make up two or three examples for each part of Problem 43, and evaluate the determinants using your graphing calculator.

10.6 3 × 3 Determinants and Systems of Three Linear Equations in Three Variables

This section will extend the concept of a determinant to include 3 × 3 determinants and then extend the use of determinants to solve systems of three linear equations in three variables.

For a 3 × 3 matrix

$$\begin{bmatrix} a_1 & b_1 & c_1 \\ a_2 & b_2 & c_2 \\ a_3 & b_3 & c_3 \end{bmatrix}$$

the determinant is written as

$$\begin{vmatrix} a_1 & b_1 & c_1 \\ a_2 & b_2 & c_2 \\ a_3 & b_3 & c_3 \end{vmatrix}$$

and defined by

$$\begin{vmatrix} a_1 & b_1 & c_1 \\ a_2 & b_2 & c_2 \\ a_3 & b_3 & c_3 \end{vmatrix} = a_1 b_2 c_3 + b_1 c_2 a_3 + c_1 a_2 b_3 - a_3 b_2 c_1 - b_3 c_2 a_1 - c_3 a_2 b_1 \qquad (1)$$

It is evident that the definition given by equation (1) is a bit complicated to be very useful in practice. Fortunately, there is a method, called **expansion of a determinant by minors,** that we can use to calculate such a determinant.

The **minor** of an element in a determinant is the determinant that remains after deleting the row and column in which the element appears. For example, consider the determinant of equation (1).

The minor of a_1 is $\begin{vmatrix} b_2 & c_2 \\ b_3 & c_3 \end{vmatrix}$.

The minor of a_2 is $\begin{vmatrix} b_1 & c_1 \\ b_3 & c_3 \end{vmatrix}$.

The minor of a_3 is $\begin{vmatrix} b_1 & c_1 \\ b_2 & c_2 \end{vmatrix}$.

Now let's consider the terms, in pairs, of the right side of equation (1) and show the tie-in with minors.

$$a_1 b_2 c_3 - b_3 c_2 a_1 = a_1 (b_2 c_3 - b_3 c_2)$$

$$= a_1 \begin{vmatrix} b_2 & c_2 \\ b_3 & c_3 \end{vmatrix}$$

$$c_1 a_2 b_3 - c_3 a_2 b_1 = -(c_3 a_2 b_1 - c_1 a_2 b_3)$$

$$= -a_2 (b_1 c_3 - b_3 c_1)$$

$$= -a_2 \begin{vmatrix} b_1 & c_1 \\ b_3 & c_3 \end{vmatrix}$$

$$b_1 c_2 a_3 - a_3 b_2 c_1 = a_3 (b_1 c_2 - b_2 c_1)$$

$$= a_3 \begin{vmatrix} b_1 & c_1 \\ b_2 & c_2 \end{vmatrix}$$

Therefore, we have

$$\begin{vmatrix} a_1 & b_1 & c_1 \\ a_2 & b_2 & c_2 \\ a_3 & b_3 & c_3 \end{vmatrix} = a_1 \begin{vmatrix} b_2 & c_2 \\ b_3 & c_3 \end{vmatrix} - a_2 \begin{vmatrix} b_1 & c_1 \\ b_3 & c_3 \end{vmatrix} + a_3 \begin{vmatrix} b_1 & c_1 \\ b_2 & c_2 \end{vmatrix}$$

and this is called the **expansion of the determinant by minors about the first column.**

E X A M P L E 1

Evaluate $\begin{vmatrix} 1 & 2 & -1 \\ 3 & 1 & -2 \\ 2 & 4 & 3 \end{vmatrix}$ by expanding by minors about the first column.

Solution

$$\begin{vmatrix} 1 & 2 & -1 \\ 3 & 1 & -2 \\ 2 & 4 & 3 \end{vmatrix} = 1 \begin{vmatrix} 1 & -2 \\ 4 & 3 \end{vmatrix} - 3 \begin{vmatrix} 2 & -1 \\ 4 & 3 \end{vmatrix} + 2 \begin{vmatrix} 2 & -1 \\ 1 & -2 \end{vmatrix}$$

$$= 1[3 - (-8)] - 3[6 - (-4)] + 2[(-4 - (-1)]$$

$$= 1(11) - 3(10) + 2(-3) = -25 \qquad \blacksquare$$

It is possible to expand a determinant by minors about any row or any column. To help determine the signs of the terms in the expansions, the following *sign array* is very useful.

$$+ \; - \; +$$
$$- \; + \; -$$
$$+ \; - \; +$$

For example, let's expand the determinant in Example 1 by minors about the second row. The second row in the sign array is $- + -$. Therefore,

$$\begin{vmatrix} 1 & 2 & -1 \\ 3 & 1 & -2 \\ 2 & 4 & 3 \end{vmatrix} = -3 \begin{vmatrix} 2 & -1 \\ 4 & 3 \end{vmatrix} + 1 \begin{vmatrix} 1 & -1 \\ 2 & 3 \end{vmatrix} - (-2) \begin{vmatrix} 1 & 2 \\ 2 & 4 \end{vmatrix}$$

$$= -3[6 - (-4)] + 1[3 - (-2)] + 2(4 - 4)$$

$$= -3(10) + 1(5) + 2(0)$$

$$= -25$$

Your decision as to which row or column to use for expanding a particular determinant by minors may depend on the numbers involved in the determinant. A row or column with one or more zeros is frequently a good choice, as the next example illustrates.

EXAMPLE 2   Evaluate $\begin{vmatrix} 3 & -1 & 4 \\ 5 & 2 & 0 \\ -2 & 6 & 0 \end{vmatrix}$.

Solution

Because the third column has two zeros, we shall expand about it.

$$\begin{vmatrix} 3 & -1 & 4 \\ 5 & 2 & 0 \\ -2 & 6 & 0 \end{vmatrix} = 4\begin{vmatrix} 5 & 2 \\ -2 & 6 \end{vmatrix} - 0\begin{vmatrix} 3 & -1 \\ -2 & 6 \end{vmatrix} + 0\begin{vmatrix} 3 & -1 \\ 5 & 2 \end{vmatrix}$$

$$= 4[30 - (-4)] - 0 + 0 = 136$$

(Note that because of the zeros, there is no need to evaluate the last two minors.)

REMARK 1: The expansion-by-minors method can be extended to determinants of size 4 × 4, 5 × 5, and so on. However, it should be obvious that it becomes increasingly tedious with bigger determinants. Fortunately, the computer handles the calculation of such determinants with a different technique.

REMARK 2: There is another method for evaluating 3 × 3 determinants. This method is demonstrated in Problem 36 of the next problem set. If you choose to use that method, keep in mind that it works *only* for 3 × 3 determinants.

Without showing all of the details, we will simply state that Cramer's rule also applies to solving systems of three linear equations in three variables. It can be stated as follows:

Cramer's Rule

Given the system

$$\begin{pmatrix} a_1x + b_1 y + c_1z = d_1 \\ a_2x + b_2 y + c_2z = d_2 \\ a_3x + b_3 y + c_3z = d_3 \end{pmatrix}$$

with

$$D = \begin{vmatrix} a_1 & b_1 & c_1 \\ a_2 & b_2 & c_2 \\ a_3 & b_3 & c_3 \end{vmatrix} \neq 0 \qquad D_x = \begin{vmatrix} d_1 & b_1 & c_1 \\ d_2 & b_2 & c_2 \\ d_3 & b_3 & c_3 \end{vmatrix}$$

$$D_y = \begin{vmatrix} a_1 & d_1 & c_1 \\ a_2 & d_2 & c_2 \\ a_3 & d_3 & c_3 \end{vmatrix} \qquad D_z = \begin{vmatrix} a_1 & b_1 & d_1 \\ a_2 & b_2 & d_2 \\ a_3 & b_3 & d_3 \end{vmatrix}$$

then $x = \dfrac{D_x}{D}$, $y = \dfrac{D_y}{D}$, and $z = \dfrac{D_z}{D}$.

Note that the elements of D are the coefficients of the variables in the given system. Then D_x, D_y, and D_z are formed by replacing the elements in the x, y, and z columns, respectively, by the constants of the system d_1, d_2, and d_3. Again, note the restriction $D \neq 0$. As before, if $D = 0$, then Cramer's rule does not apply, and you can use the elimination method to determine whether the system has no solution or infinitely many solutions.

E X A M P L E   3

Use Cramer's rule to solve the system $\begin{pmatrix} x - 2y + z = -4 \\ 2x + y - z = 5 \\ 3x + 2y + 4z = 3 \end{pmatrix}$.

Solution

To find D, let's expand about row 1.

$$D = \begin{vmatrix} 1 & -2 & 1 \\ 2 & 1 & -1 \\ 3 & 2 & 4 \end{vmatrix} = 1\begin{vmatrix} 1 & -1 \\ 2 & 4 \end{vmatrix} - (-2)\begin{vmatrix} 2 & -1 \\ 3 & 4 \end{vmatrix} + 1\begin{vmatrix} 2 & 1 \\ 3 & 2 \end{vmatrix}$$

$$= 1[4 - (-2)] + 2[8 - (-3)] + 1(4 - 3)$$

$$= 1(6) + 2(11) + 1(1) = 29$$

To find D_x, let's expand about column 3.

$$D_x = \begin{vmatrix} -4 & -2 & 1 \\ 5 & 1 & -1 \\ 3 & 2 & 4 \end{vmatrix} = 1\begin{vmatrix} 5 & 1 \\ 3 & 2 \end{vmatrix} - (-1)\begin{vmatrix} -4 & -2 \\ 3 & 2 \end{vmatrix} + 4\begin{vmatrix} -4 & -2 \\ 5 & 1 \end{vmatrix}$$

$$= 1(10 - 3) + 1[-8 - (-6)] + 4[-4 - (-10)]$$

$$= 1(7) + 1(-2) + 4(6)$$

$$= 29$$

To find D_y, let's expand about row 1.

$$D_y = \begin{vmatrix} 1 & -4 & 1 \\ 2 & 5 & -1 \\ 3 & 3 & 4 \end{vmatrix} = 1\begin{vmatrix} 5 & -1 \\ 3 & 4 \end{vmatrix} - (-4)\begin{vmatrix} 2 & -1 \\ 3 & 4 \end{vmatrix} + 1\begin{vmatrix} 2 & 5 \\ 3 & 3 \end{vmatrix}$$

$$= 1[20 - (-3)] + 4[8 - (-3)] + 1(6 - 15)$$

$$= 1(23) + 4(11) + 1(-9)$$

$$= 58$$

To find D_z, let's expand about column 1.

$$D_z = \begin{vmatrix} 1 & -2 & -4 \\ 2 & 1 & 5 \\ 3 & 2 & 3 \end{vmatrix} = 1\begin{vmatrix} 1 & 5 \\ 2 & 3 \end{vmatrix} - 2\begin{vmatrix} -2 & -4 \\ 2 & 3 \end{vmatrix} + 3\begin{vmatrix} -2 & -4 \\ 1 & 5 \end{vmatrix}$$

$$= 1(3 - 10) - 2[-6 - (-8)] + 3[-10 - (-4)]$$

$$= 1(-7) - 2(2) + 3(-6)$$

$$= -29$$

Thus

$$x = \frac{D_x}{D} = \frac{29}{29} = 1$$

$$y = \frac{D_y}{D} = \frac{58}{29} = 2$$

$$z = \frac{D_z}{D} = \frac{-29}{29} = -1$$

The solution set is $\{(1, 2, -1)\}$. (Be sure to check it!)

■

EXAMPLE 4   Use Cramer's rule to solve the system $\begin{pmatrix} 2x - y + 3z = -17 \\ 3y + z = 5 \\ x - 2y - z = -3 \end{pmatrix}$.

Solution

To find D, let's expand about column 1.

$$D = \begin{vmatrix} 2 & -1 & 3 \\ 0 & 3 & 1 \\ 1 & -2 & -1 \end{vmatrix} = 2 \begin{vmatrix} 3 & 1 \\ -2 & -1 \end{vmatrix} - 0 \begin{vmatrix} -1 & 3 \\ -2 & -1 \end{vmatrix} + 1 \begin{vmatrix} -1 & 3 \\ 3 & 1 \end{vmatrix}$$

$$= 2[-3 - (-2)] - 0 + 1(-1 - 9)$$

$$= 2(-1) - 0 - 10 = -12$$

To find D_x, let's expand about column 3.

$$D_x = \begin{vmatrix} -17 & -1 & 3 \\ 5 & 3 & 1 \\ -3 & -2 & -1 \end{vmatrix} = 3 \begin{vmatrix} 5 & 3 \\ -3 & -2 \end{vmatrix} - 1 \begin{vmatrix} -17 & -1 \\ -3 & -2 \end{vmatrix} + (-1) \begin{vmatrix} -17 & -1 \\ 5 & 3 \end{vmatrix}$$

$$= 3[-10 - (-9)] - 1(34 - 3) - 1[-51 - (-5)]$$

$$= 3(-1) - 1(31) - 1(-46) = 12$$

To find D_y, let's expand about column 1.

$$D_y = \begin{vmatrix} 2 & -17 & 3 \\ 0 & 5 & 1 \\ 1 & -3 & -1 \end{vmatrix} = 2 \begin{vmatrix} 5 & 1 \\ -3 & -1 \end{vmatrix} - 0 \begin{vmatrix} -17 & 3 \\ -3 & -1 \end{vmatrix} + 1 \begin{vmatrix} -17 & 3 \\ 5 & 1 \end{vmatrix}$$

$$= 2[-5 - (-3)] - 0 + 1(-17 - 15)$$

$$= 2(-2) - 0 + 1(-32) = -36$$

To find D_z, let's expand about column 1.

$$D_z = \begin{vmatrix} 2 & -1 & -17 \\ 0 & 3 & 5 \\ 1 & -2 & -3 \end{vmatrix} = 2 \begin{vmatrix} 3 & 5 \\ -2 & -3 \end{vmatrix} - 0 \begin{vmatrix} -1 & -17 \\ -2 & -3 \end{vmatrix} + 1 \begin{vmatrix} -1 & -17 \\ 3 & 5 \end{vmatrix}$$

$$= 2[-9 - (-10)] - 0 + 1[-5 - (-51)]$$

$$= 2(1) - 0 + 1(46) = 48$$

Thus

$$x = \frac{D_x}{D} = \frac{12}{-12} = -1 \qquad y = \frac{D_y}{D} = \frac{-36}{-12} = 3$$

$$z = \frac{D_z}{D} = \frac{48}{-12} = -4.$$

The solution set is $\{(-1, 3, -4)\}$.

PROBLEM SET 10.6

For Problems 1–10, use expansion by minors to evaluate each determinant.

1. $\begin{vmatrix} 2 & 7 & 5 \\ 1 & -1 & 1 \\ -4 & 3 & 2 \end{vmatrix}$

2. $\begin{vmatrix} 2 & 4 & 1 \\ -1 & 5 & 1 \\ -3 & 6 & 2 \end{vmatrix}$

3. $\begin{vmatrix} 3 & -2 & 1 \\ 2 & 1 & 4 \\ -1 & 3 & 5 \end{vmatrix}$

4. $\begin{vmatrix} 1 & -1 & 2 \\ 2 & 1 & 3 \\ -1 & -2 & 1 \end{vmatrix}$

5. $\begin{vmatrix} -3 & -2 & 1 \\ 5 & 0 & 6 \\ 2 & 1 & -4 \end{vmatrix}$

6. $\begin{vmatrix} -5 & 1 & -1 \\ 3 & 4 & 2 \\ 0 & 2 & -3 \end{vmatrix}$

7. $\begin{vmatrix} 3 & -4 & -2 \\ 5 & -2 & 1 \\ 1 & 0 & 0 \end{vmatrix}$

8. $\begin{vmatrix} -6 & 5 & 3 \\ 2 & 0 & -1 \\ 4 & 0 & 7 \end{vmatrix}$

9. $\begin{vmatrix} 4 & -2 & 7 \\ 1 & -1 & 6 \\ 3 & 5 & -2 \end{vmatrix}$

10. $\begin{vmatrix} -5 & 2 & 6 \\ 1 & -1 & 3 \\ 4 & -2 & -4 \end{vmatrix}$

For Problems 11–30, use Cramer's rule to find the solution set of each system.

11. $\begin{pmatrix} 2x - y + 3z = -10 \\ x + 2y - 3z = 2 \\ 3x - 2y + 5z = -16 \end{pmatrix}$

12. $\begin{pmatrix} -x + y - z = 1 \\ 2x + 3y - 4z = 10 \\ -3x - y + z = -5 \end{pmatrix}$

13. $\begin{pmatrix} x - y + 2z = -8 \\ 2x + 3y - 4z = 18 \\ -x + 2y - z = 7 \end{pmatrix}$

14. $\begin{pmatrix} x - 2y + z = 3 \\ 3x + 2y + z = -3 \\ 2x - 3y - 3z = -5 \end{pmatrix}$

15. $\begin{pmatrix} 3x - 2y - 3z = -5 \\ x + 2y + 3z = -3 \\ -x + 4y - 6z = 8 \end{pmatrix}$

16. $\begin{pmatrix} 2x - 3y + 3z = -3 \\ -2x + 5y - 3z = 5 \\ 3x - y + 6z = -1 \end{pmatrix}$

17. $\begin{pmatrix} -x + y + z = -1 \\ x - 2y + 5z = -4 \\ 3x + 4y - 6z = -1 \end{pmatrix}$

18. $\begin{pmatrix} x - 2y + 3z = 1 \\ 2x + y + z = 4 \\ 4x - 3y + 7z = 6 \end{pmatrix}$

19. $\begin{pmatrix} x - y + 2z = 4 \\ 3x - 2y + 4z = 6 \\ 2x - 2y + 4z = -1 \end{pmatrix}$

20. $\begin{pmatrix} -x - 2y + z = 8 \\ 3x + y - z = 5 \\ 5x - y + 4z = 33 \end{pmatrix}$

21. $\begin{pmatrix} 2x - y + 3z = -5 \\ 3x + 4y - 2z = -25 \\ -x + z = 6 \end{pmatrix}$

22. $\begin{pmatrix} 3x - 2y + z = 11 \\ 5x + 3y = 17 \\ x + y - 2z = 6 \end{pmatrix}$

23. $\begin{pmatrix} 2y - z = 10 \\ 3x + 4y = 6 \\ x - y + z = -9 \end{pmatrix}$

24. $\begin{pmatrix} 6x - 5y + 2z = 7 \\ 2x + 3y - 4z = -21 \\ 2y + 3z = 10 \end{pmatrix}$

25. $\begin{pmatrix} -2x + 5y - 3z = -1 \\ 2x - 7y + 3z = 1 \\ 4x - y - 6z = -6 \end{pmatrix}$

26. $\begin{pmatrix} 7x - 2y + 3z = -4 \\ 5x + 2y - 3z = 4 \\ -3x - 6y + 12z = -13 \end{pmatrix}$

27. $\begin{pmatrix} -x - y + 5z = 4 \\ x + y - 7z = -6 \\ 2x + 3y + 4z = 13 \end{pmatrix}$

28. $\begin{pmatrix} x + 7y - z = -1 \\ -x - 9y + z = 3 \\ 3x + 4y - 6z = 5 \end{pmatrix}$

29. $\begin{pmatrix} 5x - y + 2z = 10 \\ 7x + 2y - 2z = -4 \\ -3x - y + 4z = 1 \end{pmatrix}$

30. $\begin{pmatrix} 4x - y - 3z = -12 \\ 5x + y + 6z = 4 \\ 6x - y - 3z = -14 \end{pmatrix}$

■ ■ ■ **Thoughts into words**

31. How would you explain the process of evaluating 3 × 3 determinants to a friend who missed class the day it was discussed?

32. Explain how to use determinants to solve the system

$$\begin{pmatrix} x - 2y + z = 1 \\ 2x - y - z = 5 \\ 5x + 3y + 4z = -6 \end{pmatrix}$$

■ ■ ■ **Further investigations**

33. Evaluate the following determinant by expanding about the second column.

$$\begin{vmatrix} a & e & a \\ b & f & b \\ c & g & c \end{vmatrix}$$

Make a conjecture about determinants that contain two identical columns.

34. Show that $\begin{vmatrix} 1 & -1 & 2 \\ 2 & 3 & -1 \\ -1 & 2 & 4 \end{vmatrix} = -\begin{vmatrix} -1 & 1 & 2 \\ 3 & 2 & -1 \\ 2 & -1 & 4 \end{vmatrix}$.

Make a conjecture about the result of interchanging two columns of a determinant.

35. (a) Show that $\begin{vmatrix} 2 & 1 & 2 \\ 4 & -1 & -2 \\ 6 & 3 & 1 \end{vmatrix} = 2\begin{vmatrix} 1 & 1 & 2 \\ 2 & -1 & -2 \\ 3 & 3 & 1 \end{vmatrix}$.

Make a conjecture about the result of factoring a common factor from each element of a column in a determinant.

(b) Use your conjecture from part (a) to help evaluate the following determinant.

$$\begin{vmatrix} 2 & 4 & -1 \\ -3 & -4 & -2 \\ 5 & 4 & 3 \end{vmatrix}$$

36. We can describe another technique for evaluating 3 × 3 determinants as follows: First, let's write the given determinant with its first two columns repeated on the right.

$$\begin{vmatrix} a_1 & b_1 & c_1 \\ a_2 & b_2 & c_2 \\ a_3 & b_3 & c_3 \end{vmatrix} \begin{matrix} a_1 & b_1 \\ a_2 & b_2 \\ a_3 & b_3 \end{matrix}$$

Then we can add the three products shown with + and subtract the three products shown with −.

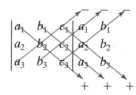

(a) Be sure that the previous description will produce equation (1) on page 529.

(b) Use this technique to do Problems 1–10.

▦ **Graphing calculator activities**

37. Use your graphing calculator to check your answers for Problems 1–10.

38. Return to Problem 43 of Problem Set 10.5. Make up two examples using 3 × 3 determinants for each part of Problem 43. Use your graphing calculator to evaluate the determinants.

10.7   Systems Involving Nonlinear Equations and Systems of Inequalities

Thus far in this chapter, we have solved systems of linear equations. In this section, we shall consider some systems of **linear inequalities** and also some systems where at least one of the equations is *nonlinear*. Let's begin by considering a system of one linear equation and one quadratic equation.

EXAMPLE 1   Solve the system $\left(\begin{array}{l} x^2 + y^2 = 17 \\ x + y = 5 \end{array} \right)$.

Solution

First, let's graph the system so that we can predict approximate solutions. From our previous graphing experiences in Chapters 8 and 9, we should recognize $x^2 + y^2 = 17$ as a circle and $x + y = 5$ as a straight line (Figure 10.8).

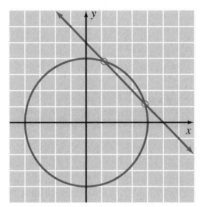

Figure 10.8

The graph indicates that there should be two ordered pairs with positive components (the points of intersection occur in the first quadrant) as solutions for this system. In fact, we could guess that these solutions are (1, 4) and (4, 1), and verify our guess by checking them in the given equations.

Let's also solve the system analytically using the substitution method as follows: Change the form of $x + y = 5$ to $y = 5 - x$ and substitute $5 - x$ for y in the first equation.

$$x^2 + y^2 = 17$$
$$x^2 + (5 - x)^2 = 17$$
$$x^2 + 25 - 10x + x^2 = 17$$
$$2x^2 - 10x + 8 = 0$$
$$x^2 - 5x + 4 = 0$$

$$(x - 4)(x - 1) = 0$$

$$x - 4 = 0 \quad \text{or} \quad x - 1 = 0$$

$$x = 4 \quad \text{or} \quad x = 1$$

Substitute 4 for x and then 1 for x in the second equation of the system to produce

$$x + y = 5 \qquad x + y = 5$$

$$4 + y = 5 \qquad 1 + y = 5$$

$$y = 1 \qquad\quad y = 4$$

Therefore, the solution set is $\{(1, 4), (4, 1)\}$. ■

 E X A M P L E   2 Solve the system $\left(\begin{array}{l} y = -x^2 + 1 \\ y = x^2 - 2 \end{array} \right)$.

Solution

Again, let's get an idea of approximate solutions by graphing the system. Both equations produce parabolas, as indicated in Figure 10.9. From the graph, we can predict two nonintegral ordered-pair solutions, one in the third quadrant and the other in the fourth quadrant.

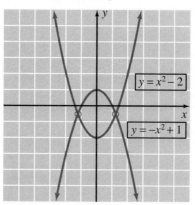

Figure 10.9

Substitute $-x^2 + 1$ for y in the second equation to obtain

$$y = x^2 - 2$$

$$-x^2 + 1 = x^2 - 2$$

$$3 = 2x^2$$

$$\frac{3}{2} = x^2$$

$$\pm\sqrt{\frac{3}{2}} = x$$

$$\pm\frac{\sqrt{6}}{2} = x$$

Substitute $\dfrac{\sqrt{6}}{2}$ for x in the second equation to yield

$$y = x^2 - 2$$

$$y = \left(\frac{\sqrt{6}}{2}\right)^2 - 2$$

$$= \frac{6}{4} - 2$$

$$= -\frac{1}{2}$$

Substitute $-\dfrac{\sqrt{6}}{2}$ for x in the second equation to yield

$$y = x^2 - 2$$

$$y = \left(-\frac{\sqrt{6}}{2}\right)^2 - 2$$

$$= \frac{6}{4} - 2 = -\frac{1}{2}$$

The solution set is $\left\{\left(-\dfrac{\sqrt{6}}{2}, -\dfrac{1}{2}\right), \left(\dfrac{\sqrt{6}}{2}, -\dfrac{1}{2}\right)\right\}$. Check it!   ■

EXAMPLE 3

Solve the system $\left(\begin{array}{l} y = x^2 + 2 \\ 6x - 4y = -5 \end{array}\right)$.

Solution

From previous graphing experiences, we recognize that $y = x^2 + 2$ is the basic parabola shifted upward 2 units and that $6x - 4y = -5$ is a straight line (see Figure 10.10). Because of the close proximity of the curves, it is difficult to tell whether they intersect. In other words, the graph does not definitely indicate any real number solutions for the system.

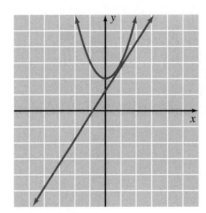

Figure 10.10

Let's solve the system using the substitution method. We can substitute $x^2 + 2$ for y in the second equation, which produces two values for x.

$$6x - 4(x^2 + 2) = -5$$

$$6x - 4x^2 - 8 = -5$$

$$-4x^2 + 6x - 3 = 0$$

$$4x^2 - 6x + 3 = 0$$

$$x = \frac{6 \pm \sqrt{36 - 48}}{8}$$

$$x = \frac{6 \pm \sqrt{-12}}{8}$$

$$x = \frac{6 \pm 2i\sqrt{3}}{8}$$

$$x = \frac{3 \pm i\sqrt{3}}{4}$$

It is now obvious that the system has no real number solutions. That is, the line and the parabola do not intersect in the real number plane. However, there will be two pairs of complex numbers in the solution set. We can substitute $\dfrac{(3 + i\sqrt{3})}{4}$ for x in the first equation.

$$y = \left(\frac{3 + i\sqrt{3}}{4}\right)^2 + 2$$

$$= \frac{6 + 6i\sqrt{3}}{16} + 2$$

$$= \frac{6 + 6i\sqrt{3} + 32}{16}$$

$$= \frac{38 + 6i\sqrt{3}}{16} = \frac{19 + 3i\sqrt{3}}{8}$$

Likewise, we can substitute $\dfrac{(3 - i\sqrt{3})}{4}$ for x in the first equation.

$$y = \left(\frac{3 - i\sqrt{3}}{4}\right)^2 + 2$$

$$= \frac{6 - 6i\sqrt{3}}{16} + 2$$

$$= \frac{6 - 6i\sqrt{3} + 32}{16}$$

$$= \frac{38 - 6i\sqrt{3}}{16}$$

$$= \frac{19 - 3i\sqrt{3}}{8}$$

The solution set is $\left\{\left(\dfrac{3 + i\sqrt{3}}{4}, \dfrac{19 + 3i\sqrt{3}}{4}\right), \left(\dfrac{3 - i\sqrt{3}}{4}, \dfrac{19 - 3i\sqrt{3}}{4}\right)\right\}$. ■

In Example 3, the use of a graphing utility may not, at first, indicate whether or not the system has any real number solutions. Suppose that we graph the system using a viewing rectangle such that $-15 \leq x \leq 15$ and $-10 \leq y \leq 10$. In Figure 10.11, we cannot tell whether or not the line and parabola intersect. However, if we change the viewing rectangle so that $0 \leq x \leq 2$ and $0 \leq y \leq 4$, as shown in Figure 10.12, then it becomes apparent that the two graphs do not intersect.

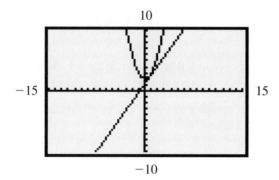

Figure 10.11

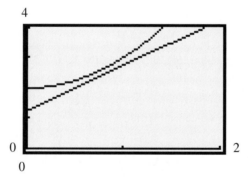

Figure 10.12

Systems of Linear Inequalities

Finding solution sets for systems of linear inequalities relies heavily on the graphing approach. The solution set of a system of linear inequalities, such as

$$\begin{pmatrix} x + y > 2 \\ x - y < 2 \end{pmatrix}$$

is the intersection of the solution sets of the individual inequalities. In Figure 10.13(a) we indicated the solution set for $x + y > 2$, and in Figure 10.13(b) we indicated the solution set for $x - y < 2$. Then, in Figure 10.13(c), we shaded the region that represents the intersection of the two solution sets from parts (a) and (b); thus it is the graph of the system. Remember that dashed lines are used to indicate that the points on the lines are not included in the solution set.

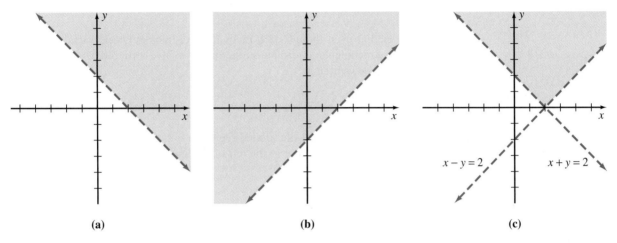

(a)   (b)   (c)

Figure 10.13

In the following examples, we indicated only the final solution set for the system.

E X A M P L E   4   Solve the following system by graphing.

$$\begin{pmatrix} 2x - y \geq 4 \\ x + 2y < 2 \end{pmatrix}$$

Solution

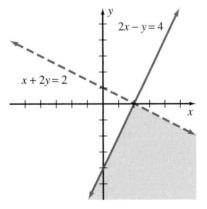

The graph of $2x - y \geq 4$ consists of all points *on or below* the line $2x - y = 4$. The graph of $x + 2y < 2$ consists of all points *below* the line $x + 2y = 2$. The graph of the system is indicated by the shaded region in Figure 10.14. Note that all points in the shaded region are on or below the line $2x - y = 4$ *and* below the line $x + 2y = 2$.

Figure 10.14

EXAMPLE 5   Solve the following system by graphing.

$$\begin{pmatrix} x \le 2 \\ y \ge -1 \end{pmatrix}$$

Solution

Remember that even though each inequality contains only one variable, we are working in a rectangular coordinate system that involves ordered pairs. That is, the system could be written as

$$\begin{pmatrix} x + 0(y) \le 2 \\ 0(x) + y \ge -1 \end{pmatrix}$$

The graph of the system is the shaded region in Figure 10.15. Note that all points in the shaded region are on or to the left of the line $x = 2$ and on or above the line $y = -1$.

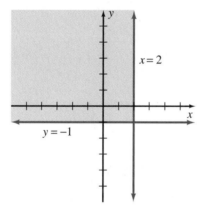

Figure 10.15

PROBLEM SET 10.7

For Problems 1–16, (a) graph each system so that approximate real number solutions (if there are any) can be predicted, and (b) solve each system using the substitution method or the elimination-by-addition method.

1. $\begin{pmatrix} y = (x + 2)^2 \\ y = -2x - 4 \end{pmatrix}$

2. $\begin{pmatrix} y = x^2 \\ y = x + 2 \end{pmatrix}$

3. $\begin{pmatrix} x^2 + y^2 = 13 \\ 3x + 2y = 0 \end{pmatrix}$

4. $\begin{pmatrix} x^2 + y^2 = 26 \\ x + y = 6 \end{pmatrix}$

5. $\begin{pmatrix} y = x^2 + 6x + 7 \\ 2x + y = -5 \end{pmatrix}$

6. $\begin{pmatrix} y = x^2 - 4x + 5 \\ -x + y = 1 \end{pmatrix}$

7. $\begin{pmatrix} y = x^2 \\ y = x^2 - 4x + 4 \end{pmatrix}$

8. $\begin{pmatrix} y = -x^2 + 3 \\ y = x^2 + 1 \end{pmatrix}$

9. $\begin{pmatrix} x + y = -8 \\ x^2 - y^2 = 16 \end{pmatrix}$

10. $\begin{pmatrix} x - y = 2 \\ x^2 - y^2 = 16 \end{pmatrix}$

11. $\begin{pmatrix} y = x^2 + 2x - 1 \\ y = x^2 + 4x + 5 \end{pmatrix}$

12. $\begin{pmatrix} 2x^2 + y^2 = 8 \\ x^2 + y^2 = 4 \end{pmatrix}$

13. $\begin{pmatrix} xy = 4 \\ y = x \end{pmatrix}$

14. $\begin{pmatrix} y = x^2 + 2 \\ y = 2x^2 + 1 \end{pmatrix}$

15. $\begin{pmatrix} x^2 + y^2 = 2 \\ x - y = 4 \end{pmatrix}$

16. $\begin{pmatrix} y = -x^2 + 1 \\ x + y = 2 \end{pmatrix}$

For Problems 17–32, indicate the solution set for each system of inequalities by shading the appropriate region.

17. $\begin{pmatrix} 3x - 4y \geq 0 \\ 2x + 3y \leq 0 \end{pmatrix}$

18. $\begin{pmatrix} 3x + 2y \leq 6 \\ 2x - 3y \geq 6 \end{pmatrix}$

19. $\begin{pmatrix} x - 3y < 6 \\ x + 2y \geq 4 \end{pmatrix}$

20. $\begin{pmatrix} 2x - y \leq 4 \\ 2x + y > 4 \end{pmatrix}$

21. $\begin{pmatrix} x + y < 4 \\ x - y > 2 \end{pmatrix}$

22. $\begin{pmatrix} x + y > 1 \\ x - y < 1 \end{pmatrix}$

23. $\begin{pmatrix} y < x + 1 \\ y \geq x \end{pmatrix}$

24. $\begin{pmatrix} y > x - 3 \\ y < x \end{pmatrix}$

25. $\begin{pmatrix} y > x \\ y > 2 \end{pmatrix}$

26. $\begin{pmatrix} 2x + y > 6 \\ 2x + y < 2 \end{pmatrix}$

27. $\begin{pmatrix} x \geq -1 \\ y < 4 \end{pmatrix}$

28. $\begin{pmatrix} x < 3 \\ y > 2 \end{pmatrix}$

29. $\begin{pmatrix} 2x - y > 4 \\ 2x - y > 0 \end{pmatrix}$

30. $\begin{pmatrix} x + y > 4 \\ x + y > 6 \end{pmatrix}$

31. $\begin{pmatrix} 3x - 2y < 6 \\ 2x - 3y < 6 \end{pmatrix}$

32. $\begin{pmatrix} 2x + 5y > 10 \\ 5x + 2y > 10 \end{pmatrix}$

■ ■ ■ Thoughts into words

33. What happens if you try to graph the system

$$\begin{pmatrix} x^2 + 4y^2 = 16 \\ 2x^2 + 5y^2 = -12 \end{pmatrix}?$$

34. Explain how you would solve the system

$$\begin{pmatrix} x^2 + y^2 = 9 \\ y^2 = x^2 + 4 \end{pmatrix}.$$

35. How do you know by inspection, without graphing, that the solution set of the system $\begin{pmatrix} 3x - 2y > 5 \\ 3x - 2y < 2 \end{pmatrix}$ is the null set?

Graphing calculator activities

36. Use a graphing calculator to graph the systems in Problems 1–16, and check the reasonableness of your answers.

37. For each of the following systems, (a) use your graphing calculator to show that there are no real number solutions, and (b) solve the system by the substitution method or the elimination-by-addition method to find the complex solutions.

(a) $\begin{pmatrix} y = x^2 + 1 \\ y = -3 \end{pmatrix}$

(b) $\begin{pmatrix} y = -x^2 + 1 \\ y = 3 \end{pmatrix}$

(c) $\begin{pmatrix} y = x^2 \\ x - y = 4 \end{pmatrix}$

(d) $\begin{pmatrix} y = x^2 + 1 \\ y = -x^2 \end{pmatrix}$

(e) $\begin{pmatrix} x^2 + y^2 = 1 \\ x + y = 2 \end{pmatrix}$

(f) $\begin{pmatrix} x^2 + y^2 = 2 \\ x^2 - y^2 = 6 \end{pmatrix}$

38. Graph the system $\begin{pmatrix} y = x^2 + 2 \\ 6x - 4y = -5 \end{pmatrix}$ and use the TRACE and ZOOM features of your calculator to demonstrate clearly that this system has no real number solutions.

CHAPTER 10

SUMMARY

(10.1) Graphing a **system of two linear equations in two variables** produces one of the following results.

1. The graphs of the two equations are two intersecting lines, which indicates that there is **one unique solution** of the system. Such a system is called a **consistent system.**

2. The graphs of the two equations are two parallel lines, which indicates that there is **no solution** for the system. It is called an **inconsistent system.**

3. The graphs of the two equations are the same line, which indicates **infinitely many solutions** for the system. The equations are called **dependent** equations.

We can describe the **substitution method** of solving a system of equations as follows:

STEP 1 Solve one of the equations for one variable in terms of the other variable if neither equation is in such a form. (If possible, make a choice that will avoid fractions.)

STEP 2 Substitute the expression obtained in Step 1 into the other equation to produce an equation with one variable.

STEP 3 Solve the equation obtained in Step 2.

STEP 4 Use the solution obtained in Step 3, along with the expression obtained in Step 1, to determine the solution of the system.

(10.2) The **elimination-by-addition method** involves the replacement of a system of equations with equivalent systems until a system is obtained whereby the solutions can be easily determined. The following operations or transformations can be performed on a system to produce an equivalent system.

1. Any two equations of the system can be interchanged.

2. Both sides of any equation of the system can be multiplied by any nonzero real number.

3. Any equation of the system can be replaced by the *sum* of that equation and a nonzero multiple of another equation.

(10.3) Solving **a system of three linear equations in three variables** produces one of the following results.

1. There is **one ordered triple** that satisfies all three equations.

2. There are **infinitely many ordered triples** in the solution set, all of which are coordinates of points on a line common to the planes.

3. There are **infinitely many ordered triples** in the solution set, all of which are coordinates of points on a plane.

4. The solution set is empty; it is $\varnothing$.

(10.4) A **matrix** is an array of numbers arranged in horizontal rows and vertical columns. A matrix of m rows and n columns is called an $m \times n$ ("m-by-n") matrix. The **augmented matrix** of the system

$$\begin{pmatrix} 5x - 2y - z = 4 \\ 3x - y - z = 7 \\ 2x + 3y + 7z = 9 \end{pmatrix}$$

is

$$\begin{bmatrix} 5 & -2 & -1 & \vdots & 4 \\ 3 & -1 & -1 & \vdots & 7 \\ 2 & 3 & 7 & \vdots & 9 \end{bmatrix}$$

The following **elementary row operations** provide the basis for transforming matrices.

1. Any two rows of an augmented matrix can be interchanged.

2. Any row can be multiplied by a nonzero constant.

3. Any row can be replaced by adding a nonzero multiple of another row to that row.

Transforming an augmented matrix to **triangular form** and then using back-substitution provides a systematic technique for solving systems of linear equations.

(10.5) A rectangular array of numbers is called a **matrix.** A **square matrix** has the same number of rows as columns. For a 2×2 matrix,

$$\begin{bmatrix} a_1 & b_1 \\ a_2 & b_2 \end{bmatrix}$$

the **determinant** of the matrix is written as

$$\begin{vmatrix} a_1 & b_1 \\ a_2 & b_2 \end{vmatrix}$$

and defined by

$$\begin{vmatrix} a_1 & b_1 \\ a_2 & b_2 \end{vmatrix} = a_1b_2 - a_2b_1$$

Cramer's rule for solving a system of two linear equations in two variables is stated as follows:

Given the system $\begin{pmatrix} a_1x + b_1y = c_1 \\ a_2x + b_2y = c_2 \end{pmatrix}$ with

$$D = \begin{vmatrix} a_1 & b_1 \\ a_2 & b_2 \end{vmatrix} \ne 0 \quad D_x = \begin{vmatrix} c_1 & b_1 \\ c_2 & b_2 \end{vmatrix} \quad D_y = \begin{vmatrix} a_1 & c_1 \\ a_2 & c_2 \end{vmatrix}$$

then

$$x = \frac{D_x}{D} \quad \text{and} \quad y = \frac{D_y}{D}$$

(10.6) A 3×3 determinant is defined by

$$\begin{vmatrix} a_1 & b_1 & c_1 \\ a_2 & b_2 & c_2 \\ a_3 & b_3 & c_3 \end{vmatrix} = a_1b_2c_3 + b_1c_2a_3 + c_1a_2b_3 \\ - a_3b_2c_1 - b_3c_2a_1 - c_3a_2b_1$$

The **minor** of an element in a determinant is the determinant that remains after deleting the row and column in which the element appears. A determinant can be evaluated by **expansion by minors** of the elements of any row or any column.

Cramer's rule for solving a system of three linear equations in three variables is stated as follows:

Given the system $\begin{pmatrix} a_1x + b_1y + c_1z = d_1 \\ a_2x + b_2y + c_2z = d_2 \\ a_3x + b_3y + c_3z = d_3 \end{pmatrix}$ with

$$D = \begin{vmatrix} a_1 & b_1 & c_1 \\ a_2 & b_2 & c_2 \\ a_3 & b_3 & c_3 \end{vmatrix} \ne 0 \quad D_x = \begin{vmatrix} d_1 & b_1 & c_1 \\ d_2 & b_2 & c_2 \\ d_3 & b_3 & c_3 \end{vmatrix}$$

$$D_y = \begin{vmatrix} a_1 & d_1 & c_1 \\ a_2 & d_2 & c_2 \\ a_3 & d_3 & c_3 \end{vmatrix} \quad D_z = \begin{vmatrix} a_1 & b_1 & d_1 \\ a_2 & b_2 & d_2 \\ a_3 & b_3 & d_3 \end{vmatrix}$$

then $x = \dfrac{D_x}{D}$, $y = \dfrac{D_y}{D}$, and $z = \dfrac{D_z}{D}$.

(10.7) The substitution and elimination methods can also be used to solve systems involving **nonlinear equations.**

The solution set of a system of **linear inequalities** is the intersection of the solution sets of the individual inequalities.

CHAPTER 10 REVIEW PROBLEM SET

For Problems 1–4, solve each system of equations using (a) the substitution method, (b) the elimination method, (c) a matrix approach, and (d) Cramer's rule.

1. $\begin{pmatrix} 3x - 2y = -6 \\ 2x + 5y = 34 \end{pmatrix}$ **2.** $\begin{pmatrix} x + 4y = 25 \\ y = -3x - 2 \end{pmatrix}$

3. $\begin{pmatrix} x = 5y - 49 \\ 4x + 3y = -12 \end{pmatrix}$ **4.** $\begin{pmatrix} x - 6y = 7 \\ 3x + 5y = 9 \end{pmatrix}$

For Problems 5–14, solve each system using the method that seems most appropriate to you.

5. $\begin{pmatrix} x - 3y = 25 \\ -3x + 2y = -26 \end{pmatrix}$ **6.** $\begin{pmatrix} 5x - 7y = -66 \\ x + 4y = 30 \end{pmatrix}$

7. $\begin{pmatrix} 4x + 3y = -9 \\ 3x - 5y = 15 \end{pmatrix}$ **8.** $\begin{pmatrix} 2x + 5y = 47 \\ 4x - 7y = -25 \end{pmatrix}$

9. $\begin{pmatrix} 7x - 3y = 25 \\ y = 3x - 9 \end{pmatrix}$ 10. $\begin{pmatrix} x = -4 - 5y \\ y = 4x + 16 \end{pmatrix}$

11. $\begin{pmatrix} \dfrac{1}{2}x + \dfrac{2}{3}y = 6 \\ \dfrac{3}{4}x - \dfrac{5}{6}y = -24 \end{pmatrix}$ 12. $\begin{pmatrix} \dfrac{3}{4}x - \dfrac{1}{2}y = 14 \\ \dfrac{5}{12}x + \dfrac{3}{4}y = 16 \end{pmatrix}$

13. $\begin{pmatrix} 6x - 4y = 7 \\ 9x + 8y = 0 \end{pmatrix}$ 14. $\begin{pmatrix} 4x - 5y = -5 \\ 6x - 10y = -9 \end{pmatrix}$

For Problems 15–18, evaluate each of the determinants.

15. $\begin{vmatrix} 3 & 5 \\ 6 & 4 \end{vmatrix}$ 16. $\begin{vmatrix} -1 & -5 \\ 4 & 9 \end{vmatrix}$

17. $\begin{vmatrix} 4 & -1 & -3 \\ 2 & 1 & 4 \\ -3 & 2 & 2 \end{vmatrix}$ 18. $\begin{vmatrix} 5 & 3 & 4 \\ -2 & 0 & 1 \\ -1 & -2 & 6 \end{vmatrix}$

For Problems 19 and 20, solve each system of equations using (a) the elimination method, (b) a matrix approach, and (c) Cramer's rule.

19. $\begin{pmatrix} x - 2y + 4z = -14 \\ 3x - 5y + z = 20 \\ -2x + y - 5z = 22 \end{pmatrix}$

20. $\begin{pmatrix} x + 3y - 2z = 28 \\ 2x - 8y + 3z = -63 \\ 3x + 8y - 5z = 72 \end{pmatrix}$

For Problems 21–24, solve each system using the method that seems most appropriate to you.

21. $\begin{pmatrix} x + y - z = -2 \\ 2x - 3y + 4z = 17 \\ -3x + 2y + 5z = -7 \end{pmatrix}$ 22. $\begin{pmatrix} -x - y + z = -3 \\ 3x + 2y - 4z = 12 \\ 5x + y + 2z = 5 \end{pmatrix}$

23. $\begin{pmatrix} 3x + y - z = -6 \\ 3x + 2y + 3z = 9 \\ 6x - 2y + 2z = 8 \end{pmatrix}$ 24. $\begin{pmatrix} x - 3y + z = 2 \\ 2x - 5y - 3z = 22 \\ -4x + 3y + 5z = -26 \end{pmatrix}$

25. Graph the following system, and then find the solution set by using either the substitution or the elimination method.

$$\begin{pmatrix} y = 2x^2 - 1 \\ 2x + y = 3 \end{pmatrix}$$

26. Indicate the solution set of the following system of inequalities by graphing the system and shading the appropriate region.

$$\begin{pmatrix} 3x + y > 6 \\ x - 2y \le 4 \end{pmatrix}$$

For Problems 27–33, solve each problem by setting up and solving a system of two equations and two unknowns.

27. The sum of the squares of two numbers is 13. If one number is 1 larger than the other number, find the numbers.

28. The sum of the squares of two numbers is 34. The difference of the squares of the same two numbers is 16. Find the numbers.

29. A number is 1 larger than the square of another number. The sum of the two numbers is 7. Find the numbers.

30. The area of a rectangular region is 54 square meters and its perimeter is 30 meters. Find the length and width of the rectangle.

31. At a local confectionery, 7 pounds of cashews and 5 pounds of Spanish peanuts cost $88, and 3 pounds of cashews and 2 pounds of Spanish peanuts cost $37. Find the price per pound for cashews and for Spanish peanuts.

32. We bought 2 cartons of pop and 4 pounds of candy for $12. The next day we bought 3 cartons of pop and 2 pounds of candy for $9. Find the price of a carton of pop and also the price of a pound of candy.

33. Suppose that a mail-order company charges a fixed fee for shipping merchandise that weighs 1 pound or less, plus an additional fee for each pound over 1 pound. If the shipping charge for 5 pounds is $2.40 and for 12 pounds is $3.10, find the fixed fee and the additional fee.

CHAPTER 10

TEST

For Problems 1–4, refer to the following systems of equations:

$$\textbf{I. } \begin{pmatrix} 5x - 2y = 12 \\ 2x + 5y = 7 \end{pmatrix} \qquad \textbf{II. } \begin{pmatrix} x - 4y = 1 \\ 2x - 8y = 2 \end{pmatrix}$$

$$\textbf{III. } \begin{pmatrix} 4x - 5y = 6 \\ 4x - 5y = 1 \end{pmatrix} \qquad \textbf{IV. } \begin{pmatrix} 2x + 3y = 9 \\ 7x - 4y = 9 \end{pmatrix}$$

1. For which of these systems are the equations said to be dependent?

2. For which of these systems does the solution set consist of a single ordered pair?

3. For which of these systems are the graphs parallel lines?

4. For which of these systems are the graphs perpendicular lines?

5. Evaluate $\begin{vmatrix} -4 & 3 \\ 2 & 8 \end{vmatrix}$.

6. Evaluate $\begin{vmatrix} 2 & -1 & 1 \\ 3 & 2 & 1 \\ 2 & 4 & -1 \end{vmatrix}$.

7. Use the elimination-by-addition method to solve the system $\begin{pmatrix} 2x - 3y = -17 \\ 5x + y = 17 \end{pmatrix}$.

8. Use the substitution method to solve the system $\begin{pmatrix} -5x + 4y = 35 \\ x - 3y = -18 \end{pmatrix}$.

9. Use Cramer's rule to solve the system $\begin{pmatrix} 6x + 7y = -40 \\ 4x - 3y = 4 \end{pmatrix}$.

10. Use a matrix approach to solve the system $\begin{pmatrix} x - 4y = 15 \\ 3x + 8y = -15 \end{pmatrix}$.

For Problems 11–14, solve each of the systems using the method that seems most appropriate to you.

11. $\begin{pmatrix} 2x - 7y = -8 \\ 4x + 5y = 3 \end{pmatrix}$

12. $\begin{pmatrix} \dfrac{2}{3}x - \dfrac{1}{2}y = 7 \\ \dfrac{1}{4}x + \dfrac{1}{3}y = 12 \end{pmatrix}$

13. $\begin{pmatrix} x - 2y - z = 2 \\ 2x + y + z = 3 \\ 3x - 4y - 2z = 5 \end{pmatrix}$

14. $\begin{pmatrix} x - 2y + z = -8 \\ 2x + y - z = 3 \\ -3x - 4y + 2z = -11 \end{pmatrix}$

15. Find the value of x in the solution for the system $\begin{pmatrix} x = -2y + 5 \\ 7x + 3y = 46 \end{pmatrix}$.

16. Find the value of y in the solution for the system $\begin{pmatrix} x - y + 4z = 25 \\ 3x + 2y - z = 5 \\ 5y + 2z = 5 \end{pmatrix}$.

17. Find the value of z in the solution for the system $\begin{pmatrix} x - y - z = -6 \\ 4x + y + 3z = -4 \\ 5x - 2y - 4z = -18 \end{pmatrix}$.

For Problems 18–20, determine how many ordered pairs of real numbers are in the solution set for each of the systems.

18. $\begin{pmatrix} y = 2x^2 + 1 \\ 3x + 4y = 12 \end{pmatrix}$

19. $\begin{pmatrix} x^2 + 2y^2 = 8 \\ 2x^2 + y^2 = 8 \end{pmatrix}$

20. $\begin{pmatrix} y = x^3 \\ y = (x - 1)^2 - 1 \end{pmatrix}$

21. Solve the system $\begin{pmatrix} y = x^2 - 4x + 7 \\ y = -2x^2 + 8x - 5 \end{pmatrix}$.

22. Graph the solution set for the system $\begin{pmatrix} x + 3y < 3 \\ 2x - y > 2 \end{pmatrix}$.

For Problems 23–25, set up and solve a system of equations to help solve each problem.

23. A box contains $7.80 in nickels, dimes, and quarters. There are 6 more dimes than nickels and three

times as many quarters as nickels. Find the number of quarters.

24. One solution contains 30% alcohol and another solution contains 80% alcohol. Some of each of the two solutions is mixed to produce 5 liters of a 60%-alcohol solution. How many liters of the 80%-alcohol solution are used?

25. The units digit of a two-digit number is 1 more than three times the tens digit. If the digits are reversed, the new number formed is 45 larger than the original number. Find the original number.

CUMULATIVE REVIEW PROBLEM SET *Chapters 1-10*

For Problems 1–5, evaluate each algebraic expression for the given values of the variables.

1. $-5(x - 1) - 3(2x + 4) + 3(3x - 1)$ for $x = -2$

2. $\dfrac{14a^3b^2}{7a^2b}$ for $a = -1$ and $b = 4$

3. $\dfrac{2}{n} - \dfrac{3}{2n} + \dfrac{5}{3n}$ for $n = 4$

4. $-4\sqrt{2x - y} + 5\sqrt{3x + y}$ for $x = 16$ and $y = 16$

5. $\dfrac{3}{x - 2} - \dfrac{5}{x + 3}$ for $x = 3$

For Problems 6–15, perform the indicated operations and express the answers in simplified form.

6. $(-5\sqrt{6})(3\sqrt{12})$

7. $(2\sqrt{x} - 3)(\sqrt{x} + 4)$

8. $(3\sqrt{2} - \sqrt{6})(\sqrt{2} + 4\sqrt{6})$

9. $(2x - 1)(x^2 + 6x - 4)$

10. $\dfrac{x^2 - x}{x + 5} \cdot \dfrac{x^2 + 5x + 4}{x^4 - x^2}$

11. $\dfrac{16x^2y}{24xy^3} \div \dfrac{9xy}{8x^2y^2}$

12. $\dfrac{x + 3}{10} + \dfrac{2x + 1}{15} - \dfrac{x - 2}{18}$

13. $\dfrac{7}{12ab} - \dfrac{11}{15a^2}$

14. $\dfrac{8}{x^2 - 4x} + \dfrac{2}{x}$

15. $(8x^3 - 6x^2 - 15x + 4) \div (4x - 1)$

For Problems 16–19, simplify each of the complex fractions.

16. $\dfrac{\dfrac{5}{x^2} - \dfrac{3}{x}}{\dfrac{1}{y} + \dfrac{2}{y^2}}$

17. $\dfrac{\dfrac{2}{x} - 3}{\dfrac{3}{y} + 4}$

18. $\dfrac{2 - \dfrac{1}{n + 2}}{3 + \dfrac{4}{n + 3}}$

19. $\dfrac{3a}{2 - \dfrac{1}{a}} - 1$

For Problems 20–25, factor each of the algebraic expressions completely.

20. $20x^2 + 7x - 6$

21. $16x^3 + 54$

22. $4x^4 - 25x^2 + 36$

23. $12x^3 - 52x^2 - 40x$

24. $xy - 6x + 3y - 18$

25. $10 + 9x - 9x^2$

For Problems 26–33, evaluate each of the numerical expressions.

26. $\left(\dfrac{2}{3}\right)^{-4}$

27. $\dfrac{3}{\left(\dfrac{4}{3}\right)^{-1}}$

28. $\sqrt[3]{-\dfrac{27}{64}}$

29. $-\sqrt{0.09}$

30. $(27)^{-\frac{4}{3}}$

31. $4^0 + 4^{-1} + 4^{-2}$

32. $\left(\dfrac{3^{-1}}{2^{-3}}\right)^{-2}$

33. $(2^{-3} - 3^{-2})^{-1}$

For Problems 34–36, find the indicated products and quotients and express the final answers with positive integral exponents only.

34. $(-3x^{-1}y^2)(4x^{-2}y^{-3})$

35. $\dfrac{48x^{-4}y^2}{6xy}$

36. $\left(\dfrac{27a^{-4}b^{-3}}{-3a^{-1}b^{-4}}\right)^{-1}$

For Problems 37–44, express each radical expression in simplest radical form.

37. $\sqrt{80}$

38. $-2\sqrt{54}$

39. $\sqrt{\dfrac{75}{81}}$

40. $\dfrac{4\sqrt{6}}{3\sqrt{8}}$

41. $\sqrt[3]{56}$

42. $\dfrac{\sqrt[3]{3}}{\sqrt[3]{4}}$

43. $4\sqrt{52x^3y^2}$ **44.** $\sqrt{\dfrac{2x}{3y}}$

For Problems 45–47, use the distributive property to help simplify each of the following.

45. $-3\sqrt{24} + 6\sqrt{54} - \sqrt{6}$ **46.** $\dfrac{\sqrt{8}}{3} - \dfrac{3\sqrt{18}}{4} - \dfrac{5\sqrt{50}}{2}$

47. $8\sqrt[3]{3} - 6\sqrt[3]{24} - 4\sqrt[3]{81}$

For Problems 48 and 49, rationalize the denominator and simplify.

48. $\dfrac{\sqrt{3}}{\sqrt{6} - 2\sqrt{2}}$ **49.** $\dfrac{3\sqrt{5} - \sqrt{3}}{2\sqrt{3} + \sqrt{7}}$

For Problems 50–52, use scientific notation to help perform the indicated operations.

50. $\dfrac{(0.00016)(300)(0.028)}{0.064}$

51. $\dfrac{0.00072}{0.0000024}$ **52.** $\sqrt{0.00000009}$

For Problems 53–56, find each of the indicated products or quotients and express the answers in standard form.

53. $(5 - 2i)(4 + 6i)$ **54.** $(-3 - i)(5 - 2i)$

55. $\dfrac{5}{4i}$ **56.** $\dfrac{-1 + 6i}{7 - 2i}$

57. Find the slope of the line determined by the points $(2, -3)$ and $(-1, 7)$.

58. Find the slope of the line determined by the equation $4x - 7y = 9$.

59. Find the length of the line segment whose endpoints are $(4, 5)$ and $(-2, 1)$.

60. Write the equation of the line that contains the points $(3, -1)$ and $(7, 4)$.

61. Write the equation of the line that is perpendicular to the line $3x - 4y = 6$ and contains the point $(-3, -2)$.

62. Find the center and the length of a radius of the circle $x^2 + 4x + y^2 - 12y + 31 = 0$.

63. Find the coordinates of the vertex of the parabola $y = x^2 + 10x + 21$.

64. Find the length of the major axis of the ellipse $x^2 + 4y^2 = 16$.

For Problems 65–70, graph each of the equations.

65. $-x + 2y = -4$ **66.** $x^2 + y^2 = 9$

67. $x^2 - y^2 = 9$ **68.** $x^2 + 2y^2 = 8$

69. $y = -3x$ **70.** $x^2y = 4$

For Problems 71–76, graph each of the functions.

71. $f(x) = -2x - 4$

72. $f(x) = -2x^2 - 2$

73. $f(x) = x^2 - 2x - 2$

74. $f(x) = \sqrt{x + 1} + 2$

75. $f(x) = 2x^2 + 8x + 9$

76. $f(x) = -|x - 2| + 1$

77. If $f(x) = x - 3$ and $g(x) = 2x^2 - x - 1$, find $(g \circ f)(x)$ and $(f \circ g)(x)$.

78. Find the inverse (f^{-1}) of $f(x) = 3x - 7$.

79. Find the inverse of $f(x) = -\dfrac{1}{2}x + \dfrac{2}{3}$.

80. Find the constant of variation if y varies directly as x, and $y = 2$ when $x = -\dfrac{2}{3}$.

81. If y is inversely proportional to the square of x, and $y = 4$ when $x = 3$, find y when $x = 6$.

82. The volume of a gas at a constant temperature varies inversely as the pressure. What is the volume of a gas under a pressure of 25 pounds if the gas occupies 15 cubic centimeters under a pressure of 20 pounds?

For Problems 83 and 84, evaluate each of the determinants.

83. $\begin{vmatrix} -2 & 4 \\ 7 & 6 \end{vmatrix}$ **84.** $\begin{vmatrix} 1 & -2 & -1 \\ 2 & 1 & 3 \\ -1 & -3 & 4 \end{vmatrix}$

For Problems 85–105, solve each of the equations.

85. $3(2x - 1) - 2(5x + 1) = 4(3x + 4)$

86. $n + \dfrac{3n - 1}{9} - 4 = \dfrac{3n + 1}{3}$

87. $0.92 + 0.9(x - 0.3) = 2x - 5.95$

88. $|4x - 1| = 11$

89. $3x^2 = 7x$

90. $x^3 - 36x = 0$

91. $30x^2 + 13x - 10 = 0$

92. $8x^3 + 12x^2 - 36x = 0$

93. $x^4 + 8x^2 - 9 = 0$

94. $(n + 4)(n - 6) = 11$

95. $2 - \dfrac{3x}{x - 4} = \dfrac{14}{x + 7}$

96. $\dfrac{2n}{6n^2 + 7n - 3} - \dfrac{n - 3}{3n^2 + 11n - 4} = \dfrac{5}{2n^2 + 11n + 12}$

97. $\sqrt{3y} - y = -6$

98. $\sqrt{x + 19} - \sqrt{x + 28} = -1$

99. $(3x - 1)^2 = 45$

100. $(2x + 5)^2 = -32$

101. $2x^2 - 3x + 4 = 0$

102. $3n^2 - 6n + 2 = 0$

103. $\dfrac{5}{n - 3} - \dfrac{3}{n + 3} = 1$

104. $12x^4 - 19x^2 + 5 = 0$

105. $2x^2 + 5x + 5 = 0$

For Problems 106–115, solve each of the inequalities.

106. $-5(y - 1) + 3 > 3y - 4 - 4y$

107. $0.06x + 0.08(250 - x) \geq 19$

108. $|5x - 2| > 13$

109. $|6x + 2| < 8$

110. $\dfrac{x - 2}{5} - \dfrac{3x - 1}{4} \leq \dfrac{3}{10}$

111. $(x - 2)(x + 4) \leq 0$

112. $(3x - 1)(x - 4) > 0$

113. $x(x + 5) < 24$

114. $\dfrac{x - 3}{x - 7} \geq 0$

115. $\dfrac{2x}{x + 3} > 4$

For Problems 116–120, solve each of the systems of equations.

116. $\begin{pmatrix} 4x - 3y = 18 \\ 3x - 2y = 15 \end{pmatrix}$

117. $\begin{pmatrix} y = \dfrac{2}{5}x - 1 \\ 3x + 5y = 4 \end{pmatrix}$

118. $\begin{pmatrix} \dfrac{x}{2} - \dfrac{y}{3} = 1 \\ \dfrac{2x}{5} + \dfrac{y}{2} = 2 \end{pmatrix}$

119. $\begin{pmatrix} 4x - y + 3z = -12 \\ 2x + 3y - z = 8 \\ 6x + y + 2z = -8 \end{pmatrix}$

120. $\begin{pmatrix} x - y + 5z = -10 \\ 5x + 2y - 3z = 6 \\ -3x + 2y - z = 12 \end{pmatrix}$

For Problems 121–135, set up an equation, an inequality, or a system of equations to help solve each problem.

121. Find three consecutive odd integers whose sum is 57.

122. Suppose that Eric has a collection of 63 coins consisting of nickels, dimes, and quarters. The number of dimes is 6 more than the number of nickels, and the number of quarters is 1 more than twice the number of nickels. How many coins of each kind are in the collection?

123. One of two supplementary angles is 4° more than one-third of the other angle. Find the measure of each of the angles.

124. If a ring costs a jeweler $300, at what price should it be sold for the jeweler to make a profit of 50% on the selling price?

125. Last year Beth invested a certain amount of money at 8% and $300 more than that amount at 9%. Her total yearly interest was $316. How much did she invest at each rate?

126. Two trains leave the same depot at the same time, one traveling east and the other traveling west. At the end of $4\frac{1}{2}$ hours, they are 639 miles apart. If the rate of the train traveling east is 10 miles per hour greater than that of the other train, find their rates.

127. Suppose that a 10-quart radiator contains a 50% solution of antifreeze. How much needs to be drained out and replaced with pure antifreeze to obtain a 70%-antifreeze solution?

128. Sam shot rounds of 70, 73, and 76 on the first three days of a golf tournament. What must he shoot on the fourth day of the tournament to average 72 or less for the four days?

129. The cube of a number equals nine times the same number. Find the number.

130. A strip of uniform width is to be cut off of both sides and both ends of a sheet of paper that is 8 inches by 14 inches to reduce the size of the paper to an area of 72 square inches. (See Figure 10.16.) Find the width of the strip.

131. A sum of $2450 is to be divided between two people in the ratio of 3 to 4. How much does each person receive?

132. Working together, Crystal and Dean can complete a task in $1\frac{1}{5}$ hours. Dean can do the task by himself in 2 hours. How long would it take Crystal to complete the task by herself?

133. Dudley bought a number of shares of stock for $300. A month later he sold all but 10 shares at a profit of $5 per share and regained his original investment of $300. How many shares did he originally buy and at what price per share?

134. The units digit of a two-digit number is 1 more than twice the tens digit. The sum of the digits is 10. Find the number.

135. The sum of the two smallest angles of a triangle is 40° less than the other angle. The sum of the smallest and largest angles is twice the other angle. Find the measures of the three angles of the triangle.

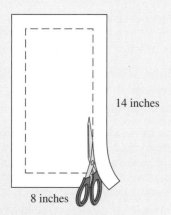

14 inches

8 inches

Figure 10.16

Because the Richter number for reporting the intensity of an earthquake is calculated from a logarithm, it is referred to as a logarithmic scale. Logarithmic scales are commonly used in science and mathematics to transform very large numbers to a smaller scale.

AP/Wide World Photos

Exponential and Logarithmic Functions

How long will it take $100 to triple itself if it is invested at 8% interest compounded continuously? We can use the formula $A = Pe^{rt}$ to generate the equation $300 = 100e^{0.08t}$, which can be solved for t by using logarithms. It will take approximately 13.7 years for the money to triple itself.

This chapter will expand the meaning of an exponent and introduce the concept of a logarithm. We will (1) work with some exponential functions, (2) work with some logarithmic functions, and (3) use the concepts of exponent and logarithm to expand our capabilities for solving problems. Your calculator will be a valuable tool throughout this chapter.

InfoTrac Project   Do a keyword search on compound interest and find the article titled "Tapping the Power of Compound Interest." Write a brief summary of the article. The formula for the amount of money (A) in an account for which interest is compounded continuously is given by $A = Pe^{rt}$, where P is the initial investment, r is the rate of interest, and t is the time (in years) the money is invested. Find the amount of money in an account after 35 years if the initial investment of $2500 was invested at 7% (compounded continuously). At what interest rate must the money be invested for there to be $35,000 in the account at the end of 35 years?

11.1 Exponents and Exponential Functions

In Chapter 1, the expression b^n was defined as n factors of b, where n is any positive integer and b is any real number. For example,

$$4^3 = 4 \cdot 4 \cdot 4 = 64$$

$$\left(\frac{1}{2}\right)^4 = \left(\frac{1}{2}\right)\left(\frac{1}{2}\right)\left(\frac{1}{2}\right)\left(\frac{1}{2}\right) = \frac{1}{16}$$

$$(-0.3)^2 = (-0.3)(-0.3) = 0.09$$

In Chapter 5, by defining $b^0 = 1$ and $b^{-n} = \dfrac{1}{b^n}$, where n is any positive integer and b is any nonzero real number, we extended the concept of an exponent to include all integers. For example,

$$2^{-3} = \frac{1}{2^3} = \frac{1}{2 \cdot 2 \cdot 2} = \frac{1}{8} \qquad \left(\frac{1}{3}\right)^{-2} = \frac{1}{\left(\frac{1}{3}\right)^2} = \frac{1}{\frac{1}{9}} = 9$$

$$(0.4)^{-1} = \frac{1}{(0.4)^1} = \frac{1}{0.4} = 2.5 \qquad (-0.98)^0 = 1$$

In Chapter 5 we also provided for the use of all rational numbers as exponents by defining $b^{\frac{m}{n}} = \sqrt[n]{b^m}$, where n is a positive integer greater than 1, and b is a real number such that $\sqrt[n]{b}$ exists. For example,

$$8^{\frac{2}{3}} = \sqrt[3]{8^2} = \sqrt[3]{64} = 4$$

$$16^{\frac{1}{4}} = \sqrt[4]{16^1} = 2$$

$$32^{-\frac{1}{5}} = \frac{1}{32^{\frac{1}{5}}} = \frac{1}{\sqrt[5]{32}} = \frac{1}{2}$$

To extend the concept of an exponent formally to include the use of irrational numbers requires some ideas from calculus and is therefore beyond the scope of this text. However, here's a glance at the general idea involved. Consider the number $2^{\sqrt{3}}$. By using the nonterminating and nonrepeating decimal representation 1.73205 . . . for $\sqrt{3}$, form the sequence of numbers 2^1, $2^{1.7}$, $2^{1.73}$, $2^{1.732}$, $2^{1.7320}$, $2^{1.73205}$. . . . It would seem reasonable that each successive power gets closer to $2^{\sqrt{3}}$. This is precisely what happens if b^n, where n is irrational, is properly defined by using the concept of a limit.

From now on, then, we can use any real number as an exponent, and the basic properties stated in Chapter 5 can be extended to include all real numbers as exponents. Let's restate those properties at this time with the restriction that the bases a and b are to be positive numbers (to avoid expressions such as $(-4)^{\frac{1}{2}}$, which do not represent real numbers).

PROPERTY 11.1

If a and b are positive real numbers and m and n are any real numbers, then

1. $b^n \cdot b^m = b^{n+m}$ Product of two powers

2. $(b^n)^m = b^{mn}$ Power of a power

3. $(ab)^n = a^n b^n$ Power of a product

4. $\left(\dfrac{a}{b}\right)^n = \dfrac{a^n}{b^n}$ Power of a quotient

5. $\dfrac{b^n}{b^m} = b^{n-m}$ Quotient of two powers

Another property that can be used to solve certain types of equations involving exponents can be stated as follows:

PROPERTY 11.2

If $b > 0$, $b \neq 1$, and m and n are real numbers, then

$$b^n = b^m \quad \text{if and only if } n = m$$

The following examples illustrate the use of Property 11.2.

EXAMPLE 1

Solve $2^x = 32$.

Solution

$$2^x = 32$$

$$2^x = 2^5 \qquad 32 = 2^5$$

$$x = 5 \qquad \text{Property 11.2}$$

The solution set is {5}.   ■

EXAMPLE 2

Solve $3^{2x} = \dfrac{1}{9}$.

Solution

$$3^{2x} = \frac{1}{9} = \frac{1}{3^2}$$

$$3^{2x} = 3^{-2}$$

$$2x = -2 \qquad \text{Property 11.2}$$

$$x = -1$$

The solution set is $\{-1\}$.

EXAMPLE 3

Solve $\left(\dfrac{1}{5}\right)^{x-2} = \dfrac{1}{125}$.

Solution

$$\left(\frac{1}{5}\right)^{x-2} = \frac{1}{125} = \left(\frac{1}{5}\right)^3$$

$$x - 2 = 3 \qquad \text{Property 11.2}$$

$$x = 5$$

The solution set is $\{5\}$.

EXAMPLE 4

Solve $8^x = 32$.

Solution

$$8^x = 32$$

$$(2^3)^x = 2^5 \qquad 8 = 2^3$$

$$2^{3x} = 2^5$$

$$3x = 5 \qquad \text{Property 11.2}$$

$$x = \frac{5}{3}$$

The solution set is $\left\{\dfrac{5}{3}\right\}$.

EXAMPLE 5

Solve $(3^{x+1})(9^{x-2}) = 27$.

Solution

$$(3^{x+1})(9^{x-2}) = 27$$

$$(3^{x+1})(3^2)^{x-2} = 3^3$$

$$(3^{x+1})(3^{2x-4}) = 3^3$$

$$3^{3x-3} = 3^3$$

$$3x - 3 = 3 \qquad \text{Property 11.2}$$

$$3x = 6$$

$$x = 2$$

The solution set is {2}.

Exponential Functions

If b is any positive number, then the expression b^x designates exactly one real number for every real value of x. Thus the equation $f(x) = b^x$ defines a function whose domain is the set of real numbers. Furthermore, if we impose the additional restriction $b \neq 1$, then any equation of the form $f(x) = b^x$ describes a one-to-one function and is called an **exponential function.** This leads to the following definition.

DEFINITION 11.1

If $b > 0$ and $b \neq 1$, then the function f defined by

$$f(x) = b^x$$

where x is any real number, is called the **exponential function with base b.**

REMARK: The function $f(x) = 1^x$ is a constant function whose graph is a horizontal line, and therefore it is not a one-to-one function. Remember from Chapter 9 that one-to-one functions have inverses; this becomes a key issue in a later section.

Now let's consider graphing some exponential functions.

EXAMPLE 6

Graph the function $f(x) = 2^x$.

Solution

First, let's set up a table of values.

x	$f(x) = 2^x$
-2	$\dfrac{1}{4}$
-1	$\dfrac{1}{2}$
0	1
1	2
2	4
3	8

Plot these points and connect them with a smooth curve to produce Figure 11.1.

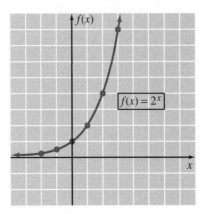

Figure 11.1

E X A M P L E   7

Graph $f(x) = \left(\dfrac{1}{2}\right)^x$.

Solution

Again, let's set up a table of values. Plot these points and connect them with a smooth curve to produce Figure 11.2.

x	$f(x) = \left(\dfrac{1}{2}\right)^x$
-2	4
-1	2
0	1
1	$\dfrac{1}{2}$
2	$\dfrac{1}{4}$
3	$\dfrac{1}{8}$

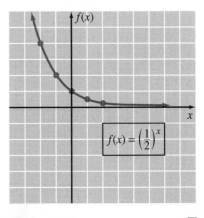

Figure 11.2

In the tables for Examples 6 and 7 we chose integral values for x to keep the computation simple. However, with the use of a calculator, we could easily acquire functional values by using nonintegral exponents. Consider the following additional values for each of the tables.

$f(x) = 2^x$	
$f(0.5) \approx 1.41$	$f(-0.5) \approx 0.71$
$f(1.7) \approx 3.25$	$f(-2.6) \approx 0.16$

$f(x) = \left(\dfrac{1}{2}\right)^x$	
$f(0.7) \approx 0.62$	$f(-0.8) \approx 1.74$
$f(2.3) \approx 0.20$	$f(-2.1) \approx 4.29$

Use your calculator to check these results. Also, it would be worthwhile for you to go back and see that the points determined do fit the graphs in Figures 11.1 and 11.2.

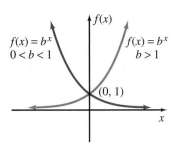

Figure 11.3

The graphs in Figures 11.1 and 11.2 illustrate a general behavior pattern of exponential functions. That is, if $b > 1$, then the graph of $f(x) = b^x$ *goes up to the right* and the function is called an **increasing function.** If $0 < b < 1$, then the graph of $f(x) = b^x$ *goes down to the right* and the function is called a **decreasing function.** These facts are illustrated in Figure 11.3. Note that because $b^0 = 1$ for any $b > 0$, all graphs of $f(x) = b^x$ contain the point $(0, 1)$.

As you graph exponential functions, don't forget your previous graphing experiences.

1. The graph of $f(x) = 2^x - 4$ is the graph of $f(x) = 2^x$ moved down 4 units.
2. The graph of $f(x) = 2^{x+3}$ is the graph of $f(x) = 2^x$ moved 3 units to the left.
3. The graph of $f(x) = -2^x$ is the graph of $f(x) = 2^x$ reflected across the x axis.

We used a graphing calculator to graph these four functions on the same set of axes, as shown in Figure 11.4.

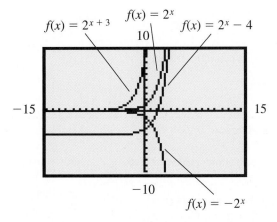

Figure 11.4

PROBLEM SET 11.1

For Problems 1–34, solve each equation.

1. $3^x = 27$

2. $2^x = 64$

3. $2^{2x} = 16$

4. $3^{2x} = 81$

5. $\left(\dfrac{1}{4}\right)^x = \dfrac{1}{256}$

6. $\left(\dfrac{1}{2}\right)^x = \dfrac{1}{128}$

7. $5^{x+2} = 125$

8. $4^{x-3} = 16$

9. $3^{-x} = \dfrac{1}{243}$

10. $5^{-x} = \dfrac{1}{25}$

11. $6^{3x-1} = 36$

12. $2^{2x+3} = 32$

13. $4^x = 8$

14. $16^x = 64$

15. $8^{2x} = 32$

16. $9^{3x} = 27$

17. $\left(\dfrac{1}{2}\right)^{2x} = 64$

18. $\left(\dfrac{1}{3}\right)^{5x} = 243$

19. $\left(\dfrac{3}{4}\right)^x = \dfrac{64}{27}$

20. $\left(\dfrac{2}{3}\right)^x = \dfrac{9}{4}$

21. $9^{4x-2} = \dfrac{1}{81}$

22. $8^{3x+2} = \dfrac{1}{16}$

23. $6^{2x} + 3 = 39$

24. $5^{2x} - 2 = 123$

25. $10^x = 0.1$

26. $10^x = 0.0001$

27. $32^x = \dfrac{1}{4}$

28. $9^x = \dfrac{1}{27}$

29. $(2^{x+1})(2^x) = 64$

30. $(2^{2x-1})(2^{x+2}) = 32$

31. $(27)(3^x) = 9^x$

32. $(3^x)(3^{5x}) = 81$

33. $(4^x)(16^{3x-1}) = 8$

34. $(8^{2x})(4^{2x-1}) = 16$

For Problems 35–52, graph each exponential function.

35. $f(x) = 4^x$

36. $f(x) = 3^x$

37. $f(x) = 6^x$

38. $f(x) = 5^x$

39. $f(x) = \left(\dfrac{1}{3}\right)^x$

40. $f(x) = \left(\dfrac{1}{4}\right)^x$

41. $f(x) = \left(\dfrac{3}{4}\right)^x$

42. $f(x) = \left(\dfrac{2}{3}\right)^x$

43. $f(x) = 2^{x-2}$

44. $f(x) = 2^{x+1}$

45. $f(x) = 3^{-x}$

46. $f(x) = 2^{-x}$

47. $f(x) = 3^{2x}$

48. $f(x) = 2^{2x}$

49. $f(x) = 3^x - 2$

50. $f(x) = 2^x + 1$

51. $f(x) = 2^{-x-2}$

52. $f(x) = 3^{-x+1}$

■ ■ ■ **Thoughts into words**

53. Explain how you would solve the equation (2^{x+1}) $(8^{2x-3}) = 64$.

54. Why is the base of an exponential function restricted to positive numbers not including 1?

55. Explain how you would graph the function

$$f(x) = -\left(\dfrac{1}{3}\right)^x.$$

▦ **Graphing calculator activities**

56. Use a graphing calculator to check your graphs for Problems 35–52.

57. Graph $f(x) = 2^x$. Where should the graphs of $f(x) = 2^{x-5}, f(x) = 2^{x-7}$, and $f(x) = 2^{x+5}$ be located? Graph all three functions on the same set of axes with $f(x) = 2^x$.

58. Graph $f(x) = 3^x$. Where should the graphs of $f(x) = 3^x + 2, f(x) = 3^x - 3$, and $f(x) = 3^x - 7$ be located? Graph all three functions on the same set of axes with $f(x) = 3^x$.

59. Graph $f(x) = \left(\frac{1}{2}\right)^x$. Where should the graphs of

$f(x) = -\left(\frac{1}{2}\right)^x$, $f(x) = \left(\frac{1}{2}\right)^{-x}$, and $f(x) = -\left(\frac{1}{2}\right)^{-x}$

be located? Graph all three functions on the same set of axes with $f(x) = \left(\frac{1}{2}\right)^x$.

60. Graph $f(x) = (1.5)^x$, $f(x) = (5.5)^x$, $f(x) = (0.3)^x$, and $f(x) = (0.7)^x$ on the same set of axes. Are these graphs consistent with Figure 11.3?

61. What is the solution for $3^x = 5$? Do you agree that it is between 1 and 2 because $3^1 = 3$ and $3^2 = 9$? Now graph $f(x) = 3^x - 5$ and use the ZOOM and TRACE features of your graphing calculator to find an approximation, to

the nearest hundredth, for the x intercept. You should get an answer of 1.46, to the nearest hundredth. Do you see that this is an approximation for the solution of $3^x = 5$? Try it; raise 3 to the 1.46 power.

Find an approximate solution, to the nearest hundredth, for each of the following equations by graphing the appropriate function and finding the x intercept.

(a) $2^x = 19$ **(b)** $3^x = 50$

(c) $4^x = 47$ **(d)** $5^x = 120$

(e) $2^x = 1500$ **(f)** $3^{x-1} = 34$

11.2   Applications of Exponential Functions

Equations that describe exponential functions can represent many real-world situations that exhibit growth or decay. For example, suppose that an economist predicts an annual inflation rate of 5% for the next 10 years. This means an item that presently costs $8 will cost $8(105\%) = 8(1.05) = \$8.40$ a year from now. The same item will cost $(105\%)[8(105\%)] = 8(1.05)^2 = \$8.82$ in 2 years. In general, the equation

$$P = P_0(1.05)^t$$

yields the predicted price P of an item in t years at the annual inflation rate of 5%, where that item presently costs P_0. By using this equation, we can look at some future prices based on the prediction of a 5% inflation rate.

A \$0.79 jar of mustard will cost $\$0.79(1.05)^3 = \$0.91$ in 3 years.

A \$2.69 bag of potato chips will cost $\$2.69(1.05)^5 = \$3.43$ in 5 years.

A \$6.69 can of coffee will cost $\$6.69(1.05)^7 = \$9.41$ in 7 years.

Compound Interest

Compound interest provides another illustration of exponential growth. Suppose that $500 (called the **principal**) is invested at an interest rate of 8% **compounded annually.** The interest earned the first year is $500(0.08) = \$40$, and this amount is added to the original $500 to form a new principal of $540 for the second year. The interest earned during the second year is $540(0.08) = \$43.20$, and this amount is added to $540 to form a new principal of $583.20 for the third year. Each year a new principal is formed by reinvesting the interest earned that year.

In general, suppose that a sum of money P (called the principal) is invested at an interest rate of r percent compounded annually. The interest earned the first year is Pr, and the new principal for the second year is $P + Pr$ or $P(1 + r)$. Note that the new principal for the second year can be found by multiplying the original principal P by $(1 + r)$. In like fashion, the new principal for the third year can be found by multiplying the previous principal $P(1 + r)$ by $(1 + r)$, thus obtaining $P(1 + r)^2$. If this process is continued, then after t years the total amount of money accumulated, A, is given by

$$A = P(1 + r)^t$$

Consider the following examples of investments made at a certain rate of interest compounded annually.

1. $750 invested for 5 years at 9% compounded annually produces

$$A = \$750(1.09)^5 = \$1153.97$$

2. $1000 invested for 10 years at 11% compounded annually produces

$$A = \$1000(1.11)^{10} = \$2839.42$$

3. $5000 invested for 20 years at 12% compounded annually produces

$$A = \$5000(1.12)^{20} = \$48{,}231.47$$

If we invest money at a certain rate of interest to be compounded more than once a year, then we can adjust the basic formula, $A = P(1 + r)^t$, according to the number of compounding periods in the year. For example, for **compounding semiannually,** the formula becomes $A = P\left(1 + \dfrac{r}{2}\right)^{2t}$ and for **compounding quarterly,** the formula becomes $A = P\left(1 + \dfrac{r}{4}\right)^{4t}$. In general, we have the following formula, where n represents the number of compounding periods in a year.

$$A = P\left(1 + \frac{r}{n}\right)^{nt}$$

The following examples should clarify the use of this formula.

1. $750 invested for 5 years at 9% compounded semiannually produces

$$A = \$750\left(1 + \frac{0.09}{2}\right)^{2(5)} = \$750(1.045)^{10} = \$1164.73$$

2. $1000 invested for 10 years at 11% compounded quarterly produces

$$A = \$1000\left(1 + \frac{0.11}{4}\right)^{4(10)} = \$1000(1.0275)^{40} = \$2959.87$$

3. $5000 invested for 20 years at 12% compounded monthly produces

$$A = \$5000\left(1 + \frac{0.12}{12}\right)^{12(20)} = \$5000(1.01)^{240} = \$54{,}462.77$$

You may find it interesting to compare these results with those obtained earlier for compounding annually.

Exponential Decay

Suppose it is estimated that the value of a car depreciates 15% per year for the first 5 years. Thus a car that costs $19,500 will be worth $19500(100\% - 15\%) = 19500(85\%) = 19500(0.85) = \$16{,}575$ in 1 year. In 2 years the value of the car will have depreciated to $19500(0.85)^2 = \$14{,}089$ (nearest dollar). The equation

$$V = V_0(0.85)^t$$

yields the value V of a car in t years at the annual depreciation rate of 15%, where the car initially cost V_0. By using this equation, we can estimate some car values, to the nearest dollar, as follows:

A $17,000 car will be worth $\$17{,}000(0.85)^5 = \$7543$ in 5 years.
A $25,000 car will be worth $\$25{,}000(0.85)^4 = \$13{,}050$ in 4 years.
A $40,000 car will be worth $\$40{,}000(0.85)^3 = \$24{,}565$ in 3 years.

Another example of exponential decay is associated with radioactive substances. We can describe the rate of decay exponentially, on the basis of the half-life of a substance. The *half-life* of a radioactive substance is the amount of time that it takes for one-half of an initial amount of the substance to disappear as the result of decay. For example, suppose that we have 200 grams of a certain substance that has a half-life of 5 days. After 5 days, $200\left(\frac{1}{2}\right) = 100$ grams remain. After 10 days, $200\left(\frac{1}{2}\right)^2 = 50$ grams remain. After 15 days, $200\left(\frac{1}{2}\right)^3 = 25$ grams remain. In general, after t days, $200\left(\frac{1}{2}\right)^{\frac{t}{5}}$ grams remain.

This discussion leads us to the following half-life formula. Suppose there is an initial amount, Q_0, of a radioactive substance with a half-life of h. The amount of substance remaining, Q, after a time period of t, is given by the formula

$$Q = Q_0\left(\frac{1}{2}\right)^{\frac{t}{h}}$$

The units of measure for t and h must be the same.

E X A M P L E   1

Barium-140 has a half-life of 13 days. If there are 500 milligrams of barium initially, how many milligrams remain after 26 days? After 100 days?

Solution

When we use $Q_0 = 500$ and $h = 13$, the half-life formula becomes

$$Q = 500\left(\frac{1}{2}\right)^{\frac{t}{13}}$$

If $t = 26$, then

$$Q = 500\left(\frac{1}{2}\right)^{\frac{26}{13}}$$

$$= 500\left(\frac{1}{2}\right)^{2}$$

$$= 500\left(\frac{1}{4}\right)$$

$$= 125$$

Thus 125 milligrams remain after 26 days. If $t = 100$, then

$$Q = 500\left(\frac{1}{2}\right)^{\frac{100}{13}}$$

$$= 500(0.5)^{\frac{100}{13}}$$

$$= 2.4, \quad \text{to the nearest tenth of a milligram}$$

Thus approximately 2.4 milligrams remain after 100 days.

REMARK: The solution to Example 1 clearly demonstrates one facet of the role of the calculator in the application of mathematics. We solved the first part of the problem easily without the calculator, but the calculator certainly was helpful for the second part of the problem.

Number e

An interesting situation occurs if we consider the compound interest formula for $P = \$1$, $r = 100\%$, and $t = 1$ year. The formula becomes $A = 1\left(1 + \dfrac{1}{n}\right)^{n}$. The accompanying table shows some values, rounded to eight decimal places, of $\left(1 + \dfrac{1}{n}\right)^{n}$ for different values of n.

n	$\left(1 + \dfrac{1}{n}\right)^{n}$
1	2.00000000
10	2.59374246
100	2.70481383
1000	2.71692393
10,000	2.71814593
100,000	2.71826824
1,000,000	2.71828047
10,000,000	2.71828169
100,000,000	2.71828181
1,000,000,000	2.71828183

The table suggests that as n increases, the value of $\left(1 + \dfrac{1}{n}\right)^{n}$ gets closer and closer to some fixed number. This does happen, and the fixed number is called e. To five decimal places, $e = 2.71828$.

The function defined by the equation $f(x) = e^{x}$ is the **natural exponential function.** It has a great many real-world applications; some we will look at in a moment. First, however, let's get a picture of the natural exponential function. Because $2 < e < 3$, the graph of $f(x) = e^{x}$ must fall between the graphs of $f(x) = 2^{x}$ and $f(x) = 3^{x}$. To be more specific, let's use our calculator to determine a table of values. Use the $\boxed{e^{x}}$ key, and round the results to the nearest tenth to obtain the following table. Plot the points determined by this table and connect them with a smooth curve to produce Figure 11.5.

x	$f(x) = e^{x}$
0	1.0
1	2.7
2	7.4
−1	0.4
−2	0.1

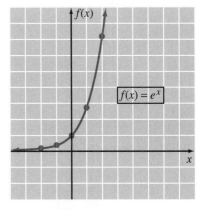

Figure 11.5

Let's return to the concept of compound interest. If the number of compounding periods in a year is increased indefinitely, we arrive at the concept of **compounding continuously.** Mathematically, this can be accomplished by applying the limit concept to the expression $P\left(1 + \dfrac{r}{n}\right)^{nt}$. We will not show the details here, but the following result is obtained. The formula

$$A = Pe^{rt}$$

yields the accumulated value, A, of a sum of money, P, that has been invested for t years at a rate of r percent compounded continuously. The following examples show the use of the formula.

1. \$750 invested for 5 years at 9% compounded continuously produces

   $$A = \$750e^{(0.09)(5)} = 750e^{0.45} = \$1176.23$$

2. \$1000 invested for 10 years at 11% compounded continuously produces

   $$A = \$1000e^{(0.11)(10)} = 1000e^{1.1} = \$3004.17$$

3. \$5000 invested for 20 years at 12% compounded continuously produces

   $$A = \$5000e^{(0.12)(20)} = 5000e^{2.4} = \$55,115.88$$

Again you may find it interesting to compare these results with those you obtained earlier when using a different number of compounding periods.

The ideas behind compounding continuously carry over to other growth situations. The law of exponential growth,

$$Q(t) = Q_0 e^{kt}$$

is used as a mathematical model for numerous growth-and-decay applications. In this equation, $Q(t)$ represents the quantity of a given substance at any time t; Q_0 is the initial amount of the substance (when $t = 0$); and k is a constant that depends on the particular application. If $k < 0$, then $Q(t)$ decreases as t increases, and we refer to the model as the **law of decay.** Let's consider some growth-and-decay applications.

E X A M P L E   2

Suppose that in a certain culture, the equation $Q(t) = 15000e^{0.3t}$ expresses the number of bacteria present as a function of the time t, where t is expressed in hours. Find (a) the initial number of bacteria, and (b) the number of bacteria after 3 hours.

Solution

(a) The initial number of bacteria is produced when $t = 0$.

$$Q(0) = 15,000e^{0.3(0)}$$
$$= 15,000e^{0}$$
$$= 15,000 \qquad e^{0} = 1$$

(b) $Q(3) = 15,000e^{0.3(3)}$

$\qquad\qquad = 15,000e^{0.9}$

$\qquad\qquad = 36,894,$ to the nearest whole number

Therefore, there should be approximately 36,894 bacteria present after 3 hours.   ■

E X A M P L E   3

Suppose the number of bacteria present in a certain culture after t minutes is given by the equation $Q(t) = Q_0 e^{0.05t}$, where Q_0 represents the initial number of bacteria. If 5000 bacteria are present after 20 minutes, how many bacteria were present initially?

Solution

If 5000 bacteria are present after 20 minutes, then $Q(20) = 5000$.

$\qquad 5000 = Q_0 e^{0.05(20)}$

$\qquad 5000 = Q_0 e^{1}$

$\qquad \dfrac{5000}{e} = Q_0$

$\qquad 1839 = Q_0,$ to the nearest whole number

Therefore, there were approximately 1839 bacteria present initially.   ■

E X A M P L E   4

The number of grams Q of a certain radioactive substance present after t seconds is given by $Q(t) = 200e^{-0.3t}$. How many grams remain after 7 seconds?

Solution

We need to evaluate $Q(7)$.

$\qquad Q(7) = 200e^{-0.3(7)}$

$\qquad\qquad = 200e^{-2.1}$

$\qquad\qquad = 24,$ to the nearest whole number

Thus approximately 24 grams remain after 7 seconds.   ■

Finally, let's use a graphing calculator to graph a special exponential function.

E X A M P L E   5

Graph the function

$$y = \frac{1}{\sqrt{2\pi}} e^{-x^2/2}$$

and find its maximum value.

Solution

If $x = 0$, then $y = \dfrac{1}{\sqrt{2\pi}}\, e^0 = \dfrac{1}{\sqrt{2\pi}} \approx 0.4$. Let's set the boundaries of the viewing rectangle so that $-5 \leq x \leq 5$ and $0 \leq y \leq 1$ with a y scale of 0.1; the graph of the function is shown in Figure 11.6. From the graph, we see that the maximum value of the function occurs at $x = 0$, which we have already determined to be approximately 0.4.

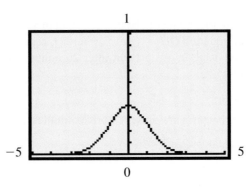

Figure 11.6

REMARK: The curve in Figure 11.6 is called the standard normal distribution curve. You may want to ask your instructor to explain what it means to assign grades on the basis of the normal distribution curve.

PROBLEM SET 11.2

1. Assuming that the rate of inflation is 4% per year, the equation $P = P_0(1.04)^t$ yields the predicted price P, in t years, of an item that presently costs P_0. Find the predicted price of each of the following items for the indicated years ahead.
   (a) $0.77 can of soup in 3 years
   (b) $3.43 container of cocoa mix in 5 years
   (c) $1.99 jar of coffee creamer in 4 years
   (d) $1.05 can of beans and bacon in 10 years
   (e) $18,000 car in 5 years (nearest dollar)
   (f) $120,000 house in 8 years (nearest dollar)
   (g) $500 TV set in 7 years (nearest dollar)

2. Suppose it is estimated that the value of a car depreciates 30% per year for the first 5 years. The equation $A = P_0(0.7)^t$ yields the value (A) of a car after t years if the original price is P_0. Find the value (to the nearest dollar) of each of the following cars after the indicated time.
   (a) $16,500 car after 4 years
   (b) $22,000 car after 2 years
   (c) $27,000 car after 5 years
   (d) $40,000 car after 3 years

For Problems 3–14, use the formula $A = P\left(1 + \dfrac{r}{n}\right)^{nt}$ to find the total amount of money accumulated at the end of the indicated time period for each of the following investments.

3. $200 for 6 years at 6% compounded annually

4. $250 for 5 years at 7% compounded annually

5. $500 for 7 years at 8% compounded semiannually

6. $750 for 8 years at 8% compounded semiannually

7. $800 for 9 years at 9% compounded quarterly

8. $1200 for 10 years at 10% compounded quarterly

9. $1500 for 5 years at 12% compounded monthly

10. $2000 for 10 years at 9% compounded monthly

11. $5000 for 15 years at 8.5% compounded annually

12. $7500 for 20 years at 9.5% compounded semiannually

13. $8000 for 10 years at 10.5% compounded quarterly

14. $10,000 for 25 years at 9.25% compounded monthly

For Problems 15–23, use the formula $A = Pe^{rt}$ to find the total amount of money accumulated at the end of the indicated time period by compounding continuously.

15. $400 for 5 years at 7%

16. $500 for 7 years at 6%

17. $750 for 8 years at 8%

18. $1000 for 10 years at 9%

19. $2000 for 15 years at 10%

20. $5000 for 20 years at 11%

21. $7500 for 10 years at 8.5%

22. $10,000 for 25 years at 9.25%

23. $15,000 for 10 years at 7.75%

24. Complete the following chart, which illustrates what happens to $1000 invested at various rates of interest for different lengths of time but always compounded continuously. Round your answers to the nearest dollar.

$1000 compounded continuously			
8%	**10%**	**12%**	**14%**
5 years			
10 years			
15 years			
20 years			
25 years			

25. Complete the following chart, which illustrates what happens to $1000 invested at 12% for different lengths of time and different numbers of compounding periods. Round all of your answers to the nearest dollar.

$1000 at 12%			
Compounded annually			
1 year	5 years	10 years	20 years
Compounded semiannually			
1 year	5 years	10 years	20 years
Compounded quarterly			
1 year	5 years	10 years	20 years
Compounded monthly			
1 year	5 years	10 years	20 years
Compounded continuously			
1 year	5 years	10 years	20 years

26. Complete the following chart, which illustrates what happens to $1000 in 10 years for different rates of interest and different numbers of compounding periods. Round your answers to the nearest dollar.

$1000 for 10 years			
Compounded annually			
8%	10%	12%	14%
Compounded semiannually			
8%	10%	12%	14%
Compounded quarterly			
8%	10%	12%	14%
Compounded monthly			
8%	10%	12%	14%
Compounded continuously			
8%	10%	12%	14%

27. Suppose that Nora invested $500 at 8.25% compounded annually for 5 years and Patti invested $500 at 8% compounded quarterly for 5 years. At the end of 5 years, who will have the most money and by how much?

28. Two years ago Daniel invested some money at 8% interest compounded annually. Today it is worth $758.16. How much did he invest two years ago?

29. What rate of interest (to the nearest hundredth of a percent) is needed so that an investment of $2500 will yield $3000 in 2 years if the money is compounded annually?

30. Suppose that a certain radioactive substance has a half-life of 20 years. If there are presently 2500 milligrams of the substance, how much, to the nearest milligram, will remain after 40 years? After 50 years?

31. Strontium-90 has a half-life of 29 years. If there are 400 grams of strontium initially, how much, to the nearest gram, will remain after 87 years? After 100 years?

32. The half-life of radium is approximately 1600 years. If the present amount of radium in a certain location is 500 grams, how much will remain after 800 years? Express your answer to the nearest gram.

For Problems 33–38, graph each of the exponential functions.

33. $f(x) = e^x + 1$

34. $f(x) = e^x - 2$

35. $f(x) = 2e^x$

36. $f(x) = -e^x$

37. $f(x) = e^{2x}$

38. $f(x) = e^{-x}$

For Problems 39–44, express your answers to the nearest whole number.

39. Suppose that in a certain culture, the equation $Q(t) = 1000e^{0.4t}$ expresses the number of bacteria present as a function of the time t, where t is expressed in hours. How many bacteria are present at the end of 2 hours? 3 hours? 5 hours?

40. The number of bacteria present at a given time under certain conditions is given by the equation $Q = 5000e^{0.05t}$,

where t is expressed in minutes. How many bacteria are present at the end of 10 minutes? 30 minutes? 1 hour?

41. The number of bacteria present in a certain culture after t hours is given by the equation $Q = Q_0e^{0.3t}$, where Q_0 represents the initial number of bacteria. If 6640 bacteria are present after 4 hours, how many bacteria were present initially?

42. The number of grams Q of a certain radioactive substance present after t seconds is given by the equation $Q = 1500e^{-0.4t}$. How many grams remain after 5 seconds? 10 seconds? 20 seconds?

43. Suppose that the present population of a city is 75,000. Using the equation $P(t) = 75,000e^{0.01t}$ to estimate future growth, estimate the population
   (a) 10 years from now
   (b) 15 years from now
   (c) 25 years from now

44. Suppose that the present population of a city is 150,000. Use the equation $P(t) = 150,000e^{0.032t}$ to estimate future growth. Estimate the population
   (a) 10 years from now
   (b) 20 years from now
   (c) 30 years from now

45. The atmospheric pressure, measured in pounds per square inch, is a function of the altitude above sea level. The equation $P(a) = 14.7e^{-0.21a}$, where a is the altitude measured in miles, can be used to approximate atmospheric pressure. Find the atmospheric pressure at each of the following locations. Express each answer to the nearest tenth of a pound per square inch.
   (a) Mount McKinley in Alaska—altitude of 3.85 miles
   (b) Denver, Colorado—the "mile-high" city
   (c) Asheville, North Carolina—altitude of 1985 feet (5280 feet = 1 mile)
   (d) Phoenix, Arizona—altitude of 1090 feet

■ ■ ■ **Thoughts into words**

46. Explain the difference between simple interest and compound interest.

47. Would it be better to invest $5000 at 6.25% interest compounded annually for 5 years or to invest $5000 at 6% interest compounded continuously for 5 years? Defend your answer.

Graphing calculator activities

48. Use a graphing calculator to check your graphs for Problems 33–38.

49. Graph $f(x) = 2^x$, $f(x) = e^x$, and $f(x) = 3^x$ on the same set of axes. Are these graphs consistent with the discussion prior to Figure 11.5?

50. Graph $f(x) = e^x$. Where should the graphs of $f(x) = e^{x-4}$, $f(x) = e^{x-6}$, and $f(x) = e^{x+5}$ be located? Graph all three functions on the same set of axes with $f(x) = e^x$.

51. Graph $f(x) = e^x$. Now predict the graphs for $f(x) = -e^x$, $f(x) = e^{-x}$, and $f(x) = -e^{-x}$. Graph all three functions on the same set of axes with $f(x) = e^x$.

52. How do you think the graphs of $f(x) = e^x$, $f(x) = e^{2x}$, and $f(x) = 2e^x$ will compare? Graph them on the same set of axes to see whether you were correct.

11.3 Logarithms

In Sections 11.1 and 11.2, (1) we learned about exponential expressions of the form b^n, where b is any positive real number and n is any real number, (2) we used exponential expressions of the form b^n to define exponential functions, and (3) we used exponential functions to help solve problems. In the next three sections we will follow the same basic pattern with respect to a new concept, that of a **logarithm.** Let's begin with the following definition.

> ### DEFINITION 11.2
>
> If r is any positive real number, then the unique exponent t such that $b^t = r$ is called the **logarithm of r with base b** and is denoted by $\log_b r$.

According to Definition 11.2, the logarithm of 8 base 2 is the exponent t such that $2^t = 8$; thus we can write $\log_2 8 = 3$. Likewise, we can write $\log_{10} 100 = 2$ because $10^2 = 100$. In general, we can remember Definition 11.2 in terms of the statement

$$\log_b r = t \quad \text{is equivalent to} \quad b^t = r$$

Thus we can easily switch back and forth between exponential and logarithmic forms of equations, as the next examples illustrate.

$$\log_3 81 = 4 \quad \text{is equivalent to} \quad 3^4 = 81$$
$$\log_{10} 100 = 2 \quad \text{is equivalent to} \quad 10^2 = 100$$
$$\log_{10} 0.001 = -3 \quad \text{is equivalent to} \quad 10^{-3} = 0.001$$
$$\log_2 128 = 7 \quad \text{is equivalent to} \quad 2^7 = 128$$
$$\log_m n = p \quad \text{is equivalent to} \quad m^p = n$$

$$2^4 = 16 \quad \text{is equivalent to } \log_2 16 = 4$$
$$5^2 = 25 \quad \text{is equivalent to } \log_5 25 = 2$$
$$\left(\frac{1}{2}\right)^4 = \frac{1}{16} \quad \text{is equivalent to } \log_{\frac{1}{2}}\left(\frac{1}{16}\right) = 4$$
$$10^{-2} = 0.01 \quad \text{is equivalent to } \log_{10} 0.01 = -2$$
$$a^b = c \quad \text{is equivalent to } \log_a c = b$$

We can conveniently calculate some logarithms by changing to exponential form, as in the next examples.

EXAMPLE 1 Evaluate $\log_4 64$.

Solution

Let $\log_4 64 = x$. Then, by switching to exponential form, we have $4^x = 64$, which we can solve as we did back in Section 11.1.

$$4^x = 64$$
$$4^x = 4^3$$
$$x = 3$$

Therefore, we can write $\log_4 64 = 3$. ∎

EXAMPLE 2 Evaluate $\log_{10} 0.1$.

Solution

Let $\log_{10} 0.1 = x$. Then, by switching to exponential form, we have $10^x = 0.1$, which we can solve as follows:

$$10^x = 0.1$$
$$10^x = \frac{1}{10}$$
$$10^x = 10^{-1}$$
$$x = -1$$

Thus we obtain $\log_{10} 0.1 = -1$ ∎

The link between logarithms and exponents also provides the basis for solving some equations that involve logarithms, as the next two examples illustrate.

E X A M P L E 3 Solve $\log_8 x = \dfrac{2}{3}$.

Solution

$$\log_8 x = \frac{2}{3}$$

$$8^{\frac{2}{3}} = x \qquad \text{By switching to exponential form}$$

$$\sqrt[3]{8^2} = x$$

$$(\sqrt[3]{8})^2 = x$$

$$4 = x$$

The solution set is {4}. ∎

E X A M P L E 4 Solve $\log_b 1000 = 3$.

Solution

$$\log_b 1000 = 3$$

$$b^3 = 1000$$

$$b = 10$$

The solution set is {10}. ∎

Properties of Logarithms

There are some properties of logarithms that are a direct consequence of Definition 11.2 and our knowledge of exponents. For example, writing the exponential equations $b^1 = b$ and $b^0 = 1$ in logarithmic form yields the following property.

> **PROPERTY 11.3**
>
> For $b > 0$ and $b \neq 1$,
>
>   **1.** $\log_b b = 1$ **2.** $\log_b 1 = 0$

Thus we can write

$$\log_{10} 10 = 1$$

$$\log_2 2 = 1$$

$$\log_{10} 1 = 0$$

$$\log_5 1 = 0$$

By Definition 11.2, $\log_b r$ is the exponent t such that $b^t = r$. Therefore, raising b to the $\log_b r$ power must produce r. We state this fact in Property 11.4.

PROPERTY 11.4

For $b > 0$, $b \neq 1$, and $r > 0$

$$b^{\log_b r} = r$$

The following examples illustrate Property 11.4.

$$10^{\,\log_{10} 19} = 19$$

$$2^{\log_2 14} = 14$$

$$e^{\log_e 5} = 5$$

Because a logarithm is by definition an exponent, it would seem reasonable to predict that there are some properties of logarithms that correspond to the basic exponential properties. This is an accurate prediction; these properties provide a basis for computational work with logarithms. Let's state the first of these properties and show how it can be verified by using our knowledge of exponents.

PROPERTY 11.5

For positive real numbers b, r, and s, where $b \neq 1$,

$$\log_b rs = \log_b r + \log_b s$$

To verify Property 11.5 we can proceed as follows: Let $m = \log_b r$ and $n = \log_b s$. Change each of these equations to exponential form.

$m = \log_b r$ becomes $r = b^m$

$n = \log_b s$ becomes $s = b^n$

Thus the product rs becomes

$$rs = b^m \cdot b^n = b^{m+n}$$

Now, by changing $rs = b^{m+n}$ back to logarithmic form, we obtain

$$\log_b rs = m + n$$

Replacing m with $\log_b r$ and n with $\log_b s$ yields

$$\log_b rs = \log_b r + \log_b s$$

The following three examples demonstrate a use of Property 11.5.

EXAMPLE 5   If $\log_2 5 = 2.3222$ and $\log_2 3 = 1.5850$, evaluate $\log_2 15$.

Solution
Because $15 = 5 \cdot 3$, we can apply Property 11.5.

$$\log_2 15 = \log_2 (5 \cdot 3)$$
$$= \log_2 5 + \log_2 3$$
$$= 2.3222 + 1.5850$$
$$= 3.9072$$

EXAMPLE 6   If $\log_{10} 178 = 2.2504$ and $\log_{10} 89 = 1.9494$, evaluate $\log_{10} (178 \cdot 89)$.

Solution

$$\log_{10} (178 \cdot 89) = \log_{10} 178 + \log_{10} 89$$
$$= 2.2504 + 1.9494$$
$$= 4.1998$$

EXAMPLE 7   If $\log_3 8 = 1.8928$, evaluate $\log_3 72$.

Solution

$$\log_3 72 = \log_3 (9 \cdot 8)$$
$$= \log_3 9 + \log_3 8$$
$$= 2 + 1.8928 \qquad \log_3 9 = 2 \text{ because } 3^2 = 9.$$
$$= 3.8928$$

Because $\dfrac{b^m}{b^n} = b^{m-n}$, we would expect a corresponding property pertaining to logarithms. There is such a property, Property 11.6.

PROPERTY 11.6

For positive numbers b, r, and s, where $b \neq 1$,

$$\log_b \left(\frac{r}{s} \right) = \log_b r - \log_b s$$

This property can be verified by using an approach similar to the one we used to verify Property 11.5. We leave it for you to do in an exercise in the next problem set.

We can use Property 11.6 to change a division problem into a subtraction problem, as in the next two examples.

EXAMPLE 8   If $\log_5 36 = 2.2265$ and $\log_5 4 = 0.8614$, evaluate $\log_5 9$.

Solution

Because $9 = \dfrac{36}{4}$, we can use Property 11.6.

$$\log_5 9 = \log_5 \left(\frac{36}{4} \right)$$

$$= \log_5 36 - \log_5 4$$

$$= 2.2265 - 0.8614$$

$$= 1.3651 \qquad \blacksquare$$

EXAMPLE 9   Evaluate $\log_{10} \left(\dfrac{379}{86} \right)$ given that $\log_{10} 379 = 2.5786$ and $\log_{10} 86 = 1.9345$.

Solution

$$\log_{10} \left(\frac{379}{86} \right) = \log_{10} 379 - \log_{10} 86$$

$$= 2.5786 - 1.9345$$

$$= 0.6441 \qquad \blacksquare$$

The next property of logarithms provides the basis for evaluating expressions such as $3^{\sqrt{2}}$, $(\sqrt{5})^{\frac{2}{3}}$, and $(0.076)^{\frac{2}{3}}$. We cite the property, consider a basis for its justification, and offer illustrations of its use.

PROPERTY 11.7

If r is a positive real number, b is a positive real number other than 1, and p is any real number, then

$$\log_b r^p = p(\log_b r)$$

As you might expect, the exponential property $(b^n)^m = b^{mn}$ plays an important role in the verification of Property 11.7. This is an exercise for you in the next problem set. Let's look at some uses of Property 11.7.

EXAMPLE 10

Evaluate $\log_2 22^{\frac{1}{3}}$ given that $\log_2 22 = 4.4598$.

Solution

$$\log_2 22^{\frac{1}{3}} = \frac{1}{3} \log_2 22 \qquad \text{Property 11.7}$$

$$= \frac{1}{3}(4.4598)$$

$$= 1.4866$$

■

EXAMPLE 11

Evaluate $\log_{10}(8540)^{\frac{3}{5}}$ given that $\log_{10} 8540 = 3.9315$.

Solution

$$\log_{10}(8540)^{\frac{3}{5}} = \frac{3}{5} \log_{10} 8540 \qquad \text{Property 11.7}$$

$$= \frac{3}{5}(3.9315)$$

$$= 2.3589$$

■

Together, the properties of logarithms enable us to change the forms of various logarithmic expressions. For example, an expression such as $\log_b \sqrt{\dfrac{xy}{z}}$ can be rewritten in terms of sums and differences of simpler logarithmic quantities as follows:

$$\log_b \sqrt{\frac{xy}{z}} = \log_b \left(\frac{xy}{z}\right)^{\frac{1}{2}}$$

$$= \frac{1}{2}\log_b \left(\frac{xy}{z}\right) \qquad \text{Property 11.7}$$

$$= \frac{1}{2}(\log_b xy - \log_b z) \qquad \text{Property 11.6}$$

$$= \frac{1}{2}(\log_b x + \log_b y - \log_b z) \qquad \text{Property 11.5}$$

Sometimes we need to change from an indicated sum or difference of logarithmic quantities to an indicated product or quotient. This is especially helpful when solving certain kinds of equations that involve logarithms. Note in these next two examples how we can use the properties, along with the process of changing from logarithmic form to exponential form, to solve some equations.

EXAMPLE 12 Solve $\log_{10} x + \log_{10}(x + 9) = 1$.

Solution

$$\log_{10} x + \log_{10}(x + 9) = 1$$

$$\log_{10}[x(x + 9)] = 1 \qquad \text{Property 11.5}$$

$$10^1 = x(x + 9) \qquad \text{Change to exponential form.}$$

$$10 = x^2 + 9x$$

$$0 = x^2 + 9x - 10$$

$$0 = (x + 10)(x - 1)$$

$$x + 10 = 0 \qquad \text{or} \qquad x - 1 = 0$$

$$x = -10 \qquad \text{or} \qquad x = 1$$

Because logarithms are defined only for positive numbers, x and $x + 9$ have to be positive. Therefore, the solution of -10 must be discarded. The solution set is $\{1\}$. ■

EXAMPLE 13 Solve $\log_5(x + 4) - \log_5 x = 2$.

Solution

$$\log_5(x + 4) - \log_5 x = 2$$

$$\log_5\left(\frac{x + 4}{x}\right) = 2 \qquad \text{Property 11.6}$$

$$5^2 = \frac{x + 4}{x} \qquad \text{Change to exponential form.}$$

$$25 = \frac{x + 4}{x}$$

$$25x = x + 4$$

$$24x = 4$$

$$x = \frac{4}{24} = \frac{1}{6}$$

The solution set is $\left\{\dfrac{1}{6}\right\}$. ■

PROBLEM SET 11.3

For Problems 1–10, write each of the following in logarithmic form. For example, $2^3 = 8$ becomes $\log_2 8 = 3$ in logarithmic form.

1. $2^7 = 128$

2. $3^3 = 27$

3. $5^3 = 125$

4. $2^6 = 64$

5. $10^3 = 1000$

6. $10^1 = 10$

7. $2^{-2} = \left(\dfrac{1}{4}\right)$ **8.** $3^{-4} = \left(\dfrac{1}{81}\right)$

9. $10^{-1} = 0.1$ **10.** $10^{-2} = 0.01$

For Problems 11–20, write each of the following in exponential form. For example, $\log_2 8 = 3$ becomes $2^3 = 8$ in exponential form.

11. $\log_3 81 = 4$ **12.** $\log_2 256 = 8$

13. $\log_4 64 = 3$ **14.** $\log_5 25 = 2$

15. $\log_{10} 10{,}000 = 4$ **16.** $\log_{10} 100{,}000 = 5$

17. $\log_2 \left(\dfrac{1}{16}\right) = -4$ **18.** $\log_5 \left(\dfrac{1}{125}\right) = -3$

19. $\log_{10} 0.001 = -3$ **20.** $\log_{10} 0.000001 = -6$

For Problems 21–40, evaluate each expression.

21. $\log_2 16$ **22.** $\log_3 9$

23. $\log_3 81$ **24.** $\log_2 512$

25. $\log_6 216$ **26.** $\log_4 256$

27. $\log_7 \sqrt{7}$ **28.** $\log_2 \sqrt[3]{2}$

29. $\log_{10} 1$ **30.** $\log_{10} 10$

31. $\log_{10} 0.1$ **32.** $\log_{10} 0.0001$

33. $10^{\log_{10} 5}$ **34.** $10^{\log_{10} 14}$

35. $\log_2 \left(\dfrac{1}{32}\right)$ **36.** $\log_5 \left(\dfrac{1}{25}\right)$

37. $\log_5 (\log_2 32)$ **38.** $\log_2 (\log_4 16)$

39. $\log_{10} (\log_7 7)$ **40.** $\log_2 (\log_5 5)$

For Problems 41–50, solve each equation.

41. $\log_7 x = 2$ **42.** $\log_2 x = 5$

43. $\log_8 x = \dfrac{4}{3}$ **44.** $\log_{16} x = \dfrac{3}{2}$

45. $\log_9 x = \dfrac{3}{2}$ **46.** $\log_8 x = -\dfrac{2}{3}$

47. $\log_4 x = -\dfrac{3}{2}$ **48.** $\log_9 x = -\dfrac{5}{2}$

49. $\log_x 2 = \dfrac{1}{2}$ **50.** $\log_x 3 = \dfrac{1}{2}$

For Problems 51–59, you are given that $\log_2 5 = 2.3219$ and $\log_2 7 = 2.8074$. Evaluate each expression using Properties 11.5–11.7.

51. $\log_2 35$ **52.** $\log_2 \left(\dfrac{7}{5}\right)$

53. $\log_2 125$ **54.** $\log_2 49$

55. $\log_2 \sqrt{7}$ **56.** $\log_2 \sqrt[3]{5}$

57. $\log_2 175$ **58.** $\log_2 56$

59. $\log_2 80$

For Problems 60–68, you are given that $\log_8 5 = 0.7740$ and $\log_8 11 = 1.1531$. Evaluate each expression using Properties 11.5–11.7.

60. $\log_8 55$ **61.** $\log_8 \left(\dfrac{5}{11}\right)$

62. $\log_8 25$ **63.** $\log_8 \sqrt{11}$

64. $\log_8 (5)^{\frac{2}{3}}$ **65.** $\log_8 88$

66. $\log_8 320$ **67.** $\log_8 \left(\dfrac{25}{11}\right)$

68. $\log_8 \left(\dfrac{121}{25}\right)$

For Problems 69–80, express each of the following as the sum or difference of simpler logarithmic quantities. Assume that all variables represent positive real numbers. For example,

$$\log_b \frac{x^3}{y^2} = \log_b x^3 - \log_b y^2$$

$$= 3 \log_b x - 2 \log_b y$$

69. $\log_b xyz$ **70.** $\log_b 5x$

71. $\log_b \left(\dfrac{y}{z}\right)$ **72.** $\log_b \left(\dfrac{x^2}{y}\right)$

73. $\log_b y^3 z^4$ **74.** $\log_b x^2 y^3$

75. $\log_b \left(\dfrac{x^{\frac{1}{2}} y^{\frac{1}{3}}}{z^4}\right)$ **76.** $\log_b x^{\frac{2}{3}} y^{\frac{3}{4}}$

77. $\log_b \sqrt[3]{x^2 z}$ **78.** $\log_b \sqrt{xy}$

79. $\log_b \left(x \sqrt{\dfrac{x}{y}} \right)$ **80.** $\log_b \sqrt{\dfrac{x}{y}}$

For Problems 81–92, solve each of the equations.

81. $\log_3 x + \log_3 4 = 2$

82. $\log_7 5 + \log_7 x = 1$

83. $\log_{10} x + \log_{10} (x - 21) = 2$

84. $\log_{10} x + \log_{10} (x - 3) = 1$

85. $\log_2 x + \log_2 (x - 3) = 2$

86. $\log_3 x + \log_3 (x - 2) = 1$

87. $\log_{10} (2x - 1) - \log_{10} (x - 2) = 1$

88. $\log_{10} (9x - 2) = 1 + \log_{10} (x - 4)$

89. $\log_5 (3x - 2) = 1 + \log_5 (x - 4)$

90. $\log_6 x + \log_6 (x + 5) = 2$

91. $\log_8 (x + 7) + \log_8 x = 1$

92. $\log_6 (x + 1) + \log_6 (x - 4) = 2$

93. Verify Property 11.6.

94. Verify Property 11.7.

■ ■ ■ **Thoughts into words**

95. Explain, without using Property 11.4, why $4^{\log_4 9}$ equals 9.

96. How would you explain the concept of a logarithm to someone who had just completed an elementary algebra course?

97. In the next section we will show that the logarithmic function $f(x) = \log_2 x$ is the inverse of the exponential function $f(x) = 2^x$. From that information, how could you sketch a graph of $f(x) = \log_2 x$?

11.4 Logarithmic Functions

We can now use the concept of a logarithm to define a new function.

> ## DEFINITION 11.3
>
> If $b > 0$ and $b \neq 1$, then the function f defined by
>
> $$f(x) = \log_b x$$
>
> where x is any positive real number, is called the **logarithmic function with base b.**

We can obtain the graph of a specific logarithmic function in various ways. For example, we can change the equation $y = \log_2 x$ to the exponential equation $2^y = x$, where we can determine a table of values. The next set of exercises asks you to graph some logarithmic functions with this approach.

We can obtain the graph of a logarithmic function by setting up a table of values directly from the logarithmic equation. Example 1 illustrates this approach.

E X A M P L E   1

Graph $f(x) = \log_2 x$.

Solution

Let's choose some values for x where the corresponding values for $\log_2 x$ are easily determined. (Remember that logarithms are defined only for the positive real numbers.)

x	$f(x)$	
$\dfrac{1}{8}$	-3	$\log_2 \dfrac{1}{8} = -3$ because $2^{-3} = \dfrac{1}{2^3} = \dfrac{1}{8}$
$\dfrac{1}{4}$	-2	
$\dfrac{1}{2}$	-1	
1	0	$\log_2 1 = 0$ because $2^0 = 1$
2	1	
4	2	
8	3	

Plot these points and connect them with a smooth curve to produce Figure 11.7.

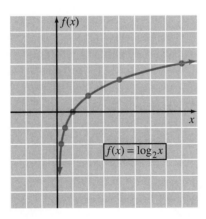

Figure 11.7

Suppose that we consider the following two functions f and g.

$$f(x) = b^x \qquad \text{Domain: all real numbers}$$
$$\text{Range: positive real numbers}$$

$$g(x) = \log_b x \qquad \text{Domain: positive real numbers}$$
$$\text{Range: all real numbers}$$

Furthermore, suppose that we consider the composition of f and g and the composition of g and f.

$$(f \circ g)(x) = f(g(x)) = f(\log_b x) = b^{\log_b x} = x$$

$$(g \circ f)(x) = g(f(x)) = g(b^x) = \log_b b^x = x \log_b b = x(1) = x$$

Therefore, because the domain of f is the range of g, the range of f is the domain of g, $f(g(x)) = x$, and $g(f(x)) = x$, the two functions f and g are inverses of each other.

Remember also from Chapter 9 that the graphs of a function and its inverse are reflections of each other through the line $y = x$. Thus the graph of a logarithmic function can also be determined by reflecting the graph of its inverse exponential function through the line $y = x$. We see this in Figure 11.8, where the graph of $y = 2^x$ has been reflected across the line $y = x$ to produce the graph of $y = \log_2 x$.

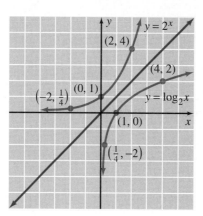

Figure 11.8

The general behavior patterns of exponential functions were illustrated by two graphs back in Figure 11.3. We can now reflect each of those graphs through the line $y = x$ and observe the general behavior patterns of logarithmic functions, as shown in Figure 11.9.

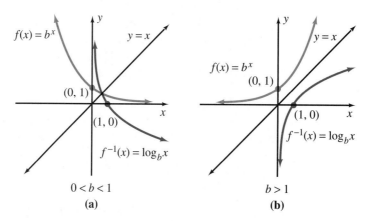

Figure 11.9

Finally, when graphing logarithmic functions, don't forget about variations of the basic curves.

1. The graph of $f(x) = 3 + \log_2 x$ is the graph of $f(x) = \log_2 x$ moved up 3 units. (Because $\log_2 x + 3$ is apt to be confused with $\log_2 (x + 3)$, we commonly write $3 + \log_2 x$.)
2. The graph of $f(x) = \log_2 (x - 4)$ is the graph of $f(x) = \log_2 x$ moved 4 units to the right.
3. The graph of $f(x) = -\log_2 x$ is the graph of $f(x) = \log_2 x$ reflected across the x axis.

Common Logarithms—Base 10

The properties of logarithms we discussed in Section 11.3 are true for any valid base. For example, because the Hindu-Arabic numeration system that we use is a base-10 system, logarithms to base 10 have historically been used for computational purposes. Base-10 logarithms are called **common logarithms.**

Originally, common logarithms were developed to assist in complicated numerical calculations that involved products, quotients, and powers of real numbers. Today they are seldom used for that purpose, because the calculator and computer can much more effectively handle the messy computational problems. However, common logarithms do still occur in applications; they are deserving of our attention.

As we know from earlier work, the definition of a logarithm provides the basis for evaluating $\log_{10} x$ for values of x that are integral powers of 10. Consider the following examples.

$$\log_{10} 1000 = 3 \quad \text{because } 10^3 = 1000$$

$$\log_{10} 100 = 2 \quad \text{because } 10^2 = 100$$

$$\log_{10} 10 = 1 \quad \text{because } 10^1 = 10$$

$$\log_{10} 1 = 0 \quad \text{because } 10^0 = 1$$

$$\log_{10} 0.1 = -1 \quad \text{because } 10^{-1} = \frac{1}{10} = 0.1$$

$$\log_{10} 0.01 = -2 \quad \text{because } 10^{-2} = \frac{1}{10^2} = 0.01$$

$$\log_{10} 0.001 = -3 \quad \text{because } 10^{-3} = \frac{1}{10^3} = 0.001$$

When working with base-10 logarithms, it is customary to omit writing the numeral 10 to designate the base. Thus the expression $\log_{10} x$ is written as $\log x$, and a statement such as $\log_{10} 1000 = 3$ becomes $\log 1000 = 3$. We will follow this practice from now on in this chapter, but don't forget that the base is understood to be 10.

$$\log_{10} x = \log x$$

To find the common logarithm of a positive number that is not an integral power of 10, we can use an appropriately equipped calculator. Using a calculator equipped with a common logarithm function (ordinarily a key labeled $\boxed{\log}$ is used), we obtained the following results, rounded to four decimal places.

$\log 1.75 = 0.2430$

$\log 23.8 = 1.3766$

$\log 134 = 2.1271$ (Be sure that you can use a calculator and obtain these results.)

$\log 0.192 = -0.7167$

$\log 0.0246 = -1.6091$

In order to use logarithms to solve problems, we sometimes need to be able to determine a number when the logarithm of the number is known. That is, we may need to determine x when $\log x$ is known. Let's consider an example.

EXAMPLE 2

Find x if $\log x = 0.2430$.

Solution

If $\log x = 0.2430$, then by changing to exponential form, we have $10^{0.2430} = x$. Therefore, using the $\boxed{10^x}$ key, we can find x.

$$x = 10^{0.2430} \approx 1.749846689$$

Therefore, $x = 1.7498$, rounded to five significant digits. ■

Be sure that you can use your calculator to obtain the following results. We have rounded the values for x to five significant digits.

If $\log x = 0.7629$, then $x = 10^{0.7629} = 5.7930$.

If $\log x = 1.4825$, then $x = 10^{1.4825} = 30.374$.

If $\log x = 4.0214$, then $x = 10^{4.0214} = 10,505$.

If $\log x = -1.5162$, then $x = 10^{-1.5162} = 0.030465$.

If $\log x = -3.8921$, then $x = 10^{-3.8921} = 0.00012820$.

The **common logarithmic function** is defined by the equation $f(x) = \log x$. It should now be a simple matter to set up a table of values and sketch the function. We will have you do this in the next set of exercises. Remember that $f(x) = 10^x$ and $g(x) = \log x$ are inverses of each other. Therefore, we could also get the graph of $g(x) = \log x$ by reflecting the exponential curve $f(x) = 10^x$ across the line $y = x$.

Natural Logarithms—Base e

In many practical applications of logarithms, the number e (remember that $e \approx 2.71828$) is used as a base. Logarithms with a base of e are called **natural logarithms,** and the symbol $\ln x$ is commonly used instead of $\log_e x$.

$$\log_e x = \ln x$$

Natural logarithms can be found with an appropriately equipped calculator. Using a calculator with a natural logarithm function (ordinarily a key labeled $\boxed{\ln x}$), we can obtain the following results, rounded to four decimal places.

$$\ln 3.21 = 1.1663$$
$$\ln 47.28 = 3.8561$$
$$\ln 842 = 6.7358$$
$$\ln 0.21 = -1.5606$$
$$\ln 0.0046 = -5.3817$$
$$\ln 10 = 2.3026$$

Be sure that you can use your calculator to obtain these results. Keep in mind the significance of a statement such as $\ln 3.21 = 1.1663$. By changing to exponential form, we are claiming that e raised to the 1.1663 power is approximately 3.21. Using a calculator, we obtain $e^{1.1663} = 3.210093293$.

Let's do a few more problems and find x when given $\ln x$. Be sure that you agree with these results.

If $\ln x = 2.4156$, then $x = e^{2.4156} = 11.196$.

If $\ln x = 0.9847$, then $x = e^{0.9847} = 2.6770$.

If $\ln x = 4.1482$, then $x = e^{4.1482} = 63.320$.

If $\ln x = -1.7654$, then $x = e^{-1.7654} = 0.17112$.

The **natural logarithmic function** is defined by the equation $f(x) = \ln x$. It is the inverse of the natural exponential function $f(x) = e^x$. Thus one way to graph $f(x) = \ln x$ is to reflect the graph of $f(x) = e^x$ across the line $y = x$. We will have you do this in the next set of problems.

PROBLEM SET 11.4

For Problems 1–10, use a calculator to find each common logarithm. Express answers to four decimal places.

1. log 7.24

2. log 2.05

3. log 52.23

4. log 825.8

5. log 3214.1

6. log 14,189

7. log 0.729

8. log 0.04376

9. log 0.00034

10. log 0.000069

For Problems 11–20, use your calculator to find x when given log x. Express answers to five significant digits.

11. $\log x = 2.6143$

12. $\log x = 1.5263$

13. $\log x = 4.9547$

14. $\log x = 3.9335$

15. $\log x = 1.9006$

16. $\log x = 0.5517$

17. $\log x = -1.3148$

18. $\log x = -0.1452$

19. $\log x = -2.1928$

20. $\log x = -2.6542$

For Problems 21–30, use your calculator to find each natural logarithm. Express answers to four decimal places.

21. ln 5

22. ln 18

23. ln 32.6

24. ln 79.5

25. ln 430

26. ln 371.8

27. ln 0.46   **28.** ln 0.524

29. ln 0.0314   **30.** ln 0.008142

For Problems 31–40, use your calculator to find x when given ln x. Express answers to five significant digits.

31. ln $x = 0.4721$ **32.** ln $x = 0.9413$

33. ln $x = 1.1425$ **34.** ln $x = 2.7619$

35. ln $x = 4.6873$ **36.** ln $x = 3.0259$

37. ln $x = -0.7284$ **38.** ln $x = -1.6246$

39. ln $x = -3.3244$ **40.** ln $x = -2.3745$

41. (a) Complete the following table and then graph $f(x) = $ log x. (Express the values for log x to the nearest tenth.)

x	0.1	0.5	1	2	4	8	10
log x							

(b) Complete the following table and express values for 10^x to the nearest tenth.

x	−1	−0.3	0	0.3	0.6	0.9	1
10^x							

Then graph $f(x) = 10^x$ and reflect it across the line $y = x$ to produce the graph for $f(x) = $ log x.

42. (a) Complete the following table and then graph $f(x) = $ ln x. (Express the values for ln x to the nearest tenth.)

x	0.1	0.5	1	2	4	8	10
ln x							

(b) Complete the following table and express values for e^x to the nearest tenth.

x	−2.3	−0.7	0	0.7	1.4	2.1	2.3
e^x							

Then graph $f(x) = e^x$ and reflect it across the line $y = x$ to produce the graph for $f(x) = $ ln x.

43. Graph $y = \log_{\frac{1}{2}} x$ by graphing $\left(\dfrac{1}{2}\right)^y = x$.

44. Graph $y = \log_2 x$ by graphing $2^y = x$.

45. Graph $f(x) = \log_3 x$ by reflecting the graph of $g(x) = 3^x$ across the line $y = x$.

46. Graph $f(x) = \log_4 x$ by reflecting the graph of $g(x) = 4^x$ across the line $y = x$.

For Problems 47–53, graph each of the functions. Remember that the graph of $f(x) = \log_2 x$ is given in Figure 11.7.

47. $f(x) = 3 + \log_2 x$ **48.** $f(x) = -2 + \log_2 x$

49. $f(x) = \log_2(x + 3)$ **50.** $f(x) = \log_2(x - 2)$

51. $f(x) = \log_2 2x$ **52.** $f(x) = -\log_2 x$

53. $f(x) = 2 \log_2 x$

For Problems 54–61, perform the following calculations and express answers to the nearest hundredth. (These calculations are in preparation for our work in the next section.)

54. $\dfrac{\log 7}{\log 3}$ **55.** $\dfrac{\ln 2}{\ln 7}$

56. $\dfrac{2 \ln 3}{\ln 8}$ **57.** $\dfrac{\ln 5}{2 \ln 3}$

58. $\dfrac{\ln 3}{0.04}$ **59.** $\dfrac{\ln 2}{0.03}$

60. $\dfrac{\log 2}{5 \log 1.02}$ **61.** $\dfrac{\log 5}{3 \log 1.07}$

■ ■ ■ **Thoughts into words**

62. Why is the number 1 excluded from being a base of a logarithmic function?

63. How do we know that $\log_2 6$ is between 2 and 3?

Graphing calculator activities

64. Graph $f(x) = x$, $f(x) = e^x$, and $f(x) = \ln x$ on the same set of axes.

65. Graph $f(x) = x$, $f(x) = 10^x$, and $f(x) = \log x$ on the same set of axes.

66. Graph $f(x) = \ln x$. How should the graphs of $f(x) = 2 \ln x$, $f(x) = 4 \ln x$, and $f(x) = 6 \ln x$ compare to the graph of $f(x) = \ln x$? Graph the three functions on the same set of axes with $f(x) = \ln x$.

67. Graph $f(x) = \log x$. Now predict the graphs for $f(x) = 2 + \log x$, $f(x) = -2 + \log x$, and $f(x) = -6 + \log x$. Graph the three functions on the same set of axes with $f(x) = \log x$.

68. Graph $\ln x$. Now predict the graphs for $f(x) = \ln (x - 2)$, $f(x) = \ln (x - 6)$, and $f(x) = \ln (x + 4)$. Graph the three functions on the same set of axes with $f(x) = \ln x$.

69. For each of the following, (a) predict the general shape and location of the graph, and (b) use your graphing calculator to graph the function and thus check your prediction.

(a) $f(x) = \log x + \ln x$

(b) $f(x) = \log x - \ln x$

(c) $f(x) = \ln x - \log x$

(d) $f(x) = \ln x^2$

11.5 Exponential Equations, Logarithmic Equations, and Problem Solving

In Section 11.1 we solved exponential equations such as $3^x = 81$ when we expressed both sides of the equation as a power of 3 and then applied the property "if $b^n = b^m$, then $n = m$." However, if we try to use this same approach with an equation such as $3^x = 5$, we face the difficulty of expressing 5 as a power of 3. We can solve this type of problem by using the properties of logarithms and the following property of equality.

> **PROPERTY 11.8**
>
> If $x > 0$, $y > 0$, and $b \neq 1$, then
>
> $$x = y \quad \text{if and only if} \quad \log_b x = \log_b y$$

Property 11.8 is stated in terms of any valid base b; however, for most applications we use either common logarithms (base 10) or natural logarithms (base e). Let's consider some examples.

EXAMPLE 1 Solve $3^x = 5$ to the nearest hundredth.

Solution

By using common logarithms, we can proceed as follows:

$$3^x = 5$$

$$\log 3^x = \log 5 \qquad \text{Property 11.8}$$

$$x \log 3 = \log 5 \qquad \log r^p = p \log r$$

$$x = \frac{\log 5}{\log 3}$$

$$x = 1.46, \quad \text{to the nearest hundredth}$$

Check

Because $3^{1.46} \approx 4.972754647$, we say that, to the nearest hundredth, the solution set for $3^x = 5$ is $\{1.46\}$. ∎

E X A M P L E   2

Solve $e^{x+1} = 5$ to the nearest hundredth.

Solution

The base e is used in the exponential expression, so let's use natural logarithms to help solve this equation.

$$e^{x+1} = 5$$

$$\ln e^{x+1} = \ln 5 \qquad \text{Property 11.8}$$

$$(x + 1)\ln e = \ln 5 \qquad \ln r^p = p \ln r$$

$$(x + 1)(1) = \ln 5 \qquad \ln e = 1$$

$$x = \ln 5 - 1$$

$$x \approx 0.609437912$$

$$x = 0.61, \quad \text{to the nearest hundredth}$$

The solution set is $\{0.61\}$. Check it! ∎

Logarithmic Equations

In Example 12 of Section 11.3 we solved the logarithmic equation

$$\log_{10} x + \log_{10}(x + 9) = 1$$

by simplifying the left side of the equation to $\log_{10}[x(x + 9)]$ and then changing the equation to exponential form to complete the solution. Now, using Property 11.8, we can solve such a logarithmic equation another way and also expand our equation-solving capabilities. Let's consider some examples.

E X A M P L E   3

Solve $\log x + \log(x - 15) = 2$.

Solution

Because $\log 100 = 2$, the given equation becomes

$$\log x + \log(x - 15) = \log 100$$

Now simplify the left side, apply Property 11.8, and proceed as follows:

$$\log(x)(x - 15) = \log 100$$

$$x(x - 15) = 100$$

$$x^2 - 15x - 100 = 0$$

$$(x - 20)(x + 5) = 0$$

$$x - 20 = 0 \quad \text{or} \quad x + 5 = 0$$

$$x = 20 \quad \text{or} \quad x = -5$$

The domain of a logarithmic function must contain only positive numbers, so x and $x - 15$ must be positive in this problem. Therefore, we discard the solution -5; the solution set is $\{20\}$. ■

EXAMPLE 4 Solve $\ln(x + 2) = \ln(x - 4) + \ln 3$.

Solution

$$\ln(x + 2) = \ln(x - 4) + \ln 3$$

$$\ln(x + 2) = \ln[3(x - 4)]$$

$$x + 2 = 3(x - 4)$$

$$x + 2 = 3x - 12$$

$$14 = 2x$$

$$7 = x$$

The solution set is $\{7\}$. ■

Problem Solving

In Section 11.2 we used the compound interest formula

$$A = P\left(1 + \frac{r}{n}\right)^{nt}$$

to determine the amount of money (A) accumulated at the end of t years if P dollars is invested at r rate of interest compounded n times per year. Now let's use this formula to solve other types of problems that deal with compound interest.

EXAMPLE 5 How long will $500 take to double itself if it is invested at 12% interest compounded quarterly?

Solution

To "double itself" means that the $500 will grow into $1000. Thus

$$1000 = 500\left(1 + \frac{0.12}{4}\right)^{4t}$$

$$= 500(1 + 0.03)^{4t}$$

$$= 500(1.03)^{4t}$$

Multiply both sides of $1000 = 500(1.03)^{4t}$ by $\dfrac{1}{500}$ to yield

$$2 = (1.03)^{4t}$$

Therefore,

$$\log 2 = \log(1.03)^{4t} \qquad \text{Property 11.8}$$

$$\log 2 = 4t \log 1.03 \qquad \log r^p = p \log r$$

Solve for t to obtain

$$\log 2 = 4t \log 1.03$$

$$\frac{\log 2}{\log 1.03} = 4t$$

$$\frac{\log 2}{4 \log 1.03} = t \qquad \text{Multiply both sides by } \frac{1}{4}.$$

$$t \approx 5.862443063$$

$$t = 5.9, \quad \text{to the nearest tenth}$$

Therefore, we are claiming that $500 invested at 12% interest compounded quarterly will double itself in approximately 5.9 years.

 Check

$500 invested at 12% compounded quarterly for 5.9 years will produce

$$A = \$500\left(1 + \frac{0.12}{4}\right)^{4(5.9)}$$

$$= \$500(1.03)^{23.6}$$

$$= \$1004.45 \qquad\qquad\qquad\qquad\qquad \blacksquare$$

In Section 11.2, we also used the formula $A = Pe^{rt}$ when money was to be compounded continuously. At this time, with the help of natural logarithms, we can extend our use of this formula.

EXAMPLE 6   How long will it take $100 to triple itself if it is invested at 8% interest compounded continuously?

Solution

To "triple itself" means that the $100 will grow into $300. Thus, using the formula for interest that is compounded continuously, we can proceed as follows:

$$A = Pe^{rt}$$

$$\$300 = \$100e^{(0.08)t}$$

$$3 = e^{0.08t}$$

$$\ln 3 = \ln e^{0.08t} \qquad \text{Property 11.8}$$

$$\ln 3 = 0.08t \ln e \qquad \ln r^p = p \ln r$$

$$\ln 3 = 0.08t \qquad \ln e = 1$$

$$\frac{\ln 3}{0.08} = t$$

$$t \approx 13.73265361$$

$$t = 13.7, \quad \text{to the nearest tenth}$$

Therefore, $100 invested at 8% interest compounded continuously will triple itself in approximately 13.7 years.

Check

$100 invested at 8% compounded continuously for 13.7 years produces

$$A = Pe^{rt}$$

$$= \$100e^{0.08(13.7)}$$

$$= \$100e^{1.096}$$

$$\approx \$299.22$$ ■

Richter Numbers

Seismologists use the Richter scale to measure and report the magnitude of earthquakes. The equation

$$R = \log \frac{I}{I_0} \qquad \text{\textit{R} is called a Richter number.}$$

compares the intensity I of an earthquake to a minimum or reference intensity I_0. The reference intensity is the smallest earth movement that can be recorded on a seismograph. Suppose that the intensity of an earthquake was determined to be 50,000 times the reference intensity. In this case, $I = 50,000I_0$ and the Richter number is calculated as follows:

$$R = \log \frac{50,000 I_0}{I_0}$$

$$R = \log 50,000$$

$$R \approx 4.698970004$$

Thus a Richter number of 4.7 would be reported. Let's consider two more examples that involve Richter numbers.

EXAMPLE 7

An earthquake in San Francisco in 1989 was reported to have a Richter number of 6.9. How did its intensity compare to the reference intensity?

Solution

$$6.9 = \log \frac{I}{I_0}$$

$$10^{6.9} = \frac{I}{I_0}$$

$$I = (10^{6.9})(I_0)$$

$$I \approx 7943282 I_0$$

Thus its intensity was a little less than 8 million times the reference intensity. ■

EXAMPLE 8

An earthquake in Iran in 1990 had a Richter number of 7.7. Compare the intensity level of that earthquake to the one in San Francisco (Example 7).

Solution

From Example 7 we have $I = (10^{6.9})(I_0)$ for the earthquake in San Francisco. Using a Richter number of 7.7, we obtain $I = (10^{7.7})(I_0)$ for the earthquake in Iran. Therefore, by comparison,

$$\frac{(10^{7.7})(I_0)}{(10^{6.9})(I_0)} = 10^{7.7-6.9} = 10^{0.8} \approx 6.3$$

The earthquake in Iran was about 6 times as intense as the one in San Francisco. ■

Change-of-Base for Logarithms with Bases Other Than 10 or e

Now let's use either common or natural logarithms to evaluate logarithms that have bases other than 10 or e. Consider the following example.

EXAMPLE 9

Evaluate $\log_3 41$.

Solution

Let $x = \log_3 41$. Change to exponential form to obtain

$$3^x = 41$$

Now we can apply Property 11.8 and proceed as follows:

$$\log 3^x = \log 41$$

$$x \log 3 = \log 41$$

$$x = \frac{\log 41}{\log 3}$$

$$x = \frac{1.6128}{0.4771} = 3.3804, \quad \text{rounded to four decimal places}$$

Using the method of Example 9 to evaluate $\log_a r$ produces the following formula, which we often refer to as the **change-of-base** formula for logarithms.

PROPERTY 11.9

If a, b, and r are positive numbers with $a \neq 1$ and $b \neq 1$, then

$$\log_a r = \frac{\log_b r}{\log_b a}$$

By using Property 11.9, we can easily determine a relationship between logarithms of different bases. For example, suppose that in Property 11.9 we let $a = 10$ and $b = e$. Then

$$\log_a r = \frac{\log_b r}{\log_b a}$$

becomes

$$\log_{10} r = \frac{\log_e r}{\log_e 10}$$

which can be written as

$$\log_e r = (\log_e 10)(\log_{10} r)$$

Because $\log_e 10 = 2.3026$, rounded to four decimal places, we have

$$\log_e r = (2.3026)(\log_{10} r)$$

Thus the natural logarithm of any positive number is approximately equal to the common logarithm of the number times 2.3026.

Now we can use a graphing utility to graph logarithmic functions such as $f(x) = \log_2 x$. Using the change-of-base formula, we can express this function as $f(x) = \dfrac{\log x}{\log 2}$ or as $f(x) = \dfrac{\ln x}{\ln 2}$. The graph of $f(x) = \log_2 x$ is shown in Figure 11.10.

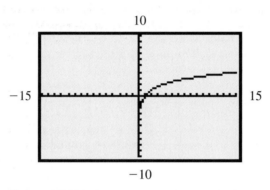

Figure 11.10

PROBLEM SET 11.5

For Problems 1–14, solve each exponential equation and express solutions to the nearest hundredth.

1. $3^x = 32$

2. $2^x = 40$

3. $4^x = 21$

4. $5^x = 73$

5. $3^{x-2} = 11$

6. $2^{x+1} = 7$

7. $5^{3x+1} = 9$

8. $7^{2x-1} = 35$

9. $e^x = 5.4$

10. $e^x = 45$

11. $e^{x-2} = 13.1$

12. $e^{x-1} = 8.2$

13. $3e^x = 35.1$

14. $4e^x - 2 = 26$

For Problems 15–22, solve each logarithmic equation.

15. $\log x + \log(x + 21) = 2$

16. $\log x + \log(x + 3) = 1$

17. $\log(3x - 1) = 1 + \log(5x - 2)$

18. $\log(2x - 1) - \log(x - 3) = 1$

19. $\log(x + 2) - \log(2x + 1) = \log x$

20. $\log(x + 1) - \log(x + 2) = \log\dfrac{1}{x}$

21. $\ln(2t + 5) = \ln 3 + \ln(t - 1)$

22. $\ln(3t - 4) - \ln(t + 1) = \ln 2$

For Problems 23–32, approximate each of the following logarithms to three decimal places.

23. $\log_2 23$

24. $\log_3 32$

25. $\log_6 0.214$

26. $\log_5 1.4$

27. $\log_7 421$

28. $\log_8 514$

29. $\log_9 0.0017$

30. $\log_4 0.00013$

31. $\log_3 720$

32. $\log_2 896$

For Problems 33–41, solve each problem and express answers to the nearest tenth.

33. How long will it take $750 to be worth $1000 if it is invested at 12% interest compounded quarterly?

34. How long will it take $1000 to double itself if it is invested at 9% interest compounded semiannually?

35. How long will it take $2000 to double itself if it is invested at 13% interest compounded continuously?

36. How long will it take $500 to triple itself if it is invested at 9% interest compounded continuously?

37. For a certain strain of bacteria, the number present after t hours is given by the equation $Q = Q_0 e^{0.34t}$, where Q_0 represents the initial number of bacteria. How long will it take 400 bacteria to increase to 4000 bacteria?

38. A piece of machinery valued at $30,000 depreciates at a rate of 10% yearly. How long will it take until it has a value of $15,000?

39. The number of grams of a certain radioactive substance present after t hours is given by the equation $Q = Q_0 e^{-0.45t}$, where Q_0 represents the initial number of grams. How long will it take 2500 grams to be reduced to 1250 grams?

40. For a certain culture the equation $Q(t) = Q_0 e^{0.4t}$, where Q_0 is an initial number of bacteria and t is time measured in hours, yields the number of bacteria as a function of time. How long will it take 500 bacteria to increase to 2000?

41. Suppose that the equation $P(t) = P_0 e^{0.02t}$, where P_0 represents an initial population and t is the time in years, is used to predict population growth. How long does this equation predict it would take a city of 50,000 to double its population?

Solve each of Problems 42–46.

42. The equation $P(a) = 14.7 e^{-0.21a}$, where a is the altitude above sea level measured in miles, yields the atmospheric pressure in pounds per square inch. If the atmospheric pressure at Cheyenne, Wyoming, is approximately 11.53 pounds per square inch, find that city's altitude above sea level. Express your answer to the nearest hundred feet.

43. An earthquake in Los Angeles in 1971 had an intensity of approximately five million times the reference intensity. What was the Richter number associated with that earthquake?

44. An earthquake in San Francisco in 1906 was reported to have a Richter number of 8.3. How did its intensity compare to the reference intensity?

45. Calculate how many times more intense an earthquake with a Richter number of 7.3 is than an earthquake with a Richter number of 6.4.

46. Calculate how many times more intense an earthquake with a Richter number of 8.9 is than an earthquake with a Richter number of 6.2.

■ ■ ■ Thoughts into words

47. Explain how to determine $\log_7 46$ without using Property 11.9.

48. Explain the concept of a Richter number.

49. Explain how you would solve the equation $2^x = 64$ and also how you would solve the equation $2^x = 53$.

50. How do logarithms with a base of 9 compare to logarithms with a base of 3? Explain how you reached this conclusion.

Graphing calculator activities

51. Graph $f(x) = x$, $f(x) = 2^x$, and $f(x) = \log_2 x$ on the same set of axes.

52. Graph $f(x) = x$, $f(x) = (0.5)^x$, and $f(x) = \log_{0.5} x$ on the same set of axes.

53. Graph $f(x) = \log_2 x$. Now predict the graphs for $f(x) = \log_3 x$, $f(x) = \log_4 x$, and $f(x) = \log_8 x$. Graph these

three functions on the same set of axes with $f(x) = \log_2 x$.

54. Graph $f(x) = \log_5 x$. Now predict the graphs for $f(x) = 2 \log_5 x$, $f(x) = -4 \log_5 x$, and $f(x) = \log_5(x + 4)$. Graph these three functions on the same set of axes with $f(x) = \log_5 x$.

SUMMARY

(11.1) If a and b are positive real numbers, and m and n are any real numbers, then

1. $b^n \cdot b^m = b^{m+n}$ Product of two powers

2. $(b^n)^m = b^{mn}$ Power of a power

3. $(ab)^n = a^n b^n$ Power of a product

4. $\left(\dfrac{a}{b}\right)^n = \dfrac{a^n}{b^n}$ Power of a quotient

5. $\dfrac{b^n}{b^m} = b^{n-m}$ Quotient of two powers

If $b > 0$, $b \neq 1$, and m and n are real numbers, then

$$b^n = b^m \quad \text{if and only if } n = m$$

A function defined by an equation of the form

$$f(x) = b^x, \qquad b > 0 \text{ and } b \neq 1$$

is called an **exponential function.**

(11.2) A general formula for any principal, P, invested for any number of years (t) at a rate of r percent compounded n times per year is

$$A = P\left(1 + \frac{r}{n}\right)^{nt}$$

where A represents the total amount of money accumulated at the end of the t years. The value of $\left(1 + \dfrac{1}{n}\right)^n$ as n gets infinitely large, approaches the number e, where e equals 2.71828 to five decimal places.

The formula

$$A = Pe^{rt}$$

yields the accumulated value, A, of a sum of money, P, that has been invested for t years at a rate of r percent **compounded continuously.**

The equation

$$Q(t) = Q_0 e^{kt}$$

is used as a mathematical model for many growth-and-decay applications.

(11.3) If r is any positive real number, then the unique exponent t such that $b^t = r$ is called the **logarithm of r with base b** and is denoted by $\log_b r$.

For $b > 0$ and $b \neq 1$, and $r > 0$,

1. $\log_b b = 1$

2. $\log_b 1 = 0$

3. $r = b^{\log_b r}$

The following properties of logarithms are derived from the definition of a logarithm and the properties of exponents. For positive real numbers b, r, and s, where $b \neq 1$,

1. $\log_b rs = \log_b r + \log_b s$

2. $\log_b\left(\dfrac{r}{s}\right) = \log_b r - \log_b s$

3. $\log_b r^p = p \log_b r$ p is any real number.

(11.4) A function defined by an equation of the form

$$f(x) = \log_b x, \qquad b > 0 \text{ and } b \neq 1$$

is called a **logarithmic function.** The equation $y = \log_b x$ is equivalent to $x = b^y$. The two functions $f(x) = b^x$ and $g(x) = \log_b x$ are inverses of each other.

Logarithms with a base of 10 are called **common logarithms.** The expression $\log_{10} x$ is usually written as $\log x$.

Natural logarithms are logarithms that have a base of e, where e is an irrational number whose decimal approximation to eight digits is 2.7182818. Natural logarithms are denoted by $\log_e x$ or **ln x.**

(11.5) The properties of equality and the properties of exponents and logarithms combine to help us solve a variety of exponential and logarithmic equations. Using these properties, we can now solve problems that deal with various applications, including compound interest and growth problems.

The formula

$$\log_a r = \frac{\log_b r}{\log_b a}$$

is often called the **change-of-base** formula.

CHAPTER 11 REVIEW PROBLEM SET

For Problems 1–8, evaluate each expression without using a calculator.

1. $\log_2 128$

2. $\log_4 64$

3. $\log 10{,}000$

4. $\log 0.001$

5. $\ln e^2$

6. $5^{\log_5 13}$

7. $\log(\log_3 3)$

8. $\log_2\left(\dfrac{1}{4}\right)$

For Problems 9–16, solve each equation without using your calculator.

9. $2^x = \dfrac{1}{16}$

10. $3^x - 4 = 23$

11. $16^x = \dfrac{1}{8}$

12. $\log_8 x = \dfrac{2}{3}$

13. $\log_x 3 = \dfrac{1}{2}$

14. $\log_5 2 + \log_5 (3x + 1) = 1$

15. $\log_2(x + 5) - \log_2 x = 1$

16. $\log_2 x + \log_2 (x - 4) = 5$

For Problems 17–20, use your calculator to find each logarithm. Express answers to four decimal places.

17. $\log 73.14$

18. $\ln 114.2$

19. $\ln 0.014$

20. $\log 0.00235$

For Problems 21–24, use your calculator to find x when given $\log x$ or $\ln x$. Express answers to five significant digits.

21. $\ln x = 0.1724$

22. $\log x = 3.4215$

23. $\log x = -1.8765$

24. $\ln x = -2.5614$

For Problems 25–28, use your calculator to help solve each equation. Express solutions to the nearest hundredth.

25. $3^x = 42$

26. $2e^x = 14$

27. $2^{x+1} = 79$

28. $e^{x-2} = 37$

For Problems 29–32, graph each of the functions.

29. $f(x) = 2^x - 3$

30. $f(x) = -3^x$

31. $f(x) = -1 + \log_2 x$

32. $f(x) = \log_2(x + 1)$

33. Approximate a value for $\log_5 97$ to three decimal places.

34. Suppose that \$800 is invested at 9% interest compounded annually. How much money has accumulated at the end of 15 years?

35. If \$2500 is invested at 10% interest compounded quarterly, how much money has accumulated at the end of 12 years?

36. If \$3500 is invested at 8% interest compounded continuously, how much money will accumulate in 6 years?

37. Suppose that a certain radioactive substance has a half-life of 40 days. If there are presently 750 grams of the substance, how much, to the nearest gram, will remain after 100 days?

38. How long will it take \$100 to double itself if it is invested at 14% interest compounded annually?

39. How long will it take \$1000 to be worth \$3500 if it is invested at 10.5% interest compounded quarterly?

40. Suppose that the present population of a city is 50,000. Use the equation $P(t) = P_0 e^{0.02t}$, where P_0 represents an initial population, to estimate future populations. Estimate the population of that city in 10 years, 15 years, and 20 years.

41. The number of bacteria present in a certain culture after t hours is given by the equation $Q = Q_0 e^{0.29t}$, where Q_0 represents the initial number of bacteria. How long will it take 500 bacteria to increase to 2000 bacteria?

42. An earthquake in Mexico City in 1985 had an intensity level about 125,000,000 times the reference intensity. Find the Richter number for that earthquake.

CHAPTER 11

TEST

For Problems 1– 4, evaluate each expression.

1. $\log_3 \sqrt{3}$

2. $\log_2(\log_2 4)$

3. $-2 + \ln e^3$

4. $\log_2(0.5)$

For Problems 5 –10, solve each equation.

5. $4^x = \dfrac{1}{64}$

6. $9^x = \dfrac{1}{27}$

7. $2^{3x-1} = 128$

8. $\log_9 x = \dfrac{5}{2}$

9. $\log x + \log(x + 48) = 2$

10. $\ln x = \ln 2 + \ln(3x - 1)$

For Problems 11–14, given that $\log_3 4 = 1.2619$ and $\log_3 5 = 1.4650$, evaluate each of the following.

11. $\log_3 100$

12. $\log_3 1.25$

13. $\log_3 \sqrt{5}$

14. $\log_3(16 \cdot 25)$

15. Solve $e^x = 176$ to the nearest hundredth.

16. Solve $2^{x-2} = 314$ to the nearest hundredth.

17. Determine $\log_5 632$ to four decimal places.

18. Express $3 \log_b x + 2 \log_b y - \log_b z$ as a single logarithm with a coefficient of 1.

19. If \$3500 is invested at 7.5% interest compounded quarterly, how much money has accumulated at the end of 8 years?

20. How long will it take \$5000 to be worth \$12,500 if it is invested at 7% compounded annually? Express your answer to the nearest tenth of a year.

21. The number of bacteria present in a certain culture after t hours is given by $Q(t) = Q_0 e^{0.23t}$, where Q_0 represents the initial number of bacteria. How long will it take 400 bacteria to increase to 2400 bacteria? Express your answer to the nearest tenth of an hour.

22. Suppose that a certain radioactive substance has a half-life of 50 years. If there are presently 7500 grams of the substance, how much will remain after 32 years? Express your answer to the nearest gram.

For Problems 23 –25, graph each of the functions.

23. $f(x) = e^x - 2$

24. $f(x) = -3^{-x}$

25. $f(x) = \log_2(x - 2)$

When objects are arranged in a sequence, the total number of objects is the sum of the terms of the sequence. This indicated sum is called a series.

Royalty-Free/CORBIS

Sequences and Series

S uppose that an auditorium has 35 seats in the first row, 40 seats in the second row, 45 seats in the third row, and so on for ten rows. The numbers 35, 40, 45, 50, . . . , 80 represent the number of seats per row from row 1 through row 10. The list of numbers has a constant difference of 5 between any two successive numbers in the list; such a list is called an arithmetic sequence.

Suppose that a fungus culture growing under controlled conditions doubles in size each day. If today the size of the culture is 6 units, then the numbers 12, 24, 48, 96, 192 represent the size of the culture for the next five days. In this list of numbers, each number after the first is two times the previous number; such a list is called a geometric sequence. Arithmetic and geometric sequences will be our focus in this chapter.

InfoTrac Project Do a keyword search on bacteria growth and find the article titled "Logarithms and Life." Write a brief summary of the article. Using the information in the article, find the total number of bacteria (assuming none die) 24 hours after the fly landed on the food.

12.1 Arithmetic Sequences

An **infinite sequence** is a function whose domain is the set of positive integers. For example, consider the function defined by the equation

$$f(n) = 3n + 2$$

where the domain is the set of positive integers. Furthermore, let's substitute the numbers of the domain, in order, starting with 1. We list the resulting ordered pairs as

$$(1, 5), (2, 8), (3, 11), (4, 14), (5, 17)$$

and so on. Because we have agreed to use the domain of positive integers in order, starting with 1, there is no need to use ordered pairs. We can simply express the infinite sequence as

$$5, 8, 11, 14, 17, \ldots .$$

Frequently, we use the letter a to represent *sequential functions,* and we write the functional value at n as a_n (read "a sub n") instead of $a(n)$. The sequence is then

$$a_1, a_2, a_3, a_4, \ldots, a_n, \ldots$$

where a_1 is the first term, a_2 the second term, a_3 the third term, and so on. The expression a_n, which defines the sequence, is called the **general term** of the sequence. Knowing the general term of a sequence allows us to find as many terms of the sequence as needed and also to find any specific terms. Consider the following examples.

E X A M P L E   1

Find the first five terms of each of the following sequences.

(a) $a_n = n^2 + 1$ **(b)** $a_n = 2^n$

Solution

(a) The first five terms are found by replacing n with 1, 2, 3, 4, and 5.

$$a_1 = 1^2 + 1 = 2 \qquad\qquad a_2 = 2^2 + 1 = 5$$

$$a_3 = 3^2 + 1 = 10 \qquad\qquad a_4 = 4^2 + 1 = 17$$

$$a_5 = 5^2 + 1 = 26$$

Thus the first five terms are 2, 5, 10, 17, and 26.

(b) $a_1 = 2^1 = 2$ $a_2 = 2^2 = 4$

$a_3 = 2^3 = 8$ $a_4 = 2^4 = 16$

$a_5 = 2^5 = 32$

The first five terms are 2, 4, 8, 16, and 32.

E X A M P L E   2   Find the 12th and 25th terms of the sequence $a_n = 5n - 1$.

Solution

$$a_{12} = 5(12) - 1 = 59$$

$$a_{25} = 5(25) - 1 = 124$$

The 12th term is 59 and the 25th term is 124.

An **arithmetic sequence** (also called an *arithmetic progression*) is a sequence in which there is a **common difference** between successive terms. The following are examples of arithmetic sequences.

(a) 1, 4, 7, 10, 13, . . .
(b) 6, 11, 16, 21, 26, . . .
(c) 14, 25, 36, 47, 58, . . .
(d) 4, 2, 0, −2, −4, . . .
(e) −1, −7, −13, −19, −25, . . .

The common difference in (a) of the previous list is 3. That is to say, $4 - 1 = 3$, $7 - 4 = 3$, $10 - 7 = 3$, $13 - 10 = 3$, and so on. The common differences for (b), (c), (d), and (e) are 5, 11, −2, and −6, respectively. It is sometimes stated that arithmetic sequences exhibit *constant growth*. This is an accurate description if we keep in mind that this "growth" may be in a negative direction, which is illustrated by (d) and (e) above.

In a more general setting, we say that the sequence

$$a_1, a_2, a_3, a_4, \ldots, a_n, \ldots$$

is an arithmetic sequence if and only if there is a real number d such that

$$a_{k+1} - a_k = d \qquad \textbf{(1)}$$

for every positive integer k. The number d is called the **common difference.**

From equation (1) we see that $a_{k+1} = a_k + d$. In other words, we can generate an arithmetic sequence that has a common difference of d by starting with a first term of a_1 and then simply adding d to each successive term as follows:

First term	a_1
Second term	$a_1 + d$
Third term	$a_1 + 2d$ $(a_1 + d) + d = a_1 + 2d$
Fourth term	$a_1 + 3d$ $(a_1 + 2d) + d = a_1 + 3d$
.	
.	
.	
nth term	$a_1 + (n - 1)d$

Thus the *general term of an arithmetic sequence* is given by

$$a_n = a_1 + (n - 1)d$$

where a_1 is the first term and d the common difference. This general-term formula provides the basis for doing a variety of problems that involve arithmetic sequences.

EXAMPLE 3   Find the general term for each of the following arithmetic sequences.

    (a) $1, 5, 9, 13, \ldots$ **(b)** $5, 2, -1, -4, \ldots$

Solution

    (a) The common difference is 4 and the first term is 1. Substituting these values into $a_n = a_1 + (n - 1)d$ and simplifying, we obtain

$$\begin{aligned} a_n &= a_1 + (n - 1)d \\ &= 1 + (n - 1)4 \\ &= 1 + 4n - 4 \\ &= 4n - 3 \end{aligned}$$

    (Perhaps you should verify that the general term $a_n = 4n - 3$ does produce the sequence $1, 5, 9, 13, \ldots$.)

    (b) Because the first term is 5 and the common difference is -3, we obtain

$$\begin{aligned} a_n &= a_1 + (n - 1)d \\ &= 5 + (n - 1)(-3) \\ &= 5 - 3n + 3 = -3n + 8 \end{aligned}$$

$\blacksquare$

EXAMPLE 4   Find the 50th term of the arithmetic sequence 2, 6, 10, 14, ...

Solution

Certainly, we could simply continue to write the terms of the given sequence until the 50th term is reached. However, let's use the general-term formula, $a_n = a_1 + (n-1)d$, to find the 50th term without determining all of the other terms.

$$a_{50} = 2 + (50 - 1)4$$
$$= 2 + 49(4) = 2 + 196 = 198$$

EXAMPLE 5   Find the first term of the arithmetic sequence where the third term is 13 and the tenth term is 62.

Solution

Using $a_n = a_1 + (n-1)d$ with $a_3 = 13$ (the third term is 13) and $a_{10} = 62$ (the tenth term is 62), we have

$$13 = a_1 + (3 - 1)d = a_1 + 2d$$
$$62 = a_1 + (10 - 1)d = a_1 + 9d$$

Solve the system of equations

$$\begin{pmatrix} a_1 + 2d = 13 \\ a_1 + 9d = 62 \end{pmatrix}$$

to yield $a_1 = -1$. Thus the first term is -1.

REMARK: Perhaps you can think of another way to solve the problem in Example 5 without using a system of equations. [*Hint:* How many "differences" are there between the third and tenth terms of an arithmetic sequence?]

Phrases such as "the set of odd whole numbers," "the set of even whole numbers," and "the set of positive multiples of 5" are commonly used in mathematical literature to refer to various subsets of the whole numbers. Though no specific ordering of the numbers is implied by these phrases, most of us would probably react with a natural ordering. For example, if we were asked to list the set of odd whole numbers, our answer probably would be 1, 3, 5, 7, ... Using such an ordering, we can think of the set of odd whole numbers as an arithmetic sequence. Therefore, we can formulate a general representation for the set of odd whole numbers by using the general-term formula. Thus

$$a_n = a_1 + (n-1)d$$
$$= 1 + (n-1)2$$
$$= 1 + 2n - 2$$
$$= 2n - 1$$

The final example of this section illustrates the use of an arithmetic sequence to solve a problem that deals with the constant growth of a man's yearly salary.

E X A M P L E   6

A man started to work in 1970 at an annual salary of $9500. He received a $700 raise each year. How much was his annual salary in 1991?

Solution

The following arithmetic sequence represents the annual salary beginning in 1970.

$$9500, 10{,}200, 10{,}900, 11{,}600, \ldots.$$

The general term of this sequence is

$$a_n = a_1 + (n - 1)d$$
$$= 9500 + (n - 1)700$$
$$= 9500 + 700n - 700$$
$$= 700n + 8800$$

The man's 1991 salary is the 22nd term of this sequence. Thus,

$$a_{22} = 700(22) + 8800 = 24{,}200$$

His salary in 1991 was $24,200. ■

PROBLEM SET 12.1

For Problems 1–14, write the first five terms of each sequence that has the indicated general term.

1. $a_n = 3n - 4$

2. $a_n = 2n + 3$

3. $a_n = -2n + 5$

4. $a_n = -3n - 2$

5. $a_n = n^2 - 2$

6. $a_n = n^2 + 3$

7. $a_n = -n^2 + 1$

8. $a_n = -n^2 - 2$

9. $a_n = 2n^2 - 3$

10. $a_n = 3n^2 + 2$

11. $a_n = 2^{n-2}$

12. $a_n = 3^{n+1}$

13. $a_n = -2(3)^{n-2}$

14. $a_n = -3(2)^{n-3}$

15. Find the 8th and 12th terms of the sequence where $a_n = n^2 - n - 2$.

16. Find the 10th and 15th terms of the sequence where $a_n = -n^2 - 2n + 3$.

17. Find the 7th and 8th terms of the sequence where $a_n = (-2)^{n-2}$.

18. Find the 6th and 7th terms of the sequence where $a_n = -(3)^{n-3}$.

For Problems 19–28, find the general term (nth term) for each of the arithmetic sequences.

19. $1, 3, 5, 7, 9, \ldots$

20. $2, 4, 6, 8, 10, \ldots$

21. $-2, 2, 6, 10, 14, \ldots$

22. $-3, 2, 7, 12, 17, \ldots$

23. $5, 3, 1, -1, -3, \ldots$

24. $2, -1, -4, -7, -10, \ldots$

25. $-7, -10, -13, -16, -19, \ldots$

26. $-3, -5, -7, -9, -11, \ldots$

27. $1, \dfrac{3}{2}, 2, \dfrac{5}{2}, 3, \ldots$

28. $\dfrac{3}{2}, 3, \dfrac{9}{2}, 6, \dfrac{15}{2}, \ldots$

For Problems 29–34, find the indicated term of the arithmetic sequence.

29. The 10th term of 7, 10, 13, 16, . . .

30. The 12th term of 9, 11, 13, 15, . . .

31. The 20th term of 2, 6, 10, 14, . . .

32. The 50th term of −1, 4, 9, 14, . . .

33. The 75th term of −7, −9, −11, −13, . . .

34. The 100th term of −7, −10, −13, −16, . . .

For Problems 35–40, find the number of terms in each of the finite arithmetic sequences.

35. 1, 3, 5, 7, . . . , 211

36. 2, 4, 6, 8, . . . , 312

37. 10, 13, 16, 19, . . . , 157

38. 9, 13, 17, 21, . . . , 849

39. −7, −9, −11, −13, . . . , −345

40. −4, −7, −10, −13, . . . , −331

41. If the 6th term of an arithmetic sequence is 24 and the 10th term is 44, find the first term.

42. If the 5th term of an arithmetic sequence is 26 and the 12th term is 75, find the first term.

43. If the 4th term of an arithmetic sequence is −9 and the 9th term is −29, find the 5th term.

44. If the 6th term of an arithmetic sequence is −4 and the 14th term is −20, find the 10th term.

45. In the arithmetic sequence 0.97, 1.00, 1.03, 1.06, . . . , which term is 5.02?

46. In the arithmetic sequence 1, 1.2, 1.4, 1.6, . . . , which term is 35.4?

For Problems 47–50, set up an arithmetic sequence and use $a_n = a_1 + (n - 1)d$ to solve each problem.

47. A woman started to work in 1975 at an annual salary of $12,500. She received a $900 raise each year. How much was her annual salary in 1992?

48. Math University had an enrollment of 8500 students in 1976. Each year the enrollment has increased by 350 students. What was the enrollment in 1990?

49. Suppose you are offered a job starting at $900 a month with a guaranteed increase of $30 a month every 6 months for the next 5 years. What will your monthly salary be for the last 6 months of the fifth year of your employment?

50. Between 1976 and 1990 a person invested $500 at 14% simple interest at the beginning of each year. By the end of 1990, how much interest had been earned by the $500 that was invested at the beginning of 1982?

■ ■ ■ **Thoughts into words**

51. How would you explain the concept of an arithmetic sequence to someone taking an elementary algebra course?

52. Is the sequence whose nth term is $a_n = 2^n$ an arithmetic sequence? Defend your answer.

12.2 Arithmetic Series

Let's begin this section with a problem to solve. Study the solution very carefully.

PROBLEM 1 Find the sum of the first one hundred positive integers.

Solution

We are being asked to find the sum of $1 + 2 + 3 + 4 + \cdots + 100$. Rather than adding in the usual way, we will find the sum in the following manner. Let's simply write the indicated sum forward and backward, and then add in a column fashion.

$$
\begin{array}{cccccccc}
1 & + & 2 & + & 3 & + & 4 & + \cdots + & 100 \\
100 & + & 99 & + & 98 & + & 97 & + \cdots + & 1 \\
\hline
101 & + & 101 & + & 101 & + & 101 & + \cdots + & 101
\end{array}
$$

In so doing, we have produced 100 sums of 101. However, this sum is double the amount we want, because we wrote the sum twice. To find the sum of just the numbers 1 to 100, we need to multiply 100 by 101 and then divide by 2.

$$
\frac{100(101)}{2} = \frac{\overset{50}{\cancel{100}}\,(101)}{2} = 5050
$$

Thus the sum of the first 100 positive integers is 5050.

Now let's introduce some terminology that applies to problems like Problem 1. The indicated sum of a sequence is called a **series.** Associated with the finite sequence

$$a_1, a_2, a_3, \ldots, a_n$$

is a **finite series**

$$a_1 + a_2 + a_3 + \cdots + a_n$$

Likewise, from the infinite sequence

$$a_1, a_2, a_3, \ldots$$

we can form the **infinite series**

$$a_1 + a_2 + a_3 + \cdots$$

In this section we will direct our attention to working with **arithmetic series**—that is, the indicated sums of arithmetic sequences.

We could have stated Problem 1 as "Find the sum of the first one hundred terms of the arithmetic series that has a general term (nth term) of $a_n = n$." In fact, the forward–backward approach used to solve the problem can be applied to the general arithmetic series

$$a_1 + a_2 + a_3 + \cdots + a_n$$

to produce a formula for finding the sum of the first n terms of any arithmetic series. Use S_n to represent the sum of the first n terms and proceed as follows:

$$S_n = a_1 + (a_1 + d) + (a_1 + 2d) + \cdots + (a_n - 2d) + (a_n - d) + a_n$$

Now write this sum in reverse as

$$S_n = a_n + (a_n - d) + (a_n - 2d) + \cdots + (a_1 + 2d) + (a_1 + d) + a_1$$

Add the two equations to produce

$$2S_n = (a_1 + a_n) + (a_1 + a_n) + (a_1 + a_n) + \cdots + (a_1 + a_n)$$
$$+ (a_1 + a_n) + (a_1 + a_n)$$

That is, we have n sums of $(a_1 + a_n)$, so

$$2S_n = n(a_1 + a_n)$$

from which we obtain

$$S_n = \frac{n(a_1 + a_n)}{2}$$

Using the nth-term formula $a_n = a_1 + (n - 1)d$ and the sum formula $\frac{n(a_1 + a_n)}{2}$, we can solve a variety of problems that involve arithmetic series.

EXAMPLE 1

Find the sum of the first 50 terms of the series

$$2 + 5 + 8 + 11 + \cdots$$

Solution

Using $a_n = a_1 + (n - 1)d$, we find the 50th term to be

$$a_{50} = 2 + 49(3) = 149$$

Then using $S_n = \frac{n(a_1 + a_n)}{2}$, we obtain

$$S_{50} = \frac{50(2 + 149)}{2} = 3775$$

EXAMPLE 2

Find the sum of all odd numbers between 7 and 433, inclusive.

Solution

We need to find the sum $7 + 9 + 11 + \cdots + 433$. To use $S_n = \frac{n(a_1 + a_n)}{2}$, we need the number of terms, n. Perhaps you could figure this out without a formula (try it), but suppose we use the nth-term formula.

$$a_n = a_1 + (n - 1)d$$
$$433 = 7 + (n - 1)2$$

$$433 = 7 + 2n - 2$$

$$433 = 2n + 5$$

$$428 = 2n$$

$$214 = n$$

Then use $n = 214$ in the sum formula to yield

$$S_{214} = \frac{214(7 + 433)}{2} = 47{,}080$$

E X A M P L E 3

Find the sum of the first 75 terms of the series that has a general term of $a_n = -5n + 9$.

Solution

Using $a_n = -5n + 9$, we can generate the series as follows:

$$a_1 = -5(1) + 9 = 4$$

$$a_2 = -5(2) + 9 = -1$$

$$a_3 = -5(3) + 9 = -6$$

$$\cdot$$
$$\cdot$$
$$\cdot$$

$$a_{75} = -5(75) + 9 = -366$$

Thus we have the series

$$4 + (-1) + (-6) + \cdots + (-366)$$

Using the sum formula, we obtain

$$S_{75} = \frac{75[4 + (-366)]}{2} = -13{,}575$$

E X A M P L E 4

Sue is saving quarters. She saves 1 quarter the first day, 2 quarters the second day, 3 quarters the third day, and so on. How much money will she have saved in 30 days?

Solution

The total number of quarters will be the sum of the series

$$1 + 2 + 3 + \cdots + 30$$

Using the sum formula yields

$$S_{30} = \frac{30(1 + 30)}{2} = 465$$

Thus Sue will have saved $(465)(0.25) = \$116.25$ at the end of 30 days.   ■

REMARK: The sum formula, $S_n = \dfrac{n(a_1 + a_n)}{2}$, was developed by using the **forward–backward** technique that we previously used on a specific problem. Now that we have the sum formula, we have two choices as we meet problems where the formula applies. We can memorize the formula and use it as it applies, or we can disregard the formula and use the forward–backward approach. Furthermore, even if you use the formula, and some day you forget it, you can use the forward–backward technique. In other words, once you understand the development of a formula, you can do some problems even if you have forgotten the formula itself.

PROBLEM SET 12.2

For Problems 1–10, find the sum of the indicated number of terms of the given series.

1. First 50 terms of $2 + 4 + 6 + 8 + \cdots$

2. First 45 terms of $1 + 3 + 5 + 7 + \cdots$

3. First 60 terms of $3 + 8 + 13 + 18 + \cdots$

4. First 80 terms of $2 + 6 + 10 + 14 + \cdots$

5. First 65 terms of $(-1) + (-3) + (-5) + (-7) + \cdots$

6. First 100 terms of $(-1) + (-4) + (-7) + (-10) + \cdots$

7. First 40 terms of $\dfrac{1}{2} + 1 + \dfrac{3}{2} + 2 + \cdots$

8. First 50 terms of $1 + \dfrac{5}{2} + 4 + \dfrac{11}{2} + \cdots$

9. First 75 terms of $7 + 10 + 13 + 16 + \cdots$

10. First 90 terms of $(-8) + (-1) + 6 + 13 + \cdots$

For Problems 11–16, find the sum of each finite arithmetic series.

11. $4 + 8 + 12 + 16 + \cdots + 212$

12. $7 + 9 + 11 + 13 + \cdots + 179$

13. $(-4) + (-1) + 2 + 5 + \cdots + 173$

14. $5 + 10 + 15 + 20 + \cdots + 495$

15. $2.5 + 3.0 + 3.5 + 4.0 + \cdots + 18.5$

16. $1 + (-6) + (-13) + (-20) + \cdots + (-202)$

For Problems 17–22, find the sum of the indicated number of terms of the series with the given nth term.

17. First 50 terms of series with $a_n = 3n - 1$

18. First 150 terms of series with $a_n = 2n - 7$

19. First 125 terms of series with $a_n = 5n + 1$

20. First 75 terms of series with $a_n = 4n + 3$

21. First 65 terms of series with $a_n = -4n - 1$

22. First 90 terms of series with $a_n = -3n - 2$

For Problems 23–34, use arithmetic sequences and series to help solve the problem.

23. Find the sum of the first 350 positive even whole numbers.

24. Find the sum of the first 200 odd whole numbers.

25. Find the sum of all odd whole numbers between 15 and 397, inclusive.

26. Find the sum of all even whole numbers between 14 and 286, inclusive.

27. An auditorium has 20 seats in the front row, 24 seats in the second row, 28 seats in the third row, and so on, for 15 rows. How many seats are there in the last row? How many seats are there in the auditorium?

28. A pile of wood has 15 logs in the bottom row, 14 logs in the row next to the bottom, and so on, with one less log in each row until the top row, which consists of 1 log. How many logs are there in the pile?

29. A raffle is organized so that the amount paid for each ticket is determined by a number on the ticket. The tickets are numbered with the consecutive odd whole numbers 1, 3, 5, 7, ... Each participant pays as many cents as the number on the ticket drawn. How much money will the raffle take in if 1000 tickets are sold?

30. A woman invests $700 at 13% simple interest at the beginning of each year for a period of 15 years. Find the

total accumulated value of all of the investments at the end of the 15-year period.

31. A man started to work in 1970 at an annual salary of $18,500. He received a $1500 raise each year through 1982. What were his total earnings for the 13-year period?

32. A well driller charges $9.00 per foot for the first 10 feet, $9.10 per foot for the next 10 feet, $9.20 per foot for the next 10 feet, and so on; he continues to increase the price by $.10 per foot for succeeding intervals of 10 feet. How much would it cost to drill a well with a depth of 150 feet?

33. A display in a grocery store has cans stacked with 25 cans in the bottom row, 23 cans in the second row from the bottom, 21 cans in the third row from the bottom, and so on until there is only 1 can in the top row. How many cans are there in the display?

34. Suppose that a person starts on the first day of August and saves a dime the first day, $.20 the second day, and $.30 the third day and continues to save $.10 more per day than the previous day. How much could be saved in the 31 days of August?

■ ■ ▨ **Thoughts into words**

35. Explain how to find the sum $1 + 2 + 3 + \cdots + 150$ without using the sum formula.

36. Explain how you would find the sum of the first 150 terms of the arithmetic series that has a general term of $a_n = -3n + 7$.

■ ■ ▨ **Further investigations**

37. We can express a series where the general term is known in a convenient and compact form using the symbol Σ along with the general-term expression. For example, consider the finite arithmetic series

$$1 + 3 + 5 + 7 + 9 + 11$$

where the general term is $a_n = 2n - 1$. This series can be expressed in *summation notation* as

$$\sum_{i=1}^{6} (2i - 1)$$

where the letter i is used as the **index of summation.** The individual terms of the series can be generated by successively replacing i in the expression $(2i - 1)$ with the numbers 1, 2, 3, 4, 5, and 6. Thus the first term is $2(1) - 1 = 1$, the second term $2(2) - 1 = 3$, the third term $2(3) - 1 = 5$, and so on. Write out the terms and find the sum of each of the following series.

(a) $\displaystyle\sum_{i=1}^{3} (5i + 2)$ **(b)** $\displaystyle\sum_{i=1}^{4} (6i - 7)$

(c) $\displaystyle\sum_{i=1}^{6} (-2i - 1)$ **(d)** $\displaystyle\sum_{i=1}^{5} (-3i + 4)$

(e) $\displaystyle\sum_{i=1}^{5} 3i$ **(f)** $\displaystyle\sum_{i=1}^{6} -4i$

38. Write each of the following in summation notation. For example, because $3 + 8 + 13 + 18 + 23 + 28$ is an arithmetic series, the general-term formula $a_n = a_1 + (n - 1)d$ yields

$$a_n = 3 + (n - 1)5$$

$$= 3 + 5n - 5$$

$$= 5n - 2$$

Now, using i as an index of summation, we can write

$$\sum_{i=1}^{6} (5i - 2)$$

(a) $2 + 5 + 8 + 11 + 14$

(b) $8 + 15 + 22 + 29 + 36 + 43$

(c) $1 + (-1) + (-3) + (-5) + (-7)$

(d) $(-5) + (-9) + (-13) + (-17) + (-21) + (-25) + (-29)$

12.3   Geometric Sequences and Series

A **geometric sequence** or **geometric progression** is a sequence in which each term after the first is obtained by multiplying the preceding term by a common multiplier. The common multiplier is called the **common ratio** of the sequence. The following geometric sequences have common ratios of $2, 3, \frac{1}{2}$, and -4, respectively.

$$1, 2, 4, 8, 16, \ldots \qquad 3, 9, 27, 81, 243, \ldots$$

$$8, 4, 2, 1, \frac{1}{2}, \ldots \qquad 1, -4, 16, -64, 256, \ldots$$

We find the common ratio of a geometric sequence by dividing a term (other than the first term) by the preceding term. In a more general setting, we say that the sequence

$$a_1, a_2, a_3, a_4, \ldots, a_n, \ldots$$

is a geometric sequence if and only if there is a nonzero real number r such that

$$a_{k+1} = ra_k \qquad \qquad \textbf{(1)}$$

for every positive integer k. The nonzero real number r is called the common ratio.

Equation (1) can be used to generate a general geometric sequence that has a_1 as a first term and r as a common ratio. We can proceed as follows:

First term	a_1	
Second term	$a_1 r$	
Third term	$a_1 r^2$	$(a_1 r)(r) = a_1 r^2$
Fourth term	$a_1 r^3$	$(a_1 r^2)(r) = a_1 r^3$
.		
.		
.		
nth term	$a_1 r^{n-1}$	

Thus the *general term of a geometric sequence* is given by

$$a_n = a_1 r^{n-1}$$

where a_1 is the first term and r is the common ratio.

EXAMPLE 1   Find the general term for the geometric sequence $2, 4, 8, 16, \ldots$

Solution

Using $a_n = a_1 r^{n-1}$, we obtain

$$a_n = 2(2)^{n-1} \qquad r = \frac{4}{2} = \frac{8}{4} = \frac{16}{8} = 2$$

$$= 2^n \qquad 2^1(2)^{n-1} = 2^{1+n-1} = 2^n$$

EXAMPLE 2   Find the tenth term of the geometric sequence $9, 3, 1, \ldots$.

Solution

Using $a_n = a_1 r^{n-1}$, we can find the tenth term as follows:

$$a_{10} = 9\left(\frac{1}{3}\right)^{10-1}$$

$$= 9\left(\frac{1}{3}\right)^9$$

$$= 9\left(\frac{1}{19{,}683}\right)$$

$$= \frac{1}{2187}$$

A **geometric series** is the indicated sum of a geometric sequence. The following are examples of geometric series.

$$1 + 2 + 4 + 8 + \cdots$$

$$3 + 9 + 27 + 81 + \cdots$$

$$8 + 4 + 2 + 1 + \cdots$$

$$1 + (-4) + 16 + (-64) + \cdots$$

Before we develop a general formula for finding the sum of a geometric series, let's consider a specific example.

EXAMPLE 3

Find the sum of $1 + 2 + 4 + 8 + \cdots + 512$.

Solution

Let S represent the sum and we can proceed as follows:

$$S = 1 + 2 + 4 + 8 + \cdots + 512, \tag{2}$$

$$2S = \quad\ 2 + 4 + 8 + \cdots + 512 + 1024 \tag{3}$$

Equation (3) is the result of multiplying both sides of equation (2) by 2. Subtract equation (2) from equation (3) to yield

$$S = 1024 - 1 = 1023 \qquad\blacksquare$$

Now let's consider the general geometric series

$$a_1 + a_1 r + a_1 r^2 + \cdots + a_1 r^{n-1}$$

By applying a procedure similar to the one used in Example 3, we can develop a formula for finding the sum of the first n terms of any geometric series.

Let S_n represent the sum of the first n terms. Thus

$$S_n = a_1 + a_1 r + a_1 r^2 + \cdots + a_1 r^{n-1} \tag{4}$$

Multiply both sides of equation (4) by the common ratio r to produce

$$r S_n = a_1 r + a_1 r^2 + a_1 r^3 + \cdots + a_1 r^{n-1} + a_1 r^n \tag{5}$$

Subtract equation (4) from equation (5) to yield

$$r S_n - S_n = a_1 r^n - a_1$$

Apply the distributive property on the left side and then solve for S_n to obtain

$$S_n(r - 1) = a_1 r^n - a_1$$

$$S_n = \frac{a_1 r^n - a_1}{r - 1}, \quad r \neq 1$$

Therefore, the sum of the first n terms of a geometric series that has a first term of a_1 and a common ratio of r is given by

$$S_n = \frac{a_1 r^n - a_1}{r - 1}, \quad r \neq 1$$

EXAMPLE 4

Find the sum of the first seven terms of the geometric series $2 + 6 + 18 + \cdots$.

Solution

Use the sum formula to obtain

$$S_7 = \frac{2(3)^7 - 2}{3 - 1}$$

$$= \frac{2(3^7 - 1)}{2}$$

$$= 3^7 - 1$$

$$= 2187 - 1$$

$$= 2186 \qquad \blacksquare$$

If the common ratio of a geometric series is less than 1, it may be more convenient to change the form of the sum formula. That is, we can change the fraction $\dfrac{a_1 r^n - a_1}{r - 1}$ to $\dfrac{a_1 - a_1 r^n}{1 - r}$ by multiplying both the numerator and the denominator by -1. Thus, if we use $S_n = \dfrac{a_1 - a_1 r^n}{1 - r}$ when $r < 1$, we can sometimes avoid unnecessary work with negative numbers, as the next example illustrates.

EXAMPLE 5

Find the sum of the geometric series $1 + \dfrac{1}{2} + \dfrac{1}{4} + \cdots + \dfrac{1}{256}$.

Solution A

To use the sum formula, we need to know the number of terms, which can be found by simply counting them or by applying the nth-term formula as follows:

$$a_n = a_1 r^{n-1}$$

$$\frac{1}{256} = 1\left(\frac{1}{2}\right)^{n-1}$$

$$\left(\frac{1}{2}\right)^8 = \left(\frac{1}{2}\right)^{n-1}$$

$$8 = n - 1 \qquad \text{Remember that if } b^n = b^m, \text{ then } n = m.$$

$$9 = n$$

Using $n = 9$, $a_1 = 1$, and $r = \dfrac{1}{2}$ in the form of the sum formula

$$S_n = \frac{a_1 - a_1 r^n}{1 - r}$$

we obtain

$$S_9 = \frac{1 - 1\left(\dfrac{1}{2}\right)^9}{1 - \dfrac{1}{2}}$$

$$= \frac{1 - \dfrac{1}{512}}{\dfrac{1}{2}}$$

$$= \frac{\dfrac{511}{512}}{\dfrac{1}{2}}$$

$$= \left(\frac{511}{512}\right)\left(\frac{2}{1}\right) = \frac{511}{256}, \text{ or } 1\frac{255}{256}$$

You should realize that a problem such as Example 5 can be done without using the sum formula; you can apply the general technique used to develop the formula. Solution B illustrates this approach.

Solution B

Let S represent the desired sum. Thus

$$S = 1 + \frac{1}{2} + \frac{1}{4} + \cdots + \frac{1}{256}$$

Multiply both sides by $\dfrac{1}{2}$ (the common ratio).

$$\frac{1}{2}S = \frac{1}{2} + \frac{1}{4} + \cdots + \frac{1}{256} + \frac{1}{512}$$

Subtract the second equation from the first equation to produce

$$\frac{1}{2}S = 1 - \frac{1}{512}$$

$$\frac{1}{2}S = \frac{511}{512}$$

$$S = \frac{511}{256}$$

$$= 1\frac{255}{256}$$

EXAMPLE 6

Suppose your employer agrees to pay you a penny for your first day's wages and then to double your pay on each succeeding day. How much will you earn on the 15th day? What will be your total earnings for the first 15 days?

Solution

The terms of the geometric series $1 + 2 + 4 + 8 + \cdots$ depict your daily wages, and the sum of the first 15 terms is your total earnings for the 15 days. The formula $a_n = a_1 r^{n-1}$ can be used to find the 15th day's wages.

$$a_{15} = (1)(2)^{14} = 16{,}384$$

Because the terms of the series are expressed in cents, your wages for the 15th day will be $163.84. Now, using the sum formula, we can find your total earnings as follows:

$$S_n = \frac{a_1 r^n - a_1}{r - 1}$$

$$S_{15} = \frac{1(2)^{15} - 1}{1} = 32{,}768 - 1 = 32{,}767$$

Thus, for the 15 days, you will earn a total of $327.67. ∎

PROBLEM SET 12.3

For Problems 1–12, find the general term (*n*th term) of each geometric sequence.

1. $1, 3, 9, 27, \ldots$

2. $1, 2, 4, 8, \ldots$

3. $2, 8, 32, 128, \ldots$

4. $3, 9, 27, 81, \ldots$

5. $1, \dfrac{1}{3}, \dfrac{1}{9}, \dfrac{1}{27}, \ldots$

6. $\dfrac{1}{2}, \dfrac{1}{4}, \dfrac{1}{8}, \dfrac{1}{16}, \ldots$

7. $0.2, 0.04, 0.008, 0.0016, \ldots$

8. $1, 0.3, 0.09, 0.027, \ldots$

9. $9, 6, 4, \dfrac{8}{3}, \ldots$

10. $6, 2, \dfrac{2}{3}, \dfrac{2}{9}, \ldots$

11. $1, -4, 16, -64, \ldots$

12. $1, -2, 4, -8, \ldots$

For Problems 13–18, find the indicated term of the geometric sequence.

13. 12th term of $\dfrac{1}{9}, \dfrac{1}{3}, 1, 3, \ldots$

14. 9th term of $2, 4, 8, 16, \ldots$

15. 10th term of $1, -2, 4, -8, \ldots$

16. 8th term of $\dfrac{1}{2}, \dfrac{1}{8}, \dfrac{1}{32}, \dfrac{1}{128}, \ldots$

17. 9th term of $-1, -\dfrac{3}{2}, -\dfrac{9}{4}, -\dfrac{27}{8}, \ldots$

18. 11th term of $1, \dfrac{2}{3}, \dfrac{4}{9}, \dfrac{8}{27}, \ldots$

For Problems 19–24, find the sum of the indicated number of terms of each geometric series.

19. First 10 terms of $\dfrac{1}{2} + \dfrac{3}{2} + \dfrac{9}{2} + \dfrac{27}{2} + \cdots$

20. First 9 terms of $1 + 2 + 4 + 8 + \cdots$

21. First 9 terms of $-2 + 6 + (-18) + 54 + \cdots$

22. First 10 terms of $-4 + 8 + (-16) + 32 + \cdots$

23. First 7 terms of $1 + 3 + 9 + 27 + \cdots$

24. First 8 terms of $4 + 2 + 1 + \dfrac{1}{2} + \cdots$

For Problems 25–30, find the sum of the indicated number of terms of the geometric series with the given nth term.

25. First 9 terms of series where $a_n = 2^{n-1}$

26. First 8 terms of series where $a_n = 3^n$

27. First 8 terms of series where $a_n = 2(3)^n$

28. First 10 terms of series where $a_n = \dfrac{1}{2^{n-4}}$

29. First 12 terms of series where $a_n = (-2)^n$

30. First 9 terms of series where $a_n = (-3)^{n-1}$

For Problems 31–36, find the sum of each finite geometric series.

31. $1 + 3 + 9 + \cdots + 729$

32. $2 + 8 + 32 + \cdots + 2048$

33. $1 + \dfrac{1}{2} + \dfrac{1}{4} + \cdots + \dfrac{1}{1024}$

34. $1 + (-2) + 4 + \cdots + (-128)$

35. $8 + 4 + 2 + \cdots + \dfrac{1}{32}$

36. $2 + 6 + 18 + \cdots + 4374$

For Problems 37–48, use geometric sequences and series to help solve the problem.

37. Find the common ratio of a geometric sequence if the second term is $\dfrac{1}{6}$ and the fifth term is $\dfrac{1}{48}$.

38. Find the first term of a geometric sequence if the fifth term is $\dfrac{32}{3}$ and the common ratio is 2.

39. Find the sum of the first 16 terms of the geometric series where $a_n = (-1)^n$. Also find the sum of the first 19 terms.

40. A fungus culture growing under controlled conditions doubles in size each day. How many units will the culture contain after 7 days if it originally contained 5 units?

41. A tank contains 16,000 liters of water. Each day one-half of the water in the tank is removed and not replaced. How much water remains in the tank at the end of the seventh day?

42. Suppose that you save 25 cents the first day of a week, 50 cents the second day, and $1 the third day and continue to double your savings each day. How much will you save on the seventh day? What will be your total savings for the week?

43. Suppose you save a nickel the first day of a month, a dime the second day, and 20 cents the third day and continue to double your savings each day. How much will you save on the 12th day of the month? What will be your total savings for the first 12 days?

44. Suppose an element has a half-life of 3 hours. This means that if n grams of it exist at a specific time, then only $\dfrac{1}{2}n$ grams remain 3 hours later. If at a particular moment we have 40 grams of the element, how much of it will remain 24 hours later?

45. A rubber ball is dropped from a height of 486 meters, and each time, it rebounds one-third of the height from which it last fell (see Figure 12.1). How far has the ball traveled by the time it strikes the ground for the seventh time?

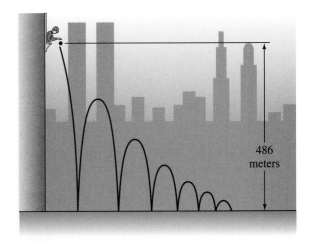

486 meters

Figure 12.1

46. A pump is attached to a container for the purpose of creating a vacuum. For each stroke of the pump, one-fourth of the air remaining in the container is removed.

(a) Form a geometric sequence where each term represents the fractional part of the air *that still remains* in the container after each stroke. Then use this sequence to find out how much of the air remains after 6 strokes.

(b) Form a geometric sequence where each term represents the fractional part of the air *being removed* from the container on each stroke of the pump.

Then use this sequence (or the associated series) to find out how much of the air remains after 6 strokes.

47. If you pay $9500 for a car and its value depreciates 10% per year, how much will it be worth in 5 years?

48. Suppose that you could get a job that pays only a penny for the first day of employment but doubles your wages each succeeding day. How much would you be earning on the 31st day of your employment?

■ ■ ■ **Thoughts into words**

49. Explain the difference between an arithmetic sequence and a geometric sequence.

50. Consider the sequence whose nth term is $a_n = 3n$. How can you determine whether this is an arithmetic sequence or a geometric sequence?

■ ■ ■ **Further investigations**

For Problems 51–56, use your calculator to help find the indicated term of each geometric sequence.

51. 20th term of $2, 4, 8, 16, \ldots$

52. 15th term of $3, 9, 27, 81, \ldots$

53. 12th term of $\dfrac{2}{3}, \dfrac{4}{9}, \dfrac{8}{27}, \dfrac{16}{81}, \ldots$

54. 10th term of $-\dfrac{3}{4}, \dfrac{9}{16}, -\dfrac{27}{64}, \dfrac{81}{256}, \ldots$

55. 11th term of $-\dfrac{3}{2}, \dfrac{9}{4}, -\dfrac{27}{8}, \dfrac{81}{16}, \ldots$

56. 6th term of the sequence where $a_n = (0.1)^n$

57. In Problem 37 of Problem Set 12.2 we introduced the summation notation, which can also be used with geometric series. For example, the series

$$2 + 4 + 8 + 16 + 32$$

can be expressed as

$$\sum_{i=1}^{5} 2^i$$

Write out the terms and find the sum of each of the following.

(a) $\displaystyle\sum_{i=1}^{6} 2^i$ **(b)** $\displaystyle\sum_{i=1}^{5} 3^i$

(c) $\displaystyle\sum_{i=1}^{5} 2^{i-1}$ **(d)** $\displaystyle\sum_{i=1}^{6} \left(\dfrac{1}{2}\right)^{i+1}$

(e) $\displaystyle\sum_{i=1}^{4} \left(\dfrac{2}{3}\right)^{i}$ **(f)** $\displaystyle\sum_{i=1}^{5} \left(-\dfrac{3}{4}\right)^{i}$

12.4   Infinite Geometric Series

In Section 12.3 we used the formula

$$S_n = \frac{a_1 - a_1 r^n}{1 - r}, \quad r \neq 1 \tag{1}$$

to find the sum of the first n terms of a geometric series. By using the property $\frac{a - b}{c} = \frac{a}{c} - \frac{b}{c}$, we can express the right side of equation (1) in terms of two fractions as follows:

$$S_n = \frac{a_1 - a_1 r^n}{1 - r} = \frac{a_1}{1 - r} - \frac{a_1 r^n}{1 - r}, \quad r \neq 1 \tag{2}$$

Now let's examine the behavior of r^n for $|r| < 1$ — that is, for $-1 < r < 1$. For example, suppose that $r = \frac{1}{3}$; then

$$r^2 = \left(\frac{1}{3}\right)^2 = \frac{1}{9} \qquad r^3 = \left(\frac{1}{3}\right)^3 = \frac{1}{27}$$

$$r^4 = \left(\frac{1}{3}\right)^4 = \frac{1}{81} \qquad r^5 = \left(\frac{1}{3}\right)^5 = \frac{1}{243}$$

and so on. We can make $\left(\frac{1}{3}\right)^n$ as close to 0 as we please by taking sufficiently large values for n. In general, for values of r such that $|r| < 1$, the expression r^n will approach 0 as n increases. Therefore, in equation (2) the fraction $\frac{a_1 r^n}{1 - r}$ will approach 0 as n increases, and we say that the *sum of an infinite geometric series* is given by

$$S_\infty = \frac{a_1}{1 - r}, \quad |r| < 1$$

EXAMPLE 1   Find the sum of the infinite geometric series

$$1 + \frac{2}{3} + \frac{4}{9} + \frac{8}{27} + \cdots$$

Solution

Because $a_1 = 1$ and $r = \dfrac{2}{3}$, we obtain

$$S_\infty = \frac{1}{1 - \dfrac{2}{3}}$$

$$= \frac{1}{\dfrac{1}{3}} = 3$$

In Example 1, by stating $S_\infty = 3$ we mean that as we add more and more terms, the sum approaches 3.

First term:	1
Sum of first 2 terms:	$1 + \dfrac{2}{3} = 1\dfrac{2}{3}$
Sum of first 3 terms:	$1 + \dfrac{2}{3} + \dfrac{4}{9} = 2\dfrac{1}{9}$
Sum of first 4 terms:	$1 + \dfrac{2}{3} + \dfrac{4}{9} + \dfrac{8}{27} = 2\dfrac{11}{27}$
Sum of first 5 terms:	$1 + \dfrac{2}{3} + \dfrac{4}{9} + \dfrac{8}{27} + \dfrac{16}{81} = 2\dfrac{49}{81}$, etc.

EXAMPLE 2 Find the sum of the infinite geometric series

$$\frac{1}{2} - \frac{1}{4} + \frac{1}{8} - \frac{1}{16} + \cdots$$

Solution

Because $a_1 = \dfrac{1}{2}$ and $r = -\dfrac{1}{2}$, we obtain

$$S_\infty = \frac{\dfrac{1}{2}}{1 - \left(-\dfrac{1}{2}\right)} = \frac{\dfrac{1}{2}}{\dfrac{3}{2}} = \frac{1}{3}$$

If $|r| > 1$, the absolute value of r^n increases without bound as n increases. Consider the following two examples and note the unbounded growth of the absolute value of r^n.

Let $r = 2$	Let $r = -3$	
$r^2 = 2^2 = 4$	$r^2 = (-3)^2 = 9$	
$r^3 = 2^3 = 8$	$r^3 = (-3)^3 = -27$	$\lvert -27 \rvert = 27$
$r^4 = 2^4 = 16$	$r^4 = (-3)^4 = 81$	
$r^5 = 2^5 = 32$	$r^5 = (-3)^5 = -243$	$\lvert -243 \rvert = 243$
etc.	etc.	

If $r = 1$, then $S_n = na_1$, and as n increases without bound, $\lvert S_n \rvert$ also increases without bound. If $r = -1$, then S_n will be either a_1 or 0. Therefore, we say that the sum of any infinite geometric sequence where $\lvert r \rvert \geq 1$ does not exist.

Repeating Decimals as Infinite Geometric Series

In Section 1.1 we learned that a rational number is a number that can be represented as either a terminating decimal or a repeating decimal. For example,

$$0.23, \qquad 0.147, \qquad 0.\overline{3}, \qquad 0.\overline{14}, \qquad \text{and} \qquad 0.5\overline{81}$$

are rational numbers. (Remember that $0.\overline{3}$ means $0.333\ldots$.) Our knowledge of place value provides the basis for changing terminating decimals such as 0.23 and 0.147 to $\dfrac{a}{b}$ form, where a and b are integers, $b \neq 0$.

$$0.23 = \frac{23}{100}$$

$$0.147 = \frac{147}{1000}$$

However, changing repeating decimals to $\dfrac{a}{b}$ form requires a different technique, and our work with infinite geometric series provides the basis for one such approach. Consider the following examples.

EXAMPLE 3   Change $0.\overline{3}$ to $\dfrac{a}{b}$ form, where a and b are integers, $b \neq 0$.

Solution

We can write the repeating decimal $0.\overline{3}$ as the infinite geometric series

$$0.3 + 0.03 + 0.003 + 0.0003 + \cdots$$

with $a_1 = 0.3$ and $r = 0.1$. Therefore, we can use the sum formula and obtain

$$S_\infty = \frac{a_1}{1 - r} = \frac{0.3}{1 - 0.1} = \frac{0.3}{0.9} = \frac{3}{9} = \frac{1}{3}$$

Thus $0.\overline{3} = \dfrac{1}{3}$.

EXAMPLE 4   Change $0.\overline{14}$ to $\dfrac{a}{b}$ form, where a and b are integers, $b \neq 0$.

Solution

We can write the repeating decimal $0.\overline{14}$ as the infinite geometric series

$$0.14 + 0.0014 + 0.000014 + \cdots$$

with $a_1 = 0.14$ and $r = 0.01$. The sum formula produces

$$S_\infty = \frac{0.14}{1 - 0.01} = \frac{0.14}{0.99} = \frac{14}{99}$$

Thus $0.\overline{14} = \dfrac{14}{99}$. ■

If the repeating block of digits does not begin immediately after the decimal point, we can make a slight adjustment, as the final example illustrates.

EXAMPLE 5   Change $0.5\overline{81}$ to $\dfrac{a}{b}$ form, where a and b are integers, $b \neq 0$.

Solution

We can write the repeating decimal $0.5\overline{81}$ as

$$[0.5] + [0.081 + 0.00081 + 0.0000081 + \cdots]$$

where

$$0.081 + 0.00081 + 0.0000081 + \cdots$$

is an infinite geometric series with $a_1 = 0.081$ and $r = 0.01$. Thus

$$S_\infty = \frac{0.081}{1 - 0.01} = \frac{0.081}{0.99} = \frac{81}{990} = \frac{9}{110}$$

Therefore,

$$0.5\overline{81} = 0.5 + \frac{9}{110}$$

$$= \frac{5}{10} + \frac{9}{110}$$

$$= \frac{55}{110} + \frac{9}{110}$$

$$= \frac{64}{110}$$

$$= \frac{32}{55}$$

■

PROBLEM SET 12.4

For Problems 1–20, find the sum of the infinite geometric series. If the series has no sum, so state.

1. $1 + \dfrac{3}{4} + \dfrac{9}{16} + \dfrac{27}{64} + \cdots$ **2.** $\dfrac{2}{3} + \dfrac{2}{9} + \dfrac{2}{27} + \dfrac{2}{81} + \cdots$

3. $\dfrac{1}{2} + \dfrac{1}{4} + \dfrac{1}{8} + \dfrac{1}{16} + \cdots$ **4.** $1 + \dfrac{1}{2} + \dfrac{1}{4} + \dfrac{1}{8} + \cdots$

5. $\dfrac{2}{3} + \dfrac{4}{9} + \dfrac{8}{27} + \dfrac{16}{81} + \cdots$ **6.** $\dfrac{1}{3} + \dfrac{1}{9} + \dfrac{1}{27} + \dfrac{1}{81} + \cdots$

7. $1 - \dfrac{1}{2} + \dfrac{1}{4} - \dfrac{1}{8} + \cdots$ **8.** $1 + 2 + 4 + 8 + \cdots$

9. $6 + 2 + \dfrac{2}{3} + \dfrac{2}{9} + \cdots$

10. $4 + (-2) + 1 + \left(-\dfrac{1}{2}\right) + \cdots$

11. $2 + (-6) + 18 + (-54) + \cdots$

12. $4 + 2 + 1 + \dfrac{1}{2} + \cdots$

13. $1 + \left(-\dfrac{3}{4}\right) + \dfrac{9}{16} + \left(-\dfrac{27}{64}\right) + \cdots$

14. $9 - 3 + 1 - \dfrac{1}{3} + \cdots$ **15.** $8 - 4 + 2 - 1 + \cdots$

16. $5 + 3 + \dfrac{9}{5} + \dfrac{27}{5} + \cdots$ **17.** $1 + \dfrac{3}{2} + \dfrac{9}{4} + \dfrac{27}{8} + \cdots$

18. $1 - \dfrac{4}{3} + \dfrac{16}{9} - \dfrac{64}{27} + \cdots$ **19.** $27 + 9 + 3 + 1 + \cdots$

20. $9 + 3 + 1 + \dfrac{1}{3} + \cdots$

For Problems 21–34, change each repeating decimal to $\dfrac{a}{b}$ form, where a and b are integers, $b \neq 0$. Express $\dfrac{a}{b}$ in reduced form.

21. $0.\overline{4}$ **22.** $0.\overline{7}$ **23.** $0.\overline{47}$

24. $0.\overline{23}$ **25.** $0.\overline{45}$ **26.** $0.\overline{72}$

27. $0.\overline{427}$ **28.** $0.\overline{129}$ **29.** $0.4\overline{6}$

30. $0.8\overline{6}$ **31.** $2.1\overline{8}$ **32.** $2.9\overline{6}$

33. $0.42\overline{7}$ **34.** $0.23\overline{6}$

■ ■ ■ **Thoughts into words**

35. What does it mean to say that the sum of the infinite geometric series $1 + \dfrac{1}{2} + \dfrac{1}{4} + \dfrac{1}{8} + \cdots$ is 2?

36. What do we mean when we say that the infinite geometric sequence $1 + 2 + 4 + 8 + \cdots$ has no sum?

37. Why don't we discuss the sum of an infinite arithmetic series?

12.5 Binomial Expansions

In Chapter 3, we used the pattern $(x + y)^2 = x^2 + 2xy + y^2$ to square binomials and the pattern $(x + y)^3 = x^3 + 3x^2y + 3xy^2 + y^3$ to cube binomials. At this time, we can extend those ideas to arrive at a pattern that will allow us to write the expansion of $(x + y)^n$, where n is *any* positive integer. Let's begin by looking at some specific expansions that we can verify by direct multiplication.

$$(x + y)^1 = x + y$$
$$(x + y)^2 = x^2 + 2xy + y^2$$
$$(x + y)^3 = x^3 + 3x^2y + 3xy^2 + y^3$$
$$(x + y)^4 = x^4 + 4x^3y + 6x^2y^2 + 4xy^3 + y^4$$
$$(x + y)^5 = x^5 + 5x^4y + 10x^3y^2 + 10x^2y^3 + 5xy^4 + y^5$$

First, note the patterns of the exponents for x and y on a term-by-term basis. The exponents of x begin with the exponent of the binomial and term-by-term decrease by 1 until the last term has x^0, which is 1. The exponents of y begin with $0(y^0 = 1)$ and term-by-term increase by 1 until the last term contains y to the power of the original binomial. In other words, the variables in the expansion of $(x + y)^n$ exhibit the following pattern.

$$x^n, \qquad x^{n-1}y, \qquad x^{n-2}y^2, \qquad x^{n-3}y^3, \qquad \ldots, \qquad xy^{n-1}, \qquad y^n$$

Note that the sum of the exponents of x and y for each term is n.

Next, let's arrange the **coefficients** in the following triangular formation that yields an easy-to-remember pattern.

$$
\begin{array}{ccccccccccc}
 & & & & 1 & & 1 & & & & \\
 & & & 1 & & 2 & & 1 & & & \\
 & & 1 & & 3 & & 3 & & 1 & & \\
 & 1 & & 4 & & 6 & & 4 & & 1 & \\
1 & & 5 & & 10 & & 10 & & 5 & & 1 \\
\end{array}
$$

The number of the row of the formation contains the coefficients of the expansion of $(x + y)$ to that power. For example, the fifth row contains 1   5   10   10   5   1, which are the coefficients of the terms of the expansion of $(x + y)^5$. Furthermore, each row can be formed from the previous row as follows:

1. Start and end each row with 1.
2. All other entries result from adding the two numbers in the row immediately above, one number to the left and one number to the right.

Thus, from row 5 we can form row 6 as follows:

Row 5:   1   5   10   10   5   1
           Add   Add   Add   Add   Add

Row 6:   1   6   15   20   15   6   1

We can use the row-6 coefficients and our previous discussion relative to the exponents and write out the expansion for $(x + y)^6$.

$$(x + y)^6 = x^6 + 6x^5y + 15x^4y^2 + 20x^3y^3 + 15x^2y^4 + 6xy^5 + y^6$$

REMARK: We often refer to the triangular formation of numbers that we have been discussing as Pascal's triangle. This is in honor of Blaise Pascal, a 17th-century mathematician, to whom the discovery of this pattern is attributed.

Although Pascal's triangle will work for any positive integral power of a binomial, it does become somewhat impractical for large powers, so we need another technique for determining the coefficients. Let's look at the following notational agreements. $n!$ (read "n factorial") means $n(n - 1)(n - 2) \ldots 1$, where n is any positive integer. For example,

$3!$ means $3 \cdot 2 \cdot 1 = 6$

$5!$ means $5 \cdot 4 \cdot 3 \cdot 2 \cdot 1 = 120$

We also agree that $0! = 1$. (Note that both $0!$ and $1!$ equal 1.)

Let us now use the factorial notation and state the expression of the general case $(x + y)^n$, where n is any positive integer.

$$(x + y)^n = x^n + nx^{n-1}y + \frac{n(n - 1)}{2!}x^{n-2}y^2 + \frac{n(n - 1)(n - 2)}{3!}x^{n-3}y^3$$
$$+ \cdots + y^n$$

The binomial expansion for the general case may look a little confusing, but actually it is quite easy to apply once you try it a few times on some specific examples. Remember the decreasing pattern for the exponents of x and the increasing pattern for the exponents of y. Furthermore, note the pattern of the coefficients:

$$1, \quad n, \quad \frac{n(n - 1)}{2!}, \quad \frac{n(n - 1)(n - 2)}{3!}, \quad \text{etc.}$$

Keep these ideas in mind as you study the following examples.

E X A M P L E   1

Expand $(x + y)^7$.

Solution

We can expand as follows:

$$(x + y)^7 = x^7 + 7x^6y + \frac{7 \cdot 6}{2!}x^5y^2 + \frac{7 \cdot 6 \cdot 5}{3!}x^4y^3 + \frac{7 \cdot 6 \cdot 5 \cdot 4}{4!}x^3y^4 + \frac{7 \cdot 6 \cdot 5 \cdot 4 \cdot 3}{5!}x^2y^5$$

$$+ \frac{7 \cdot 6 \cdot 5 \cdot 4 \cdot 3 \cdot 2}{6!}xy^6 + y^7$$

$$= x^7 + 7x^6y + 21x^5y^2 + 35x^4y^3 + 35x^3y^4 + 21x^2y^5 + 7xy^6 + y^7 \quad \blacksquare$$

Expand $(x - y)^5$.

Solution

We shall treat $(x - y)^5$ as $[x + (-y)]^5$.

$$[x + (-y)]^5 = x^5 + 5x^4(-y) + \frac{5 \cdot 4}{2!}x^3(-y)^2 + \frac{5 \cdot 4 \cdot 3}{3!}x^2(-y)^3$$

$$+ \frac{5 \cdot 4 \cdot 3 \cdot 2}{4!}x(-y)^4 + (-y)^5$$

$$= x^5 - 5x^4y + 10x^3y^2 - 10x^2y^3 + 5xy^4 - y^5 \qquad \blacksquare$$

EXAMPLE 3

Expand and simplify $(2a + 3b)^4$.

Solution

Let $x = 2a$ and $y = 3b$.

$$(2a + 3b)^4 = (2a)^4 + 4(2a)^3(3b) + \frac{4 \cdot 3}{2!}(2a)^2(3b)^2 + \frac{4 \cdot 3 \cdot 2}{3!}(2a)(3b)^3 + (3b)^4$$

$$= 16a^4 + 96a^3b + 216a^2b^2 + 216ab^3 + 81b^4 \qquad \blacksquare$$

Finding Specific Terms

Sometimes it is convenient to find a specific term of a binomial expansion without writing out the entire expansion. For example, suppose that we need the sixth term of the expansion $(x + y)^{12}$. We could proceed as follows:

The sixth term will contain y^5. (Note in the general expansion that the **exponent of y is always 1 less than the number of the term**.) Because the sum of the exponents for x and y must be 12 (the exponent of the binomial), the sixth term will also contain x^7. Again looking back at the general binomial expansion, note that the **denominators of the coefficients** are of the form $r!$, where the value of r agrees with the exponent of y for each term. Thus, if we have y^5, the denominator of the coefficient is 5!. In the general expansion, each **numerator of a coefficient** contains r factors, where the first factor is the exponent of the binomial and each succeeding factor is 1 less than the preceding one. Thus the sixth term of $(x + y)^{12}$ is $\dfrac{12 \cdot 11 \cdot 10 \cdot 9 \cdot 8}{5!}x^7y^5$, which simplifies to $792x^7y^5$.

EXAMPLE 4

Find the fourth term of $(3a + 2b)^7$.

Solution

The fourth term will contain $(2b)^3$ and thus $(3a)^4$. The coefficient is $\dfrac{7 \cdot 6 \cdot 5}{3!}$. Therefore, the fourth term is

$$\frac{7 \cdot 6 \cdot 5}{3!}(3a)^4(2b)^3$$

which simplifies to $22,680a^4b^3$. $\qquad \blacksquare$

PROBLEM SET 12.5

For Problems 1–6, use Pascal's triangle to help expand each of the following.

1. $(x + y)^8$

2. $(x + y)^7$

3. $(3x + y)^4$

4. $(x + 2y)^4$

5. $(x - y)^5$

6. $(x - y)^4$

For Problems 7–20, expand and simplify.

7. $(x + y)^{10}$

8. $(x + y)^9$

9. $(2x + y)^6$

10. $(x + 3y)^5$

11. $(x - 3y)^5$

12. $(2x - y)^6$

13. $(3a - 2b)^5$

14. $(2a - 3b)^4$

15. $(x + y^3)^6$

16. $(x^2 + y)^5$

17. $(x + 2)^7$

18. $(x + 3)^6$

19. $(x - 3)^4$

20. $(x - 1)^9$

For Problems 21–24, write the first four terms of the expansion.

21. $(x + y)^{15}$

22. $(x + y)^{12}$

23. $(a - 2b)^{13}$

24. $(x - y)^{20}$

For Problems 25–30, find the indicated term of the expansion.

25. Seventh term of $(x + y)^{11}$

26. Fourth term of $(x + y)^8$

27. Fourth term of $(x - 2y)^6$

28. Fifth term of $(x - y)^9$

29. Third term of $(2x - 5y)^5$

30. Sixth term of $(3a + b)^7$

■ ■ ■ **Thoughts into words**

31. How would you explain binomial expansions to an elementary algebra student?

32. Explain how to find the fifth term of the expansion of $(2x + 3y)^9$ without writing out the entire expansion.

SUMMARY

(12.1) An **infinite sequence** is a function whose domain is the set of positive integers. We frequently express a general infinite sequence as

$$a_1, a_2, a_3, \ldots, a_n, \ldots$$

where a_1 is the first term, a_2 the second term, and so on, and a_n represents the general, or nth, term.

An **arithmetic sequence** is a sequence where there is a **common difference** between successive terms.

The **general term of an arithmetic sequence** is given by

$$a_n = a_1 + (n-1)d$$

where a_1 is the first term and d is the common difference.

(12.2) The indicated sum of a sequence is called a **series**. The sum of the first n terms of an arithmetic series is given by

$$S_n = \frac{n(a_1 + a_n)}{2}$$

(12.3) A **geometric sequence** is a sequence in which each term after the first is obtained by multiplying the preceding term by a common multiplier. The common multiplier is called the **common ratio** of the sequence.

The **general term of a geometric sequence** is given by

$$a_n = a_1 r^{n-1}$$

where a_1 is the first term and r is the common ratio.

The sum of the first n terms of a geometric series is given by

$$S_n = \frac{a_1 r^n - a_1}{r - 1}, \qquad r \neq 1$$

(12.4) The sum of an infinite geometric series is given by

$$S_\infty = \frac{a_1}{1 - r}, \qquad |r| < 1$$

Any infinite geometric series where $|r| \geq 1$ has no sum.

This sum formula can be used to change repeating decimals to $\dfrac{a}{b}$ form.

(12.5) The expansion of $(x + y)^n$, where n is a positive integer, is given by

$$(x + y)^n = x^n + nx^{n-1}y + \frac{n(n-1)}{2!}x^{n-2}y^2$$

$$+ \frac{n(n-1)(n-2)}{3!}x^{n-3}y^3 + \cdots + y^n$$

To find a specific term of a binomial expansion, review Example 4 of Section 12.5.

CHAPTER 12 REVIEW PROBLEM SET

For Problems 1–10, find the general term (nth term) for each of the following sequences. These problems contain a mixture of arithmetic and geometric sequences.

1. $3, 9, 15, 21, \ldots$

2. $\dfrac{1}{3}, 1, 3, 9, \ldots$

3. $10, 20, 40, 80, \ldots$

4. $5, 2, -1, -4, \ldots$

5. $-5, -3, -1, 1, \ldots$

6. $9, 3, 1, \dfrac{1}{3}, \ldots$

7. $-1, 2, -4, 8, \ldots$

8. $12, 15, 18, 21, \ldots$

9. $\dfrac{2}{3}, 1, \dfrac{4}{3}, \dfrac{5}{3}, \ldots$

10. $1, 4, 16, 64, \ldots$

For Problems 11–16, find the indicated term of each of the sequences.

11. The 19th term of $1, 5, 9, 13, \ldots$

12. The 28th term of $-2, 2, 6, 10, \ldots$

13. The 9th term of $8, 4, 2, 1, \ldots$

14. The 8th term of $\dfrac{243}{32}, \dfrac{81}{16}, \dfrac{27}{8}, \dfrac{9}{4}, \ldots$

15. The 34th term of $7, 4, 1, -2, \ldots$

16. The 10th term of $-32, 16, -8, 4, \ldots$

17. If the 5th term of an arithmetic sequence is -19 and the 8th term is -34, find the common difference of the sequence.

18. If the 8th term of an arithmetic sequence is 37 and the 13th term is 57, find the 20th term.

19. Find the first term of a geometric sequence if the 3rd term is 5 and the 6th term is 135.

20. Find the common ratio of a geometric sequence if the 2nd term is $\dfrac{1}{2}$ and the 6th term is 8.

21. Find the sum of the first 9 terms of the series $81 + 27 + 9 + 3 + \cdots$.

22. Find the sum of the first 70 terms of the series $-3 + 0 + 3 + 6 + \cdots$.

23. Find the sum of the first 75 terms of the series $5 + 1 + (-3) + (-7) + \cdots$.

24. Find the sum of the first 10 terms of the series for which $a_n = 2^{5-n}$.

25. Find the sum of the first 95 terms of the series for which $a_n = 7n + 1$.

26. Find the sum $5 + 7 + 9 + \cdots + 137$.

27. Find the sum $64 + 16 + 4 + \cdots + \dfrac{1}{64}$.

28. Find the sum of all even numbers between 8 and 384, inclusive.

29. Find the sum of all multiples of 3 between 27 and 276, inclusive.

30. Find the sum of the infinite geometric series $64 + 16 + 4 + 1 + \cdots$.

31. Change $0.\overline{36}$ to reduced $\dfrac{a}{b}$ form, where a and b are integers, $b \neq 0$.

32. Change $0.4\overline{5}$ to reduced $\dfrac{a}{b}$ form, where a and b are integers, $b \neq 0$.

Solve Problems 33–37 by using your knowledge of arithmetic and geometric sequences.

33. Suppose that at the beginning of the year your savings account contains $3750. If you withdraw $250 per month from the account, how much will it contain at the end of the year?

34. Sonya has decided to start saving dimes. She plans to save 1 dime the first day of April, 2 dimes the second day, 3 dimes the third day, 4 dimes the fourth day, and so on, for the 30 days of April. How much money will she save in April?

35. Nancy has decided to start saving dimes. She plans to save 1 dime the first day of April, 2 dimes the second day, 4 dimes the third day, 8 dimes the fourth day, and so on, for the first 15 days of April. How much will she save in 15 days?

36. A tank contains 61,440 gallons of water. Each day one-fourth of the water is to be drained out. How much will remain in the tank at the end of 6 days?

37. An object, falling from rest in a vacuum, falls 16 feet the first second, 48 feet the second second, 80 feet the third second, 112 feet the fourth second, and so on. How far will the object fall in 15 seconds?

38. Expand $(2x + y)^7$.

39. Expand $(x - 3y)^4$.

40. Find the 5th term of the expansion of $(x + 2y)^8$.

TEST

1. Find the 15th term of the sequence for which $a_n = -3n - 1$.

2. Find the 5th term of the sequence for which $a_n = 3(2)^{n-1}$.

3. Find the general term of the sequence $-3, 1, 5, 9, \ldots$.

4. Find the general term of the sequence
$$5, \frac{5}{2}, \frac{5}{4}, \frac{5}{8}, \ldots.$$

5. Find the general term of the sequence $6, 3, 0, -3, \ldots$.

6. Find the 7th term of the sequence $8, 12, 18, 27, \ldots$.

7. Find the 75th term of the sequence $1, 4, 7, 10, \ldots$.

8. Find the number of terms in the sequence $7, 11, 15, \ldots, 243$.

9. If the 4th term of an arithmetic sequence is 13 and the 7th term is 22, find the 15th term.

10. Find the sum of the first 40 terms of the series $1 + 4 + 7 + 10 + \cdots$.

11. Find the sum of the first 8 terms of the series $3 + 6 + 12 + 24 + \cdots$.

12. Find the sum of the series $3 + 1 + (-1) + \cdots + (-55)$.

13. Find the sum of the series $3 + 9 + 27 + \cdots + 2187$.

14. Find the sum of the first 45 terms of the series for which $a_n = 7n - 2$.

15. Find the sum of the first 10 terms of the series for which $a_n = 3(2)^n$.

16. Find the sum of the first 150 positive even whole numbers.

17. Find the sum of the odd numbers between 11 and 193, inclusive.

18. A woman invests \$350 at 12% simple interest at the beginning of each year for a period of 10 years. Find the total accumulated value of all the investments at the end of the 10-year period.

19. Suppose you save a dime the first day of a month, \$.20 the second day, and \$.40 the third day and continue to double how much you save per day for 15 days. Find the total amount that you will have saved at the end of 15 days.

20. Find the sum of the infinite geometric series $9 + 3 + 1 + \dfrac{1}{3} + \cdots$.

21. Change the repeating decimal $0.\overline{37}$ to $\dfrac{a}{b}$ form, where a and b are integers and $b \neq 0$.

22. Change the repeating decimal $0.2\overline{6}$ to $\dfrac{a}{b}$ form, where a and b are integers and $b \neq 0$.

23. Expand $(x - 3y)^5$.

24. Write the first three terms of the expansion of $(2x + y)^7$.

25. Find the 5th term of the expansion of $(a + b)^{12}$.

Appendix

A   Prime Numbers and Operations with Fractions

This appendix reviews the operations with rational numbers in common fraction form. Throughout this section, we will speak of "multiplying fractions." Be aware that this phrase means multiplying rational numbers in common fraction form. A strong foundation here will simplify your later work in rational expressions. Because prime numbers and prime factorization play an important role in the operations with fractions, let's begin by considering two special kinds of whole numbers, prime numbers and composite numbers.

> ### DEFINITION A.1
>
> A **prime number** is a whole number greater than 1 that has no factors (divisors) other than itself and 1. Whole numbers greater than 1 that are not prime numbers are called **composite numbers.**

The prime numbers less than 50 are 2, 3, 5, 7, 11, 13, 17, 19, 23, 29, 31, 37, 41, 43, and 47. Note that each of these has no factors other than itself and 1. We can express every composite number as the indicated product of prime numbers. Consider the following examples:

$$4 = 2 \cdot 2 \qquad 6 = 2 \cdot 3 \qquad 8 = 2 \cdot 2 \cdot 2 \qquad 10 = 2 \cdot 5 \qquad 12 = 2 \cdot 2 \cdot 3$$

In each case we express a composite number as the indicated product of prime numbers. The indicated-product form is called the prime-factored form of the number. There are various procedures to find the prime factors of a given composite number. For our purposes, the simplest technique is to factor the given composite number into any two easily recognized factors and then continue to factor each of these until we obtain only prime factors. Consider these examples:

$$18 = 2 \cdot 9 = 2 \cdot 3 \cdot 3 \qquad\qquad\qquad 27 = 3 \cdot 9 = 3 \cdot 3 \cdot 3$$

$$24 = 4 \cdot 6 = 2 \cdot 2 \cdot 2 \cdot 3 \qquad\qquad 150 = 10 \cdot 15 = 2 \cdot 5 \cdot 3 \cdot 5$$

It does not matter which two factors we choose first. For example, we might start by expressing 18 as $3 \cdot 6$ and then factor 6 into $2 \cdot 3$, which produces a final result of $18 = 3 \cdot 2 \cdot 3$. Either way, 18 contains two prime factors of 3 and one prime factor of 2. The order in which we write the prime factors is not important.

Least Common Multiple

It is sometimes necessary to determine the smallest common nonzero multiple of two or more whole numbers. We call this nonzero number the **least common multiple.** In our work with fractions, there will be problems where it will be necessary to find the least common multiple of some numbers, usually the denominators of fractions. So let's review the concepts of multiples. We know that 35 is a multiple of 5 because $5 \cdot 7 = 35$. The set of all whole numbers that are multiples of 5 consists of 0, 5, 10, 15, 20, 25, and so on. In other words, 5 times each successive whole number ($5 \cdot 0 = 0, 5 \cdot 1 = 5, 5 \cdot 2 = 10, 5 \cdot 3 = 15$, etc.) produces the multiples of 5. In a like manner, the set of multiples of 4 consists of 0, 4, 8, 12, 16, and so on. We can illustrate the concept of least common multiple and find the least common multiple of 5 and 4 by using a simple listing of the multiples of 5 and the multiples of 4.

Multiples of 5 are 0, 5, 10, 15, 20, 25, 30, 35, 40, 45, . . .

Multiples of 4 are 0, 4, 8, 12, 16, 20, 24, 28, 32, 36, 40, 44, 48, . . .

The nonzero numbers in common on the lists are 20 and 40. The least of these, 20, is the least common multiple. Stated another way, 20 is the smallest nonzero whole number that is divisible by both 4 and 5.

Often, from your knowledge of arithmetic, you will be able to determine the least common multiple by inspection. For instance, the least common multiple of 6 and 8 is 24. Therefore, 24 is the smallest nonzero whole number that is divisible by both 6 and 8. If we cannot determine the least common multiple by inspection, then using the prime-factorized form of composite numbers is helpful. The procedure is as follows.

STEP 1   Express each number as a product of prime factors.

STEP 2   The least common multiple contains each different prime factor as many times as the most times it appears in any one of the factorizations from Step 1.

The following examples illustrate this technique for finding the least common multiple of two or more numbers.

E X A M P L E   1   Find the least common multiple of 24 and 36.

Solution

Let's first express each number as a product of prime factors.

$$24 = 2 \cdot 2 \cdot 2 \cdot 3$$

$$36 = 2 \cdot 2 \cdot 3 \cdot 3$$

The prime factor 2 occurs the most times (three times) in the factorization of 24. Because the factorization of 24 contains three 2s, the least common multiple must have three 2s. The prime factor 3 occurs the most times (two times) in the factorization of 36. Because the factorization of 36 contains two 3s, the least common multiple must have two 3s. The least common multiple of 24 and 36 is therefore $2 \cdot 2 \cdot 2 \cdot 3 \cdot 3 = 72$. ∎

EXAMPLE 2   Find the least common multiple of 48 and 84.

Solution

$$48 = 2 \cdot 2 \cdot 2 \cdot 2 \cdot 3$$
$$84 = 2 \cdot 2 \cdot 3 \cdot 7$$

We need four 2s in the least common multiple because of the four 2s in 48. We need one 3 because of the 3 in each of the numbers, and we need one 7 because of the 7 in 84. The least common multiple of 48 and 84 is $2 \cdot 2 \cdot 2 \cdot 2 \cdot 3 \cdot 7 = 336$. ∎

EXAMPLE 3   Find the least common multiple of 12, 18, and 28.

Solution

$$28 = 2 \cdot 2 \cdot 7$$
$$18 = 2 \cdot 3 \cdot 3$$
$$12 = 2 \cdot 2 \cdot 3$$

The least common multiple is $2 \cdot 2 \cdot 3 \cdot 3 \cdot 7 = 252$. ∎

EXAMPLE 4   Find the least common multiple of 8 and 9.

Solution

$$9 = 3 \cdot 3$$
$$8 = 2 \cdot 2 \cdot 2$$

The least common multiple is $2 \cdot 2 \cdot 2 \cdot 3 \cdot 3 = 72$. ∎

Multiplying Fractions

We can define the multiplication of fractions in common fractional form as follows:

Multiplying Fractions

If a, b, c, and d are integers, with b and d not equal to zero, then $\dfrac{a}{b} \cdot \dfrac{c}{d} = \dfrac{a \cdot c}{b \cdot d}$.

To multiply fractions in common fractional form, we simply multiply numerators and multiply denominators. The following examples illustrate the multiplying of fractions.

$$\frac{1}{3} \cdot \frac{2}{5} = \frac{1 \cdot 2}{3 \cdot 5} = \frac{2}{15}$$

$$\frac{3}{4} \cdot \frac{5}{7} = \frac{3 \cdot 5}{4 \cdot 7} = \frac{15}{28}$$

$$\frac{3}{5} \cdot \frac{5}{3} = \frac{15}{15} = 1$$

The last of these examples is a very special case. If the product of two numbers is 1, then the numbers are said to be reciprocals of each other.

Before we proceed too far with multiplying fractions, we need to learn about reducing fractions. The following property is applied throughout our work with fractions. We call this property the fundamental property of fractions.

Fundamental Property of Fractions

If b and k are nonzero integers, and a is any integer, then $\dfrac{a \cdot k}{b \cdot k} = \dfrac{a}{b}$.

The fundamental property of fractions provides the basis for what is often called reducing fractions to lowest terms, or expressing fractions in simplest or reduced form. Let's apply the property to a few examples.

EXAMPLE 5 Reduce $\dfrac{12}{18}$ to lowest terms.

Solution

$$\frac{12}{18} = \frac{2 \cdot 6}{3 \cdot 6} = \frac{2}{3} \qquad \text{A common factor of 6 has been divided out of both numerator and denominator.}$$

EXAMPLE 6 Change $\dfrac{14}{35}$ to simplest form.

Solution

$$\frac{14}{35} = \frac{2 \cdot 7}{5 \cdot 7} = \frac{2}{5} \qquad \text{A common factor of 7 has been divided out of both numerator and denominator.}$$

EXAMPLE 7

Reduce $\dfrac{72}{90}$.

Solution

$$\frac{72}{90} = \frac{2 \cdot 2 \cdot 2 \cdot 3 \cdot 3}{2 \cdot 3 \cdot 3 \cdot 5} = \frac{4}{5}$$ The prime-factored forms of the numerator and denominator may be used to find common factors.   ■

We are now ready to consider multiplication problems with the understanding that the final answer should be expressed in reduced form. Study the following examples carefully; we use different methods to simplify the problems.

EXAMPLE 8

Multiply $\left(\dfrac{9}{4}\right)\left(\dfrac{14}{15}\right)$.

Solution

$$\left(\frac{9}{4}\right)\left(\frac{14}{15}\right) = \frac{3 \cdot 3 \cdot 2 \cdot 7}{2 \cdot 2 \cdot 3 \cdot 5} = \frac{21}{10}$$ ■

EXAMPLE 9

Find the product of $\dfrac{8}{9}$ and $\dfrac{18}{24}$.

Solution

$$\frac{\overset{1}{8}}{\underset{1}{9}} \cdot \frac{\overset{2}{18}}{\underset{3}{24}} = \frac{2}{3}$$ A common factor of 8 has been divided out of 8 and 24, and a common factor of 9 has been divided out of 9 and 18.   ■

Dividing Fractions

The next example motivates a definition for division of rational numbers in fractional form:

$$\frac{\frac{3}{4}}{\frac{2}{3}} = \frac{\left(\frac{3}{4}\right)\left(\frac{3}{2}\right)}{\left(\frac{2}{3}\right)\left(\frac{3}{2}\right)} = \frac{\left(\frac{3}{4}\right)\left(\frac{3}{2}\right)}{1} = \left(\frac{3}{4}\right)\left(\frac{3}{2}\right) = \frac{9}{8}$$

Note that $\left(\dfrac{\frac{3}{2}}{\frac{3}{2}}\right)$ is a form of 1, and $\dfrac{3}{2}$ is the reciprocal of $\dfrac{2}{3}$. In other words,

$\dfrac{3}{4}$ divided by $\dfrac{2}{3}$ is equivalent to $\dfrac{3}{4}$ times $\dfrac{3}{2}$. The following definition for division now should seem reasonable.

Division of Fractions

If b, c, and d are nonzero integers, and a is any integer, then $\dfrac{a}{b} \div \dfrac{c}{d} = \dfrac{a}{b} \cdot \dfrac{d}{c}$.

Note that to divide $\dfrac{a}{b}$ by $\dfrac{c}{d}$, we multiply $\dfrac{a}{b}$ times the reciprocal of $\dfrac{c}{d}$, which is $\dfrac{d}{c}$. The next examples demonstrate the important steps of a division problem.

$$\frac{2}{3} \div \frac{1}{2} = \frac{2}{3} \cdot \frac{2}{1} = \frac{4}{3}$$

$$\frac{5}{6} \div \frac{3}{4} = \frac{5}{6} \cdot \frac{4}{3} = \frac{5 \cdot 4}{6 \cdot 3} = \frac{5 \cdot 2 \cdot 2}{2 \cdot 3 \cdot 3} = \frac{10}{9}$$

$$\frac{\frac{6}{7}}{\frac{2}{2}} = \frac{\overset{3}{\cancel{6}}}{7} \cdot \frac{1}{\underset{1}{\cancel{2}}} = \frac{3}{7}$$

Adding and Subtracting Fractions

Suppose that it is one-fifth of a mile between your dorm and the union and two-fifths of a mile between the union and the library along a straight line as indicated in Figure A.1. The total distance between your dorm and the library is three-fifths of a mile, and we write $\dfrac{1}{5} + \dfrac{2}{5} = \dfrac{3}{5}$.

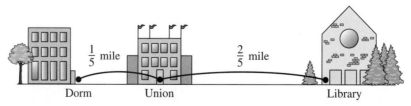

Figure A.1

A pizza is cut into seven equal pieces and you eat two of the pieces (see Figure A.2). How much of the pizza remains? We represent the whole pizza by $\frac{7}{7}$ and conclude that $\frac{7}{7} - \frac{2}{7} = \frac{5}{7}$ of the pizza remains.

Figure A.2

These examples motivate the following definition for addition and subtraction of rational numbers in $\frac{a}{b}$ form.

Addition and Subtraction of Fractions

If a, b, and c are integers, and b is not zero, then

$$\frac{a}{b} + \frac{c}{b} = \frac{a + c}{b} \qquad \text{Addition}$$

$$\frac{a}{b} - \frac{c}{b} = \frac{a - c}{b} \qquad \text{Subtraction}$$

We say that fractions with common denominators can be added or subtracted by adding or subtracting the numerators and placing the results over the common denominator. Consider the following examples:

$$\frac{3}{7} + \frac{2}{7} = \frac{3 + 2}{7} = \frac{5}{7}$$

$$\frac{7}{8} - \frac{2}{8} = \frac{7 - 2}{8} = \frac{5}{8}$$

$$\frac{5}{6} - \frac{1}{6} = \frac{5 - 1}{6} = \frac{4}{6} = \frac{2}{3} \qquad \text{We agree to reduce the final answer.}$$

How do we add or subtract if the fractions do not have a common denominator? We use the fundamental principle of fractions, $\frac{a \cdot k}{b \cdot k} = \frac{a}{b}$, to get equivalent fractions that have a common denominator. **Equivalent fractions** are fractions that name the same number. Consider the next example, which shows the details.

EXAMPLE 10 Add $\dfrac{1}{4} + \dfrac{2}{5}$.

Solution

$$\dfrac{1}{4} = \dfrac{1 \cdot 5}{4 \cdot 5} = \dfrac{5}{20} \qquad \dfrac{1}{4} \text{ and } \dfrac{5}{20} \text{ are equivalent fractions.}$$

$$\dfrac{2}{5} = \dfrac{2 \cdot 4}{5 \cdot 4} = \dfrac{8}{20} \qquad \dfrac{2}{5} \text{ and } \dfrac{8}{20} \text{ are equivalent fractions.}$$

$$\dfrac{5}{20} + \dfrac{8}{20} = \dfrac{13}{20}$$

$\blacksquare$

Note that in Example 10 we chose 20 as the common denominator, and 20 is the least common multiple of the original denominators 4 and 5. (Recall that the least common multiple is the smallest nonzero whole number divisible by the given numbers.) In general, we use the least common multiple of the denominators of the fractions to be added or subtracted as a **least common denominator** (LCD).

Recall that the least common multiple may be found either by inspection or by using prime factorization forms of the numbers. Consider some examples involving these procedures.

EXAMPLE 11 Subtract $\dfrac{5}{8} - \dfrac{7}{12}$.

Solution

By inspection the LCD is 24.

$$\dfrac{5}{8} - \dfrac{7}{12} = \dfrac{5 \cdot 3}{8 \cdot 3} - \dfrac{7 \cdot 2}{12 \cdot 2} = \dfrac{15}{24} - \dfrac{14}{24} = \dfrac{1}{24}$$

$\blacksquare$

If the LCD is not obvious by inspection, then we can use the technique of prime factorization to find the least common multiple.

EXAMPLE 12 Add $\dfrac{5}{18} + \dfrac{7}{24}$.

Solution

If we cannot find the LCD by inspection, then we can use the prime-factorized forms.

$$\left. \begin{array}{l} 18 = 2 \cdot 3 \cdot 3 \\ 24 = 2 \cdot 2 \cdot 2 \cdot 3 \end{array} \right\} \longrightarrow \text{LCD} = 2 \cdot 2 \cdot 2 \cdot 3 \cdot 3 = 72$$

$$\dfrac{5}{18} + \dfrac{7}{24} = \dfrac{5 \cdot 4}{18 \cdot 4} + \dfrac{7 \cdot 3}{24 \cdot 3} = \dfrac{20}{72} + \dfrac{21}{72} = \dfrac{41}{72}$$

$\blacksquare$

EXAMPLE 13

Marcey put $\frac{5}{8}$ pound of chemicals in the spa to adjust the water quality. Michael, not realizing Marcey had already put in chemicals, put $\frac{3}{14}$ pound of chemicals in the spa. The chemical manufacturer states that you should never add more than 1 pound of chemicals. Have Marcey and Michael together put in more than 1 pound of chemicals?

Solution

Add $\frac{5}{8} + \frac{3}{14}$.

$$\left.\begin{array}{l} 8 = 2 \cdot 2 \cdot 2 \\ \\ 14 = 2 \cdot 7 \end{array}\right\} \longrightarrow \text{LCD} = 2 \cdot 2 \cdot 2 \cdot 7 = 56$$

$$\frac{5}{8} + \frac{3}{14} = \frac{5 \cdot 7}{8 \cdot 7} + \frac{3 \cdot 4}{14 \cdot 4} = \frac{35}{56} + \frac{12}{56} = \frac{47}{56}$$

No, Marcey and Michael have not added more than 1 pound of chemicals. ■

Simplifying Numerical Expressions

We now consider simplifying numerical expressions that contain fractions. In agreement with the order of operations, first multiplications and divisions are done as they appear from left to right, and then additions and subtractions are performed as they appear from left to right. In these next examples, we show only the major steps. Be sure you can fill in all the details.

EXAMPLE 14

Simplify $\frac{3}{4} + \frac{2}{3} \cdot \frac{3}{5} - \frac{1}{2} \cdot \frac{1}{5}$.

Solution

$$\frac{3}{4} + \frac{2}{3} \cdot \frac{3}{5} - \frac{1}{2} \cdot \frac{1}{5} = \frac{3}{4} + \frac{2}{5} - \frac{1}{10}$$

$$= \frac{15}{20} + \frac{8}{20} - \frac{2}{20} = \frac{15 + 8 - 2}{20} = \frac{21}{20}$$ ■

EXAMPLE 15

Simplify $\frac{5}{8}\left(\frac{1}{2} + \frac{1}{3}\right)$.

Solution

$$\frac{5}{8}\left(\frac{1}{2} + \frac{1}{3}\right) = \frac{5}{8}\left(\frac{3}{6} + \frac{2}{6}\right) = \frac{5}{8}\left(\frac{5}{6}\right) = \frac{25}{48}$$ ■

PRACTICE EXERCISES

For Problems 1–12, factor each composite number into a product of prime numbers; for example, $18 = 2 \cdot 3 \cdot 3$.

1. 26

2. 16

3. 36

4. 80

5. 49

6. 92

7. 56

8. 144

9. 120

10. 84

11. 135

12. 98

For Problems 13–24, find the least common multiple of the given numbers.

13. 6 and 8

14. 8 and 12

15. 12 and 16

16. 9 and 12

17. 28 and 35

18. 42 and 66

19. 49 and 56

20. 18 and 24

21. 8, 12, and 28

22. 6, 10, and 12

23. 9, 15, and 18

24. 8, 14, and 24

For Problems 25–30, reduce each fraction to lowest terms.

25. $\dfrac{8}{12}$

26. $\dfrac{12}{16}$

27. $\dfrac{16}{24}$

28. $\dfrac{18}{32}$

29. $\dfrac{15}{9}$

30. $\dfrac{48}{36}$

For Problems 31–36, multiply or divide as indicated, and express answers in reduced form.

31. $\dfrac{3}{4} \cdot \dfrac{5}{7}$

32. $\dfrac{4}{5} \cdot \dfrac{3}{11}$

33. $\dfrac{2}{7} \div \dfrac{3}{5}$

34. $\dfrac{5}{6} \div \dfrac{11}{13}$

35. $\dfrac{3}{8} \cdot \dfrac{12}{15}$

36. $\dfrac{4}{9} \cdot \dfrac{3}{2}$

37. A certain recipe calls for $\dfrac{3}{4}$ cup of milk. To make half of the recipe, how much milk is needed?

38. John is adding a diesel fuel additive to his fuel tank, which is half full. The directions say to add $\dfrac{1}{3}$ of the bottle to a full fuel tank. What portion of the bottle should he add to the fuel tank?

39. Mark shares a computer with his roommates. He has partitioned the hard drive in such a way that he gets $\dfrac{1}{3}$ of the disk space. His part of the hard drive is currently $\dfrac{2}{3}$ full. What portion of the computer's hard drive space is he currently taking up?

40. Angelina teaches $\dfrac{2}{3}$ of the deaf children in her local school. Her local school educates $\dfrac{1}{2}$ of the deaf children in the school district. What portion of the school district's deaf children is Angelina teaching?

For Problems 41–57, add or subtract as indicated and express answers in lowest terms.

41. $\dfrac{2}{7} + \dfrac{3}{7}$

42. $\dfrac{3}{11} + \dfrac{5}{11}$

43. $\dfrac{7}{9} - \dfrac{2}{9}$

44. $\dfrac{11}{13} - \dfrac{6}{13}$

45. $\dfrac{3}{4} + \dfrac{9}{4}$

46. $\dfrac{5}{6} + \dfrac{7}{6}$

47. $\dfrac{11}{12} - \dfrac{3}{12}$

48. $\dfrac{13}{16} - \dfrac{7}{16}$

49. $\dfrac{5}{24} + \dfrac{11}{24}$

50. $\dfrac{7}{36} + \dfrac{13}{36}$

51. $\dfrac{1}{3} + \dfrac{1}{5}$

52. $\dfrac{1}{6} + \dfrac{1}{8}$

53. $\dfrac{15}{16} - \dfrac{3}{8}$

54. $\dfrac{13}{12} - \dfrac{1}{6}$

55. $\dfrac{7}{10} + \dfrac{8}{15}$

56. $\dfrac{7}{12} + \dfrac{5}{8}$

57. $\dfrac{11}{24} + \dfrac{5}{32}$

58. Alicia and her brother Jeff shared a pizza. Alicia ate $\dfrac{1}{8}$ of the pizza, while Jeff ate $\dfrac{2}{3}$ of the pizza. How much of the pizza has been eaten?

59. Rosa has $\dfrac{1}{3}$ pound of blueberries, $\dfrac{1}{4}$ pound of strawberries, and $\dfrac{1}{2}$ pound of raspberries. If she combines these for a fruit salad, how many pounds of these berries will be in the salad?

60. A chemist has $\dfrac{11}{16}$ of an ounce of dirt residue to perform crime lab tests. He needs $\dfrac{3}{8}$ of an ounce to perform a test for iron content. How much of the dirt residue will be left for the chemist to use in other testing?

For Problems 61–68, simplify each numerical expression, expressing answers in reduced form.

61. $\dfrac{1}{4} - \dfrac{3}{8} + \dfrac{5}{12} - \dfrac{1}{24}$

62. $\dfrac{3}{4} + \dfrac{2}{3} - \dfrac{1}{6} + \dfrac{5}{12}$

63. $\dfrac{5}{6} + \dfrac{2}{3} \cdot \dfrac{3}{4} - \dfrac{1}{4} \cdot \dfrac{2}{5}$

64. $\dfrac{2}{3} + \dfrac{1}{2} \cdot \dfrac{2}{5} - \dfrac{1}{3} \cdot \dfrac{1}{5}$

65. $\dfrac{3}{4} \cdot \dfrac{6}{9} - \dfrac{5}{6} \cdot \dfrac{8}{10} + \dfrac{2}{3} \cdot \dfrac{6}{8}$

66. $\dfrac{3}{5} \cdot \dfrac{5}{7} + \dfrac{2}{3} \cdot \dfrac{3}{5} - \dfrac{1}{7} \cdot \dfrac{2}{5}$

67. $\dfrac{7}{13}\left(\dfrac{2}{3} - \dfrac{1}{6}\right)$

68. $48\left(\dfrac{5}{12} - \dfrac{1}{6} + \dfrac{3}{8}\right)$

69. Blake Scott leaves $\dfrac{1}{4}$ of his estate to the Boy Scouts, $\dfrac{2}{5}$ to the local cancer fund, and the rest to his church. What fractional part of the estate does the church receive?

70. Franco has $\dfrac{7}{8}$ of an ounce of gold. He wants to give $\dfrac{3}{16}$ of an ounce to his friend Julie. He plans to divide the remaining amount of his gold in half to make two rings. How much gold will he have for each ring?

Photo Credits

Answers to Odd-Numbered Problems and All Chapter Review, Chapter Test, and Cumulative Review Problems

Problem Set 1.1 (page 9)
1. True **3.** False **5.** True
7. False **9.** True **11.** 0 and 14
13. $0, 14, \frac{2}{3}, -\frac{11}{14}, 2.34, 3.2\overline{1}, 6\frac{7}{8}, -19,$ and -2.6
15. 0 and 14 **17.** All of them **19.** $\not\subseteq$
21. $\subseteq$ **23.** $\not\subseteq$ **25.** $\subseteq$ **27.** $\not\subseteq$
29. Real, rational, an integer, and negative
31. Real, irrational, and negative
33. $\{1, 2\}$ **35.** $\{0, 1, 2, 3, 4, 5\}$
37. $\{\ldots, -1, 0, 1, 2\}$ **39.** $\varnothing$ **41.** $\{0, 1, 2, 3, 4\}$
43. -6 **45.** 2 **47.** $3x + 1$ **49.** $5x$
51. 26 **53.** 84 **55.** 23 **57.** 65
59. 60 **61.** 33 **63.** 1320 **65.** 20
67. 119 **69.** 18 **71.** 4 **73.** 31

Problem Set 1.2 (page 19)
1. -7 **3.** -19 **5.** -22 **7.** -7 **9.** 108
11. -70 **13.** 14 **15.** -7 **17.** $3\frac{1}{2}$ **19.** $5\frac{1}{2}$
21. $-\frac{2}{15}$ **23.** -4 **25.** 0 **27.** Undefined
29. -60 **31.** -4.8 **33.** 14.13
35. -6.5 **37.** -38.88 **39.** 0.2

41. $-\frac{13}{12}$ **43.** $-\frac{3}{4}$ **45.** $-\frac{13}{9}$ **47.** $-\frac{3}{5}$
49. $-\frac{3}{2}$ **51.** -12 **53.** -24 **55.** $\frac{35}{4}$
57. 15 **59.** -17 **61.** $\frac{47}{12}$ **63.** 5
65. 0 **67.** 26 **69.** 6 **71.** 25
73. 78 **75.** -10 **77.** 5 **79.** -5
81. 10.5 **83.** -3.3 **85.** 19.5 **87.** $\frac{3}{4}$
89. $\frac{5}{2}$ **93.** 10 over par **95.** Lost $16.50
97. A gain of 0.88 dollar
99. No; they made it 49.1 pounds lighter.

Problem Set 1.3 (page 28)
1. Associative property of addition
3. Commutative property of addition
5. Additive inverse property
7. Multiplication property of negative one
9. Commutative property of multiplication
11. Distributive property
13. Associative property of multiplication
15. 18 **17.** 2 **19.** -1300 **21.** 1700
23. -47 **25.** 3200 **27.** -19 **29.** -41
31. -17 **33.** -39 **35.** 24 **37.** 20
39. 55 **41.** 16 **43.** 49 **45.** -216

47. -14 **49.** -8 **51.** $\dfrac{3}{16}$ **53.** $-\dfrac{10}{9}$
57. 2187 **59.** -2048
61. $-15{,}625$ **63.** 3.9525416

Problem Set 1.4 (page 36)

1. $4x$ **3.** $-a^2$ **5.** $-6n$ **7.** $-5x + 2y$
9. $6a^2 + 5b^2$ **11.** $21x - 13$ **13.** $-2a^2b - ab^2$
15. $8x + 21$ **17.** $-5a + 2$ **19.** $-5n^2 + 11$
21. $-7x^2 + 32$ **23.** $22x - 3$ **25.** $-14x - 7$
27. $-10n^2 + 4$ **29.** $4x - 30y$ **31.** $-13x - 31$
33. $-21x - 9$ **35.** -17 **37.** 12 **39.** 4
41. 3 **43.** -38 **45.** -14 **47.** 64
49. 104 **51.** 5 **53.** 4 **55.** $-\dfrac{22}{3}$ **57.** $\dfrac{29}{4}$
59. 221.6 **61.** 1092.4 **63.** 1420.5
65. $n + 12$ **67.** $n - 5$ **69.** $50n$ **71.** $\dfrac{1}{2}n - 4$
73. $\dfrac{n}{8}$ **75.** $2n - 9$ **77.** $10(n - 6)$
79. $n + 20$ **81.** $2t - 3$ **83.** $n + 47$ **85.** $8y$
87. 25 cm **89.** $\dfrac{c}{25}$ **91.** $n + 2$ **93.** $\dfrac{c}{5}$
95. $12d$ **97.** $3y + f$ **99.** $5280m$

Chapter 1 Review Problem Set (page 40)

1. (a) 67 **(b)** $0, -8,$ and 67 **(c)** 0 and 67
 (d) $0, \dfrac{3}{4}, -\dfrac{5}{6}, 8\dfrac{1}{3}, -8, 0.34, 0.2\overline{3}, 67,$ and $\dfrac{9}{7}$
 (e) $\sqrt{2}$ and $-\sqrt{3}$
2. Associative property for addition
3. Substitution property of equality
4. Multiplication property of negative one
5. Distributive property
6. Associative property for multiplication
7. Commutative property for addition
8. Distributive property
9. Multiplicative inverse property
10. Symmetric property of equality
11. $-6\dfrac{1}{2}$ **12.** $-6\dfrac{1}{6}$ **13.** -8 **14.** -15
15. 20 **16.** 49 **17.** -56 **18.** -24 **19.** 6
20. 4 **21.** 100 **22.** 8 **23.** $-4a^2 - 5b^2$
24. $3x - 2$ **25.** ab^2 **26.** $-\dfrac{7}{3}x^2y$
27. $10n^2 - 17$ **28.** $-13a + 4$ **29.** $-2n + 2$
30. $-7x - 29y$ **31.** $-7a - 9$ **32.** $-9x^2 + 7$
33. $-6\dfrac{1}{2}$ **34.** $-\dfrac{5}{16}$ **35.** -55 **36.** 144

37. -16 **38.** -44 **39.** 19.4 **40.** 59.6
41. $-\dfrac{59}{3}$ **42.** $\dfrac{9}{2}$ **43.** $4 + 2n$ **44.** $3n - 50$
45. $\dfrac{2}{3}n - 6$ **46.** $10(n - 14)$ **47.** $5n - 8$
48. $\dfrac{n}{n - 3}$ **49.** $5(n + 2) - 3$ **50.** $\dfrac{3}{4}(n + 12)$
51. $37 - n$ **52.** $\dfrac{w}{60}$ **53.** $2y - 7$ **54.** $n + 3$
55. $p + 5n + 25q$ **56.** $\dfrac{i}{48}$ **57.** $24f + 72y$
58. $10d$ **59.** $12f + i$ **60.** $25 - c$

Chapter 1 Test (page 42)

1. Symmetric property **2.** Distributive property
3. -3 **4.** -23 **5.** $-\dfrac{23}{6}$ **6.** 11 **7.** 8
8. -94 **9.** -4 **10.** 960 **11.** -32
12. $-x^2 - 8x - 2$ **13.** $-19n - 20$ **14.** 27
15. $\dfrac{11}{16}$ **16.** $\dfrac{2}{3}$ **17.** 77 **18.** -22.5
19. 93 **20.** -5 **21.** $6n - 30$
22. $3n + 28$ or $3(n + 8) + 4$
23. $\dfrac{72}{n}$ **24.** $5n + 10d + 25q$ **25.** $6x + 2y$

CHAPTER 2

Problem Set 2.1 (page 50)

1. $\{4\}$ **3.** $\{-3\}$ **5.** $\{-14\}$ **7.** $\{6\}$ **9.** $\left\{\dfrac{19}{3}\right\}$
11. $\{1\}$ **13.** $\left\{-\dfrac{10}{3}\right\}$ **15.** $\{4\}$ **17.** $\left\{-\dfrac{13}{3}\right\}$
19. $\{3\}$ **21.** $\{8\}$ **23.** $\{-9\}$ **25.** $\{-3\}$
27. $\{0\}$ **29.** $\left\{-\dfrac{7}{2}\right\}$ **31.** $\{-2\}$ **33.** $\left\{-\dfrac{5}{3}\right\}$
35. $\left\{\dfrac{33}{2}\right\}$ **37.** $\{-35\}$ **39.** $\left\{\dfrac{1}{2}\right\}$ **41.** $\left\{\dfrac{1}{6}\right\}$
43. $\{5\}$ **45.** $\{-1\}$ **47.** $\left\{-\dfrac{21}{16}\right\}$ **49.** $\left\{\dfrac{12}{7}\right\}$
51. 14 **53.** 13, 14, and 15 **55.** 9, 11, and 13
57. 14 and 81 **59.** \$11 per hour
61. 30 pennies, 50 nickels, and 70 dimes **63.** \$300
65. 20 three-bedroom, 70 two-bedroom, and 140 one-bedroom
73. (a) $\varnothing$ **(c)** $\{0\}$ **(e)** $\varnothing$

Problem Set 2.2 (page 58)

1. {12} **3.** $\left\{-\dfrac{3}{5}\right\}$ **5.** {3} **7.** {−2}

9. {−36} **11.** $\left\{\dfrac{20}{9}\right\}$ **13.** {3} **15.** {3}

17. {−2} **19.** $\left\{\dfrac{8}{5}\right\}$ **21.** {−3} **23.** $\left\{\dfrac{48}{17}\right\}$

25. $\left\{\dfrac{103}{6}\right\}$ **27.** {3} **29.** $\left\{\dfrac{40}{3}\right\}$ **31.** $\left\{-\dfrac{20}{7}\right\}$

33. $\left\{\dfrac{24}{5}\right\}$ **35.** {−10} **37.** $\left\{-\dfrac{25}{4}\right\}$ **39.** {0}

41. 18 **43.** 16 inches long and 5 inches wide
45. 14, 15, and 16 **47.** 8 feet
49. Angie is 22 and her mother is 42.
51. Sydney is 18 and Marcus is 36. **53.** 80, 90, and 94
55. 48° and 132° **57.** 78°

Problem Set 2.3 (page 66)

1. {20} **3.** {50} **5.** {40} **7.** {12}
9. {6} **11.** {400} **13.** {400} **15.** {38}
17. {6} **19.** {3000} **21.** {3000} **23.** {400}
25. {14} **27.** {15} **29.** $90 **31.** $54.40
33. $48 **35.** $400 **37.** 65% **39.** 62.5%
41. $32,500 **43.** $3000 at 10% and $4500 at 11%
45. $53,000 **47.** 8 pennies, 15 nickels, and 18 dimes
49. 15 dimes, 45 quarters, and 10 half-dollars
55. {7.5} **57.** {−4775} **59.** {8.7}
61. {17.1} **63.** {13.5}

Problem Set 2.4 (page 77)

1. $120 **3.** 3 years **5.** 6% **7.** $800
9. $1600 **11.** 8% **13.** $200
15. 6 feet; 14 feet; 10 feet; 20 feet; 7 feet; 2 feet

17. $h = \dfrac{V}{B}$ **19.** $h = \dfrac{V}{\pi r^2}$ **21.** $r = \dfrac{C}{2\pi}$

23. $C = \dfrac{100M}{I}$ **25.** $C = \dfrac{5}{9}(F - 32)$ or $C = \dfrac{5F - 160}{9}$

27. $x = \dfrac{y - b}{m}$ **29.** $x = \dfrac{y - y_1 + mx_1}{m}$

31. $x = \dfrac{ab + bc}{b - a}$ **33.** $x = a + bc$

35. $x = \dfrac{3b - 6a}{2}$ **37.** $x = \dfrac{5y + 7}{2}$

39. $y = -7x - 4$ **41.** $x = \dfrac{6y + 4}{3}$

43. $x = \dfrac{cy - ac - b^2}{b}$ **45.** $y = \dfrac{x - a + 1}{a - 3}$

47. 22 meters long and 6 meters wide **49.** $11\dfrac{1}{9}$ years

51. $11\dfrac{1}{9}$ years **53.** 4 hours **55.** 3 hours

57. 40 miles
59. 15 quarts of 30% solution and 5 quarts of
70% solution
61. 25 milliliters **67.** $596.25 **69.** 1.5 years
71. 14.5% **73.** $1850

Problem Set 2.5 (page 85)

1. $(1, \infty)$

3. $[-1, \infty)$

5. $(-\infty, -2)$

7. $(-\infty, 2]$

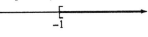

9. $x < 4$ **11.** $x \leq -7$ **13.** $x > 8$ **15.** $x \geq -7$
17. $(1, \infty)$

19. $(-\infty, -4]$

21. $(-\infty, -2]$

23. $(-\infty, 2)$

25. $(-1, \infty)$

27. $[-1, \infty)$

29. $(-2, \infty)$

31. $(-2, \infty)$

33. $(-\infty, -2)$

35. $[-3, \infty)$

37. $(0, \infty)$

39. $[4, \infty)$

41. $\left(\dfrac{7}{2}, \infty\right)$ **43.** $\left(\dfrac{12}{5}, \infty\right)$ **45.** $\left(-\infty, -\dfrac{5}{2}\right]$

47. $\left[\dfrac{5}{12}, \infty\right)$ **49.** $(-6, \infty)$ **51.** $(-5, \infty)$

53. $\left(-\infty, \dfrac{5}{3}\right]$ **55.** $(-36, \infty)$ **57.** $\left(-\infty, -\dfrac{8}{17}\right]$

59. $\left(-\dfrac{11}{2}, \infty\right)$ **61.** $(23, \infty)$ **63.** $(-\infty, 3)$

65. $\left(-\infty, -\dfrac{1}{7}\right]$ **67.** $(-22, \infty)$ **69.** $\left(-\infty, \dfrac{6}{5}\right)$

Problem Set 2.6 (page 94)

1. $(4, \infty)$ **3.** $\left(-\infty, \dfrac{23}{3}\right)$ **5.** $[5, \infty)$

7. $[-9, \infty)$ **9.** $\left(-\infty, -\dfrac{37}{3}\right]$ **11.** $\left(-\infty, -\dfrac{19}{6}\right)$

13. $(-\infty, 50]$ **15.** $(300, \infty)$ **17.** $[4, \infty)$

19. $(-1, 2)$

21. $(-1, 2]$

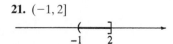

23. $(-\infty, -1) \cup (2, \infty)$

25. $(-\infty, 1] \cup (3, \infty)$

27. $(0, \infty)$

29. $\varnothing$

31. $(-\infty, \infty)$

33. $(-1, \infty)$

35. $(1, 3)$

37. $(-\infty, -5) \cup (1, \infty)$

39. $[3, \infty)$

41. $\left(\dfrac{1}{3}, \dfrac{2}{5}\right)$

43. $(-\infty, -1) \cup \left(-\dfrac{1}{3}, \infty\right)$

45. $(-2, 2)$ **47.** $[-5, 4]$ **49.** $\left(-\dfrac{1}{2}, \dfrac{3}{2}\right)$

51. $\left(-\dfrac{1}{4}, \dfrac{11}{4}\right)$ **53.** $[-11, 13]$ **55.** $(-1, 5)$

57. More than 10%   **59.** 5 feet and 10 inches or better

61. 168 or better   **63.** 77 or less

65. $163°F \le C \le 218°F$ **67.** $6.3 \le M \le 11.25$

Problem Set 2.7 (page 101)

1. $(-5, 5)$

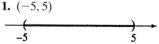

3. $[-2, 2]$

5. $(-\infty, -2) \cup (2, \infty)$

7. $(-1, 3)$

9. $[-6, 2]$

11. $(-\infty, -3) \cup (-1, \infty)$

13. $(-\infty, 1] \cup [5, \infty)$

15. $\{-7, 9\}$ **17.** $(-\infty, -4) \cup (8, \infty)$ **19.** $(-8, 2)$
21. $\{-1, 5\}$ **23.** $[-4, 5]$
25. $\left(-\infty, -\dfrac{7}{2}\right] \cup \left[\dfrac{5}{2}, \infty\right)$ **27.** $\left\{-5, \dfrac{7}{3}\right\}$
29. $\{-1, 5\}$ **31.** $(-\infty, -2) \cup (6, \infty)$
33. $\left(-\dfrac{1}{2}, \dfrac{3}{2}\right)$ **35.** $\left[-5, \dfrac{7}{5}\right]$ **37.** $\left\{\dfrac{1}{12}, \dfrac{17}{12}\right\}$
39. $[-3, 10]$ **41.** $(-5, 11)$
43. $\left(-\infty, -\dfrac{3}{2}\right) \cup \left(\dfrac{1}{2}, \infty\right)$ **45.** $\{0, 3\}$
47. $(-\infty, -14] \cup [0, \infty)$ **49.** $[-2, 3]$
51. $\varnothing$ **53.** $(-\infty, \infty)$ **55.** $\left\{\dfrac{2}{5}\right\}$ **57.** $\varnothing$
59. $\varnothing$ **65.** $\left\{-2, -\dfrac{4}{3}\right\}$ **67.** $\{-2\}$ **69.** $\{0\}$

Chapter 2 Review Problem Set (page 104)

1. $\{18\}$ **2.** $\{-14\}$ **3.** $\{0\}$ **4.** $\left\{\dfrac{1}{2}\right\}$
5. $\{10\}$ **6.** $\left\{\dfrac{7}{3}\right\}$ **7.** $\left\{\dfrac{28}{17}\right\}$ **8.** $\left\{-\dfrac{1}{38}\right\}$
9. $\left\{\dfrac{27}{17}\right\}$ **10.** $\left\{-\dfrac{10}{3}, 4\right\}$ **11.** $\{50\}$
12. $\left\{-\dfrac{39}{2}\right\}$ **13.** $\{200\}$

14. $\{-8\}$ **15.** $\left\{-\dfrac{7}{2}, \dfrac{1}{2}\right\}$
16. $x = \dfrac{2b + 2}{a}$ **17.** $x = \dfrac{c}{a - b}$
18. $x = \dfrac{pb - ma}{m - p}$ **19.** $x = \dfrac{11 + 7y}{5}$
20. $x = \dfrac{by + b + ac}{c}$ **21.** $s = \dfrac{A - \pi r^2}{\pi r}$
22. $b_2 = \dfrac{2A - hb_1}{h}$ **23.** $n = \dfrac{2S_n}{a_1 + a_2}$
24. $R = \dfrac{R_1 R_2}{R_1 + R_2}$ **25.** $[-5, \infty)$ **26.** $(4, \infty)$
27. $\left(-\dfrac{7}{3}, \infty\right)$ **28.** $\left[\dfrac{17}{2}, \infty\right)$ **29.** $\left(-\infty, \dfrac{1}{3}\right)$
30. $\left(\dfrac{53}{11}, \infty\right)$ **31.** $[6, \infty)$ **32.** $(-\infty, 100]$
33. $(-5, 6)$ **34.** $\left(-\infty, -\dfrac{11}{3}\right) \cup (3, \infty)$
35. $(-\infty, -17)$ **36.** $\left(-\infty, -\dfrac{15}{4}\right)$

37.

38.

39.

40.

41.

42.

43.

44. $\varnothing$
45. The length is 15 meters and the width is 7 meters.
46. \$200 at 7% and \$300 at 8%   **47.** 88 or better
48. 4, 5, and 6   **49.** \$10.50 per hour
50. 20 nickels, 50 dimes, and 75 quarters   **51.** $80°$
52. \$45.60   **53.** $6\dfrac{2}{3}$ pints   **54.** 55 miles per hour
55. Sonya for $3\dfrac{1}{4}$ hours and Rita for $4\dfrac{1}{2}$ hours
56. $6\dfrac{1}{4}$ cups

Chapter 2 Test (page 107)

1. $\{-3\}$ **2.** $\{5\}$ **3.** $\left\{\dfrac{1}{2}\right\}$ **4.** $\left\{\dfrac{16}{5}\right\}$

5. $\left\{-\dfrac{14}{5}\right\}$ **6.** $\{-1\}$ **7.** $\left\{-\dfrac{3}{2}, 3\right\}$ **8.** $\{3\}$

9. $\left\{\dfrac{31}{3}\right\}$ **10.** $\{650\}$ **11.** $y = \dfrac{8x - 24}{9}$

12. $h = \dfrac{S - 2\pi r^2}{2\pi r}$ **13.** $(-2, \infty)$ **14.** $[-4, \infty)$

15. $(-\infty, -35]$ **16.** $(-\infty, 10)$ **17.** $(3, \infty)$

18. $(-\infty, 200]$ **19.** $\left(-1, \dfrac{7}{3}\right)$

20. $\left(-\infty, -\dfrac{11}{4}\right] \cup \left[\dfrac{1}{4}, \infty\right)$ **21.** $72

22. 19 centimeters   **23.** $\dfrac{2}{3}$ of a cup

24. 97 or better   **25.** $70°$

CHAPTER 3

Problem Set 3.1 (page 113)

1. 2   **3.** 3   **5.** 2   **7.** 6   **9.** 0
11. $10x - 3$ **13.** $-11t + 5$ **15.** $-x^2 + 2x - 2$
17. $17a^2b^2 - 5ab$ **19.** $-9x + 7$ **21.** $-2x + 6$
23. $10a + 7$ **25.** $4x^2 + 10x + 6$
27. $-6a^2 + 12a + 14$ **29.** $3x^3 + x^2 + 13x - 11$
31. $7x + 8$ **33.** $-3x - 16$ **35.** $2x^2 - 2x - 8$
37. $-3x^2 + 5x^2 - 2x + 9$ **39.** $5x^2 - 4x + 11$
41. $-6x^2 + 9x + 7$ **43.** $-2x^2 + 9x + 4$
45. $-10n^2 + n + 9$ **47.** $8x - 2$ **49.** $8x - 14$
51. $-9x^2 - 12x + 4$ **53.** $10x^2 + 13x - 18$
55. $-n^2 - 4n - 4$ **57.** $-x + 6$ **59.** $6x^2 - 4$
61. $-7n^2 + n + 6$ **63.** $t^2 - 4t + 8$
65. $4n^2 - n - 12$ **67.** $-4x - 2y$
69. $-x^3 - x^2 + 3x$ **71. (a)** $8x + 4$ **(c)** $12x + 6$
73. $8\pi h + 32\pi$ **(a)** 226.1   **(c)** 452.2

Problem Set 3.2 (page 120)

1. $36x^4$ **3.** $-12x^5$ **5.** $4a^3b^4$ **7.** $-3x^3y^2z^6$
9. $-30xy^4$ **11.** $27a^4b^5$ **13.** $-m^3n^3$
15. $\dfrac{3}{10}x^3y^6$ **17.** $-\dfrac{3}{20}a^3b^4$ **19.** $-\dfrac{1}{6}x^3y^4$
21. $30x^6$ **23.** $-18x^9$ **25.** $-3x^6y^6$ **27.** $-24y^9$
29. $-56a^4b^2$ **31.** $-18a^3b^3$ **33.** $-10x^7y^7$
35. $50x^5y^2$ **37.** $27x^3y^6$ **39.** $-32x^{10}y^5$
41. $x^{16}y^{20}$ **43.** $a^6b^{12}c^{18}$ **45.** $64a^{12}b^{18}$
47. $81x^2y^8$ **49.** $81a^4b^{12}$ **51.** $-16a^4b^4$

53. $-x^6y^{12}z^{18}$ **55.** $-125a^6b^6c^3$ **57.** $-x^7y^{28}z^{14}$
59. $3x^3y^3$ **61.** $-5x^3y^2$ **63.** $9bc^2$
65. $-18xyz^4$ **67.** $-a^2b^3c^2$ **69.** 9   **71.** $-b^2$
73. $-18x^3$ **75.** $6x^{3n}$ **77.** a^{5n+3} **79.** x^{4n}
81. a^{5n+1} **83.** $-10x^{2n}$ **85.** $12a^{n+4}$ **87.** $6x^{3n+2}$
89. $12x^{n+2}$ **91.** $22x^2; 6x^3$ **93.** $\pi r^2 - 36\pi$

Problem Set 3.3 (page 127)

1. $10x^2y^3 + 6x^3y^4$ **3.** $-12a^3b^3 + 15a^5b$
5. $24a^4b^5 - 16a^4b^6 + 32a^5b^6$
7. $-6x^3y^3 - 3x^4y^4 + x^5y^2$ **9.** $ax + ay + 2bx + 2by$
11. $ac + 4ad - 3bc - 12bd$ **13.** $x^2 + 16x + 60$
15. $y^2 + 6y - 55$ **17.** $n^2 - 5n - 14$ **19.** $x^2 - 36$
21. $x^2 - 12x + 36$ **23.** $x^2 - 14x + 48$
25. $x^3 - 4x^2 + x + 6$ **27.** $x^3 - x^2 - 9x + 9$
29. $t^2 + 18t + 81$ **31.** $y^2 - 14y + 49$
33. $4x^2 + 33x + 35$ **35.** $9y^2 - 1$
37. $14x^2 + 3x - 2$ **39.** $5 + 3t - 2t^2$
41. $9t^2 + 42t + 49$ **43.** $4 - 25x^2$
45. $49x^2 - 56x + 16$ **47.** $18x^2 - 39x - 70$
49. $2x^2 + xy - 15y^2$ **51.** $25x^2 - 4a^2$
53. $t^3 - 14t - 15$ **55.** $x^3 + x^2 - 24x + 16$
57. $2x^3 + 9x^2 + 2x - 30$ **59.** $12x^3 - 7x^2 + 25x - 6$
61. $x^4 + 5x^3 + 11x^2 + 11x + 4$
63. $2x^4 - x^3 - 12x^2 + 5x + 4$
65. $x^3 + 6x^2 + 12x + 8$ **67.** $x^3 - 12x^2 + 48x - 64$
69. $8x^3 + 36x^2 + 54x + 27$
71. $64x^3 - 48x^2 + 12x - 1$
73. $125x^3 + 150x^2 + 60x + 8$ **75.** $x^{2n} - 16$
77. $x^{2a} + 4x^a - 12$ **79.** $6x^{2n} + x^n - 35$
81. $x^{4a} - 10x^{2a} + 21$ **83.** $4x^{2n} + 20x^n + 25$
87. $2x^2 + 6$ **89.** $4x^3 - 64x^2 + 256x; 256 - 4x^2$
93. (a) $a^6 + 6a^5b + 15a^4b^2 + 20a^3b^3 + 15a^2b^4 + 6ab^5 + b^6$
   (c) $a^8 + 8a^7b + 28a^6b^2 + 56a^5b^3 + 70a^4b^4 + 56a^3b^5 +$
   $28a^2b^6 + 8ab^7 + b^8$

Problem Set 3.4 (page 135)

1. Composite   **3.** Prime   **5.** Composite
7. Composite   **9.** Prime   **11.** $2 \cdot 2 \cdot 7$
13. $2 \cdot 2 \cdot 11$ **15.** $2 \cdot 2 \cdot 2 \cdot 7$ **17.** $2 \cdot 2 \cdot 2 \cdot 3 \cdot 3$
19. $3 \cdot 29$ **21.** $3(2x + y)$ **23.** $2x(3x + 7)$
25. $4y(7y - 1)$ **27.** $5x(4y - 3)$ **29.** $x^2(7x + 10)$
31. $9ab(2a + 3b)$ **33.** $3x^3y^3(4y - 13x)$
35. $4x^2(2x^2 + 3x - 6)$ **37.** $x(5 + 7x + 9x^3)$
39. $5xy^2(3xy + 4 + 7x^2y^2)$ **41.** $(y + 2)(x + 3)$
43. $(2a + b)(3x - 2y)$ **45.** $(x + 2)(x + 5)$
47. $(a + 4)(x + y)$ **49.** $(a - 2b)(x + y)$

51. $(a - b)(3x - y)$ **53.** $(a + 1)(2x + y)$
55. $(a - 1)(x^2 + 2)$ **57.** $(a + b)(2c + 3d)$
59. $(a + b)(x - y)$ **61.** $(x + 9)(x + 6)$
63. $(x + 4)(2x + 1)$ **65.** $\{-7, 0\}$ **67.** $\{0, 1\}$

69. $\{0, 5\}$ **71.** $\left\{-\dfrac{1}{2}, 0\right\}$ **73.** $\left\{-\dfrac{7}{3}, 0\right\}$

75. $\left\{0, \dfrac{5}{4}\right\}$ **77.** $\left\{0, \dfrac{1}{4}\right\}$ **79.** $\{-12, 0\}$

81. $\left\{0, \dfrac{3a}{5b}\right\}$ **83.** $\left\{-\dfrac{3a}{2b}, 0\right\}$ **85.** $\{a, -2b\}$

87. 0 or 7   **89.** 6 units   **91.** $\dfrac{4}{\pi}$ units

93. The square is 100 feet by 100 feet, and the rectangle is 50 feet by 100 feet.

95. 6 units   **101.** $x^a(2x^a - 3)$ **103.** $y^{2m}(y^m + 5)$
105. $x^{4a}(2x^{2a} - 3x^a + 7)$

Problem Set 3.5 (page 142)
1. $(x + 1)(x - 1)$ **3.** $(4x + 5)(4x - 5)$
5. $(3x + 5y)(3x - 5y)$ **7.** $(5xy + 6)(5xy - 6)$
9. $(2x + y^2)(2x - y^2)$ **11.** $(1 + 12n)(1 - 12n)$
13. $(x + 2 + y)(x + 2 - y)$
15. $(2x + y + 1)(2x - y - 1)$
17. $(3a + 2b + 3)(3a - 2b - 3)$
19. $-5(2x + 9)$ **21.** $9(x + 2)(x - 2)$
23. $5(x^2 + 1)$ **25.** $8(y + 2)(y - 2)$
27. $ab(a + 3)(a - 3)$ **29.** Not factorable
31. $(n + 3)(n - 3)(n^2 + 9)$ **33.** $3x(x^2 + 9)$
35. $4xy(x + 4y)(x - 4y)$ **37.** $6x(1 + x)(1 - x)$
39. $(1 + xy)(1 - xy)(1 + x^2y^2)$ **41.** $4(x + 4y)(x - 4y)$
43. $3(x + 2)(x - 2)(x^2 + 4)$ **45.** $(a - 4)(a^2 + 4a + 16)$
47. $(x + 1)(x^2 - x + 1)$
49. $(3x + 4y)(9x^2 - 12xy + 16y^2)$
51. $(1 - 3a)(1 + 3a + 9a^2)$
53. $(xy - 1)(x^2y^2 + xy + 1)$
55. $(x + y)(x - y)(x^2 - xy + y^2)(x^2 + xy + y^2)$

57. $\{-5, 5\}$ **59.** $\left\{-\dfrac{7}{3}, \dfrac{7}{3}\right\}$ **61.** $\{-2, 2\}$

63. $\{-1, 0, 1\}$ **65.** $\{-2, 2\}$ **67.** $\{-3, 3\}$
69. $\{0\}$ **71.** $-3, 0,$ or 3 **73.** 4 and 8
75. 10 meters long and 5 meters wide   **77.** 6 inches
79. 8 yards

Problem Set 3.6 (page 151)
1. $(x + 5)(x + 4)$ **3.** $(x - 4)(x - 7)$
5. $(a + 9)(a - 4)$ **7.** $(y + 6)(y + 14)$
9. $(x - 7)(x + 2)$ **11.** Not factorable
13. $(6 - x)(1 + x)$ **15.** $(x + 3y)(x + 12y)$

17. $(a - 8b)(a + 7b)$ **19.** $(3x + 1)(5x + 6)$
21. $(4x - 3)(3x + 2)$ **23.** $(a + 3)(4a - 9)$
25. $(n - 4)(3n + 5)$ **27.** Not factorable
29. $(2n - 7)(5n + 3)$ **31.** $(4x - 5)(2x + 9)$
33. $(1 - 6x)(6 + x)$ **35.** $(5y + 9)(4y - 1)$
37. $(12n + 5)(2n - 1)$ **39.** $(5n + 3)(n + 6)$
41. $(x + 10)(x + 15)$ **43.** $(n - 16)(n - 20)$
45. $(t + 15)(t - 12)$ **47.** $(t^2 - 3)(t^2 - 2)$
49. $(2x^2 - 1)(5x^2 + 4)$ **51.** $(x + 1)(x - 1)(x^2 - 8)$
53. $(3n + 1)(3n - 1)(2n^2 + 3)$
55. $(x + 1)(x - 1)(x + 4)(x - 4)$
57. $2(t + 2)(t - 2)$ **59.** $(4x + 5y)(3x - 2y)$
61. $3n(2n + 5)(3n - 1)$ **63.** $(n - 12)(n - 5)$
65. $(6a - 1)^2$ **67.** $6(x^2 + 9)$ **69.** Not factorable
71. $(x + y - 7)(x - y + 7)$
73. $(1 + 4x^2)(1 + 2x)(1 - 2x)$ **75.** $(4n + 9)(n + 4)$
77. $n(n + 7)(n - 7)$ **79.** $(x - 8)(x + 1)$
81. $3x(x - 3)(x^2 + 3x + 9)$ **83.** $(x^2 + 3)^2$
85. $(x + 3)(x - 3)(x^2 + 4)$ **87.** $(2w - 7)(3w + 5)$
89. Not factorable   **91.** $2n(n^2 + 7n - 10)$
93. $(2x + 1)(y + 3)$ **99.** $(x^a + 3)(x^a + 7)$
101. $(2x^a + 5)^2$ **103.** $(5x^n - 1)(4x^n + 5)$
105. $(x - 4)(x - 2)$ **107.** $(3x - 11)(3x + 2)$
109. $(3x + 4)(5x + 9)$

Problem Set 3.7 (page 157)
1. $\{-3, -1\}$ **3.** $\{-12, -6\}$ **5.** $\{4, 9\}$
7. $\{-6, 2\}$ **9.** $\{-1, 5\}$ **11.** $\{-13, -12\}$
13. $\left\{-5, \dfrac{1}{3}\right\}$ **15.** $\left\{-\dfrac{7}{2}, -\dfrac{2}{3}\right\}$ **17.** $\{0, 4\}$
19. $\left\{\dfrac{1}{6}, 2\right\}$ **21.** $\{-6, 0, 6\}$ **23.** $\{-4, 6\}$
25. $\{-4, 4\}$ **27.** $\{-11, 4\}$ **29.** $\{-5, 5\}$
31. $\left\{-\dfrac{5}{3}, -\dfrac{3}{5}\right\}$ **33.** $\left\{-\dfrac{1}{8}, 6\right\}$ **35.** $\left\{\dfrac{3}{7}, \dfrac{5}{4}\right\}$
37. $\left\{-\dfrac{2}{7}, \dfrac{4}{5}\right\}$ **39.** $\left\{-7, \dfrac{2}{3}\right\}$ **41.** $\{-20, 18\}$
43. $\left\{-2, -\dfrac{1}{3}, \dfrac{1}{3}, 2\right\}$ **45.** $\left\{-\dfrac{2}{3}, 16\right\}$ **47.** $\left\{-\dfrac{3}{2}, 1\right\}$
49. $\left\{-\dfrac{5}{2}, -\dfrac{4}{3}, 0\right\}$ **51.** $\left\{-1, \dfrac{5}{3}\right\}$ **53.** $\left\{-\dfrac{3}{2}, \dfrac{1}{2}\right\}$
55. 8 and 9 or -9 and -8 **57.** 7 and 15
59. 10 inches by 6 inches   **61.** -7 and -6 or 6 and 7
63. 4 centimeters by 4 centimeters and 6 centimeters by 8 centimeters   **65.** 3, 4, and 5 units
67. 9 inches and 12 inches
69. An altitude of 4 inches and a side 14 inches long
77. (a) 0.28 and 3.73   **(c)** 2.27 and 5.76   **(e)** 0.71

Chapter 3 Review Problem Set (page 161)

1. $5x - 3$ **2.** $3x^2 + 12x - 2$ **3.** $12x^2 - x + 5$

4. $-20x^5y^7$ **5.** $-6a^5b^5$ **6.** $15a^4 - 10a^3 - 5a^2$

7. $24x^2 + 2xy - 15y^2$ **8.** $3x^3 + 7x^2 - 21x - 4$

9. $256x^8y^{12}$ **10.** $9x^2 - 12xy + 4y^2$

11. $-8x^6y^9z^3$ **12.** $-13x^2y$ **13.** $2x + y - 2$

14. $x^4 + x^3 - 18x^2 - x + 35$ **15.** $21 + 26x - 15x^2$

16. $-12a^5b^7$ **17.** $-8a^7b^3$ **18.** $7x^2 + 19x - 36$

19. $6x^3 - 11x^2 - 7x + 2$ **20.** $6x^{4n}$

21. $4x^2 + 20xy + 25y^2$ **22.** $x^3 - 6x^2 + 12x - 8$

23. $8x^3 + 60x^2 + 150x + 125$ **24.** $(x + 7)(x - 4)$

25. $2(t + 3)(t - 3)$ **26.** Not factorable

27. $(4n - 1)(3n - 1)$ **28.** $x^2(x^2 + 1)(x + 1)(x - 1)$

29. $x(x - 12)(x + 6)$ **30.** $2a^2b(3a + 2b - c)$

31. $(x - y + 1)(x + y - 1)$ **32.** $4(2x^2 + 3)$

33. $(4x + 7)(3x - 5)$ **34.** $(4n - 5)^2$

35. $4n(n - 2)$ **36.** $3w(w^2 + 6w - 8)$

37. $(5x + 2y)(4x - y)$ **38.** $16a(a - 4)$

39. $3x(x + 1)(x - 6)$ **40.** $(n + 8)(n - 16)$

41. $(t + 5)(t - 5)(t^2 + 3)$ **42.** $(5x - 3)(7x + 2)$

43. $(3 - x)(5 - 3x)$ **44.** $(4n - 3)(16n^2 + 12n + 9)$

45. $2(2x + 5)(4x^2 - 10x + 25)$ **46.** $\{-3, 3\}$

47. $\{-6, 1\}$ **48.** $\left\{\dfrac{2}{7}\right\}$ **49.** $\left\{-\dfrac{2}{5}, \dfrac{1}{3}\right\}$

50. $\left\{-\dfrac{1}{3}, 3\right\}$ **51.** $\{-3, 0, 3\}$ **52.** $\{-1, 0, 1\}$

53. $\{-7, 9\}$ **54.** $\left\{-\dfrac{4}{7}, \dfrac{2}{7}\right\}$ **55.** $\left\{-\dfrac{4}{5}, \dfrac{5}{6}\right\}$

56. $\{-2, 2\}$ **57.** $\left\{\dfrac{5}{3}\right\}$ **58.** $\{-8, 6\}$

59. $\left\{-5, \dfrac{2}{7}\right\}$ **60.** $\{-8, 5\}$ **61.** $\{-12, 1\}$

62. $\varnothing$ **63.** $\left\{-5, \dfrac{6}{5}\right\}$ **64.** $\{0, 1, 8\}$

65. $\left\{-10, \dfrac{1}{4}\right\}$ **66.** 8, 9, and 10 or -1, 0, and 1

67. -6 and 8   **68.** 13 and 15

69. 12 miles and 16 miles   **70.** 4 meters by 12 meters

71. 9 rows and 16 chairs per row

72. The side is 13 feet long and the altitude is 6 feet.

73. 3 feet

74. 5 centimeters by 5 centimeters and 8 centimeters by 8 centimeters   **75.** 6 inches

Chapter 3 Test (page 163)

1. $2x - 11$ **2.** $-48x^4y^4$ **3.** $-27x^6y^{12}$

4. $20x^2 + 17x - 63$ **5.** $6n^2 - 13n + 6$

6. $x^3 - 12x^2y + 48xy^2 - 64y^3$

7. $2x^3 + 11x^2 - 11x - 30$ **8.** $-14x^3y$

9. $(6x - 5)(x + 4)$ **10.** $3(2x + 1)(2x - 1)$

11. $(4 + t)(16 - 4t + t^2)$ **12.** $2x(3 - 2x)(5 + 4x)$

13. $(x - y)(x + 4)$ **14.** $(3n + 8)(8n - 3)$

15. $\{-12, 4\}$ **16.** $\left\{0, \dfrac{1}{4}\right\}$ **17.** $\left\{\dfrac{3}{2}\right\}$ **18.** $\{-4, -1\}$

19. $\{-9, 0, 2\}$ **20.** $\left\{-\dfrac{3}{7}, \dfrac{4}{5}\right\}$ **21.** $\left\{-\dfrac{1}{3}, 2\right\}$

22. $\{-2, 2\}$ **23.** 9 inches   **24.** 15 rows

25. 8 feet

Cumulative Review Problem Set (page 164)

1. 4   **2.** -19 **3.** 9   **4.** 21   **5.** -78

6. -33 **7.** -43 **8.** -11 **9.** -39 **10.** 57

11. $2x - 11$ **12.** $36a^2b^6$ **13.** $30x^2 - 37x - 7$

14. $-2x^2 - 11x - 12$ **15.** $-64a^6b^9$

16. $5x^3 - 6x^2 - 20x + 24$ **17.** $x^3 - 4x^2 - x + 12$

18. $2x^4 - x^3 - 2x^2 - 19x - 28$ **19.** $7(x + 1)(x - 1)$

20. $(2a - b)^2$ **21.** $(3x + 7)(x - 8)$

22. $(1 - x)(1 + x + x^2)$ **23.** $(y - 5)(x + 2)$

24. $3(x - 4)^2$ **25.** $(4n^2 + 3)(n + 1)(n - 1)$

26. $4x(2x + 3)(4x^2 - 6x + 9)$ **27.** $4(x^2 + 9)$

28. $(3x + 4)(2x - 1)$ **29.** $(3x - 5)^2$

30. $(x + 3y)(2x + 1)$ **31.** $(2a + 3b)(4a^2 - 6ab + 9b^2)$

32. $(x^2 + 4)(x + 2)(x - 2)$

33. $2m^2n^2(5m^2 - mn - 2n^2)$

34. $(2y + 7z)(5x - 12)$ **35.** $(3x + 5)(x - 2)$

36. $(5 - 2a)(5 + 2a)$ **37.** $(6x + 5)^2$

38. $(4y + 1)(16y^2 - 4y + 1)$

39. $x = \dfrac{2y + 6}{5}$ **40.** $y = \dfrac{12 - 3x}{4}$

41. $h = \dfrac{V - 2\pi r^2}{2\pi r}$ **42.** $R_1 = \dfrac{RR_2}{R_2 - R}$

43. 10.5%   **44.** $-15°$ **45.** $\{-6, 3\}$

46. $\left\{-\dfrac{7}{3}, \dfrac{2}{5}\right\}$ **47.** $\{-4\}$ **48.** $\{-9, 2\}$

49. $\{-1, 1\}$ **50.** $\{-10\}$ **51.** $\{15\}$

52. $\left\{-\dfrac{25}{4}\right\}$ **53.** $\left\{-\dfrac{5}{3}, 3\right\}$ **54.** $\{-1, 5\}$

55. $\{400\}$ **56.** $\{-4, 10\}$ **57.** $\{-4, 0, 4\}$

58. $\{-2, 3\}$ **59.** $\left\{-\dfrac{1}{4}, \dfrac{2}{3}\right\}$ **60.** $\{-6, 5\}$

61. $\{-5, 0, 2\}$ **62.** $\left\{\dfrac{17}{12}\right\}$ **63.** $(-22, \infty)$

64. $(23, \infty)$ **65.** $(-\infty, -3) \cup (4, \infty)$

66. $\left(-7, \dfrac{7}{3}\right)$ **67.** $(300, \infty)$ **68.** $\left(-\infty, \dfrac{7}{8}\right]$

69. $\left[\dfrac{5}{2}, \infty\right)$ **70.** $\left(-\infty, \dfrac{32}{31}\right)$

71. 7, 9, and 11
72. 8 nickels, 15 dimes, 25 quarters
73. 12 and 34   **74.** 62° and 118°
75. $400 at 8% and $600 at 9%
76. 35 pennies, 40 nickels, 70 dimes
77. 1 hour and 40 minutes
78. 25 milliliters   **79.** 40%
80. Better than 88   **81.** 4 inches
82. 7 meters by 14 meters
83. 8 rows and 12 chairs per row
84. 9 feet, 12 feet, and 15 feet

C H A P T E R 4

Problem Set 4.1 (page 171)

1. $\dfrac{3}{4}$ **3.** $\dfrac{5}{6}$ **5.** $-\dfrac{2}{5}$ **7.** $\dfrac{2}{7}$ **9.** $\dfrac{2x}{7}$

11. $\dfrac{2a}{5b}$ **13.** $-\dfrac{y}{4x}$ **15.** $-\dfrac{9c}{13d}$ **17.** $\dfrac{5x^2}{3y^3}$

19. $\dfrac{x-2}{x}$ **21.** $\dfrac{3x+2}{2x-1}$ **23.** $\dfrac{a+5}{a-9}$

25. $\dfrac{n-3}{5n-1}$ **27.** $\dfrac{5x^2+7}{10x}$ **29.** $\dfrac{3x+5}{4x+1}$

31. $\dfrac{3x}{x^2+4x+16}$ **33.** $\dfrac{x+6}{3x-1}$ **35.** $\dfrac{x(2x+7)}{y(x+9)}$

37. $\dfrac{y+4}{5y-2}$ **39.** $\dfrac{3x(x-1)}{x^2+1}$ **41.** $\dfrac{2(x+3y)}{3x(3x+y)}$

43. $\dfrac{3n-2}{7n+2}$ **45.** $\dfrac{4-x}{5+3x}$ **47.** $\dfrac{9x^2+3x+1}{2(x+2)}$

49. $\dfrac{-2(x-1)}{x+1}$ **51.** $\dfrac{y+b}{y+c}$ **53.** $\dfrac{x+2y}{2x+y}$

55. $\dfrac{x+1}{x-6}$ **57.** $\dfrac{2s+5}{3s+1}$ **59.** -1 **61.** $-n-7$

63. $-\dfrac{2}{x+1}$ **65.** -2 **67.** $-\dfrac{n+3}{n+5}$

Problem Set 4.2 (page 177)

1. $\dfrac{1}{10}$ **3.** $-\dfrac{4}{15}$ **5.** $\dfrac{3}{16}$ **7.** $-\dfrac{5}{6}$ **9.** $-\dfrac{2}{3}$

11. $\dfrac{10}{11}$ **13.** $-\dfrac{5x^3}{12y^2}$ **15.** $\dfrac{2a^3}{3b}$ **17.** $\dfrac{3x^3}{4}$

19. $\dfrac{25x^3}{108y^2}$ **21.** $\dfrac{ac^2}{2b^2}$ **23.** $\dfrac{3x}{4y}$ **25.** $\dfrac{3(x^2+4)}{5y(x+8)}$

27. $\dfrac{5(a+3)}{a(a-2)}$ **29.** $\dfrac{3}{2}$ **31.** $\dfrac{3xy}{4(x+6)}$

33. $\dfrac{5(x-2y)}{7y}$ **35.** $\dfrac{5+n}{3-n}$ **37.** $\dfrac{x^2+1}{x^2-10}$

39. $\dfrac{6x+5}{3x+4}$ **41.** $\dfrac{2t^2+5}{2(t^2+1)(t+1)}$ **43.** $\dfrac{t(t+6)}{4t+5}$

45. $\dfrac{n+3}{n(n-2)}$ **47.** $\dfrac{25x^3y^3}{4(x+1)}$ **49.** $\dfrac{2(a-2b)}{a(3a-2b)}$

Problem Set 4.3 (page 185)

1. $\dfrac{13}{12}$ **3.** $\dfrac{11}{40}$ **5.** $\dfrac{19}{20}$ **7.** $\dfrac{49}{75}$ **9.** $\dfrac{17}{30}$

11. $-\dfrac{11}{84}$ **13.** $\dfrac{2x+4}{x-1}$ **15.** 4   **17.** $\dfrac{7y-10}{7y}$

19. $\dfrac{5x+3}{6}$ **21.** $\dfrac{12a+1}{12}$ **23.** $\dfrac{n+14}{18}$

25. $-\dfrac{11}{15}$ **27.** $\dfrac{3x-25}{30}$ **29.** $\dfrac{43}{40x}$

31. $\dfrac{20y-77x}{28xy}$ **33.** $\dfrac{16y+15x-12xy}{12xy}$

35. $\dfrac{21+22x}{30x^2}$ **37.** $\dfrac{10n-21}{7n^2}$ **39.** $\dfrac{45-6n+20n^2}{15n^2}$

41. $\dfrac{11x-10}{6x^2}$ **43.** $\dfrac{42t+43}{35t^3}$ **45.** $\dfrac{20b^2-33a^3}{96a^2b}$

47. $\dfrac{14-24y^3+45xy}{18xy^3}$ **49.** $\dfrac{2x^2+3x-3}{x(x-1)}$

51. $\dfrac{a^2-a-8}{a(a+4)}$ **53.** $\dfrac{-41n-55}{(4n+5)(3n+5)}$

55. $\dfrac{-3x+17}{(x+4)(7x-1)}$ **57.** $\dfrac{-x+74}{(3x-5)(2x+7)}$

59. $\dfrac{38x+13}{(3x-2)(4x+5)}$ **61.** $\dfrac{5x+5}{2x+5}$ **63.** $\dfrac{x+15}{x-5}$

65. $\dfrac{-2x-4}{2x+1}$ **67. (a)** -1 **(c)** 0

Problem Set 4.4 (page 194)

1. $\dfrac{7x+20}{x(x+4)}$ **3.** $\dfrac{-x-3}{x(x+7)}$ **5.** $\dfrac{6x-5}{(x+1)(x-1)}$

7. $\dfrac{1}{a+1}$ **9.** $\dfrac{5n+15}{4(n+5)(n-5)}$ **11.** $\dfrac{x^2+60}{x(x+6)}$

13. $\dfrac{11x+13}{(x+2)(x+7)(2x+1)}$

15. $\dfrac{-3a+1}{(a-5)(a+2)(a+9)}$

17. $\dfrac{3a^2 + 14a + 1}{(4a - 3)(2a + 1)(a + 4)}$

19. $\dfrac{3x^2 + 20x - 111}{(x^2 + 3)(x + 7)(x - 3)}$

21. $\dfrac{-7y - 14}{(y + 8)(y - 2)}$ **23.** $\dfrac{-2x^2 - 4x + 3}{(x + 2)(x - 2)}$

25. $\dfrac{2x^2 + 14x - 19}{(x + 10)(x - 2)}$ **27.** $\dfrac{2n + 1}{n - 6}$

29. $\dfrac{2x^2 - 32x + 16}{(x + 1)(2x - 1)(3x - 2)}$ **31.** $\dfrac{1}{(n^2 + 1)(n + 1)}$

33. $\dfrac{-16x}{(5x - 2)(x - 1)}$ **35.** $\dfrac{t + 1}{t - 2}$ **37.** $\dfrac{2}{11}$

39. $-\dfrac{7}{27}$ **41.** $\dfrac{x}{4}$ **43.** $\dfrac{3y - 2x}{4x - 7}$

45. $\dfrac{6ab^2 - 5a^2}{12b^2 + 2a^2b}$ **47.** $\dfrac{2y - 3xy}{3x + 4xy}$ **49.** $\dfrac{3n + 14}{5n + 19}$

51. $\dfrac{5n - 17}{4n - 13}$ **53.** $\dfrac{-x + 5y - 10}{3y - 10}$

55. $\dfrac{-x + 15}{-2x - 1}$ **57.** $\dfrac{3a^2 - 2a + 1}{2a - 1}$

59. $\dfrac{-x^2 + 6x - 4}{3x - 2}$

Problem Set 4.5 (page 201)

1. $3x^3 + 6x^2$ **3.** $-6x^4 + 9x^6$ **5.** $3a^2 - 5a - 8$
7. $-13x^2 + 17x - 28$ **9.** $-3xy + 4x^2y - 8xy^2$

11. $x - 13$ **13.** $x + 20$ **15.** $2x + 1 - \dfrac{3}{x - 1}$

17. $5x - 1$ **19.** $3x^2 - 2x - 7$ **21.** $x^2 + 5x - 6$

23. $4x^2 + 7x + 12 + \dfrac{30}{x - 2}$ **25.** $x^3 - 4x^2 - 5x + 3$

27. $x^2 + 5x + 25$ **29.** $x^2 - x + 1 + \dfrac{63}{x + 1}$

31. $2x^2 - 4x + 7 - \dfrac{20}{x + 2}$ **33.** $4a - 4b$

35. $4x + 7 + \dfrac{23x - 6}{x^2 - 3x}$ **37.** $8y - 9 + \dfrac{8y + 5}{y^2 + y}$

39. $2x - 1$ **41.** $x - 3$ **43.** $5a - 8 + \dfrac{42a - 41}{a^2 + 3a - 4}$

45. $2n^2 + 3n - 4$ **47.** $x^4 + x^3 + x^2 + x + 1$

49. $x^3 - x^2 + x - 1$ **51.** $3x^2 + x + 1 + \dfrac{7}{x^2 - 1}$

53. $(x - 6)$ **55.** $x + 6, R = 14$ **57.** $x^2 - 1$
59. $x^2 - 2x - 3$ **61.** $2x^2 - x - 6, R = -6$
63. $x^3 - 7x^2 + 21x + 56, R = 167$

Problem Set 4.6 (page 209)

1. $\{2\}$ **3.** $\{-3\}$ **5.** $\{6\}$ **7.** $\left\{-\dfrac{85}{18}\right\}$

9. $\left\{\dfrac{7}{10}\right\}$ **11.** $\{5\}$ **13.** $\{58\}$

15. $\left\{\dfrac{1}{4}, 4\right\}$ **17.** $\left\{-\dfrac{2}{5}, 5\right\}$

19. $\{-16\}$ **21.** $\left\{-\dfrac{13}{3}\right\}$ **23.** $\{-3, 1\}$

25. $\left\{-\dfrac{5}{2}\right\}$ **27.** $\{-51\}$ **29.** $\left\{-\dfrac{5}{3}, 4\right\}$ **31.** $\varnothing$

33. $\left\{-\dfrac{11}{8}, 2\right\}$ **35.** $\{-29, 0\}$ **37.** $\{-9, 3\}$

39. $\left\{-2, \dfrac{23}{8}\right\}$ **41.** $\left\{\dfrac{11}{23}\right\}$ **43.** $\left\{3, \dfrac{7}{2}\right\}$

45. \$750 and \$1000   **47.** $48°$ and $72°$ **49.** $\dfrac{2}{7}$ or $\dfrac{7}{2}$

51. \$1080   **53.** \$69 for Tammy and \$51.75 for Laura
55. 8 and 82   **57.** 14 feet and 6 feet
59. 690 females and 460 males

Problem Set 4.7 (page 218)

1. $\{-21\}$ **3.** $\{-1, 2\}$ **5.** $\{2\}$ **7.** $\left\{\dfrac{37}{15}\right\}$

9. $\{-1\}$ **11.** $\{-1\}$ **13.** $\left\{0, \dfrac{13}{2}\right\}$

15. $\left\{-2, \dfrac{19}{2}\right\}$ **17.** $\{-2\}$ **19.** $\left\{-\dfrac{1}{5}\right\}$ **21.** $\varnothing$

23. $\left\{\dfrac{7}{2}\right\}$ **25.** $\{-3\}$ **27.** $\left\{-\dfrac{7}{9}\right\}$ **29.** $\left\{-\dfrac{7}{6}\right\}$

31. $x = \dfrac{18y - 4}{15}$ **33.** $y = \dfrac{-5x + 22}{2}$

35. $M = \dfrac{IC}{100}$ **37.** $R = \dfrac{ST}{S + T}$

39. $y = \dfrac{bx - x - 3b + a}{a - 3}$ **41.** $y = \dfrac{ab - bx}{a}$

43. $y = \dfrac{-2x - 9}{3}$

45. 50 miles per hour for Dave and 54 miles per hour for Kent
47. 60 minutes
49. 60 words per minute for Connie and 40 words per minute for Katie
51. Plane B could travel at 400 miles per hour for 5 hours and plane A at 350 miles per hour for 4 hours, or

plane B could travel at 250 miles per hour for 8 hours and plane A at 200 miles per hour for 7 hours.

53. 60 minutes for Nancy and 120 minutes for Amy

55. 3 hours

57. 16 miles per hour on the way out and 12 miles per hour on the way back, or 12 miles per hour out and 8 miles per hour back

Chapter 4 Review Problem Set (page 222)

1. $\dfrac{2y}{3x^2}$ **2.** $\dfrac{a-3}{a}$ **3.** $\dfrac{n-5}{n-1}$ **4.** $\dfrac{x^2+1}{x}$

5. $\dfrac{2x+1}{3}$ **6.** $\dfrac{x^2-10}{2x^2+1}$ **7.** $\dfrac{3}{22}$ **8.** $\dfrac{18y+20x}{48y-9x}$

9. $\dfrac{3x+2}{3x-2}$ **10.** $\dfrac{x-1}{2x-1}$ **11.** $\dfrac{2x}{7y^2}$ **12.** $3b$

13. $\dfrac{n(n+5)}{n-1}$ **14.** $\dfrac{x(x-3y)}{x^2+9y^2}$ **15.** $\dfrac{23x-6}{20}$

16. $\dfrac{57-2n}{18n}$ **17.** $\dfrac{3x^2-2x-14}{x(x+7)}$ **18.** $\dfrac{2}{x-5}$

19. $\dfrac{5n-21}{(n-9)(n+4)(n-1)}$ **20.** $\dfrac{6y-23}{(2y+3)(y-6)}$

21. $6x-1$ **22.** $3x^2-7x+22-\dfrac{90}{x+4}$ **23.** $\left\{\dfrac{4}{13}\right\}$

24. $\left\{\dfrac{3}{16}\right\}$ **25.** $\varnothing$ **26.** $\{-17\}$ **27.** $\left\{\dfrac{2}{7},\dfrac{7}{2}\right\}$

28. $\{22\}$ **29.** $\left\{-\dfrac{6}{7},3\right\}$ **30.** $\left\{\dfrac{3}{4},\dfrac{5}{2}\right\}$ **31.** $\left\{\dfrac{9}{7}\right\}$

32. $\left\{-\dfrac{5}{4}\right\}$ **33.** $y=\dfrac{3x+27}{4}$ **34.** $y=\dfrac{bx-ab}{a}$

35. \$525 and \$875

36. 20 minutes for Julio and 30 minutes for Dan

37. 50 miles per hour and 55 miles per hour or $8\tfrac{1}{3}$ miles per hour and $13\tfrac{1}{3}$ miles per hour

38. 9 hours   **39.** 80 hours   **40.** 13 miles per hour

Chapter 4 Test (page 224)

1. $\dfrac{13y^2}{24x}$ **2.** $\dfrac{3x-1}{x(x-6)}$ **3.** $\dfrac{2n-3}{n+4}$ **4.** $-\dfrac{2x}{x+1}$

5. $\dfrac{3y^2}{8}$ **6.** $\dfrac{a-b}{4(2a+b)}$ **7.** $\dfrac{x+4}{5x-1}$ **8.** $\dfrac{13x+7}{12}$

9. $\dfrac{3x}{2}$ **10.** $\dfrac{10n-26}{15n}$ **11.** $\dfrac{3x^2+2x-12}{x(x-6)}$

12. $\dfrac{11-2x}{x(x-1)}$ **13.** $\dfrac{13n+46}{(2n+5)(n-2)(n+7)}$

14. $3x^2-2x-1$ **15.** $\dfrac{18-2x}{8+9x}$ **16.** $y=\dfrac{4x+20}{3}$

17. $\{1\}$ **18.** $\left\{\dfrac{1}{10}\right\}$ **19.** $\{-35\}$ **20.** $\{-1,5\}$

21. $\left\{\dfrac{5}{3}\right\}$ **22.** $\left\{-\dfrac{9}{13}\right\}$ **23.** $\dfrac{27}{72}$

24. 1 hour   **25.** 15 miles per hour

CHAPTER 5

Problem Set 5.1 (page 232)

1. $\dfrac{1}{27}$ **3.** $-\dfrac{1}{100}$ **5.** 81 **7.** -27 **9.** -8

11. 1 **13.** $\dfrac{9}{49}$ **15.** 16 **17.** $\dfrac{1}{1000}$

19. $\dfrac{1}{1000}$ **21.** 27 **23.** $\dfrac{1}{125}$ **25.** $\dfrac{9}{8}$

27. $\dfrac{256}{25}$ **29.** $\dfrac{2}{25}$ **31.** $\dfrac{81}{4}$ **33.** 81

35. $\dfrac{1}{10,000}$ **37.** $\dfrac{13}{36}$ **39.** $\dfrac{1}{2}$ **41.** $\dfrac{72}{17}$

43. $\dfrac{1}{x^6}$ **45.** $\dfrac{1}{a^3}$ **47.** $\dfrac{1}{a^8}$ **49.** $\dfrac{y^6}{x^2}$ **51.** $\dfrac{c^8}{a^4b^{12}}$

53. $\dfrac{y^{12}}{8x^9}$ **55.** $\dfrac{x^3}{y^{12}}$ **57.** $\dfrac{4a^4}{9b^2}$ **59.** $\dfrac{1}{x^2}$

61. a^5b^2 **63.** $\dfrac{6y^3}{x}$ **65.** $7b^2$ **67.** $\dfrac{7x}{y^2}$

69. $-\dfrac{12b^3}{a}$ **71.** $\dfrac{x^5y^5}{5}$ **73.** $\dfrac{b^{20}}{81}$ **75.** $\dfrac{x+1}{x^3}$

77. $\dfrac{y-x^3}{x^3y}$ **79.** $\dfrac{3b+4a^2}{a^2b}$

81. $\dfrac{1-x^2y}{xy^2}$ **83.** $\dfrac{2x-3}{x^2}$

Problem Set 5.2 (page 243)

1. 8 **3.** -10 **5.** 3 **7.** -4 **9.** 3

11. $\dfrac{4}{5}$ **13.** $-\dfrac{6}{7}$ **15.** $\dfrac{1}{2}$ **17.** $\dfrac{3}{4}$ **19.** 8

21. $3\sqrt{3}$ **23.** $4\sqrt{2}$ **25.** $4\sqrt{5}$ **27.** $4\sqrt{10}$

29. $12\sqrt{2}$ **31.** $-12\sqrt{5}$ **33.** $2\sqrt{3}$ **35.** $3\sqrt{6}$

37. $-\dfrac{5}{3}\sqrt{7}$ **39.** $\dfrac{\sqrt{19}}{2}$ **41.** $\dfrac{3\sqrt{3}}{4}$ **43.** $\dfrac{5\sqrt{3}}{9}$

45. $\dfrac{\sqrt{14}}{7}$ **47.** $\dfrac{\sqrt{6}}{3}$ **49.** $\dfrac{\sqrt{15}}{6}$ **51.** $\dfrac{\sqrt{66}}{12}$

53. $\dfrac{\sqrt{6}}{3}$ **55.** $\sqrt{5}$ **57.** $\dfrac{2\sqrt{21}}{7}$ **59.** $-\dfrac{8\sqrt{15}}{5}$

61. $\dfrac{\sqrt{6}}{4}$ **63.** $-\dfrac{12}{25}$ **65.** $2\sqrt[3]{2}$ **67.** $6\sqrt[3]{3}$

69. $\dfrac{2\sqrt[3]{3}}{3}$ **71.** $\dfrac{3\sqrt[3]{2}}{2}$ **73.** $\dfrac{\sqrt[3]{12}}{2}$

75. 42 miles per hour; 49 miles per hour; 65 miles per hour

77. 107 square centimeters   **79.** 140 square inches

85. (a) 1.414   **(c)** 12.490   **(e)** 57.000
(g) 0.374   **(i)** 0.930

Problem Set 5.3 (page 249)

1. $13\sqrt{2}$ **3.** $54\sqrt{3}$ **5.** $-30\sqrt{2}$ **7.** $-\sqrt{5}$

9. $-21\sqrt{6}$ **11.** $-\dfrac{7\sqrt{7}}{12}$ **13.** $\dfrac{37\sqrt{10}}{10}$

15. $\dfrac{41\sqrt{2}}{20}$ **17.** $-9\sqrt[3]{3}$ **19.** $10\sqrt[3]{2}$ **21.** $4\sqrt{2x}$

23. $5x\sqrt{3}$ **25.** $2x\sqrt{5y}$ **27.** $8xy^3\sqrt{xy}$

29. $3a^2b\sqrt{6b}$ **31.** $3x^3y^4\sqrt{7}$ **33.** $4a\sqrt{10a}$

35. $\dfrac{8y}{3}\sqrt{6xy}$ **37.** $\dfrac{\sqrt{10xy}}{5y}$ **39.** $\dfrac{\sqrt{15}}{6x^2}$

41. $\dfrac{5\sqrt{2y}}{6y}$ **43.** $\dfrac{\sqrt{14xy}}{4y^3}$ **45.** $\dfrac{3y\sqrt{2xy}}{4x}$

47. $\dfrac{2\sqrt{42ab}}{7b^2}$ **49.** $2\sqrt[3]{3y}$ **51.** $2x\sqrt[3]{2x}$

53. $2x^2y^2\sqrt[3]{7y^2}$ **55.** $\dfrac{\sqrt[3]{21x}}{3x}$ **57.** $\dfrac{\sqrt[3]{12x^2y}}{4x^2}$

59. $\dfrac{\sqrt[3]{4x^2y^2}}{xy^2}$ **61.** $2\sqrt{2x+3y}$ **63.** $4\sqrt{x+3y}$

65. $33\sqrt{x}$ **67.** $-30\sqrt{2x}$ **69.** $7\sqrt{3n}$

71. $-40\sqrt{ab}$ **73.** $-7x\sqrt{2x}$

79. (a) $5|x|\sqrt{5}$ **(b)** $4x^2$ **(c)** $2b\sqrt{2b}$
(d) $y^2\sqrt{3y}$ **(e)** $12|x^3|\sqrt{2}$ **(f)** $2m^4\sqrt{7}$
(g) $8|c^5|\sqrt{2}$ **(h)** $3d^3\sqrt{2d}$ **(i)** $7|x|$
(j) $4n^{10}\sqrt{5}$ **(k)** $9h\sqrt{h}$

Problem Set 5.4 (page 256)

1. $6\sqrt{2}$ **3.** $18\sqrt{2}$ **5.** $-24\sqrt{10}$ **7.** $24\sqrt{6}$
9. 120   **11.** 24   **13.** $56\sqrt[3]{3}$ **15.** $\sqrt{6}+\sqrt{10}$
17. $6\sqrt{10}-3\sqrt{35}$ **19.** $24\sqrt{3}-60\sqrt{2}$
21. $-40-32\sqrt{15}$ **23.** $15\sqrt{2x}+3\sqrt{xy}$
25. $5xy-6x\sqrt{y}$ **27.** $2\sqrt{10xy}+2y\sqrt{15xy}$
29. $-25\sqrt{6}$ **31.** $-25-3\sqrt{3}$ **33.** $23-9\sqrt{5}$
35. $6\sqrt{35}+3\sqrt{10}-4\sqrt{21}-2\sqrt{6}$
37. $8\sqrt{3}-36\sqrt{2}+6\sqrt{10}-18\sqrt{15}$
39. $11+13\sqrt{30}$ **41.** $141-51\sqrt{6}$ **43.** -10
45. -8 **47.** $2x-3y$ **49.** $10\sqrt[3]{12}+2\sqrt[3]{18}$

51. $12-36\sqrt[3]{2}$ **53.** $\dfrac{\sqrt{7}-1}{3}$ **55.** $\dfrac{-3\sqrt{2}-15}{23}$

57. $\dfrac{\sqrt{7}-\sqrt{2}}{5}$ **59.** $\dfrac{2\sqrt{5}+\sqrt{6}}{7}$

61. $\dfrac{\sqrt{15}-2\sqrt{3}}{2}$ **63.** $\dfrac{6\sqrt{7}+4\sqrt{6}}{13}$

65. $\sqrt{3}-\sqrt{2}$ **67.** $\dfrac{2\sqrt{x}-8}{x-16}$ **69.** $\dfrac{x+5\sqrt{x}}{x-25}$

71. $\dfrac{x-8\sqrt{x}+12}{x-36}$ **73.** $\dfrac{x-2\sqrt{xy}}{x-4y}$

75. $\dfrac{6\sqrt{xy}+9y}{4x-9y}$

Problem Set 5.5 (page 262)

1. $\{20\}$ **3.** $\varnothing$ **5.** $\left\{\dfrac{25}{4}\right\}$ **7.** $\left\{\dfrac{4}{9}\right\}$ **9.** $\{5\}$

11. $\left\{\dfrac{39}{4}\right\}$ **13.** $\varnothing$ **15.** $\{1\}$ **17.** $\left\{\dfrac{3}{2}\right\}$

19. $\{3\}$ **21.** $\left\{\dfrac{61}{25}\right\}$ **23.** $\{-3, 3\}$ **25.** $\{-9, -4\}$

27. $\{0\}$ **29.** $\{3\}$ **31.** $\{4\}$ **33.** $\{-4, -3\}$
35. $\{12\}$ **37.** $\{25\}$ **39.** $\{29\}$ **41.** $\{-15\}$

43. $\left\{-\dfrac{1}{3}\right\}$ **45.** $\{-3\}$ **47.** $\{0\}$ **49.** $\{5\}$

51. $\{2, 6\}$ **53.** 56 feet; 106 feet; 148 feet
55. 3.2 feet; 5.1 feet; 7.3 feet

Problem Set 5.6 (page 267)

1. 9   **3.** 3   **5.** -2 **7.** -5 **9.** $\dfrac{1}{6}$

11. 3   **13.** 8   **15.** 81   **17.** -1

19. -32 **21.** $\dfrac{81}{16}$ **23.** 4   **25.** $\dfrac{1}{128}$

27. -125 **29.** 625   **31.** $\sqrt[3]{x^4}$ **33.** $3\sqrt{x}$

35. $\sqrt[3]{2y}$ **37.** $\sqrt{2x-3y}$ **39.** $\sqrt[3]{(2a-3b)^2}$

41. $\sqrt[3]{x^2y}$ **43.** $-3\sqrt[5]{xy^2}$ **45.** $5^{\frac{1}{2}}y^{\frac{1}{2}}$ **47.** $3y^{\frac{1}{2}}$

49. $x^{\frac{1}{3}}y^{\frac{2}{3}}$ **51.** $a^{\frac{1}{2}}b^{\frac{3}{4}}$ **53.** $(2x-y)^{\frac{3}{5}}$ **55.** $5xy^{\frac{1}{2}}$

57. $-(x+y)^{\frac{1}{3}}$ **59.** $12x^{\frac{13}{20}}$ **61.** $y^{\frac{5}{12}}$ **63.** $\dfrac{4}{x^{\frac{1}{10}}}$

65. $16xy^2$ **67.** $2x^2y$ **69.** $4x^{\frac{4}{15}}$

71. $\dfrac{4}{b^{\frac{5}{12}}}$ **73.** $\dfrac{36x^{\frac{4}{5}}}{49y^{\frac{4}{3}}}$ **75.** $\dfrac{y^{\frac{3}{2}}}{x}$ **77.** $4x^{\frac{1}{6}}$

79. $\dfrac{16}{a^{\frac{11}{10}}}$ **81.** $\sqrt[6]{243}$ **83.** $\sqrt[4]{216}$ **85.** $\sqrt[12]{3}$

87. $\sqrt{2}$ **89.** $\sqrt[4]{3}$

93. (a) 12   **(c)** 7   **(e)** 11

95. (a) 1024   **(c)** 512   **(e)** 49

Problem Set 5.7 (page 273)

1. $(8.9)(10)^1$ **3.** $(4.29)(10)^3$ **5.** $(6.12)(10)^6$

7. $(4)(10)^7$ **9.** $(3.764)(10)^2$ **11.** $(3.47)(10)^{-1}$

13. $(2.14)(10)^{-2}$ **15.** $(5)(10)^{-5}$ **17.** $(1.94)(10)^{-9}$

19. 23   **21.** 4190   **23.** 500,000,000

25. 31,400,000,000   **27.** 0.43   **29.** 0.000914

31. 0.00000005123   **33.** 0.000000074   **35.** 0.77

37. 300,000,000,000   **39.** 0.000000004   **41.** 1000

43. 1000   **45.** 3000   **47.** 20   **49.** 27,000,000

51. $(6.02)(10^{23})$ **53.** 831   **55.** $(2.07)(10^4)$ dollars

57. $(1.99)(10^{-26})$ kg   **59.** 1833

63. (a) 7000   **(c)** 120   **(e)** 30

65. (a) $(4.385)(10)^{14}$ **(c)** $(2.322)(10)^{17}$

    (e) $(3.052)(10)^{12}$

Chapter 5 Review Problem Set (page 276)

1. $\dfrac{1}{64}$ **2.** $\dfrac{9}{4}$ **3.** 3   **4.** -2 **5.** $\dfrac{2}{3}$ **6.** 32

7. 1   **8.** $\dfrac{4}{9}$ **9.** -64 **10.** 32   **11.** 1

12. 27   **13.** $3\sqrt{6}$ **14.** $4x\sqrt{3xy}$ **15.** $2\sqrt{2}$

16. $\dfrac{\sqrt{15x}}{6x^2}$ **17.** $2\sqrt[3]{7}$ **18.** $\dfrac{\sqrt[3]{6}}{3}$ **19.** $\dfrac{3\sqrt{5}}{5}$

20. $\dfrac{x\sqrt{21x}}{7}$ **21.** $3xy^2\sqrt[3]{4xy^2}$ **22.** $\dfrac{15\sqrt{6}}{4}$

23. $2y\sqrt{5xy}$ **24.** $2\sqrt{x}$ **25.** $24\sqrt{10}$ **26.** 60

27. $24\sqrt{3}-6\sqrt{14}$ **28.** $x-2\sqrt{x}-15$ **29.** 17

30. $12-8\sqrt{3}$ **31.** $6a-5\sqrt{ab}-4b$ **32.** 70

33. $\dfrac{2(\sqrt{7}+1)}{3}$ **34.** $\dfrac{2\sqrt{6}-\sqrt{15}}{3}$

35. $\dfrac{3\sqrt{5}-2\sqrt{3}}{11}$ **36.** $\dfrac{6\sqrt{3}+3\sqrt{5}}{7}$ **37.** $\dfrac{x^6}{y^8}$

38. $\dfrac{27a^3b^{12}}{8}$ **39.** $20x^{\frac{7}{10}}$ **40.** $7a^{\frac{5}{12}}$ **41.** $\dfrac{y^{\frac{4}{3}}}{x}$

42. $\dfrac{x^{12}}{9}$ **43.** $\sqrt{5}$ **44.** $5\sqrt[3]{3}$ **45.** $\dfrac{29\sqrt{6}}{5}$

46. $-15\sqrt{3x}$ **47.** $\dfrac{y+x^2}{x^2y}$ **48.** $\dfrac{b-2a}{a^2b}$

49. $\left\{\dfrac{19}{7}\right\}$ **50.** $\{4\}$ **51.** $\{8\}$ **52.** $\varnothing$

53. $\{14\}$ **54.** $\{-10,1\}$ **55.** $\{2\}$ **56.** $\{8\}$

57. 0.000000006   **58.** 36,000,000,000   **59.** 6

60. 0.15   **61.** 0.000028   **62.** 0.002   **63.** 0.002

64. 8,000,000,000

Chapter 5 Test (page 278)

1. $\dfrac{1}{32}$ **2.** -32 **3.** $\dfrac{81}{16}$ **4.** $\dfrac{1}{4}$ **5.** $3\sqrt{7}$

6. $3\sqrt[3]{4}$ **7.** $2x^2y\sqrt{13y}$ **8.** $\dfrac{5\sqrt{6}}{6}$ **9.** $\dfrac{\sqrt{42x}}{12x^2}$

10. $72\sqrt{2}$ **11.** $-5\sqrt{6}$ **12.** $-38\sqrt{2}$

13. $\dfrac{3\sqrt{6}+3}{10}$ **14.** $\dfrac{9x^2y^2}{4}$ **15.** $-\dfrac{12}{a^{\frac{3}{10}}}$

16. $\dfrac{y^3+x}{xy^3}$ **17.** $-12x^{\frac{1}{4}}$ **18.** 33   **19.** 600

20. 0.003   **21.** $\left\{\dfrac{8}{3}\right\}$ **22.** $\{2\}$ **23.** $\{4\}$

24. $\{5\}$ **25.** $\{4,6\}$

CHAPTER 6

Problem Set 6.1 (page 286)

1. False   **3.** True   **5.** True   **7.** True

9. $10+8i$ **11.** $-6+10i$ **13.** $-2-5i$

15. $-12+5i$ **17.** $-1-23i$ **19.** $-4-5i$

21. $1+3i$ **23.** $\dfrac{5}{3}-\dfrac{5}{12}i$ **25.** $-\dfrac{17}{9}+\dfrac{23}{30}i$

27. $9i$ **29.** $i\sqrt{14}$ **31.** $\dfrac{4}{5}i$ **33.** $3i\sqrt{2}$

35. $5i\sqrt{3}$ **37.** $6i\sqrt{7}$ **39.** $-8i\sqrt{5}$

41. $36i\sqrt{10}$ **43.** -8 **45.** $-\sqrt{15}$ **47.** $-3\sqrt{6}$

49. $-5\sqrt{3}$ **51.** $-3\sqrt{6}$ **53.** $4i\sqrt{3}$ **55.** $\dfrac{5}{2}$

57. $2\sqrt{2}$ **59.** $2i$ **61.** $-20+0i$ **63.** $42+0i$

65. $15+6i$ **67.** $-42+12i$ **69.** $7+22i$

71. $40-20i$ **73.** $-3-28i$ **75.** $-3-15i$

77. $-9+40i$ **79.** $-12+16i$ **81.** $85+0i$

83. $5+0i$ **85.** $\dfrac{3}{5}+\dfrac{3}{10}i$ **87.** $\dfrac{5}{17}-\dfrac{3}{17}i$

89. $2+\dfrac{2}{3}i$ **91.** $0-\dfrac{2}{7}i$ **93.** $\dfrac{22}{25}-\dfrac{4}{25}i$

95. $-\dfrac{18}{41}+\dfrac{39}{41}i$ **97.** $\dfrac{9}{2}-\dfrac{5}{2}i$ **99.** $\dfrac{4}{13}-\dfrac{1}{26}i$

Problem Set 6.2 (page 294)

1. $\{0,9\}$ **3.** $\{-3,0\}$ **5.** $\{-4,0\}$

7. $\left\{0, \dfrac{9}{5}\right\}$ **9.** $\{-6, 5\}$ **11.** $\{7, 12\}$

13. $\left\{-8, -\dfrac{3}{2}\right\}$ **15.** $\left\{-\dfrac{7}{3}, \dfrac{2}{5}\right\}$ **17.** $\left\{\dfrac{3}{5}\right\}$

19. $\left\{-\dfrac{3}{2}, \dfrac{7}{3}\right\}$ **21.** $\{1, 4\}$ **23.** $\{8\}$ **25.** $\{12\}$

27. $\{0, 5k\}$ **29.** $\{0, 16k^2\}$ **31.** $\{5k, 7k\}$

33. $\left\{\dfrac{k}{2}, -3k\right\}$ **35.** $\{\pm 1\}$ **37.** $\{\pm 6i\}$

39. $\{\pm\sqrt{14}\}$ **41.** $\{\pm 2\sqrt{7}\}$ **43.** $\{\pm 3\sqrt{2}\}$

45. $\left\{\pm\dfrac{\sqrt{14}}{2}\right\}$ **47.** $\left\{\pm\dfrac{2\sqrt{3}}{3}\right\}$ **49.** $\left\{\pm\dfrac{2i\sqrt{30}}{5}\right\}$

51. $\left\{\pm\dfrac{\sqrt{6}}{2}\right\}$ **53.** $\{-1, 5\}$ **55.** $\{-8, 2\}$

57. $\{-6 \pm 2i\}$ **59.** $\{1, 2\}$ **61.** $\{4 \pm \sqrt{5}\}$

63. $\{-5 \pm 2\sqrt{3}\}$ **65.** $\left\{\dfrac{2 \pm 3i\sqrt{3}}{3}\right\}$ **67.** $\{-12, -2\}$

69. $\left\{\dfrac{2 \pm \sqrt{10}}{5}\right\}$ **71.** $2\sqrt{13}$ centimeters

73. $4\sqrt{5}$ inches   **75.** 8 yards   **77.** $6\sqrt{2}$ inches
79. $a = b = 4\sqrt{2}$ meters
81. $b = 3\sqrt{3}$ inches and $c = 6$ inches
83. $a = 7$ centimeters and $b = 7\sqrt{3}$ centimeters
85. $a = \dfrac{10\sqrt{3}}{3}$ feet and $c = \dfrac{20\sqrt{3}}{3}$ feet
87. 17.9 feet   **89.** 38 meters
91. 53 meters   **95.** 10.8 centimeters   **97.** $h = s\sqrt{2}$

Problem Set 6.3 (page 300)

1. $\{-6, 10\}$ **3.** $\{4, 10\}$ **5.** $\{-5, 10\}$ **7.** $\{-8, 1\}$

9. $\left\{-\dfrac{5}{2}, 3\right\}$ **11.** $\left\{-3, \dfrac{2}{3}\right\}$ **13.** $\{-16, 10\}$

15. $\{-2 \pm \sqrt{6}\}$ **17.** $-3 \pm 2\sqrt{3}\}$
19. $\{5 \pm \sqrt{26}\}$ **21.** $\{4 \pm i\}$ **23.** $\{-6 \pm 3\sqrt{3}\}$

25. $\{-1 \pm i\sqrt{5}\}$ **27.** $\left\{\dfrac{-3 \pm \sqrt{17}}{2}\right\}$

29. $\left\{\dfrac{-5 \pm \sqrt{21}}{2}\right\}$ **31.** $\left\{\dfrac{7 \pm \sqrt{37}}{2}\right\}$

33. $\left\{\dfrac{-2 \pm \sqrt{10}}{2}\right\}$ **35.** $\left\{\dfrac{3 \pm i\sqrt{6}}{3}\right\}$

37. $\left\{\dfrac{-5 \pm \sqrt{37}}{6}\right\}$ **39.** $\{-12, 4\}$ **41.** $\left\{\dfrac{4 \pm \sqrt{10}}{2}\right\}$

43. $\left\{-\dfrac{9}{2}, \dfrac{1}{3}\right\}$ **45.** $\{-3, 8\}$ **47.** $\{3 \pm 2\sqrt{3}\}$

49. $\left\{\dfrac{3 \pm i\sqrt{3}}{3}\right\}$ **51.** $\{-20, 12\}$ **53.** $\left\{-1, -\dfrac{2}{3}\right\}$

55. $\left\{\dfrac{1}{2}, \dfrac{3}{2}\right\}$ **57.** $\{-6 \pm 2\sqrt{10}\}$ **59.** $\left\{\dfrac{-1 \pm \sqrt{3}}{2}\right\}$

61. $\left\{\dfrac{-b \pm \sqrt{b^2 - 4ac}}{2a}\right\}$ **65.** $x = \dfrac{a\sqrt{b^2 - y^2}}{b}$

67. $r = \dfrac{\sqrt{A\pi}}{\pi}$ **69.** $\{2a, 3a\}$ **71.** $\left\{\dfrac{a}{2}, -\dfrac{2a}{3}\right\}$

73. $\left\{\dfrac{2b}{3}\right\}$

Problem Set 6.4 (page 308)
1. Two real solutions; $\{-7, 3\}$

3. One real solution; $\left\{\dfrac{1}{3}\right\}$

5. Two complex solutions; $\left\{\dfrac{7 \pm i\sqrt{3}}{2}\right\}$

7. Two real solutions; $\left\{-\dfrac{4}{3}, \dfrac{1}{5}\right\}$

9. Two real solutions; $\left\{\dfrac{-2 \pm \sqrt{10}}{3}\right\}$

11. $\{-1 \pm \sqrt{2}\}$ **13.** $\left\{\dfrac{-5 \pm \sqrt{37}}{2}\right\}$ **15.** $\{4 \pm 2\sqrt{5}\}$

17. $\left\{\dfrac{-5 \pm i\sqrt{7}}{2}\right\}$ **19.** $\{8, 10\}$ **21.** $\left\{\dfrac{9 \pm \sqrt{61}}{2}\right\}$

23. $\left\{\dfrac{-1 \pm \sqrt{33}}{4}\right\}$ **25.** $\left\{\dfrac{-1 \pm i\sqrt{3}}{4}\right\}$

27. $\left\{\dfrac{4 \pm \sqrt{10}}{3}\right\}$ **29.** $\left\{-1, \dfrac{5}{2}\right\}$ **31.** $\left\{-5, -\dfrac{4}{3}\right\}$

33. $\left\{\dfrac{5}{6}\right\}$ **35.** $\left\{\dfrac{1 \pm \sqrt{13}}{4}\right\}$ **37.** $\left\{0, \dfrac{13}{5}\right\}$

39. $\left\{\pm\dfrac{\sqrt{15}}{3}\right\}$ **41.** $\left\{\dfrac{-1 \pm \sqrt{73}}{12}\right\}$

43. $\{-18, -14\}$ **45.** $\left\{\dfrac{11}{4}, \dfrac{10}{3}\right\}$

47. $\left\{\dfrac{2 \pm i\sqrt{2}}{2}\right\}$ **49.** $\left\{\dfrac{1 \pm \sqrt{7}}{6}\right\}$

55. $\{-1.381, 17.381\}$ **57.** $\{-13.426, 3.426\}$
59. $\{-0.347, -8.653\}$ **61.** $\{0.119, 1.681\}$
63. $\{-0.708, 4.708\}$ **65.** $k = 4$ or $k = -4$

Problem Set 6.5 (page 318)

1. $\{2 \pm \sqrt{10}\}$ **3.** $\left\{-9, \dfrac{4}{3}\right\}$ **5.** $\{9 \pm 3\sqrt{10}\}$

7. $\left\{\dfrac{3 \pm i\sqrt{23}}{4}\right\}$ **9.** $\{-15, -9\}$ **11.** $\{-8, 1\}$

13. $\left\{\dfrac{2 \pm i\sqrt{10}}{2}\right\}$ **15.** $\{9 \pm \sqrt{66}\}$ **17.** $\left\{-\dfrac{5}{4}, \dfrac{2}{5}\right\}$

19. $\left\{\dfrac{-1 \pm \sqrt{2}}{2}\right\}$ **21.** $\left\{\dfrac{3}{4}, 4\right\}$ **23.** $\left\{\dfrac{11 \pm \sqrt{109}}{2}\right\}$

25. $\left\{\dfrac{3}{7}, 4\right\}$ **27.** $\left\{\dfrac{7 \pm \sqrt{129}}{10}\right\}$ **29.** $\left\{-\dfrac{10}{7}, 3\right\}$

31. $\{1 \pm \sqrt{34}\}$ **33.** $\{\pm\sqrt{6}, \pm 2\sqrt{3}\}$

35. $\left\{\pm 3, \pm\dfrac{2\sqrt{6}}{3}\right\}$ **37.** $\left\{\pm\dfrac{i\sqrt{15}}{3}, \pm 2i\right\}$

39. $\left\{\pm\dfrac{\sqrt{14}}{2}, \pm\dfrac{2\sqrt{3}}{3}\right\}$ **41.** 8 and 9 **43.** 9 and 12

45. $5 + \sqrt{3}$ and $5 - \sqrt{3}$ **47.** 3 and 6

49. 9 inches and 12 inches **51.** 1 meter

53. 8 inches by 14 inches **55.** 20 miles per hour for Lorraine and 25 miles per hour for Charlotte, or 45 miles per hour for Lorraine and 50 miles per hour for Charlotte **57.** 55 miles per hour

59. 6 hours for Tom and 8 hours for Terry **61.** 2 hours

63. 8 students **65.** 40 shares at $20 per share

67. 50 numbers **69.** 9%

75. $\{9, 36\}$ **77.** $\{1\}$ **79.** $\left\{-\dfrac{8}{27}, \dfrac{27}{8}\right\}$

81. $\left\{-4, \dfrac{3}{5}\right\}$ **83.** $\{4\}$ **85.** $\{\pm 4\sqrt{2}\}$

87. $\left\{\dfrac{3}{2}, \dfrac{5}{2}\right\}$ **89.** $\{-1, 3\}$

Problem Set 6.6 (page 326)

1. $(-\infty, -2) \cup (1, \infty)$

3. $(-4, -1)$

5. $\left(-\infty, -\dfrac{7}{3}\right] \cup \left[\dfrac{1}{2}, \infty\right)$

7. $\left[-2, \dfrac{3}{4}\right]$

9. $(-1, 1) \cup (3, \infty)$

11. $(-\infty, -2] \cup [0, 4]$

13. $(-\infty, -1) \cup (2, \infty)$

15. $(-2, 3)$

17. $(-\infty, 0) \cup \left[\dfrac{1}{2}, \infty\right)$

19. $(-\infty, 1) \cup [2, \infty)$

21. $(-7, 5)$ **23.** $(-\infty, 4) \cup (7, \infty)$ **25.** $\left[-5, \dfrac{2}{3}\right]$

27. $\left(-\infty, -\dfrac{5}{2}\right] \cup \left[-\dfrac{1}{4}, \infty\right)$ **29.** $\left(-\infty, -\dfrac{4}{5}\right) \cup (8, \infty)$

31. $(-\infty, \infty)$ **33.** $\left\{-\dfrac{5}{2}\right\}$ **35.** $(-1, 3) \cup (3, \infty)$

37. $(-6, -3)$ **39.** $(-\infty, 5) \cup [9, \infty)$

41. $\left(-\infty, \dfrac{4}{3}\right) \cup (3, \infty)$ **43.** $(-4, 6]$ **45.** $(-\infty, 2)$

55. (a) $(-\infty, -2] \cup [1, \infty)$

(c) $(-1, 1) \cup (3, \infty)$ (e) $(-\infty, -4) \cup (-1, 5)$

Chapter 6 Review Problem Set (page 329)

1. $2 - 2i$ **2.** $-3 - i$ **3.** $30 + 15i$ **4.** $86 - 2i$

5. $-32 + 4i$ **6.** $25 + 0i$ **7.** $\dfrac{9}{20} + \dfrac{13}{20}i$ **8.** $-\dfrac{3}{29} + \dfrac{7}{29}i$

9. One real solution with a multiplicity of 2.

10. Two nonreal complex solutions

11. Two unequal real solutions

12. Two unequal real solutions **13.** $\{0, 17\}$

14. $\{-4, 8\}$ **15.** $\left\{\dfrac{1 \pm 8i}{2}\right\}$ **16.** $\{-3, 7\}$

17. $\{-1 + \sqrt{10}\}$ **18.** $\{3 \pm 5i\}$ **19.** $\{25\}$

20. $\left\{-4, \dfrac{2}{3}\right\}$ **21.** $\{-10, 20\}$ **22.** $\left\{\dfrac{-1 \pm \sqrt{61}}{6}\right\}$

23. $\left\{\dfrac{1 \pm i\sqrt{11}}{2}\right\}$ **24.** $\left\{\dfrac{5 \pm i\sqrt{23}}{4}\right\}$

25. $\left\{\dfrac{-2 \pm \sqrt{14}}{2}\right\}$ **26.** $\{-9, 4\}$ **27.** $\{-2 \pm i\sqrt{5}\}$

28. $\{-6, 12\}$　　**29.** $\{1 \pm \sqrt{10}\}$

30. $\left\{\pm\dfrac{\sqrt{14}}{2}, \pm 2\sqrt{2}\right\}$　　**31.** $\left\{\dfrac{-3 \pm \sqrt{97}}{2}\right\}$

32. $(-\infty, -5) \cup (2, \infty)$　　**33.** $\left[\pm\dfrac{7}{2}, 3\right]$

34. $(-\infty, -6) \cup [4, \infty)$　　**35.** $\left(-\dfrac{5}{2}, -1\right)$

36. $3 + \sqrt{7}$ and $3 - \sqrt{7}$　　**37.** 20 shares at \$15 per share　**38.** 45 miles per hour and 52 miles per hour

39. 8 units　　**40.** 8 and 10　　**41.** 7 inches by 12 inches

42. 4 hours for Reena and 6 hours for Billy

43. 10 meters

Chapter 6 Test (page 331)

1. $39 - 2i$　　**2.** $-\dfrac{6}{25} - \dfrac{17}{25}i$　　**3.** $\{0, 7\}$

4. $\{-1, 7\}$　　**5.** $\{-6, 3\}$　　**6.** $\{1 - \sqrt{2}, 1 + \sqrt{2}\}$

7. $\left\{\dfrac{1 - 2i}{5}, \dfrac{1 + 2i}{5}\right\}$　　**8.** $\{-16, -14\}$

9. $\left\{\dfrac{1 - 6i}{3}, \dfrac{1 + 6i}{3}\right\}$　　**10.** $\left\{-\dfrac{7}{4}, \dfrac{6}{5}\right\}$　　**11.** $\left\{-3, \dfrac{19}{6}\right\}$

12. $\left\{-\dfrac{10}{3}, 4\right\}$　　**13.** $\{-2, 2, -4i, 4i\}$　　**14.** $\left\{-\dfrac{3}{4}, 1\right\}$

15. $\left\{\dfrac{1 - \sqrt{10}}{3}, \dfrac{1 + \sqrt{10}}{3}\right\}$　　**16.** Two equal real solutions　**17.** Two nonreal complex solutions

18. $[-6, 9]$　　**19.** $(-\infty, -2) \cup \left(\dfrac{1}{3}, \infty\right)$

20. $[-10, -6)$　　**21.** 20.8 feet

22. 29 meters　　**23.** 150 shares

24. $6\dfrac{1}{2}$ inches　　**25.** $3 + \sqrt{5}$

Cumulative Review Problem Set (page 332)

1. $\dfrac{64}{15}$　　**2.** $\dfrac{11}{3}$　　**3.** $\dfrac{1}{6}$　　**4.** $-\dfrac{44}{5}$　　**5.** -7

6. $-24a^4b^5$　　**7.** $2x^3 + 5x^2 - 7x - 12$　　**8.** $\dfrac{3x^2y^2}{8}$

9. $\dfrac{a(a + 1)}{2a - 1}$　　**10.** $\dfrac{-x + 14}{18}$　　**11.** $\dfrac{5x + 19}{x(x + 3)}$

12. $\dfrac{2}{n + 8}$　　**13.** $\dfrac{x - 14}{(5x - 2)(x + 1)(x - 4)}$

14. $y^2 - 5y + 6$　　**15.** $x^2 - 3x - 2$　　**16.** $20 + 7\sqrt{10}$

17. $2x - 2\sqrt{xy} - 12y$　　**18.** $-\dfrac{3}{8}$　　**19.** $-\dfrac{2}{3}$

20. 0.2　　**21.** $\dfrac{1}{2}$　　**22.** $\dfrac{13}{9}$　　**23.** -27

24. $\dfrac{16}{9}$　　**25.** $\dfrac{8}{27}$　　**26.** $3x(x + 3)(x^2 - 3x + 9)$

27. $(6x - 5)(x + 4)$　　**28.** $(4 + 7x)(3 - 2x)$

29. $(3x + 2)(3x - 2)(x^2 + 8)$　　**30.** $(2x - y)(a - b)$

31. $(3x - 2y)(9x^2 + 6xy + 4y^2)$　　**32.** $\left\{-\dfrac{12}{7}\right\}$

33. $\{150\}$　　**34.** $\{25\}$　　**35.** $\{0\}$　　**36.** $\{-2, 2\}$

37. $\{-7\}$　　**38.** $\left\{-6, \dfrac{4}{3}\right\}$　　**39.** $\left\{\dfrac{5}{4}\right\}$　　**40.** $\{3\}$

41. $\left\{\dfrac{4}{5}, 1\right\}$　　**42.** $\left\{-\dfrac{10}{3}, 4\right\}$　　**43.** $\left\{\dfrac{1}{4}, \dfrac{2}{3}\right\}$

44. $\left\{-\dfrac{3}{2}, 3\right\}$　　**45.** $\left\{\dfrac{1}{5}\right\}$　　**46.** $\left\{\dfrac{5}{7}\right\}$　　**47.** $\{-2, 2\}$

48. $\{0\}$　　**49.** $\{-6, 19\}$

50. $\left\{-\dfrac{3}{4}, \dfrac{2}{3}\right\}$　　**51.** $\{1 \pm 5i\}$　　**52.** $\{1, 3\}$

53. $\left\{-4, \dfrac{1}{3}\right\}$　　**54.** $\{-2 \pm 4i\}$　　**55.** $\left\{\dfrac{1 \pm \sqrt{33}}{4}\right\}$

56. $(-\infty, -2]$　　**57.** $\left(-\infty, \dfrac{19}{5}\right)$　　**58.** $\left(\dfrac{1}{4}, \infty\right)$

59. $(-2, 3)$　　**60.** $\left(-\infty, -\dfrac{13}{3}\right) \cup (3, \infty)$

61. $(-\infty, 29]$　　**62.** $[-2, 4]$　　**63.** $(-\infty, -5) \cup \left(\dfrac{1}{3}, \infty\right)$

64. $(-\infty, -2] \cup (7, \infty)$　　**65.** $(-3, 4)$　　**66.** 6 liters

67. \$900 and \$1350　　**68.** 12 inches by 17 inches

69. 5 hours　　**70.** 7 golf balls　　**71.** 12 minutes

72. 7%　　**73.** 15 chairs per row　　**74.** 140 shares

CHAPTER 7

Problem Set 7.1 (page 348)

1. (a) III　　**(c)** II　　**3. (a)** I and III　　**(c)** II and III

5. $(-3, -1); (3, 1); (3, -1)$　　**7.** $(7, 2); (-7, -2); (-7, 2)$

9. $(5, 0); (-5, 0); (-5, 0)$　　**11.** x axis　　**13.** y axis

15. x axis, y axis, and origin　　**17.** x axis　　**19.** None

21. Origin　　**23.** y axis

25.　　**27.**

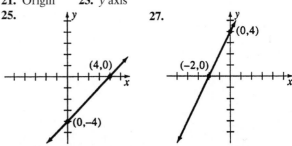

29.

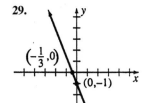

$\left(-\frac{1}{3},0\right)$ (0,−1)

31.

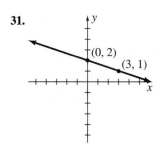

(0, 2) (3, 1)

45.

(−1,−3) (1,−3)

47.

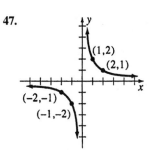

(1,2) (2,1) (−2,−1) (−1,−2)

33.

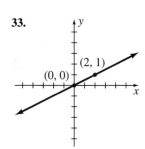

(2, 1) (0, 0)

35.

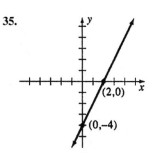

(2,0) (0,−4)

49.

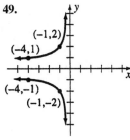

(−1,2) (−4,1) (−4,−1) (−1,−2)

51.

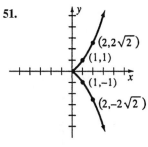

(2,2√2) (1,1) (1,−1) (2,−2√2)

37.

(2, 0) (0, −1)

39.

(−1,3) (1,3) (0,2)

53.

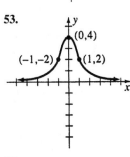

(0,4) (−1,−2) (1,2)

55.

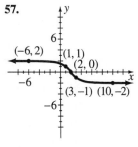

(−1, 1) (1, 1)

57.

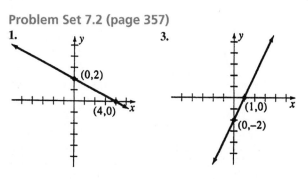

6 (−6, 2) (1, 1) (2, 0) −6 (3,−1) (10,−2) −6

Problem Set 7.2 (page 357)

1.

(0,2) (4,0)

3.

(1,0) (0,−2)

41.

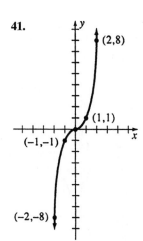

(2,8) (1,1) (−1,−1) (−2,−8)

43.

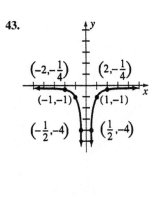

$\left(-2,-\frac{1}{4}\right)$ $\left(2,-\frac{1}{4}\right)$ (−1,−1) (1,−1) $\left(-\frac{1}{2},-4\right)$ $\left(\frac{1}{2},-4\right)$

5.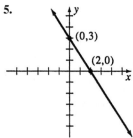

7.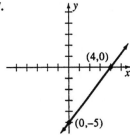

25.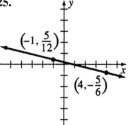

27.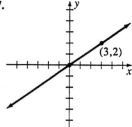

9.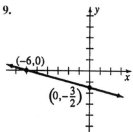

11.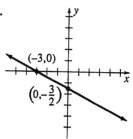

29.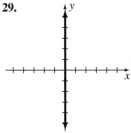

31.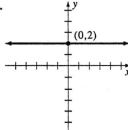

13.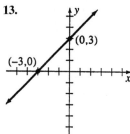

15.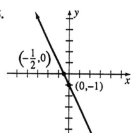

33.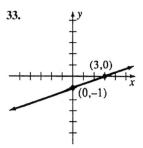

35.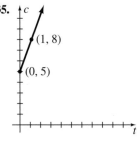

17.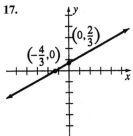

19.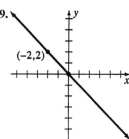

37.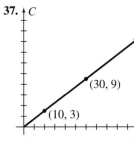

39.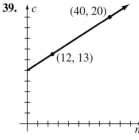

21.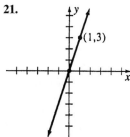

23.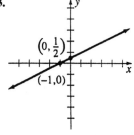

45.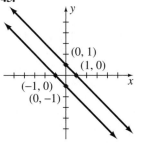

47.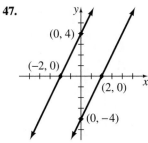

Problem Set 7.3 (page 364)

1.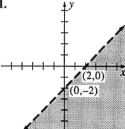

3.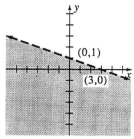

21.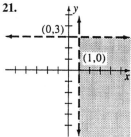

23.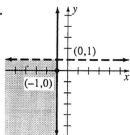

5.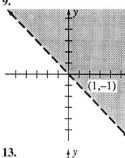

7.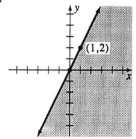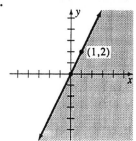

27.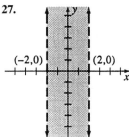

29.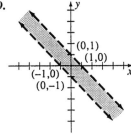

9.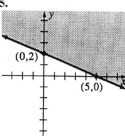

11.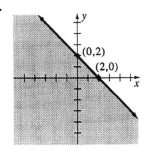

Problem Set 7.4 (page 374)

1. 15   **3.** $\sqrt{13}$ **5.** $3\sqrt{2}$ **7.** $3\sqrt{5}$

9. 6   **11.** $3\sqrt{10}$

13. The lengths of the sides are 10, $5\sqrt{5}$, and 5. Because $10^2 + 5^2 = (5\sqrt{5})^2$, it is a right triangle.

15. The distances between (3, 6), and (7, 12), between (7, 12) and (11, 18), and between (11, 18) and (15, 24) are all $2\sqrt{13}$ units.

17. $\dfrac{4}{3}$ **19.** $-\dfrac{7}{3}$ **21.** -2 **23.** $\dfrac{3}{5}$ **25.** 0

27. $\dfrac{1}{2}$ **29.** 7   **31.** -2

33–39. Answers will vary.

13.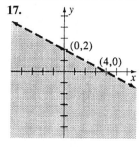

15.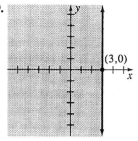

17.

19.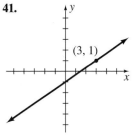

41.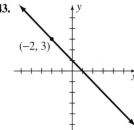

43.

45.

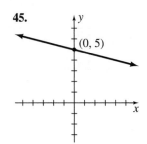

47.

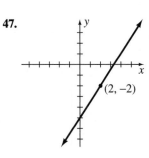

53.

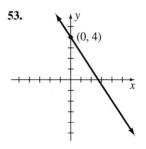

55.

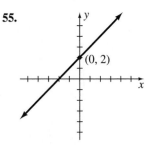

49. $-\dfrac{2}{3}$ **51.** $\dfrac{1}{2}$ **53.** $\dfrac{4}{7}$

55. 0   **57.** -5 **59.** 105.6 feet

61. 8.1%   **63.** 19 centimeters

69. (a) $(3, 5)$ **(c)** $(2, 5)$ **(e)** $\left(\dfrac{17}{8}, -7\right)$

57.

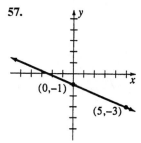

59.

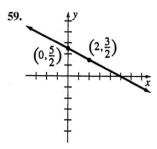

Problem Set 7.5 (page 386)

1. $x - 2y = -7$ **3.** $3x - y = -10$

5. $3x + 4y = -15$ **7.** $5x - 4y = 28$ **9.** $x - y = 1$

11. $5x - 2y = -4$ **13.** $x + 7y = 11$

15. $x + 2y = -9$ **17.** $7x - 5y = 0$

19. $y = \dfrac{3}{7}x + 4$ **21.** $y = 2x - 3$

23. $y = -\dfrac{2}{5}x + 1$ **25.** $y = 0(x) - 4$

27. $2x - y = 4$ **29.** $5x + 8y = -15$

31. $x + 0(y) = 2$ **33.** $0(x) + y = 6$

35. $x + 5y = 16$ **37.** $4x - 7y = 0$

39. $x + 2y = 5$ **41.** $3x + 2y = 0$

43. $m = -3$ and $b = 7$ **45.** $m = -\dfrac{3}{2}$ and $b = \dfrac{9}{2}$

47. $m = \dfrac{1}{5}$ and $b = -\dfrac{12}{5}$

61.

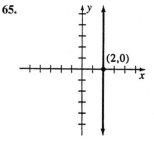

63.

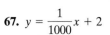

65.

49.

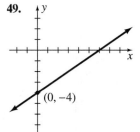

51.

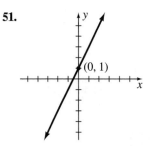

67. $y = \dfrac{1}{1000}x + 2$ **69.** $y = \dfrac{9}{5}x + 32$

77. (a) $2x - y = 1$ **(b)** $5x - 6y = 29$

(c) $x + y = 2$ **(d)** $3x - 2y = 18$

Chapter 7 Review Problem Set (page 391)

1. (a) $\dfrac{6}{5}$ **(b)** $-\dfrac{2}{3}$ **2.** 5   **3.** -1

4. (a) $m = -4$ **(b)** $m = \dfrac{2}{7}$

5. 5, 10, and $\sqrt{97}$ **6. (a)** $2\sqrt{10}$ **(b)** $\sqrt{58}$

8. $7x + 4y = 1$ **9.** $3x + 7y = 28$ **10.** $2x - 3y = 16$

11. $x - 2y = -8$ **12.** $2x - 3y = 14$

13. $x - y = -4$ **14.** $x + y = -2$ **15.** $4x + y = -29$

16.

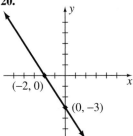

17.

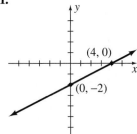

18.

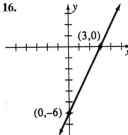

19.

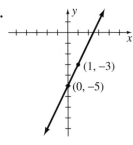

20.

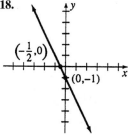

21.

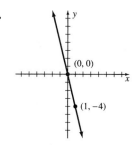

22.

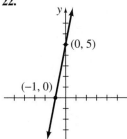

23.

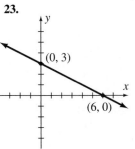

24.

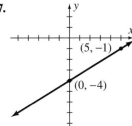

25.

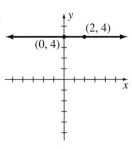

26.

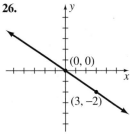

27.

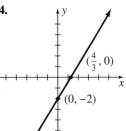

28.

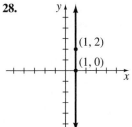

29.

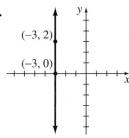

30.

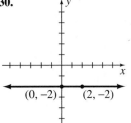

31.

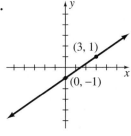

32.

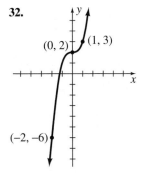

33.

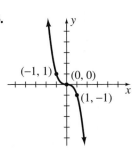

34.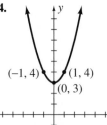

35.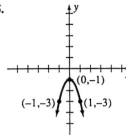

5. $y = -\dfrac{1}{6}x + \dfrac{4}{3}$ **6.** $5x + 2y = -18$

7. $6x + y = 31$ **8.** Origin symmetry

9. x axis, y axis, and origin symmetry

10. x axis symmetry   **11.** $\dfrac{7}{2}$ **12.** $\dfrac{9}{4}$ **13.** $\dfrac{10}{9}$

14. $-\dfrac{5}{8}$ **15.** 480 feet   **16.** 6.7%

17. 43 centimeters

36.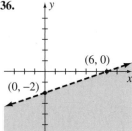

37.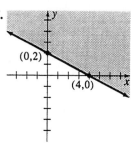

18.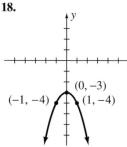

19.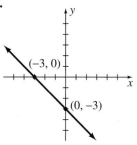

38.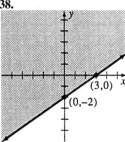

39.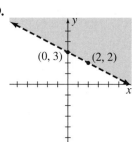

20.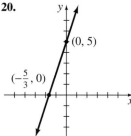

21.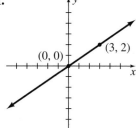

40.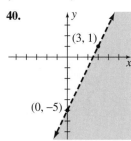

41.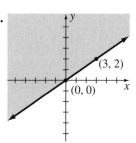

22.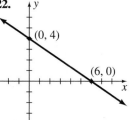

23.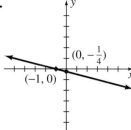

42. 316.8 feet   **43.** 8 inches   **44.** $-\dfrac{5}{3}$ **45.** $-\dfrac{4}{5}$

46. $y = \dfrac{3}{200}x - 600$ **47.** $y = \dfrac{1}{5}x - 20$

48. $y = 8x$ **49.** $y = 300x - 150$

50. (a) y axis   **(b)** Origin   **(c)** Origin   **(d)** x axis

Chapter 7 Test (page 393)

1. $-\dfrac{6}{5}$ **2.** $\dfrac{3}{7}$ **3.** $\sqrt{58}$ **4.** $3x + 2y = 2$

24.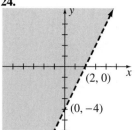

25.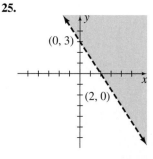

CHAPTER 8

Problem Set 8.1 (page 404)

1.
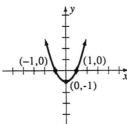

3.
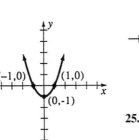

5.
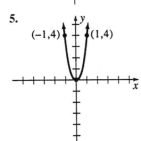

7.
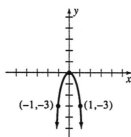

9.
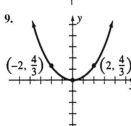

11.
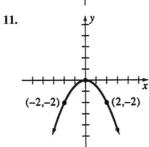

13.
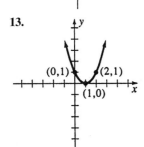

15.
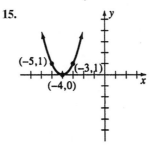

17.
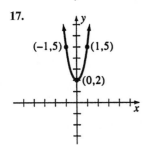

19.
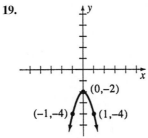

21.
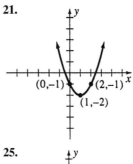

23.
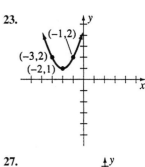

25.
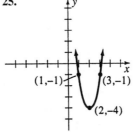

27.
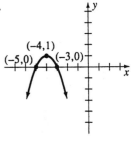

29.
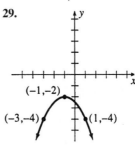

Problem Set 8.2 (page 412)

1.

3.
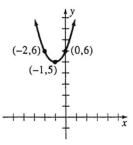

5.
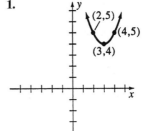

7.
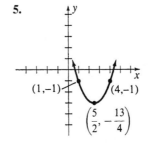

9.
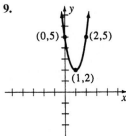

11.
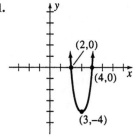

35.
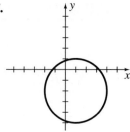

37.
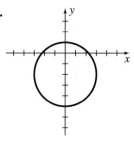

13.
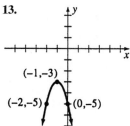

15.
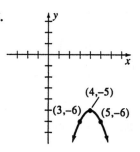

39.
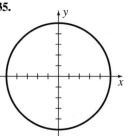

41.
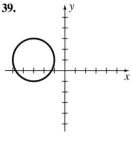

17.
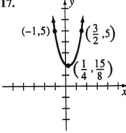

19.
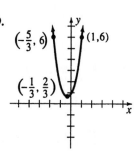

43.
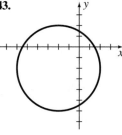

45. $x^2 + y^2 - 6x - 10y + 9 = 0$

47. $x^2 + y^2 + 8x - 2y - 47 = 0$

49. $x^2 + y^2 + 4x + 12y + 22 = 0$ **51.** $x^2 + y^2 - 20 = 0$

53. $x^2 + y^2 - 10x + 16y - 7 = 0$

55. $x^2 + y^2 - 8y = 0$ **57.** $x^2 + y^2 + 8x - 6y = 0$

63. (a) $(1, 4), r = 3$ **(c)** $(-6, -4), r = 8$
    (e) $(0, 6), r = 9$

21.
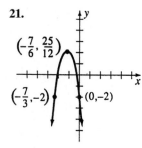

23. $(1, 3), r = 4$

25. $(-3, -5), r = 4$ **27.** $(0, 0), r = \sqrt{10}$

29. $(8, -3), r = \sqrt{2}$ **31.** $(-3, 4), r = 5$

33. $\left(-\dfrac{1}{2}, 4\right), r = 2\sqrt{2}$

Problem Set 8.3 (page 418)

1.

3.

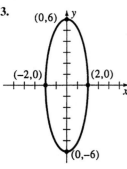

(0,6)

(−2,0)   (2,0)

(0,−6)

5.

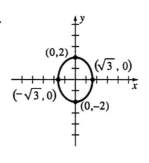

(0,2)

($\sqrt{3}$, 0)

(−$\sqrt{3}$, 0)

(0,−2)

Problem Set 8.4 (page 426)

1. $(4, 0), (-4, 0), \quad y = \frac{1}{3}x$ and $y = -\frac{1}{3}x$

3. $(0, 6), (0, -6)$ $y = 3x$ and $y = -3x$

5. $\left(\frac{2}{5}, 0\right), \left(-\frac{2}{5}, 0\right), \quad y = \frac{5}{3}x$ and $y = -\frac{5}{3}x$

7. $y = -x - 5$ and $y = x - 1$

9. $y = \frac{3}{2}x - \frac{9}{2}$ and $y = -\frac{3}{2}x - \frac{3}{2}$

7.

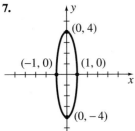

(0, 4)

(−1, 0)   (1, 0)

(0, −4)

9.

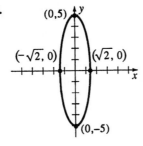

(0,5)

(−$\sqrt{2}$, 0)   ($\sqrt{2}$, 0)

(0,−5)

11.

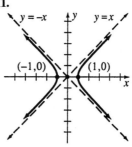

$y = -x$ $y = x$

(−1,0)   (1,0)

13.

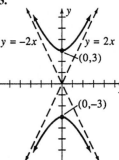

$y = -2x$ $y = 2x$

(0,3)

(0,−3)

11.

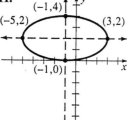

(−1,4)

(−5,2)   (3,2)

(−1,0)

13.

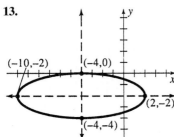

(−10,−2)   (−4,0)

(2,−2)

(−4,−4)

15.

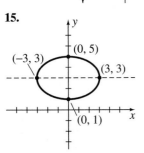

(0, 5)

(−3, 3)   (3, 3)

(0, 1)

15.

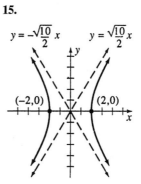

$y = -\frac{\sqrt{10}}{2}x$ $y = \frac{\sqrt{10}}{2}x$

(−2,0)   (2,0)

17.

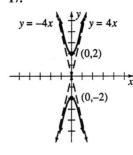

$y = -4x$ $y = 4x$

(0,2)

(0,−2)

19.

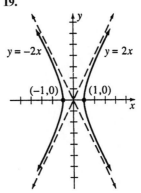

$y = -2x$ $y = 2x$

(−1,0)   (1,0)

21.

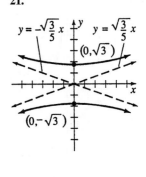

$y = -\sqrt{\frac{3}{5}}x$ $y = \sqrt{\frac{3}{5}}x$

(0,$\sqrt{3}$)

(0,−$\sqrt{3}$)

23.

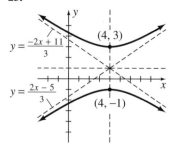

$y = \dfrac{-2x + 11}{3}$

(4, 3)

$y = \dfrac{2x - 5}{3}$

(4, −1)

25.

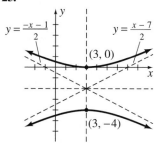

$y = \dfrac{-x - 1}{2}$

$y = \dfrac{x - 7}{2}$

(3, 0)

(3, −4)

27.

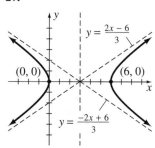

$y = \dfrac{2x - 6}{3}$

(0, 0)   (6, 0)

$y = \dfrac{-2x + 6}{3}$

29.

(a)

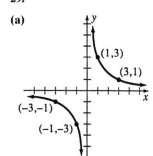

(1, 3)

(3, 1)

(−3, −1)

(−1, −3)

(c)

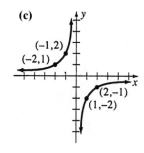

(−1, 2)

(−2, 1)

(2, −1)

(1, −2)

31. (a) Origin

(c)

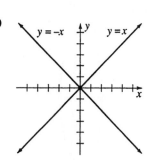

$y = -x$ $y = x$

Chapter 8 Review Problem Set (page 428)

1. $(0, 6)$ **2.** $(0, -8)$ **3.** $(-3, -1)$
4. $(7, 5)$ **5.** $(6, -8)$ **6.** $(-4, -9)$
7. $x^2 - 4x + y^2 + 12y + 15 = 0$
8. $x^2 + 8x + y^2 + 16y + 68 = 0$
9. $x^2 + y^2 - 10y = 0$
10. $(-7, 4)$ and $r = 7$
11. $(-8, 0)$ and $r = 5$
12. $(6, -8)$ and $r = 10$
13. $(0, 0)$ and $r = 2\sqrt{6}$ **14.** 16 and 4
15. 8 and 6   **16.** 10 and 4
17. $2\sqrt{14}$ and 4   **18.** 6 and 2

19. 6 and 4   **20.** $y = \pm\dfrac{1}{3}x$ **21.** $y = \pm 2x$

22. $y = \pm\dfrac{5}{3}x$ **23.** $y = \pm\dfrac{1}{2}x$

24. $y = -\dfrac{5}{2}x - 2$ and $y = \dfrac{5}{2}x + 8$

25. $y = -\dfrac{1}{6}x + \dfrac{25}{6}$ and $y = \dfrac{1}{6}x + \dfrac{23}{6}$

26.

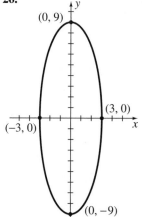

(0, 9)

(−3, 0)   (3, 0)

(0, −9)

27.

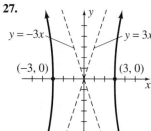

$y = -3x$ $y = 3x$

(−3, 0)   (3, 0)

28.

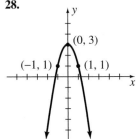

29.

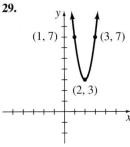

30.

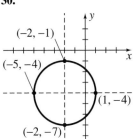

31.

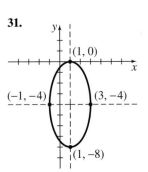

32.

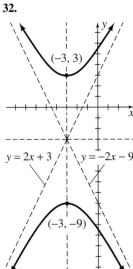

33.

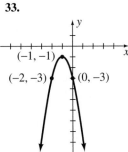

34.

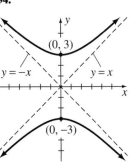

35.

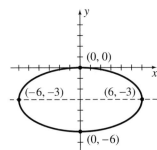

36.

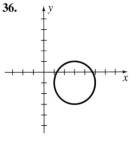

37.

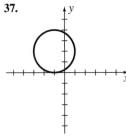

38.

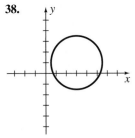

39.

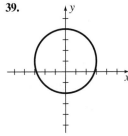

Chapter 8 Test (page 430)

1. $(0, 9)$ **2.** $(1, 7)$ **3.** $(-4, -2)$

4. $(3, 0)$ **5.** $x^2 + 8x + y^2 - 29 = 0$

6. $x^2 - 4x + y^2 - 16y + 59 = 0$

7. $x^2 + 6x + y^2 + 8y = 0$

8. $(0, 0)$ and $r = 4\sqrt{2}$

9. $(6, -4)$ and $r = 7$ **10.** $(-5, -1)$ and $r = 8$

11. 8 units   **12.** 6 units   **13.** 6 units

14. 4 units   **15.** $y = \pm 4x$ **16.** $y = \pm\dfrac{5}{4}x$

17. $y = \dfrac{1}{5}x - \dfrac{6}{5}$ and $y = -\dfrac{1}{5}x - \dfrac{4}{5}$

18.

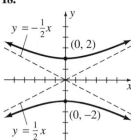

19.

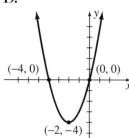

20.

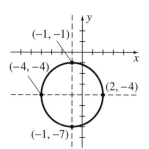

21.

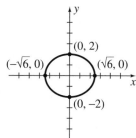

22.

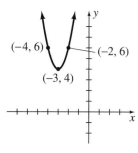

23.

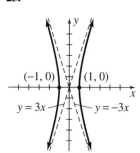

24.

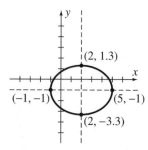

25.

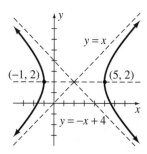

CHAPTER 9

Problem Set 9.1 (page 438)

1. $D = \{1, 2, 3, 4\}$, $R = \{5, 8, 11, 14\}$ It is a function.

3. $D = \{0, 1\}$, $R = \{-2\sqrt{6}, -5, 5, 2\sqrt{6}\}$ It is not a function.

5. $D = \{1, 2, 3, 4, 5\}$, $R = \{2, 5, 10, 17, 26\}$ It is a function.

7. $D = \{$all reals$\}$ or D: $(-\infty, \infty)$, $R = \{$all reals$\}$ or R: $(-\infty, \infty)$, yes

9. $D = \{$all reals$\}$ or D: $(-\infty, \infty)$, $R = \{y|y \geq 0\}$ or R: $[0, \infty)$, yes

11. $\{$all reals$\}$ or $(-\infty, \infty)$

13. $\{x|x \neq 1\}$ or $(-\infty, 1) \cup (1, \infty)$

15. $\left\{x|x \neq \dfrac{3}{4}\right\}$ or $\left(-\infty, \dfrac{3}{4}\right) \cup \left(\dfrac{3}{4}, \infty\right)$

17. $\{x|x \neq -1, x \neq 4\}$ or $(-\infty, -1) \cup (-1, 4) \cup (4, \infty)$

19. $\{x|x \neq -8, x \neq 5\}$ or $(-\infty, -8) \cup (-8, 5) \cup (5, \infty)$

21. $\{x|x \neq -6, x \neq 0\}$ or $(-\infty, -6) \cup (-6, 0) \cup (0, \infty)$

23. $\{$all reals$\}$ or $(-\infty, \infty)$

25. $\{t|t \neq -2, t \neq 2\}$ or $(-\infty, -2) \cup (-2, 2) \cup (2, \infty)$

27. $\{x|x \geq -4\}$ or $[-4, \infty)$

29. $\left\{s|s \geq \dfrac{5}{4}\right\}$ or $\left[\dfrac{5}{4}, \infty\right)$

31. $\{x|x \leq -4$ or $x \geq 4\}$ or $(-\infty, -4] \cup [4, \infty)$

33. $\{x|x \leq -3$ or $x \geq 6\}$ or $(-\infty, -3] \cup [6, \infty)$

35. $\{x|-1 \leq x \leq 1\}$ or $[-1, 1]$

37. $f(0) = -2, f(2) = 8, f(-1) = -7, f(-4) = -22$

39. $f(-2) = -\dfrac{7}{4}, f(0) = -\dfrac{3}{4}, f\left(\dfrac{1}{2}\right) = -\dfrac{1}{2}, f\left(\dfrac{2}{3}\right) = -\dfrac{5}{12}$

41. $g(-1) = 0$; $g(2) = -9$; $g(-3) = 26$; $g(4) = 5$

43. $h(-2) = -2$; $h(-3) = -11$; $h(4) = -32$; $h(5) = -51$

45. $f(3) = \sqrt{7}$; $f(4) = 3$; $f(10) = \sqrt{21}$; $f(12) = 5$

47. $f(1) = -1$; $f(-1) = -2$; $f(3) = -\dfrac{2}{3}$; $f(-6) = \dfrac{4}{3}$

49. $f(-2) = 27$; $f(3) = 42$; $g(-4) = -37$; $g(6) = -17$

51. $f(-2) = 5$; $f(3) = 8$; $g(-4) = -3$; $g(5) = -4$

53. -3 **55.** $-2a - h$ **57.** $4a - 1 + 2h$

59. $-8a - 7 - 4h$

61. $h(1) = 48$; $h(2) = 64$; $h(3) = 48$; $h(4) = 0$

63. $C(75) = \$74$; $C(150) = \$98$; $C(225) = \$122$; $C(650) = \$258$

65. $I(0.11) = 55$; $I(0.12) = 60$; $I(0.135) = 67.5$; $I(0.15) = 75$

Problem Set 9.2 (page 451)

1.

3.

5.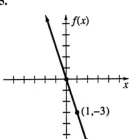

7.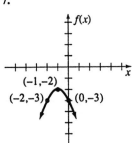

9.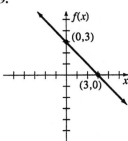

11.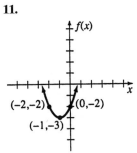

13.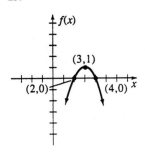

15.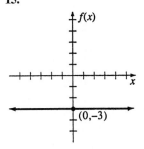

17.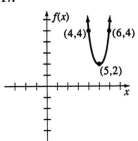

19.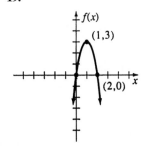

21.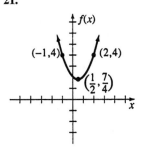

23.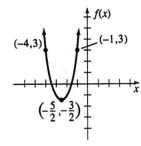

25.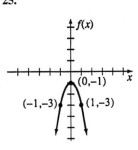

27.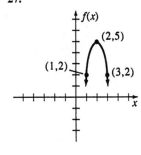

29.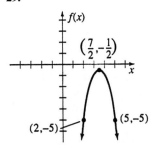

31. (a) $.42   **(c)** Answers will vary.
33. $26; $30.50; $50; $60.50
35. $2.10; $4.55; $20.72; $29.40; $33.88
37. $f(p) = 0.8p;$ $7.60; $12; $60; $10; $600
39. 80 items   **41.** 5 and 25
43. 60 meters by 60 meters
45. 1100 subscribers at $13.75 per month

Problem Set 9.3 (page 461)

1.

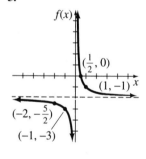

3.
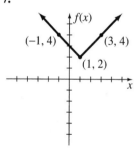

5.
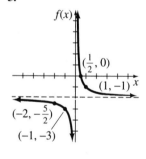

7.
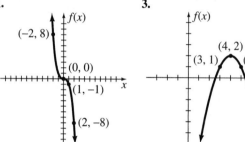

9.
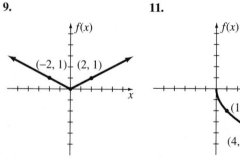

11.
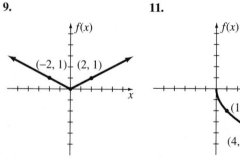

13.
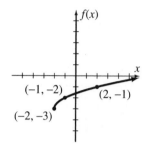

15.
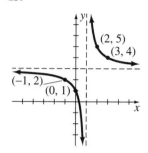

17.
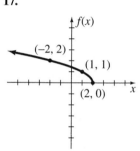

19.
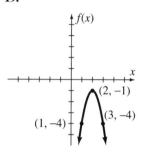

21.
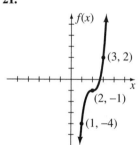

23.
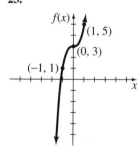

25.
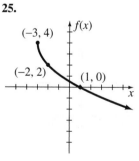

27.
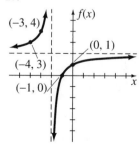

29.

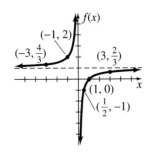

31.

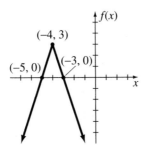

33.

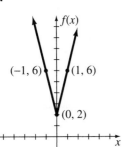

35. (a)

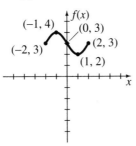

35. (c)

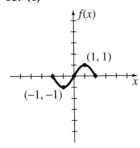

19. $(f \circ g)(x) = 2x^2 + 11x + 17, D = \{\text{all reals}\}$ or $D: (-\infty, \infty)$
$(g \circ f)(x) = -2x^2 + x + 1, D = \{\text{all reals}\}$ or $D: (-\infty, \infty)$

21. $(f \circ g)(x) = \dfrac{3}{4x - 9}, D = \left\{x \mid x \neq \dfrac{9}{4}\right\}$ or

$D: \left(-\infty, \dfrac{9}{4}\right) \cup \left(\dfrac{9}{4}, \infty\right)$

$(g \circ f)(x) = \dfrac{12 - 9x}{x}, D = \{x \mid x \neq 0\}$ or

$D: (-\infty, 0) \cup (0, \infty)$

23. $(f \circ g)(x) = \sqrt{5x + 4}, D = \left\{x \mid x \geq -\dfrac{4}{5}\right\}$

or $D: \left[-\dfrac{4}{5}, \infty\right)$

$(g \circ f)(x) = 5\sqrt{x + 1} + 3, D = \{x \mid x \geq -1\}$ or
$D: [-1, \infty)$

25. $(f \circ g)(x) = x - 4, D = \{x \mid x \neq 4\}$ or
$D: (-\infty, 4) \cup (4, \infty)$

$(g \circ f)(x) = \dfrac{x}{1 - 4x}, D = \left\{x \mid x \neq 0 \text{ and } x \neq \dfrac{1}{4}\right\}$ or

$D: (-\infty, 0) \cup \left(0, \dfrac{1}{4}\right) \cup \left(\dfrac{1}{4}, \infty\right)$

27. $(f \circ g)(x) = \dfrac{2\sqrt{x}}{x}, D = \{x \mid x > 0\}$ or $D: (0, \infty)$

$(g \circ f)(x) = \dfrac{4\sqrt{x}}{x}, D = \{x \mid x > 0\}$ or $D: (0, \infty)$

29. $(f \circ g)(x) = \dfrac{3x + 3}{2}, D = \{x \mid x \neq -1\}$ or

$D: (-\infty, -1) \cup (-1, \infty)$

$(g \circ f)(x) = \dfrac{2x}{2x + 3}, D = \left\{x \mid x \neq 0 \text{ and } x \neq -\dfrac{3}{2}\right\}$ or

$D: \left(-\infty, -\dfrac{3}{2}\right) \cup \left(-\dfrac{3}{2}, 0\right) \cup (0, \infty)$

Problem Set 9.4 (page 467)

1. 124 and -130 **3.** 323 and 257   **5.** $\dfrac{1}{2}$ and -1

7. Undefined and undefined   **9.** $\sqrt{7}$ and 0
11. 37 and 27   **13.** $(f \circ g)(x) = 15x - 3,$
$D = \{\text{all reals}\}$ or $D: (-\infty, \infty)$ $(g \circ f)(x) = -15x - 1,$
$D = \{\text{all reals}\}$ or $D: (-\infty, \infty)$
15. $(f \circ g)(x) = -14x - 7, D = \{\text{all reals}\}$ or $D: (-\infty, \infty)$
$(g \circ f)(x) = -14x + 11, D = \{\text{all reals}\}$ or $D: (-\infty, \infty)$
17. $(f \circ g)(x) = 3x^2 + 11, D = \{\text{all reals}\}$ or $D: (-\infty, \infty)$
$(g \circ f)(x) = 9x^2 + 12x + 7, D = \{\text{all reals}\}$ or $D: (-\infty, \infty)$

Problem Set 9.5 (page 474)

1. Not a function   **3.** Function   **5.** Function
7. Function   **9.** One-to-one function
11. Not a one-to-one function
13. Not a one-to-one function
15. One-to-one function
17. Domain of f: $\{1, 2, 3, 4\}$
Range of f: $\{3, 6, 11, 18\}$
f^{-1}: $\{(3, 1), (6, 2), (11, 3), (18, 4)\}$
Domain of f^{-1}: $\{3, 6, 11, 18\}$
Range of f^{-1}: $\{1, 2, 3, 4\}$

19. Domain of f: $\{-2, -1, 0, 5\}$
Range of f: $\{-1, 1, 5, 10\}$
f^{-1}: $\{(-1, -2), (1, -1), (5, 0), (10, 5)\}$
Domain of f^{-1}: $\{-1, 1, 5, 10\}$
Range of f^{-1}: $\{-2, -1, 0, 5\}$

21. $f^{-1}(x) = \dfrac{x + 4}{5}$ **23.** $f^{-1}(x) = \dfrac{1 - x}{2}$

25. $f^{-1}(x) = \dfrac{5}{4}x$ **27.** $f^{-1}(x) = 2x - 8$

29. $f^{-1}(x) = \dfrac{15x + 6}{5}$ **31.** $f^{-1}(x) = \dfrac{x - 4}{9}$

33. $f^{-1}(x) = \dfrac{-x - 4}{5}$ **35.** $f^{-1}(x) = \dfrac{-3x + 21}{2}$

37. $f^{-1}(x) = \dfrac{3}{4}x + \dfrac{3}{16}$ **39.** $f^{-1}(x) = -\dfrac{7}{3}x - \dfrac{14}{9}$

41. $f^{-1}(x) = \dfrac{1}{4}x$ **43.** $f^{-1}(x) = -3x$

45. $f^{-1}(x) = \dfrac{x + 3}{3}$ **47.** $f^{-1}(x) = \dfrac{-x - 4}{2}$

49. $f^{-1}(x) = \sqrt{x}, x \geq 0$

53. (a) $f^{-1}(x) = \dfrac{x - 1}{2}$ **(c)** $f^{-1}(x) = \dfrac{-x + 5}{4}$

(e) $f^{-1}(x) = \dfrac{1}{2}x$

Problem Set 9.6 (page 482)

1. $y = \dfrac{k}{x^2}$ **3.** $C = \dfrac{kg}{t^3}$ **5.** $V = kr^3$ **7.** $S = ke^2$

9. $V = khr^2$ **11.** $\dfrac{2}{3}$ **13.** -4 **15.** $\dfrac{1}{3}$

17. -2 **19.** 2   **21.** 5   **23.** 9   **25.** 9

27. $\dfrac{1}{6}$ **29.** 112   **31.** 12 cubic centimeters

33. 28   **35.** 2 seconds   **37.** 12 ohms

39. (a) \$210   **(c)** \$1050   **41.** 3560.76 cubic meters

43. 0.048

Chapter 9 Review Problem Set (page 486)

1. $D = \{1, 2, 4\}$ **2.** $D = \{x \mid x \neq 5\}$ or
D: $(-\infty, 5) \cup (5, \infty)$

3. $D = \{x \mid x \neq 0 \text{ and } x \neq -4\}$ or
D: $(-\infty, -4) \cup (-4, 0) \cup (0, \infty)$

4. $D = \{x \mid x \geq 5 \text{ or } x \leq -5\}$ or D: $(-\infty, -5] \cup [5, \infty)$

5. $f(2) = -1, f(-3) = 14; f(a) = a^2 - 2a - 1$

6. $4a + 2h + 1$

7.
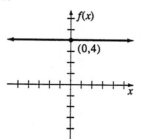

8.
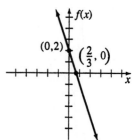

9.
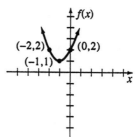

10.
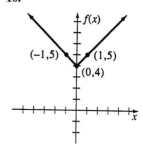

11.
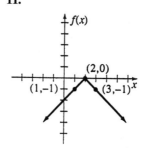

12.
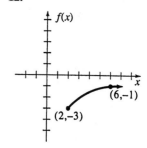

13.
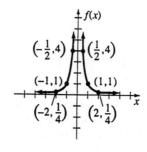

14.
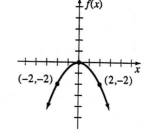

15.

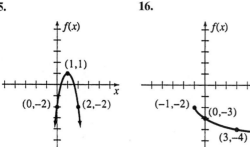

16.

17. (a) $(-5, -28); x = -5$ **(b)** $\left(-\dfrac{7}{6}, \dfrac{67}{2}\right); x = -\dfrac{7}{2}$

18. $(f \circ g)(x) = 6x - 11$ and $(g \circ f)(x) = 6x - 13$
19. $(f \circ g)(x) = x^2 - 2x - 1$ and $(g \circ f)(x) = x^2 - 10x + 27$
20. $(f \circ g)(x) = 4x^2 - 20x + 20$ and $(g \circ f)(x) = -2x^2 + 15$
21. $f^{-1}(x) = \dfrac{x + 1}{6}$ **22.** $f^{-1}(x) = \dfrac{3x - 21}{2}$
23. $f^{-1}(x) = \dfrac{-35x - 10}{21}$ **24.** $k = 9$ **25.** $y = 120$
26. 128 pounds **27.** 20 and 20 **28.** 3 and 47
29. 25 students **30.** 600 square inches
31. \$.72 **32.** $f(x) = 0.7x$; \$45.50; \$33.60; \$10.85

Chapter 9 Test (page 488)

1. $D = \left\{x \mid x \ne -4 \text{ and } x \ne \dfrac{1}{2}\right\}$ or

$D: (-\infty, -4) \cup \left(-4, \dfrac{1}{2}\right) \cup \left(\dfrac{1}{2}, \infty\right)$

2. $D = \left\{x \mid x \le \dfrac{5}{3}\right\}$ or $D: \left(-\infty, \dfrac{5}{3}\right]$ **3.** $\dfrac{11}{6}$

4. 11 **5.** $(-6, 3)$ **6.** $6a + 3h + 2$
7. $(f \circ g)(x) = -21x - 2$

8. $(g \circ f)(x) = 8x^2 + 38x + 48$ **9.** $(f \circ g)(x) = \dfrac{3x}{2 - 2x}$

10. $f^{-1}(x) = \dfrac{x + 9}{5}$ **11.** $f^{-1}(x) = \dfrac{-x - 6}{3}$

12. $f^{-1}(x) = \dfrac{15x + 9}{10}$ **13.** -4 **14.** 15

15. 6 and 54 **16.** \$96
17. The graph of $f(x) = (x - 6)^3 - 4$ is the graph of $f(x) = x^3$ translated 6 units to the right and 4 units downward.
18. The graph of $f(x) = -|x| + 8$ is the graph of $f(x) = |x|$ reflected across the x axis and translated 8 units upward.

19. The graph of $f(x) = -\sqrt{x + 5} + 7$ is the graph of $f(x) = \sqrt{x}$ reflected across the x axis and translated 5 units to the left and 7 units upward.

20.

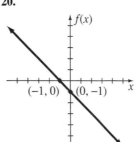

21.

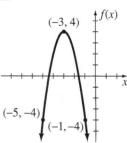

22.

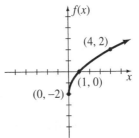

23.

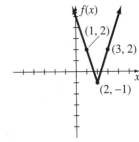

24.

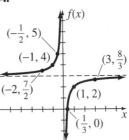

25.

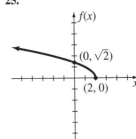

CHAPTER 10

Problem Set 10.1 (page 497)
1. $\{(3, 2)\}$ **3.** $\{(-2, 1)\}$ **5.** Dependent
7. $\{(4, -3)\}$ **9.** Inconsistent **11.** $\{(8, 12)\}$
13. $\{(-4, -6)\}$ **15.** $\{(-9, 3)\}$ **17.** $\{(1, 3)\}$

19. $\left\{\left(5, \dfrac{3}{2}\right)\right\}$ **21.** $\left\{\left(\dfrac{9}{5}, -\dfrac{7}{25}\right)\right\}$ **23.** $\{(-2, -4)\}$

25. $\{(5, 2)\}$ **27.** $\{(-4, -8)\}$ **29.** $\left\{\left(\dfrac{11}{20}, \dfrac{7}{20}\right)\right\}$

31. $\{(-1, 5)\}$ **33.** $\left\{\left(-\dfrac{3}{4}, -\dfrac{6}{5}\right)\right\}$

35. $\left\{\left(\dfrac{5}{27}, -\dfrac{26}{27}\right)\right\}$

37. $2000 at 7% and $8000 at 8%   **39.** 92
41. 34 and 97   **43.** 42 females
45. 20 inches by 27 inches
47. 60 five-dollar bills and 40 ten-dollar bills
49. 2500 student tickets and 500 nonstudent tickets

Problem Set 10.2 (page 506)
1. $\{(4, -3)\}$ **3.** $\{(-1, -3)\}$ **5.** $\{(-8, 2)\}$
7. $\{(-4, 0)\}$ **9.** $\{(1, -1)\}$ **11.** Inconsistent
13. $\left\{\left(-\dfrac{1}{11}, \dfrac{4}{11}\right)\right\}$ **15.** $\left\{\left(\dfrac{3}{2}, -\dfrac{1}{3}\right)\right\}$ **17.** $\{(4, -9)\}$
19. $\{(7, 0)\}$ **21.** $\{(7, 12)\}$ **23.** $\left\{\left(\dfrac{7}{11}, \dfrac{2}{11}\right)\right\}$
25. Inconsistent   **27.** $\left\{\left(\dfrac{51}{31}, -\dfrac{32}{31}\right)\right\}$ **29.** $\{(-2, -4)\}$
31. $\left\{\left(-1, -\dfrac{14}{3}\right)\right\}$ **33.** $\{(-6, 12)\}$ **35.** $\{(2, 8)\}$
37. $\{(-1, 3)\}$ **39.** $\{(16, -12)\}$ **41.** $\left\{\left(-\dfrac{3}{4}, \dfrac{3}{2}\right)\right\}$
43. $\{(5, -5)\}$
45. 5 gallons of 10% solution and 15 gallons of 20% solution
47. $1 for a tennis ball and $2 for a golf ball
49. 40 double rooms and 15 single rooms   **51.** 9 feet
53. $\dfrac{3}{4}$ **55.** 18 centimeters by 24 centimeters
57. 8 feet   **61. (a)** Consistent   **(c)** Consistent
(e) Dependent   **(g)** Inconsistent

Problem Set 10.3 (page 516)
1. $\{(-2, 5, 2)\}$ **3.** $\{(4, -1, -2)\}$ **5.** $\{(-1, 3, 5)\}$
7. Infinitely many solutions   **9.** $\varnothing$ **11.** $\left\{\left(-2, \dfrac{3}{2}, 1\right)\right\}$
13. $\left\{\left(\dfrac{1}{3}, -\dfrac{1}{2}, 1\right)\right\}$ **15.** $\left\{\left(\dfrac{2}{3}, -4, \dfrac{3}{4}\right)\right\}$
17. $\{(-2, 4, 0)\}$ **19.** $\left\{\left(\dfrac{1}{2}, \dfrac{1}{3}, \dfrac{1}{6}\right)\right\}$ **21.** 194
23. $.70 per bottle of catsup, $1 per jar of peanut butter, and $.80 per jar of pickles

25. $-2, 6$, and 16 **27.** $40°, 60°$, and $80°$
29. $500 at 12%, $1000 at 13%, and $1500 at 14%

Problem Set 10.4 (page 522)
1. $\{(6, -4)\}$ **3.** $\{(-4, -4)\}$ **5.** $\left\{\left(-\dfrac{11}{7}, -\dfrac{13}{7}\right)\right\}$
7. $\{(1, -3)\}$ **9.** $\{(-1, -2, -3)\}$ **11.** $\{(3, 1, 4)\}$
13. $\{(4, 3, -2)\}$ **15.** $\{(-5, 2, 0)\}$ **17.** $\{(-2, -1, 1)\}$
19. $\{(-1, 4, -1)\}$ **21.** $\{(2, 0, -3)\}$
25. $\{(1, -1, 2, -3)\}$ **27.** $\varnothing$

Problem Set 10.5 (page 527)
1. 10   **3.** -48 **5.** -29 **7.** 50   **9.** 8
11. 6   **13.** -32 **15.** -2 **17.** 5
19. $-\dfrac{7}{20}$ **21.** $\{(3, 8)\}$ **23.** $\{(-2, 5)\}$
25. $\{(-2, 2)\}$ **27.** $\varnothing$
29. $\{(1, 6)\}$ **31.** $\left\{\left(-\dfrac{1}{4}, \dfrac{2}{3}\right)\right\}$ **33.** $\left\{\left(\dfrac{17}{62}, \dfrac{4}{31}\right)\right\}$
35. $\{(-1, 3)\}$ **37.** $\{(9, -2)\}$ **39.** $\{(-4, -3)\}$

Problem Set 10.6 (page 534)
1. -57 **3.** 14   **5.** -41 **7.** -8
9. -96 **11.** $\{(-3, 1, -1)\}$ **13.** $\{(0, 2, -3)\}$
15. $\left\{\left(-2, \dfrac{1}{2}, -\dfrac{2}{3}\right)\right\}$ **17.** $\{(-1, -1, -1)\}$ **19.** $\varnothing$
21. $\{(-5, -2, 1)\}$ **23.** $\{(-2, 3, -4)\}$
25. $\left\{\left(-\dfrac{1}{2}, 0, \dfrac{2}{3}\right)\right\}$ **27.** $\{(-6, 7, 1)\}$
29. $\left\{\left(1, -6, -\dfrac{1}{2}\right)\right\}$ **33.** 0   **35. (b)** -20

Problem Set 10.7 (page 542)
1. $\{(-2, 0), (-4, 4)\}$ **3.** $\{(2, -3), (-2, 3)\}$
5. $\{(-6, 7), (-2, -1)\}$ **7.** $\{(1, 1)\}$ **9.** $\{(-5, -3)\}$
11. $\{(-3, 2)\}$ **13.** $\{(2, 2), (-2, -2)\}$
15. $\{(2 + i\sqrt{3}, -2 + i\sqrt{3}), (2 - i\sqrt{3}, -2 - i\sqrt{3})\}$
17.   **19.**

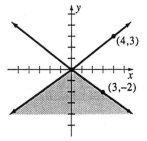

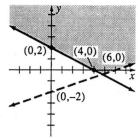

21.

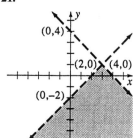

23.

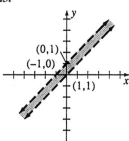

22. $\{(2, -1, -2)\}$ **23.** $\left\{\left(-\dfrac{1}{3}, -1, 4\right)\right\}$

24. $\{(0, -2, -4)\}$
25. $\{(1, 1), (-2, 7)\}$ **26.**

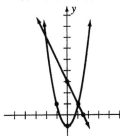

25.

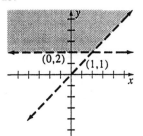

27.

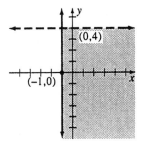

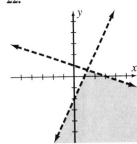

27. 2 and 3 or -3 and -2
28. 5 and 3 or 5 and -3 or -5 and 3 or -5 and -3
29. 2 and 5 or -3 and 10   **30.** 6 meters by 9 meters
31. \$9 per pound for cashews and \$5 per pound for Spanish peanuts
32. \$1.50 for a carton of pop and \$2.25 for a pound of candy
33. The fixed fee is \$2, and the additional fee is \$.10 per pound.

29.

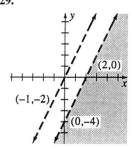

31.

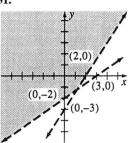

Chapter 10 Test (page 547)
1. II   **2.** I and IV   **3.** III   **4.** I   **5.** -38
6. -9 **7.** $\{(2, 7)\}$ **8.** $\{(-3, 5)\}$ **9.** $\{(-2, -4)\}$
10. $\{(3, -3)\}$ **11.** $\left\{\left(-\dfrac{1}{2}, 1\right)\right\}$ **12.** $\{(24, 18)\}$

13. $\{(1, -2, 3)\}$ **14.** $\{(-1, 2, -3)\}$ **15.** $x = 7$
16. $y = -1$ **17.** $z = 0$ **18.** Two
19. Four   **20.** One   **21.** $\{(2, 3)\}$

22.

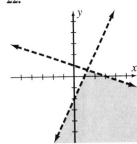

Chapter 10 Review Problem Set (page 545)
1. $\{(2, 6)\}$ **2.** $\{(-3, 7)\}$ **3.** $\{(-9, 8)\}$
4. $\left\{\left(\dfrac{89}{23}, -\dfrac{12}{23}\right)\right\}$ **5.** $\{(4, -7)\}$ **6.** $\{(-2, 8)\}$
7. $\{(0, 3)\}$ **8.** $\{(6, 7)\}$ **9.** $\{(1, -6)\}$
10. $\{(-4, 0)\}$ **11.** $\{(-12, 18)\}$ **12.** $\{(24, 8)\}$
13. $\left\{\left(\dfrac{2}{3}, -\dfrac{3}{4}\right)\right\}$ **14.** $\left\{\left(-\dfrac{1}{2}, \dfrac{3}{5}\right)\right\}$ **15.** -18
16. 11   **17.** -29 **18.** 59   **19.** $\{(2, -4, -6)\}$
20. $\{(-1, 5, -7)\}$ **21.** $\{(2, -3, 1)\}$

23. 24 quarters   **24.** 3 liters   **25.** 27

Cumulative Review Problem Set (page 549)

1. -6 **2.** -8 **3.** $\dfrac{13}{24}$ **4.** 24 **5.** $\dfrac{13}{6}$

6. $-90\sqrt{2}$ **7.** $2x + 5\sqrt{x} - 12$ **8.** $-18 + 22\sqrt{3}$

9. $2x^3 + 11x^2 - 14x + 4$ **10.** $\dfrac{x + 4}{x(x + 5)}$

11. $\dfrac{16x^2}{27y}$ **12.** $\dfrac{16x + 43}{90}$ **13.** $\dfrac{35a - 44b}{60a^2b}$

14. $\dfrac{2}{x - 4}$ **15.** $2x^2 - x - 4$ **16.** $\dfrac{5y^2 - 3xy^2}{x^2y + 2x^2}$

17. $\dfrac{2y - 3xy}{3x + 4xy}$ **18.** $\dfrac{(2n - 5)(n + 3)}{(n - 2)(3n + 13)}$

19. $\dfrac{3a^2 - 2a + 1}{2a - 1}$ **20.** $(5x - 2)(4x + 3)$

21. $2(2x + 3)(4x^2 - 6x + 9)$

22. $(2x + 3)(2x - 3)(x + 2)(x - 2)$

23. $4x(3x + 2)(x - 5)$ **24.** $(y - 6)(x + 3)$

25. $(5 - 3x)(2 + 3x)$ **26.** $\dfrac{81}{16}$ **27.** 4 **28.** $-\dfrac{3}{4}$

29. -0.3 **30.** $\dfrac{1}{81}$ **31.** $\dfrac{21}{16}$ **32.** $\dfrac{9}{64}$ **33.** 72

34. $-\dfrac{12}{x^3y}$ **35.** $\dfrac{8y}{x^5}$ **36.** $-\dfrac{a^3}{9b}$ **37.** $4\sqrt{5}$

38. $-6\sqrt{6}$ **39.** $\dfrac{5\sqrt{3}}{9}$ **40.** $\dfrac{2\sqrt{3}}{3}$ **41.** $2\sqrt[3]{7}$

42. $\dfrac{\sqrt[3]{6}}{2}$ **43.** $8xy\sqrt{13x}$ **44.** $\dfrac{\sqrt{6xy}}{3y}$

45. $11\sqrt{6}$ **46.** $-\dfrac{169\sqrt{2}}{12}$ **47.** $-16\sqrt[3]{3}$

48. $\dfrac{-3\sqrt{2} - 2\sqrt{6}}{2}$ **49.** $\dfrac{6\sqrt{15} - 3\sqrt{35} - 6 + \sqrt{21}}{5}$

50. 0.021 **51.** 300 **52.** 0.0003 **53.** $32 + 22i$

54. $-17 + i$ **55.** $0 - \dfrac{5}{4}i$ **56.** $-\dfrac{19}{53} + \dfrac{40}{53}i$

57. $-\dfrac{10}{3}$ **58.** $\dfrac{4}{7}$ **59.** $2\sqrt{13}$

60. $5x - 4y = 19$ **61.** $4x + 3y = -18$

62. $(-2, 6)$ and $r = 3$ **63.** $(-5, -4)$ **64.** 8 units

65.

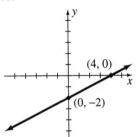

66.

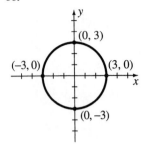

67.

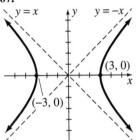

68.

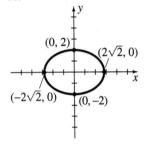

69.

70.

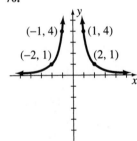

71.

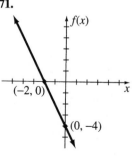

72.

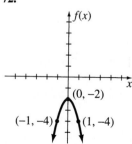

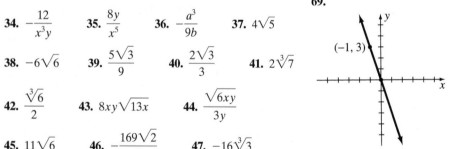

73.
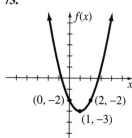

74.
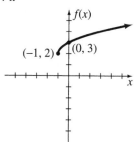

75.   **76.**
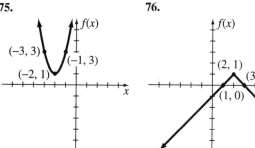

77. $(g \circ f)(x) = 2x^2 - 13x + 20$

    $(f \circ g)(x) = 2x^2 - x - 4$ **78.** $f^{-1}(x) = \dfrac{x + 7}{3}$

79. $f^{-1}(x) = -2x + \dfrac{4}{3}$ **80.** $k = -3$ **81.** $y = 1$

82. 12 cubic centimeters   **83.** -40 **84.** 40

85. $\left\{-\dfrac{21}{16}\right\}$ **86.** $\left\{\dfrac{40}{3}\right\}$ **87.** $\{6\}$

88. $\left\{-\dfrac{5}{2}, 3\right\}$ **89.** $\left\{0, \dfrac{7}{3}\right\}$ **90.** $\{-6, 0, 6\}$

91. $\left\{-\dfrac{5}{6}, \dfrac{2}{5}\right\}$ **92.** $\left\{-3, 0, \dfrac{3}{2}\right\}$ **93.** $\{\pm 1, \pm 3i\}$

94. $\{-5, 7\}$ **95.** $\{-29, 0\}$ **96.** $\left\{\dfrac{7}{2}\right\}$

97. $\{12\}$ **98.** $\{-3\}$ **99.** $\left\{\dfrac{1 \pm 3\sqrt{5}}{3}\right\}$

100. $\left\{\dfrac{-5 \pm 4i\sqrt{2}}{2}\right\}$ **101.** $\left\{\dfrac{3 \pm i\sqrt{23}}{4}\right\}$

102. $\left\{\dfrac{3 \pm \sqrt{3}}{3}\right\}$ **103.** $\{1 \pm \sqrt{34}\}$

104. $\left\{\pm\dfrac{\sqrt{5}}{2}, \pm\dfrac{\sqrt{3}}{3}\right\}$ **105.** $\left\{\dfrac{-5 \pm i\sqrt{15}}{4}\right\}$

106. $(-\infty, 3)$ **107.** $(-\infty, 50]$

108. $\left(-\infty -\dfrac{11}{5}\right) \cup (3, \infty)$ **109.** $\left(-\dfrac{5}{3}, 1\right)$

110. $\left[-\dfrac{9}{11}, \infty\right)$ **111.** $[-4, 2]$

112. $\left(-\infty, \dfrac{1}{3}\right) \cup (4, \infty)$ **113.** $(-8, 3)$

114. $(-\infty, 3] \cup (7, \infty)$ **115.** $(-6, -3)$ **116.** $\{(9, 6)\}$

117. $\left\{\left(\dfrac{9}{5}, -\dfrac{7}{25}\right)\right\}$ **118.** $\left\{\left(\dfrac{70}{23}, \dfrac{36}{23}\right)\right\}$

119. $\{(-2, 4, 0)\}$ **120.** $\{(-1, 4, -1)\}$

121. 17, 19, and 21

122. 14 nickels, 20 dimes, and 29 quarters

123. $48°$ and $132°$ **124.** \$600

125. \$1700 at 8% and \$2000 at 9%

126. 66 miles per hour and 76 miles per hour

127. 4 quarts   **128.** 69 or less   **129.** $-3, 0,$ or 3

130. 1 inch   **131.** \$1050 and \$1400

132. 3 hours   **133.** 30 shares at \$10 per share

134. 37   **135.** $10°, 60°,$ and $110°$

CHAPTER 11

Problem Set 11.1 (page 560)

1. $\{3\}$ **3.** $\{2\}$ **5.** $\{4\}$ **7.** $\{1\}$ **9.** $\{5\}$

11. $\{1\}$ **13.** $\left\{\dfrac{3}{2}\right\}$ **15.** $\left\{\dfrac{5}{6}\right\}$ **17.** $\{-3\}$

19. $\{-3\}$ **21.** $\{0\}$ **23.** $\{1\}$ **25.** $\{-1\}$

27. $\left\{-\dfrac{2}{5}\right\}$ **29.** $\left\{\dfrac{5}{2}\right\}$ **31.** $\{3\}$ **33.** $\left\{\dfrac{1}{2}\right\}$

35.   **37.**
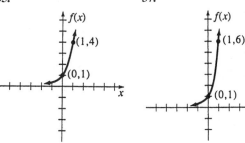

39.

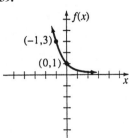

41.

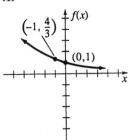

43.

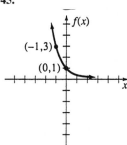

45.

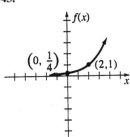

47.

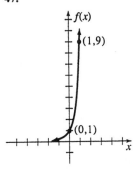

49.

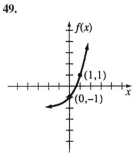

51.

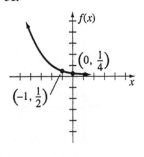

Problem Set 11.2 (page 568)

1. (a) $0.87   **(c)** $2.33   **(e)** $21,900   **(g)** $658
3. $283.70   **5.** $865.84   **7.** $1782.25
9. $2725.05   **11.** $16,998.71   **13.** $22,553.65
15. $567.63   **17.** $1422.36   **19.** $8963.38
21. $17,547.35   **23.** $32,558.88
25.

	1 yr	5 yr	10 yr	20 yr
Compounded annually	$1120	1762	3106	9,646
Compounded semiannually	1124	1791	3207	10,286
Compounded quarterly	1126	1806	3262	10,641
Compounded monthly	1127	1817	3300	10,893
Compounded continuously	1127	1822	3320	11,023

27. Nora will have $.24 more.   **29.** 9.54%
31. 50 grams; 37 grams

33.

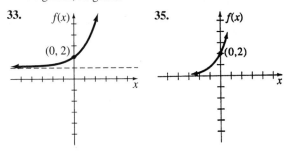

35.

37.

39. 2226; 3320; 7389   **41.** 2000
43. (a) 82,888   **(c)** 96,302
45. (a) 6.5 pounds per square inch
    (c) 13.6 pounds per square inch

Problem Set 11.3 (page 578)

1. $\log_2 128 = 7$ **3.** $\log_5 125 = 3$

5. $\log_{10} 1000 = 3$ **7.** $\log_2 \left(\dfrac{1}{4}\right) = -2$

9. $\log_{10} 0.1 = -1$ **11.** $3^4 = 81$ **13.** $4^3 = 64$

15. $10^4 = 10,000$ **17.** $2^{-4} = \dfrac{1}{16}$ **19.** $10^{-3} = 0.001$

21. 4   **23.** 4   **25.** 3   **27.** $\dfrac{1}{2}$ **29.** 0

31. −1 **33.** 5 **35.** −5 **37.** 1 **39.** 0

41. {49} **43.** {16} **45.** {27} **47.** $\left\{\dfrac{1}{8}\right\}$

49. {4} **51.** 5.1293 **53.** 6.9657 **55.** 1.4037
57. 7.4512 **59.** 6.3219 **61.** −0.3791
63. 0.5766 **65.** 2.1531 **67.** 0.3949
69. $\log_b x + \log_b y + \log_b z$ **71.** $\log_b y - \log_b z$

73. $3\log_b y + 4\log_b z$ **75.** $\dfrac{1}{2}\log_b x + \dfrac{1}{3}\log_b y - 4\log_b z$

77. $\dfrac{2}{3}\log_b x + \dfrac{1}{3}\log_b z$ **79.** $\dfrac{3}{2}\log_b x - \dfrac{1}{2}\log_b y$

81. $\left\{\dfrac{9}{4}\right\}$ **83.** {25} **85.** {4} **87.** $\left\{\dfrac{19}{8}\right\}$

89. {9} **91.** {1}

Problem Set 11.4 (page 585)

1. 0.8597 **3.** 1.7179 **5.** 3.5071 **7.** −0.1373
9. −3.4685 **11.** 411.43 **13.** 90,095
15. 79.543 **17.** 0.048440 **19.** 0.0064150
21. 1.6094 **23.** 3.4843 **25.** 6.0638
27. −0.7765 **29.** −3.4609 **31.** 1.6034
33. 3.1346 **35.** 108.56 **37.** 0.48268
39. 0.035994

41.

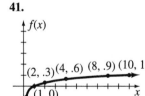

43.

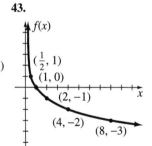

45.

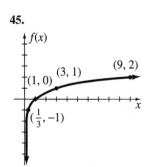

47.

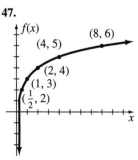

49.

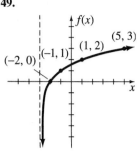

51.

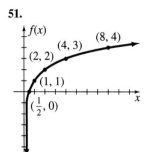

53.

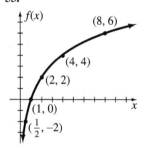

55. 0.36 **57.** 0.73 **59.** 23.10
61. 7.93

Problem Set 11.5 (page 594)

1. {3.15} **3.** {2.20} **5.** {4.18} **7.** {0.12}
9. {1.69} **11.** {4.57} **13.** {2.46} **15.** {4}

17. $\left\{\dfrac{19}{47}\right\}$ **19.** {1} **21.** {8} **23.** 4.524

25. −0.860 **27.** 3.105 **29.** −2.902
31. 5.989 **33.** 2.4 years **35.** 5.3 years
37. 6.8 hours **39.** 1.5 hours
41. 34.7 years **43.** 6.7
45. Approximately 8 times

Chapter 11 Review Problem Set (page 597)

1. 7 **2.** 3 **3.** 4 **4.** −3 **5.** 2 **6.** 13

7. 0 **8.** −2 **9.** {−4} **10.** {3} **11.** $\left\{-\dfrac{3}{4}\right\}$

12. {4} **13.** {9} **14.** $\left\{\dfrac{1}{2}\right\}$ **15.** {5} **16.** {8}

17. 1.8642 **18.** 4.7380 **19.** −4.2687
20. −2.6289 **21.** 1.1882 **22.** 2639.4
23. 0.013289 **24.** 0.077197 **25.** {3.40}
26. {1.95} **27.** {5.30} **28.** {5.61}

29.

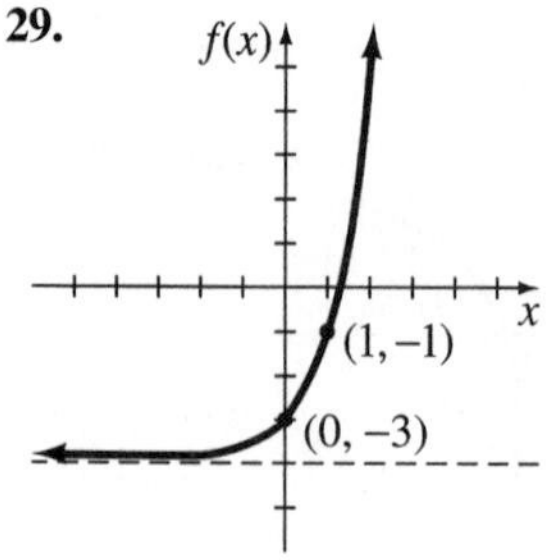

30.

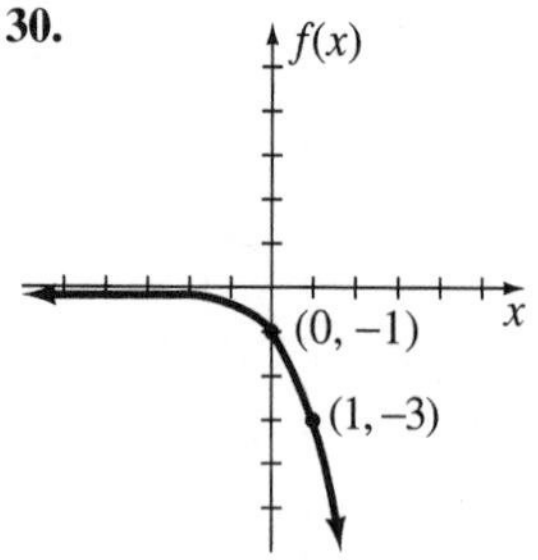

31.

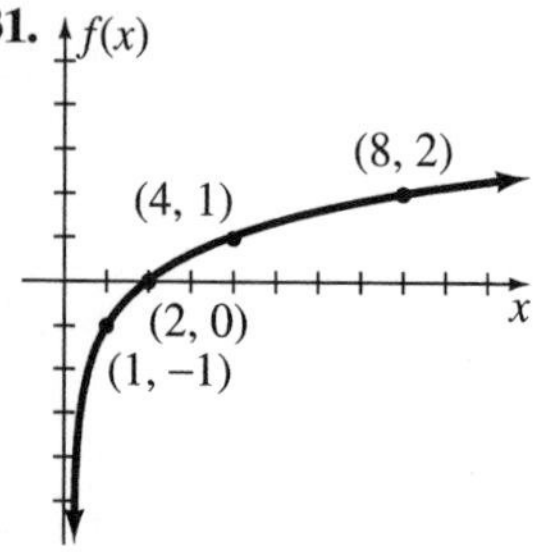

32.

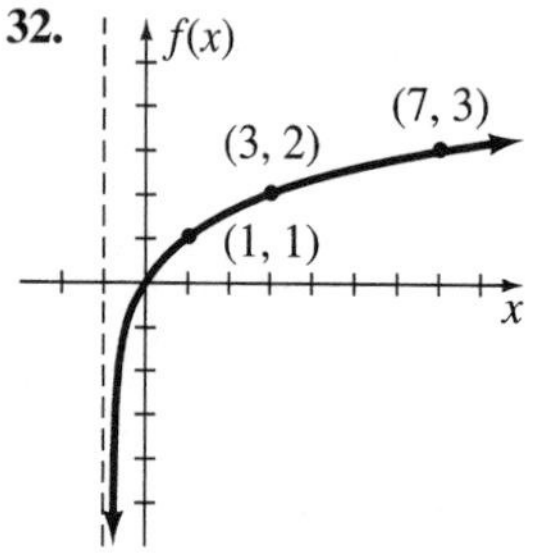

33. 2.842 **34.** \$2913.99 **35.** \$8178.72
36. \$5656.26 **37.** 133 grams
38. Approximately 5.3 years
39. Approximately 12.1 years
40. 61,070; 67,493; 74,591
41. Approximately 4.8 hours **42.** 8.1

Chapter 11 Test (page 598)

1. $\frac{1}{2}$ **2.** 1 **3.** 1 **4.** −1 **5.** $\{-3\}$
6. $\left\{-\frac{3}{2}\right\}$ **7.** $\left\{\frac{8}{3}\right\}$ **8.** $\{243\}$ **9.** $\{2\}$
10. $\left\{\frac{2}{5}\right\}$ **11.** 4.1919 **12.** 0.2031 **13.** 0.7325
14. 5.4538 **15.** $\{5.17\}$ **16.** $\{10.29\}$
17. 4.0069 **18.** $\log_b\left(\frac{x^3y^2}{z}\right)$ **19.** \$6342.08
20. 13.5 years **21.** 7.8 hours **22.** 4813 grams

23.

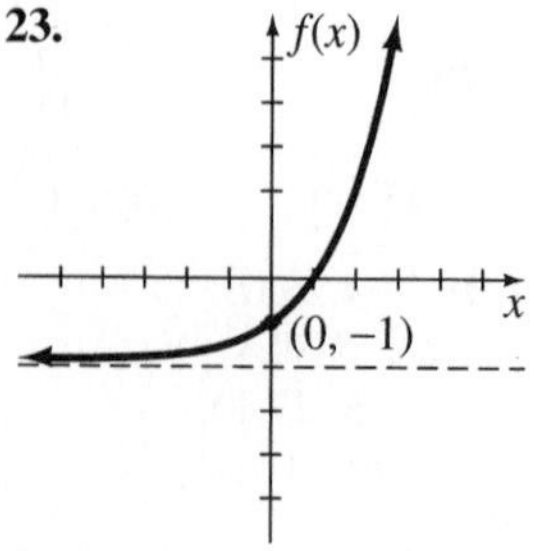

24.

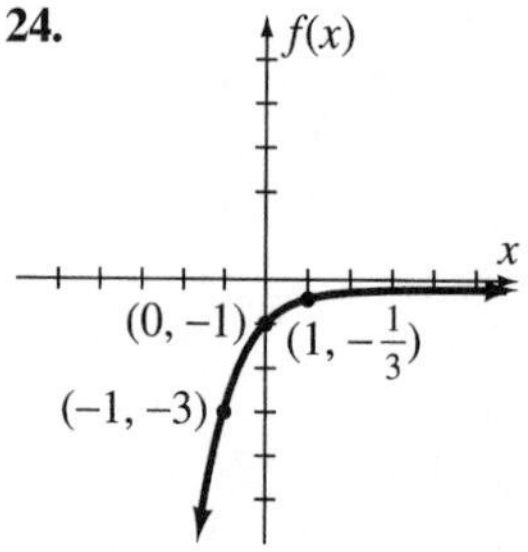

25.

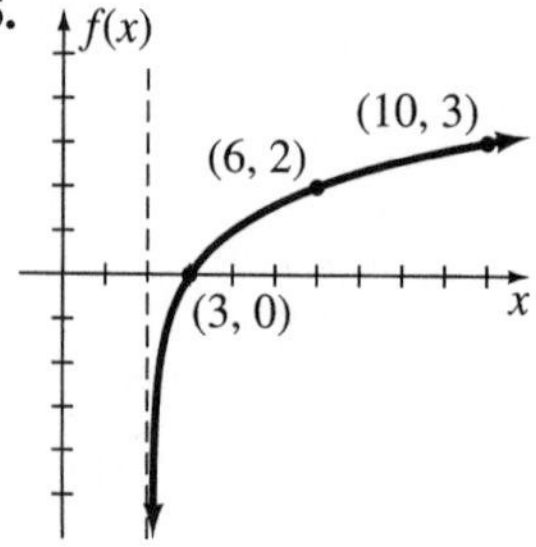

CHAPTER 12

Problem Set 12.1 (page 604)

1. −1, 2, 5, 8, 11 **3.** 3, 1, −1, −3, −5
5. −1, 2, 7, 14, 23 **7.** 0, −3, −8, −15, −24
9. −1, 5, 15, 29, 47 **11.** $\frac{1}{2}$, 1, 2, 4, 8
13. $-\frac{2}{3}$, −2, −6, −18, −54 **15.** $a_8 = 54$ and $a_{12} = 130$
17. $a_7 = -32$ and $a_8 = 64$ **19.** $a_n = 2n - 1$
21. $a_n = 4n - 6$ **23.** $a_n = -2n + 7$
25. $a_n = -3n - 4$ **27.** $a_n = \frac{1}{2}(n + 1)$
29. 34 **31.** 78 **33.** −155 **35.** 106
37. 50 **39.** 170 **41.** −1 **43.** −13
45. 136 **47.** \$27,800 **49.** \$1170

Problem Set 12.2 (page 609)

1. 2550 **3.** 9030 **5.** −4225 **7.** 410
9. 8850 **11.** 5724 **13.** 5070 **15.** 346.5
17. 3775 **19.** 39,500 **21.** −8645
23. 122,850 **25.** 39,552 **27.** 76 seats, 720 seats
29. \$10,000 **31.** \$357,500 **33.** 169
37. **(a)** 7 + 12 + 17; 36
(c) (−3) + (−5) + (−7) + (−9) + (−11) + (−13); −48
(e) 3 + 6 + 9 + 12 + 15; 45

Problem Set 12.3 (page 616)

1. $a_n = 3^{n-1}$ **3.** $a_n = 2(4)^{n-1}$
5. $a_n = \left(\frac{1}{3}\right)^{n-1}$ or $a_n = \frac{1}{3^{n-1}}$
7. $a_n = 0.2(0.2)^{n-1}$ or $a_n = (0.2)^n$
9. $a_n = 9\left(\frac{2}{3}\right)^{n-1}$ or $a_n = (3^{-n+3})(2)^{n-1}$
11. $a_n = (-4)^{n-1}$
13. 19,683 **15.** −512 **17.** $-\frac{6561}{256}$
19. 14,762 **21.** −9842 **23.** 1093 **25.** 511

27. 19,680　　**29.** 2730　　**31.** 1093　　**33.** $1\frac{1023}{1024}$

35. $15\frac{31}{32}$　　**37.** $\frac{1}{2}$　　**39.** 0; −1　　**41.** 125 liters

43. \$102.40; \$204.75　　**45.** $971.\overline{3}$ meters

47. \$5609.66　　**51.** 1,048,576　　**53.** $\frac{4096}{531,441}$

55. $-\frac{177,147}{2048}$　　**57. (a)** 126　　**(c)** 31　　**(e)** $1\frac{49}{81}$

Problem Set 12.4 (page 623)

1. 4　　**3.** 1　　**5.** 2　　**7.** $\frac{2}{3}$　　**9.** 9

11. No sum　　**13.** $\frac{4}{7}$　　**15.** $\frac{16}{3}$　　**17.** No sum

19. $\frac{81}{2}$　　**21.** $\frac{4}{9}$　　**23.** $\frac{47}{99}$　　**25.** $\frac{5}{11}$

27. $\frac{427}{999}$　　**29.** $\frac{7}{15}$　　**31.** $\frac{24}{11}$　　**33.** $\frac{47}{110}$

Problem Set 12.5 (page 627)

1. $x^8 + 8x^7y + 28x^6y^2 + 56x^5y^3 + 70x^4y^4 + 56x^3y^5 + 28x^2y^6 + 8xy^7 + y^8$

3. $81x^4 + 108x^3y + 54x^2y^2 + 12xy^3 + y^4$

5. $x^5 - 5x^4y + 10x^3y^2 - 10x^2y^3 + 5xy^4 - y^5$

7. $x^{10} + 10x^9y + 45x^8y^2 + 120x^7y^3 + 210x^6y^4 + 252x^5y^5 + 210x^4y^6 + 120x^3y^7 + 45x^2y^8 + 10xy^9 + y^{10}$

9. $64x^6 + 192x^5y + 240x^4y^2 + 160x^3y^3 + 60x^2y^4 + 12xy^5 + y^6$

11. $x^5 - 15x^4y + 90x^3y^2 - 270x^2y^3 + 405xy^4 + 243y^5$

13. $243a^5 - 810a^4b + 1080a^3b^2 - 720a^2b^3 + 240ab^4 - 32b^5$

15. $x^6 + 6x^5y^3 + 15x^4y^6 + 20x^3y^9 + 15x^2y^{12} + 6xy^{15} + y^{18}$

17. $x^7 + 14x^6 + 84x^5 + 280x^4 + 560x^3 + 672x^2 + 448x + 128$　　**19.** $x^4 - 12x^3 + 54x^2 - 108x + 81$

21. $x^{15} + 15x^{14}y + 105x^{13}y^2 + 455x^{12}y^3$

23. $a^{13} - 26a^{12}b + 312a^{11}b^2 - 2288a^{10}b^3$　　**25.** $462x^5y^6$

27. $-160x^3y^3$　　**29.** $2000x^3y^2$

Chapter 12 Review Problem Set (page 628)

1. $a_n = 6n - 3$　　**2.** $a_n = 3^{n-2}$　　**3.** $a_n = 5 \cdot 2^n$

4. $a_n = -3n + 8$　　**5.** $a_n = 2n - 7$　　**6.** $a_n = 3^{3-n}$

7. $a_n = -(-2)^{n-1}$　　**8.** $a_n = 3n + 9$　　**9.** $a_n = \frac{n+1}{3}$

10. $a_n = 4^{n-1}$　　**11.** 73　　**12.** 106　　**13.** $\frac{1}{32}$

14. $\frac{4}{9}$　　**15.** −92　　**16.** $\frac{1}{16}$　　**17.** −5

18. 85　　**19.** $\frac{5}{9}$　　**20.** 2 or −2　　**21.** $121\frac{40}{81}$

22. 7035　　**23.** −10,725　　**24.** $31\frac{31}{32}$　　**25.** 32,015

26. 4757　　**27.** $85\frac{21}{64}$　　**28.** 37,044　　**29.** 12,726

30. $85\frac{1}{3}$　　**31.** $\frac{4}{11}$　　**32.** $\frac{41}{90}$　　**33.** \$750

34. \$46.50　　**35.** \$3276.70　　**36.** 10,935 gallons

37. 3600 feet

38. $128x^7 + 448x^6y + 672x^5y^2 + 560x^4y^3 + 280x^3y^4 + 84x^2y^5 + 14xy^6 + y^7$

39. $x^4 - 12x^3y + 54x^2y^2 - 108xy^3 + 81y^4$

40. $1120x^4y^4$

Chapter 12 Test (page 630)

1. −46　　**2.** 48　　**3.** $a_n = 4n - 7$

4. $a_n = 5(2)^{1-n}$　　**5.** $a_n = -3n + 9$　　**6.** $\frac{729}{8}$ or $91\frac{1}{8}$

7. 223　　**8.** 60 terms　　**9.** 46　　**10.** 2380

11. 765　　**12.** −780　　**13.** 3279　　**14.** 7155

15. 6138　　**16.** 22,650　　**17.** 9384　　**18.** \$5810

19. \$3276.70　　**20.** $13\frac{1}{2}$　　**21.** $\frac{37}{99}$　　**22.** $\frac{4}{15}$

23. $x^5 - 15x^4y + 90x^3y^2 - 270x^2y^3 + 405xy^4 - 243y^5$

24. $128x^7 + 448x^6y + 672x^5y^2$　　**25.** $495a^8b^4$

APPENDIX A

Practice Exercises (page 640)

1. $2 \cdot 13$　　**2.** $2 \cdot 2 \cdot 2 \cdot 2$　　**3.** $2 \cdot 2 \cdot 3 \cdot 3$

4. $2 \cdot 2 \cdot 2 \cdot 2 \cdot 5$　　**5.** $7 \cdot 7$　　**6.** $2 \cdot 2 \cdot 23$

7. $2 \cdot 2 \cdot 2 \cdot 7$　　**8.** $2 \cdot 2 \cdot 2 \cdot 2 \cdot 3 \cdot 3$

9. $2 \cdot 2 \cdot 2 \cdot 3 \cdot 5$　　**10.** $2 \cdot 2 \cdot 3 \cdot 7$

11. $3 \cdot 3 \cdot 3 \cdot 5$　　**12.** $2 \cdot 7 \cdot 7$　　**13.** 24

14. 24　　**15.** 48　　**16.** 36　　**17.** 140

18. 462　　**19.** 392　　**20.** 72　　**21.** 168

22. 60　　**23.** 90　　**24.** 168　　**25.** $\frac{2}{3}$　　**26.** $\frac{3}{4}$

27. $\frac{2}{3}$　　**28.** $\frac{9}{16}$　　**29.** $\frac{5}{3}$　　**30.** $\frac{4}{3}$　　**31.** $\frac{15}{28}$

32. $\frac{12}{55}$　　**33.** $\frac{10}{21}$　　**34.** $\frac{65}{66}$　　**35.** $\frac{3}{10}$　　**36.** $\frac{2}{3}$

37. $\frac{3}{8}$ cup　　**38.** $\frac{1}{6}$ of the bottle

39. $\frac{2}{9}$ of the disk space   **40.** $\frac{1}{3}$ **41.** $\frac{5}{7}$

42. $\frac{8}{11}$ **43.** $\frac{5}{9}$ **44.** $\frac{5}{13}$ **45.** 3

46. 2   **47.** $\frac{2}{3}$ **48.** $\frac{3}{8}$ **49.** $\frac{2}{3}$

50. $\frac{5}{9}$ **51.** $\frac{8}{15}$ **52.** $\frac{7}{24}$ **53.** $\frac{9}{16}$

54. $\frac{11}{12}$ **55.** $\frac{37}{30}$ **56.** $\frac{29}{24}$ **57.** $\frac{59}{96}$

58. $\frac{19}{24}$ **59.** $\frac{13}{12}$ **60.** $\frac{5}{16}$ **61.** $\frac{1}{4}$

62. $\frac{5}{3}$ **63.** $\frac{37}{30}$ **64.** $\frac{4}{5}$ **65.** $\frac{1}{3}$ **66.** $\frac{27}{35}$

67. $\frac{7}{26}$ **68.** 30   **69.** $\frac{7}{20}$ **70.** $\frac{11}{32}$

Index

Properties of Absolute Value

$|a| \geq 0$

$|a| = |-a|$

$|a - b| = |b - a|$

$|a^2| = |a|^2 = a^2$

Multiplication Patterns

$(a + b)^2 = a^2 + 2ab + b^2$

$(a - b)^2 = a^2 - 2ab + b^2$

$(a + b)(a - b) = a^2 - b^2$

$(a + b)^3 = a^3 + 3a^2b + 3ab^2 + b^3$

$(a - b)^3 = a^3 - 3a^2b + 3ab^2 - b^3$

Properties of Exponents

$b^n \cdot b^m = b^{n+m}$ $\dfrac{b^n}{b^m} = b^{n-m}$

$(b^n)^m = b^{mn}$

$(ab)^n = a^n b^n$

$\left(\dfrac{a}{b}\right)^n = \dfrac{a^n}{b^n}$

Equations Determining Functions

Linear function: $f(x) = ax + b$

Quadratic function: $f(x) = ax^2 + bx + c$

Polynomial function: $f(x) = a_n x^n + a_{n-1} x^{n-1} + \ldots + a_1 x + a_0$

Rational function: $f(x) = \dfrac{g(x)}{h(x)}$,   where g and h are polynomials; $h(x) \neq 0$

Exponential function: $f(x) = b^x$,   where $b > 0$ and $b \neq 1$

Logarithmic function: $f(x) = \log_b x$,   where $b > 0$ and $b \neq 1$

Interval Notation

(a, ∞)

$(-\infty, b)$

(a, b)

$[a, \infty)$

$(-\infty, b]$

$(a, b]$

$[a, b)$

$[a, b]$

Set Notation

$\{x \mid x > a\}$

$\{x \mid x < b\}$

$\{x \mid a < x < b\}$

$\{x \mid x \geq a\}$

$\{x \mid x \leq b\}$

$\{x \mid a < x \leq b\}$

$\{x \mid a \leq x < b\}$

$\{x \mid a \leq x \leq b\}$

Properties of Logarithms

$\log_b b = 0$

$\log_b 1 = 0$

$\log_b rs = \log_b r + \log_b s$

$\log_b \left(\dfrac{r}{s}\right) = \log_b r - \log_b s$

$\log_b r^p = p(\log_b r)$

Factoring Patterns

$a^2 - b^2 = (a + b)(a - b)$

$a^3 - b^3 = (a - b)(a^2 + ab + b^2)$

$a^3 + b^3 = (a + b)(a^2 - ab + b^2)$